D1321447

Leo J. Ritter, Jr., is Senior Editor of *Engineering News-Record* and is a registered professional engineer. He has taught at Mississippi State University, Texas A&M University, University of Florida, New York University, and Northwestern University. He has also held a Fulbright Professorship at the University of the Philippines. He is active in various professional and technical societies and is a past President of the Construction Writers Association.

Radnor J. Paquette is Professor of Civil Engineering at Georgia Institute of Technology. He has also taught at the University of Florida. Professor Paquette has been, at various times, a mining engineer, highway engineer, construction engineer, and highway planning engineer for a number of companies and state agencies and has served as President of the Educational Division of the American Road Builders' Association. He is currently Chairman of the Annual Georgia Highway Conference; Technical Advisor, National Association of County Engineers; and a member of various technical boards and committees. He has written numerous articles and books in the field of highway engineering.

HIGHWAY ENGINEERING

LEO J. RITTER, JR.
Senior Editor
Engineering News-Record

AND

RADNOR J. PAQUETTE
Georgia Institute of Technology

THIRD EDITION

THE RONALD PRESS COMPANY · NEW YORK

3MP

Library of Congress Catalog Card Number: 67-13195
PRINTED IN THE UNITED STATES OF AMERICA

PREFACE

This book is designed as a basic textbook for the course in highway engineering. It will also be useful as a professional reference for the practicing highway engineer.

Since the Second Edition was published, highway travel and highway expenditures have soared in the United States and in the rest of the free world. During this period progress has been substantial on the greatest roadbuilding project in world history—construction of the 41,000-mile National System of Interstate and Defense Highways, now estimated to cost over fifty billion dollars. And the demand for improved highway transportation shows no signs of slackening.

The practice of highway engineering has changed substantially in the intervening years, to keep pace with a growing volume of work and an ever-increasing public demand for higher levels of traffic service. New concepts of administration, planning, design, and construction are speeding work, increasing engineering productivity, and providing a better end product. New methods also are being applied to traffic operations and maintenance.

The original purpose of this book—to give an integrated picture of the whole broad field of highway engineering—remains unchanged. Its arrangement corresponds to the order in which factors in the development of a highway or highway system commonly occur. More than the usual amount of space has been devoted to certain of the more general aspects of highway engineering in the chapters on administration, economics and financing, and planning. Thus the student does not lose his perspective in the maze of detail involved in design and construction. The chapter arrangement leads from these more general factors, through design and the preparation of plans and contracts, to construction and maintenance, in a logical, orderly manner, with stress on fundamental principles.

Extensive changes have been made in the text to bring it up to date and to incorporate major technological advances. For example, highway engineering applications of electronic computers are described in several chapters, reflecting the greatly expanded use of these tools in modern highway practice. Throughout the book there are extensive changes that reflect advances in materials, equipment, and construction procedures. All numerical problems are new or have been revised.

Several chapters and major sections have been rewritten completely. The chapter on planning has been reorganized and expanded to reflect new planning concepts and techniques, particularly those used in urban areas. Chapter 5 contains important new data from the 1966 edition of the "Blue Book" (*Policy on Geometric Design of Rural Highways*) of the American Association of State Highway Officials; the previous edition of this basic design guide was issued in 1954.

Latest available data on the hydraulic design of culverts have been incorporated in the chapter on drainage; these procedures, developed by the Bureau of Public Roads, were in the formative stages at the time of the Second Edition. The description of major bridge types is completely revised and profusely illustrated. Other new information includes the critical path method, the new specifications for asphalts, and new methods for the design of both flexible and rigid pavements.

Sections of the book dealing with thickness design of flexible and rigid pavements have been revised drastically to reflect changes in practice brought about by the $27,000,000 AASHO Road Test, on which traffic stopped rolling in 1962.

Credit is due many organizations and individuals for their cooperation in supplying information for this Third Edition. Direct credit has been given in the text to these sources. A complete bibliography is included at the end of the book.

LEO J. RITTER, JR.
RADNOR J. PAQUETTE

New York, N.Y.
Atlanta, Georgia
March, 1967

CONTENTS

HIGHWAY ENGINEERING

CHAPTER 1

INTRODUCTION

One of the most striking features of our modern American economy and way of life is our dependence upon highway transportation. In fact, good highways are so interwoven with every phase of our daily activities that it is almost impossible to imagine what life would be like without them. We depend upon highways for the movement of goods, for travel to and from work, for services, for social and recreational purposes, and for many other functions necessary to the functioning of our complex society. Despite our concern with air travel and the exploration of space, we are still living in the age of the motor vehicle—and will be for many years to come.

The United States is covered by a tremendous network of roads, ranging in type from high-speed, controlled-access freeways connecting major centers of population and complex urban expressways to local rural roads and residential streets, and reaching into virtually every corner of the nation. There are more than 3.6 million miles of highways and streets within the limits of the continental United States. Soon, more than 110 million vehicles of all sorts—more than in all the rest of the world put together—will be operating over this vast highway system. Approximately one in every seven workers in this country is employed in highway transportation and its allied fields. These few statistics serve to focus attention upon the importance of highways in the United States and, thus, upon highway engineering.

The Development of Highways

Before beginning the study of highway engineering as an important phase of the general field of highway transportation, it is interesting and informative briefly to trace the development of highways and highway systems from early historical periods to the modern era.

1–1. Ancient Roads. The great highway systems of our modern civilization have their origin in the period before the dawn of recorded history. Even before the invention of the wheel, which is popularly supposed to have occurred some ten thousand years ago, individual and mass movements of people undoubtedly took place. The earliest travel was on foot; later, pack animals were utilized, crude sleds were developed, and simple wheeled vehicles came into being. Many of the migrations of the early historical period involved large numbers of people and covered relatively great distances. More or less regularly traveled routes developed, extending to the limits of the then-known world.

As various civilizations reached a higher level, many of the ancient peoples came to a realization of the importance of improved roads. The streets of the city of Babylon were paved as early as 2000 B.C. History also records the construction of a magnificent road to aid in the building

FIG. 1–1. The Appian Way in the Roman Campagna, with the Claudian Aqueduct in the background. (Courtesy Bureau of Public Roads.)

of the Great Pyramid in Egypt nearly 3000 years before the birth of Christ. Traces of early roads have been found on the island of Crete, and it is known that the early civilizations of the Chinese, Carthaginians, and Incas also led to extensive road building.

By far the most advanced highway system of the ancient world was that of the Romans. (See Fig. 1–1.) When Roman civilization was at its peak, a great system of military roads reached to the limits of the empire. Many of these roads were built of stone and were 3 feet or more in thickness. Traces of this magnificent system are still in existence on the European Continent; in fact, some of these roads still serve as bases for sections of modern highways.

After the decline and fall of the Roman Empire, road building, along with virtually all other forms of scientific activity, practically ceased for a period of a thousand years. Even as late as the early portion of the eighteenth century, the only convenient means of travel between cities was on foot or on horseback. Stagecoaches were introduced in 1659, but travel in them proved exceedingly difficult in most instances because of the extremely poor condition of rural roads.

1–2. Later European Development. Interest in the art of road building was revived in Europe in the late eighteenth century. During this period, Trésaguet, a noted French engineer, advocated a method of road construction utilizing a broken-stone base covered with smaller stones. The regime of Napoleon in France (1800–1814) gave a great impetus to road construction, chiefly for military purposes, and led to the establishment of a national system of highways in that country.

At about the same time in England, two Scottish engineers, Thomas Telford and John L. McAdam, developed similar types of construction. Telford urged the use of large pieces of ledge stone to form a base with smaller stones for the wearing surface. McAdam advocated the use of smaller broken stone throughout. This latter type of construction is still in extensive use, being the forerunner of various types of modern macadam bases and pavements.

Development of Highways in the United States

During the early history of the American colonies, travel was primarily local in character, and rural roads were generally little more than trails or cleared paths through the forests. Toward the end of the eighteenth century, public demand led to the improvement of various roads by private enterprise. These improvements generally took the form of toll roads or "turnpikes," and were principally located in areas adjacent to the larger cities. The first American turnpike made use of existing roads in Virginia to connect Alexandria with settlements in the Blue Ridge Mountains. Another important toll road of this period was the Philadelphia and Lancaster Turnpike Road, from Philadelphia to Lancaster, Pennsylvania.

Another famous early road in American history was the Wilderness Road, blazed by Daniel Boone in 1775, which led from the Shenandoah Valley in Virginia through the Cumberland Gap into Kentucky.

In 1806 the federal government entered the field of highway construction for the first time with the authorization of construction of the National Pike, or Cumberland Road. The first contract on this route was let in 1811, and by 1816 the road extended from Cumberland, Maryland, the terminus of the Chesapeake and Ohio Canal, to Wheeling, West Virginia. When construction was completed (1841), it had reached Vandalia, Illinois. This route, which was surfaced largely with macadam, was nearly 800 miles in length and was built at a cost of approximately $7 million (see Fig. 1–2).

FIG. 1–2. A stone bridge on the National Pike or Cumberland Road (circa 1840). (Courtesy Bureau of Public Roads.)

During the middle of the nineteenth century the great western expansion of the country began, and many roads or "trails" figured prominently in this development. Probably the most famous of the great trails was the Oregon Trail. This route began at Independence, Missouri, went overland to the Platte River, near Grand Island, Nebraska, and thence along the Platte River and its North Fork to Fort Laramie, Wyoming. Continuing westward, the trail followed the Sweetwater and crossed the Continental Divide through South Pass, in Wyoming. West of this point travelers followed the path of the Snake River into Oregon. Other famous western roads, each of which has a fascinating history, included the Santa Fe Trail, the Mormon Trail, the California Trail, and the Overland Trail.

In 1830, however, the steam locomotive demonstrated its superiority over horse-drawn vehicles, and interest in road building began to wane. By the time the first transcontinental railroad (the Union Pacific) was completed in 1869, road-building activities outside the cities virtually ceased until near the beginning of the twentieth century.

By 1900 a strong popular demand for highways again existed. The principal demand came from farmers, who clamored for farm-to-market roads, so that they might more readily move their agricultural products to the nearest railhead. Some demand was also being felt for improved routes connecting the larger centers of population. During this period certain states began to recognize the need for state financial aid for road construction. The first state-aid law was enacted by New Jersey in 1891, and by 1900 six other states had enacted similar legislation.

1–3. Advent of the Motor Vehicle. The year 1904 marked the beginning of a new era in highway transportation in America with the advent of motor vehicles in considerable numbers. Almost overnight a tremendous demand was created for improved highways, not only for farm-to-market roads but also for through routes connecting the metropolitan areas. Additional state-aid laws were enacted, and by 1917 every state participated in highway construction in some fashion. By this time, also, most states had established some sort of highway agency and had delegated to these bodies the responsibility for the construction and maintenance of the principal state routes. Figure 1–3 shows the evolution of one United States road from those early days to the present time.

This period was also marked by radical changes in road-construction methods, particularly with regard to wearing surfaces. The early roads of American history were largely natural earth. Since timber was readily available, plank and corduroy roads were numerous, while later, wood blocks were used. Some gravel was used on early surfaces, while many city streets were paved with cobblestones. The invention of the power stone crusher and the steam roller led to the construction of a considerable mileage of broken-stone surfaces. Development of this type of surface generally paralleled that in Europe.

In the cities, relatively large concentrations of wheeled vehicles and the need for abatement of noise and dust brought various improved surfaces into being. The first brick pavement in this country is supposed to have been built in Charleston, West Virginia, in 1871, and asphalt was used for paving Pennsylvania Avenue, in Washington, D. C., in 1867. Concrete pavements were introduced about 1893, and the first rural road of concrete was built in 1909 in Wayne County, Michigan.

Fig. 1-3. Two views of the Pacific Highway (Interstate 5) in Oregon. Top view is a newly completed section of the highway in 1920; at the right is the old wagon road that served as the highway in still earlier times. Bottom view is the present-day highway near Grants Pass. (Oregon State Highway Department photo.)

1-4. Development of the Federal-Aid Program. World War I intensified highway problems in the United States and led the federal government again actively to enter the field of highway construction. Actually, federal participation in highway affairs on a continuing basis began in 1893, when Congress established the Office of Road Inquiry, an agency whose work was primarily educational in nature. This agency later became the Bureau of Public Roads of the Department of Agriculture, and in 1939 it became the Public Roads Administration under the Federal Works Agency. In 1949, this agency was transferred to the Department of Commerce and its name again changed to the Bureau of Public Roads. In 1967, B.P.R. was transferred to the Department of Transportation, as a part of the Federal Highway Administration.

The modern era of federal aid for highways began with the passage of the Federal Road Act of 1916, which authorized the expenditure of $75 million over a period of five years for improvement of rural highways. The funds appropriated under this Act were apportioned to the individual states on the basis of area, population, and mileage of rural roads, the apportioning ratios being based upon the ratio of the amount of each of these items within the individual state to similar totals for the country as a whole. States were required to match the federal funds on a 50–50 basis, with sliding scale in the public land states of the West.

The Federal-Aid Highway Act of 1921 extended the principle of federal aid in highway construction and strengthened it in two important respects. First, it required that each state designate a connected system of interstate and intrastate routes, not to exceed 7 per cent of the total rural mileage then existing within the state, and it further directed that federal funds be expended upon this designated system. Second, the Act placed the responsibility for maintenance of these routes upon the individual states.

During this period many states, experiencing difficulty in securing the necessary funds to match the federal appropriations, looked for new sources of revenue. This fact led to the enactment of gasoline taxes by several states in 1919, and other states rapidly followed suit.

During the ensuing years, appropriations for federal aid were steadily increased, while the basis of participation remained practically the same. Funds were provided (as now) for the improvement of roads in national parks, national forests, Indian reservations, and other public lands. The principle of federal aid was broadened to allow the use of federal funds for the improvement of extensions of the federal-aid system into and through urban areas and for the construction of secondary roads.

Another important step taken during this period was the authorization by Congress, in 1934, of the expenditure of not more than 1.5 per cent of the annual federal funds by the states in making highway planning surveys and other important investigations. The first planning survey was inaugurated in 1935, and work of this nature was being conducted by all the states by 1940.

World War II focused attention upon the role of highways in national defense and funds were provided for the construction of access roads to military establishments and for the performance of various other activities geared to the war effort. Normal highway development ceased during the war years.

The Federal-Aid Highway Act of 1944 provided funds for highway improvements in the postwar years. Basically, the Act provided $500 million for each of three years. Funds were earmarked as follows: $225 million for projects on the federal-aid system; $150 million for projects on the principal secondary and farm-to-market roads; and $125 million for projects on the federal-aid system in urban areas. The law also required the designation of two new highway systems. One is the National System of Interstate Highways, which will be discussed later in this chapter, and the other is one composed of the principal secondary routes (Federal-Aid Secondary System).

The 1952 Act, for the first time, contained specific authorization of funds for the National System of Interstate Highways. In 1954, funds for the Interstate System were greatly increased, with apportionment based one-half on the traditional formula for the primary system and one-half on population. The Secretary of Commerce was directed to make studies of highway financing, with special attention to toll roads and to problems of relocating public utilities services. In 1955, the President's Advisory Committee on a National Highway Program (Clay Committee) reported the need for a greatly increased program for modernization of the Interstate System. Intensive efforts by all interested groups in 1955 and 1956 led to the passage of the Federal-Aid Highway Act of 1956.

1–5. Federal-Aid Highway Act of 1956. The Federal-Aid Highway Act of 1956 was a milestone in the development of highway transportation in the United States. It marked the beginning of the largest peacetime public works program in the history of the world. The Act contained many significant provisions—among the most important of which were the following:

1. It authorized completion, within a period of 13 to 15 years, of the 41,000-mile National System of Interstate and Defense Highways (see Art. 1–6). Federal funds to be expended on this system were set at $24.825 billion, beginning July 1, 1956, and ending June 30, 1969.

Federal funds were made available on a 90–10 matching basis; state expenditures were expected to be approximately $2.6 billion, making a total of more than $27 billion available for the program. Subsequent increases in costs have led to upward revision of estimates of the cost of completing the system. An estimate submitted to Congress in 1965 showed a total cost of $46.8 billion; by 1966 unofficial estimates put the cost over $50 billion.

2. Apportionment of funds for the Interstate System during the first three years was based upon the existing formula (two-thirds population, one-sixth area, and one-sixth rural road mileage). Subsequent state shares were allocated on the basis of remaining interstate needs, as determined by periodic studies made jointly by the Bureau of Public Roads and the various states.

3. The law also provided increased funds for the regular federal-aid systems (primary, secondary, and urban extensions). ABC funds for 1957 were increased from $700 million (authorized under the 1954 Act) to $825 million. The amount for 1958 was stepped up to $875 million. By 1964, ABC fund authorizations had risen to $1 billion per year. These funds are distributed on the traditional 50–50 matching basis.

4. The Act made apportioned federal-aid funds available for the acquisition of right-of-way on any of the federal-aid systems in anticipation of construction to take place within 5 years (later modified to 7 years).

5. Several new provisions were included which are applicable only to the Interstate System. One of these provisions denies federal funds to any state which permits vehicles with excessive weight and length to use the Interstate System. Approved design standards, including control of access, must be used, and the design was to be adequate to accommodate traffic expected in 1975. (In 1963, Congress amended the highway act to require designs to be based on 20-year traffic forecasts, dating from the time when the plans for each project are approved.) Toll roads could be included in the Interstate System, if suitably located and adequately designed. Congress indicated its intent to consider at a future time whether or not, and how, the states should be reimbursed for previously constructed toll and free portions of the Interstate System.

6. Title II of the 1956 Act, cited as the Highway Revenue Act of 1956, required revenues derived from taxes levied by the federal government to be earmarked and put into a special trust fund (the Highway Trust Fund).

Increases in the estimated cost of completing the Interstate System and a desire to complete the system within the time period originally established led Congress to increase highway user taxes in subsequent years. (See Art. 3–18 for detailed information on the financing provisions of the various highway acts.)

Meanwhile, the biennial federal-aid highway acts provided for improvement of the regular federal-aid systems and for roads on public lands.

An important provision of the 1962 Act was a requirement that highway agencies in all United States metropolitan areas with a population of 50,000 or more be engaged in a "comprehensive, cooperative and continuing transportation planning process" as a prerequisite to the initiation of federal-aid highway projects in those areas after July 1, 1965.

The 1964 Highway Act raised the level of authorizations for the ABC systems to $1 billion for each of two fiscal years (1966 and 1967). Of this amount, 45 per cent goes for federal-aid primary highways, 30 per cent for secondary roads, and 25 per cent for urban highways. The Act also provided $358 million for roads on public lands for the 2-year period. The 1966 Act authorizes the same level of ABC funds for fiscal 1968 and 1969.

In 1965, Congress authorized two additional highway improvement programs, both financed from the General Treasury and both outside the regular federal-aid program. One of these is a $1.2 billion, 5-year program to build 2350 miles of "developmental highways" in the 11-state Appalachian area, plus another $50 million to build access roads. The purpose of the program, in which the federal government pays 70 per cent of the cost of the developmental highways, is to spur economic development of the region. Congress also established a $320 million highway beautification program; the money is to be used to control billboards and junkyards adjacent to Interstate and primary highways, and for landscaping areas adjacent to highway rights-of-way.

Highway Systems in the United States

1–6. National System of Interstate and Defense Highways. There is no federal system of highways as such in the United States, i.e., no system of highways built and maintained by the federal government, except for relatively small mileages in national parks and forests, Indian reservations, etc. The closest thing to a federal road system is the Interstate System; however, selection, construction, and maintenance of these roads are prerogatives of the individual states.

Designation of the National System of Interstate Highways was authorized by the Federal-Aid Highway Act of 1944, with a limitation of 40,000 miles. The 1956 Act authorized an additional 1000 miles and amended the name to the National System of Interstate and Defense Highways (see Fig. 1–4).

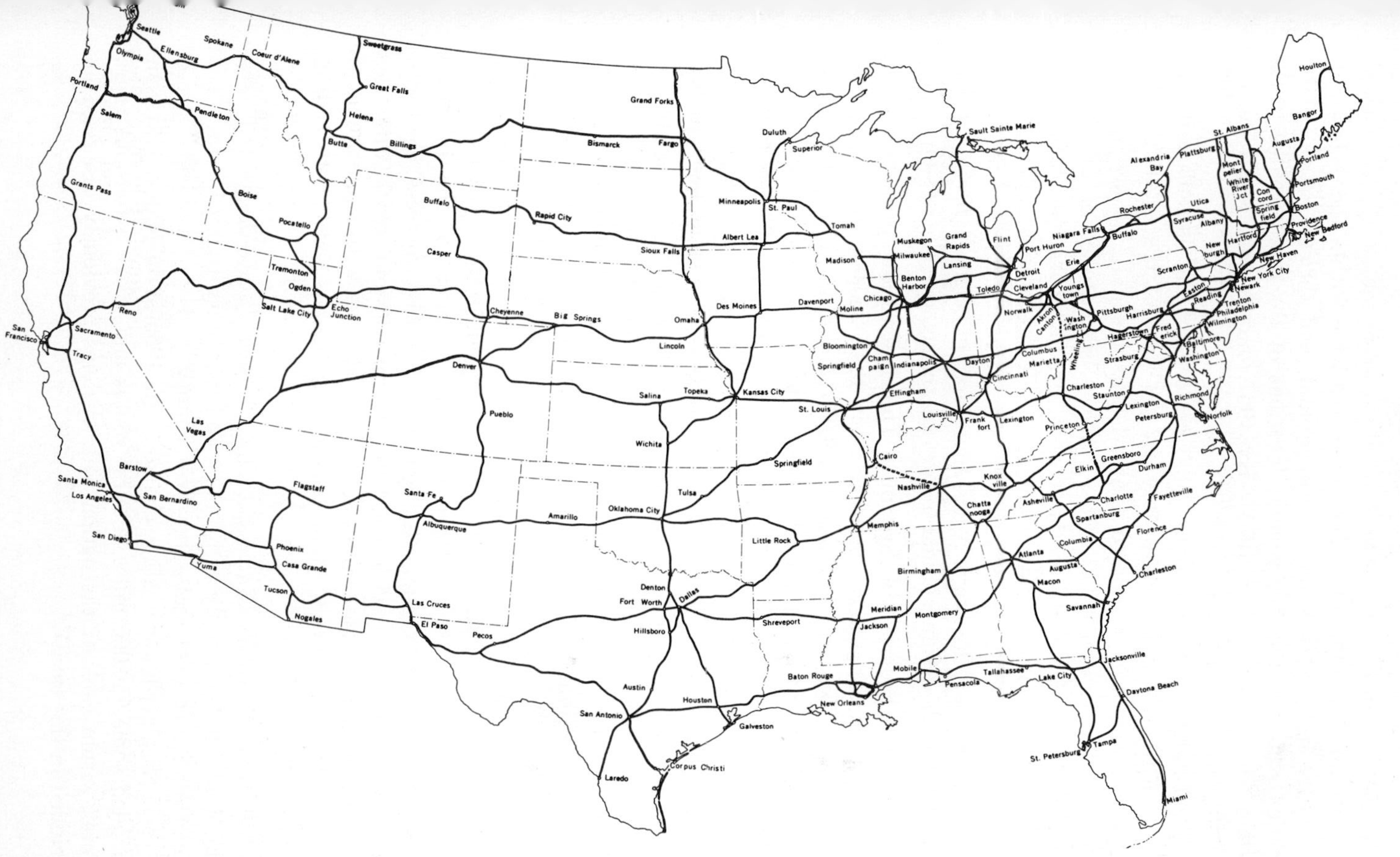

FIG. 1-4. National System of Interstate and Defense Highways. (Courtesy Bureau of Public Roads.)

In 1947, a nationwide network of city-to-city routes was designated. At that time, the remaining 2300 miles available within the 40,000-mile limit were reserved for routes into, around, and through cities. After detailed consideration, the general locations of the urban routes in the system were designated in 1955. By 1967, only a very small mileage of undesignated routes remained.

The Interstate System connects 90 per cent of the nation's principal cities. Although it comprises only 1.2 per cent of the total road and street mileage, it is estimated that when completed this key network will carry 20 per cent of all traffic. Access is controlled on all sections of the system. On most of the system, grade separations carry cross traffic over or under the interstate route. There are no railroad grade crossings, no traffic lights, and no stop signs. Many miles of the system comprise divided highways with four, six, or eight lanes of traffic. Uniformly high design standards prevail throughout the system.

1–7. Other Federal-Aid Systems. The federal-aid primary system with its urban extensions comprises more than 250,000 miles, including the Interstate System and nearly 28,000 miles in urban areas. The federal-aid secondary system, which is composed of feeder roads under state, county, and other local control, is made up of approximately 630,000 miles of roads, including urban extensions.

1–8. U.S.-numbered Highways. Every tourist is familiar with the comprehensive network of interstate highways which are designated as "U.S.-numbered Highways." These routes are marked by characteristic shield-shaped emblems bearing the initials "U.S.," the name of the state, and a route number. They have been designated by joint action of the various state highway departments, acting through the American Association of State Highway Officials. Highways included in this system are normally a part of the federal-aid and state systems. Some have been incorporated in or replaced by elements of the Interstate System. Routes traversing the country from east to west bear even numbers, while those with characteristic north and south directions have been given odd numbers. Among the more famous of these routes are U.S. 1, "from Maine to Miami," and U.S. 30, the "Lincoln Highway."

1–9. State Systems. Each state has its own "state-designated" system, made up principally of primary and secondary routes of state-wide importance. In some states, however, county roads and municipal streets have been placed under state control. In 1965 state highway systems included some 750,000 miles of rural routes and urban extensions (rural routes within each of the nation-wide systems previously discussed are included in this total).

1–10. Local Road Systems. By far the greatest percentage of mileage of roads in the United States is under the control of local governmental units. Included in this classification are more than 2,300,000 miles of county, township, town, and village roads. County systems are predominant, with approximately two-thirds of the above mileage being under county jurisdiction. Principal areas in which the county is not the primary unit of local road administration include the New England States and Pennsylvania.

Four states, North Carolina, West Virginia, Virginia (except for two counties), and Delaware, have assumed responsibility for all local roads. In a few other states some counties have transferred responsibility for local roads to the state highway department.

Certain routes under local control are included in the Federal-Aid Secondary System, as previously described.

1–11. City Streets. The picture of highway systems in the United States is completed by recognition of the fact that there exist more than 441,000 miles of city streets and alleys which are constructed and maintained by the individual municipalities.

1–12. Toll Roads. During the period 1950 to 1958, many toll roads or turnpikes were built and placed in service in various sections of the country. The second "toll-road era" of American history was brought about by the necessity for providing high-type facilities capable of relieving pressing traffic problems and corresponding inability to construct such facilities from current revenue. Toll roads were, in general, built by special authorities created by the various state governments. Revenue bonds were sold, and the income from tolls collected used to retire the bonds and interest charges.

The initial toll road of this era was the original Pennsylvania Turnpike, completed in 1940. Toll roads were built and placed in operation in Colorado, Connecticut, Florida, Illinois, Indiana, Kansas, Kentucky, Maine, Massachusetts, New Hampshire, New Jersey, New York, Ohio, Oklahoma, Pennsylvania, Texas, Virginia, and West Virginia. Total mileage of toll roads in operation was 3210 miles, built at an estimated cost of more than $5 billion.

With passage of the Federal-Aid Highway Act of 1956, interest in the construction of new toll roads waned in most states. However, toll road mileage has been built in the intervening years in several states, including Delaware and Maryland in addition to the original list. The purpose generally has been to speed construction of intrastate routes not on the Interstate System. By 1965, there were more than 3500 miles of toll roads in operation in the United States. About 2300 miles of these routes had been incorporated in the Interstate System.

Present Status of United States Highways

Highway systems in the United States have achieved a level of development without parallel in the modern world. The period from 1920 to 1941 was marked by the construction of an extensive system of two-lane rural roads, practically all of which were constructed with adequate, all-weather surfaces. The postwar years saw additional sizable improvements in the rural road systems. At the present time, only a very small percentage of the Federal-Aid Primary System remains unsurfaced. In the cities a large proportion of streets has been surfaced in one fashion or another. It is estimated that less than 8 per cent of the total mileage of city streets given earlier is not unsurfaced.

With regard to the vast mileage of county, township, town, and village roads, approximately two-thirds of the total 2,300,000 miles have been surfaced, generally with gravel or other low-type improvement. Many of these routes that are still unsurfaced are so remote and serve such a small volume of traffic that their improvement to any considerable extent will not be economically feasible or justifiable for many years to come.

In spite of the tremendous strides which have been made in the development of highways during the last several decades and the recent construction of thousands of miles of multiple-lane, limited-access routes, the country as a whole still faces a tremendous highway problem. Many thousands of miles of primary routes are inadequate and fail to meet the demands of modern traffic. Many roads have become obsolete due to inadequate traffic capacities, narrow pavements, undesirable curvatures, excessive grade, restricted sight distances, unprotected railroad crossings, dangerous intersections, or other faults. Thousands of highway bridges have their design capacities exceeded every day, while others are narrow or in dangerous locations. With increasing urbanization of the entire nation, United States cities have become focal points of the problem. There, traffic congestion and the parking problem have created a situation of far-reaching social and economic consequences. Highway construction and increasing reliance upon the private automobile for individual transportation have resulted in decentralization and new patterns of urban-suburban development in many United States cities. Difficulties encountered in maintaining the character and function of the central business district have led to increasing pressures for more urban highway construction and, at the same time in some cities, for development or improvement of rail transit systems.

Traffic fatalities on the nation's highways reached the shocking total of nearly 50,000 in 1965, and this carnage shows little sign of slackening. The Federal-Aid Highway Act of 1956 was a long step forward in the

. A. M. Fallenbuchl takes on e subject of "What's Wrong th the Polish Economy?" om 7-9 p.m. Feb. 12 in the ternational Institute, 111 E. rby.

DETROIT PUPPETEERS UILD — The Guild sponsors a cture by Dr. Robinson of the stitute of Arts on the life and orks of the late Paul Mcnarlin, the man who travelled e world collecting puppets nd who donated them to the stitute. Fri. Feb. 12 at 8 p.m. oom 156 Art Education Bldg., ayne St. Campus. Free.

Folk/Rock

ICTHUS COFFEEHOUSE — s live or die for Icthus. And in der to live, they've got to aise $400 this weekend. So ey've brought in **Michael Cooey**, the young man who stages one-man folk festival just by cking up his guitar, and who as, in addition, picked up the ritics' applause all the way om the New York Times and illage Voice to the Kentish mes of London. At 8 p.m. eb. 12 and 13. Adm. $2.50. 50 E. Jefferson.

BENEFIT DANCE CONCERT - The Muscular Dystrophy Asociation receives the proceeds the concert and with Iron orse Exchange, Nobody's ildren and Bornagain.

Livernois, on the corn McNichols.

PERCULATOR PIT — Banning, Jan Broughto others perform a kind Kind to Your Mind sort on Friday. Saturday, **Moore and Rita Cole** a happy foot stompers. F p.m. til 12 a.m. Adn 22395 Eureka.

AMANI COFFEEHOU **Dave Reske, John Roun** **Mary Foley** give you slants on the folk sound might just come away c all three. Feb. 13 fror p.m. til 12 a.m. Adm. $1 Mile and Gratiot.

history of United States highway development. Highway problems, however, will continue to challenge the best efforts of highway administrators, planners, and engineers for many years to come.

Role of the Highway Engineer

What is the role of the highway engineer in the highway program? The development of the modern highway system is largely dependent upon the efforts of the highway engineer, who must translate the desires of the people for better highway transportation into physical being. The engineer is concerned with and generally responsible for the planning and administration of the highway program, for the design and construction of highway improvements, and for maintenance of the highway system after it is built.

Modern highway development calls for the utilization of many specialists, men who are trained and experienced in such fields as traffic, roadway design, materials, bridge design, aerial photogrammetry, cost analysis, administration, and a host of others. The highway engineer still must perform surveys, prepare plans, and supervise construction. Yet, in a broader sense, he must be a man of vision, engaged in planning and construction on a tremendous scale, aware of his function in modern society and of the importance of highway improvement in the over-all march of modern civilization.

Outline of Highway Engineering

In this book an attempt has been made to consider the various aspects of highway engineering in the approximate order in which they might appear as factors in the improvement of a highway or a highway system, including those in urban areas. Attention has also been given to some of the problems encountered in urban areas. The principal divisions of the book include Administration, Planning, Design, Location, Construction, and Maintenance.

Thus, administration of the various United States highway systems is discussed first, followed by a chapter on economics and finance. These chapters serve as a background for the study of highway planning. Considerable space is then devoted to a discussion of the physical design of highways, including general and miscellaneous design features, soil engineering, and drainage. In the chapters immediately following are discussions relative to location surveys, the preparation of plans, estimates, and other necessary contractual documents.

In normal highway operations these steps lead to a fairly clear-cut dividing line, the line where location and design are complete and construction is ready to begin. The first construction processes discussed are those relative to earthwork, the performance of which must necessarily precede the construction of any wearing surface. Bituminous materials and the thickness design of flexible pavements are then considered as a background for the presentation of practice relative to the construction of various types of flexible pavements and bases. Chapters are then devoted to various types of flexible surfaces, ranging from low-cost treatments to high-type bituminous pavements. Attention is then given to the design and construction of rigid pavements (portland cement concrete). Finally, a chapter is devoted to the very important subject of maintenance, including snow removal and ice control.

CHAPTER 2

HIGHWAY ADMINISTRATION

Administration of public highways in the United States is a governmental function, responsibility for which is delegated, in whole or in part, to appropriate agencies of the federal government, to the various state governments, and to numerous local governmental units. Although there is and must be a separation of function and responsibility among the governmental units mentioned, close contact and cooperation among all governmental units concerned with road and street construction is a necessity if the development of the over-all highway system is to proceed in an intelligent, adequate, and economical fashion.

The Bureau of Public Roads

The general nature and extent of the federal-aid program for highways have been discussed in the preceding chapter. The Bureau of Public Roads, a division of the Department of Transportation, acts as the representative of the federal government in all matters relating to public highways. The principal functions of this organization are the administration of the federal-aid program and of highway construction projects on federally held lands. The Bureau of Public Roads also engages in extensive programs of highway planning, research and development, and highway safety. The agency represents the United States government in various programs of technical assistance to foreign countries and administers construction of the Inter-American Highway in Central America.

2–1. Organization. The Bureau of Public Roads is headed by the Federal Highway Administrator, under whom the Director of the Bureau of Public Roads serves, as do the chief engineer and the heads of the offices of the general counsel, audits and investigations, and administration. The Washington headquarters staff is responsible for policy formulation and general direction of public roads operations in engineering, finance, management, and legal fields.

The Bureau of Public Roads has nine regional offices located across the country, each supervising the federal-aid program in from four to eight states. The regions have division offices in every state of the Union

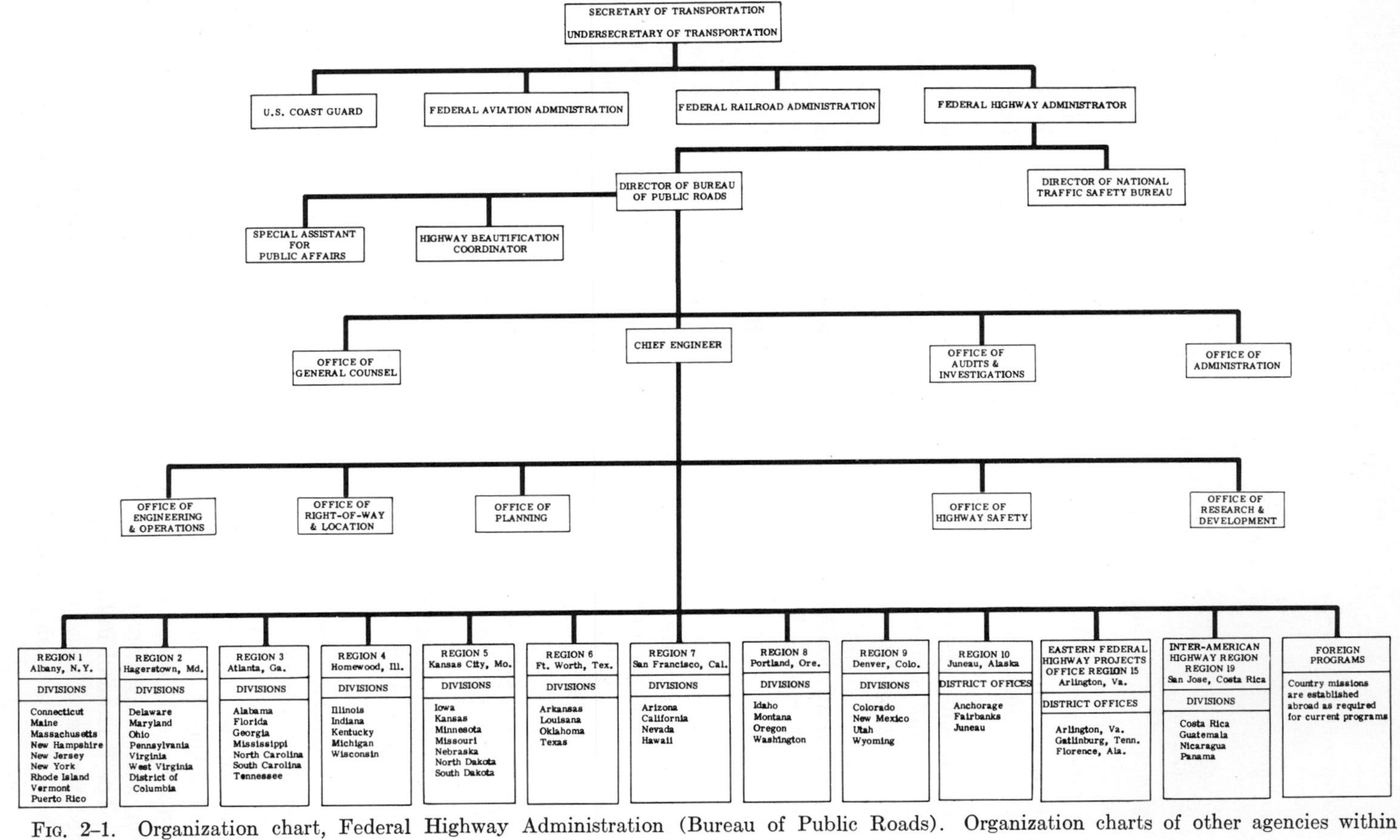

FIG. 2–1. Organization chart, Federal Highway Administration (Bureau of Public Roads). Organization charts of other agencies within the Department of Transportation not shown. (Courtesy Bureau of Public Roads.)

(except Alaska), Puerto Rico, and the District of Columbia. It is through this field organization that most relations with the state highway departments are carried on. A regional office handles federal-aid work in Alaska, where the work differs somewhat from federal-aid work in the continental United States. Organization of the headquarters staff of the Bureau of Public Roads is shown in Fig. 2–1.

2–2. Administration of the Federal-Aid Program. Some of the principal functions of the Bureau of Public Roads in its administration of federal aid to the states are described in the following paragraphs:

AUTHORIZATION AND APPORTIONMENT. The history of a federal-aid project begins with the authorization by Congress of federal funds for highway construction, generally made on a biennial basis for the ABC programs and at irregular intervals for the Interstate System. The federal funds for each year are apportioned among the states, usually from six to twelve months before the beginning of the indicated fiscal year, and are available for that fiscal year and for two years thereafter. Within this time limit the funds apportioned to a state must be committed to specific construction work by formal agreement between the state and the Bureau of Public Roads; once such commitments are made, the actual expenditures may continue beyond the three-year limit. Apportionments to the states are made according to formulas provided by law. Amounts available to the states also are limited by mandatory operation of the program on a pay-as-you-go basis within the limits of revenue available to the Highway Trust Fund.

PROGRAMING AND SUBSEQUENT STEPS. Knowing the amount of federal aid available to it in each fiscal year, each state draws up programs of projects to be built with those funds, based on preliminary surveys and cost estimates. Since there is close contact between the Bureau of Public Roads and the states, questions as to the selection of projects and their location, design, and cost are settled in the early stages of the program planning.

After acceptance of the programs by the Bureau of Public Roads, the state makes detailed surveys for each project and prepares plans, specifications, and final cost estimates. These must be approved by the Public Roads division engineer. The state then advertises for bids and awards a contract for construction, subject to concurrence by the Bureau of Public Roads. The day-to-day supervision of the work is performed by state engineers resident on the job, but a Public Roads engineer or investigating team makes periodic inspections on the site.

Public Roads does not prescribe detailed design standards and construction specifications for federal-aid work. The general policies of design standards adopted by the American Association of State Highway

Officials (AASHO) are used by most of the states and are endorsed by the Bureau of Public Roads. Details of design and construction specifications are prepared by each state to fit its individual needs, and are subject to review and approval by the federal bureau.

PAYMENTS. As the work progresses on a project, the state periodically pays the contractor and claims reimbursement from Public Roads for the allowable federal share of the cost. When the project has been completed, inspected, and accepted, a final detailed accounting of costs is made by the state and reviewed by Public Roads auditors, after which final payment of the federal share is made. Thus the federal-aid funds are needed over a period of time rather than all in advance, and the federal aid for each fiscal year is initially made available only as an authorization. After completion of a federal-aid project, Public Roads engineers continue to inspect it at intervals to see that it is being properly maintained.

2–3. Road Building by the Bureau of Public Roads. In addition to its work in the supervision of the federal-aid program to the states, the Bureau of Public Roads cooperates with other federal agencies in the construction of roads in national parks and forests, Indian reservations, and other federally controlled lands. These projects are generally constructed entirely with federal funds.

Many routes of outstanding engineering accomplishment and scenic beauty have resulted from this work of the federal agency. Included among these are the Blue Ridge Parkway in Virginia and North Carolina; the Trail Ridge Road in Rocky Mountain National Park, Colorado; and the Natchez Trace Parkway, now under construction from Nashville, Tennessee, to Natchez, Mississippi.

2–4. Other Activities of the Bureau of Public Roads. Research was one of the purposes for which the Office of Road Inquiry was created in 1893 and is one of the important functions of the Bureau of Public Roads today. In its own offices and laboratories, and in cooperation with the state highway departments, universities, and other organizations, both public and private, Public Roads conducts research and development programs in five major areas: economics and fiscal requirements, engineering systems, materials, structures and applied mechanics, and traffic systems.

Highway research and development programs were stimulated by the Federal Aid Highway Act of 1962, which provided for mandatory, rather than permissive, use of 1½ per cent of the funds apportioned to each state for ABC and Interstate projects, for planning and for research and development projects. An additional ½ per cent of ABC funds may be used by the states for such projects. In 1962, also, BPR,

AASHO, and the Highway Research Board inaugurated the National Cooperative Highway Research Program, which involves the expenditure of about $2.5 million a year in a planned program.

Research activities of the Bureau of Public Roads are very broad in scope, involving transportation and economic studies, physical research of many kinds, design investigations, and many other activities. Work of the Bureau has led to many advances in design, construction, and management. Some of the results of this research are disseminated through the magazine *Public Roads*.

Planning activities of the Bureau of Public Roads were accelerated by the highway acts of 1956 and 1962, which thrust the agency headlong into the field of urban transportation planning. The agency's activities include future planning (with the states, BPR in 1965 began planning for a national highway program after 1972, the planned date of completion of the Interstate System); execution of the highway cost allocation study, which recommended (in 1961 and 1965) the manner in which highway costs should be allocated among the various classes of highway users; studies of planning techniques; and collection and dissemination of highway statistics.

In 1964 and 1965, President Johnson placed emphasis on three aspects of the highway program: reduction in traffic accidents, more aesthetics in highway location and design, and development of a scenic highway system. In all of these activities, most of which are carried on through the regular federal-aid highway program, the Bureau acts as the agent of the federal government.

State Highway Departments

State governments, through the various state highway departments, occupy the key position in the development of highway systems in the United States at the present time. The relation of the Bureau of Public Roads to highway development has already been discussed, but it must be emphasized that federal-aid programs are undertaken at the option of the individual states and that the states are responsible for the design, construction, and maintenance of routes constructed with federal participation, subject only to review and approval by the Bureau of Public Roads.

In the majority of states, the state highway department (or some comparable agency) has the responsibility for the development of roads included within the state-designated system, which is generally made up of primary and secondary routes of state-wide importance, including urban extensions. As has previously been mentioned, certain states have assumed responsibility for all local roads.

2–5. Functions of State Highway Departments. Functions of the various state highway departments vary widely because of differences in the laws, directives, and precedents under which they are organized and operated. However, the principal function of any state highway department is the construction and maintenance of highways contained within the state-designated system, including those built with federal participation. The departments are also usually responsible for the planning and programing of improvements to be made on the various components of the state highway system and for the allocation of state funds to the various local units of government. They are also usually charged with the duty of distributing information regarding highways to the general public, and they may be responsible for motor vehicle registration. In some states the highway department exercises police functions relative to safety on the highways, usually through the state highway patrol.

2–6. Organization of State Highway Departments. As might be expected, state highway departments vary a great deal in their administrative organization. A comprehensive study (*68*)* of state highway administrative bodies showed that at that time in 12 states the state highway agency was administered by a single executive; 7 others had the single-executive type of organization, with boards or commissions acting in advisory or coordinate capacities. The remaining state highway departments were administered by commissions. In 9 of these states the commissioners exercise full administrative control, while in the others the commissioners had limited administrative control (generally limited to determination of policy). Changes occur in individual state organizations from time to time. For example, in 1965, Michigan went from a single-executive form, in which the state highway commissioner was an elected official, to a four-man, bipartisan commission appointed by the governor.

Almost without exception the members of the administrative bodies of the various state highway departments, including those of the single-executive form, are appointed by the governor of the state concerned. In many states such appointments are subject to approval by the state senate or some similar legislative body.

Detailed organization charts of the different state highway departments also vary greatly. Some departments have changed their administrative organizations with the passage of time, while others have retained practically the same organizational form for many years. Organization charts of two state highway departments, California and Texas, are shown in Figs. 2–2 and 2–3. The California form of organiza-

* References are to items in the Bibliography at the end of the book.

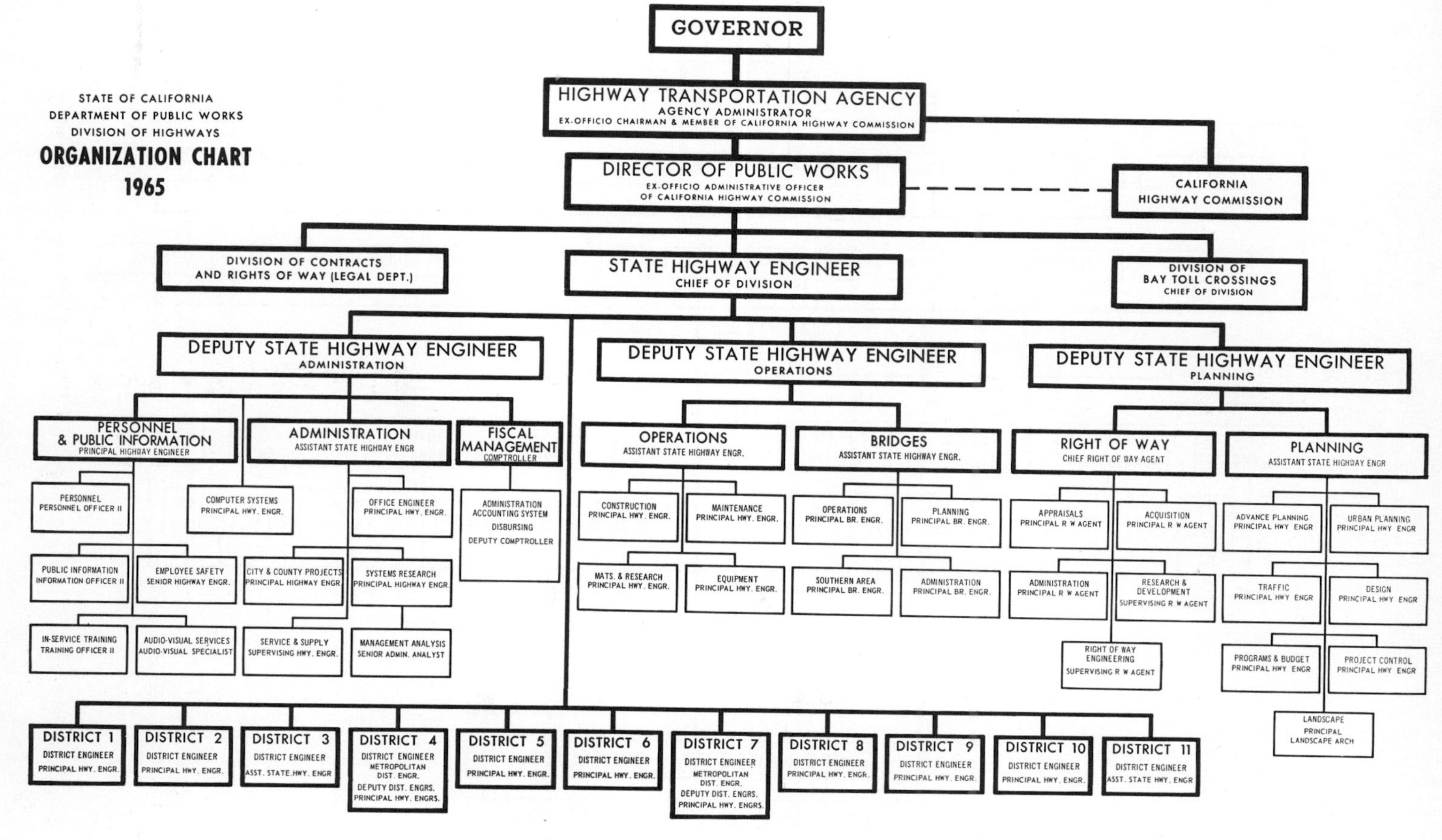

FIG. 2–2. Organization chart of the California Division of Highways. (Courtesy California Division of Highways.)

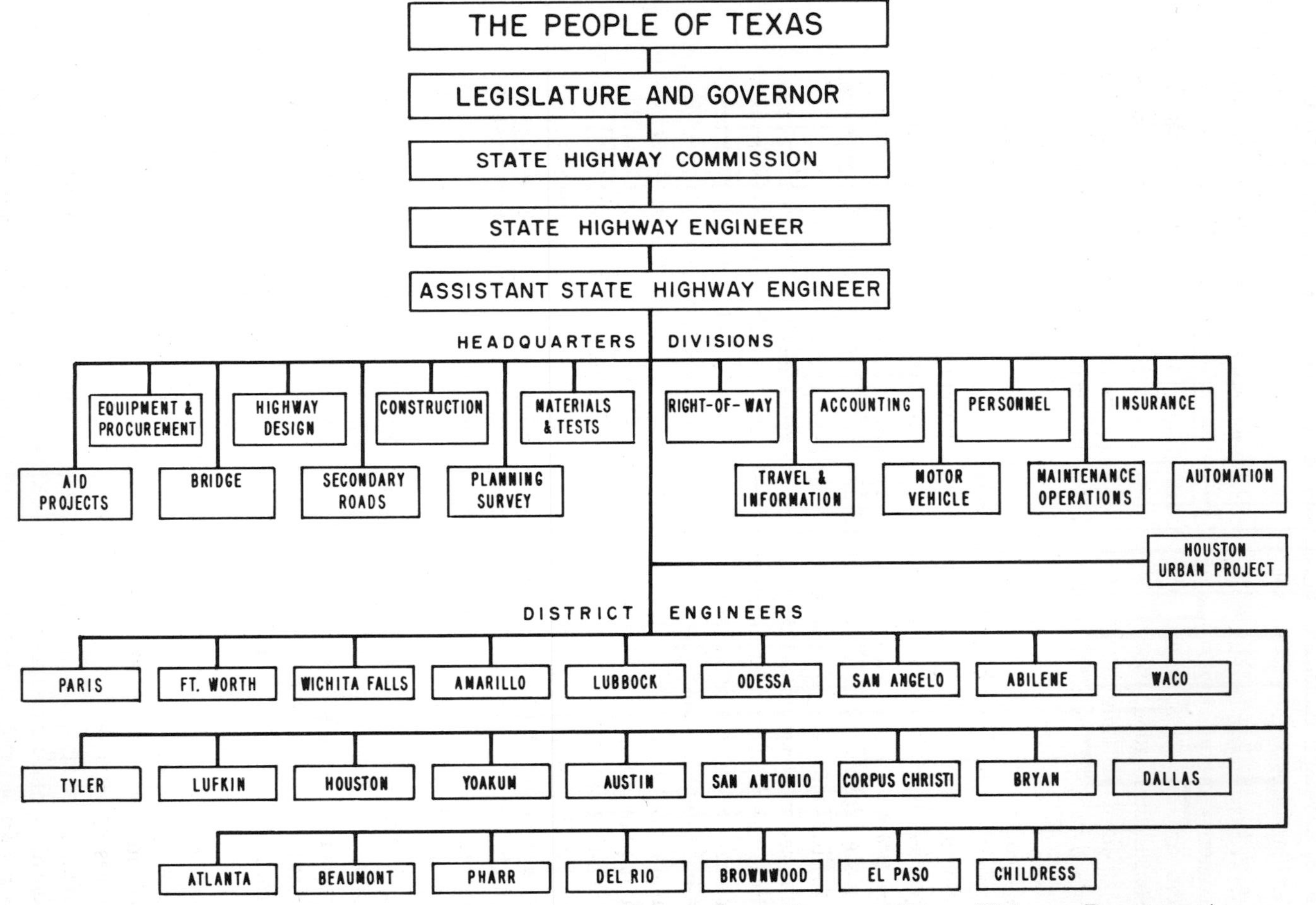

Fig. 2–3. Organization chart of the Texas Highway Department. (Courtesy Texas Highway Department.)

tion is a "single executive with coordinate commission"; that of Texas is an "administrative commission."

2–7. Recommendations for Good Administration. State highway departments are currently engaged in truly "big business," as most of them are operating under budgets of many millions of dollars. It is essential that good, modern organizational and management procedures be utilized in their administration; that sound business practices be applied in their operation; that modern engineering standards be applied to the design and location of highways within their jurisdiction; and that construction methods and materials utilized in road building be properly selected and controlled. Only through application of these principles will the public receive the maximum benefit from the funds expended on highways. In addition, it is axiomatic that, insofar as it is practicable, political influences must be removed from the conduct of state highway affairs in order to ensure the continuing improvement of the highway system on a sound, well-planned basis.

In recent years, state highway departments have made extensive use of business machines and electronic computers to aid in the management process. Such machines are used for payroll and accounting purposes, for bid tabulation and analysis, and for systematic analysis of the information needed by top administrators in making management decisions about planning and programing.

Such analytical approaches as the critical path method (CPM, see Art. 10–40) are being used as management tools. An administrator can use such a program to develop a plan for the best utilization of available funds, manpower, and time, taking into consideration such basic factors as project descriptions and costs, project priorities, and revenue schedules.

2-8. Personnel Needs of State Highway Departments. In order to conduct the highway program properly, it is essential that highway departments be staffed with an adequate number of skilled professional engineers. Many state highway departments have found it difficult to attract and hold qualified engineers during the postwar years because of various factors. Factors which continue to hamper recruitment of young engineers in some states include political interference with engineering personnel, low salaries, lack of security, and lack of a planned program of advancement. With industry actively competing for the services of engineers already employed in highway work and for engineering graduates, many state highway departments have altered and modernized their personnel policies. Efforts have been made to increase salaries to a level where they are commensurate with responsibility; to provide a civil service or merit system for the protection of engineering employees; to establish definite job classifications; and to provide for the regular advancement of engineers both in responsibility and pay.

Realizing the necessity for direct action to attract engineering graduates to highway work, the Bureau of Public Roads and several state highway departments have moved to establish definite training programs designed to give the young engineer a background in all phases of their work and to prepare him for a position of responsibility in the organization. Under these plans the engineering graduate is systematically moved from job to job and given additional training within the ranks of the organization until he is ready to assume additional responsibility in the particular branch of highway engineering for which he is best suited. The program of the Bureau of Public Roads has been an outstanding success; since World War II, more than 90 per cent of the trainees who completed the program are still employed by the Bureau. Partially to offset engineering shortages, many highway departments are making extensive use of consulting engineers.

Local Road Administration

As has been previously pointed out, some 2,300,000 miles of highways are currently under the administration of local authorities. Administration of this tremendous mileage of local roads presents an extremely varied pattern.

2–9. Types and General Features. In most sections of the country, the county is the prevailing unit of local road administration. Exceptions to this exist principally in the New England states and Pennsylvania where the township continues to be the most important unit having responsibility for local roads. In some of these states there are township systems of "feeder" roads, tributary to the main county routes.

In considering the administration of local roads by the important county unit, considerable differences in practice are also to be noted. In some counties road affairs are administered for the county as a unit; this is the so-called "county-unit" plan. In others, responsibility for roads is divided among various precincts or commissioners' districts, each with its own funds, personnel, and equipment. Extreme variations in area (from a few square miles up to more than 20,000 square miles in one county in California) and population (from a few hundred or less up to more than 5 million in Cook County, Illinois, and over 6 million in the City and County of Los Angeles) further account for the various types of county road administrations.

It is generally believed that the county-unit plan is the most efficient unit of local road administration. Advantages claimed for the county-unit plan over smaller units of local road administration include the creation of a more efficient and capable organization, a more equitable distribution of the tax burden, higher standards of engineering and construction, and increased ability to cooperate with state highway departments. There has also been some indication that consolidation of the road-building activities of adjacent counties may be advisable in certain instances where the individual counties do not possess sufficient resources to purchase and maintain adequate equipment or to secure necessary engineering services.

County engineering organizations are as varied in character and quality as the county administrative organizations. Generally speaking, the larger and more populous counties have entirely adequate engineering organizations, many of which are on a par with those of the state highway departments. Many other counties have inadequate engineering forces, while hundreds of counties operate with little or no engineering aid. For the country as a whole, only about 900 of the more than 3000 counties have county engineers.

The Federal-Aid Secondary Program, with its large allocations of funds for the secondary road program, has stimulated local road affairs and led many counties to establish engineering organizations capable of cooperating with the state and federal agencies in the planning, design, and construction of local roads.

The following criteria have been enumerated for good highway administration at the county level (*247*):

A road commission whose functions are limited to policy formulation, with an engineer responsible for all management functions.

Maintenance of adequate and uniform records for guidance in management, for developing and carrying out a long-range road improvement program, and for use in making such reports as the state legislature may direct.

Cooperation with other county and city agencies in the development of integrated regional highway programs.

Operation under an annual improvement program, based on a long-range plan and related to a definite annual budget.

Preparation of construction plans and specifications for all contractual work.

Maintenance of roads and bridges to uniform standards.

Extension of contractual highway services to small cities and villages upon request.

Installation of traffic-control and safety measures in accordance with uniform standards.

Adoption of uniform accounting practices.

Adoption of standard specifications for purchasing, in accordance with sound business practice.

Municipal Street Administration

Administration of the thousands of miles of city streets in this country is also quite variable in nature. This is not surprising when it is realized that this mileage exists in thousands of incorporated communities, the large majority of which have populations of less than 5000.

Generally speaking, responsibility for city streets is vested in the city engineer, director of public works, city manager, or some similar official, depending on the form of government used. Functions of the city street department, or similar body, include design and construction of street improvements, formulation of long-range plans for street improvement, the making of traffic surveys, installation of traffic-control and safety devices, and the supervision of maintenance.

As in the case of the counties, street departments in the larger cities operate at a high level of efficiency, performing all necessary engineering functions. In many smaller cities, engineering organizations are inadequate or nonexistent. However, it may be noted that there is an increasing appreciation of the desirability of providing engineering control in the planning, design, and construction of city streets, with the result that many municipal engineering organizations have been created or expanded in recent years.

Necessity for Interagency Cooperation

It is recognized that the various components of the highway system are not separate entities but are, or must become, integrated parts of a coordinated system if highway development is to keep pace with the demands of the modern motorist. In recognition of this fact there must be close cooperation among all the units of government concerned with highways and streets, including the federal government, the states, counties, and cities.

The federal-aid program has been, and is continuing to be, a potent force working toward this end. Construction of the Federal-Aid Primary System has brought the federal government into close contact with the states. Later federal-aid acts have necessitated close cooperation among the federal, state, and county governments in the designation and improvement of the Federal Aid Secondary System. Likewise, activities conducted under the Federal-Aid Urban Program have brought about closer relationships among representatives of federal, state, county, and city governments. Close cooperation of all agencies is involved in the planning and construction of the National System of Interstate and Defense Highways. Interagency cooperation in urban transportation has been stimulated by the planning requirements of the 1962 Highway Act, as explained previously.

In all these activities the state highway departments are the key agencies. They are usually responsible for the distribution of state funds to the various units of local government; they are in a position to supply the technical knowledge and engineering skill so badly needed in many local agencies; they possess facilities for performing special surveys and analyses, such as urban traffic studies; and they possess a wealth of information which should be made available to local units of government for planning and programing purposes. In view of these and other facts, several state highway departments have established special county (secondary road) and urban divisions to cooperate more closely with the appropriate local governments and to further the development of an integrated highway system. In many urban places, cooperative transportation planning agencies have been established.

Related Organizations

In addition to the governmental agencies described, many other organizations contribute to the development of the highway system. One of the most important of these groups is the American Association of State Highway Officials (AASHO). The membership of this group consists of the principal executive and engineering officers of the various state highway departments and the Bureau of Public Roads. The principal objective of the Association is the fostering of "the development, operation, and maintenance of a nation-wide system of highways to adequately serve the transportation needs of our country." Work of the various committees of the Association has resulted in a large number of publications which are widely used by other groups.

The American Road Builders' Association (ARBA) is another large and powerful group in the highway field in this country. The membership of ARBA comprises seven divisions, including persons and organizations from every branch of the highway industry and representatives from many foreign countries. As a result of the work of the divisions and the committees included therein, the organization issues technical bulletins and reports at intervals; it also publishes a monthly magazine, *The American Road Builder*. ARBA is active in furthering highway legislation and in disseminating information about highways to the general public. ARBA holds annual and regional meetings, with a "Road Show" that is held at periodic intervals.

The Highway Research Board, which has been a vital factor in the advancement of technical knowledge in the last 50 years, is a division of the National Academy of Sciences–National Research Council. The board does not, in general, perform research itself, but has its primary function in the correlation of research efforts of educational institutions, governmental agencies, and industry. The board holds an annual meeting and issues various publications.

Many other groups are engaged in activities related to highway development, including organizations of highway users, materials producers, equipment manufacturers and dealers, automotive manufacturers, engineering groups, colleges and universities, and many others. Many such groups could be named and publications of some of them have been cited at various points in this book.

PROBLEMS

2–1. Determine how state highway affairs are administered in your state. Is the highway department administered by a single executive or by a commission? Draw an organization chart for the headquarters organization, showing lines of responsibility for the various primary functions of the department. (Consult the annual or biennial reports of the highway department in your college library.)

2–2. How are local road affairs administered in your state? Is the county, township, or some other form of local government responsible for roads? Is there an engineer responsible for such functions as design, construction, and maintenance in the place in which you live or attend school? If so, draw a typical organization chart for this engineering organization.

2–3. Has your state highway department set up special divisions to cooperative with local governmental units in the improvement of secondary roads and urban extensions of state routes? If so, briefly describe the functions of these divisions.

2–4. Prepare a brief report describing an important project now under construction (or recently completed) in your state which required close cooperation among representatives of federal, state, and local highway agencies.

2–5. Prepare a report outlining the organization and functions of the Highway Research Board.

2–6. Prepare a report describing the organization, membership, and activities of the American Association of State Highway Officials. List the current books of standards and policies adopted by this association. What do you think is this group's most valuable contribution to the highway program?

2–7. Determine who is responsible for the planning, design, construction, and maintenance of streets in your community. How much money is expended in a typical year? What are the sources of street funds? Is there a long-range program of street improvement? If so, how did the program come into being? Briefly outline the main points of the program.

2–8. Determine what your state highway department is doing to attract and hold engineering graduates. Is there a formal training program for the young engineer? A cooperative program of summer employment with contractor groups? Some other type of cooperative program? Write a report on your findings.

2–9. In many states, extensive use has been made of consulting engineers in the expanded highway program. Has this been done in your state? If it has, describe one major project in which consulting engineers played a major role, and explain the functions which they performed on that project.

2–10. Prepare a brief report outlining the duties and responsibilities of one of the following persons:

(a) A project engineer employed by a state highway department on a project which is being carried out by contract.

(b) A construction superintendent or project manager for a contractor on state highway department work.

(c) A county engineer.

(d) A city engineer or city manager.

2–11. Investigate and report on the use that your state highway department is making of modern business machines and methods. Are such programs as CPM or PERT used in planning and programing? If so, how?

CHAPTER 3

HIGHWAY ECONOMICS AND FINANCE

Among the most important fields of the many-sided subject of highway engineering are those of highway economics and highway finance, including the field of taxation for highway purposes. These two fields, which are of essentially equal importance, may be treated as separate subjects, but are here discussed together because of their fundamental similarity and close relationship. Generally speaking, this relationship may be visualized by the statement that the principles of highway economics must be employed in order to ensure that public funds, obtained by highway financing methods, are expended in a fashion which will guarantee the maximum possible benefit to all concerned.

As is the case with many of the subjects covered in later sections of this text, space will not permit the detailed discussion of all the principles and factors involved in the application of engineering economics to highway improvement or those relating to highway finance and taxation. This fact will necessitate some simplification of the material presented, with emphasis being placed upon certain principles of fundamental importance.

3–1. Economic Studies. Although costs have long been a factor in various decisions made by highway engineers and officials, formal economic studies are of comparatively recent vintage in most highway agencies.

Some of the data needed for reasonably complete analyses have been lacking. And data now available are by no means complete. For example, much more information is needed about the operating costs of trucks and buses, particularly in urban areas. Intensified research is filling some of the gaps, but there is still a long way to go.

However, increasing amounts of money available for highways; increasing complexity of highway location and design problems, particularly in urban areas; greater awareness of the socioeconomic impact of highway improvements; and a growing necessity for comparing various forms of transportation in order to develop balanced urban transportation systems have focused attention on economic studies.

Economic studies for highway purposes principally are done for one or more of these reasons:

1. To compare alternate locations
2. To evaluate various features of highway design, e.g., the type of surface to be used
3. To determine priority of improvement
4. To allocate responsibility for the costs of highway improvement among the various classes of highway users (and nonusers, in some cases)
5. Occasionally, to compare proposals for highway improvement with proposals for other public projects such as education

Economists disagree as to both general concepts and detailed methods of computation to be applied to highway problems. Most of the state highway departments in this country use the concepts and methods recommended by the American Association of State Highway Officials (*105*). The AASHO "Red Book" advocates use of the benefit-cost ratio (Art. 3–12) in economic calculations designed to compare design alternates or alternate locations of the same general highway facility; the method is not suitable for direct comparison of projects that have "dissimilar traffic, terrain and design conditions" or for priority determination of projects on an area- or state-wide basis.

Two other methods of making economic calculations are considered briefly in this chapter: the "present worth" method and the "rate of return" method (Art. 3–17).

Costs of Highway Transportation

The costs of highway transportation may be divided into two broad groups: (1) road costs and (2) road user costs. Both are usually expressed on an annual basis. Thus, road costs may be generally defined as the annual costs accruing from the construction, maintenance, and operation of an individual highway or highway system. Similarly, road user costs are those annual costs accruing from the ownership and operation of vehicles over the highway system or individual units thereof. Both road and road user costs are of primary importance to the highway engineer. This is true because it is an axiom of highway economics that, generally speaking, since the same people pay both the road costs and the road user costs, maximum economy is secured only when the sum of these two costs is a minimum consistent with convenience and safety.

3–2. Annual Road Cost. The annual road cost is the sum of the annual capital cost plus the annual cost of maintenance and operation of the highway and its pertinent structures.

The annual capital cost is the yearly amount required to amortize the total cost of the improvement plus interest. The capital cost for the highway improvement is the total first cost or initial investment, including the cost of engineering, right-of-way, grading and drainage, pavement, and major structures (such as culverts and bridges). The total cost of an improvement is usually an engineer's estimate based upon preliminary plans.

The total annual road cost may be computed as follows (*105*):

$$R = [(C_1 \cdot K_1) + (C_2 \cdot K_2) + (C_3 \cdot K_3) + (C_4 \cdot K_4)] + M \quad (3\text{–}1)$$

where R = total annual road cost

C_1, C_2, C_3, C_4 = the capital costs of the various items which make up the improvement

A common breakdown is costs for (1) right-of-way, (2) grading, drainage, and minor structures, (3) major structures, and (4) pavement and appurtenances. Breakdown of this type is necessary because of the variation in useful lives of the various elements.

K_1, K_2, K_3, K_4 are capital recovery factors for a given rate of interest and amortization of total cost of each element based upon its average life. The factor may be determined from Table 3–1 for commonly used rates of interest and amortization periods. The selection of a proper rate of interest is a debatable point, with various organizations and economists utilizing rates from 0 to 8 per cent. The 1960 version of the AASHO Red Book contains examples with interest rates of from 3.5 to 6 per cent. E. L. Grant, of Stanford University, suggests an interest rate of 7 per cent for highway calculations (*19*). Robley Winfrey, of the Bureau of Public Roads, recommends 5 to 8 per cent (*279*). By contrast, an interest rate of 0 per cent is used by the State of California.

The question of salvage value (the value of a depreciable facility at the end of its estimated useful life) sometimes arises in economic calculations for highway purposes. In general, a salvage value of zero may be assumed for calculations of this type. Often an existing highway has no real value at the time it is improved or replaced. In comparing alternate locations of new routes there is no existing facility to be salvaged. If an existing facility does have a salvage value (such as the value of the land in the right-of-way), this fact can be considered in the calculations.

M equals the annual cost for maintenance and operation of the improved highway. This item is estimated by study of actual costs for similar highways and conditions.

The computation of total annual road costs based on the summation of the annual costs of the individual elements of improvement and their average lives is the proper and accurate method. For quick analyses and for comparisons of projects which have similar ratios of costs of individual items to total cost, an estimate of total over-all cost and over-all average life may provide sufficient accuracy. In this case, Eq. 3–1 reduces to $R = C \cdot K + M$, in which C is the total cost of all elements of improvement and K is the capital recovery factor corresponding to the over-all average life of the improvement.

TABLE 3–1

CAPITAL RECOVERY FACTORS TO BE USED IN COMPUTING ANNUAL ROAD COSTS

Years	Rate of Interest, r, Per Cent								
n	0	2.0	3.0	4.0	5.0	6.0	7.0	8.0	10.0
5	0.2000	0.2121	0.2183	0.2246	0.2310	0.2374	0.24389	0.25046	0.26380
10	0.1000	0.1113	0.1172	0.1233	0.1295	0.1359	0.14238	0.14903	0.16275
15	0.0667	0.0778	0.0838	0.0899	0.0963	0.1030	0.10979	0.11683	0.13147
20	0.0500	0.0611	0.0672	0.0736	0.0802	0.0872	0.09439	0.10185	0.11746
25	0.0400	0.0512	0.0574	0.0640	0.0709	0.0782	0.08581	0.09368	0.11017
30	0.0333	0.0446	0.0510	0.0578	0.0651	0.0726	0.08059	0.08883	0.10608
40	0.0250	0.0365	0.0433	0.0505	0.0583	0.0665	0.07501	0.08386	0.10226
50	0.0200	0.0318	0.0389	0.0465	0.0548	0.0634	0.07246	0.08174	0.10086
60	0.0167	0.0283	0.0361	0.0442	0.0528	0.0619	0.07123	0.08080	0.10033
70	0.0143	0.0267	0.0343	0.0427	0.0517	0.0610	0.07062	0.08037	0.10013
80	0.0125	0.0252	0.0331	0.0418	0.0510	0.0606	0.07031	0.08017	0.10005
90	0.0111	0.0240	0.0323	0.0412	0.0506	0.0603	0.07016	0.08008	0.10002
100	0.0100	0.0232	0.0316	0.0408	0.0504	0.0602	0.07008	0.08004	0.10001

K = capital recovery factor, the annuity for n years which 1 will buy or the annuity needed to discharge a debt of 1 in n years with interest.

$$K = \frac{r(1 + r)^n}{(1 + r)^n - 1}$$

where r = rate of interest,
n = number of years.

In most cases the annual capital costs may be determined directly for the entire length of the highway, without dividing it into sections. An exception occurs when economic calculation involves a highway which is to be brought to a level of partial improvement (stage construction) over a portion of its length.

It must be noted that the periods of time used in computing annual road costs usually are not the same as those used in traffic forecasts upon which road user costs and benefit-cost analyses are based. In many cases, average life periods for capital costs are longer than traffic forecast periods. For road user benefit analyses, such as those covered in a later section of this chapter, it is not necessary that these periods be the same. Use of a shorter traffic forecast period for the analysis makes the resulting benefit-cost ratio correct insofar as the traffic data can be established. At the end of the period of analysis, the improvement still has value since the complete capital cost items have not yet been amortized. Expression of the capital cost as the sum of the annual costs of construction and maintenance gives a proper value for any period of time. If a highway with a low-type surface is involved, where the life of the surface may be less than the traffic forecast period, the shorter period of time should be used.

The period of time to be used for the computation of annual costs for each element of capital expenditure is a matter for decision by the agency involved. Many organizations use a time of 50 or 60 years for right-of-way costs and 40 years for costs of grading, drainage, and structures. For surfaces a period of 20 years is frequently used; the more accurate determination of the useful life of road surfaces is one of the major objectives of the road-life studies described in the next paragraph.

3–3. Estimates of Service Lives of Road Surfaces. An estimate of the average life of the road surface (pavement) is necessary to the

TABLE 3–2

Average Lives of Highway Surfaces

Surface Type	Average Life (Years)
Soil surfaced	4.0
Gravel or stone	7.5
Bituminous surface treated	11.7
Mixed bituminous	12.5
Bituminous penetration	18.5
Bituminous concrete	20.3
Portland cement concrete	27.0
Brick or block	19.6

economic calculations of Art. 3–2. So far as surfaces are concerned, estimates are largely based upon data collected in the road-life studies conducted by the various state highway departments for surfaces of various types. Information of this type has been exhaustively analyzed by the Bureau of Public Roads (*42*). In Table 3–2 are shown the average service lives of a large mileage of primary roads in service during 1953.

At the end of its useful life, the surface is "retired." Retirement may be accomplished by resurfacing in some cases, by construction of a new surface in others, or by reconstruction, relocation, or abandonment of the highway structure. The date of actual retirement does not, of course, always coincide with that of economic retirement. The economic life of a highway surface may be defined as the period of time that the surface must be utilized in order that its annual costs may have their minimum value. The increased coincidence of actual and economic retirement dates is one of the long-range objectives of the road-life studies.

Information contained in Table 3–2 was obtained by the statistical analysis of road-life data. The mechanics of this and related analyses will not be discussed here.

Benefits Resulting from Highway Improvements

It is obvious that many benefits result from highway improvement or, to put it more broadly, from improved highway transportation. Some of these benefits are direct and readily apparent; others are indirect and more difficult of discernment. Likewise, some benefits may be readily evaluated in terms of dollars and cents; others defy evaluation in this fashion, although they are nonetheless as real and lasting as monetary returns. Sometimes, of course, the consequences of a given highway improvement are not entirely beneficial; for example, construction of an urban expressway may produce a decline in ridership on a commuter railroad; and public funds spent from the general treasury for highways can not be used, say, for education. Such consequences should be considered in comprehensive economic analyses.

For the purpose of discussion, it is convenient to divide benefits into three broad classes: (1) reduction in road user costs, including savings in time; (2) reduction in accident losses; and (3) general benefits, including those to adjacent property and the general public.

Road User Costs

The discussion of road user costs given here is based upon a comprehensive report of the committee on Planning and Design Policies of the American Association of State Highway Officials (*105*).

Road user costs may be defined as vehicular operating costs, usually expressed in cents per mile, covering all items involved in vehicle ownership and operation. The value of time is included as one element of road user costs.

The AASHO Committee established road user costs in terms of the following major variables.

1. Type of vehicle
   a. Passenger car
   b. Truck or bus
2. Type of area
   a. Rural
   b. Urban
3. Type of highway
   a. Two-lane
   b. Divided
4. Type of operation
   a. Free
   b. Normal
   c. Restricted
5. Running speed
   (Range of likely values)
6. Gradient class
   a. 0 to 3 per cent
   b. 3 to 5 per cent
   c. 5 to 7 per cent
   d. 7 to 9 per cent
7. Type of surface
   a. Paved
   b. Loose surface
   c. Unsurfaced
8. Alignment
   a. Tangent or flat curves
   b. Curved

Data in the necessary detail and form were not available to the committee for trucks and buses in rural areas or for passenger cars, trucks, and buses in urban areas. Such information will eventually be available as a result of continuing investigations in this field. Material fully treated by the committee covers only the operational costs of passenger cars in rural areas. Approximations for trucks and buses in rural areas and urban areas are available and are included in the discussion.

3–4. Road User Cost for Continuous Operation. For any highway, or section of highway, the annual road user cost—including both vehicular operating cost and time cost—can be determined by the use of the following expression:

$$\text{Annual road user cost} = 365 \cdot A \cdot L \cdot U \qquad (3\text{–}2)$$

where A = annual average daily traffic volume for the period involved in the economic analysis
L = section length in miles
U = combined unit operating and time cost per vehicle mile for the type of highway and anticipated operating conditions

3–5. Determination of Traffic Volume. Refer to Chapter 4 for a more complete coverage of the methods of estimating traffic volumes. For the calculation of road user cost, the committee suggests the following steps.

1. Estimate the annual average daily traffic which will use the section upon its completion.
2. Determine the number of years in the future for which the analysis is to be made and the expansion factor for traffic on the section during this period.
3. Calculate an expanded annual average daily traffic volume which is a

representative or average value for the period of analysis. This is the value to be used for "*A*" in Eq. 3–2.

A period of 15 to 25 years in the future is the normal maximum for which traffic forecasts can be made with accuracy. In most cases, it will be necessary to separate the traffic data by vehicle types, since different unit road user costs must be used for each. Light trucks, such as pick-ups, delivery vans, etc., can be classed with passenger cars. All heavier and larger trucks and buses (generally, with gross vehicle weight above 9000 pounds) should be considered separately.

3–6. Determination of Section Length. The highway should be divided into sections of convenient length. The AASHO Committee suggests that there should be a separate section for each significant variation in traffic volume. Also, there should be a separate section where a significant variation occurs in either the physical conditions (such as number of lanes, profile conditions, or type of surface) or vehicle operating conditions, as determined jointly by highway conditions and volumes. In general, the use of short sections should be avoided.

3–7. Determination of Unit Vehicle Operating and Time Cost (Tangent Roadways). Figs. 3–1 through 3–4 show values for combined unit vehicle operating and time costs for tangent roadways as based on national average prices current at the time of the committee report. They are in a form for direct use in Eq. 3–2. The proper "*U*" value for any section is read from the appropriate figure after the following conditions are established for that section:

(1) Number and arrangement of lanes (type of highway).
(2) Type of surface.
(3) Grade or profile type (gradient class).
(4) Running speed (the speed over a specified section of highway, being the distance divided by the time the vehicle is in motion).
(5) Type of operation.
(6) Alignment features (determines correction factor [Fig. 3–5] to be applied to tangent alignment costs).

In addition to the charts shown here the committee prepared detailed tables of road user costs. Use of the tables is more precise than the use of the charts.

Number and Arrangement of Lanes. For the same terrain and geometric design details a somewhat higher running speed and a somewhat different effect from other variables can be expected on a divided highway than on a two-lane highway with pavements in good condition. Unit cost values for these two types of rural highways are shown separately in Figs. 3–1 and 3–2. Running speeds on three-lane and four-lane

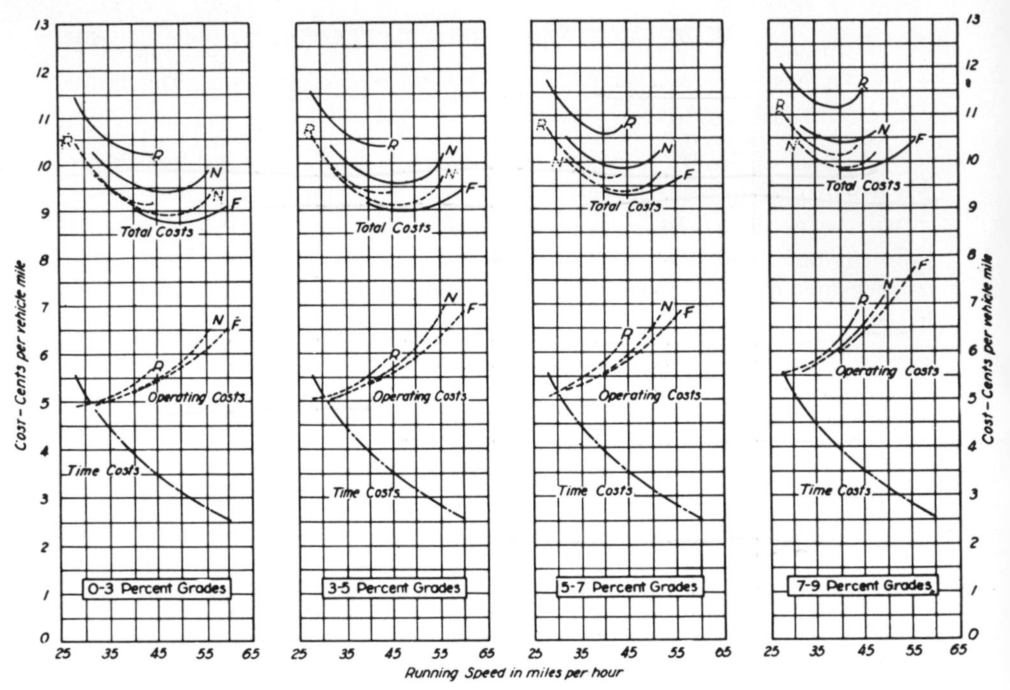

FIG. 3-1. Unit road user costs versus running speed, for tangent divided highways. Passenger cars in rural areas. Pavements in good condition. (Courtesy AASHO.)

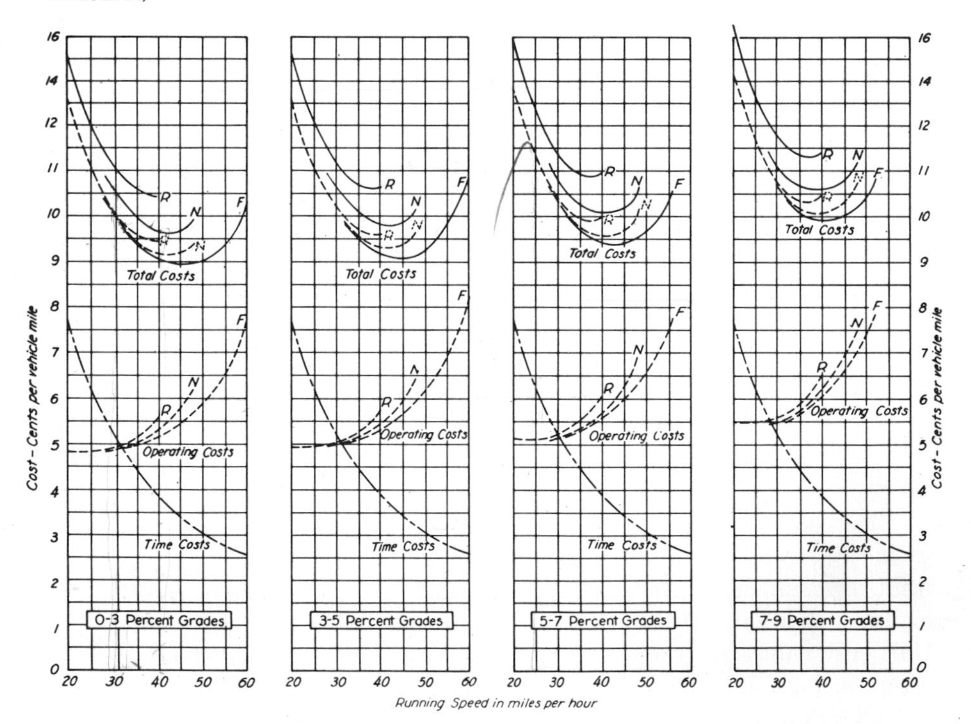

FIG. 3-2. Unit road user costs versus running speed, for tangent 2-lane highways. Passenger cars in rural areas. Pavements in good condition. (Courtesy AASHO.)

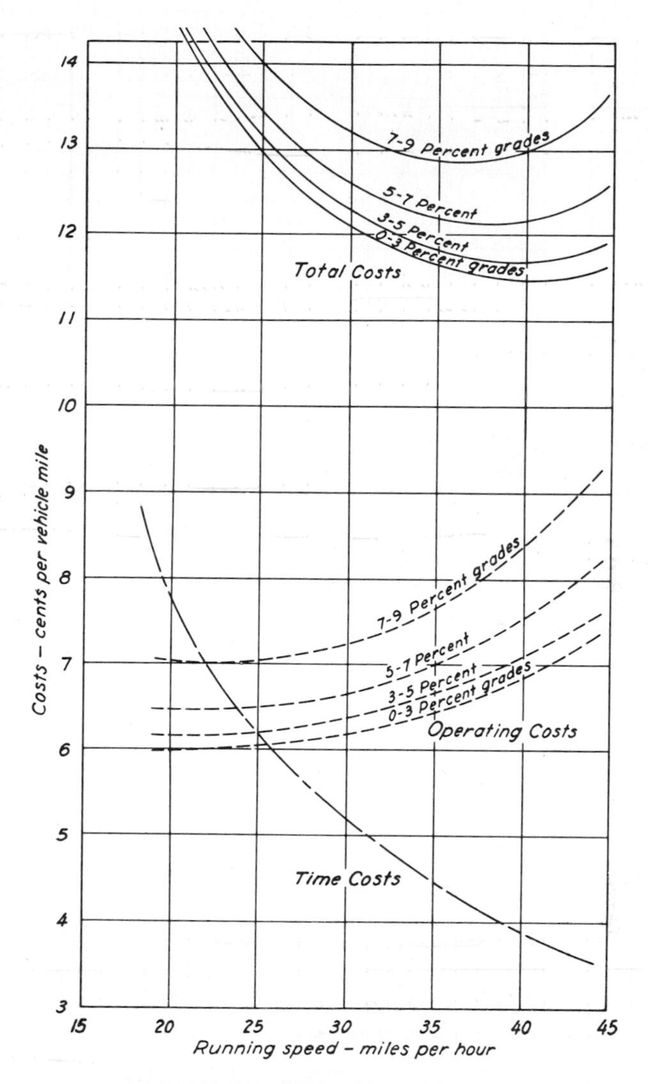

UNIT ROAD USER COSTS VS. RUNNING SPEED
PASSENGER CARS IN RURAL AREAS

TANGENT LOOSE SURFACE HIGHWAYS
IN GOOD CONDITION

FIG. 3–3. Unit road user costs versus running speed, for tangent loose-surface highways. Passenger cars in rural areas. Highways in good condition. (Courtesy AASHO.)

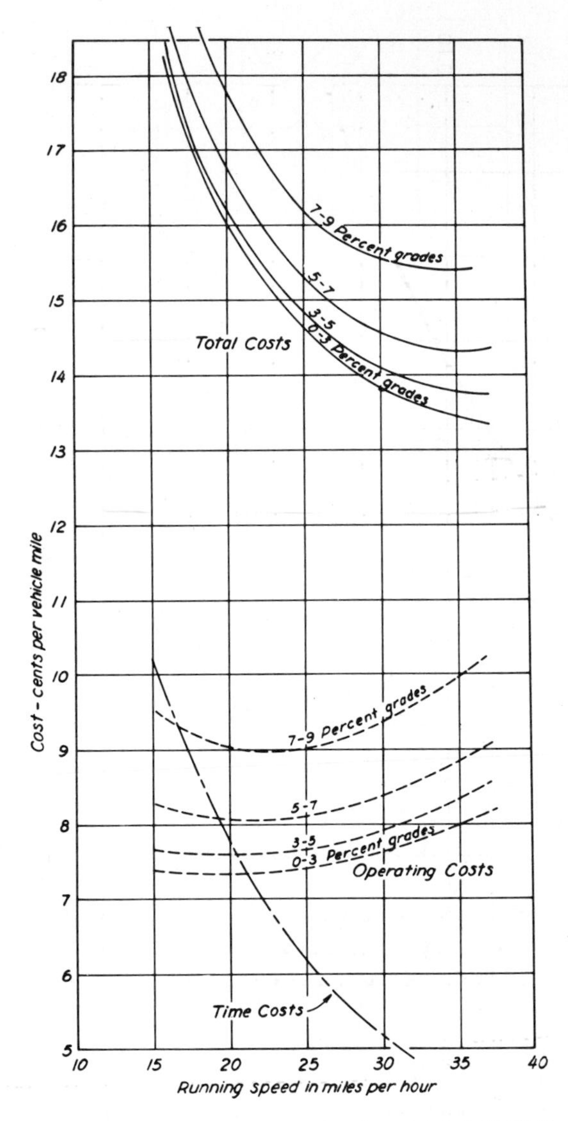

Fig. 3-4. Unit road user costs versus running speed for tangent unsurfaced roads. Passenger cars in rural areas. (Courtesy AASHO.)

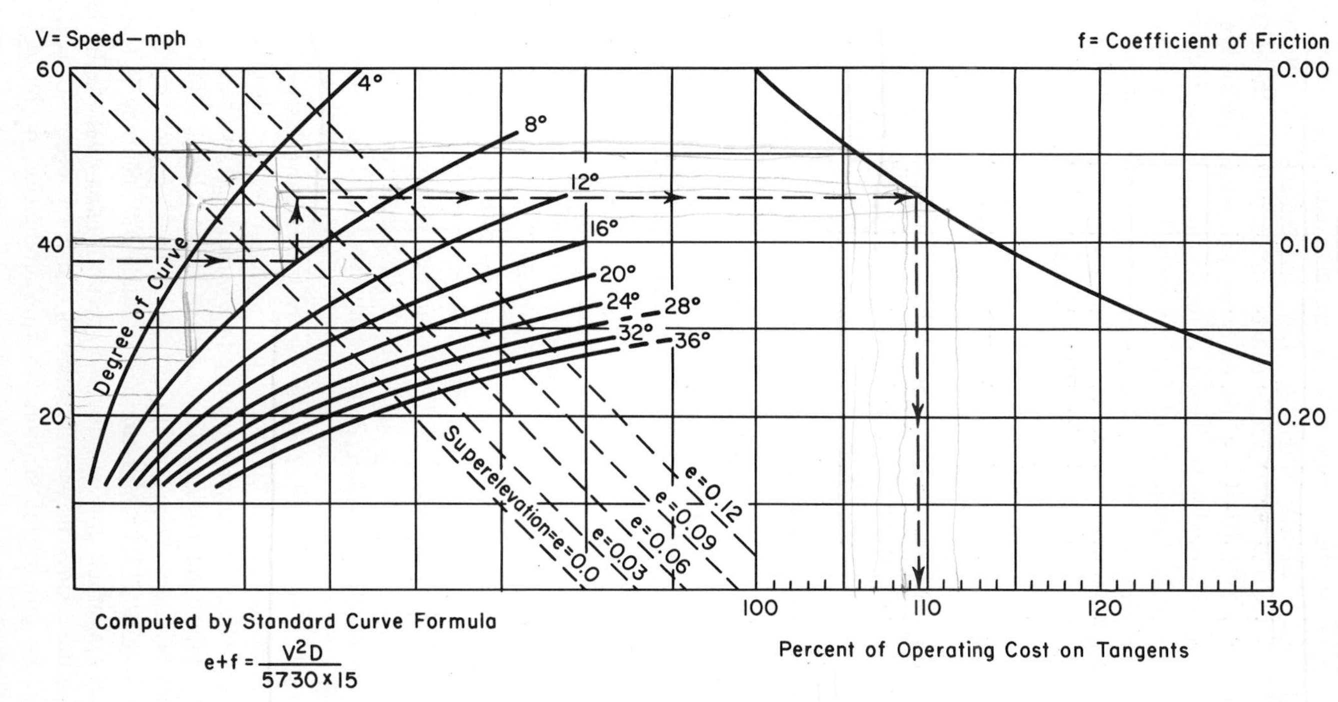

Fig. 3-5. Relation between operating costs on curves and on tangents. *Example:* Assume 38 mph operation on 8° curve, 0.06 superelevation. Follow arrows and read 109.50% as the factor to apply to correct the tangent road user cost values. (Courtesy AASHO.)

undivided highways are about equal to or only slightly greater than on two-lane highways for comparable conditions; effects of other variables are also only slightly different. For this reason, the committee felt that the preparation of separate charts and tables for these types of highways was not justified. Unit costs on three-lane highways can be assumed to be practically the same as costs on two-lane highways; unit costs on undivided highways of four or more lanes can be approximated as values between those for two-lane and divided highways for running speeds of 40 mph or less.

Type of Surface. It is readily apparent that the type of surface is a major factor in determining road user costs, since such items as time, fuel consumption, and others are directly affected. Differences in road user costs are marked between paved surfaces, loose surfaces (such as gravel), and unsurfaced roads. The committee separated unit road cost for three categories of surfaces: (1) paved surfaces, either rigid or flexible; (2) loose surfaces, principally all-weather gravel; and (3) unsurfaced.

The data contained in Figs. 3–1 through 3–3 are representative for surfaces in good condition. Unit costs for paved surfaces in fair to poor condition can be obtained by interpolation between the "paved" and "loose" values. Likewise, values for gravel surfaces in poor condition can be estimated by interpretation between "loose" and "unsurfaced."

Gradient Class. Running speeds and resultant unit costs are affected by the profile type and gradients involved. Various studies have shown that the effect of changes in grade is dependent upon the over-all resultant efficiency of vehicle operation over a section of highway. The operating efficiency of vehicles is found to be dependent upon such factors as rate of grade, over-all rise and fall, combination of grades, approach conditions, lengths of individual grades of any rate, power of vehicle, dispersion of energy by braking, and so on.

The committee separated unit cost data into four gradient classes: 0 to 3, 3 to 5, 5 to 7, and 7 to 9 per cent. These classes indicate an average grade along the highway; unit cost values for them include the momentum effect for operation in rolling terrain. The gradient class must be determined for each section used in the analyses. This determination can be made from profile data by one of several methods of comparable accuracy. A single method of sufficient accuracy is the calculation of the average gradient for the section under analysis by summation of actual grades at each station, or at every other station, and dividing by the number of stations in the summation or by dividing the total rise and fall by the length of the section. On long sections with considerable variation in profile, gradient determinations may be necessary for two or more subsections.

Running Speed. The speed of travel is obviously an important consideration in determining road user costs because a number of the elements of vehicle operating costs vary considerably with speed and the value of time is directly related to it. The running speed must be representative of the whole length of the section, of all vehicles of the type to which it applies, and of the whole of the period used in the analysis. The running speed must be representative of the terrain, curvature, gradient, sight distance, traffic volume, and other conditions of consequence on the section under analysis. The best source of speed data is the actual measurement of speeds on existing facilities to be analyzed. Where comparable conditions exist, it is most easily determined by dividing the total length by the total running time for representative operation. To be truly representative, the speed should be adjusted to include both dark and daylight hours, typical inclement weather conditions, and anticipated future operating conditions. Where comparable measurement is not feasible, data are available from which estimates can be made from design speed and traffic volume data.

Type of Operation. Road user costs are affected, in a very complex fashion, by the manner in which vehicles are operated. On highways of essentially the same geometric design characteristics, the general manner of operation may be markedly different because of different traffic volumes. Or where traffic volumes are about the same, the operations may be distinctly different because of differences in geometric design or in the traffic patterns of the two highways. The committee felt that a complete expression for type of operations in its relation to unit costs would include the complex combination of running speed, traffic volumes and patterns, type of vehicles, and geometric design features.

In a simplified approach, the general combined effects of terrain, type of highway, control of access, and traffic volume are grouped into three types of operation: (1) free, (2) normal, and (3) restricted. The proper type of operation is determined by the relation of the thirtieth highest traffic volume to the capacity of the highway. See Chapter 5 for a complete explanation of the terms "thirtieth highest hourly traffic volume" and "capacity." While these operating conditions are general rather than precise, the committee feels that they could serve as a practical means of distinguishing a complex combination of factors. To use this method, it is necessary to rate operating conditions as being in one of the following classes.

Type of Operation	Ratio of the Thirtieth Highest Hourly Traffic Volume to Capacity
Restricted	More than 1.25
Normal	0.75 to 1.25
Free	Less than 0.75

Alignment Features. Curvature has an effect upon operating costs, but the effect is very hard to evaluate. The committee evaluated the factors involved and proposed that a correction for conditions of curved alignment be made by increasing costs on tangent sections in accordance with the percentage obtained from Fig. 3–5. To use this figure, it is necessary to determine both the sharpness and the percentage of length of significant curvature. Little error results in using the average degree of curvature and average superelevation for the section being analyzed, weighted on the basis of length.

Components of Road User Operating Costs. A highway improvement may benefit the road user in many ways. It may reduce operating expenses for fuel, oil, tires, vehicle maintenance, and so on. It may reduce the time spent in travel. It may increase the safety of operation by reducing accidents. It may increase the service received from a vehicle for a fixed total cost of ownership. It may provide increased comfort, convenience, and pleasure in travel. To achieve a quantitative measure of road user benefits, all of these items should be considered in monetary terms. The items which can now be evaluated in monetary terms are vehicle operating costs, the value of time, and comfort and convenience.

Fuel Costs. The most important single item of vehicle operating cost is that of the fuel or gasoline used. The amount of fuel consumed varies with each of the major factors previously discussed in this article. The price of fuel varies over the country. The unit cost values in Figs. 3–1 through 3–4 are based on an average nationwide cost of 32 cents per gallon.

Nonfuel Operating Costs. This term designates the major elements of vehicle operating costs other than gasoline. As used in the committee report, these elements include costs of oil, tires, maintenance and repairs, and depreciation. The following nationwide average costs have been used for these items:

Oil—45 cents per quart.

Tires—$100 initial cost per set.

Car repairs—on good pavements, 1.2 cents per vehicle mile; on loose surfaces, 1.8 cents per vehicle mile; on unimproved surfaces, 2.4 cents per vehicle mile.

Car depreciation—1.5 cents per vehicle mile.

TIME VALUE. One of the greatest benefits from many highway improvements is the saving in travel time. There is general acceptance of the belief that saving in time for trucks and buses has value in direct ratio to those costs of operation directly related to time, such as wages paid to drivers and hourly rental of the equipment. It has value also for fixed daily or monthly costs, such as overhead, because saving in time of travel results in greater usage for a given period of time.

The acceptance of values for all savings in time of passenger vehicles is not as general, although it is admitted that assignment of some value usually is justified. Where cars are used for business, as in the case of salesmen, the cars are no different in this regard from trucks. For other uses, opinion is divided as to the value of time saved. There is agreement that drivers of passenger cars for nonbusiness purposes are attaching increasing importance to and are willing to pay for travel on facilities that result in a saving of time.

The value of time chosen by the committee for computations involving passenger cars is $1.55 per hour.

COMFORT AND CONVENIENCE. Benefits such as comfort and convenience are intangibles, but they are none the less real and should be evaluated. Certainly an expressway which allows uninterrupted travel into the center of a city renders a service which has value over and above the value of time saved and savings in operating costs. There is value in being able to drive without frequent brake applications, stops and starts, or unexpected interference to travel. Positive identification of values for assignment to various degrees of comfort and convenience is not possible. However, assumed values are included in the charts for operating costs, as follows:

Paved highways—1.0 cents per vehicle mile for restricted operation, 0.5 cent for normal operation, and 0.0 cent for free operation.

Loose surfaced highways—0.75 cent per vehicle mile.

Unsurfaced highways—1.0 cent per vehicle mile.

DERIVATION OF OPERATING COSTS. The committee report cited herein contains additional detailed information related to unit operating costs and their derivation. The student is referred to that report for a comprehensive treatment of the subject.

3–8. Calculations of Road User Costs with Intersection Delay. The road user costs discussed in Art. 3–7 are set up for continuous operation, and do not allow for the element of delays due to stops at intersections. Where stops are necessary, they result in an additional cost above that expressed in Eq. 3–2 and the curves of Figs. 3–1 through 3–4. Where intersections are infrequent and stops are only required occasionally, their cost may be an insignificant part of the total road user cost and may be ignored. Where large volumes of crossing and turning traffic force a considerable number of stops at frequent intersections, the cost increase may be substantial. Another situation in which stopping costs are important is that of intersection improvement.

The AASHO Committee suggested that the following expression could be used for total annual road user cost including intersection delay:

$$\text{Annual road user cost} = 365 \cdot A \cdot L \cdot U + 365 \cdot B \cdot S \qquad (3\text{–}3)$$

where A, L, and U = same as in Eq. 3–2

B = average daily traffic required to stop at a given intersection

S = average additional cost per stop delay per vehicle at that intersection

The annual road user cost per stop for each intersection is determined by evaluating B and S and then the quantity $365 \cdot B \cdot S$. The summation of this cost for all the intersections in the section under analysis provides the cost increment for that section of the highway.

TABLE 3–3

EXTRA COST PER VEHICLE STOP ABOVE THAT FOR CONSTANT SPEED OPERATION

Standing, Delay Period, Seconds	Extra Cost per Vehicle Stop, Cents, for an Approach Speed (mph) of					
	10	20	30	40	50	60
0	0.2	0.4	0.7	1.3	2.1	2.9
20	1.2	1.4	1.8	2.3	3.1	4.0
40	2.2	2.4	2.8	3.4	4.1	5.0
60	3.2	3.4	3.8	4.4	5.1	6.0

The evaluation of B and the "standing delay period" (average number of seconds these vehicles are required to stand after stopping), necessary for determining S, can be done by measuring these items at a typical existing intersection and then adjusting them to be representative for the period of analysis. The value of S from Table 3–3 is the additional cost per vehicle stop above that for operation of the vehicle through the intersection at the approach speed.

The value of B may consist of traffic on the highway section under study, on a road crossing it, or both.

3–9. Approximate Road User Costs for Trucks and Buses in Rural Areas. Sufficient data were not available to permit construction of charts like Figs. 3–1 through 3–4 for trucks and buses. However, approximations may be made which will permit complete analyses to be made for all types of vehicles.

In Eqs. 3–2 and 3–3, the annual average daily traffic volume was designated as "A." This quantity can be divided into two parts—A_p for passenger cars and light trucks having essentially the same characteristics, and A_t for all other trucks and buses. Thus

$$A = A_p + A_t, \qquad \text{and} \qquad A_e = A_p + F(A_t) \qquad (3\text{–}4)$$

where A_e = the equivalent passenger vehicles, including the effect of trucks

F = the ratio of the unit operating cost for trucks to that of passenger vehicles

TABLE 3–4

RATIO OF TRUCK OPERATING COSTS TO THOSE OF PASSENGER CARS

Type of Vehicle	Approximate Per Cent of Total on Main Highways	Ratio (F) of Truck Operating Costs to Those of Passenger Cars
Single-unit trucks	70	2 to 4
Truck combinations	29	4 to 6
Buses	1	2 to 4
Composite truck and bus	100	2.5 to 4.5

The value of F may be chosen from Table 3–4 for a single type of vehicle, for the indicated composite, or calculated as a weighted average where the volume of each type of vehicle is known.

3–10. Approximations for Urban Areas. The AASHO Committee suggested that approximate analyses in urban areas may be made in the following way.

(1) On major streets and expressways where traffic flow is reasonably continuous, determine the representative running speed, preferably by actual observations properly weighted for traffic volumes. Figs. 3–1 through 3–5 may then be used to determine U, as for rural areas.

(2) On major streets where traffic flow is not continuous but has only a moderate amount of interruption, determine the representative running speed and the corresponding value of U from Figs. 3–1 through

3–5. Increase this value by 10 to 30 per cent to allow for the additional cost of stops before use in Eq. 3–2 or 3–3.

(3) With stop-and-go conditions, make special field investigations for conditions and values. It is suggested that this might be done by proceeding as in (1) above and using Eq. 3–3 to include the cost of stops. The cost of stops would have to be obtained from observations on existing streets, with adjustments to be representative of future conditions.

(4) For conditions intermediate between those described, use intermediate values determined by measurement or interpolation, or both, as judgment dictates.

3–11. Accident Costs. The accidents which occur every year on the streets and highways of America represent a staggering economic loss which is quite aside from the pain and suffering which they cause. Total economic loss resulting from motor vehicle accidents in 1966 was estimated by the Insurance Information Institute to be $13 billion. This total included losses incurred in fatal accidents, accidents involving nonfatal personal injury, and those involving only property damage. It follows from an economic standpoint that highway improvements which result in decreased accident rates result in a monetary saving to the road user and the public in general.

For example, the incorporation of controlled access into the design results in an average saving of about 1/4 cent per mile when compared with other types of highways (*105*). This figure assumes considerable proportions when examined with respect to a normal volume of traffic. For example, with an ADT volume of 15,000 and an accident rate and cost conforming to these averages, the annual saving accruing to this traffic in 1 mile would be about $13,700. This amount would amortize a total sum of $157,000 at 6 per cent interest in 20 years. If this reduction in accidents could be assayed, it would be economically justified to spend an additional $157,000 per mile on a four-lane divided highway to obtain controlled access. This sum does not include additional benefits that may also accrue to the road user by the higher type of design.

A great quantity of information on highway accidents is available, but the available statistics do not, in general, establish direct relationships among elements of highway design and accident occurrence. This fact, coupled with a general lack of comprehensive knowledge of the average economic consequences of the average accident, makes the incorporation of accident costs in economic calculations difficult.

To some economists, the lack of satisfactory data is complete. In 1964, Robley Winfrey of the Bureau of Public Roads said, "The extensive literature on accident costs and highway safety does not contain a single publication which puts accident costs in the proper light or proper units for application to highway studies" (*20*).

A comprehensive study of direct costs of traffic accidents in Illinois yielded information of some usefulness (*141*). In 1958, about 1.3 million Illinois passenger cars were involved in traffic accidents on Illinois highways, at a total direct cost of $258.8 million, or an average of $199 per event. For trucks, 128,000 were involved in accidents costing $18.1 million, or an average of $141 per event. Costs amounted to 0.97 cent per passenger-car mile and 0.36 cent per truck-mile. Median cost values were $2280 for fatal-injury involvements, $310 for nonfatal injuries, and $50 for property-damage-only involvements.

Earlier studies in Massachusetts and Utah (*18*) gave these direct costs for accidents involving passenger cars: Massachusetts (1953)—fatal injury, $5213; nonfatal injury, $862; and property damage only, $203. Utah (1955)—fatal injury, $3690; nonfatal injury, $1277; and property damage only, $299.

In 1967 a report prepared by consulting engineers Wilbur Smith and Associates for the Bureau of Public Roads gave the average cost of an involvement in a fatal injury accident in the Washington metropolitan area as $47,000. Reported nonfatal injury involvements averaged $860, and those involving property damage only averaged almost $200. All reported and unreported motor vehicle accidents cost the area's residents an average of almost $100 per family per year, or about 1.1 cents per vehicle-mile of in-area travel.

The above discussion indicates the importance of reductions in accident rates to the road user, and includes source data that might be used in economic analyses. Use of these data is not widespread for the following reasons (*105*):

(1) Accident statistics which are readily available frequently do not provide the desired correlation with highway design and operational elements for unquestioned direct use in road user benefit analyses.

(2) For any particular section of highway, the number of accidents occurring is usually so small as to make difficult the computation of reliable accident rates. For example, on a 5-mile section of a two-lane highway with an average daily traffic of 2200, and using the national average of 5.6 fatalities per 100 million vehicle miles, one death could be expected every 3.4 years. It can be seen that a considerable period of time would be required to collect statistics of this order on any given section of highway. Correlation of accident data with design or operational elements requires even more time. It is also apparent that one freak accident experience could so distort the picture that any computed value would be meaningless.

(3) The number of personal injury accidents is much greater than fatal accidents, and it would seem a rate for these types could be computed with more assurance. However, this is not necessarily true, because

they are subject to greater error due to incomplete reporting not only of the damage but of the accident itself.

(4) Usually there are large benefits in the form of fewer accidents because of the replacement of a substandard facility with one of modern design. Such is not always the case, however, and in some cases the accident rate has proved to be higher on the new facility. This apparently is due to excessive speed and driver inattention. Thus, a greater understanding of the causes of accidents is needed to permit an accurate appraisal of accident rates upon new facilities.

Benefits to Adjacent Property

Those benefits which accrue to the owner of property located adjacent to or in the vicinity of improved highways are fairly simple to observe and delineate but are somewhat more difficult to evaluate. In the vast majority of cases, highway improvements result in an increase in the value of land located adjacent to or served by the highway involved. A simple case of direct property benefit may be seen in that of a farm or a number of farms located on an unimproved earth road, rough, muddy in wet weather and dusty in dry. Any degree of improvement of a road of this sort would be of direct benefit to the property owners located along it since they would be provided with easier access to market, schools, recreational facilities, and such. Similar direct benefits may accrue to city dwellers who live on an improved residential street. In many other cases property benefits may be more difficult to discern and to evaluate. The lack of adequate improvements for city streets and the problems associated with traffic congestion and parking have been factors in the decentralization that has occurred in some United States cities, the rapid growth of suburban areas, decreased property values, and the creation of blighted areas in central business districts.

In 1964, the Bureau of Public Roads published (*154*) a comprehensive report on the relationships between highway improvement and economic and social changes. The results of more than 100 economic and social impact studies were analyzed and reported. Many more such studies are currently under way.

Results of the economic impact studies contained in the B.P.R. report are too detailed to be presented here. However, results obtained in some of the more well-known studies include the following:

1. Land and improvements on it located on the Dallas Central Expressway rose about 780 per cent in value during a 10-year period that spanned the construction and opening of the highway. Land in a two-block-wide strip adjacent to property abutting the Expressway had an

increase in value, including improvemnets on it, of only 104 per cent; land beyond this strip showed a gain of about 47 per cent.

2. Before U.S. 50 was relocated and improved near Lawrenceville, Illinois, land that now faces the expressway's frontage roads abutted a local road. This land was developed to single-family residences and increased in value more than nine times during an 18-year period; similar land located along a secondary road increased in value only 1½ times.

3. The impact of the New York State Thruway on land values has been dramatic (the over-all impact of the Thruway has been compared with the impact of the long-ago construction of the Erie Canal, in the same state). Land values at Rochester Interchange have risen 700 per cent and more. Some land acquired by the state for rights-of-way at $500 per acre later was worth ten times that amount. Near Syracuse, a 21-acre site near the Thruway sold for $15,000 in 1951; in 1955, slightly over half that acreage sold for ten times the former total value.

4. Construction of Massachusetts Route 128, the circumferential highway located about 10 miles from the center of Boston, spurred industrial development. From 1951 through 1957, investment in new plants along Route 128 totaled about $85 million. In 1957, investment along Route 128 comprised almost 40 per cent of all investment of this type in the Boston metropolitan area.

5. In a 3-year period, 156 (44 per cent) of the 354 new industrial plants built in Indiana were located within a 45-mile-wide band straddling the Indiana Turnpike.

Benefits to the General Public

Benefits of improved highway transportation which are enjoyed by the general public may well be regarded as the most important of the three classes of benefits which have been listed. These benefits include many of the advantages commonly associated with modern civilization, including such things as efficient fire and police protection, rapid postal deliveries, improved access to educational, recreational, and social facilities, decreased costs of commodities, improved national defense, and so on ad infinitum. The general advantages of improved highways are so interwoven with every phase of modern life in this country that it is virtually impossible to imagine what existence would be like without them. Thus, although it is apparent that this class of benefits is of extreme importance, these benefits cannot usually be evaluated except in a general sense and are therefore seldom used in economic calculations. In highway improvements involving large areas and expenditures these benefits may be listed and defined (*159*). The B.P.R. report described in the preceding paragraphs contains a quantity of information about the social effects of highways, over and above their economic impact.

Benefit-Cost Calculations

3–12. Benefit Theory. From considerations previously expressed, it has been determined that economical highway improvement is effected when the total cost of highway transportation is a minimum consistent with convenience and safety. It is also axiomatic that improvements must be effected in such a way as to provide the greatest possible benefits to those who pay for them. If this is done, the taxpayer will be assured of receiving the greatest possible return on his investment. As will be indicated later when financing methods are studied, funds for highway improvement, and particularly for rural highways, are in large measure provided by highway users. Thus, for this class of highway at least, it is axiomatic that improvements should be made in such a way as to yield maximum benefit to the highway user, consistent with the other benefits listed above.

It also follows that, if the highway system is to be improved so as to provide benefits of various sorts to various segments of the population, then those who benefit most from the improvement should pay the most for it. Thus it seems entirely logical in many cases that the highway user should pay the largest proportion of the costs of highway improvement and, furthermore, that if certain classes of users benefit more than others, then these users should pay a proportionally higher share of the cost. An accurate and fair allocation of highway costs is, of course, difficult to determine and even more difficult to accomplish. Methods of allocating highway costs are described in Art. 3–20.

3–13. Benefit-Cost Ratio. Economic calculations intended to demonstrate economic justification for a particular project, permit comparison of alternative schemes or locations, determine priority of improvement, and so on, may be carried out by one of several methods. Methods used in engineering economy studies include a comparison of annual costs, determination of the interest rate at which alternatives are equally attractive, a comparison of the present value of all present and future expenditures, and a comparison of benefit-cost ratios. The latter is the method commonly used by highway engineers and administrators. Other methods are described briefly in Art. 3–17.

In the benefit-cost ratio method, analysis is made by a comparison of the relations of annual road user costs to the annual highway costs for logical alternatives in location and design which have the same overall traffic movements as the highway or connecting highways affected.

As presented here and developed by the AASHO Committee (*105*) the method is applicable only to alternatives of location and design of the same general highway facility. The benefit-cost ratio determined by this method is not suitable for use directly in comparison of highway projects that have dissimilar traffic, terrain, and design conditions. It is not suitable for priority determinations of projects on an area or state-wide basis. It has value as a guide factor for priority determinations, but only when used jointly with other criteria.

The benefit-cost ratio method is a comparison of the difference in annual costs to road users, when vehicles use an existing road in one case and an improved road in another, with the annual cost of making the improvement. The applicable equation is

$$\text{Benefit ratio} = \frac{R - R_1}{H_1 - H} \tag{3-5}$$

where R = the total annual road user cost for the basic condition, usually the existing road

R_1 = the total annual road user cost for the improvement, including travel on existing highways by the traffic directly involved or affected

H = the total annual highway cost for the basic condition, usually the existing road

H_1 = the total annual highway cost for the improvement.

The basic condition is usually the existing highway. Calculations then provide a comparison of the proposed improvement with the existing facility. According to the AASHO Committee, if the existing highway is used as the basic condition it is not necessary to include interest and amortization on its worth in computing annual cost. The "H" value is thus equal to the cost of maintenance (see Eq. 3–1). When an existing facility logically cannot be used as the basic condition, it is desirable to use instead the lowest type of the alternatives considered.

Benefit ratios are calculated for all logical alternatives, each compared with the basic condition. By doing this, the several ratios can be compared directly as to their relative value. A benefit ratio less than one indicates that, in a road user benefit sense, the basic condition is to be preferred over the alternative improvement.

3–14. Example of Benefit-Cost Ratio Calculations. The following example is typical of the simpler benefit ratio calculations which are frequently necessary in highway economy studies. It is taken directly from the AASHO Committee report.

Assume a relocation project where by heavy grading work on new alignment it is possible to reduce the length between two points on an existing highway. The existing highway is to be abandoned. Present traffic on the existing route is 1500 vpd and it is estimated that the average traffic for the next 20 years, the analysis period, will approximate 2500 vpd. This traffic is composed of passenger cars with a very small proportion of trucks. Due to the character of the trucks, a distinction between them and the passenger cars is not considered necessary. Assignment of type of operation is made on the basis of the indicated ratios of thirtieth highest hourly traffic volume to capacity. The proposed facility contemplates a pavement width of 20 feet compared with 18 feet on the existing facility. This factor, together with improved alignment and grades, permits distinction as to type of operation as follows:

Route	Thirtieth Highest Hour	Capacity	Ratio	Type of Operation
Existing	375	450	0.83	Normal
Proposed	375	600	0.63	Free

Other essential data are:

Route	Design Speed (mph)	Running Speed (mph)	No. Lanes	Length (Miles)	Grade Class (Per cent)	Type of Operation	Surface and Condition	Curvature
Existing	50	37	2	2.0	0–3	N	Paved–Good	50%–4°
Existing	50	34	2	0.4	3–5	N	Paved–Good	50%–4°
Proposed	60	42	2	1.5	0–3	F	Paved–Good	Negligible

The curvature proposed on the new facility is generally flat, with a minor portion approaching the design speed limit. This is to be properly superelevated so a correction for curvature can be ignored. However, the existing highway has a large number of curves approaching the maximum for a 50 mph design speed. It is estimated that 50 per cent of its length will have an average curvature of about 4 degrees with superelevation negligible. From Fig. 3–5 a correction of between 7 and 8 per cent is read for the speeds considered. Since only one-half of the roadway is curved the correction is halved to 4 per cent and applied for the whole of the route to the values from Fig. 3–2. The unit costs are as follows:

Route	Grade	Unit Costs
Existing	0–3	9.85 × 1.04 = 10.24 cents
Existing	3–5	10.20 × 1.04 = 10.61 cents
Proposed	0–3	9.01 × 1.00 = 9.01 cents

From Eq. 3–2 the average annual road user costs are calculated as follows for each section and added:

Days × A × L × U	= $ Costs
Existing Route	
365 × 2500 × 2.0 × $.1024	= $186,900
365 × 2500 × 0.4 × .1061	= 38,700
Total	$ 225,600
Proposed Route	
365 × 2500 × 1.5 × $.0901	= $123,300

The average annual unit maintenance cost on the existing route is estimated to be $1100 per mile or a total of 2.4 × $1100 = $2640, and that of the proposed route to be $880 per mile or a total of 1.5 × $880 = $1320.

The total estimated cost of the proposed improvement is $550,000, and the prevailing local interest rate is 5 per cent. This interest rate is applied with reference to the cost and life expectancy of the individual items of improvement to compute the annual cost. Capital recovery factors, K, are selected from Table 3–1.

Item	Estimated Life	Cost	K	Annual Cost
Pavement	20 years	$ 66,000	.0802	$ 5,290
ROW	60 years	33,000	.0528	1,740
Grading, drainage, and structures	40 years	451,000	.0583	26,290
			Total annual cost =	$33,320

The benefit ratio for the proposed improvement is computed by Eq. 3–5 as follows:

$$\text{Benefit ratio} = \frac{\$225{,}600 - 123{,}300}{\$34{,}640 - 2640} = 3.20$$

The analysis indicates that the annual benefits are over three times the annual project costs. From this it appears to be a worthwhile project as far as road user benefits are concerned and one that should be slated for early construction.

In order to illustrate the approximations made to include trucks in an analysis of this type, assume that 10 per cent of the average annual traffic volume consists of dual-tired light to medium trucks. The total average traffic for the analysis period consists of 2250 passenger cars and 250 trucks. It is estimated that truck operating and time costs, as compared to that of passenger cars, are in the ratio of 3 to 1. Then the total equivalent passenger cars for analysis purposes is computed by Eq. 3–4 as follows:

$$A_e = 2250 + (3 \times 250) = 3000 \text{ vpd}$$

The assumption of 10 per cent trucks (or about 7 per cent of the peak hour volume) in the traffic stream lowers the capacity, but the truck influence is insufficient to change the type of operation designation. The revised traffic volume, $A_e = 3000$, is used with the same unit costs previously given to compute annual road user costs as follows:

Days ×	A ×	L ×	U	= $ Costs
		Existing Route		
365 ×	3000 ×	2.0 ×	$.1042	= $224,300
365 ×	3000 ×	0.4 ×	.1061	= 46,500
			Total	$270,800
		Proposed Route		
365 ×	3000 ×	1.5 ×	$.0901	= $148,000

These revised total road user costs are used together with previous highway cost and maintenance figures in Eq. 3–5 to compute the benefit ratio as follows:

$$\text{Benefit ratio} = \frac{\$270{,}800 - 148{,}000}{\$34{,}640 - 2{,}640} = 3.84$$

3–15. Second Benefit Ratio. In cases in which several alternate locations or designs are being compared, says the AASHO Committee, the calculation is not always conclusive regarding the merits of the alternates as to benefits obtained by expenditure beyond that of the preferable alternate (the highest benefit ratio). Thus, a second-step benefit ratio calculation may be valuable to determine the additional benefits gained from additional increments of capital expenditure. This is done by eliminating the original basic condition, then calculating a second benefit ratio with the preferable alternate as the basic condition.

Such calculations are shown in the tabulations below, for the design of a major interchange in which six alternates were developed.

First Benefit Ratio

Alternate Plan	Total Annual Costs		Differences in Costs from Alternate 1		Benefit Ratio
	Highway	Road User	Highway	Road User	
1 (Basic)	$19,800	$411,600	–	–	–
2	22,200	356,700	$2400	$54,900	22.8
3	23,100	354,800	3300	56,800	17.2
4	25,000	336,200	5200	75,400	14.5
5	29,200	352,100	9400	59,500	6.3
6	29,700	352,600	9900	59,000	6.0

Second Benefit Ratio

Alternate Plan	Differences in Costs from Alternate 2		Benefit Ratio
	Highway	Road User	
2	–	–	–
3	$900	$1,900	2.1
4	2800	20,500	7.3
5	7000	4,600	0.66
6	7500	4,100	0.55

From the first ratios, alternate plans 2, 3, and 4 show high benefit ratios and plan 2 is preferable. In the second benefit ratio calculation, plans 3 and 4 show justification, the latter to a high degree. If funds were available, it would appear justified to build plan 4 at an extra cost of $2800 per year, rather than plan 2. Neither plan 5 nor plan 6 deserves further consideration, since additional road costs produce comparatively little in the way of road user benefits.

3–16. Effect of Interest and Other Factors. As noted previously, the rate of interest, the salvage value, and the time period have an effect on economic calculations. Estimating the salvage value is risky for periods far in the future; the assumption of zero salvage value has much to recommend it. Where salvage values can be estimated, they can be included in the calculations of Eq. 3–1. Time periods also are hard to pin down precisely, and most highway engineers use the average periods given in the preceding examples.

Interest rates have a considerable effect on benefit-cost ratios. To illustrate, suppose that the costs of the example of Art. 13–14 are computed at 8 per cent interest. They then become

Item	Estimated Life	Cost	K	Annual Cost
Pavement	20 years	\$ 66,000	0.10185	\$ 6,722
Right of way	60 years	\$ 33,000	0.08080	2,666
Grading, drainage, and structures	40 years	\$451,000	0.08386	37,822
				\$47,210

Assuming the same reduction in average annual road user costs and the same maintenance costs as in the previous example, the benefit-cost ratio becomes

$$\frac{\$225{,}600 - 123{,}300}{\$48{,}510 - 2640} = \frac{102{,}300}{45{,}870} = 2.23$$

as compared with the previously computed benefit-cost ratio of 3.20 with an interest rate of 5 per cent. Such a variation can have a marked effect on the comparison of alternatives.

3–17. Other Methods of Economic Comparison. Besides the benefit-cost ratio method, three other concepts are sometimes used for economic comparison of highway projects; they are minimum annual cost, rate of return, and present worth. Each method has its vocal adherents for particular purposes.

The minimum annual cost method is the embodiment of the earlier statement that maximum economy is obtained when the sum of annual road costs and annual user costs is a minimum consistent with convenience and safety. Numerically, the annual cost is the sum of the annual road costs of Eq. 3–1 and the annual user costs of Eq. 3–2. The calculation is made for the existing route and each of the alternatives; the alternative with the lowest annual cost is regarded as the "best."

Proponents of this method say its advantages include its ease of understanding by the layman and the fact that it categorizes the costs by public sector (road costs) and private sector (user costs). The method is, of course, very sensitive to the rate of interest chosen.

In the conventional *rate of return* method, the economist calculates the relationship between savings in annual costs and the capital investment, then expresses the relationship as a rate of interest (return on invested capital). The rates of return are calculated for the various alternatives; the capital investment that yields the greatest rate of return generally is regarded as the most desirable. A minimum acceptable rate generally is established before the calculations are made.

Numerically, the procedure is to add together, for each alternative, the annual road user costs and the maintenance costs; these constitute the numerator. The capital investment is the denominator, and the numerical value of the fraction is the capital recovery factor. With the value of *n* known, the rate of interest can be determined by interpolation from a table of capital recovery factors (such as Table 3–1). The procedure is awkward if the computations must be made for different time periods for the various road components, but not difficult.

Grant and Oglesby recommend an incremental rate of return method for application to highway projects (*70*). The method is the same as that described above, except that every alternative is compared with every other alternative, not just with the existing condition (as in Art. 3–15 for the benefit-cost ratio method).

Proponents of the rate of return method state that the method is readily understandable, since it is the method used most often by private business when considering capital investment, and that it is independent of the rate of interest. The latter claim actually is only partially true, since the "cut-off" rate of interest still must be determined in some manner; however, this decision can be postponed to the end of the process.

In the *present worth* method, the procedure is to reduce all costs to their present worth, i.e., the sum of money which, if invested at an assumed rate of interest, would be equal to the costs at the time they will be incurred. If alternatives have the same life, the alternative with the lowest present worth would be the most attractive.

Annual costs and present worths are convertible one to the other by means of the capital recovery factor. Carstens and Csanyi say that this method tends to minimize the effect of large future benefits arising from predicted increases in traffic and tends to place more emphasis on reductions in user costs that take place in the near future (*223*).

Highway Finance

The second of the two major topics which are being considered in this chapter is highway finance. Previously in this chapter we have considered certain principles applicable to the economic improvement of highways; that is, we have considered ways in which the money available for highway purposes may be spent to provide a satisfactory return from the investment made by any one of a number of different governmental units. This section is concerned with the sources and distribution of the money which is obtained for highway purposes; in other words, with the financing of highway systems and their improvement. Attention will be principally focused upon current methods of providing money for highway expenditures and the distribution of these sums. Certain general principles of highway finance and taxation will also be presented.

There are two general methods of financing highways. The first of these is the "pay-as-you-go" method, which involves paying for all highway improvements and the costs of maintaining and operating the highway system from current revenue. This is the method which is currently in use by many governmental agencies, and rightly so in many cases because of constitutional prohibitions or similar legislative provisions against borrowing for highway purposes. The second method is, of course, that of borrowing money to pay the costs of highway improvement, with the borrowed sums plus interest being repaid over a period of time from future income. This latter method may be generally termed "credit financing." In most cases it is also "bond financing." Both methods have merit under certain circumstances, and strong advocates of either method may be found among the various authorities in the highway field.

Attention will first be given to the various sources of revenue for highway purposes. These sources may in many cases be regarded as sources of current revenue, although they may also be used for debt service, either on obligations incurred in the past or for obligations incurred at the present time and against which some of these revenues may be pledged for future payment. Bond financing is considered separately in a later article.

3–18. Sources of Revenue for Highway Expenditures. There are four principal sources from which the funds necessary to the operation and improvement of the nation's highway system are obtained. These principal sources are as follows:

1. Highway user taxes
2. Federal funds
3. Property taxes
4. Tolls
5. Miscellaneous sources

HIGHWAY USER TAXES. Levies which here are classed as highway user taxes are of three general types: (a) fuel taxes, (b) registration taxes and related fees, and (c) special taxes upon commercial vehicles. These taxes are in large measure imposed by the governments of the various states, although they are also imposed by local governments in a relatively small number of instances. Income from state highway user imposts reached an all-time high of approximately $7.5 billion in

1966 out of a total income for all roads and streets of approximately $14.3 billion, not including bond issues of about $1.5 billion. Nearly two-thirds of the total obtained from state highway user taxes was derived from fuel taxes.

(a) *Fuel Taxes.* Taxes were first levied upon motor fuels in this country in 1919 by Oregon and three other states. By 1929 all states levied a gasoline tax, that is, a stated tax upon each gallon of gasoline sold as motor-vehicle fuel. At the present time gasoline taxes range from 5 to 7 cents per gallon in the continental United States, and are slightly higher in Alaska and Hawaii. A large number of states increased gasoline tax rates in the period immediately following the end of World War II. As has been indicated, this tax is by far the greatest revenue producer of all the highway user taxes, and in most cases it is far and away the biggest single source of income for the operation and improvement of the state highway system.

The gasoline tax is, of course, a direct motor-vehicle tax, being levied in direct proportion to the amount of use of the vehicle, since gasoline consumption is directly proportional to the mileage driven. The federal government also collects a tax on each gallon of gasoline sold for highway purposes. Its method of application and impact are the same as those of the state gasoline taxes, although it is technically classed as an excise tax. This federal tax will be further discussed under the heading of federal funds. A small number of local units of government also levy gallonage taxes on gasoline, but the total amount of revenue derived from these taxes is insignificant when compared with that collected by the state governments.

(b) *Registration Taxes and Allied Fees.* Laws pertaining to the registration of motor vehicles have a somewhat longer history than those relating to the imposition of a fuel tax, since the first registration law became effective in New York in 1901. By 1917 all states had enacted laws of this type. Although the procedure of registration was first adopted as a regulatory device, it soon became recognized as a ready source of highway revenue. The practice of graduating registration fees with the weight or capacity of the vehicle also began early in the modern era of highway development in this country, being first occasioned by the great increase in the number of trucks using the highway system after World War I. This practice is now followed in nearly all states. As is the case with gasoline taxes, registration fees are principally collected by the state governments. Several other taxes and fees are closely associated with registration fees and are levied by the various state governments. These special fees include such things as operators' and chauffeurs' license fees, fees for certificates of title, title-transfer fees, and fees charged for duplicate license plates. In many states income derived from these latter sources is used primarily for the operation of the highway system rather than for actual construction and maintenance. For example, in some states money derived from operators' and chauffeurs' license fees goes wholly or in part to pay the expense of maintaining a highway police organization.

As in the case of gasoline taxes, registration fees, or what are sometimes called "wheel taxes," are levied by a comparatively few local governmental units. In some 10 states local units of government are permitted to levy this form of highway user tax, with such levies being principally made in cities. Chicago is an outstanding example of a city which levies a tax of this type, with several million dollars per year being derived from this source.

(c) *Special Taxes on Commercial Vehicles.* The practice of assessing special fees against motor vehicles used for commercial purposes, including both passenger and freight carriers, has developed somewhat more gradually and on a less uniform basis than those taxes which have been previously discussed in this section. Special fees are levied against this class of motor vehicles in proportion to their use of the highway system and, roughly at least, in proportion to their effect upon the physical design of the highway. These imposts have generally taken the form of graduated fees computed upon a total mileage basis, on the basis of ton-miles or passenger-miles of travel, or taxes upon gross receipts. The special nature of these vehicles is also taken into account by the graduated registration fees previously described.

The general basis upon which these taxes are levied seems to be sound as there is little question that certain classes of commercial vehicles, for example heavy trucks and tractor-trailer combinations, have a disproportionate effect upon the design of the highway. The thickness of a pavement which is required to carry heavy trucks may be considerably greater than that required for the adequate support of passenger automobiles. Similarly, a direct effect may be seen in the necessity of providing extra lanes for heavy trucks on long grades. They are often found necessary in highways designed to carry both passenger cars and heavy commercial units in order to provide some relief from congestion. A long and continuing argument has been going on for many years between those who represent the interests of commercial haulers, on the one hand, and highway administrators and engineers, on the other, relative to the imposition and collection of these special taxes. A detailed study (*170*) showed that average user-tax payments per vehicle, per vehicle mile, and per ton-mile were $50, 0.54 cent, and 0.27 cent for passenger cars; $131, 1.21 cents, and 0.18 cent for trucks and truck combinations; and buses $470, 1.85 cents, and 0.20 cent, respectively.

Truck combinations alone contributed $881, 2.25 cents, and 0.12 cent, respectively. Representatives of the trucking industry felt that imposts on the heavier vehicles were excessive.

Methods of allocating cost responsibility among various classes of highway users are discussed in Art. 3–19.

FEDERAL FUNDS. For many years, funds collected from certain federal excise taxes (principally on gasoline) went into the general revenue fund of the federal government. Appropriations were then made from the general fund for federal aid to highways and the construction, maintenance, and operation of roads in national parks, national forests, and so on. However, the Highway Act of 1956 changed this system and established the Highway Trust Fund, from which expenditures are made directly for federal aid and for roads on federally controlled lands. Into the Trust Fund go monies received from federal taxes on gasoline and diesel fuel, tires and inner tubes, tread rubber, new trucks, truck-trailers, and buses; there is also a highway use tax on heavy vehicles. Tax rates established by 1961 amendments to the 1956 Act were as follows:

Motor fuels—4 cents per gallon.
Tires—10 cents per pound (highway); 5 cents, others.
Inner tubes—10 cents per pound.
Tread rubber—5 cents per pound.
New trucks, truck-trailers, and buses—10 per cent of sale price.
Highway use tax on vehicles over 26,000 pounds—$3.00 per 1000 pounds of gross weight.

In 1965, Congress voted to place the revenue from the 6-cent-per-gallon tax on lubricating oil and the 8 per cent tax on truck and bus parts in the Highway Trust Fund, beginning Jan. 1, 1966. In 1966, these combined levies produced nearly $4.2 billion.

A provision in the 1956 Act (the Byrd Amendment) provided that expenditures from the fund could not exceed receipts at any time. In other words, the Trust Fund had to be administered on a "pay-as-you-go" basis. (Actually, the Trust Fund can be "in the red" during a fiscal year, but must be in balance at the end—June 30—of each fiscal year.)

PROPERTY TAXES. There are two types of property taxes which are of importance when sources of funds for highways are being considered. These are (1) ad valorem property taxes and (2) special assessments against real estate lying contiguous or adjacent to a highway improvement. The first of these types of property taxes is extensively used in the financing of highway improvements by local units of government, including counties, townships, incorporated places, and similar administrative units. The second type is now principally used in towns and cities. Neither of these types is an important source of highway revenue to state governments at the present time.

The common concept of the use of general property taxes for highway improvements, which may be levied upon real property or upon both real and personal property, is that such a procedure is justifiable if the improvements which are to be effected are of general benefit to the community. The same line of reasoning is applicable if the improvements are expected to enhance the value of property located throughout the area. In practice, of course, in many local governmental units, property taxes are not levied specifically for highway improvements and are not earmarked for this purpose. Money derived from this source goes into the general revenue fund and is then budgeted, generally on an annual basis, for highways, education, public health, welfare, and such.

General property taxes are, of course, levied upon the assessed or "fair" value of the land and improvements located thereon, or on a similar value of personal property. General property taxes have numerous faults, one of the principal ones being that income from this source is likely to drop sharply in depression years. It is also claimed that taxes of this type are frequently unfairly distributed, expensive, and even confiscatory. Nevertheless it seems quite certain that many local governmental units will continue to depend upon general property taxes as a source of income for highways and other public services for many years to come.

Special assessments are simply what the name implies. They are special assessments which are made directly upon the owners of property which lies on or adjacent to a highway or street improvement in accordance with the benefits which are expected to accrue directly to the parcels of land thus affected. As has been indicated, this form of property tax is now used principally in cities and is not in widespread use for the financing of rural roads. It has been most generally used in recent years in the provision of street improvements in new residential areas and is less generally used for reconstruction. Special assessments are usually made on a frontage basis rather than on the property value. The assessments may be designed to recover the entire cost of the improvement over a period of time, or the initial cost may be shared among the property owners and the city government. Assessments may be made on abutting property only, or on property located on adjacent side streets and even parallel streets, depending largely on the nature and amount of the benefits which are expected to accrue from the proposed improvement. Assessments against property located on adjacent streets would generally be made on a reduced or sliding scale. As is the case with general property taxes, special assessments have some serious inherent disadvantages. It seems quite certain, however, that this method of financing street improvements will be continued.

TOLLS. Funds derived from the assessment of tolls for the use of a particular facility have become an important source of revenue as a result of the construction of a large number of toll roads following World War II. Prior to that time, tolls were generally limited to assessments for particular structures, such as a bridge or tunnel. Tolls are levied by public agencies, including special authorities of various types, to repay costs incurred in the construction of the facility (see Art. 3–21 for a discussion of bond financing). As soon as the debt incurred by the construction of the facility has been paid, the facility is then made "free." In 1966, toll receipts totaled approximately $744 million.

MISCELLANEOUS FUNDS. Brief attention may be given at this point to general revenue funds as a source of money for the financing of highway expenditures. Generally speaking, funds from general revenue have not been of major importance in the improvement of highways by the state governments in recent years. In a few states, however, income derived from taxes levied for highway improvement, consisting chiefly of highway user taxes, goes into the general fund from which the funds required for operation and improvement of the state highway system are budgeted periodically (this system is followed in New York, one of the nation's top roadbuilding states). In states where this system is used, the amount of money appropriated from the general fund would usually be substantially the same as that which would normally accrue to the state highway department from highway user taxes, so that the end result is about the same. Students of taxation are frequently strong advocates of this method of putting income derived from all tax sources into the general fund, with appropriations being made at intervals for all public services on the basis of need. From the viewpoint of the highway administrator, this concept has a serious disadvantage in that highway improvements are best made upon a planned, orderly basis with the assurance of a steady and predictable annual income. In 1966, combined receipts from property taxes, general revenues, and miscellaneous sources totaled approximately $2.6 billion.

Another source of revenue which is worthy of mention here is that derived from parking meters, as used in many towns and cities. Although parking meters were originally envisaged as regulatory devices, the income derived from them has proved to be very substantial and has become quite important in many urban areas. Many thousands of parking meters are now in use in cities and towns over the country; in 1966, revenues from parking fees totaled $66 million.

General Considerations Relative to Highway Revenues. Several general considerations relative to the securing of needed revenues for highway expenditures may be briefly discussed at this point. Even though highway user taxes are at a high level, a relatively low proportion of total highway transportation cost actually goes for the construction, maintenance, and operation of the highway system. For example, in New York in 1966 the total direct cost of state highway user taxes to the owner of a private passenger car which was driven 10,000 miles was $79.52. This sum was determined on the basis of 15 miles to a gallon of gasoline, a state gasoline tax of 6 cents per gallon and a license fee of $39.50. Thus, in New York highway user taxes on this vehicle total slightly less than 0.80 cent per mile of travel. Thus if it is assumed that the cost of operation of this vehicle is approximately 10 cents per mile, it is apparent that only about 10 per cent of the cost of owning and operating the average vehicle actually goes to the improvement and operation of the highway system, including the federal taxes on gasoline. In fact, in many cases the total contributed by the owner of a passenger vehicle is less than the cost of insurance on the vehicle. On this basis it seems apparent that further increases in highway user taxes may be justifiable as well as necessary if a modern highway system is to be provided and maintained.

There is, however, another aspect to this general situation. This lies in the fact that the general tax burden carried by the individual has risen to such high levels in recent years that some people, at least, believe that tax increases of any sort are unwise and undesirable. Still another aspect of the general picture of highway financing is that in many instances, both in the past and at the present time, funds available for highway improvement have not been expended either wisely or well. Attention may then be given to the proper and efficient expenditure of funds already available before new sources of revenue are considered.

3–19. Highway Cost Allocation. One underlying concept should govern all levies for highway purposes. This concept is that the cost of highway improvement should be borne on an equitable basis by the three principal groups who benefit from highway improvement, namely, (1) the highway user, (2) property owners, and (3) the general public. Every effort should be made to assess the cost of improved highways in an equitable manner against these groups. The effort should be made at every level of government. The accomplishment of the equitable distribution of the tax burden in proportion to benefits received is, of course, much more difficult than the statement of this underlying objective.

Highway cost allocation is a thorny fiscal (and philosophical) problem that has occupied the attention of highway economists for many years.

First consideration usually is the division of cost responsibility among highway users and non-users. The comprehensive allocation cost study prepared by the Bureau of Public Roads and submitted to Congress in 1961 recommended that about 8 per cent of the cost of the federal-aid highway program be charged to revenue sources other than motor vehicle taxes (i.e., to nonhighway users) (*151*). This recommendation was ignored by the Congress.

The excise tax schedule enacted by Congress in 1961 and given in Art. 3–18 was a compromise solution, which did not follow precisely any of the recommendations of the Bureau of Public Roads; the bureau had made recommendations based on four different methods of allocating cost responsibility, including the two described below.

In 1965, in conjunction with its upward revision of the estimate of the cost of completing the Interstate System (and an upward revision of tax schedules, if the program was to be completed on schedule), the Bureau of Public Roads presented another analysis of highway cost allocation to Congress. This report covered only two methods of cost allocation: the incremental cost method and the differential benefit method.

The incremental cost method is the traditional way of allocating highway user charges. In this approach, each element of highway design affected by the size or weight of vehicles is broken down into a series of additions (increments); the cost of providing each increment is assigned only to those vehicles whose size and weight require them. Thus, all vehicles share in the cost of the basic design, but the cost of each succeeding increment is borne only by the larger vehicles.

The differential benefit method assigns cost responsibility to the various users in direct proportion to the vehicular benefits they receive by their use of the highway system. Four types of vehicular benefits were considered: reductions in operating costs, in time costs, in accident costs, and in the strains and discomforts of driving. The last (impedance costs) were calculated only for passenger cars.

The following are the estimated annual cost responsibilities for federal-aid highways for various classes of vehicles under 1961 legislation, the incremental cost method, and the differential benefit method:

Automobile, 4400 lb.	$30, $31, $36
Pickup truck, 2-axle, 4-tire, 6000 lb.	$40, $24, $25
Stake truck, 2-axle, 6-tire, 14,000 lb.	$66, $41, $31
Van truck, 2-axle, 6-tire, 18,000 lb.	$80, $61, $42
Tractor-semitrailer, 3-axle, gasoline, 45,000 lb.	$466, $462, $344
Tractor-semitrailer, 4-axle, gasoline, 55,000 lb.	$676, $771, $543
Tractor-semitrailer, 4-axle, diesel, 55,000 lb.	$897, $1189, $910
Tractor-semitrailer, 5-axle, diesel, 66,000 lb.	$923, $1369, $782

3–20. Distribution of Highway Revenues. A problem which is closely akin to the one discussed in the previous article is that of equitably distributing the revenues which are available for highway purposes. It is considered fundamental that revenues which are available for highway purposes be distributed (and expended) in such a way as to provide maximum benefit to those who actually pay the greatest share of the cost. The difficulty of applying this concept is fairly obvious when the nature of the motor vehicle, the manifold uses to which it is put, and the extreme variation in driver behavior are considered. The difficulty of applying this concept may be illustrated by a simple example. Consider, first, a primary state highway which carries a very heavy volume of traffic. Contributions of those who use this facility, or what are sometimes termed the "earnings" of the highway, are likely to be very large. In other words, the income derived from highway user taxes, as determined on a vehicle-mile, ton-mile, or passenger-mile basis, is almost certain to be more than the cost of the improvement when considered over a period of years. Thus it may be said that this class of highway "pays its own way" in terms of the revenue obtained from those who use this class of facility. On the other hand, a county highway may carry only a few hundred vehicles per day and the "earnings" from this type of road may be much less than its cost. Nonetheless, the latter road may be an important link in the county highway system and may play a vital part in the welfare of the population as a whole. Improvement of this latter type of road may have to be justified by considerations relative to benefits accruing to property owners and the general public, who may or may not pay an equitable share of the cost of its improvement and operation.

Table 3–5 presents information relative to the distribution made of highway revenues in 1964 and predicted for 1970. Table 3–6 is a breakdown of estimated capital expenditures for highways for the same years.

Two of the items given in Table 3–5 deserve additional explanation. These are the items listed as "Interest on debt" and "Debt retirement." These items reflect the money expended in the payment of obligations previously incurred for highway purposes. The total highway debt of the state governments, counties, and municipalities was estimated by the Bureau of Public Roads to be somewhat more than $14.4 billion at the end of 1962; much of this was for toll facilities. Although the sums required for debt service of these obligations represents a sizable portion of the total income for highway purposes, it is generally felt that the various governmental units, taken as a whole, are still in a favorable position with regard to the relation between the existing highway debt and the total capital investment which has been made in the various units of the highway system. Several of the toll roads built since World War II have been highly successful from a financial point of view, as are many toll bridges and tunnels.

Another factor in the distribution of highway revenues is diversion. By "diversion" is meant the expenditure of funds collected from highway user taxes for nonhighway purposes. Although these diverted funds are

TABLE 3–5

DISTRIBUTION OF HIGHWAY REVENUES, ALL ROADS AND STREETS, BY FUNCTION, IN MILLIONS OF DOLLARS (ESTIMATED)

	1964	1970
Capital outlay	$8,082	$9,115
Maintenance	3,205	4,127
Administration and research[1]	625	809
Highway police and safety	428	522
Interest on debt	505	613
Total current expenditures	$12,845	$15,186
Debt retirement (par value)[2]	744	1,058
Grand total	$13,619	$16,244

[1] Includes engineering and equipment costs not charged to capital outlay and maintenance and miscellaneous expenditures.
[2] Redemptions by refundings not included.

TABLE 3–6

ESTIMATED CAPITAL EXPENDITURES FOR HIGHWAYS, IN MILLIONS OF DOLLARS

	1964	1970
State highway departments	$6117	$7050
State toll facilities	459	325
Local toll facilities	72	7
Municipalities	730	971
Federal government	189	216
Total	$8082	$9115

in general expended for very worthwhile purposes, with expenditures of this sort being principally made for education at the present time, the diversion of highway funds is regarded in a very unfavorable light by the large majority of highway administrators and engineers. The principal basis for condemnation of diversion arises from the concept that since highway-user tax schedules are predicated upon benefits received by the taxpayer in the form of highway improvements, then these taxes become unjust and inequitable if funds derived from this source are expended for purposes other than highway improvement. The amount of diversion, expressed in terms of percentage of highway user tax receipts, has decreased substantially in recent years. There is every reason to believe that this trend will continue. A large number of states now have constitutional provisions which prevent the diversion of highway funds. In 1964, diversion was estimated at about $626 million, slightly more than 9 per cent of the United States totals of highway user taxes (*208*).

3–21. Bond Financing for Highways. It sometimes becomes necessary or advisable for a governmental unit to obtain funds for highway improvements by credit financing. That is, the government must make use of its credit in order to borrow money which is then repaid, with interest, usually over a period of years. Necessary improvements of a highway system or a specific facility may be financed in this manner. The principle involved is much the same as when a private individual needs to borrow money in order to build a new home, buy a new car, or for some other purpose.

Funds borrowed for highway purposes may, in some cases, be obtained by short-term loans. Loans of this type can usually be secured from banks by means of short-term notes. Funds obtained in this manner might, for example, be used to provide means for highway expenditures when a large sum of money was needed during the construction season and the date of tax collection was somewhat later in the year. This type of borrowing is comparatively rare in modern highway finance, and by far the greater amount of credit financing for highways is done through the issue of bonds.

One form of bond which may be used for the financing of highway improvements is the general-obligation bond. This type is backed by the entire faith and credit of the issuing agency. In recent years, general-obligation bonds have been regarded with less favor than in the early period of highway improvement, and the trend has been toward the use of the so-called "limited-obligation bond." Limited-obligation bonds are usually secured by the anticipated revenues from a particular source of income. For example, many limited-obligation bonds are now secured by the pledging of future income from gasoline tax receipts to their repayment. Still another form of bond is what may be termed here a "revenue bond." A bond issue of this type may be employed, for example, in the construction of a toll road or bridge, and the bond issue secured by the pledging of future tolls to repayment of the bonds. Limited-obligation bonds are also sometimes termed, somewhat loosely perhaps, "revenue bonds."

The mechanics of a bond issue and its subsequent repayment may be generally described as follows. The total amount of the bond issue is decided upon by the governmental agency involved, and then approved, by referendum or legislative action. Bonds are then issued and are sold, frequently on a competitive bid basis, to banks, investment companies, or similar organizations, or in some instances to the general public. The amount of interest carried on the bond issue will generally be determined by the type of bond issue, the term of issue, and the credit standing of the issuing agency. The total amount of the bond issue, plus interest, is then repaid by the issuing agency in any one of several ways, as briefly described later. The bonds are said to have been "retired" when they are repaid.

Highway bonds have been issued for varying periods, with typical periods ranging from 10 to 50 years. Ten years is probably somewhat too short a period of issue for most highway bonds, while a period of 50 years might be regarded as proper under some circumstances in the financing of some relatively long-lived structure, such as a road, bridge, or tunnel.

Any one of several general methods may be employed to retire bonds which have been issued for highway improvements. Two of these general methods will be described here. One of these may be termed the "sinking-fund" method and the other the "serial" method. Bonds which are to be retired by the use of a sinking fund are generally term-issue obligations. That is, the entire amount borrowed falls due at the end of a period of time which is fixed when the bonds are issued. Money which will be required to pay the obligation when it falls due or when the bonds "mature" is provided by setting aside an annual sum toward the payment of the borrowed amount. Payments which are made to the sinking fund must be large enough so that their sum, plus whatever income may be derived from their investment, will be sufficient to retire the bonds when they mature. The sinking-fund method has several inherent disadvantages, chief among which is the temptation which the sinking fund continually presents to public officials over a period of years as it gradually accumulates. This and other factors have generally led to the abandonment of the sinking-fund method in favor of serial bonds. Serial bonds are retired by periodic (annual) payments upon the principal plus periodic payments of interest. In their most common form, constant payments are made upon the principal each year, with annual interest payments which decrease as time passes. No sinking fund is required when this method of payment is used, and the bonds are completely retired at the end of their term of issue.

There is little doubt but that bond financing is entirely proper and desirable as a financing method for highways, at least under certain circumstances. Bond issues of this type may generally be justified by application of the benefit theory which has been previously emphasized throughout this chapter. An example of this reasoning lies in the major arterial improvements which are now being planned and executed in many urban areas. Benefits derived from these improvements in the form of decreased operating costs, relief from congestion, time saving, and decreased accident rates are so large as to dwarf into relative insignificance the added cost of financing involved in bond issues. Thus, properly planned projects of this general type may be wisely and logically financed by bond issues. The improvement of an entire highway system by bond financing may also be justifiable in some instances. Benefits resulting from broad improvements may again make credit financing desirable. Credit financing may also be resorted to in an effort to bolster the general financial condition of a state through increased employment resulting from construction activity.

The chief disadvantage accruing from bond financing is the added cost occasioned by the payment of interest on the obligation. This consideration may frequently be eliminated by the benefits resulting from the improvement and the resulting reduced costs of highway transportation. Bond issues should not, however, be used for the payment of current obligations. This is an unsound policy in public finance as well as in private finance where an individual borrows money to pay current bills. Other disadvantages which may be associated with bond issues include the generation of a volume of work beyond the capacity of the contractors in the area, with a resultant lack of competitive bidding with accompanying price increases, and the fact that a volume of work may be involved which cannot properly be handled by the engineering force of the agency issuing the bonds.

PROBLEMS

3–1. A governmental agency is contemplating the improvement of a section of primary state highway by relocation between two major control points. A survey is made and it is determined that the distance of 4 miles between these two points on the old road may be shortened to 3 miles in the new location. It may be considered that the old road has reached the limit of its economic usefulness and that its abandonment need not be considered in relation to the new location. Traffic on the existing route is 1000 vpd and it may be assumed that the average traffic over the next 20 years (the design period) will be 1750 vpd. Traffic is predominantly made up of passenger vehicles. The proposed route will have a pavement width of 22 feet, compared to an existing 18 feet. Superelevation on the curves of the existing route is generally inadequate (0.05 foot/foot); the new route will be superelevated properly, so no correction for curvature is needed.

The following traffic data are of consequence in determining the type of operation.

Route	Thirtieth Highest Hour	Capacity
Existing	350	450
Proposed	375	600

Other essential data are:

Route	Design Speed	Running Speed	Lanes	Length (miles)	Grade Class	Surface and Condition	Curvature
Existing	50	37	2	3.0	0–3	Paved–good	50%–8°
Existing	50	34	2	1.0	3–5	Paved–good	50%–8°
Proposed	60	42	2	3.0	0–3	Paved–good	Negligible

The average annual unit maintenance cost on the existing route is assumed to be $1200 per mile, while on the new route it is expected to be $900 per mile. The total estimated cost of the improvement is $1 million and the local rate of interest is 7.0 per cent. Cost of the project is divided as follows:

Item	Estimated Life (Years)	Cost
Pavement	20	$100,000
ROW	60	75,000
Grading, drainage, and structures	40	825,000

Compute the benefit cost ratio for the proposed improvement.

3–2. Using the other data of Problem 3–1, assume that 10 per cent of the traffic is made up of single-unit trucks and 5 per cent is composed of truck combinations. Assume that this quantity of truck traffic does not affect the type of operation on either the existing or proposed facility. Compute the benefit-cost ratio for the changed conditions.

3–3. For the data of Problem 3–2, compute the benefit-cost ratio for an interest rate of 0 per cent, 4 per cent, and 10 per cent. Compare these values with the answer of Problem 3–2.

3–4. It is proposed to surface a section of existing gravel road on which the current traffic volume is 500 vpd; the new pavement is to be a plant-mix bituminous surface. It is estimated that the average traffic will be 600 vpd over a 12-year period, which is the estimated life of the bituminous surface. Estimated traffic will include 15 per cent light trucks, for which operating costs may be taken to be twice those of passenger cars. Minor sight distance improvements will also be made on the route.

Both the existing and improved routes have two lanes, and a design speed of 40 mph. Assume no superelevation of the existing surface and superelevation of 0.03 foot per foot on the improvement. The following are other essential data:

Route	Running Speed (mph.)	Length (miles)	Grade Class (%)	Type of Operation	Surface and Condition	Curvature
Existing	27	3.5	0–3	N	Loose–good	50%–8°
Existing	23	1.5	5–7	N	Loose–good	50%–8°
Proposed	32	3.5	0–3	F	Paved–good	50%–8°
Proposed	27	1.5	5–7	F	Paved–good	50%–8°

The total capital cost of the proposed surface is $50,000 with maintenance cost estimated at $700 per year per mile. The maintenance of the existing gravel road is estimated to cost $900 per mile per year. Interest rate is 6.0 per cent.

Compute the benefit cost ratio for the proposed improvement.

3–5. There are several possible alternative sources of action that might be followed in improving an existing route. Analysis of the alternatives yields the cost data tabulated below. Alternate No. 1 is the existing condition and the life of the improvement is estimated at 30 years.

Alternate	Capital Cost	Annual Maintenance Cost	Annual Road User Cost
1	–	$8500	$1,000,000
2	$1,000,000	7000	800,000
3	1,250,000	6000	700,000
4	1,400,000	6000	750,000

Determine the most desirable of these alternatives by:

(a) The benefit-cost ratio, using an interest rate of 7 per cent.

(b) The minimum annual cost method.

3–6. Determine the annual income of the government of your state from direct highway user taxes. Consult the annual report of your state highway department, or some other source, and determine how this money is distributed to the various governmental units in your state.

3–7. Estimate the annual contribution which the owner of an average light passenger car makes in the form of direct highway-user taxes levied by your state government. Assume 10,000 miles of travel and 15 miles to a gallon of gasoline. How much does this motorist contribute to the Federal Highway Trust Fund each year?

3–8. Prepare a brief report on a major highway (or street) improvement project which is now under way or planned in your state, for which bond financing is to be used. Describe the nature of the project, the total amounts of money involved, and the financing arrangements.

3–9. Of the taxes now levied by the federal government upon gasoline and other highway products, determine those earmarked for the Highway Trust Fund established by the Federal-Aid Highway Act of 1956. From current literature, determine the present status of the Highway Trust Fund. Draw a chart showing the relationship between estimated income and expenditures from the Trust Fund during the period of construction of the National System of Interstate and Defense Highways.

3–10. From current literature, prepare a report showing the effect of a modern highway improvement (such as the construction of an expressway) upon the value of adjacent land.

3–11. Prepare a report abstracting the major conclusions and recommendations of the Highway Cost Allocation Study authorized by the Federal-Aid Highway Act of 1956. The report was presented to Congress in 1961 and supplemented in 1965. What has been the major effect of the study?

CHAPTER 4

HIGHWAY PLANNING

Highway planning is a vital function which has as its purpose the orderly development of an adequate and economical highway system to meet present and future needs. It is often part of a larger transportation planning process, in which all forms of transport are considered. It also is a key portion of any comprehensive urban planning process. Most of the discussion presented in this chapter is directly concerned with the planning activities of state agencies, but many of the facts presented are also applicable to highway and street planning by local governmental authorities. Sound planning is necessary to the successful execution of a county highway program or city street program, just as it is to state highway programs.

In the period before 1930, most states had a primary or trunk-line system of roads, part of which had been incorporated into the federal-aid highway system and which consisted mainly of rural highways. The rapid growth of motor-vehicle registration had considerably increased the number of vehicles in use and the volume of traffic had increased at an even faster rate. Indications were that the main rural highways would soon become inadequate for future needs, and that the secondary rural roads and city streets were becoming inadequate for the increased volume of traffic they were being required to serve. In order to make a study and to develop policies as an approach to these problems, the so-called "planning surveys" were started in 1935 to gather the factual information necessary as a basis for analysis.

An act of Congress adopted June 18, 1934, and known as the Hayden-Cartwright Act authorized the expenditure of funds not to exceed 1.5 per cent of the federal-aid apportionment to any state for the making of surveys, plans, and engineering investigations of projects for future construction. Funds needed for the operation of the planning surveys were largely derived from this source. By 1940 practically all state highway departments, in cooperation with the Bureau of Public Roads, were doing such surveys. Recent federal-aid highway acts have continued to permit expenditure of funds for planning purposes; the percentage of available funds has been increased substantially beyond that of the original Hayden-Cartwright Act.

At the time of the passage of the Hayden-Cartwright Act, the philosophy of the Federal Aid Highway program was to assist the rural areas, but with the increase in motor vehicle registration and traffic volumes, signs of distress soon became apparent in urban and suburban areas. Thus it was logical that planning activities should be extended to these areas. With the passage of the Federal Aid Act of 1944, funds became available for federal-aid highway construction in urban areas.

The Federal Aid Highway Act of 1958 repealed all obsolete prior provisions of such acts and codified existing provisions into Title 23 Highways of the United States Code. The 1962 Act further defined the requirements for planning processes in urban areas over 50,000 population effective July 1, 1965. The law required the establishment of a "cooperative, comprehensive and continuing" transportation planning process in each of these areas, as a prerequisite for federal aid. By the deadline, nearly all of the more than 200 such urban areas in the country had met the requirement.

Intelligent planning requires the use of much factual information pertinent to the highway system and its use. This information can be obtained by means of various studies, including studies to determine the condition of existing facilities, the volume and nature of traffic using the highway system, the use and life expectancy of our roads, and the needs of the future, as well as fiscal studies necessary for the development of a well-coordinated program. These studies are described in the sections that follow.

4–1. Road Inventory. A road inventory consists of a complete on-the-ground check of the location and physical condition of all rural roads within a state. The width, type, and condition of all roadway surfaces are recorded. Classification of road types may be made on the following basis: (1) primitive roads, (2) unimproved roads, (3) graded and drained earth roads, (4) soil surfaced roads, (5) slag, gravel, or stone roads, (6) bituminous surface treated roads, (7) mixed bituminous roads, (8) bituminous penetration roads, (9) bituminous concrete, sheet asphalt, or rock asphalt roads, (10) portland cement concrete roads, (11) brick roads, (12) block roads, and (13) combination type roads. This classification has been recommended by the Bureau of Public Roads for the purposes of the planning surveys, and it differs somewhat from that used in later sections of this book. In the annual "mileage tables" published by the Bureau of Public Roads, similar surface types are grouped together.

The type, dimensions, and condition of all structures are recorded. This includes all bridges and culverts, overpasses, underpasses, and such.

The locations of all farms, rural dwellings, schools, churches, and other cultural features are also noted. To these items may be added

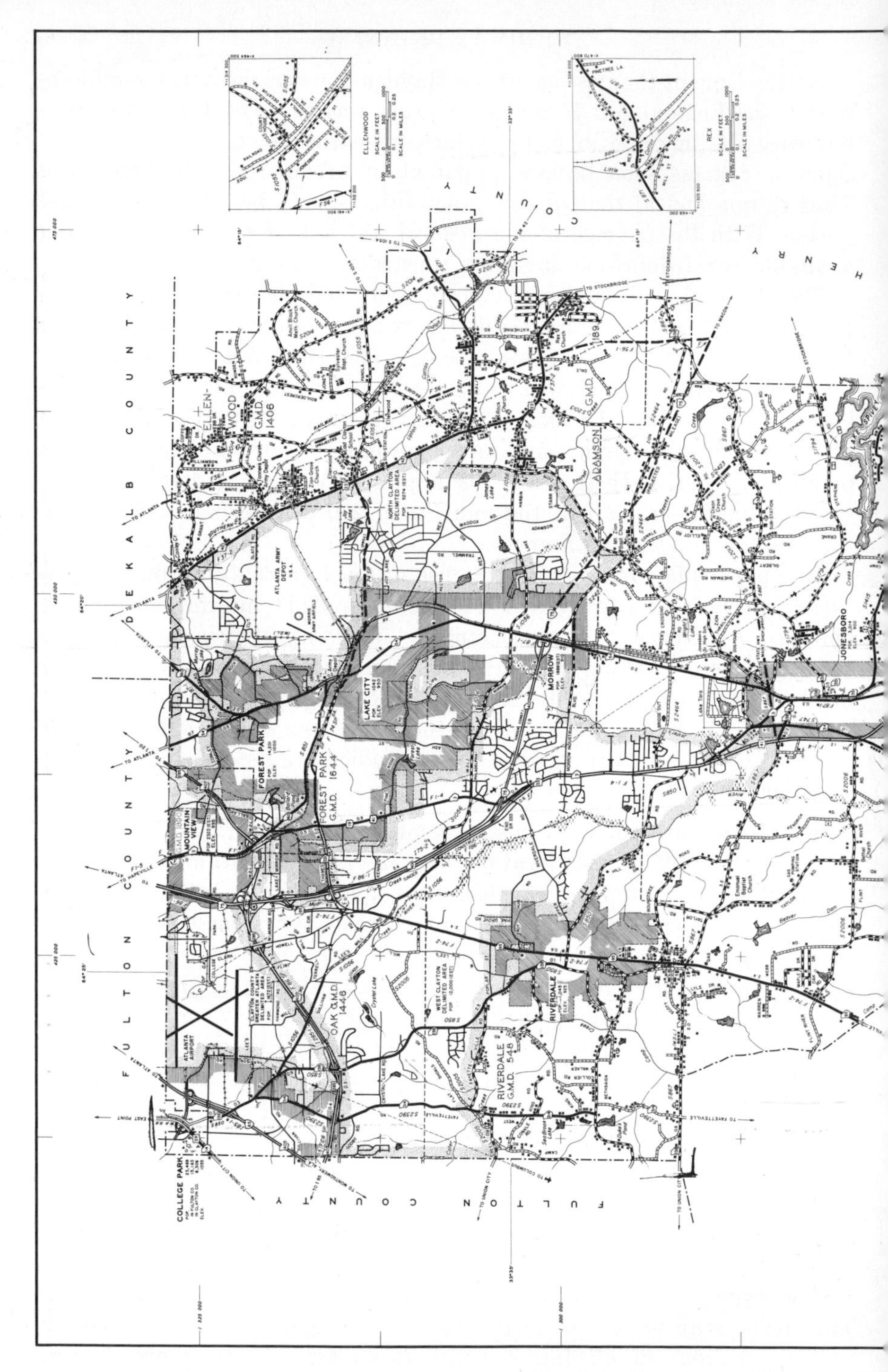

Fig. 4-1. Base map. (Courtesy Sta

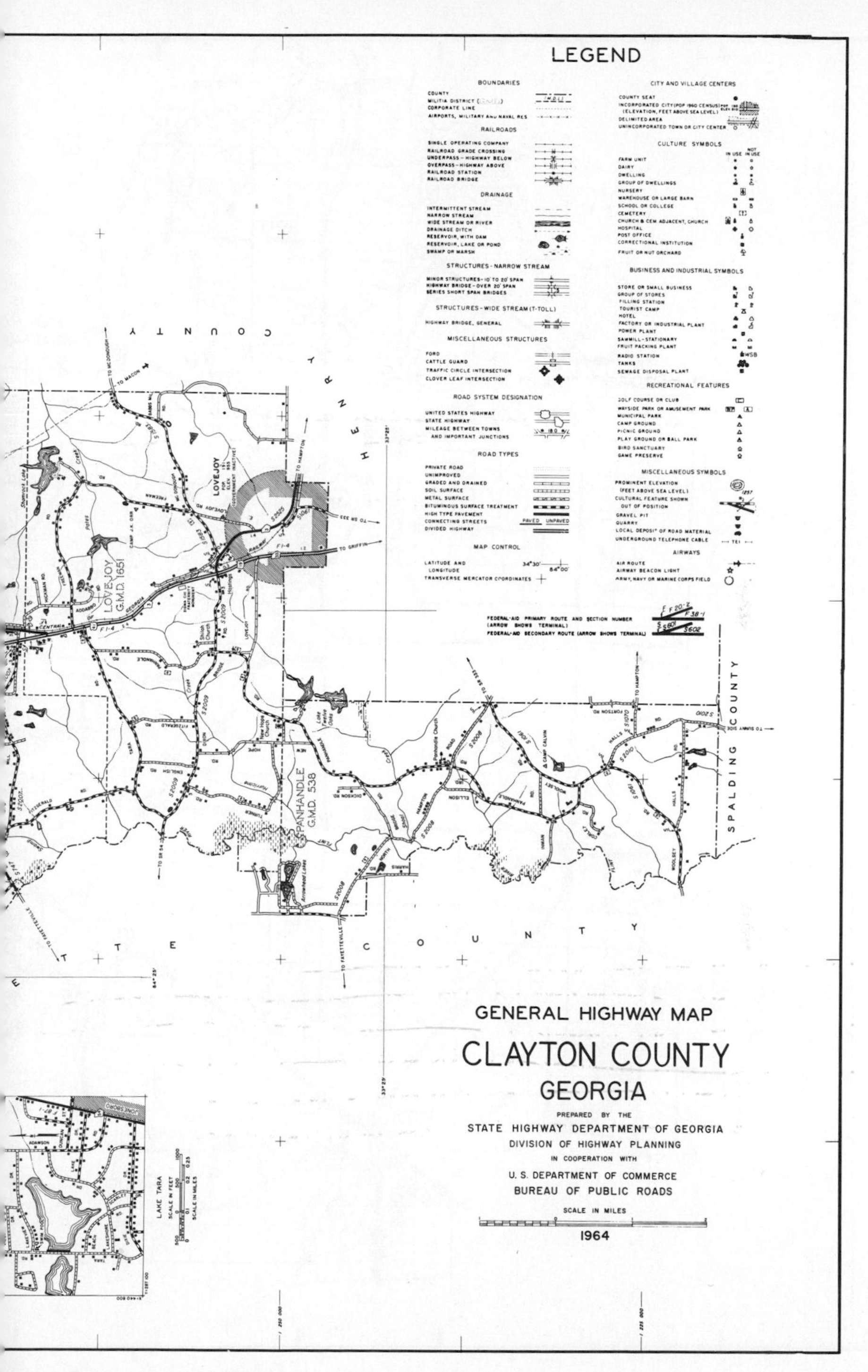

ghway Department of Georgia.)

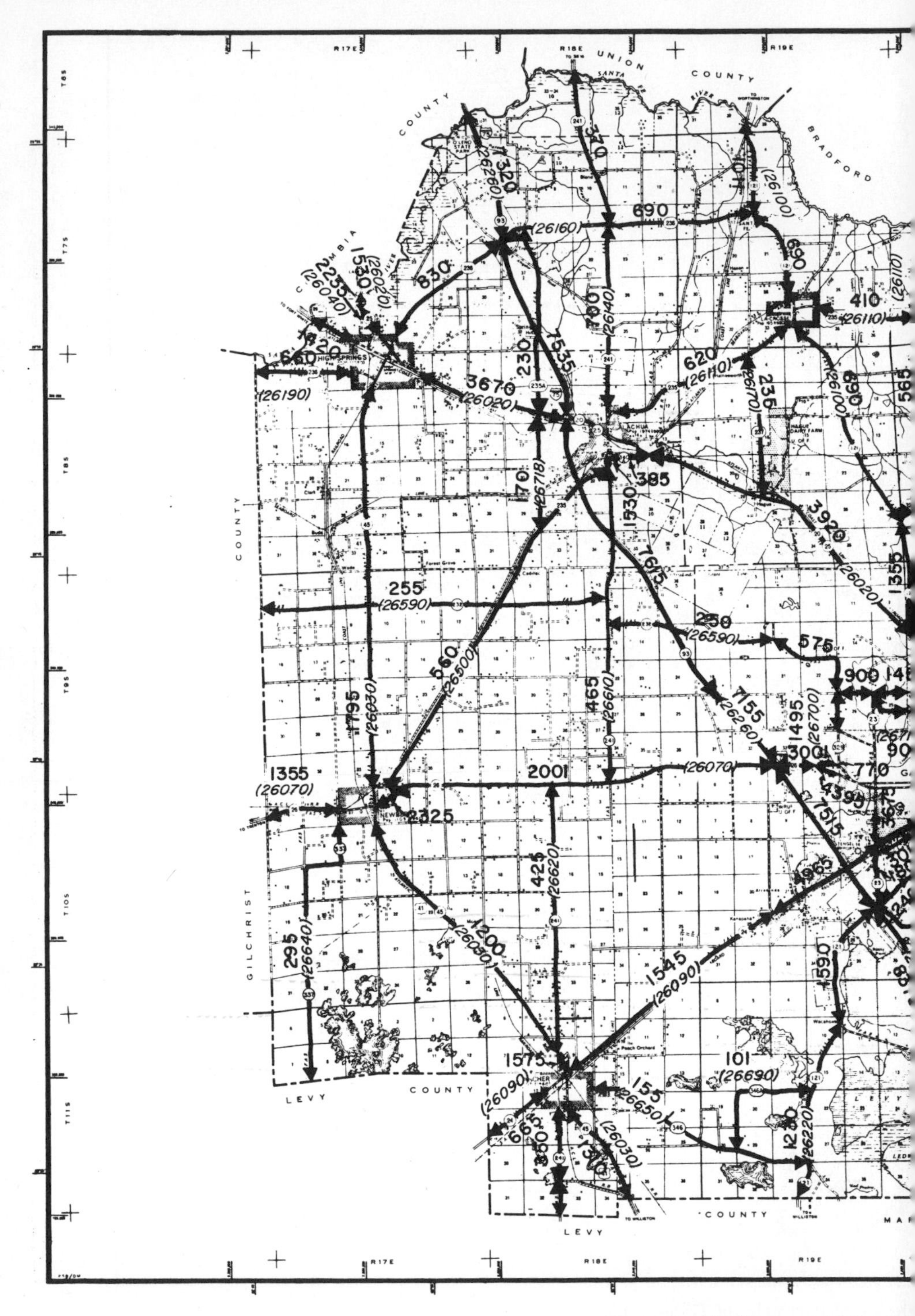

FIG. 4–2. Traffic map. (Courtes

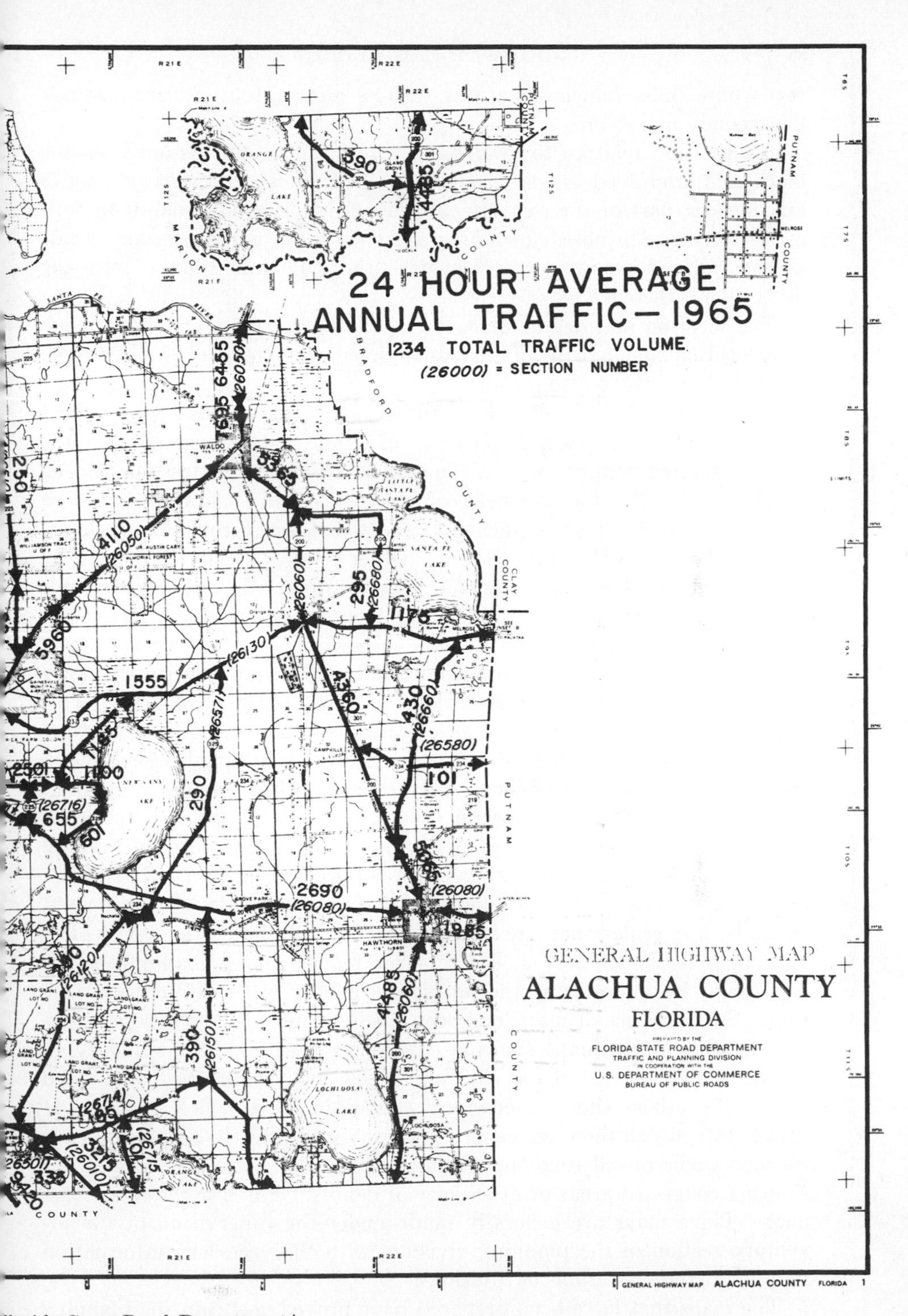
24 HOUR AVERAGE
ANNUAL TRAFFIC—1965
1234 TOTAL TRAFFIC VOLUME
(26000) = SECTION NUMBER
GENERAL HIGHWAY MAP
ALACHUA COUNTY
FLORIDA
PREPARED BY THE
FLORIDA STATE ROAD DEPARTMENT
TRAFFIC AND PLANNING DIVISION
IN COOPERATION WITH THE
U.S. DEPARTMENT OF COMMERCE
BUREAU OF PUBLIC ROADS
GENERAL HIGHWAY MAP ALACHUA COUNTY FLORIDA 1

'lorida State Road Department.)

many other miscellaneous features such as resorts, tourist camps, mines, fairgrounds, and so on.

Information relative to grades, curvature, and sight distances on the main and improved roads is also obtained by special survey parties and made a part of the road inventory. Existing conditions at all railroad crossings are noted, including sight distance, curvature, and grade at the railroad crossing, together with the type and kind of crossing protection if there is any.

The location and degree of curve of all sharp curves are determined. The location and nature of all restrictions which might be considered as traffic hazards are also noted.

Measurements are made along the roadway by means of odometers which register distances in hundredths of a mile and are usually mounted on the steering column of a car or truck. Various devices have been developed for the measurement of other necessary data, including a gradometer for the measurement of grades and a superometer for determining the rate of superelevation. Other special devices are also in use, though no attempt will be made here to discuss the detailed methods involved in securing the desired road inventory information. Tape recorders are used frequently in the field survey.

The large amount of information collected in an inventory survey must be tabulated and summarized, usually by machine methods. This information is then used in the preparation of inventory maps. A base map is first prepared which shows every mile of rural public road, every farmhouse and other residence, every church, school, store, mill, mine, and other place that is an origin or a destination of traffic that moves on the highway. These base maps also show every mile of railroad and navigable stream, and every railroad station, airport, wharf, city, village, and town. All these maps are drawn with uniform conventions, symbols, and scales that are agreed upon by all the states. A typical base map is shown in Fig. 4–1. Other maps can be made from the base maps by reproducing the latter and adding other desired information. Such a series of maps has been made to include (1) general highway transportation maps showing existing surface conditions and types of road surfaces, (2) maps showing school bus routes and post road series, (3) others showing common carrier bus and truck routes, (4) traffic maps which show the general highway map and the annual average 24-hour traffic on all roads appearing on the map, and (5) special maps showing congested areas or other special details. Fig. 4–2 shows a traffic map. These maps are generally made under the supervision of the inventory section of the planning division, with the necessary information supplied by other sections or divisions.

The maps that have been prepared have proved useful in the planning and coordinating of highway programs and improvements of the various federal, state, and local agencies. They have also proved beneficial to other agencies. For example, power companies can use them for studies of potential demand, and they can also be used by industrial concerns for charting sales territories and campaigns.

To be of value, any inventory must be taken at periodic intervals or kept current, as obsolescence, deterioration, and new construction are continually changing existing conditions.

Traffic Studies

The primary purpose of the early comprehensive traffic surveys carried out in all the states as a part of the planning surveys was to determine the traffic volume or traffic density on rural highways and to obtain information concerning the weights, heights, lengths, and other characteristics of trucks and buses. The information thus obtained was intended to serve as a basis for the determination of various features of design, such as the width and thickness of pavements, the priority of improvement on the basis of need, and the allocation of funds for the scheduling of construction operations.

To better understand the present-day traffic investigations, it will prove beneficial to study the elements of these original surveys as they constitute the basis for gathering and analyzing the factual data obtained at the present time. The benefit of experience and knowledge is reflected in the continuation and expansion of these early investigations. A description of the conduct of the original traffic surveys is given in the paragraphs immediately following.

The original traffic survey, like the inventory survey, was on a statewide basis. As a first step in determining traffic density and the characteristics of trucks using the state highway systems, a method of sampling was devised. Traffic was counted at a number of continuous counting stations, key stations, and coverage stations. The counting of all traffic was accomplished by human observers; counts of this type are referred to as manual counts.

4–2. Continuous Counting Stations. Continuous counting stations were (and are) located in various parts of the state. Some were located near urban areas, while others were in lightly populated districts. The locations of these stations were determined as a result of studies which enabled the investigators to classify each area as industrial, agricultural, recreational, or as a combination of these. The travel habits of highway users in these areas could then be determined from the traffic counts at these various stations. Consequently, the continuous stations would show practically every variation in travel habit that would be expected to occur on any route or section of the state road system.

Manual counts were made at these stations for 24 hours a day and continued for the period of the survey, which in the majority of states was for one year. The number of continuous count stations varied. Michigan had 14 such continuous stations in its original traffic survey, while South Carolina had 10.

4–3. Key Stations. Key stations, sometimes called "control stations," were located on the main trunk lines of the state road system. The location of each station was selected with care, preference being given to (1) routes carrying through traffic, (2) routes whose primary purpose was to serve as connectors between two towns, and (3) at intersections of farm-to-market routes with the state road system.

Eighteen eight-hour counts were made at each key station during the one-year period at different times of the day on different days of the week and in all seasons of the year. Fourteen of these eight-hour counts were made from 6:00 A.M. to 2:00 P.M. and from 2:00 P.M. to 10:00 P.M. They were known as "day density counts." The remaining four eight-hour counts were made from 10:00 P.M. to 6:00 A.M. and were called "night density counts."

The number of key stations operated varied according to the apparent need and the available funds. North Carolina, which may be considered as representative of the average, had 216 key stations, while Michigan had 396 key stations.

4–4. Coverage Stations. Coverage stations, which were located on the secondary road system, generally included all the main roads of the county system. At each coverage station a single eight-hour count was made from 8:00 A.M. to 4:00 P.M. The number of coverage stations was approximately one-fourth the total mileage of the state. For example, North Carolina had 14,000 coverage stations for 68,000 miles of road.

4–5. Weighing or Loadmeter Stations. As previously mentioned, in addition to determining the volume of traffic, information was desired on bus and truck characteristics and movements. Weighing or loadometer stations were established for obtaining this information. Axle loads were determined by loadometers or portable scales, and other information was obtained, including the type of vehicle, manufacturer's rated capacity, width, height, and length. Additional information as to the commodity carried, its origin, destination, and other pertinent facts, was also noted. These weighing or loadometer stations were identical with the key stations and had a similar cycle of operations. They were referred to as day loadometer and night loadometer survey stations.

Information obtained from the various key, coverage, and loadometer stations was supplemented with information obtained from toll bridges and ferries. These data proved of value in analyzing traffic habits and traffic patterns.

4–6. Traffic Analysis. One of the purposes served in taking density counts is to determine the annual average 24-hour traffic over the entire road system. The only stations from which this information can be accurately obtained are the continuous stations where the count is taken for the period of one year. It has been previously mentioned that 18 eight-hour counts were made at each key station throughout the year. This schedule was designed so as to permit determination of the average daily 24-hour traffic. These counts were so spaced as to take care of any hourly, daily, or seasonal variation in traffic volume. From these key station counts, the average daily 24-hour traffic is determined.

The next step in the analysis is to expand the eight-hour coverage station counts to average daily 24-hour traffic. It has been determined by a careful study of density counts that the total average daily 24-hour traffic volume in any area may vary considerably in total volume, but the percentage of the total volume that is recorded for any particular hour of the day is likely to be similar at all the stations in that area. This hourly variation indicates that the same travel habit prevails in the same territory. Consequently, if an eight-hour density count is made at a coverage station, it is only necessary to identify the key station that has a similar traffic habit and, as the ratio of the eight-hour count to the average daily 24-hour count is known, this factor is used to expand the eight-hour count at the coverage station to the average daily 24-hour traffic.

A further study of traffic densities of continuous count stations reveals that the total volume of traffic does not change materially from Monday through Friday, and that Saturday and Sunday traffic usually show different characteristics from that of a typical weekday. Saturday traffic is usually heavier both in passenger and commercial vehicles, whereas Sunday traffic usually shows only a few commercial vehicles. Some studies indicate that counts made on Tuesday, Wednesday, and Thursday can be expanded to average daily 24-hour traffic with a minimum of error. Eight-hour counts taken on Saturday or Sunday would have to be adjusted with factors developed by the ratio of Sunday to weekday traffic or Saturday to weekday traffic.

The average daily 24-hour traffic obtained from an eight-hour coverage count may have to be adjusted to take care of any seasonal variation that may exist in a particular area.

In general, seasonal variation in traffic volume is less marked in certain areas than was formerly assumed. This is due to improvements in snow removal and ice control, and general improvement of motor vehicles which permit operation under severe conditions. However, certain areas will continue to be recreational in character and factors will have to be developed from continuous count stations in those areas so that adjustments can be made.

The average daily 24-hour traffic was determined in this manner for each section of the state highway system under the original comprehensive traffic survey.

4–7. Short-Count Method. A fair degree of accuracy in determining traffic density can be obtained by the use of a short-count method. In the study of many continuous count stations, it has been shown that the percentage of the daily 24-hour traffic for any given hour is approximately constant at different points along the same route or on routes of the same character in the same region. Therefore, if the daily habit can be determined for a route or a section thereof, it is possible to expand a one-hour count into daily average 24-hour traffic by the use of factors developed from a continuous station. The average daily 24-hour traffic can further be expanded with seasonal factors of a continuous count station having similar seasonal variation and the average annual 24-hour traffic can thus be determined.

This method is similar to the expansion of an eight-hour coverage station to the average daily 24-hour traffic, but it is based on a one-hour count. If a number of one-hour counts are expanded and their results averaged, the resultant average would be fairly accurate. One-hour counts should be made on days when the traffic is fairly consistent. This usually occurs on Tuesday, Wednesday, and Thursday. The most favorable hours in most localities have been found to be from 9:00 to 10:00 and 10:00 to 11:00 A.M., and from 2:00 to 3:00 and 3:00 to 4:00 P.M.

4–8. Mechanical Counters. Mechanical counters were used before many of the states had completed their comprehensive traffic surveys that were started in 1935. The first type to be developed were the more expensive fixed-type photoelectric machines. These recorded the time each vehicle passed the station and were useful in estimating hourly and daily variation. According to a recent estimate, there are approximately 1000 of these fixed-type photoelectric devices in use throughout the United States. Many of these have been operated continuously at the same locations for the last 15 years or so. Information from these stations serves as a basis for determining traffic trends from year to year and permits the perfecting of schedules for counting traffic and determining the length and spacing of traffic counts.

Two other types of automatic recorders are used at the present time for carrying on traffic surveys. Both are of the movable type and operate from a battery. One, which is called a recording meter, records upon a tape the passage of each vehicle, together with the time at which the vehicle passed the meter. The other type, which is called a nonrecording meter, records the cumulative traffic for the period of its operation. Both types of recorder use a rubber tube laid in the vehicle lane which registers the passing of each axle (or pair of axles).

With the use of these mechanical recorders it is possible to make complete comprehensive traffic surveys with a minimum of manual counts. It is necessary, however, to make some manual counts for classification purposes. It is only through these manual counts that factors can be developed for determining the number of passenger cars, trucks, and buses that make up the total counts of the automatic recorders.

4–9. Traffic Surveys with Automatic Recorders. Much use is made today of automatic recorders for conducting traffic surveys. The method used in obtaining samples of traffic and making the necessary expansions has not changed materially. The annual gathering of this information supplies the basic data necessary for planning, programming, traffic control, design and maintenance and general administration of a state highway program.

The following summary of mechanical counts obtained in the state of Georgia for 1964 indicates the scope of this continuing operation needed for traffic analysis (*177*).

Rural continuous count stations	26
Urban continuous count stations	3
Seasonal control station counts	6,200
County coverage counts	16,500
City coverage counts	6,100
Special coverage counts	3,800
Total continuous	29
Total short-term	32,600

4–10. Automatic Traffic Data Acquisition. As discussed previously, the purpose of the traffic counting program is the determination of the amount and kind of traffic utilizing the streets and highways throughout the state. As traffic counting locations have become more numerous, older methods or data acquisition and analysis have grown too costly and time-consuming. Computers are now used widely for automatic acquisition and processing of traffic data.

In a report to the State Highway Board, Emory Parrish, Executive Assistant Director, Georgia State Highway Department, described the approach being used in that state. The process involved is a team effort on behalf of International Business Machines Corporation, Southern

Bell Telephone and Telegraph Company, Streeter-Amet Division of Goodman Manufacturing Company, and the State Highway Department of Georgia in cooperation with the U.S. Bureau of Public Roads. The following is abstracted from that report (*177*).

The traffic data acquisition system comprises an IBM 1710 control system consisting of a 1620 central processing unit, two disk storage drivers and supplementary equipment to convert telephone tones into a form compatible with the computer, a Bell Telphone Automatic Dialing Unit and a Data Receiving set. One Wide Area Telephone Service (WATS) line is leased for the system. At each remote counting station now on the operating system is located some type of vehicle detecting unit, an electronic storage device, a telephone data transmitter, and a local telephone business line. The telephone business line has a telephone number just as any other line would have.

At the beginning of each hour as indicated by an internal clock to the computer, the 1710 control system interrupts any program currently being processed by the computer and seizes the WATS line. When connection of the WATS line is complete and it returns a dial tone, the computer searches its memory and furnishes the first telephone number to the automatic dialing unit. The automatic dialing unit dials the number of the first traffic counting station. The telephone connection is then made through the WATS line to the local switched telephone network and local business line to the traffic counting station. The data transmitter at the traffic counting station answers the call and sends back a tone signal to the data receiving set in the general office which alerts the computer to be ready to receive the traffic count. The electronic scanner and storage device then reads out volume to the data transmitter that transmits it to the data receiving set as a voice tone where it is converted to computer language and sent to the computer. The computer automatically examines the count for reasonableness and stores it for future analysis. The control system disconnects the call and is ready to poll the next station. The total polling sequence requires approximately 15 seconds.

Georgia has 30 continuous counting stations, four of which are on the automatic system. Upon the completion of this pilot test the remaining stations will be added to the system as funds permit.

In connection with the above study, statistical information is being obtained for the High Hour data for the 30 continuous count traffic-recording stations in the state of Georgia.

4–11. Traffic Flow Maps. When all the information of the traffic survey is assembled, when the expansion factors have been determined and the annual average 24-hour traffic has been computed, traffic maps and flow maps can be prepared. A base map previously prepared by the road inventory survey is used for the traffic map. The numerical traffic densities for each segment of the road are usually placed on this map. For more rapid visualization, the volume of traffic may be shown on the traffic flow map by bands proportional in width to the volume of traffic passing through the various points at which base station counts were taken. Figure 4–3 shows a typical traffic flow map.

4–12. Estimating Future Traffic Volume. Traffic flow maps provide an invaluable guide in determining the selection of a route to be improved. They also form a basis for comparison with future traffic volume studies. It is usually assumed that a highway improvement will have a life of 20 years or more. It therefore follows that a route improved today will more than likely carry a larger volume of traffic before the expiration of its useful life. It is therefore necessary to estimate probable future traffic volume so that the highway improvement can be designed in accordance with future needs.

There are many methods of forecasting traffic volume increases. Many are based on economic factors and trends which require an intimate knowledge of economics and involve complicated computations. A simple method of approximating future traffic is based on two factors which contribute to this increase in volume. These are population and automobile registration. Population curves are plotted and extended into the future by what appears to be a normal trend. The percentage of increase in population and the percentage of increase of automobile registration for the future 20-year period may then be averaged and the resulting factor used to convert present traffic volumes to those which may be expected at the end of 20 years.

With the aid of the more than 1000 continuous counting stations that have been operating continuously for many years, forecasting methods have been improved. These stations give actual counts and indicate traffic volume trends which can be correlated with population and automobile registration trends. These forecasting methods give a fairly accurate estimate of future traffic volumes but consideration must be given to local conditions. For example, the installation of a new industrial plant or the removal of an old one from a locality will affect conditions appreciably over a large area. Other factors may be of importance in a given case. In addition to normal traffic growth, traffic projection factors should reflect the effects of generated and development traffic for new routes. Generated traffic is made up of motor-vehicle trips which would not have been made if the facility had not been provided. Development traffic is that due to improvements on adjacent land over and above that which would have taken place had the new facility not been built. Both these types of traffic are difficult to predict.

4–13. Annual Traffic Volumes and Hourly Capacities. In developing criteria for the design of a highway, use is made of the annual average daily traffic in determining required highway capacities. Highway capacity is usually expressed as the number of vehicles per hour that can be accommodated on a given highway. Traffic volumes are much heavier during certain hours of the day or year, and it is for these peak hours that the highway is designed.

It has been found that, for the United States as a whole, traffic on

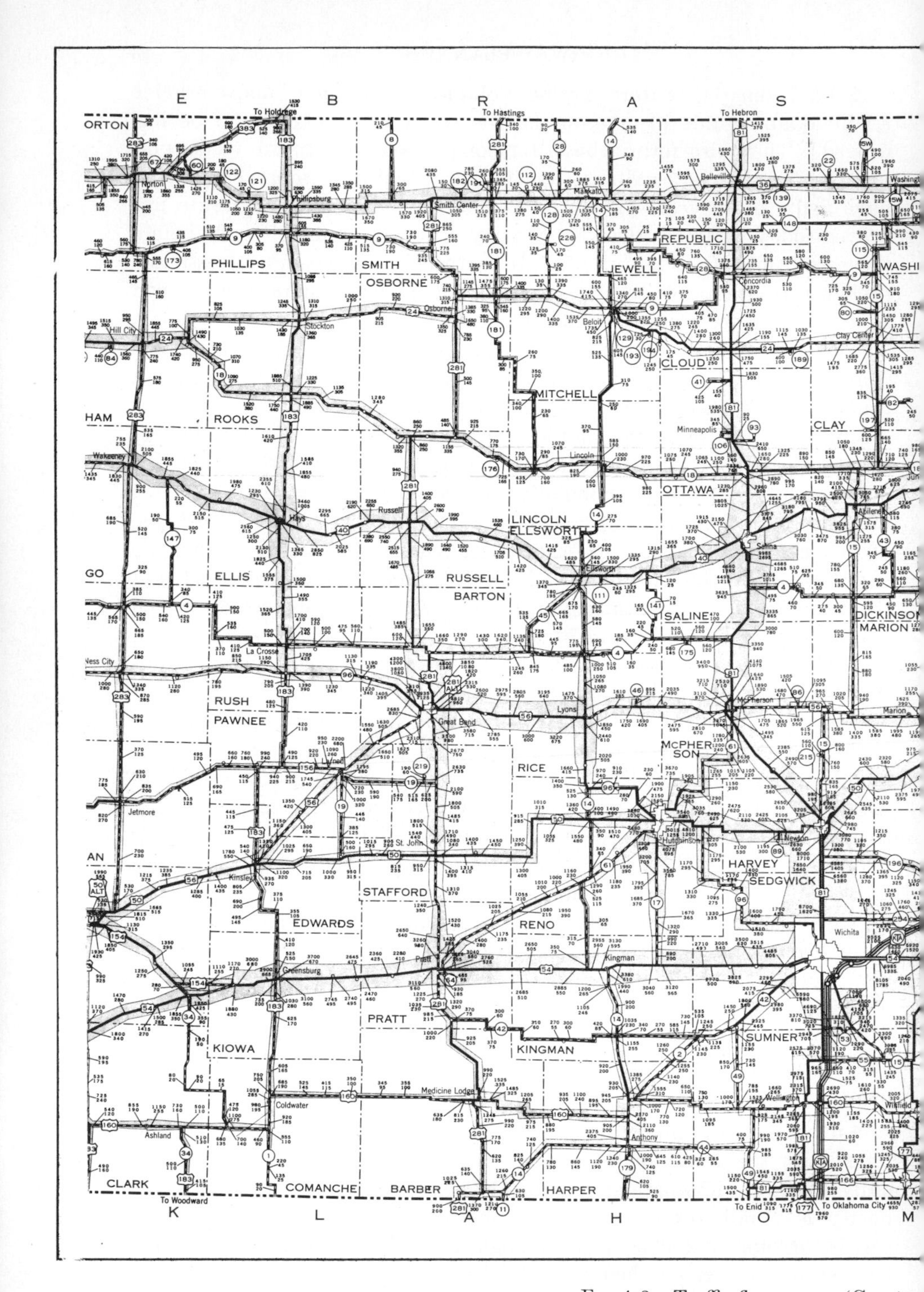

Fig. 4-3. Traffic flow map. (Courtes

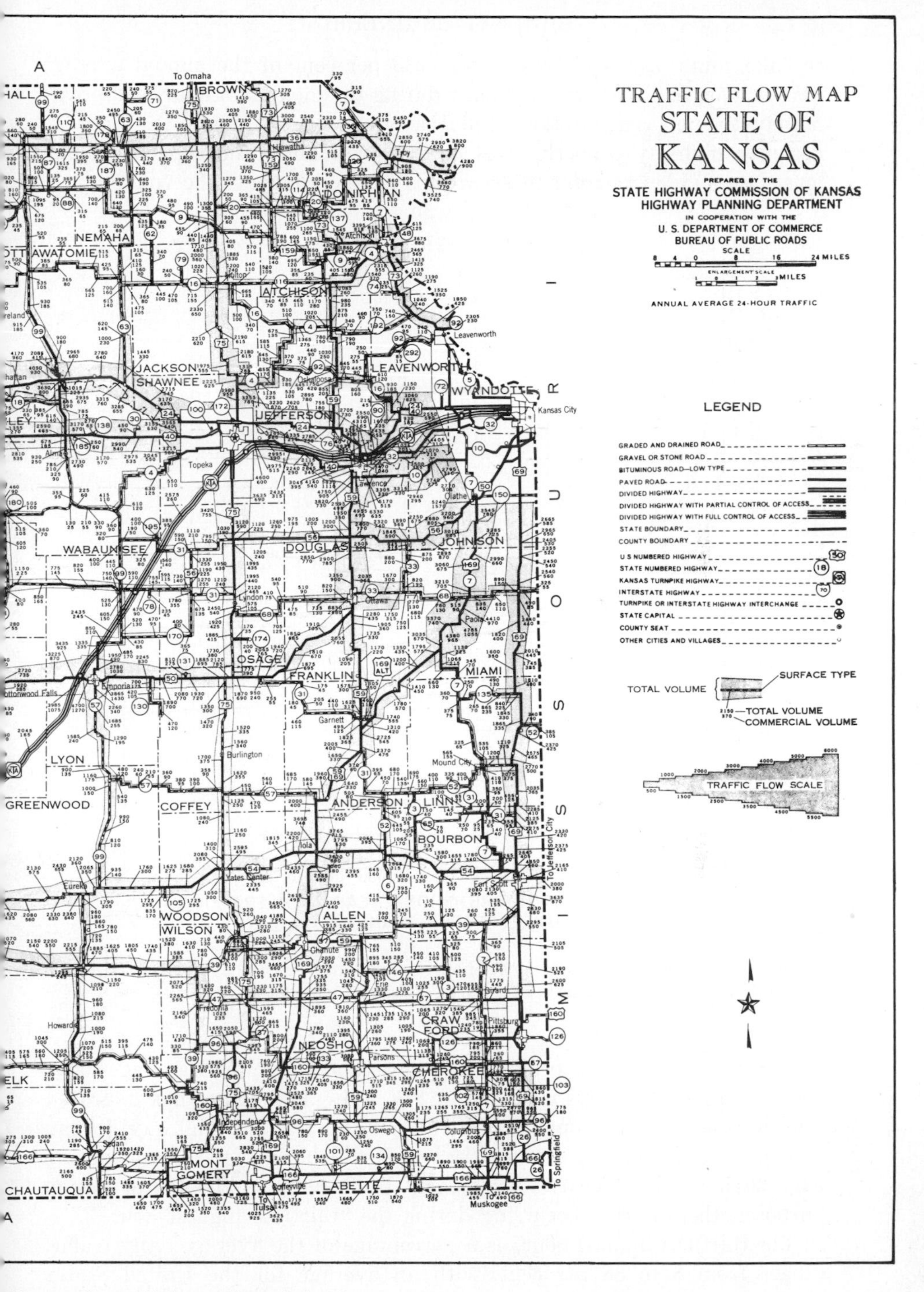

Kansas State Highway Commission.)

the maximum day is approximately 233 per cent of the annual average daily traffic, and that traffic volume during the maximum hour is approximately 25 per cent of the annual average daily traffic. In order to design a highway properly it is necessary to know the capacity which must be provided in order to accommodate the known traffic volume.

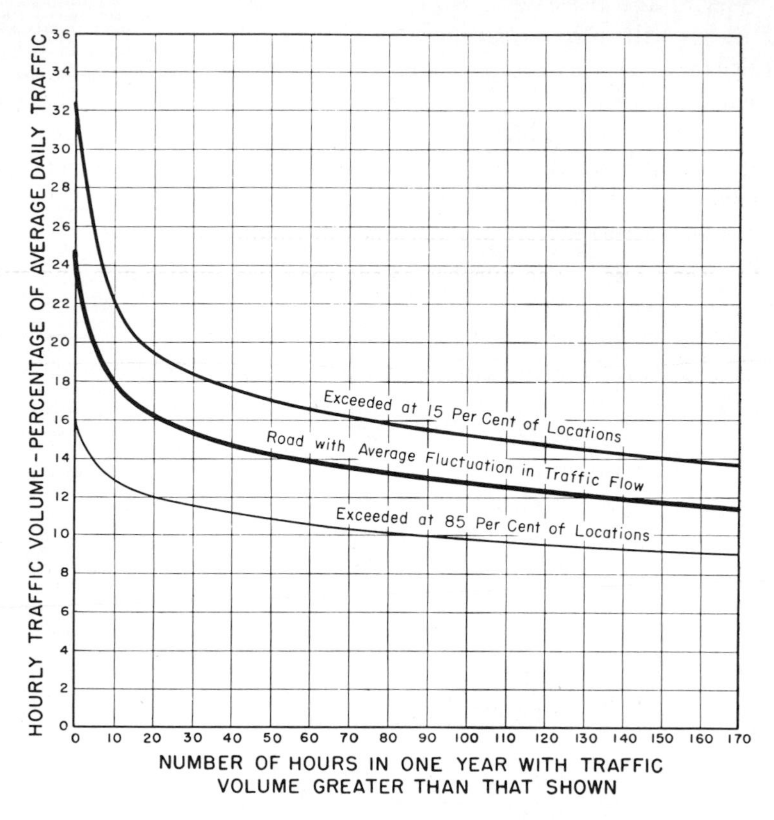

Fig. 4-4. Relation between peak hourly flows and annual average daily traffic on rural highways. (Courtesy Bureau of Public Roads.)

The relation between peak hourly flows and the annual average daily traffic on rural highways is shown in Fig. 4-4. Experience has indicated that it would be uneconomical to design the average highway for an hourly volume which is greater than that which will be exceeded during only 29 hours in a year. The hourly traffic volume chosen for design purposes, then, is that occurring during the "thirtieth highest hour."

The thirtieth highest hour, as a percentage of the average daily traffic, ranges from 8 to 38 per cent, with an average for the United States of 15 per cent for rural locations and 12 per cent for urban locations. In a study made in 1940, it was found in the comparison of 23 selected continuous counting stations in the United States that for any location, the relationship between the thirtieth highest hour and the annual average daily traffic remained unchanged from year to year. Present studies, however, indicate that the thirtieth hour factor has a tendency to decline slightly with the passing of time; if this trend continues, appropriate adjustments will have to be made in the design hourly volume for any future year.

When use is made of highway capacities for design purposes, consideration must be given to other factors which influence highway capacity. Among those are narrow lanes, restricted lateral clearances, narrow shoulders, commercial vehicles, imperfect alignment, signal control, and cross movements. These factors are discussed in Chapter 5.

4–14. Road Use. The purpose of the road-use studies was and still is to determine the relative use of the various parts of the entire highway system in a given state. Information was obtained by personal interview from a representative sample of the registered motor-vehicle operators within the state. Information requested of each was the total mileage driven in one year and the proportion of that mileage that was driven on the main trunk lines, secondary roads, and city streets. An analysis of results revealed the proportionate use made of the various roads by the residents of city, town, and country. This information was correlated with that of the financial survey which determined the amount of revenue received from each of these units. A comparison was then made of the benefits received and contributions made by the various groups. Information thus gained has been utilized in various states to more equitably distribute the costs of highway transportation among the various benefiting groups in accordance with the concepts presented in Chapter 3.

4–15. Road Life. Road-life studies were started in some states soon after the state-wide planning surveys were in operation. Others started later, and a majority of states have continued these studies up to the present time.

The primary objectives of a road-life study are to determine the average rate of retirement and the estimated average service life for each type of highway surface. Roads are usually retired by resurfacing, reconstructing, abandoning, or by transfer. Annual costs of construction, maintenance, and depreciation are included.

With this information it is possible to estimate the amount and cost of replacements that will be required in the future. Programming of construction and reconstruction operations can be made on the basis of corresponding revenue to be received in the future. As indicated in the preceding chapter, data derived from road-life studies are utilized in various economic calculations.

4–16. Highway Sufficiency Studies. With the rapid growth of motor-vehicle transportation in past decades administration of the public highways has become increasingly important. The revenue for building and maintaining the various highway systems has not kept pace with the increasing demands. In the past, and still in some locales today, improvement of highways was governed by the whims and desires of certain pressure groups. This has caused many problems in trying to improve highway systems on an equitable basis.

In order to provide a method for equitable distribution of highway funds and improvements where the need is the greatest, the use of sufficiency ratings was adopted. A sufficiency rating attempts to classify the segments of a highway system in as unprejudiced, objective, and uniform a manner as possible. This method, to be successful, must have every factor and element used in the rating adequately defined and measured. The element of personal judgment, although it can not be completely eliminated, must be minimized in order for the public to accept the method as impartial and unbiased.

Sufficiency rating is a comparison of one roadway to another roadway, or to certain standards. The ratings evaluate a particular section of roadway as to its ability to carry its quota of traffic in a safe, rapid, and economical manner. Sufficiency ratings are primarily used for long-range planning of future improvements, but they are also of importance when highway financing measures are before legislative bodies. The ratings are a good measure of the over-all adequacy of a highway system and may be used in long-range studies to measure the improvement and progress of the highway system.

There are several methods in use in the various states for determining the numerical sufficiency rating for a section of highway. The basic principles are the same in each case. Various elements are evaluated to get a measure of condition or structural adequacy, safety, and service. These elements are then combined to give an over-all rating which is indicative of how well any section of highway is performing its intended function. For example, a section of highway which is completely adequate for the traffic which it carries may have a total numerical rating of 100; a section which has a sufficiency rating of only 60 is obviously deficient in several respects. In arriving at the numerical ratings, some states assign numerical values to the factors involved and compare each section with every other section; others compare all sections against the same set of established standards.

No matter what system is used, the sufficiency ratings focus attention upon sections of the system which need improvement and are an aid in establishing priorities of improvement. Sufficiency ratings are used by more than half the state highway departments. In at least one state, Colorado, state law requires the development and publication of an annual list of highway improvements in order of priority; this requirement is met by using sufficiency ratings.

4–17. Highway Needs Studies. Highway planning surveys provided much valuable information relative to the physical plant of the highway system, the demands made upon it by the motorist, and the financial means by which it was built and is being maintained. It also provided means for determining what should be done in order to provide adequate facilities for serving present needs and for estimating the probable future demands of traffic. During the years following World War II it became evident, however, that if future programs were to be planned intelligently, the work of the original planning surveys had to be supplemented and brought up to date, and that a long-range program for the entire network of highways in a given state was necessary in order to provide for the correction of existing deficiencies and for anticipated requirements of the future.

Studies for such long-range programs have been undertaken by various committees. These committees are for the most part composed of representatives of road using and road building groups as well as public officials. The highway needs study is a very thorough and complex study of the administration of the highway system, need for a system of roads, classification and analysis of the present roads, improvements required for future traffic, and highway income and expenditures.

The study is usually made to include all deficiencies now existing and those expected to accrue within a certain period. Usual studies include needs on a basis of 5-, 10-, 15-, and 20-year programs. For instance, a 10-year program would include correction of present deficiencies, improvement of the system to accommodate the traffic expected in 10 years, and the cost of annual maintenance. The long-range program must include the anticipated needs for that entire period to take care of increased traffic and expansion of facilities, or the highway system will be as deficient at the end of that period as it is today. It will not suffice to include only maintenance and improvement of the present system; a study of the anticipated needs for that period must also be included.

Highway needs studies may be made by the state's highway department or an outside agency. The use of an independent organization eliminates the psychological barrier caused by a highway department investigating its own shortcomings. The approach to such a study usually includes an investigation of (1) the economic importance of the highway transportation system; (2) the historical development of highways with present and expected future traffic demands; (3) the existing deficiencies of highways, roads, and streets; and (4) the administration of the highway system.

The results of these studies are presented in a final report which includes a summation of all factors for an accumulated highway needs program and recommendations for accomplishing the desired results. Many states have completed such studies. Many of these reports have been presented to various legislative bodies and have been acted upon favorably; in fact, several states have had additional needs studies for an evaluation of their original needs program made previously. The Automotive Safety Foundation and the Bureau of Public Roads have cooperated with the various committees in carrying out the studies.

4–18. Fiscal Studies. The purpose of the financial survey conducted as a part of the state-wide planning survey was to obtain information relative to the financial structure of all the governmental units within a given state. From this information estimates were made of the ability to finance and maintain an adequate system of highways and streets. The work was accomplished by a staff of accountants whose source of information was published financial reports of the various governmental agencies. In the absence of reports, the information was obtained from the original financial records. The financial survey covered the financial transactions of all governmental units for a period of one year, and consisted of the following:

(1) All forms and rates of taxation for all purposes.
(2) All forms and rates of taxation for highway purposes.
(3) The total revenue received from each source of taxation for all purposes.
(4) The total revenue received from each source of taxation for highway purposes.
(5) The amount of the highway tax burden falling upon the road user and other beneficiaries in each of the taxable subdivisions.
(6) The amounts, terms, and interest rates of all existing debts for highway purposes, for each governmental agency.
(7) The amount of all expenditures for highway purposes by all governmental agencies. This included construction, maintenance, and administration.

The financial inventory revealed the existing relationship between the sum of all highway revenues and expenditures and the grand total of all revenues of the state for purposes of government and other public services. It provided factual data for a more equitable distribution of the tax burden to the various groups that were benefited. The information also provided means of determining on what basis future programs could be built and financed. The need was also found to continue

of traffic. As a result of many origin and destination and other studies in urban areas, the Federal-Aid Highway Act of 1944 provided for an annual appropriation of $125,000,000 for three years for improving the federal-aid system in urban areas. The provision of funds for urban improvement projects was continued in subsequent Federal-Aid Highway Acts, as indicated in a previous chapter.

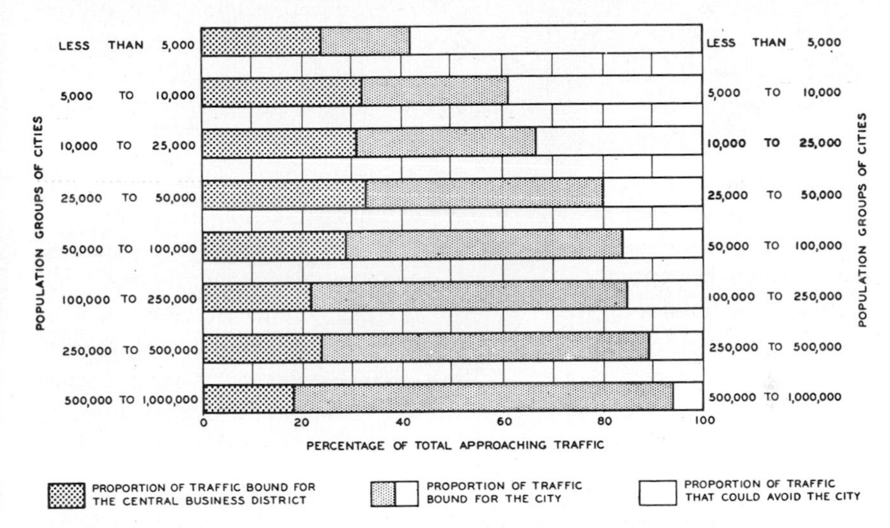

FIG. 4-5. Proportions of traffic approaching cities of various population groups which are bound beyond the city, to the city, and to the central business district. (Courtesy Bureau of Public Roads.)

Fig. 4–6 shows a flow diagram resulting from an origin and destination study made on an interstate route through a typical city of 50,000 to 100,000 population. It can be readily seen that the problem in urban areas is not one of providing bypasses, but of providing better facilities in the way of free-flowing arteries of ample capacity within the urban areas, as previously mentioned. With this in mind, attention was focused upon the problem of determining traffic volumes and characteristics within the cities, as indicated in the following articles.

4–20. Analysis of Origin and Destination Studies. The problem of analyzing city traffic in order to determine which of several routes should be improved or to serve as a basis for an over-all plan to relieve congestion is more complex than in rural areas. Between the rural and urban area is a transitional or suburban area that must be considered in the analysis of traffic problems. The large cities influence traffic for a distance up to 35 miles from their limits, while cities with a population of 10,000, more or less, influence traffic up to 5 or 6 miles from their boundaries. The magnitude of the city traffic problem is therefore due

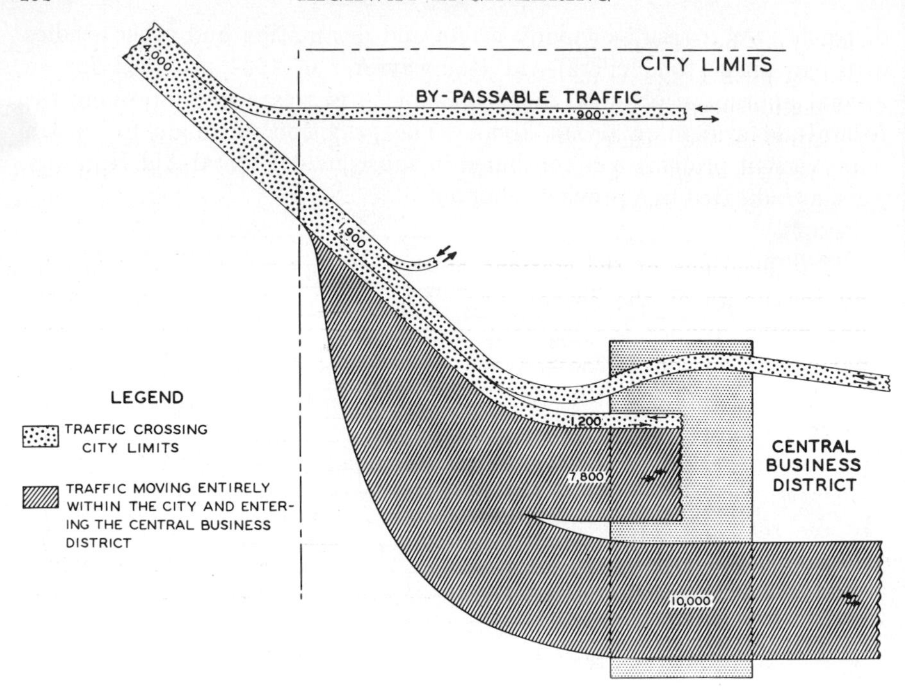

FIG. 4-6. Origins and destinations of traffic on an interstate route through a typical city of 50,000–100,000 population (based on data from 7 cities). (Courtesy Bureau of Public Roads.)

to a combination of the internal movement within a city of the volume of traffic attracted to the city from the rural and suburban areas.

Urban origin and destination surveys are now being conducted in many cities with the cooperation of the state highway departments and the Bureau of Public Roads. The basic method of making these surveys, which has been developed by the Bureau of Public Roads in cooperation with the Bureau of the Census, consists of two parts, an external survey and an internal survey. The data of both surveys must be combined for the proper evaluation of results.

4-21. External Survey. The purposes of an external survey are to determine the total number and character of all vehicles entering and leaving the metropolitan area and to obtain origin and destination information from the drivers of a large number of these vehicles. The total number of vehicles and their classification is determined by manual counts at each external station during specified periods and supplemented by automatic traffic recorder counts for longer periods. Origin and destination information is obtained by direct interviews. Traffic in both directions is stopped and the vehicle operators are questioned about the trip into, out of, and through the city. All cars going in both directions may be stopped, although random sampling methods are also used. During peak hours as many cars as possible are stopped without causing congestion because of the interviews. The information required usually pertains to the origin, destination, principal streets followed, purpose of the trip, purpose of any stops, etc. Instead of the direct interview, a mail questionnaire may be used, with analysis based on statistical methods.

The locations of the stations of the external survey are determined by the limits of the cordon line. This cordon line is a hypothetical line which divides the internal survey area from the external. Ordinarily the cordon is located far enough out to intersect a minimum number of roads and still not include much rural territory in the internal survey. The cordon line should be further located so that its course around the city is fairly uniform without much deviation from the central business area either inward or outward. Consideration should be given to the inclusion of outlying ends of transportation systems. Natural barriers, such as streams, rivers, railroads, and such, may be used as portions of the cordon line. In effect, the cordon line creates a screen through which all vehicles coming into or out of the internal area must pass and be accounted for.

4–22. Internal Survey. In order to determine the traffic habits and needs for traffic facilities of those who live in urban areas, it is necessary to know where people go, the time of day they go, the type of transportation used, and other pertinent facts relative to their daily travel. It is not necessary to interview all residents in an area in order to obtain this information. It has been well established by the Bureau of the Census and other poll groups that if a representative sample is interviewed, the habits and needs of this sample will be indicative of those of all the residents of the area.

The size of the sample needed usually depends upon the total population, the degree of accuracy desired, and to some extent upon the density of the survey area. A small city generally requires a larger sample proportionally than a large, more densely populated city. Experience has shown that for cities of various populations, the size of the home interview sample should be as follows:

Population of Area	Recommended Size of Sample
Under 50,000	20 per cent
50,000 to 300,000	10 per cent
300,000 to 1,000,000	5 per cent
Over 1,000,000	4 per cent

The internal survey consists of the following subdivisions: (1) home interviews (trips by all individuals), (2) truck studies, (3) taxi studies, and (4) public transit information.

Interviews are made on a predetermined schedule and forms are provided for tabulating the necessary information. The information required is for the day preceding the home interview. This information includes the number of occupants in a household, occupation, number making the trip, use of private or public transportation, origin and destination, purpose of the trip, etc.

In selecting the sample, use is made of certain available basic records. For example these may include: (1) Sanborn maps, which are a series of copyrighted maps prepared for the use of insurance companies, assessors, etc. These show all structures, together with other information. (2) City directories. (3) The Bureau of the Census "Block Statistics" which show blocks and block numbers. (4) Land use maps which may be available.

Truck and taxi information is generally available from the records of the companies which operate them. Usually 20 per cent of the trucks and from 50 to 100 per cent of the taxis registered in an area are sampled. Public transit information is usually obtained directly from the transit companies. This should include location of existing routes, schedule of operations, and the total number of passengers carried on an average weekday during the period of the survey.

The completeness with which trips are reported in the internal survey may be checked by three methods: (1) use of control points for comparison, (2) a screen-line comparison, and (3) a cordon-line comparison. Not more than three or four control points should be selected for use in any one study. Control points are usually viaducts, bridges, underpasses, or other points of constriction through which large volumes of traffic pass. The average traffic volume and classification at each control point must be determined by manual counts made during the survey so that the expanded counts obtained from home interviews can be compared with the actual ground counts at these points.

Screen lines are natural barriers, such as rivers or railroads, which usually have a limited number of crossings at which ground counts can be made at a minimum of expense. The purpose of the screen line is to divide the area of the internal survey into two parts in order to determine the number of vehicles moving from one part to the other. A comparison can then be made between the number of trips having origin on one side and destination on the other, as actually counted, to the number of trips as determined from the expanded interviews.

A cordon-line comparison deals with passenger car trips by residents of the internal area and truck trips by trucks registered in the area, but only those trips which cross the cordon line. The total number of such trips recorded in the external survey can be compared with the total number recorded in the internal survey.

In many recent surveys conducted in various municipalities by the method just described, about 90 per cent of the traffic passing control points and screen lines during 16 hours of operation of external stations has been accounted for by the expanded interview data. The remaining 10 per cent has been assumed to be made up of cars circulating in search of a parking place or of unimportant short trips not reported in interviews, and of trips by persons living outside the area and not

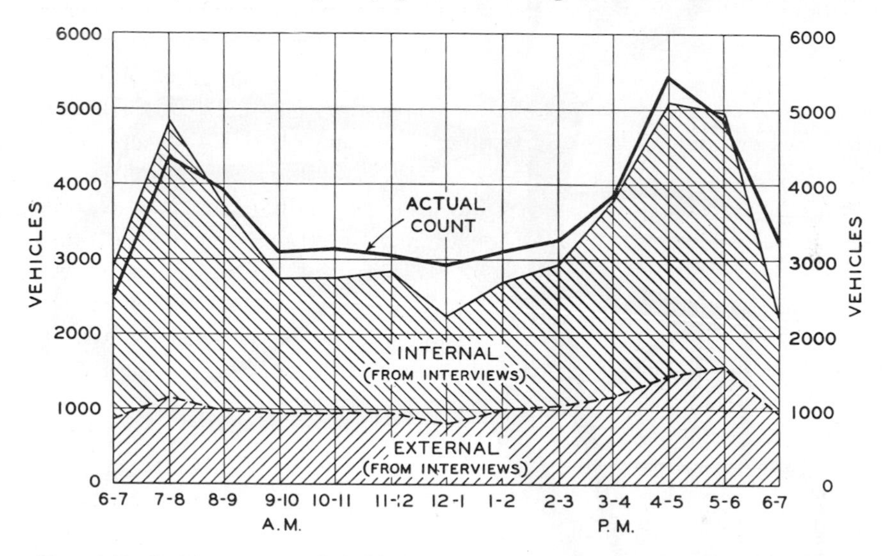

FIG. 4–7. Baltimore travel habit survey. Comparison between traffic passing three control points as reproduced from interviews and that actually counted at those points. (Courtesy Bureau of Public Roads.)

intercepted at the external cordon stations. These trips are considered to be of minor importance so far as the principal purposes of the survey are concerned.

Fig. 4–7 shows a comparison of the number of vehicles counted at control points with the number passing those points as determined from the interviews.

4–23. Desire Lines. Desire lines of course indicate the actual desires of vehicle users as direct lines of travel from one point to another within a given area, assuming that direct routes are available. The information which these lines represent is of incalculable value to the planning and design engineer whenever routes for new arterial improvements are to be selected.

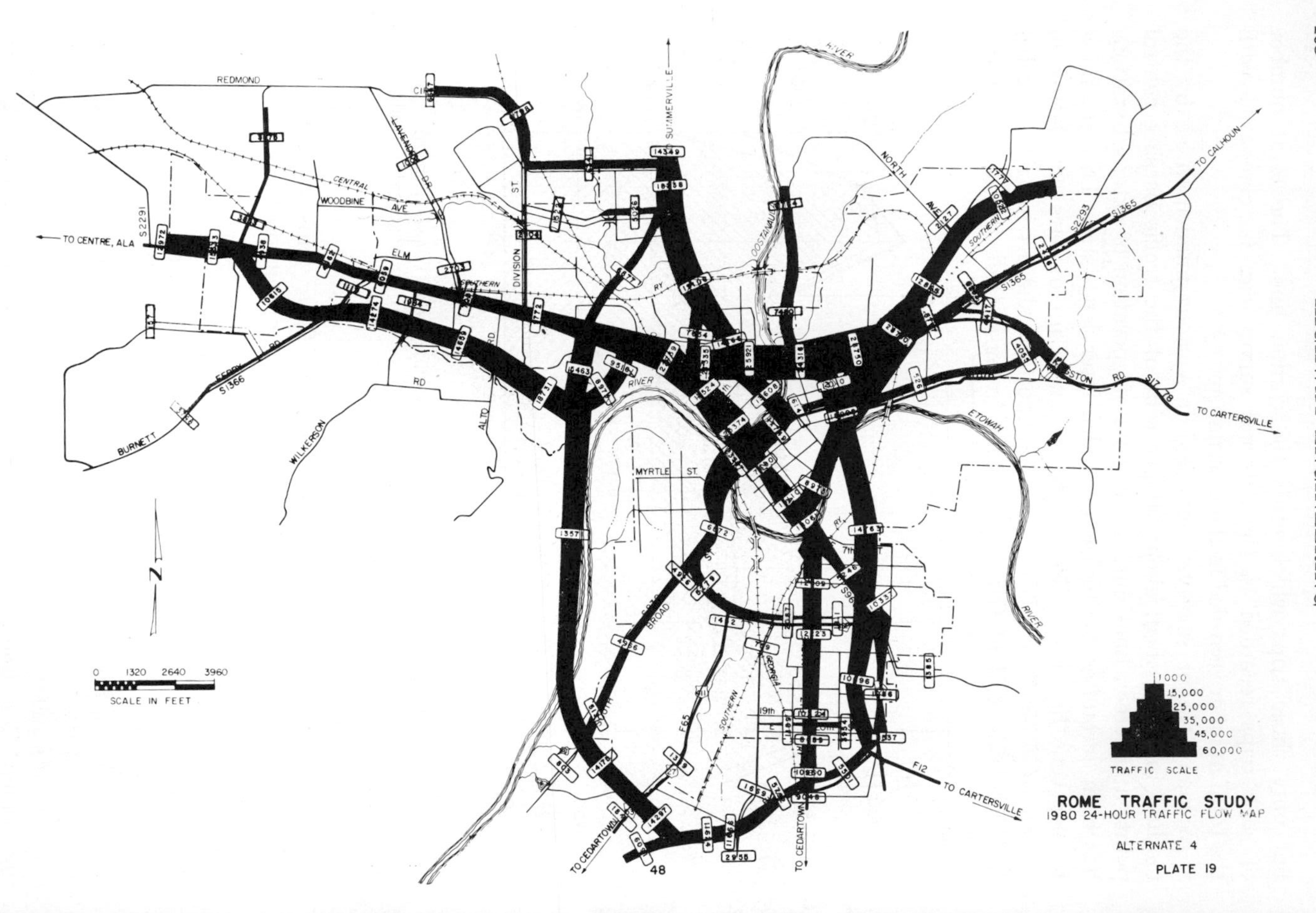
ROME TRAFFIC STUDY
1980 24-HOUR TRAFFIC FLOW MAP
ALTERNATE 4
PLATE 19
TRAFFIC SCALE
1000
15,000
25,000
35,000
45,000
60,000
0 1320 2640 3960
SCALE IN FEET
N
TO CENTRE, ALA
TO SUMMERVILLE
TO CALHOUN
TO CARTERSVILLE
TO CARTERSVILLE
TO CEDARTOWN
TO CEDARTOWN
48
REDMOND
CENTRAL
WOODBINE
AVE
ELM
SOUTHERN
DIVISION
ST
BURNETT
WILKERSON
ALTO
RD
RIVER
OOSTANAULA
NORTH
AVE
KINGSTON
RD
ETOWAH
MYRTLE ST.
BROAD
GEORGIA
19th
7th
S2291
S1366
S1365
S2293
S17 78
F12
F65

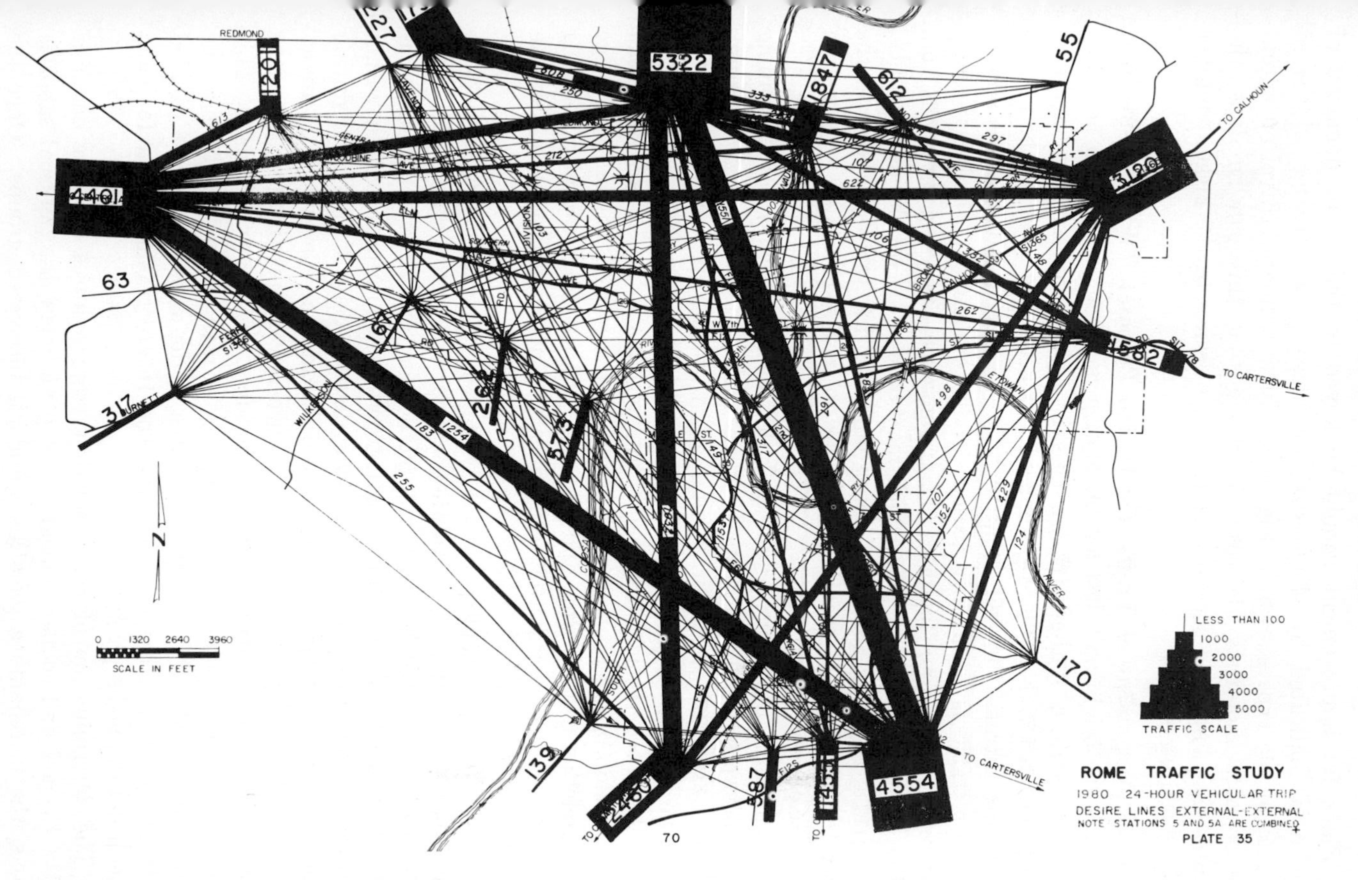

FIG. 4–8. Comparison of 1980 traffic on actual routes (above) and desire lines (below) of travel, for Rome, Georgia. (Courtesy Georgia State Highway Department.)

Plans for improvement should be keyed to the major demands of traffic, as indicated by the desire lines having the greater widths in a plot of the type shown in Fig. 4–8, which compares travel on actual routes with motorists' desires. Assuming that this can be done, the routes thus improved should best meet the desires of the largest number of vehicle users.

4–24. Assignment of Traffic for Design Purposes. While desire lines indicate the origin and destination of trip movements, they do not indicate the routing of these trips over existing or proposed new facilities. When new arterial routes are proposed it is necessary to know the volume and type of traffic that will be attracted to the new facility. This allocation of traffic is commonly called "traffic assignment" and is usually dependent upon four factors: (1) traffic which may be diverted from alternative routes, (2) traffic which may be created by the new facility, (3) traffic resulting from additional land use due to easy access to the new facility, and (4) traffic growth due to increased vehicle registration and use of vehicles.

Analyzing data from origin and destination surveys by combining trips between zones has been a tedious and time-consuming operation. With the use of electronic computers, various trip combinations can be obtained with a minimum of time. Much research has been done to make these traffic assignments by mechanical methods. California has developed a set of curves showing the per cent of freeway usage as a function of both time and distance. Other states are also carrying on studies to develop better techniques and machine procedures for the analysis of urban traffic studies.

S. E. Ridge (*93*) has discussed the use of electronic computers in the analysis and use of origin-and-destination survey data. His comments are abstracted herewith. A method of forecasted traffic distribution designed to handle differential growth (nonuniform growth within various sections of an urban area) was developed by T. J. Fratar during the course of the Cleveland origin and destination survey. It is analogous to the Hardy Cross method of successive approximations for moment distribution in indeterminate structures, and is very efficient. This method, and variations of it, are called "growth factor" methods. The name comes from the fact that the area under consideration is divided into many small areas or zones and the future traffic growth for each factor determined separately.

The determination of zone growth factors requires that the engineer have the aid and cooperation of the city planners, the political subdivisions, the merchants associations, and all interested groups. Once the growth factors are determined, an electronic computer can produce a forecast of the zone-to-zone traffic volumes in few hours. The Bureau of Public Roads has developed programs for this purpose using the largest and fastest computers available, because of the tremendous volumes of data to be handled.

After the future zone-to-zone traffic volumes are obtained, the job of selecting an integrated system of expressways to handle the anticipated volumes remains. In order to do this, the forecasted zone-to-zone volumes must each be assigned to the most logical existing or proposed facility and these results accumulated citywide in order to arrive at the traffic volume data required for geometric design.

The AASHO Policy on Arterial Highways in Urban Areas (*99*) recommends that for each zone-to-zone movement two best time paths should be determined. First, the total travel time of the movement when routed over the best existing facilities and second, the travel time of the movement over the best route of the proposed system of expressways. The ratio of the two travel times for the movement is then calculated. Standard traffic diversion curves can be consulted to determine what proportion of the movement would be diverted to the proposed facility if built and what proportion would remain on existing city streets. The volume of traffic for the particular zone-to-zone movement is then split and the proper proportion assigned to each of the two best routes. When all of the zone-to-zone movements have been assigned in this manner, forecasted traffic volumes for every major street and proposed expressway have been determined.

To be properly done, the whole process should be repeated many times to find which configuration of expressways will provide the most service for the forecasted traffic volumes at the least cost in much the same manner as several alternative locations for a rural highway would be investigated. Electronic computer programs under development will permit this to be done, with resultant improvement in the development of urban expressway systems.

4–25. Parking Surveys. The parking problem (*268*) today is one of the most important confronting most municipalities. In the larger cities, the lack of proper parking facilities is resulting in decentralization of business, with the resulting loss of land values in the central shopping district. The parking problem usually begins not at the point of destination but at the point of origin, when the motorists decide where and how to go. The purpose of a parking survey, therefore, is to determine the parking habits and requirements of motorists and the relation of these factors to other uses of existing parking facilities.

A parking survey should be designed so that the information collected will provide needed data for the evaluation of the several factors in the parking problem. The information secured should include (1) the location, kind, and capacity of existing parking facilities; (2) the amount of parking space needed to serve present demands; (3) the approximate location of possible additional parking facilities; and (4) the legal, administrative, financial, and economic aspects of parking facilities (*41*).

The first phase of the parking survey is to make an inventory of all available parking facilities. This includes curb and off-street parking areas such as parking lots, garages, and filling stations. Any physical and legal restrictions are also to be noted. The theoretical capacity of a parking area is usually determined by dividing the gross area by that needed by one car, which is assumed to be 200 square feet for car lots and garages. Twenty lineal feet is usually used in determining theoretical curb capacities. Calculation of the theoretical capacity does not imply that more cars cannot be accommodated. Some cars are parked for short periods of time and their space then becomes available to others. Cooperation of the owners of off-street parking facilities must be obtained in order to determine the actual capacities or number of cars that are accommodated for periods of the survey so that the results can be correlated with capacities of other facilities. Additional information in regard to the duration of parking is also to be obtained.

The determination of actual capacities of curb parking facilities is more readily done. Parking, when permitted, is usually for one- or two-hour periods. Therefore the number of space hours available is determined from the inventory of the curb space available. It is assumed that parking is for the legal time specified. Studies are also made in order to ascertain the extent of illegal parking and its effect on the total effective capacity of curb parking facilities.

In order to determine the number of cars which may utilize available parking spaces in a central area, a cordon line is usually drawn around the area to be studied and cordon counts are made of all vehicles entering and leaving it. The counts are made manually so that a motor-vehicle classification can be made. Automatic recorders are often used to supplement the manual counts and to serve as a basis of control for abnormal conditions.

Motorists who are interviewed at all parking facilities within the cordon area are asked for their home address, destination, and the purpose of the trip. Fig. 4–9 shows the results of a parking survey in the city of Baltimore.

A parking survey should also include a study of the area for further off-street parking facilities. This should include the investigation of vacant lots, obsolete buildings, and blighted residential or business areas within the area of the parking survey. Mention may be made of other parking areas outside of the limits of the central shopping area which may also be used. These areas are usually referred to as fringe parking areas, the idea being that the motorist can park his car near the central

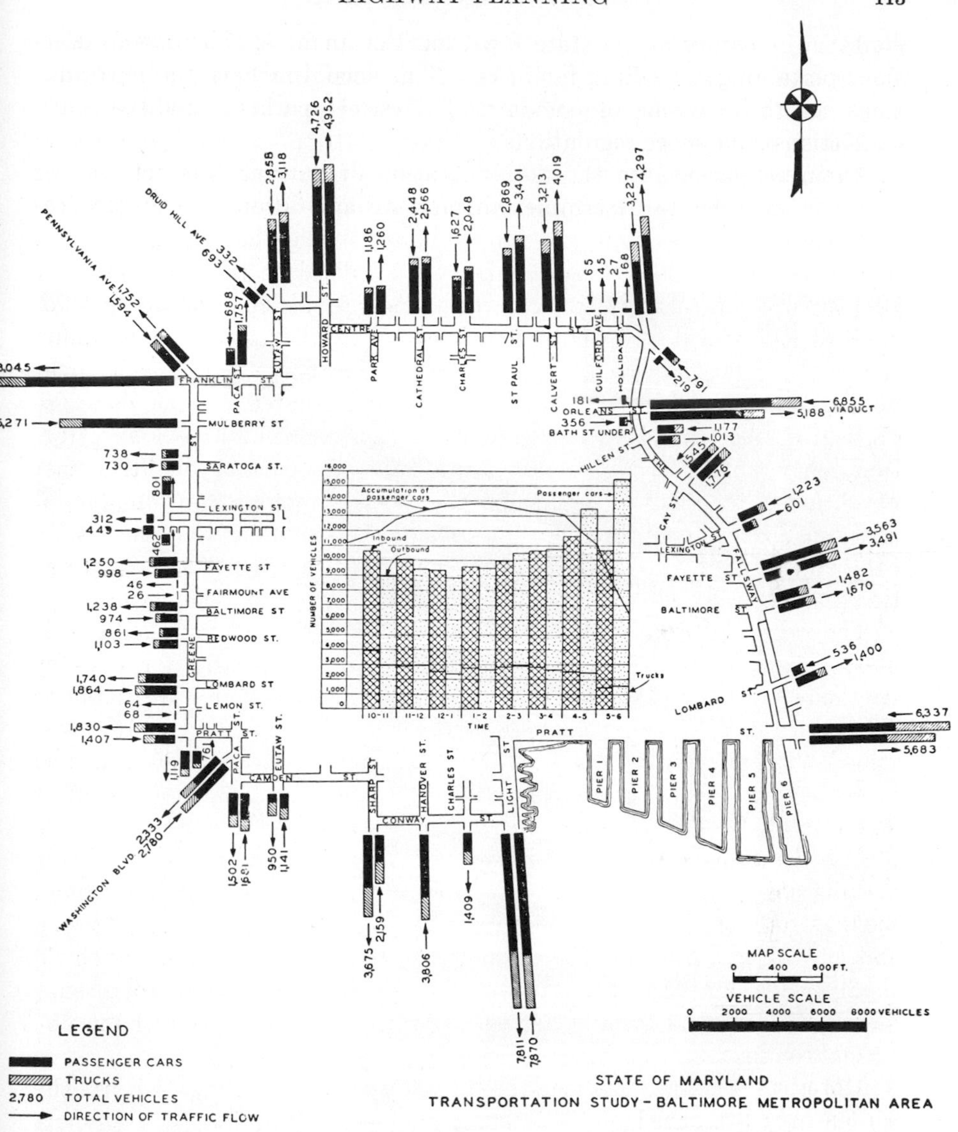

FIG. 4-9. Parking survey of number of passenger cars and trucks entering and leaving the downtown area of Baltimore City on a week-day between 10:00 a.m. and 6:00 p.m.

business district and make use of available mass transportation facilities as a means of arriving at his destination. Many municipalities have tried this type of facility, but as a general rule it has not proved as successful as desired.

In order to be properly evaluated, a parking survey should also include an inventory of the state laws, local ordinances, and judicial decisions pertaining to parking facilities. This should include public regulations for the licensing of commercial off-street parking facilities, curb restrictions, and police regulations.

An investigation into the administration of parking facilities should be made in order to determine which state and local agencies are re-

Fig. 4–10. This city-owned 12-level parking garage is one of several built in the downtown area of Chicago. (Courtesy National Highway Users Conference.)

sponsible for the planning, location, financing, construction, operation, and maintenance of parking facilities within the municipality that is being surveyed. For example, in a given municipality the police may have the responsibility for enforcing curb parking restrictions, while a traffic commission may have the authority to establish such parking regulations, including meter parking. The zoning commission may require parking provisions under zoning laws for various property uses. All administrative facts should be included so that in the final analysis recommendations may be made to remedy any deficiencies that may exist.

In matters pertaining to the financing of parking services, studies should include the aggregate revenue from parking meters and its disposition as well as the assignment of costs of parking facilities for off-street parking according to the benefits received. Benefits may be apportioned to property owners, business establishment, the motorists, the general community, and the municipality itself.

A complete comprehensive survey should do much in the way of solving the parking problem. The information received in a parking survey usually results in many recommendations for the relief of congestion and for general improvement in traffic facilities. These recommendations, for example, may be for the provision of additional parking facilities, either publicly or privately owned, better enforcement of parking regulations to make available space at the curb now being used by the illegal parking motorist, recommendations for changes in zoning restrictions, and so on. A modern parking garage is shown in Fig. 4–10. The importance of continuing parking studies and noting trends cannot be over-emphasized.

4–16. Accident Studies. Accidents resulting from the movement of vehicles and persons are one of the more important factors that must be considered in the design, construction, and use of highways. In the analysis of accidents it is necessary to study driver behavior, the physical structure of the highway, and other factors that may be the cause of accidents in order that means may be found to provide safe highways and to keep accidents to a minimum.

Accident information is usually obtained from the records of police files. Cooperation between enforcement agencies and state highway departments has largely resulted in the uniform reporting and filing of information in regard to accidents. This has facilitated the work of making accident studies.

Information on an accident report for highway traffic studies should include the location, time, and date of the accident; the condition of the driver, vehicle, and road surface; any fatalities, personal injuries, and property damage; and other vital data pertinent to the accident

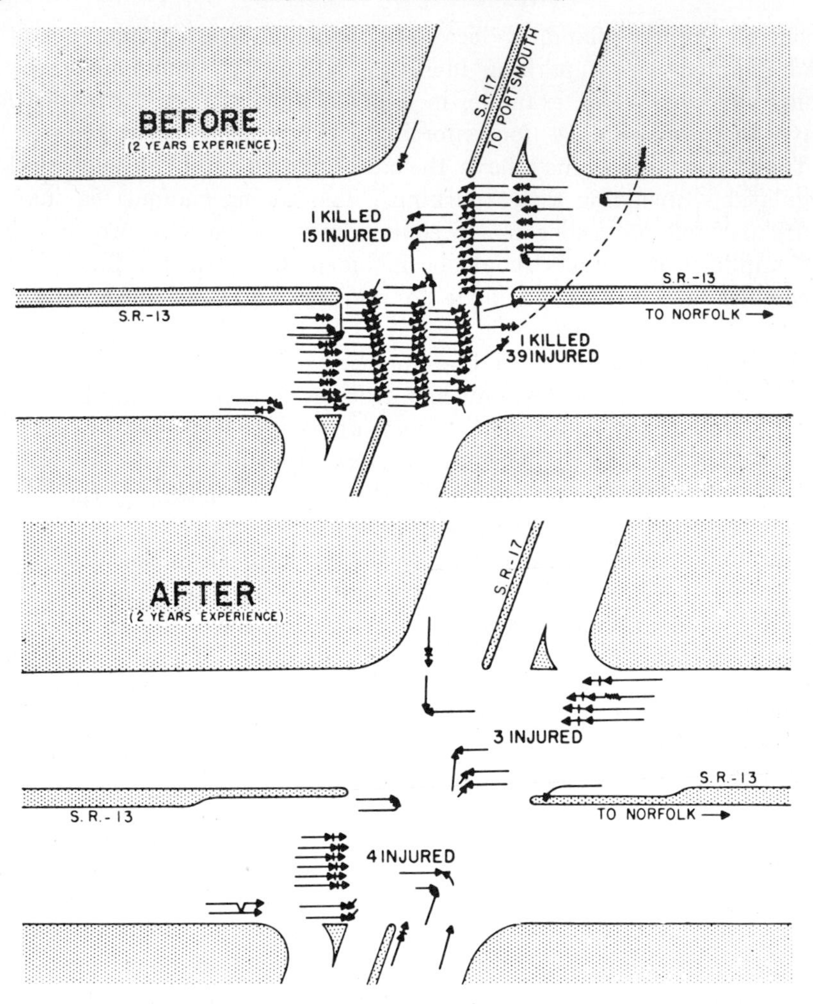

Correction of high accident location achieved near Norfolk, Va., by installation of left turn lanes and new traffic signal system, including left-turn arrow. Project cost $16,700, according to U.S. Bureau of Public Roads. After two years experience before and after, the result:

Before		After
159.4	Accident Rate	26.2
80	Total Accidents	24
2	Fatal Accidents	0
24	Injury Accidents	8
54	Property Damage Accidents	16
2	Persons Killed	0
54	Persons Injured	8
$33,500	Amount of Property Damage	$8,600

Fig. 4–11. Intersection study diagrams showing effectiveness of measures undertaken to reduce accidents. (Courtesy Bureau of Public Roads.)

necessary for completing the report, such as names, addresses, persons involved, witnesses, disposition of the injured, etc. Information gained from individual accident reports may thus be summarized and analyzed in various ways.

In most studies, maps are used for plotting the location of accidents. Large-size maps may be employed, and a dot or a pin may be used to mark the location of each accident. In this manner the frequency of accidents occurring at or near the same location becomes quite evident and will indicate the need for an investigation of the particular location. Special studies at intersections where accidents occur are usually made on a larger map or on a drawing of the intersection. Sketches showing the accident diagram, a speed diagram, existing physical conditions, and recommended changes are of much value in comparing the results after the changes have been made. Fig. 4–11 shows intersection study diagrams at one location, before and after remedial measures were taken.

A study of accident locations may reveal poor alignment, poor sight distance, or a combination of both; the presence of roadside vendors; and a lack of proper warning signs or control devices. At intersections where the frequency of accidents is high, detail studies are often necessary. These may include density counts, turning movements, and the like.

The correct reporting of accidents and the proper use of data thus obtained constitute an important phase of highway planning in both rural and urban areas. Information thus gained provides a basis for the correction of existing deficiencies in the highway or street system and for the more adequate design of new facilities. A principal objective of highway improvement is to provide for safer highway transportation and for the reduction of accident rates. Information gained from accident studies should prove to be invaluable in attaining this important objective.

The Bureau of Public Roads and the states are concentrating on a "spot improvement program" designed to eliminate high-hazard locations on the federal-aid highway systems by 1969. The program has been highly successful.

4–27. Traffic Engineering. The work of the original state-wide planning surveys has been continued and expanded in practically all the states. The importance of the work in relation to the design, construction, and maintenance of highways has been recognized, and it has resulted in the creation of a traffic division in many of the state highway departments. Some state highway departments have combined the planning divisions and traffic division.

The work of the planning surveys having been previously explained, together with the manner in which it has been carried on up to the present, an explanation of the many functions that are being performed by a traffic engineering or a planning and traffic division of a highway department seems in order. Some of these functions, which are necessary for the proper operation of an integrated highway system, are not further discussed in the text.

Although once again emphasis has been placed primarily upon the work of the state highway departments in this field, it should be noted that many of these same functions are performed by divisions of local governmental units. For example, the traffic engineering division is an important part of the engineering organization in any large city. Similarly, traffic engineering is a vital function of large county engineering organizations. In many states, traffic engineering functions required in the smaller governmental units are performed by the state agencies on a cooperative basis.

TRAFFIC SURVEYS. The work of the original surveys has been continued. Field studies are made relative to traffic volumes, traffic movements, cordon counts, parking, speed and delay, and origin and destination studies. Studies in law observance, speed zoning, and no-passing zones are also included.

TRAFFIC CONTROL. Traffic control consists of the supervision of the design, fabrication, erection, and maintenance of all traffic control devices. These include signal devices, signs, and pavement and curb markings. The establishment of loading zones, cab stands, truck routes, detours, and other means of control in municipalities are also a function of traffic control. The use of one-way streets and signal control are two of the measures that are undertaken in order to improve existing facilities.

TRAFFIC DESIGN. All plans for construction or reconstruction are required to be checked for features that affect traffic and safety. This includes a check of the many aspects of design in regard to intersection treatment, channelization, private drives and entrances, parking facilities, and so on.

TRAFFIC ACCIDENT RECORDS. These records are collected, tabulated, and summarized. Studies are made of hazardous locations and points of congestion. Complaints in regard to traffic conditions are collected and studied. Spot maps, collision, and condition diagrams are made of the above conditions.

TRAFFIC PLANNING AND RESEARCH. It is necessary to continue studies and carry on research in matters pertaining to traffic. These cover investigations relating to traffic demands, the need for traffic routing on one-way streets, turn controls, curb and off-street parking, and other methods for the improvement of traffic conditions.

TRAFFIC SAFETY AND EDUCATION. Cooperation with the police, press, radio, schools and other civic groups in the promotion of traffic safety is one of the functions that must be stressed. The preparation and distribution of highway safety literature is also important.

Although little coverage is given to the specific functions of traffic engineering listed above, primarily because of lack of space, the importance of these functions cannot be overemphasized. Traffic engineering (and traffic planning) has become a highly specialized field that requires the services of trained and experienced engineering personnel. The importance of the work performed by traffic engineers has been recognized in recent years, and many attractive careers are now being fashioned in this relatively new branch of highway engineering.

PROBLEMS

4–1. From current literature or by direct contact, prepare a report on the methods presently being used by your own state highway department in developing and maintaining the state-wide road inventory.

4–2. It is necessary to conduct a traffic survey in order to determine traffic density on the main and local roads of your county. Prepare a sketch map of the county, with the main and local roads, and mark the location of the various stations. Prepare a schedule for gathering the necessary information.

4–3. A total count of 2500 cars was obtained for an 8-hour count at a certain station in a rural area. From a continuous count station in the area having the same travel habits and characteristics, an 8-hour count represents 42.8 per cent of the 24-hour annual average traffic. Determine the 24-hour annual average traffic. Present information based on a study of traffic trends, gasoline consumption, and population increases indicate that a factor of 1.62 may be used to project a rural 24-hour annual average traffic count to a time 15 years later. On this basis, what traffic volume would be expected at this particular station? Using the nation-wide average relationships given in the next chapter, roughly estimate the traffic volume during the 30th highest hour.

4–4. How many cars could be placed in a commercial parking lot 210 feet by 300 feet in size, allowing a space of 20 feet by 8 feet for each car? Make a sketch showing the arrangement. Assume two driveways for entrance and two for exit.

4–5. Obtain a copy of the form used by the local police agency for recording traffic accident data. Make a report explaining the routine followed in reporting accidents, and the use made of the reports.

4–6. Make a report explaining how the functions of traffic engineering and traffic safety are carried out in your community, emphasizing the organizations concerned with these functions, their duties, and relation to one another.

4–7. Using a recent publication of the Highway Research Board, the Institute of Traffic Engineers, the Eno Foundation, or a similar organization, prepare a report abstracting a paper or article dealing with the planning and conduct of an urban traffic survey. Explain in some detail the technique used in conducting the origin and destination part of the survey.

4–8. Present a brief report summarizing one of the comprehensive state highway needs studies conducted under the auspices of the Automotive Safety Foundation or other group.

4–9. Detail the steps which have been taken, or are planned, to alleviate the parking problem on your campus or in your community. Comment on the plans which have been developed. Give any constructive suggestions you may have for solving the problem.

CHAPTER 5

GEOMETRIC DESIGN OF HIGHWAYS

Geometric design is the design or proportioning of the visible elements of a highway or street; it does not include "structural design" as the term is commonly used in civil engineering. Geometric design practices by the state highway departments and other designing agencies are not entirely uniform. A considerable variation exists in the laws of the various states limiting the size and weight of motor vehicles. Differences also exist in the financing ability of the various governmental agencies, and these influence the designers' decisions and modify their design standards. There are also differences in experience and the interpretation of research, which are reflected in a variety of design practices.

These differences are tolerated by the Bureau of Public Roads as unavoidable and are accepted when approval is requested on plans developed by the various states for federal-aid improvements. The strongest force tending to standardize these differences lies in the various technical committees of the American Association of State Highway Officials. All state highway departments and the Bureau of Public Roads hold membership in this association and join in the deliberations of its technical committees. Upon approval by a required majority, a standard submitted is declared adopted and becomes, in effect, a guide for all members of the association.

Geometric design standards relate to horizontal alignment and profile, intersections, clearances, and the dimensions of the highway cross-section. These features of road design are directly affected by traffic volume and traffic speeds. Studies in the dynamics of highway movements by planning surveys, along with other factual data pertaining to traffic, form the basis of the newer geometric standards in highway design.

For purposes of design, the American Association of State Highway Officials has suggested a classification of highways on the basis of (1) traffic volume, (2) character of traffic, (3) assumed designed speed, and (4) weight of traffic. The latter, which chiefly affects the design of pavements for weight-carrying capacity, is considered in other chapters of this book. In connection with the above classification, the elements of bridge design for the various types of crossing encountered must be considered.

5–1. Traffic Volume. Traffic volume, one of the factors which influence the design of a highway, is indicated by a numerical figure representing the number of vehicles expected to use the highway in a fixed period of time (1 hour).

Traffic volume varies greatly from hour to hour and day to day. It is for the peak hours that highway standards should be developed. It

TABLE 5–1 (*101*)

TRAFFIC ELEMENTS AND THEIR RELATION—RURAL HIGHWAYS

Traffic Element	Explanation and Nationwide Percentage or Factor
Average daily traffic: ADT	Average 24-hour volume for a given year; total for both directions of travel, unless otherwise specified. Directional or one-way ADT is an average 24-hour volume in one direction of travel only.
Current traffic	ADT composed of existing trips, including attracted traffic, that would use the improvement if opened to traffic today (current year specified).
Future traffic	ADT that would use a highway in the future (future year specified). Future traffic may be obtained by adding generated traffic, normal traffic growth, and development traffic to current traffic, or by multiplying current traffic by the traffic projection factor.
Traffic projection factor	Future traffic divided by current traffic. General range, 1.5 to 2.5 for 20-year period. (Freeways may be up to 20 per cent greater or 1.8 to 3.0.)
Design hour volume: DHV	Future hourly volume for use in design (two-way unless otherwise specified), usually the 30th highest hourly volume of the design year (30 HV) or equivalent, the approximate value of which can be obtained by the application of appropriate percentages to future traffic (ADT). The design hour volume, when expressed in terms of all types of vehicles, should be accompanied by factor T, the percentage of trucks during peak hours. Or, the design hour volume may be broken down to the number of passenger vehicles and the number of trucks.
Relation between DHV and ADT: *K*	DHV expressed as a percentage of ADT, both two-way; normal range, 12 to 18. Or, DHV expressed as a percentage of ADT, both one-way; normal range, 16 to 24.
Directional distribution: *D*	One-way volume in predominant direction of travel expressed as a percentage of two-way DHV. General range during design hour 50 to 80. Average, 67.
Composition of traffic: *T*	Trucks (exclusive of light delivery trucks) expressed as a percentage of DHV. Average 7 to 9. Where week-end peaks govern, average may be 5 to 8.

has been found that the thirtieth highest hourly traffic in one year generally is the most suitable value to use for design purposes. Studies made in the United States show that for any particular location the relationship between the thirtieth hour and the annual average daily traffic remains substantially unchanged from year to year.

The value most often used for traffic volume for design purposes then is the thirtieth highest hourly volume for the year for which the road is designed. On minor low-volume roads, design may be based on the average daily traffic. On highly seasonal (resort) highways, the eightieth to one hundredth highest hourly volume may be used. The selected value is also dependent upon the probable life of the roadway. Table 5–1 is a summary of the traffic elements pertinent to design. It also lists related values or factors generally representative of highways throughout the country.

5–2. Character of Traffic. The character of traffic affects the choice of the type of highway and also many details of design. Highways may be divided into three general groups with respect to the character of traffic using them.

The symbol *"P"* denotes highways on which the traffic is predominantly composed of passenger vehicles, the percentage of trucks being such that they have little or no effect on the movement of passenger vehicles. The symbol *"T"* denotes highways on which the traffic consists of a relatively high percentage of trucks and buses. The character of this traffic is such as to hinder the speed of traffic to such an extent that other vehicles are slowed appreciably, or a condition is created where there is continual passing of the slower moving vehicles with accompanying danger to the faster moving vehicles. *"M"* designates mixed traffic and includes such a proportionate amount of *P* and *T* vehicles that the slow-moving vehicles impede traffic only occasionally. It is evident that highways of a *T* classification will require a design treatment in regard to width of traffic lanes, shoulder widths, grades, and pavement thickness different from those of a *P* classification.

The percentage of trucks required to change a highway into a *T* classification cannot be fixed, but it is a variable that depends upon other factors such as alignment and grade.

5–3. Assumed Design Speed. The assumed design speed for a highway, according to the American Association of State Highway Officials (*101*), may be considered as "the maximum safe speed that can be maintained over a specified section of highway when conditions are so favorable that the design features of the highway govern."

The main factor that affects the choice of a design speed is the character of the terrain through which the highway is to pass. Choice of design speed should also be logical with respect to the type of highway being designed. Under normal conditions, a highway in gently rolling country would justify a higher design speed than one in mountainous terrain. Other considerations determining the selection of an assumed design speed are economic factors based on traffic volume, traffic characteristics, costs of rights of way, and other factors which may be of an aesthetic nature. Consideration must also be given to the speed capacity of the motor vehicle. Approved design speeds, as adopted by the American Association of State Highway Officials, are 30, 40, 50, 60, 65, 70, 75, and 80 miles per hour.

The design speed chosen is not necessarily the speed attained when the facility is constructed. Speed of operation will be dependent upon the group characteristics of the drivers under prevailing traffic conditions. Running speeds on a given highway will vary during the day depending upon the traffic volume. The relationship between the design speed and the running speed will also vary. This is illustrated in Table 5–2.

TABLE 5–2 (*101*)

Assumed Relation Between Design Speed and Average Running Speed—Main Highways

Design Speed, mph	Average Running Speed, mph—Main Highways		
	Low Volume	Intermediate Volume	Approaching Possible Capacity
30	28	26	25
40	36	34	31
50	44	41	35
60	52	47	37
65	55	50	
70	58	54	
75	61	56	
80	64	59	

5–4. Design Designation. The design designation indicates the major controls for which a highway is designed. As discussed above, these include in a broad sense traffic volume, character or composition of traffic, and design speed. Traffic volume should be designated in terms of annual daily traffic (ADT) for the current year and for the future design year. The design hourly volume (DHV), a two-way value, should be given along with the directional distribution of traffic expressed as a percentage (D) of the DHV. The character or composition of traffic is indicated by the percentage of trucks (T) of the DHV. Design-speed designation is basic to over-all standards, and with the DHV and T it gives the necessary information to be incorporated in the final plans.

The ratio (K) of the design hour volume (DHV) to the annual daily traffic (ADT), the directional distribution (D) of the design hour volume (DHV) expressed as a percentage, and the percentage of trucks (T) should be shown on the plans.

Where access is fully or partially controlled, this should be shown in the design designation; otherwise, no control of access is assumed.

An example of such designations is given below; the tabulation on the left is for a two-lane highway; the tabulation on the right is for a multilane highway:

			Control of access = full		
ADT (1966)	=	2500	ADT (1966)	=	10,200
ADT (1986)	=	5200	ADT (1986)	=	22,000
DHV	=	720	DHV	=	2950
D	=	65%	D	=	60%
T	=	12%	T	=	8%
V	=	60 mph	V	=	70 mph

5–5. Vehicle Design. The dimensions of the motor vehicle also influence design practice. The width of the vehicle naturally affects the width of the traffic lane; length has a bearing on roadway capacity and affects the turning radius; the height of the vehicle affects the clearance of the various structures. Weight, as mentioned previously, affects the structural design of the roadway.

The American Association of State Highway Officials recommends (*101*) that four design vehicles be used for determining the controls for the geometric design of the highway; these vehicles range in size from the passenger car to the large combination semi-trailer. It is usual for the engineer to select for design the largest vehicle that is anticipated in significant numbers in the design year. The dimensions of these design vehicles are given in Table 5–3.

5–6. Highway Capacity. A knowledge of highway capacity is necessary in the determination of standards to provide adequate facilities for various volumes of traffic. With a low traffic volume, the vehicle operator has a wide latitude in the selection of the speed at which he wishes to travel. As traffic volume increases, the speed of each vehicle is influenced in a large measure by the speed of the slower vehicles. As traffic density increases, a point is finally reached where all vehicles are traveling at the speed of the slower vehicles. This condition indicates that the ultimate capacity has been reached. The capacity of a highway is therefore measured by its ability to accommodate traffic,

TABLE 5–3

DIMENSIONS OF RECOMMENDED DESIGN VEHICLES

Design Vehicle		Dimension in Feet					
			Overhang		Over-all	Over-all	
Type	Symbol	Wheelbase	Front	Rear	Length	Width	Height
Passenger car	P	11	3	5	19	7	–
Single-unit truck	SU	20	4	6	30	8.5	13.5
Semi-trailer combination, intermediate	WB-40	13 + 27 = 40	4	6	50	8.5	13.5
Semi-trailer combination, large	WB-50	20 + 30 = 50	3	2	55	8.5	13.5

and is usually expressed as the number of vehicles that can pass a given point in a certain period of time at a given speed.

When highway capacity is expressed in numbers of vehicles per hour, this should not be interpreted as meaning that the annual average daily traffic would be 24 times the design value for hourly traffic. Traffic volumes are much heavier during certain hours of the day and during certain periods of the year; thus many peak hours may be obtained (see Figs. 5–1 and 5–2) (*32*). As previously indicated, it has been found uneconomical to design the average highway for a greater hourly volume than that which is exceeded during only 29 hours each year. It

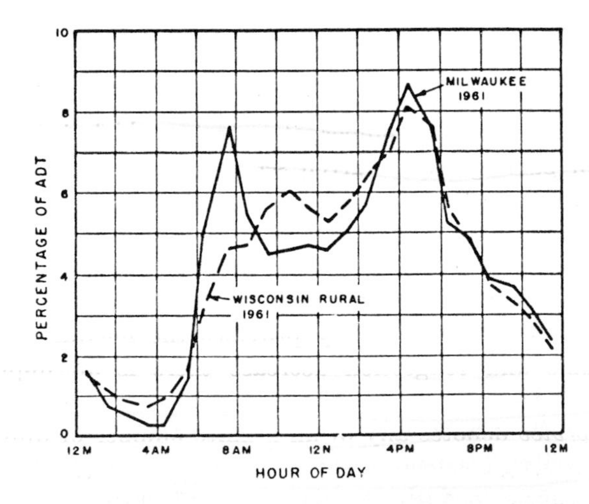

FIG. 5–1. Chart showing hourly variations in traffic by per cent of traffic in the average 24-hour day. (Courtesy Highway Research Board.)

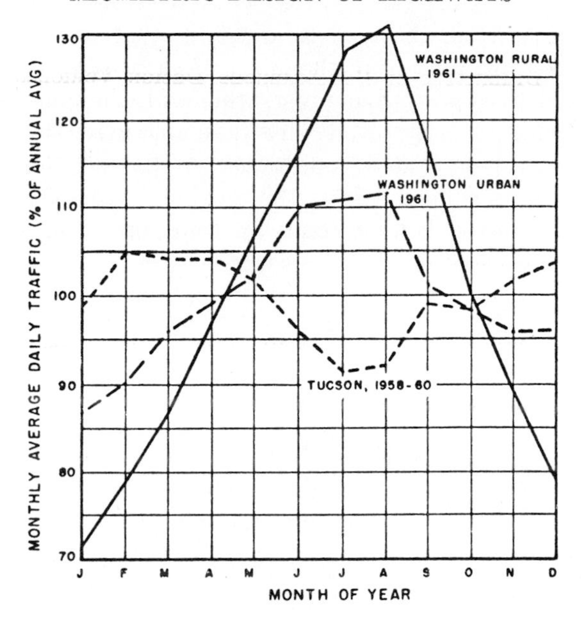

FIG. 5-2. Chart showing the seasonal variations in weekday 24-hour traffic. (Courtesy Highway Research Board.)

has been found by the use of traffic counts over a period of years that approximately 15 per cent of the annual average daily traffic would equal the volume observed during the thirtieth highest hour for rural locations. For urban areas it is 12 per cent.

Capacity is a function of the prevailing conditions. These conditions are those that are determined by the physical features of the roadway and those that are dependent upon the traffic using the roadway. The *capacity* of a given section of roadway, stated either as unidirectional or in both directions for a two-lane or three-lane roadway, may be defined as the maximum number of vehicles which has a reasonable expectation of passing over a given section of roadway during a given time period under prevailing roadway and traffic conditions (*32*).

While the maximum number of vehicles that can be accommodated remains fixed under similar roadway and traffic conditions, there is a range of lesser volumes which can be handled under differing operating conditions. Operation at capacity provides the maximum volume, but as both volume and congestion decrease there is an improvement in the level of service.

Level of service denotes any of an infinite number of differing combinations of operating conditions which can occur on a given lane or roadway when it is accommodating various traffic volumes.

Six levels of service, levels A through F, define the full range of driving conditions from the best to worst, in that order. These levels of service qualitatively measure the effect of such factors as travel time, speed, cost, and freedom to maneuver, which, in combination with other factors, determine the type of service that any given facility provides to the user under the stated conditions. With each level of service, a *service volume* is defined. It is the maximum volume that can pass over a given section of lane or roadway while operating conditions are maintained at the specified level of service.

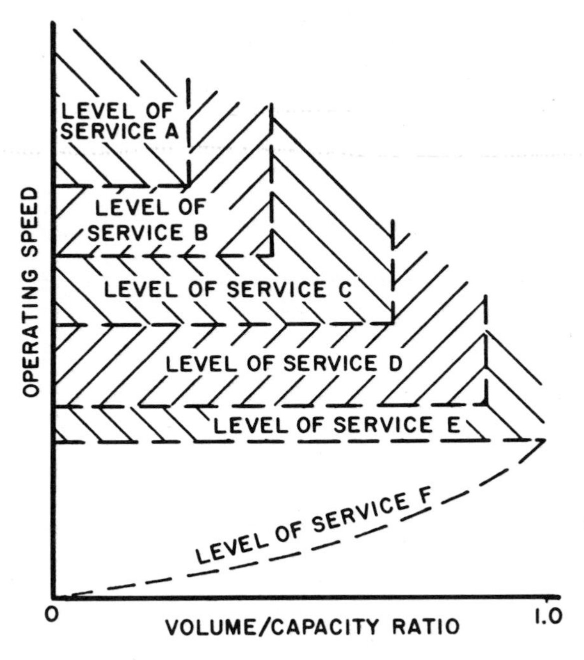

FIG. 5–3. General concept of relationship of level of service to operating speed and volume/capacity ratio (not to scale). (Courtesy Highway Research Board.)

The levels of service are based on relationships of operating speed and volume-to-capacity ratio. Fig. 5–3 (*32*), a graphical representation of the level-of-service concept, shows that, as the operating speed on a facility increases, the higher levels of service are attained; on the other hand, as the volume-to-capacity ratio increases, the facility tends to operate at a lower level of service.

Level A represents free flow, at low densities with no restrictions due to traffic conditions. Level B, the lower limit of which is often used for the design of rural highways, is the zone of stable flow with some slight restriction of driver freedom. Level C denotes the zone of stable flow with more marked restriction on the driver's selection of speed and with reduced ability to pass. The conditions of level D reflect little freedom for driver maneuverability, and, while the operating speeds are still tolerable, this region approaches the condition of unstable flow. Low operating speeds and volumes near or at capacity indicate level of service E, which is the area of unstable flow. Level F is the level of service provided by the familiar traffic jam, with frequent interruptions and breakdown of flow, as well as volumes below capacity, coupled with low operating speeds.

The individual elements used to determine the level of service differ with the type of facility. Table 5–4 sets out the individual elements used in this evaluation.

TABLE 5–4

ELEMENTS USED TO EVALUATE LEVEL OF SERVICE (32)

Element	Freeways	Multi-lane Highways	Two- and Three-lane Highways	Urban Arterials	Down-town Streets
Basic Elements					
Operating speed for section	x	x	x		
Average over-all travel speed				x	x
Volume-to-capacity ratio:					
(a) Most critical point	x	x	x	x	
(b) Each subsection	x	x	x	x	
(c) Entire section	x	x	x	x	
Related Elements					
(a) Average highway speed	x	x	x		
(b) Number of lanes	x				
(c) Sight distance			x		

TABLE 5–5

UNINTERRUPTED FLOW CAPACITIES UNDER IDEAL CONDITIONS (32)

Highway Type	Capacity (Pass. VPH)
Multilane	2000 per lane
Two-lane, two-way	2000 total both directions
Three-lane, two-way	4000 total both directions

CAPACITY UNDER UNINTERRUPTED FLOW CONDITIONS. The capacity of a roadway under uninterrupted flow conditions can be obtained by modifying the capacity of the roadway under ideal conditions, by factors that make allowance for the departure of the prevailing conditions from ideal conditions. Table 5–5 shows the flow capacities of different highway types under ideal conditions. These ideal conditions comprise uninterrupted flow, with no interference by side traffic or obstructions, a vehicle stream composed solely of passenger vehicles, 12-foot-wide traffic lanes, and an alignment capable of providing operation at 70 miles per hour, with no restriction for passing sight distance.

The adjustment factors used in modifying the capacity under ideal conditions to give capacity and service volumes under prevailing conditions may be grouped under the following:

1. Roadway factors
2. Traffic factors

The roadway factors make allowance for the effects of such design elements as lane widths, lateral clearances at the edges of lanes, alignment, and grades. Table 5–6 shows the effect of reduced lane widths on capacity, and Table 5–7 shows the effect of edge clearance. Under both conditions, with limited clearance, the driver has a feeling of restriction of freedom of action, resulting in lowered capacities and service volumes.

TABLE 5–6

EFFECT OF LANE WIDTH ON CAPACITY FOR UNINTERRUPTED FLOW CONDITIONS (*32*)

Lane Width (feet)	Capacity (Per Cent of 12-Ft Lane Cap.)	
	Two-Lane Highways	Multilane Highways
12	100	100
11	88	97
10	81	91
9	76	81

TABLE 5–7

EFFECTIVE ROADWAY WIDTH DUE TO RESTRICTED LATERAL CLEARANCES UNDER UNINTERRUPTED FLOW CONDITIONS (*32*)

Clearance from Pavement Edge to Obstruction, Both Sides (feet)	Effective Width of Two 12-Foot Lanes (feet)	Capacity of Two 12-Foot Lanes (per cent of ideal)
6	24	100
4	22	92
2	20	83
0	17	72

A similar restriction of action is found where alignment conditions prevent complete freedom of choice of speed and ability to pass. There is a resultant decrease in the average highway speed, as the faster vehicles are restricted by the speeds of slower vehicles, with a decrease in capacity as indicated in Table 5–8.

TABLE 5–8

APPARENT EFFECT OF QUALITY OF ALIGNMENT (AS REPRESENTED BY AVERAGE HIGHWAY SPEED) ON CAPACITY (*32*)

Average Highway Speed (mph)	Capacity (per cent of Ideal Alignment)	
	Multilane Highways	Two-Lane Highways
70	100	100
60	100	98
50	96	96
40	–	95
30	–	94

The presence of long adverse grades is important where truck traffic is anticipated. Under normal load, truck traffic travels at a considerably lower speed upgrade than on the level. On two-lane highways where the presence of excessive grades often implies sections with restricted passing sight distance, queues of vehicles form behind the slow-moving truck traffic, waiting for an opportunity to pass. The effect of grades of up to 7 per cent on passenger car traffic is not discernible.

In addition to roadway factors, *traffic factors* (such as many trucks and buses in the traffic stream and variation of flow) affect the capacity and service volumes of a highway. Trucks and buses, because of their restricted maneuverability, reduce the number of vehicles that a facility can handle. This reduction in vehicles is represented by the term *passenger car equivalent,* which indicates the equivalent number of passenger cars that have been displaced by the presence of each truck or bus. Fig. 5–4 shows the relation that has been observed between average truck speeds and passenger car equivalents (*32*).

CAPACITY FOR INTERRUPTED FLOW CONDITIONS. It is beyond the scope of this book to deal with capacity considerations for *interrupted flow conditions,* which require detailed analysis of such factors as interruptions to traffic, land use, traffic character, turning movements, type of traffic flow, level of service required, volumes, and intersection geometry.

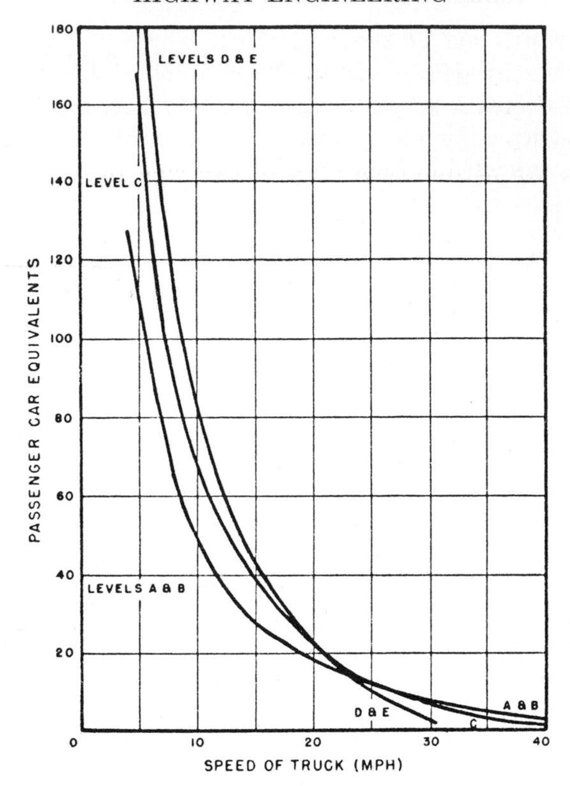

FIG. 5-4. Passenger car equivalents for various average truck speeds on two-lane highways. (Courtesy Highway Research Board.)

The reader is referred to the *Highway Capacity Manual—1965* (*32*), which deals with the recommended procedures in great detail.

5-7. The Highway Cross-Section. The width of the surfaced road should be adequate to accommodate the type and volume of traffic anticipated and the assumed design speed of vehicles. Roads presently in use include those described as single-lane, two-lane, three-lane, multilane undivided, and multilane divided. Modern practice no longer recognizes the single-lane road as a practical element of the improved rural road system, and, as money becomes available, existing one-lane roads are being replaced by two-lane facilities.

Many three-lane roads have been built in previous years and are still in use. Their great advantage seemed to stem from the large improvement in capacity over the two-lane road, with only a moderate increase in construction and right-of-way costs. The three-lane road, however, does appear to have a high accident rate, and the construction of these roads is no longer recommended.

Highways can be classified according to the number of lanes available for vehicle traffic. The classification will normally be two-lane, multilane (undivided four or more lanes), and multilane divided.

Two-Lane Highways. The bulk of our highway system is composed of two-lane highways, and over 90 per cent of the rural roads in the United States are of this type. These two-lane roads vary from low-type roads, which follow the natural ground surface, to high-speed primary highways with paved surfaces and stabilized shoulders. As traffic density, vehicle speeds, and truck widths have increased, two-lane highways have also increased in width from 16 feet to the current recommended value of 24 feet, with 10-foot shoulders on either side for primary routes. Where lighter traffic volumes are expected, the American Association of State Highway Officials recommends minimum surfacing widths based on design speed and traffic volume. Table 5–9 gives the AASHO minimums for assumed conditions (*101*).

TABLE 5–9

Minimum Widths of Surfacing for Two-Lane Highways

Design Speed, mph	Minimum Widths of Surfacing in Feet for Design Volume of:				
	Current ADT 50–250	Current ADT 250–400	Current ADT 400–750 DHV 100–200	DHV 200–400	DHV 400 and over
30	20	20	20	22	24
40	20	20	22	22	24
50	20	20	22	24	24
60	20	22	22	24	24
65	20	22	24	24	24
70	20	22	24	24	24
75	24	24	24	24	24
80	24	24	24	24	24

Four-Lane Highways. Four traffic lanes are considered ideal for purposes of design. It is assumed that traffic is flowing in opposite directions on each pair of lanes, and that passing is accomplished within the lanes of forward movement of traffic and not in the lanes of opposing traffic. This feeling of freedom from opposing traffic results generally is smoother operation and increase in the capacity per lane over that of two- and three-lane pavements. With respect to traffic capacity, the four-lane highway is the most efficient, having at least four times the capacity of a two-lane highway for the same assumed design speed.

Although the four-lane highway is the basic multi-lane type, traffic volumes sometimes warrant the use of highways having six or even eight lanes, particularly in urban areas.

The undivided multilane highway does, however, appear to have an accident rate higher than that of the two-lane highway. This would appear to be due to higher traffic volumes with traffic frictions from cars traveling both in the same direction and in opposing directions.

It is generally agreed that, for rural roads and where design speeds are high, where traffic volumes are sufficient to require multilane construction, traffic separation is desirable.

5–8. Divided Highways. In order to provide positive protection against conflict of opposing traffic, highways are frequently divided by a median strip (see Fig. 5–5). On such highways, lane widths should

Fig. 5–5. Four-lane divided highway in a rural area. (Courtesy California Division of Highways.)

be a minimum of 12 feet, with 13-foot lanes provided where many large truck combinations are likely to occur. It is highly desirable that all multilane highways should be divided. The width of these median strips varies from 4 to 60 feet, or more. A median strip less than 4 to 6 feet in width is considered to be little more than a centerline stripe and its use, except for special conditions, should be discouraged. The narrower the median, the longer must be the opening in the median to give protection to vehicles making left turns at points other than intersections. Where narrow medians must be used, many agencies install median barriers to physically separate opposing streams of traffic and minimize the number of head-on collisions. Medians of 14 to 16 feet have been used and are sufficient to provide most of the separation advantages of opposing traffic while permitting the inclusion of a median lane at crossroads;

however, medians 16 to 60 feet wide and more are now recommended. The median should also be of sufficient width to maintain vegetation and to support low-growing shrubs that reduce the headlight glare of opposing traffic. Median strips at intersections should receive careful consideration and should be designed to permit necessary turning movements.

Divided highways need not be of constant cross-section. The median strip may vary in width; the roads may be at different elevations; and the superelevation may be applied separately on each pavement (Fig. 5–6). In rolling terrain, substantial savings may be effected in construction and maintenance costs by this variation in design. This type of design also tends to eliminate the monotony of a constant width and equal grade.

Where it is necessary to narrow the median strip, or where intersections make it desirable to widen the median strip in tangent alignment, the change may be effected by reverse curves of one degree or less which may be without superelevations or transitions. Where such changes in width on curves are desirable, they should be accomplished if possible by changing the curvature of one or both roads.

A very important feature of the design of multi-lane highways is the control of access from adjacent property. Limited-access highways are discussed in Art. 5–43.

5–9. Parking Lanes. A parking lane is a lane separate and distinct from the traffic lane. Parking should be prohibited on rural highways, but in some rural areas where parking adjacent to the traffic lane cannot be avoided, parallel parking should be permitted and extra lanes provided for this purpose. Where it is desired to provide parking facilities in parks, scenic outlooks, or other points of interest, off-the-road parking space should be provided. In urban and suburban locations the parking area often includes the gutter section of the roadway, which may vary in width from 6 to 8 feet.

The minimum width of the parking lane for parallel parking is 8 feet, with 10 feet preferred. For angle parking, the width of the lane for parking purposes increases with the angle. When the angle of parking exceeds 45 degrees it is necessary to use two moving traffic lanes for maneuvering the vehicle into position. Angle parking should be used only in low-speed urban areas where parking requirements take precedence over the smooth flow of traffic. Parking at the approaches to intersections should be prohibited.

5–10. Pavement Crowns. Another element of the highway cross-section is the pavement crown, which is the raising of the centerline of the roadway above its edges. Pavement crowns have varied greatly throughout the years. On the early low-type roads, high crowns were

FIG. 5-6. Widely separated roadways make up four-lane divided high-way in Michigan.

necessary for good drainage and were ½ inch or more per foot. With the improvement of construction materials, techniques, and equipment, which permit closer control, crowns have been decreased. Present-day high-type pavements with good control of drainage now have crowns as low as ⅛ inch per foot. Low crowns are satisfactory when little or no settlement of the pavement is to be expected and when the drainage system is of sufficient capacity to quickly remove the water from the traffic lane. When four or more traffic lanes are used, it is desirable to provide a higher rate of crown on the outer lanes in order to expedite the flow of water.

Crowns may be formed by intersecting tangent lines or by curved lines that emanate from the road centerline. In the latter case, circular arcs of long radii are used, as well as parabolic arcs. It makes little difference which is employed, but the parabolic arc lends itself better to making computations for the initial offsets or ordinates in constructing templates or in setting grades.

5–11. Shoulders. Closely related to the lane width is the width of the shoulders. It is necessary to provide shoulders for safe operation and to develop full traffic capacity. Well-maintained, smooth, firm shoulders increase the effective width of the traffic lane as much as 2 feet, as most vehicle operators drive closer to the edge of the pavement in the presence of adequate shoulders. To accomplish their purpose, shoulders should be wide enough to permit and encourage vehicles to leave the pavement when stopping. The greater the traffic density the more is the likelihood of the shoulders being put to emergency use. A usable outside shoulder width of at least 10, and preferably 12 feet, clear of all obstructions is desirable for all heavily traveled and high-speed highways. Inside shoulders often are not as wide. In mountainous areas where the extra cost of providing shoulders of this width may be prohibitive or on low-type highways, a minimum width of 4 feet may be provided; width of 6 to 8 feet is preferable. Under these conditions, however, emergency parking strips should be provided at proper intervals. In terrain where guardrails or other vertical elements (such as retaining walls) are required an additional 2 feet of shoulder should be provided..

The slope of the shoulder should be greater than that of the pavement. A shoulder with a high-type surfacing should have a slope of at least ⅜ inch per foot. Sodded shoulders may have a slope as high as 1 inch per foot in order to carry water away from the pavement.

5–12. Guardrails. Guardrails should be provided at points where fills are over 8 feet in height, when shoulder slopes are greater than 1 on 4, and at locations where there is a sudden change in alignment and where a great reduction in speed is necessary. Deep roadside ditches with steep banks, or right-of-way limitations, often make it necessary to steepen the side slopes and require the use of guardrails. Where guardrails are used, the width of the shoulders is increased approximately 2 feet to allow space for the placing of the posts.

Various types of guardrail are in use at the present time. The most important of these are the ordinary wooden guardrail, the cable guardrail, wire-mesh railing, and steel plate railing.

The wooden railing may consist of a series of posts used alone or connected with wooden rails. Rustic railing, which is made entirely of logs, is used in parks or scenic areas.

The cable guardrail consists of one or two steel cables mounted on posts. The posts are usually of wood, but steel or reinforced concrete posts are also used. Expansion take-up devices are used with the cable to take care of the expansion and contraction caused by temperature changes.

In place of using steel cable, fabricated steel mesh may be employed. The wire mesh may be from 12 to 24 inches in width. Various types of steel plate rails are also used and have proved effective in preventing excessive damage to vehicles because of the deflecting effect of the plate rail.

Aluminum and masonry railings are also used.

Care must be taken when selecting the type of guardrail to be used. Cable guardrail gives more protection than a wooden guardrail, but is less visible than the wooden guardrail or the steel or aluminum plate and wire mesh. On the other hand, the steel-plate and wire-mesh type act like snow fences in certain areas and hinder snow removal considerably. A great deal of research has been done in recent years to develop safer and more effective guardrails (and median barriers). This work is continuing.

5–13. Curbs, Curb and Gutter, and Drainage Ditches. The use of curbs is generally confined to urban and suburban roadways. The design of curbs varies from a low, flat, lip-type to a nearly vertical barrier-type curb. Curbs adjacent to traffic lanes, where sidewalks are not used, should be low and very flat. The face of the curb should be no steeper than 45 degrees so that vehicles may mount them without difficulty. Curbs at parking areas and adjacent to sidewalks should be 6 to 8 inches in height, with faces nearly vertical. Clearance should be sufficient to clear fenders and bumpers and to permit the opening of car doors. From the pedestrian viewpoint, curbs should be one step in height. When a barrier or nonmountable curb is used, it should be offset a minimum of 10 feet from the edge of the traffic lane. Fig. 5–7 shows some typical curb sections (101).

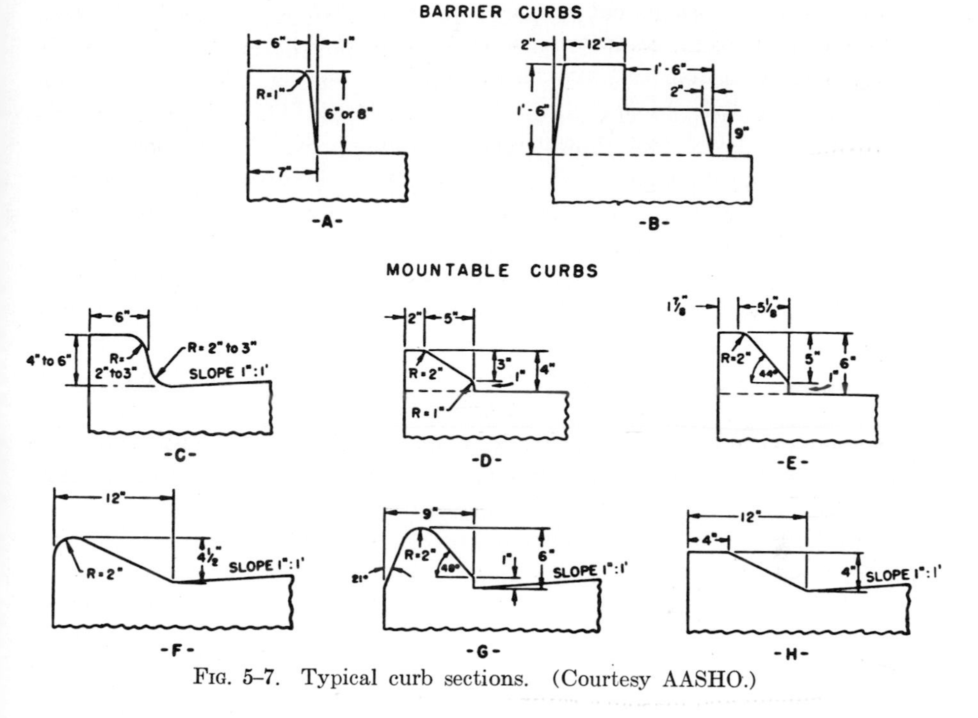

FIG. 5–7. Typical curb sections. (Courtesy AASHO.)

Drainage ditches, in their relation to the highway cross-section, should be located within or beyond the limits of the shoulder, and under normal conditions be low enough to drain the water from under the pavement. The profile gradient of the ditch may vary greatly from that of the adjacent pavement. A rounded ditch section has been found to be safer than a **V**-type ditch, which also may be subject to severe washing action. Ditch maintenance is also less on the rounded ditch sections.

5–14. Slopes. Side slopes and back slopes may vary a great deal, depending upon the type of material and the geographical location. Well-rounded flat slopes present a pleasing appearance and are most economical to build and maintain. Side slopes of 1 on 4 are used a great deal in both cut and fill sections up to about 10 feet in depth or height, but where the height of cut or fill does not exceed 6 feet a maximum side slope of 6:1 is recommended. Extremely flat slopes are sometimes used in swamp sections. As previously mentioned, where guardrail is used, slopes may be as high as 1 on 1.5. Slopes as high as 1 on 1 are generally not satisfactory and present an expensive maintenance problem. In certain fill sections, special slopes may be built with riprap, dry masonry rubble, reinforced concrete cribbing, and various types of retaining wall.

The back slopes in cut areas may vary from a 6 on 1 to vertical in rock sections or loess formations, to 1 on 1.5 in normal soil. It is sometimes advisable to have back slopes as flat as 1 on 4 when side borrow is needed. Slope transitions from cuts to fills should be gradual and should extend over a considerable length of the roadway.

5–15. Sidewalks. Another feature of the highway cross-section is sidewalks, which are accepted as an integral part of city streets. In rural areas, however, little consideration has been given to their construction, since pedestrian traffic is usually very light. Motor vehicle speeds in rural areas are high and only a few seconds elapse from the time the approaching vehicle is sighted until it is upon the pedestrian. Serious considerations should be given to the construction of sidewalks in all areas where the number of pedestrians using the highway warrants it. Justification for the construction of sidewalks is governed by the density of pedestrian and vehicular traffic, their relative timing, the traffic speeds. The decision to construct sidewalks will in general be dictated by regular usage and density of community development. In urban areas, sidewalks are traditionally used, even though the surfacing may be inferior. A typical highway cross-section of an urban project is shown in Fig. 5–8. Sidewalk surfaces should be equal in quality to the roadway surface in order to encourage their use.

5–16. Special Cross-Sectional Elements. It is sometimes necessary to furnish additional accommodations for traffic that will affect the various cross-sectional features which have been discussed so far. Special consideration for off-the-road parking, mailbox turnouts, or additional lanes for truck traffic on long grades causes corresponding changes in the cross-sections. In the design of multi-lane highways and expressways, it is necessary to separate the through traffic from the adjacent service or frontage roads. The width and type of separator is controlled by the width of the available right-of-way, location and type of overpasses or underpasses, and many other factors. The border strip, which is that portion of the highway between the curbs of the through highway and the frontage or service road, is generally used for the location of utilities, either subsurface or overhead. Highways involving these elements of the cross-section are usually treated as special cases and receive consideration at the time of their design.

5–17. Right-of-Way. The right-of-way width for a two-lane highway on secondary roads with an annual average daily traffic volume of 400 to 1000 vehicles, recommended by the American Association of State Highway Officials, is 66 feet minimum and 80 feet desirable. On the Interstate system minimum widths will vary depending on conditions from 150 feet (without frontage roads) and 250 feet (with frontage

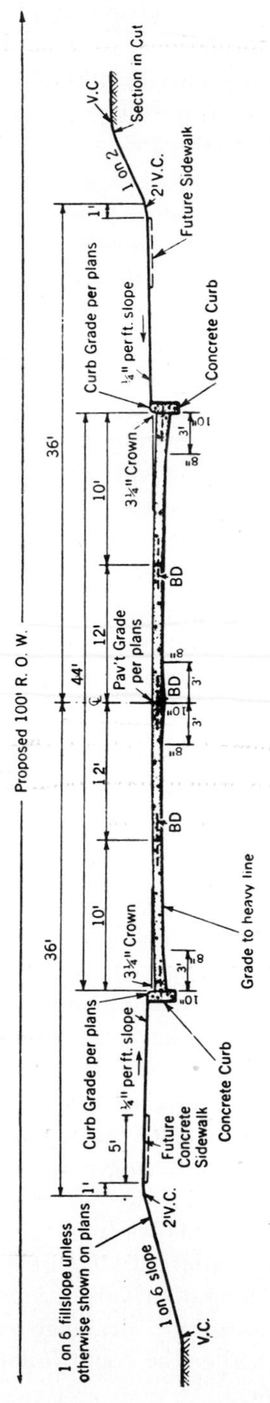

FIG. 5-8. Typical cross-section—urban project.

roads) to 200 and 300 feet for an eight-lane divided highway without and with frontage roads, respectively. On high-type two-lane highways in rural areas a minimum width of 100 feet, with 120 feet desirable, is recommended. A minimum of 150 feet and a desirable width of 250 feet is recommended for divided highways.

Right-of-way should be purchased outright or placed under control by easement or other means. When this is done, sufficient right-of-way is available when needed, and the expense of purchasing developed property or the removal of other encroachments from the highway right-of-way is eliminated. A wide section of right-of-way provides a safer highway, permits gentle rounded slopes, and, in general, lowers maintenance and snow removal costs.

In summing up the various elements of the cross-section, it can readily be seen that the opportunity for wide variation in design presents itself in the selection of crowns, pavement and shoulder widths, slopes, and other features. Every feature, from the centerline to the extremity of the right-of-way, must be given careful consideration for a balanced design. Typical designs are illustrated in Fig. 5–9.

5–18. Alignment. An ideal and most interesting roadway is one that generally follows the existing natural topography of the country. This is the most economical to construct, but there are certain aspects of design that must be adhered to which may prevent the designer from following this undulating surface without making certain adjustments in a vertical and horizontal direction.

The designer must produce an alignment in which conditions are consistent. Sudden changes in alignment should be avoided as much as possible. For example, long tangents should be connected with long sweeping curves, and short sharp curves should not be interspersed with long curves of small curvature. The ideal location is one with consistent alignment where both grade and curvature receive consideration and satisfy limiting criteria. The final alignment will be that in which the best balance between grade and curvature is achieved.

Terrain has considerable influence on the final choice of alignment. Generally, the topography of an area is fitted into one of the following three classifications: level, rolling, or mountainous.

In level country, the alignment, in general, is limited by considerations other than grade, i.e., cost of right-of-way, land use, waterways requiring expensive bridging, existing roads, railroads, canals and power lines, and subgrade conditions or the availability of suitable borrow.

In rolling country, grade and curvature must be considered carefully. Depths of cut and heights of fill, drainage structures, and number of bridges will depend on whether the route follows the ridges, the valleys, or a cross-drainage alignment.

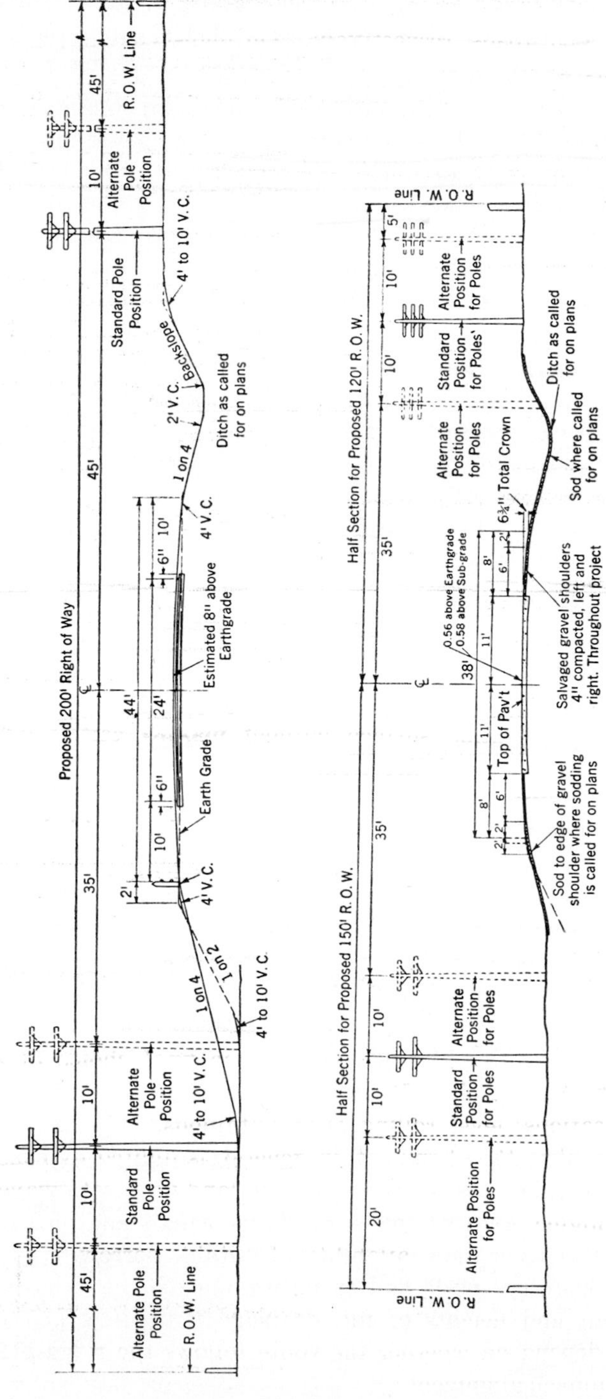

FIG. 5–9. Typical cross-sections.

In mountainous country, grades provide the greatest problem, and, in general, the horizontal alignment (curvature) is conditioned by maximum grade criteria.

5–19. Circular Curves. Circular curves may be described by giving either the radius or the "degree of curve." In highway design, the degree of curve is defined as the central angle subtended by an arc of 100 feet. This is known as the "arc definition." Books on railroad location define the degree of curve as the central angle subtended by a chord of 100 feet, and some highway departments follow this procedure. Arc or chord measurements can be considered alike for all curves less than 4 degrees without appreciable error. An examination of tables will show that the following chords may be assumed to be equal to the arcs without appreciable error:

100-foot chords up to 4 degrees
50-foot chords up to 10 degrees
25-foot chords up to 25 degrees
10-foot chords up to 100 degrees

Fig. 5–10 shows a simple highway curve and its component parts, with the necessary formulas for finding the values of the various elements. A combination of simple curves can be arranged to produce compound or reverse curves.

In general, the maximum desirable degree of curvature is from 5 to 7 degrees in open country and not over 10 degrees in mountainous areas.

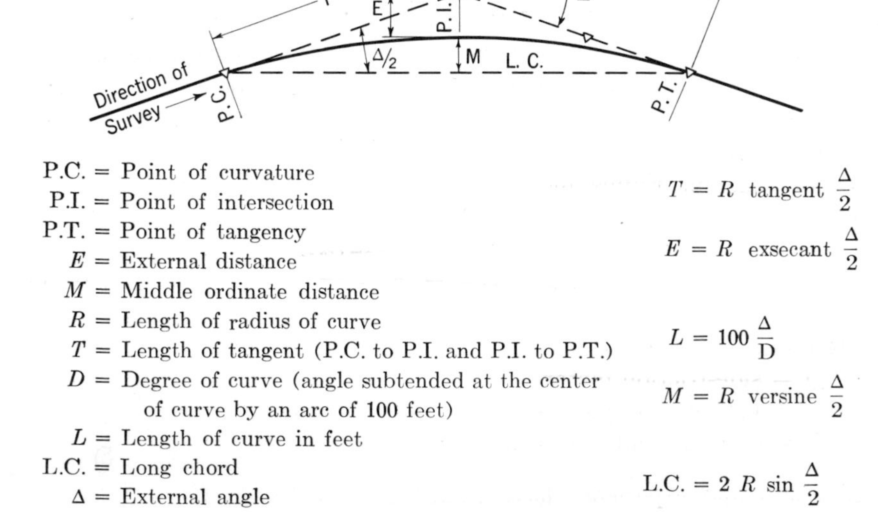

FIG. 5–10. A simple highway curve.

Many states limit curvature to 3 degrees on principal highways. However, exceptions to these limits will occur.

Two curves in the same direction connected with a short tangent known as "broken back" curves, should be combined into one continuous curve. This may be accomplished by compounding the curve, provided that the difference between the two branches of the curve does not exceed 5 degrees.

Where reversed curves are necessary, they should be separated by a tangent distance of at least 200 feet in order to allow proper easement from one curve to another.

For small deflections a curve should be at least 500 feet long for a 5-degree central angle, with an increase of 100 feet for each decrease of one degree in the central angle.

Other factors which have to be considered with a curve alignment, and which materially affect the design, are superelevations, transitions, and widening.

5–20. Superelevation of Curves. On rural highways, most drivers adopt a more or less uniform speed when traffic conditions will permit them to do so. When making a transition from a tangent section to a curved section, if the sections are not properly designed, the vehicle must be driven at reduced speed for safety as well as for the comfort of the occupants. This is due to the fact that a force is acting on the vehicle which tends to cause an outward skidding away from the center of the curve. Most highways have a slight crowned surface to take care of drainage. It can be readily seen that when these crowns are carried along the curve, the tendency to slip is retarded on the inside of the curve because of the banking effect of the crown. The hazard of slipping is increased on the outside of the curve, however, due to the outward sloping of the crown. In order to overcome this tendency to slip and to maintain average speeds, it is necessary to "superelevate" the roadway section; i.e., raise the outside edge or "bank" the curve.

Analysis of the forces acting on the vehicle as it moves around a curve of constant radius indicates that the theoretical superelevation is equal to

$$e + f = \frac{V^2}{15R} = \frac{0.067V^2}{R} \tag{5–1}$$

where e = rate of superelevation in feet per foot
f = side-friction factor
V = velocity in miles per hour
R = radius of curve in feet

Research and experience have established limiting values for e and f. Use of the maximum e with the safe f value in the formula permits determination of minimum curve radii for various design speeds. Safe side-friction factors to be used in design range from 0.16 at 30 miles per hour to 0.11 at 80 miles per hour. Present practice suggests a maximum superelevation rate of 0.12 feet per foot. Where snow and ice conditions prevail, the maximum superelevation should not exceed 0.08 foot per foot. Some states have adopted a maximum superelevation rate of 0.10; however, other rates may have application on some types of highways in certain areas.

The maximum safe degree of curvature for a given design speed can be determined from the rate of superelevation and the side-friction factor. The minimum safe radius R can be calculated from the formula given above:

$$R = \frac{V^2}{15(e + f)}$$

using D as the degree of curve (arc definition)

$$D = \frac{5729.6}{R} \quad \text{or} \quad D = \frac{85{,}900\ (e + f)}{V^2} \tag{5–2}$$

The relationship between superelevation and degree of curve is illustrated in Fig. 5–11 and Table 5–10. Fig. 5–12 is used to determine superelevation rates in Michigan.

5–21. Spirals or Transition Curves. Transition curves serve the purpose of providing a gradual change from the tangent section to the circular curve, and vice versa. Spiral curves give more comfort and safety at high operating speeds than other methods, and should be used on all curves except those which are very flat; when used in combination with superelevated sections, superelevation should be attained within the limits of the transition.

The minimum length of the transition curve is given as *(174)*:

$$L_s = 1.6 \frac{V^3}{R} \tag{5–3}$$

where L_s = length of the transition in feet
V = speed in miles per hour
R = Radius in feet

5–22. Attainment of Superelevation. The transition from the tangent section to a curved superelevated section must be accomplished without any appreciable reduction in speed and in such a manner as to insure safety and comfort of the vehicle and occupants.

In order to effect this change, it will be readily seen that the normal road cross-section will have to be tilted to the superelevated cross-section. This tilting usually is accomplished by rotating the section about the centerline axis. The effect of this rotation is to lower the inside

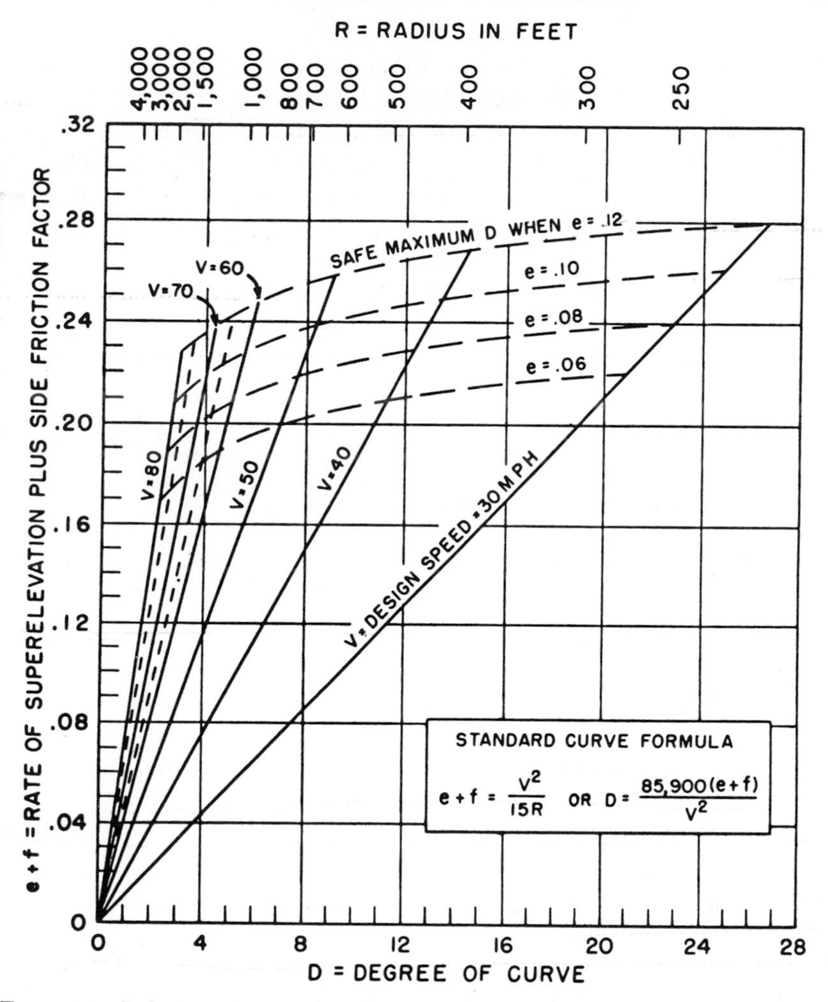

FIG. 5-11. Relation of superelevation (e) plus side-friction factor (f) to degree of curve for different design speeds. (Courtesy of AASHO.)

edge of the pavement and raise the outside edge without changing the centerline grade. Another method is to rotate about the inner edge of the pavement as an axis so that the inner edge retains its normal grade but the centerline grade is varied or rotation may be about the outside edge. Rotation about the centerline is used by a majority of the states, but for flat grades too much sag is created in the ditch grades by this method. On grades below 2 per cent, rotation about the inside edge is preferred. Regardless of which method is used, care should be exercised to provide for drainage in the ditch sections of the superelevated areas.

The roadway on full superelevated sections should be a straight in-

TABLE 5–10 (*101*)

MAXIMUM DEGREE OF CURVE AND MINIMUM RADIUS DETERMINED FOR LIMITING VALUES OF e AND f

Design Speed	Maximum e	Maximum f	Total $(e + f)$	Minimum Radius, ft.	Max. Degree of Curve	Max. Degree of Curve, Rounded
30	.06	.16	.22	273	21.0	21.0
40	.06	.15	.21	508	11.3	11.5
50	.06	.14	.20	833	6.9	7.0
60	.06	.13	.19	1263	4.5	4.5
65	.06	.13	.19	1483	3.9	4.0
70	.06	.12	.18	1815	3.2	3.0
75	.06	.11	.17	2206	2.6	2.5
80	.06	.11	.17	2510	2.3	2.5
30	.08	.16	.24	250	22.9	23.0
40	.08	.15	.23	464	12.4	12.5
50	.08	.14	.22	758	7.6	7.5
60	.08	.13	.21	1143	5.0	5.0
65	.08	.13	.21	1341	4.3	4.5
70	.08	.12	.20	1633	3.5	3.5
75	.08	.11	.19	1974	2.9	3.0
80	.08	.11	.19	2246	2.5	2.5
30	.10	.16	.26	231	24.8	25.0
40	.10	.15	.25	427	13.4	13.5
50	.10	.14	.24	694	8.3	8.5
60	.10	.13	.23	1043	5.5	5.5
65	.10	.13	.23	1225	4.7	4.5
70	.10	.12	.22	1485	3.9	4.0
75	.10	.11	.21	1786	3.2	3.0
80	.10	.11	.21	2032	2.8	3.0
30	.12	.16	.28	214	26.7	26.5
40	.12	.15	.27	395	14.5	14.5
50	.12	.14	.26	641	8.9	9.0
60	.12	.13	.25	960	6.0	6.0
65	.12	.13	.25	1127	5.1	5.0
70	.12	.12	.24	1361	4.2	4.0
75	.12	.11	.23	1630	3.5	3.5
80	.12	.11	.23	1855	3.1	3.0

clined section. When a crowned surface is rotated to the desired superelevation, the change from a crowned section to a straight inclined section should be accomplished gradually. This may be done by first changing the section from the centerline to the outside edge to a level section. Second, raise the outside edge to an amount equal to one-half the desired superelevation, and at the same time change the inner section to a straight section and put the whole section in one inclined plane. Third, continue rotation about the centerline until full superelevation is reached.

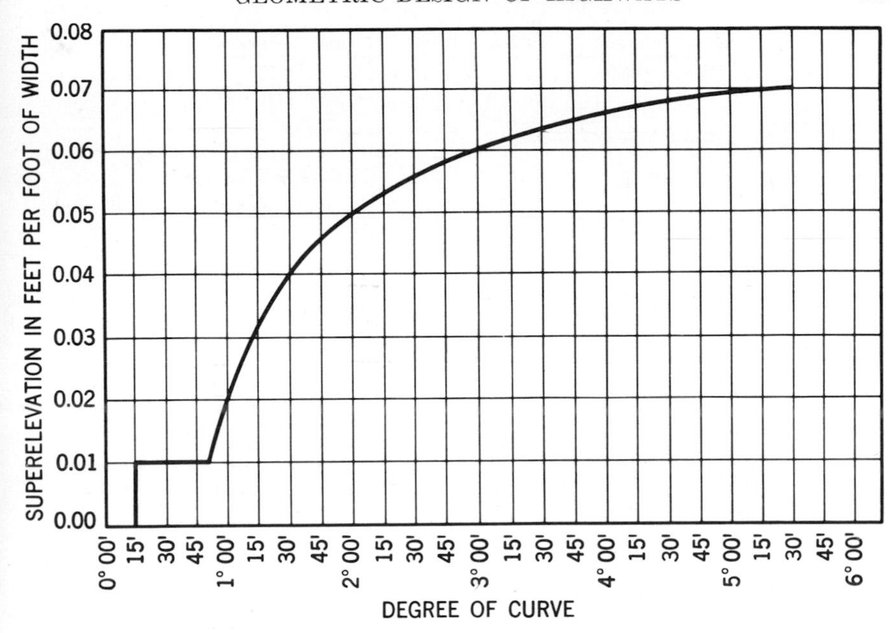

Fig. 5–12. Relationship between superelevation and degree of curve used for design in one state.

The distance required for accomplishing the transition from a normal to a superelevated section, sometimes called the "transition runoff," is a function of the design speed, degree of curvature, and the rate of superelevation. However, all roadways are not superelevated at the same rate. The American Association of State Highway Officials has recommended maximum relative slopes between the centerline and edge profiles of a two-lane highway (*101*). These recommended values, shown in Table 5–11, are dependent upon the design speed.

TABLE 5–11

Superelevation Design Data (101)

Design Speed	Maximum Relative Slope Between Profiles of Edge of Two-Lane Pavement and Centerline
30	0.66%
40	.58
50	.50
60	.45
65	.41
70	.40
75	.38
80	0.36%

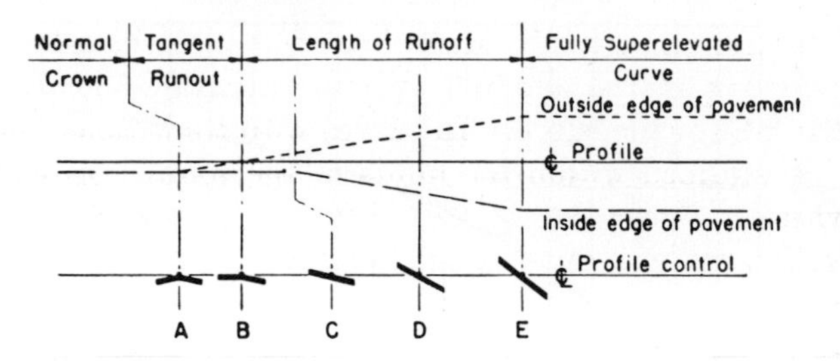

A-PAVEMENT REVOLVED ABOUT CENTER LINE

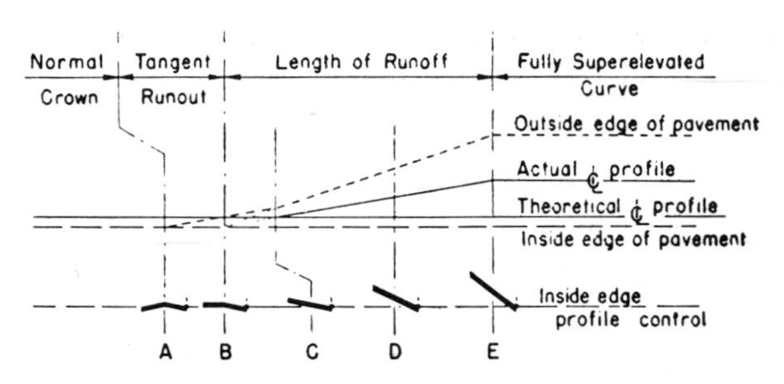

B-PAVEMENT REVOLVED ABOUT INSIDE EDGE

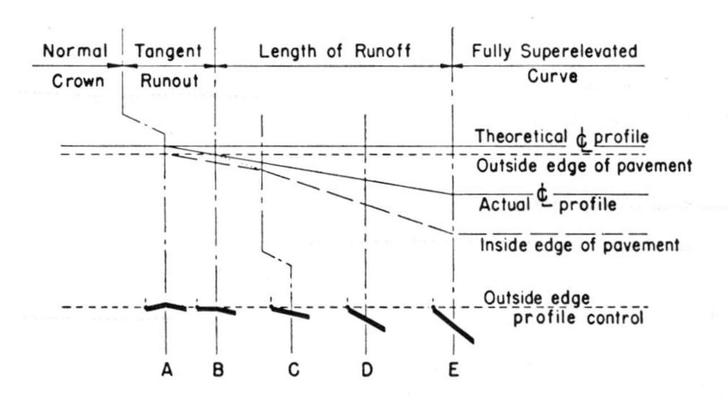

C-PAVEMENT REVOLVED ABOUT OUTSIDE EDGE

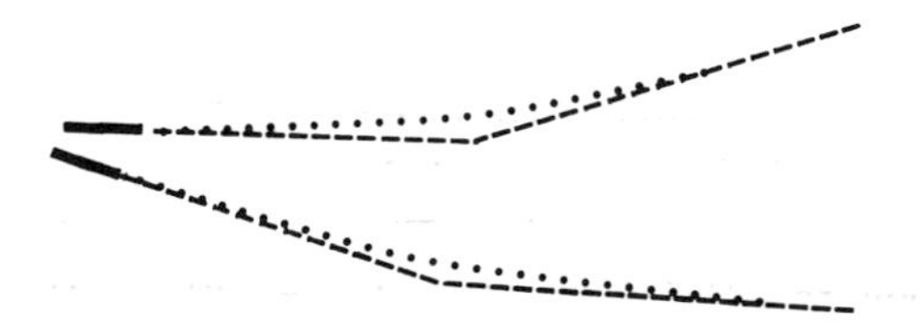

Note: Angular breaks to be appropriately rounded as shown by dotted line.

Fig. 5–13. Diagrammatic profiles showing methods of attaining superelevation. (Courtesy AASHO.)

Superelevation is usually started on the tangent at some distance before the curve starts, and full superelevation is generally reached at beyond the P.C. of the curve. In curves with transitions, the superelevation can be attained within the limits of the spiral. In curves of small degree where no transition is used, between 60 and 80 per cent of the superelevation runoff is put into the tangent.

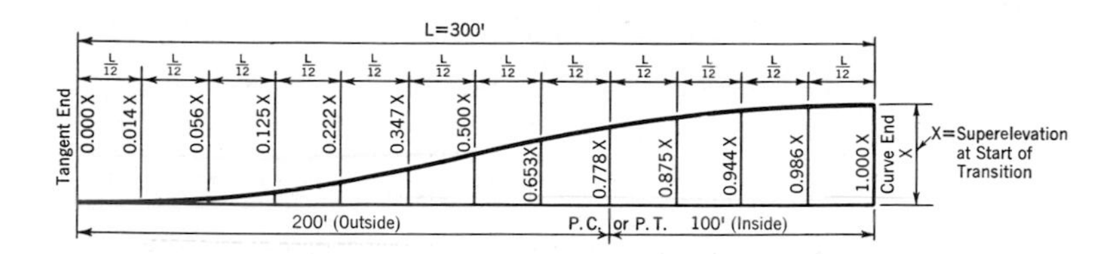

To be used on curves where the rate of superelevation is less than .04 ft. per ft. unless otherwise shown on plan.

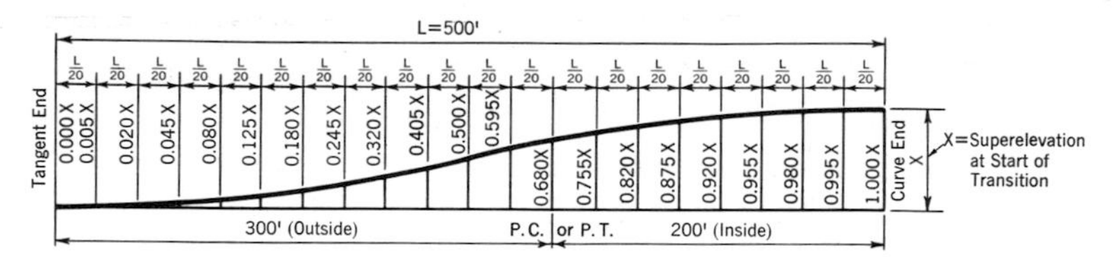

To be used on curves where the rate of superelevation is .04 foot per foot or greater unless otherwise shown on plans.

TRANSITION VERTICAL REVERSE CURVES

The transition from normal section of metal to the superelevated section shall be made by means of the proper reverse vertical curve shown above. This curve shall be applied to all grade lines which are to be depressed or raised in the superelevation transition.

For spiraled curves, the length of the superelevation transition equals the length of the inside spiral.

FIG. 5–14. Superelevation using reversed parabolic curves. (Courtesy Michigan Department of State Highways.)

On multilane undivided roads or roads with wide medians, increased runoff lengths are necessary. In order to obtain smooth profiles for the pavement edges, it is recommended that the breaks at cross-sections be replaced by smooth curves (see Fig. 5–13). Different states adopt differing methods to attain smooth edge profiles. Fig. 5–14 indicates the method adopted by one state.

5–23. Widening of Curves. Extra width of pavement may be necessary on curves. As a vehicle turns, the rear wheels follow the front wheels on a shorter radius, and this has the effect of increasing the width of the vehicle in relation to the lane width of the roadway. In the study of traffic behavior of a large number of drivers, it has also been found that they do not drive as close to the edge of curved sections as when on a tangent. This fact is hard to evaluate, and justification for widening on this basis is questionable.

Present practice requires no widening when the degree of curvature is less than 10 degrees on a two-lane pavement 24 feet wide. For greater curvatures and/or narrower pavements widening normally is from 2 to 4 feet, depending on design speed and pavement width. It is suggested that no pavement be widened less than 2 feet.

5–24. Grades and Grade Control. The vertical alignment of the roadway and its effect on the safe and economical operation of the motor vehicle constitutes one of the most important features of road design. The vertical alignment, which consists of a series of straight lines connected by vertical parabolic or circular curves, is known as the "grade line." When the grade line is increasing from the horizontal, it is known as a "plus grade," and when it is decreasing from the horizontal it is known as a "minus grade." In analyzing grade and grade controls, the designer usually studies the effect of change in grade on the centerline profile.

In the establishment of a grade, an ideal situation is one in which the cut is balanced against the fill without a great deal of borrow or an excess of cut to be wasted. All hauls should be downhill if possible, and not too long. Ideal grades have long distances between points of intersection, with long vertical curves between grade tangents to provide smooth riding qualities and good visibility. The grade should follow the general terrain and rise and fall in the direction of the existing drainage. In rock cuts and in flat swampy areas it is necessary to maintain higher grades. Future possible construction and the presence of grade separations and bridge structures also control grades.

Changes of grade from plus to minus should be placed in cuts, and changes from a minus grade to a plus grade should be placed in fills. This will generally give a good design, and many times it will avoid the appearance of building hills and producing depressions contrary to the general existing contours of the land. Other considerations for determining the grade line may be of more importance than the balancing of cuts and fills.

Urban projects usually require a more detailed study of the controls and a finer adjustment of elevations than do rural projects. It is often best to adjust the grade to meet existing conditions because of additional expense when doing otherwise.

In the analysis of grade and grade control, one of the most important considerations is the effect of grades upon the operating costs of the motor vehicle. An increase in the gasoline consumption and a reduction of speed are apparent when grades are increased. An economical approach would be to balance the added annual cost of grade reduction against the added annual cost of vehicle operation without grade reduction. An accurate solution to the problem depends on the knowledge of traffic volume and type, which can be obtained only by means of a traffic survey.

While maximum grades vary a great deal in various states, AASHO recommendations make maximum grades dependent on design speed and topography (*101*). Present practice limits grades in flat country to 6 per cent at 30-mph design speed and 3 per cent for 80 mph. In mountainous terrain 9 per cent at 30 mph and 5 per cent at 70 mph should be used. Wherever long sustained grades are used, the designer should not substantially exceed the critical length of grade without the provision of climbing lanes for slow moving vehicles. Critical grade lengths vary from 1700 feet for a 3 per cent grade to 500 feet for an 8 per cent grade.

Long-sustained grades should be less than the maximum grade used on any particular section of a highway. It is often preferred to break the long-sustained uniform grade by placing steeper grades at the bottom and lightening the grades near the top of·the ascent. Dips in the profile grade in which vehicles may be hidden from view should also be avoided.

Minimum grades are governed by drainage conditions. Level grades may be used in fill sections in rural areas when crowned pavements and sloping shoulders can take care of the pavement surface drainage. It is preferred, however, to have a minimum grade of at least 0.3 per cent under most conditions in order to secure adequate drainage.

5–25. Vertical Curves. The parabolic curve is used almost exclusively in connecting grade tangents because of the convenient manner in which the vertical offsets can be computed. A typical vertical curve is shown in Fig. 5–15.

Offsets for vertical curves may be computed from the formulas given in Fig. 5–15. It is usually necessary to make calculations at 50-foot stations, while some paving operations require elevations at 25-foot intervals or less. It is often necessary to compute other critical points on the vertical curve in order to ensure proper drainage or clearances, such as at sags and crests. The low point or high point of a parabolic curve is not usually vertically above or below the vertex of the intersecting grade tangents, but is to the left or right of this point. The distance from the P.C. of the curve to the low or high point is given as:

$$x = \frac{Lg_1}{g_1 - g_2} \tag{5-4}$$

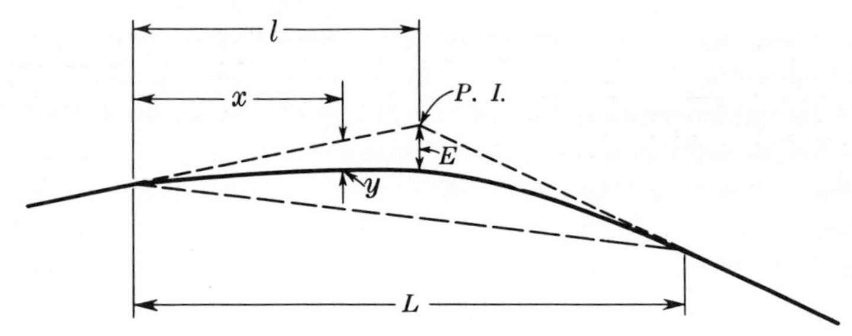

The distance from the P.I. to the middle of the parabolic curve may be found from the following formula:

$$E = \frac{gL}{8}$$

where:

E = the external distance in feet
g = the algebraic difference in grade in per cent of the intersecting grades
L = the length of the curve in stations

The vertical offset at any point on the curve may be computed by the following:

$$y = \left(\frac{x}{l}\right)^2 E$$

where:

y = offset in feet
x = any distance from the P.C. in feet
$l = \frac{L}{2} \times 100$ ft or half the length of the curve
E = external distance in feet

FIG. 5–15. A typical vertical curve.

and the difference in elevation is given as:

$$y = \frac{Lg_1^2}{2(g_1 - g_2)} \tag{5–5}$$

where x = distance from the P.C. to the turning point in stations
L = length of the curve in stations
g_1 = grade tangent from the P.C. in per cent
g_2 = grade tangent from the P.T. in per cent
y = the difference in elevation between the P.C. and the turning point in feet

The above formulas apply only for the symmetrical curve, that is, one in which the tangents are of equal length. The unequal tangent or unsymmetrical vertical curve is a compound parabolic curve. Its use is generally warranted only where a symmetrical curve cannot meet imposed alignment conditions.

5–26. Sight Distance. Safe highways must be designed to give the driver a sufficient distance of clear vision ahead so that he can avoid hitting unexpected obstacles and can pass slower vehicles without danger.

Sight distance is the length of highway visible ahead to the driver of a vehicle. When this distance is not long enough to permit passing an overtaken vehicle, it is termed "stopping (or nonpassing) sight distance." The stopping distance is the minimum distance required for stopping a vehicle traveling at or near the design speed before reaching an object in its path. This stationary object may be a vehicle or some other object on the roadway. When the sight distance is long enough to enable a vehicle to overtake and pass another vehicle on a two-lane highway without interference from an oncoming vehicle, it is termed "passing sight distance."

5–27. Minimum Stopping Sight Distance. Sight distance at every point should be as long as possible, but never less than the minimum stopping sight distance. The minimum stopping sight distance is based upon the sum of two distances: one, the distance traveled from the time the object is sighted to the instant that the brakes are applied, and two, the distance required for stopping the vehicle after the brakes are applied. The first of these two distances is dependent upon the speed of the vehicle and the perception time and brake reaction time of the operator. The second distance depends upon the speed of the vehicle; condition of brakes, tires, and roadway surface; and the alignment and grade of the highway.

5–28. Braking Distance. There is a wide variation among vehicle operators as to the time that it takes to react and to apply the brakes after an obstruction is sighted. Investigations seem to indicate that the minimum value of perception time can be assumed to vary from 2 seconds at 30 miles per hour to 1 second at 70 miles per hour. Brake reaction time is less than perception time, and tests indicate that the average brake reaction time is about ½ second. To provide a factor of safety for operators whose brake reaction time is above the average, a full second is assumed as the total brake reaction time. Some investigators feel that perception and brake reaction time should be combined and have assigned 2½ seconds for this value.

The approximate braking distance of a vehicle on a level highway is determined by:

$$d = \frac{v^2}{2fg} \qquad (5\text{–}6)$$

where d = braking distance in feet
v = velocity of the vehicle in feet per second when the brakes are applied
g = acceleration due to gravity
f = coefficient of friction between tires and roadway

Changing v in feet per second to V in miles per hour, and substituting 32.2 for g, we have:

$$d = \frac{V^2}{30f} \qquad (5\text{–}7)$$

It is assumed that the friction force is uniform throughout the braking period. This is not strictly true; it varies as some power of the velocity. Other physical factors affecting the coefficient of friction are the condition and pressure of tires, type and condition of the surface, and climatic conditions such as rain, snow, and ice. Friction factors for skidding are assumed to vary from 0.62 at 30 miles per hour to 0.55 at 70 miles per hour for dry pavements. For wet pavements these are lower, as shown in Table 5–12. Recommended minimum stopping sight distances are shown in Table 5–12. In this table, perception and brake reaction time are combined.

TABLE 5–12 (*101*)

MINIMUM STOPPING SIGHT DISTANCE

Design Speed (mph)	Assumed Speed for Condition (mph)	Perception and Brake Reaction		Coefficient of Friction (f)	Braking Distance on Level (feet)	Stopping Sight Distance	
		Time (sec.)	Distance (feet)			Computed (feet)	Rounded for Design (feet)
			Design Criteria—Wet Pavements				
30	28	2.5	103	.36	73	176	200
40	36	2.5	132	.33	131	263	275
50	44	2.5	161	.31	208	369	350
60	52	2.5	191	.30	300	491	475
65	55	2.5	202	.30	336	538	550
70	58	2.5	213	.29	387	600	600
75	61	2.5	224	.28	443	667	675
80	64	2.5	235	.27	506	741	750
			Comparative Values—Dry Pavements				
30	30	2.5	110	.62	48	158	
40	40	2.5	147	.60	89	236	
50	50	2.5	183	.58	144	327	
60	60	2.5	220	.56	214	434	
65	65	2.5	238	.56	251	489	
70	70	2.5	257	.55	297	554	
75	75	2.5	275	.54	347	622	
80	80	2.5	293	.53	403	696	

5–29. Effect of Grade on Stopping Distance. When a highway is on a grade, the formula for braking distance is modified as follows:

$$d = \frac{V^2}{30(f \pm g)} \tag{5-8}$$

in which g is the percentage of grade divided by 100. The safe stopping distances on upgrades are shorter and on downgrades longer than horizontal stopping distances. Table 5–13 shows the effect of grade on stopping distance. Where an unusual combination of steep grades and high speed occurs the minimum stopping sight distance should be adjusted to provide for this factor.

TABLE 5–13

EFFECT OF GRADE ON STOPPING SIGHT DISTANCE: WET CONDITIONS (*101*)

Design Speed, mph	Assumed Speed for Condition, mph	Correction in Stopping Distance, Feet					
		Decrease for Upgrades			Increase for Downgrades		
		3%	6%	9%	3%	6%	9%
30	28	–	10	20	10	20	30
40	36	10	20	30	10	30	50
50	44	20	30	–	20	50	–
60	52	30	50	–	30	80	–
65	55	30	60	–	40	90	–
70	58	40	70	–	50	100	–
75	61	50	80	–	60	120	–
80	64	60	90	–	70	150	–

5–30. Measuring Minimum Stopping Sight Distance. It is assumed that the height of eye of the average driver, except drivers of buses and trucks, is about 3.75 feet above the pavement. The height of the stationary object may vary, but a height of 6 inches is assumed in determining stopping sight distance. Fig. 5–16 shows a standard plan used for determining non-passing (stopping) sight distances over a crest.

5–31. Minimum Passing Sight Distance. The majority of our highways carry two lanes of traffic moving in opposite directions. In order to pass slower moving vehicles, it is necessary to use the lane of opposing traffic. If passing is to be accomplished safely, the vehicle driver must be able to see enough of the highway ahead in the opposing traffic lane to permit him to have sufficient time to pass and then return to the right traffic lane without cutting off the passed vehicle and before meeting

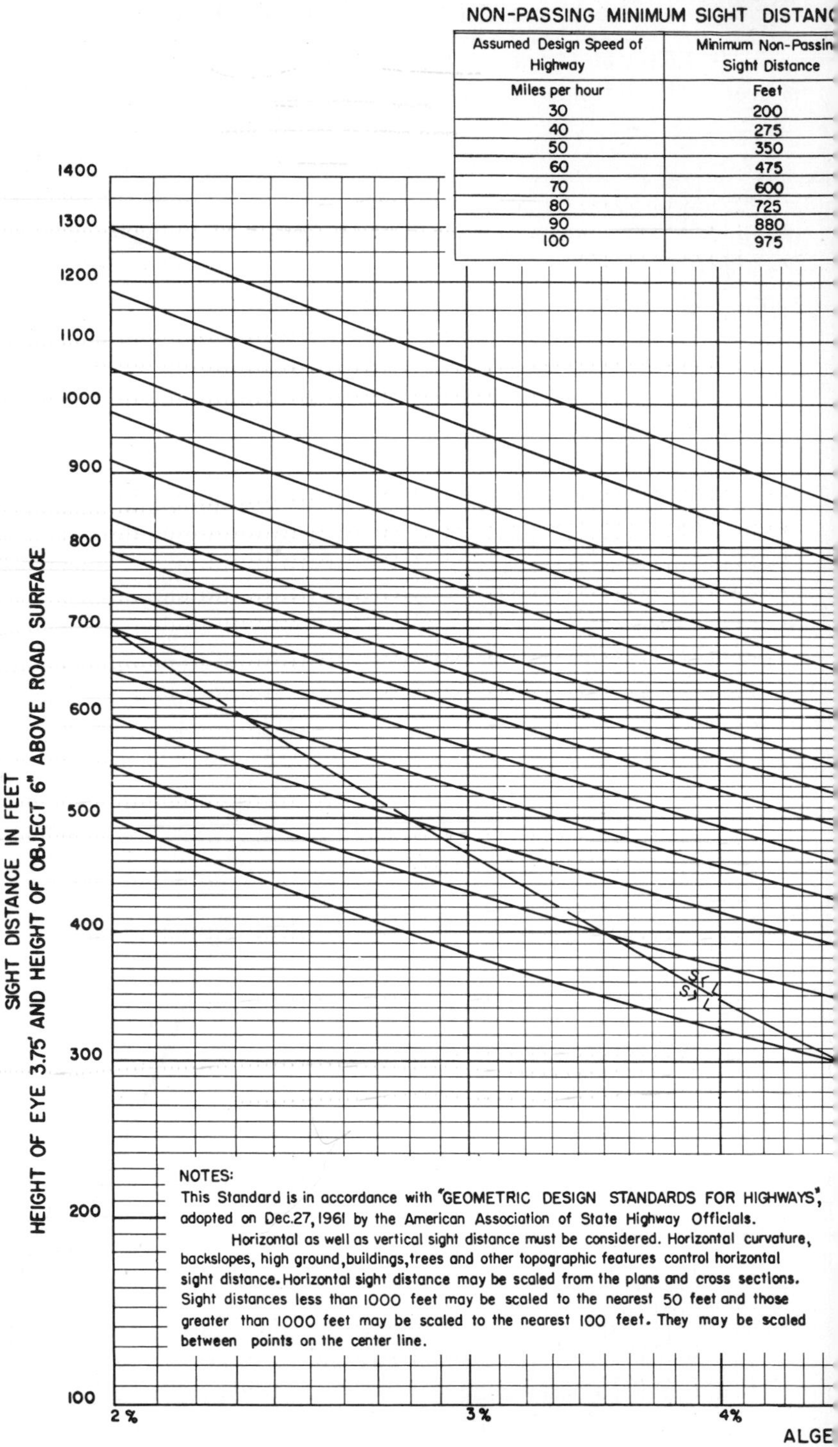

NON-PASSING MINIMUM SIGHT DISTAN

Assumed Design Speed of Highway	Minimum Non-Passin Sight Distance
Miles per hour	Feet
30	200
40	275
50	350
60	475
70	600
80	725
90	880
100	975

SIGHT DISTANCE IN FEET
HEIGHT OF EYE 3.75' AND HEIGHT OF OBJECT 6" ABOVE ROAD SURFACE

1400
1300
1200
1100
1000
900
800
700
600
500
400
300
200
100

S L
S L

NOTES:
This Standard is in accordance with "GEOMETRIC DESIGN STANDARDS FOR HIGHWAYS", adopted on Dec.27, 1961 by the American Association of State Highway Officials.

Horizontal as well as vertical sight distance must be considered. Horizontal curvature, backslopes, high ground, buildings, trees and other topographic features control horizontal sight distance. Horizontal sight distance may be scaled from the plans and cross sections. Sight distances less than 1000 feet may be scaled to the nearest 50 feet and those greater than 1000 feet may be scaled to the nearest 100 feet. They may be scaled between points on the center line.

2% 3% 4%

ALGE

FIG. 5-16. Chart for determining nonpassing sight distan

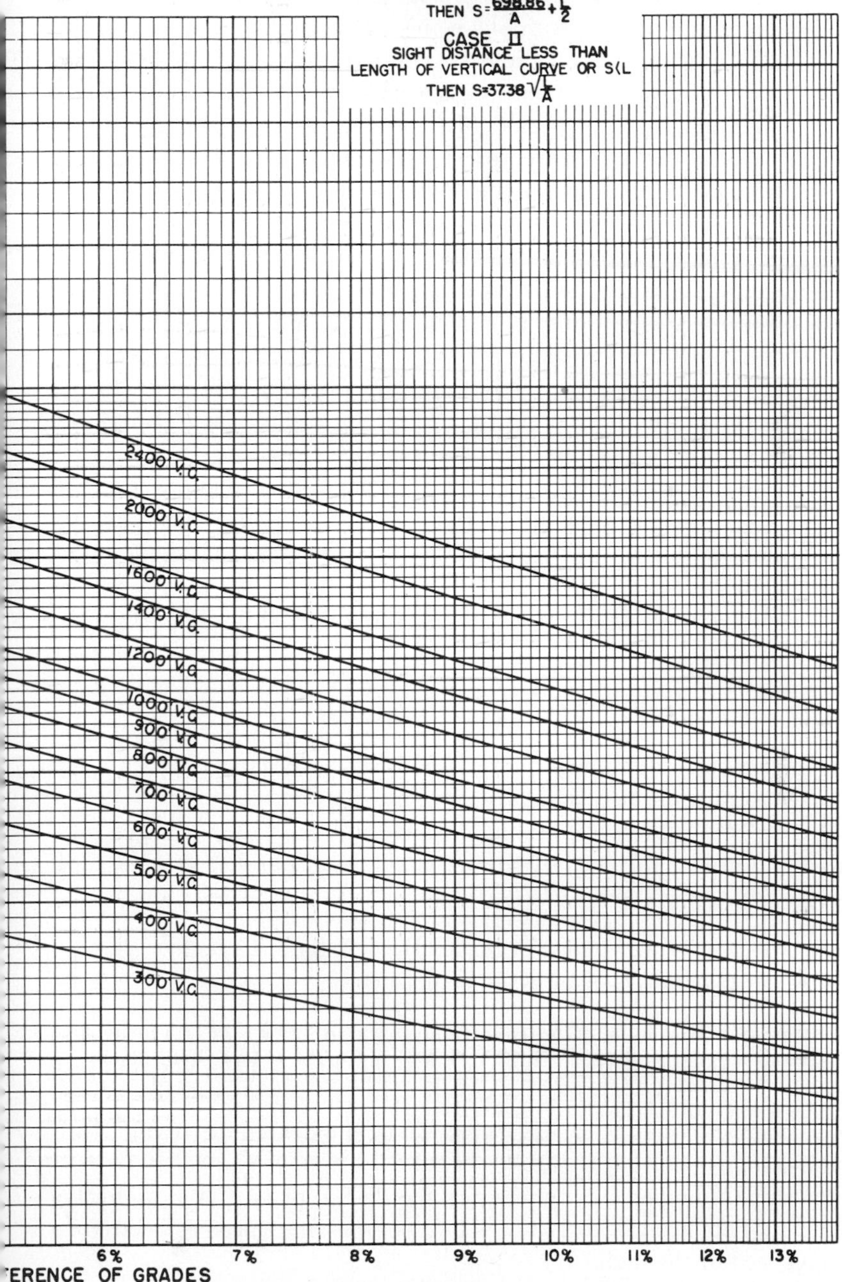

crest. (Courtesy Michigan Department of State Highways.)

the oncoming traffic. The total distance required for completing this maneuver is the passing sight distance.

When computing minimum passing sight distances, various assumptions must be made relative to traffic behavior. On two-lane highways it may be assumed that the vehicle being passed travels at a uniform speed, and that the passing vehicle is required to travel at this same speed when the sight distance is unsafe for passing. When a safe passing section is reached, a certain period of time elapses in which the driver decides whether or not it is safe to pass. When he decides to pass, it is assumed that he accelerates his speed during the entire passing operation. It is also assumed that the opposing traffic appears the instant the passing maneuver starts and arrives alongside the passing vehicle when the maneuver is completed.

Accepting these assumptions, it can be shown that the passing minimum sight distance for a two-lane highway is the sum of four distances: d_1 = distance traveled during perception and reaction time and during the initial acceleration to the point where the vehicle will turn into the opposite lane; d_2 = distance traveled while the passing vehicle occupies the left lane; d_3 = distance between the passing vehicle at the end of its maneuver and the opposing vehicle; d_4 = distance traveled by the oncoming vehicle for two-thirds of the time the passing vehicle occupies the left lane.

The preliminary delay distance d_1 is computed from the following formula:

$$d_1 = 1.47t_1\left(v - m + \frac{at_1}{2}\right) \tag{5-9}$$

where t_1 = time of preliminary delay, seconds
a = average acceleration rate, mphps
v = average speed of passing vehicle, mph
m = difference in speed of passed vehicle and passing vehicle, mph

The distance $d_2 = 1.47vt_2$

where t_2 = time passing vehicle occupies the left lane, seconds
v = average speed of passing vehicle, mph

The clearance distance d_3 varies from 110 to 300 feet. These distances are adjusted as shown in Table 5–14 and Fig. 5–17. The distance $d_4 = 2d_2/3$.

The relationship of d_1, d_2, d_3, and d_4 is illustrated in Table 5–14 and Fig. 5–17.

Minimum passing sight distances for the design of two-lane highways are shown in Table 5–15 (p. 163). Here the passing speed is related to the design speed of the highway.

TABLE 5–14

ELEMENTS OF SAFE PASSING SIGHT DISTANCE—TWO-LANE HIGHWAYS (*101*)

Speed group, mph	30–40	40–50	50–60	60–70
Average passing speed, mph	34.9	43.8	52.6	62.0
Initial maneuver:				
a = average acceleration, mphps*	1.40	1.43	1.47	1.50
t_1 = time, seconds*	3.6	4.0	4.3	4.5
d_1 = distance traveled, feet	145	215	290	370
Occupation of left lane:				
t_2 = time, seconds*	9.3	10.0	10.7	11.3
d_2 = distance traveled, feet	475	640	825	1030
Clearance length:				
d_3 = distance traveled, feet*	100	180	250	300
Opposing vehicle:				
d_4 = distance traveled, feet	315	425	550	680
Total distance, $d_1 + d_2 + d_3 + d_4$, feet	1035	1460	1915	2380

* For consistent speed relation, observed values adjusted slightly.

5–32. Sight Distances for Four-Lane Highways. A four-lane highway should be so designed that the sight distance at all points is greater than the stopping minimum. The crossing of vehicles into the opposing traffic lane on a four-lane highway should be prevented by a median barrier and prohibited by enforcement agencies.

5–33. Extra Lanes for Passing. Inability to pass indicates that the traffic density is greater than the capacity of the road at the assumed design speed and that a wider road is indicated. The capacity of the road may be increased by adding an extra lane at safe passing sections. The added lane should be at least as long as the minimum length of safe passing section to which it is added. Under such conditions it must be remembered that the primary purpose of the added lane is to provide a passing lane at the safe passing section. When additional lanes for passing are used on steep grades, the length of the additional lane should have as great a sight distance as possible.

5–34. Measuring Minimum Passing Sight Distance. As previously stated, the height of eye of the average driver is assumed to be about 3.75 feet above the pavement. As vehicles are the objects that must be seen when passing, it is assumed that the height of object for passing sight distance is the same as the height of eye. Fig. 5–18 shows a chart plan used for determining passing sight distance over a crest.

5–35. Horizontal Sight Distances. Horizontal sight distances may be scaled directly from the plans with a fair degree of accuracy. However, where the distance from the highway centerline to the obstruction is

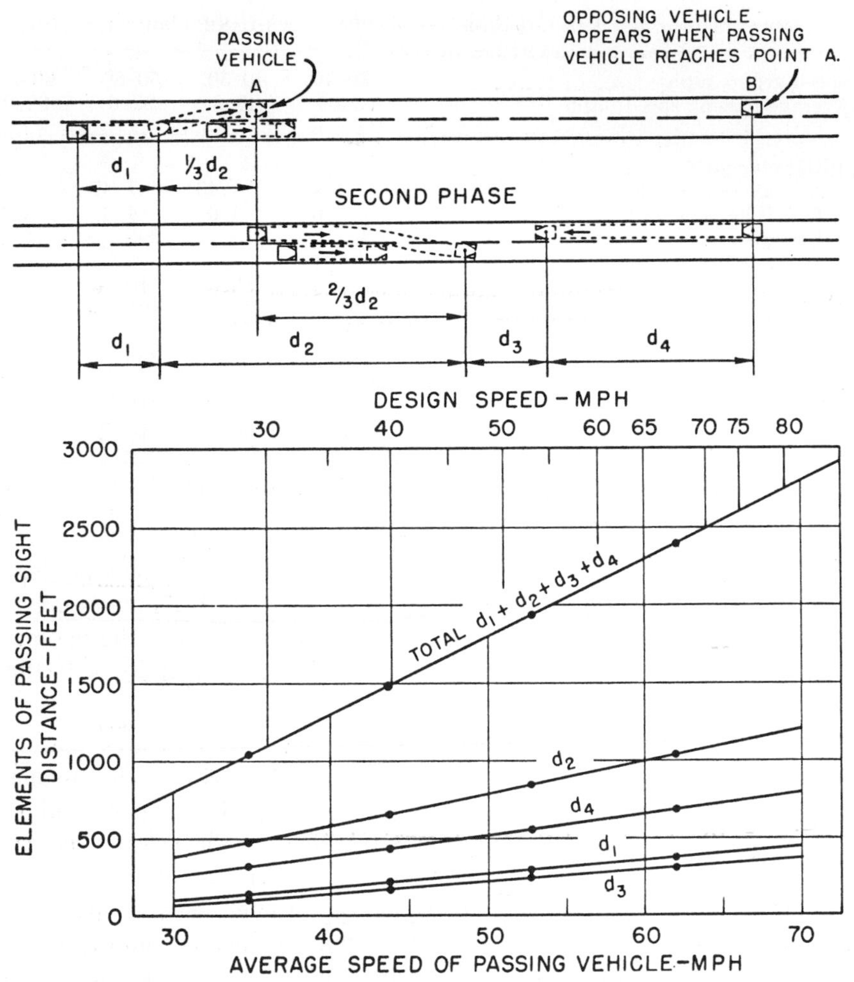

FIG. 5-17. Elements of and total passing sight distance, two-lane highways. (Courtesy AASHO.)

known, it is possible to determine the sight distance for a known radius or degree of curve from the following relations:

$$R = \frac{m}{1 - \cos \dfrac{SD}{200}} \tag{5-10}$$

$$S = \frac{200}{D} \cos^{-1} \left(\frac{R - m}{R} \right) \tag{5-11}$$

where m = distance from center of roadway to edge of obstruction in feet
S = sight distance along the center of the road in feet
R = radius of curvature in feet
D = degree of curvature

The formulas do not apply when the length of the circular curve is less than the sight distance, or when the radius of curvature is not constant throughout.

TABLE 5–15

MINIMUM PASSING SIGHT DISTANCE FOR DESIGN OF 2-LANE HIGHWAYS (*101*)

Design Speed, mph	Assumed speeds		Minimum passing sight distance, feet	
	Passed Vehicle, mph	Passing Vehicle, mph	Fig 5–17	Rounded
30	26	36	1090	1100
40	34	44	1480	1500
50	41	51	1840	1800
60	47	57	2140	2100
65	50	60	2310	2300
70	54	64	2490	2500
75	56	66	2600	2600
80	59	69	2740	2700

Fig. 5–19 shows a horizontal sight distance chart developed from the above formula.

5–36. Marking and Signing Nonpassing Zones. The marking of pavements for nonpassing zones is intended to show the motorist that sight distance is restricted. This restriction may be caused by vertical or horizontal alignment or by a combination of both.

The American Association of State Highway Officials has adopted standards for the marking of pavements. A nonpassing zone for two- and three-lane pavements is one in which the sight distance is less than 500, 600, 800, 1000, and 1200 feet for assumed design speeds of 30, 40, 50, 60, and 70 miles per hour, respectively. An example of marking and signing nonpassing zones is shown in Fig. 5–20 (*98*).

5–37. Highway Intersections. In the design of highways, consideration must be given to the conditions imposed by cross traffic in both rural and urban areas.

Design standards for highway intersections may be divided into two

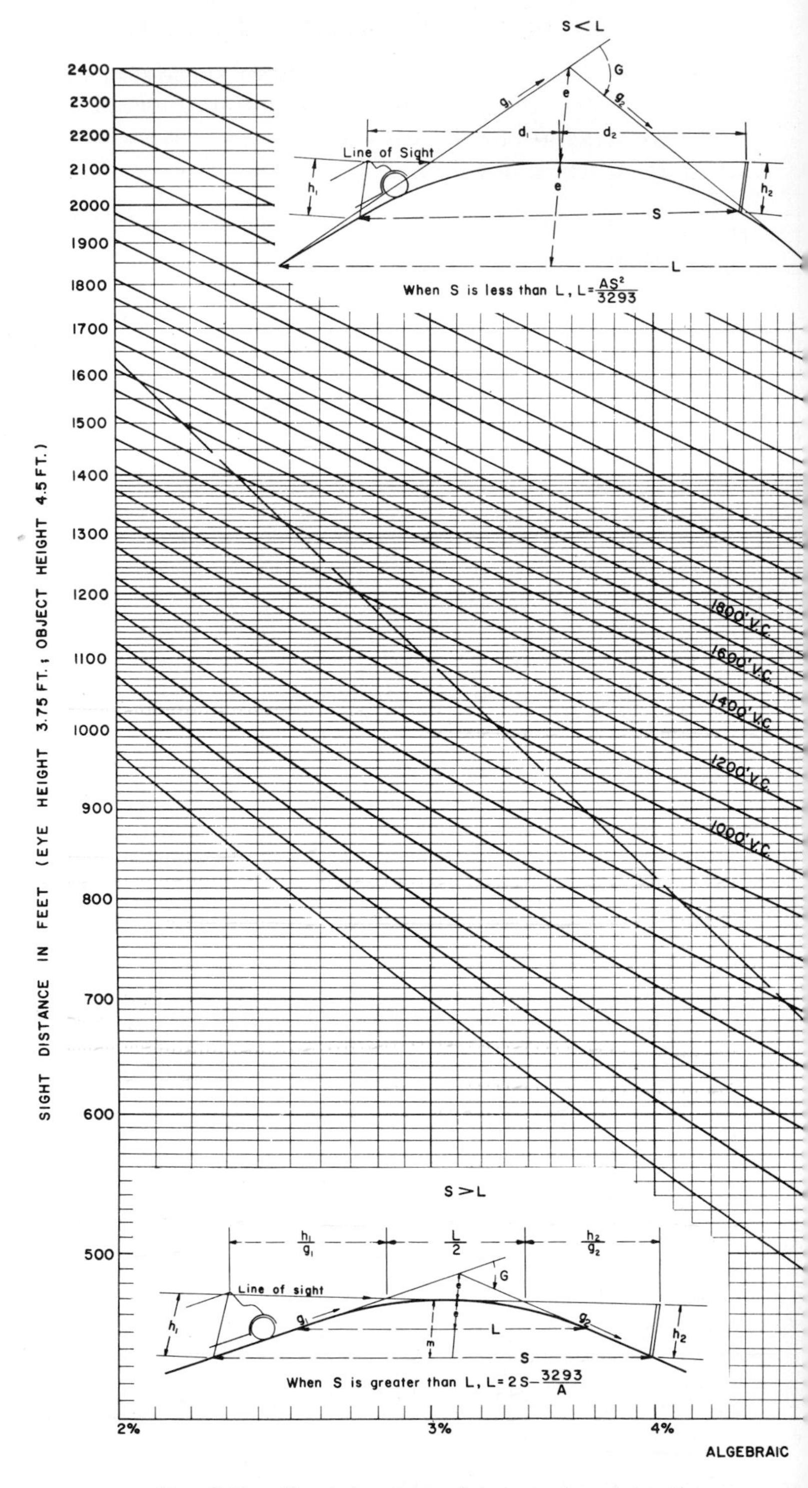

Fig. 5–18. Chart for determining passing sight distance over

NOTES: This Standard is in accordance with "A Policy on Geometric Design Standards for Highways Other Than Freeways," adopted in 1961 by the American Association of State Highway Officials.

Horizontal as well as vertical sight distance must be considered. Horizontal curvature, backslopes, high ground, buildings, trees and other topographic features control horizontal sight distance. Horizontal sight distance may be scaled from the plans and cross sections. Sight distances less than 1,000 feet may be scaled to the nearest 50 feet and those greater than 1,000 feet may be scaled to the nearest 100 feet. They may be scaled between points on the center line.

MINIMUM PASSING SIGHT DISTANCE FOR TWO LANE HIGHWAYS

Based on passing one vehicle with 10 m.p.h. difference between V and speed of passed vehicle.

Assumed Design Speed of Highway - V	Minimum Passing Sight Distance	*Maximum Sight Distance Warranting "No Passing" Signs
Miles per hour	Feet	Feet
30	1100	500
40	1500	600
50	1800	800
60	2100	1000
70	2500	1200
80	2900	1500

* According to 1963 Michigan Manual of Uniform Traffic Control Devices.

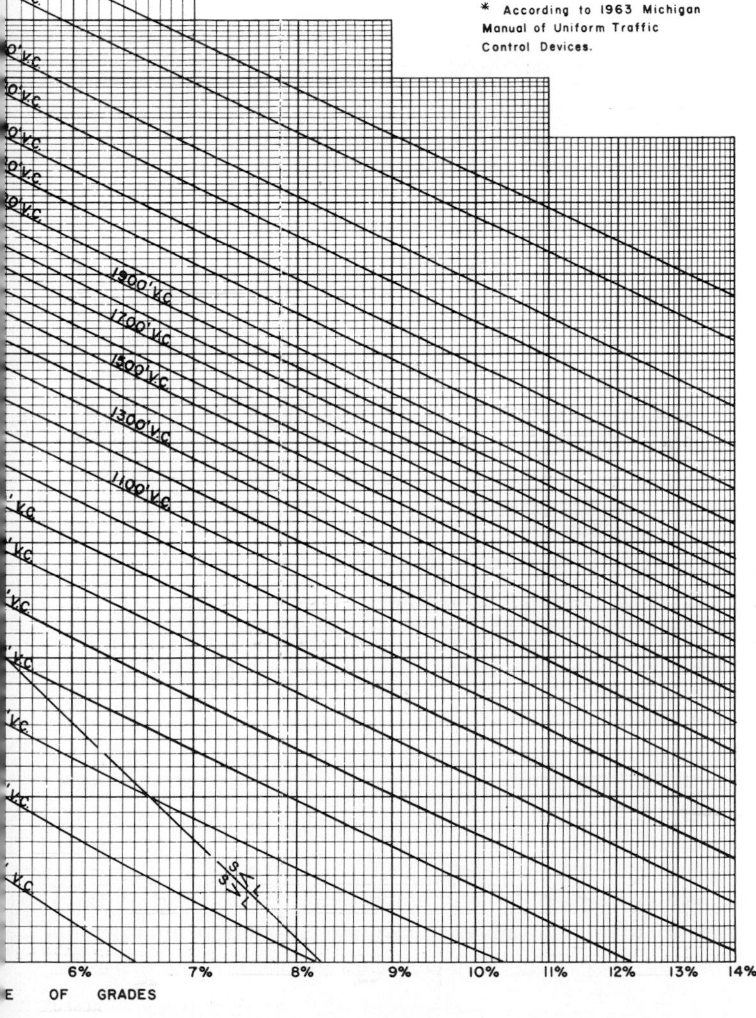

rest. (Courtesy Michigan Department of State Highways.)

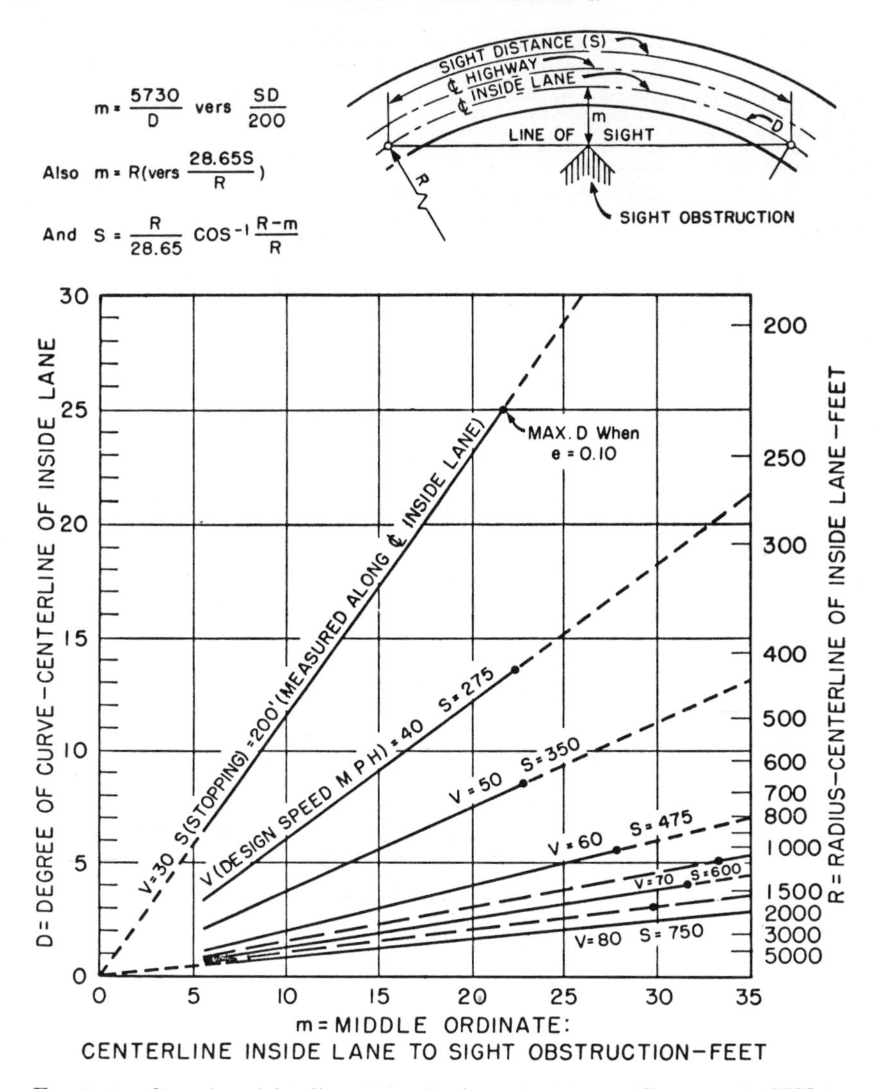

FIG. 5–19. Stopping sight distance on horizontal curves. (Courtesy AASHO.)

groups: (1) intersections at grade which include rotary intersections and (2) grade separations. Standards for these groups have been developed in detail by the American Association of State Highway Officials (*101*).

5–38. Intersections at Grade. Most highways intersect at grade and the intersection area should be designed to provide adequately for turning and crossing movements, with due consideration given to sight distance, signs, grades, and alignment.

Simple intersections at grade may consist of two intersecting highways or a junction of four roads, and a multiple intersection, which is a junction of five or more roads. A junction of three roads is indicated as a "branch," **T**, or **Y**. A branch may be defined as an offshoot of a main-traveled highway, and it usually has a small deflection angle. A **T** intersection is one in which two roads intersect to form a continuous highway, and the third road intersects at, or nearly at, right angles. A **Y** intersection is one in which three roads intersect at nearly equal angles. In addition to the above types of intersection, the flared intersection may be used. This consists of additional pavement width or

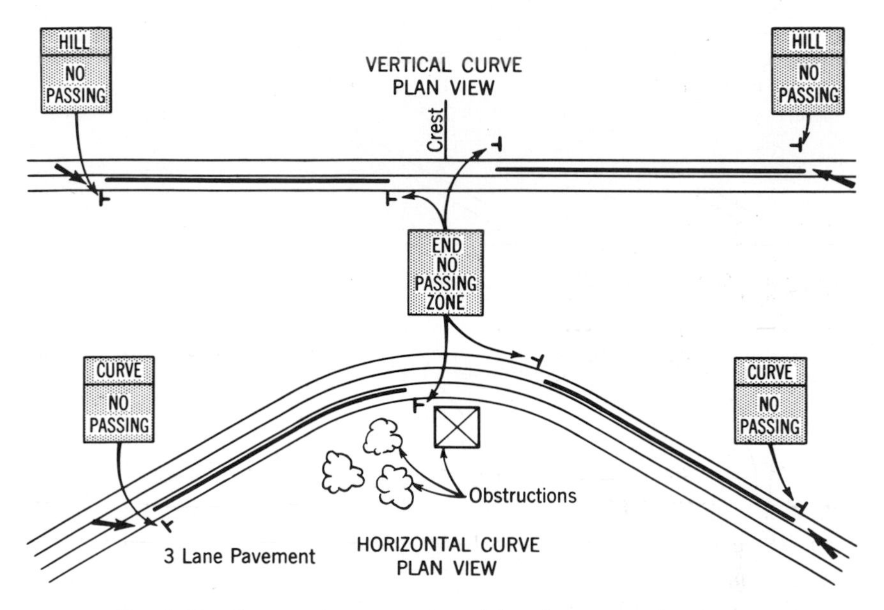

FIG. 5–20. Signs and pavement markings for nonpassing zones.

additional traffic lanes at the intersection area. A few examples of these types of intersections are given in Fig. 5–21.

The design of the edge of the pavement for a simple intersection should provide sufficient clearance between the vehicle and the other traffic lanes. It is frequently assumed that all turning movements at intersections are accomplished at speeds of less than 20 miles per hour, and the design is based on the physical characteristics of the assumed design vehicles. The radius needed for the turning movement of a passenger car is less than that needed for trucks and greatest for semitrailer-truck combinations. A minimum curb radius on 90-degree turns of 30 feet for P traffic and 50 feet for T traffic is recommended. Table 5–16 gives the minimum design for the edge of the pavement.

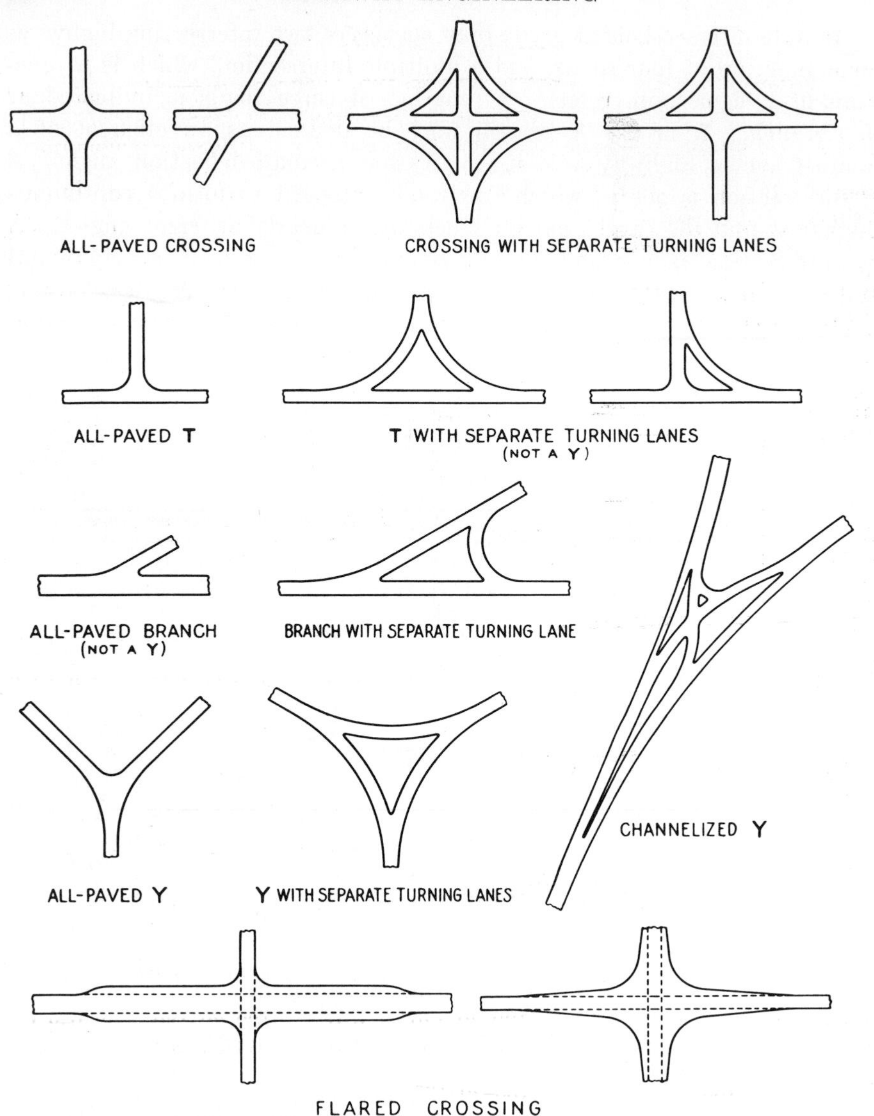

FIG. 5-21. Types of intersections at grade. (Courtesy Bureau of Public Roads.)

5-39. Islands and Channels. The movement of traffic at intersections is controlled by the use of islands. These may be in the form of pavement markings, traffic separators, median strips, and signs. The use of islands with the resulting channelization of traffic aids considerably in directing the flow of traffic and in protecting pedestrians by providing a place of refuge. Capacity can be increased and the number of accidents decreased.

TABLE 5–16 (*101*)

MINIMUM EDGE OF PAVEMENT DESIGNS FOR TURNS AT INTERSECTIONS

Design Vehicle	Angle of Turn (degrees)	Simple Curve Radius (feet)	3-Centered Compound Curve, Symmetric Radii (feet)	3-Centered Compound Curve, Symmetric Offset (feet)	3-Centered Compound Curve, Asymmetric Radii (feet)	3-Centered Compound Curve, Asymmetric Offset (feet)
P	30	60	–	–	–	–
SU		100	–	–	–	–
WB-40		150	–	–	–	–
WB-50		200	–	–	–	–
P	45	50	–	–	–	–
SU		75	–	–	–	–
WB-40		120	–	–	–	–
WB-50		170	200–100–200	3.0	–	–
P	60	40	–	–	–	–
SU		60	–	–	–	–
WB-40		90	–	–	–	–
WB-50		–	200–75–200	5.5	200–75–275	2.0–6.0
P	75	35	100–25–100	2.0	–	–
SU		55	120–45–120	2.0	–	–
WB-40		85	120–45–120	5.0	120–45–200	2.0–6.5
WB-50		–	150–50–150	6.0	150–50–225	2.0–10.0
P	90	30	100–20–100	2.5	–	–
SU		50	120–40–120	2.0	–	–
WB-40		–	120–40–120	5.0	120–40–200	2.0–6.0
WB-50		–	180–60–180	6.0	120–40–200	2.0–10.0
P	105	–	100–20–100	2.5	–	–
SU		–	100–35–100	3.0	–	–
WB-40		–	100–35–100	5.0	100–35–200	2.0–8.0
WB-50		–	180–45–180	8.0	150–40–210	2.0–10.0
P	120	–	100–20–100	2.0	–	–
SU		–	100–30–100	3.0	–	–
WB-40		–	120–30–120	6.0	100–30–180	2.0–9.0
WB-50		–	180–40–180	8.5	150–35–220	2.0–12.0
P	135	–	100–20–100	1.5	–	–
SU		–	100–30–100	4.0	–	–
WB-40		–	120–30–120	6.5	100–25–180	3.0–13.0
WB-50		–	160–35–160	9.0	130–30–185	3.0–14.0
P	150	–	75–18–75	2.0	–	–
SU		–	100–30–100	4.0	–	–
WB-40		–	100–30–100	6.0	90–25–160	3.0–11.0
WB-50		–	160–35–160	7.0	120–30–180	3.0–14.0
P	180	–	50–15–50	0.5	–	–
SU	U-Turn	–	100–30–100	1.5	–	–
WB-40		–	100–20–100	9.5	85–20–150	6.0–13.0
WB-50		–	130–25–130	9.5	100–25–180	6.0–13.0

Islands in an intersection can separate conflicting movements or control the angle at which a conflict may occur. They can regulate traffic, indicate the proper use of the intersection, and often reduce the amount of pavement required. They provide for the protection of the motorists, protection and storage of turning and crossing vehicles, and space for traffic control devices.

Islands are generally grouped into three major classes: directional, divisional, and refuge. General types and shapes of islands are shown in Fig. 5–22.

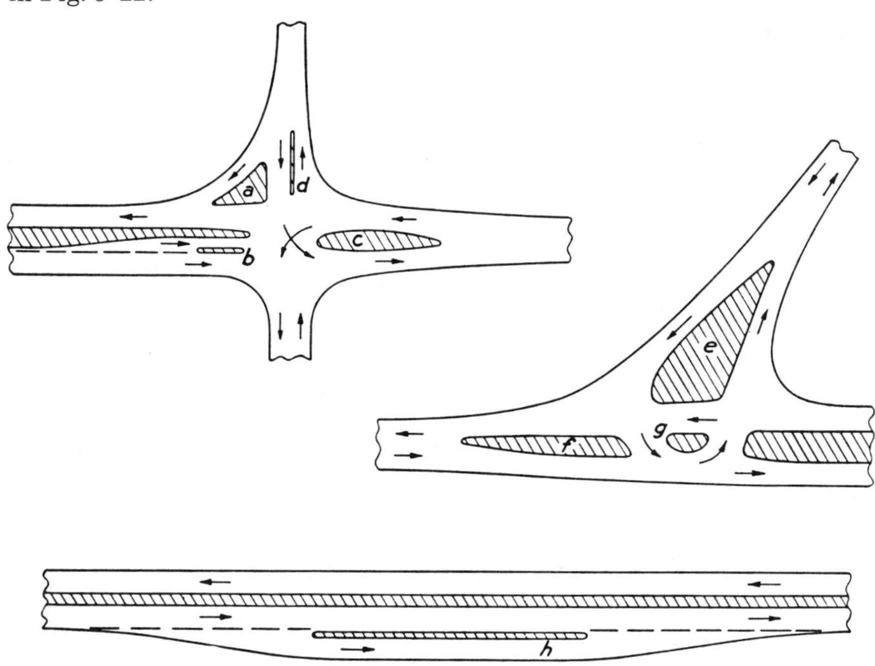

Fig. 5–22. General types and shapes of islands. (Courtesy AASHO.)

Directional islands are designed primarily to guide the motorist through the intersection by indicating the intended route. Where spacious area exists at an intersection and leaves much to the discretion of the driver, islands may be used to channel the motorist into the desired lane by placing a channeling island in the little-used portion of the intersection.

The placing of directional islands should be such that the proper course of travel is immediately evident and easy to follow. A complicated system of islands where the desired course of travel is not immediately evident may result in confusion of the motorist and may be of more hindrance than help in maintaining a steady traffic flow. Islands should be so placed that crossing streams of traffic will pass at approximately right angles and merging streams of traffic will converge at flat angles. By use of such angles there will be less hindrance to traffic on the thoroughfare, and the possibility of accidents in the intersection will be decreased.

Divisional islands are most frequently used on undivided highways approaching intersections. They serve to alert the driver to the intersection and regulate the flow of traffic into and out of the intersection. Their use is particularly advantageous for controlling left-turning traffic at skewed intersections.

A refuge island is located at or near cross walks to aid and protect pedestrians crossing the roadway. Refuge islands are most generally used on wide streets in urban areas for loading and unloading of transit riders. The design of refuge islands is the same as that of other types of islands, except that a higher barrier curb is necessary.

5-40. Rotary Intersections. A rotary intersection is one in which all traffic merges into and emerges from a one-way road around a central island (Fig. 5-23). Advantages of a rotary intersection include:

1. Continuous traffic movement from all legs at reduced speed when operating at low volumes.
2. Since crossing movements are eliminated, accidents are likely to be less serious.
3. Where more than four intersection legs are involved, the design layout may be simplified. In such a case, a channelized layout may become extremely complicated.
4. The cost of an at-grade intersection may be considerably less than that of grade separation structures.

Disadvantages include:

1. A rotary can handle no more traffic than an adequately designed channelized layout.
2. It has been found that unsatisfactory functioning occurs when two or more legs approach design capacity.
3. The area involved to satisfy proper geometric design of a rotary is extensive. This requirement for large flat areas may render the cost prohibitive in urban areas or areas of difficult topography.
4. Channelization will often prove more acceptable where large pedestrian traffic is expected.
5. Most rotaries are designed to function at low speeds. Where high speeds are anticipated the large lengths of required weaving sections may cause prohibitive land costs.
6. For proper functioning, it has been found that adequate lighting must be provided and controlled access is required on the intersecting legs. Failing these provisions, the intersection may become a serious safety hazard.

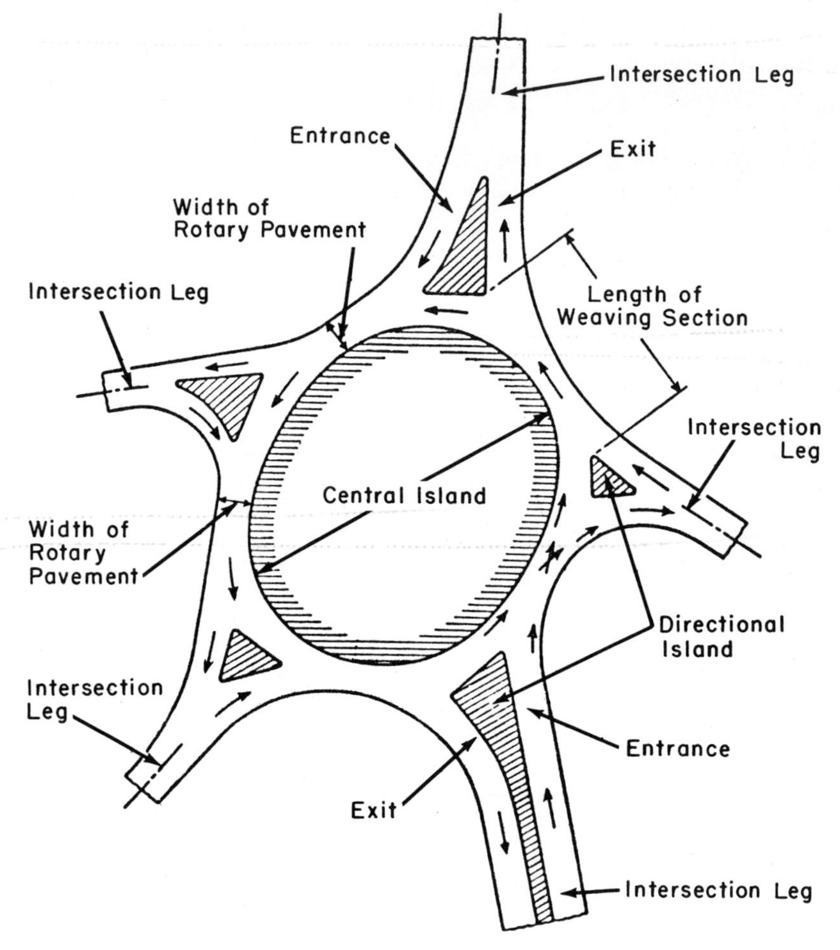

FIG. 5–23. A rotary intersection at grade. (Courtesy AASHO.)

5–41. Grade Separations and Interchanges. Intersections at grade can be eliminated by the use of grade-separation structures which permit the cross flow of traffic at different levels without interruption. The advantage of such separation is the freedom from cross interference with resultant saving of time and increase in safety for traffic movements.

Grade separations and interchanges may be warranted (1) as a part of an express highway system designed to carry heavy volumes of traffic, (2) to eliminate bottlenecks, (3) to prevent accidents, (4) where the topography is such that other types of design are not feasible, and (5) where the volumes to be catered for would require the design of an intersection, at grade, of unreasonable size (*101*).

An interchange is a grade separation in which vehicles moving in one direction of flow may transfer direction by the use of connecting roadways. These connecting roadways at interchanges are called ramps.

Many types and forms of interchanges and ramp layouts are used in the United States. These general forms may be classified into four main types:

1. T- and Y-interchanges
2. Diamond interchanges
3. Partial and full cloverleafs
4. Directional interchanges

T- AND Y-INTERCHANGES. Fig. 5–24 shows typical layouts of interchanges at three-legged junctions. The geometry of the interchange can be altered to favor certain movements by the provision of large turning radii, and to suit the topography of the site. The trumpet interchange has been found suitable for orthogonal or skewed intersections. Fig. 5–24(a) favors the left turn on the freeway by the provision of a semi-direct connecting ramp. Fig. 5–24(c) indicates an intersection where all turning movements are facilitated in this way.

DIAMOND INTERCHANGES. The diamond interchange is adaptable to both urban and rural use. The major flow is grade-separated, with turning movements to and from the minor flow achieved by diverging and merging movements with through traffic on the minor flow. Only the minor flow directions have intersections at grade. In rural areas, this is generally acceptable, owing to the light traffic on the minor flow. In urban areas, the at-grade intersections generally will require signalized control to prevent serious interference of ramp traffic and the crossing arterial street. The design of the intersection should be such that the signalization required does not impair the capacity of the arterial street. To achieve this, widening of the arterial may be necessary in the area of the interchange. Care must also be taken in the design of the ramps, so that traffic waiting to leave the ramp will not back up into through lanes of the major flow.

One disadvantage of the diamond interchange is the possibility of illegal wrong-way turns, which can cause severe accidents. Where the geometry of the intersection may lead to these turns, the designer can use channelization devices and additional signing and pavement marking. Wrong-way movements are, in general, precluded by the use of cloverleaf designs.

Fig. 5–24(d) shows the conventional diamond interchange. Increased capacity of the minor flows can be attained by means of the arrangement

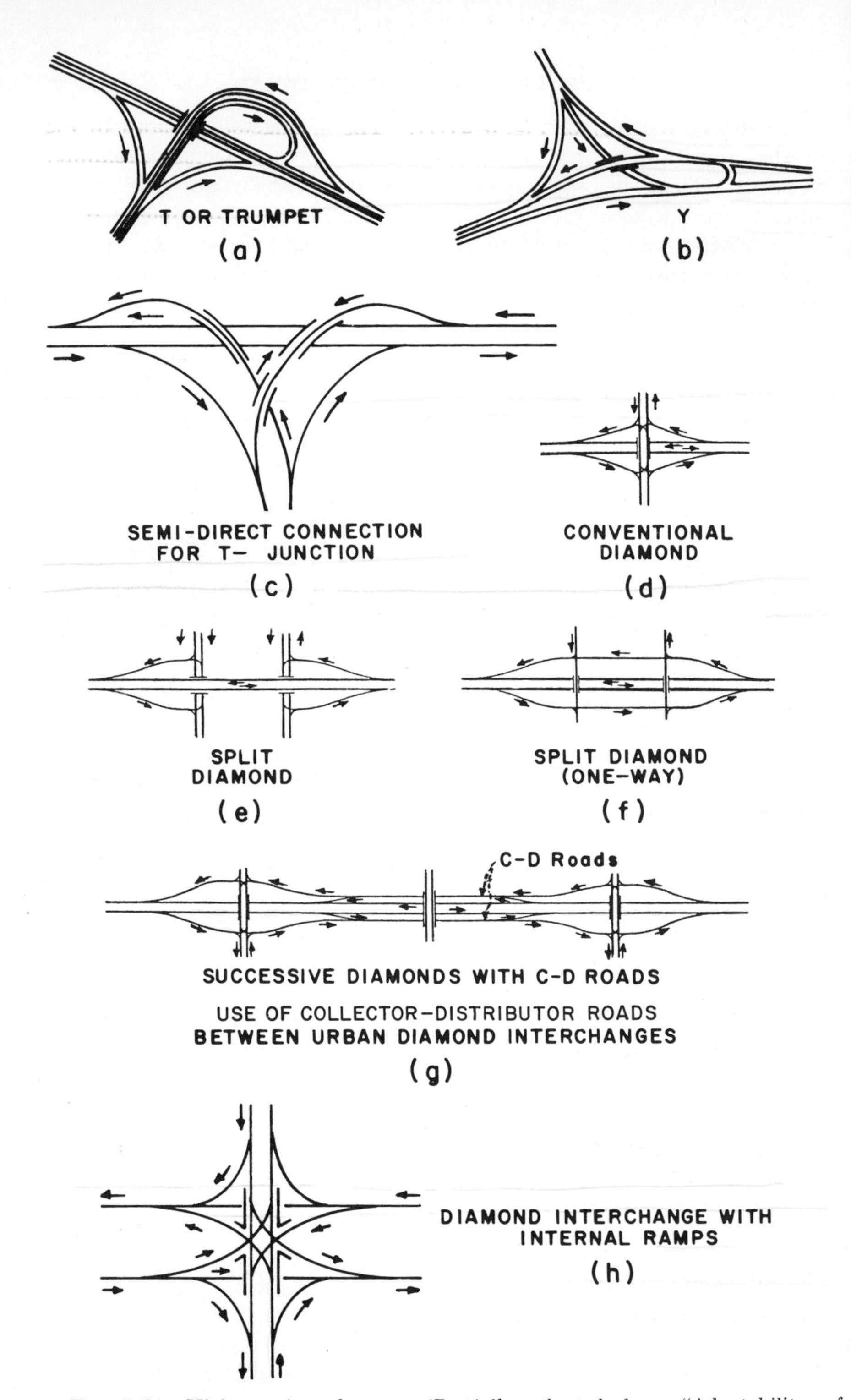

FIG. 5–24. Highway interchanges. (Partially adapted from "Adaptability of Interchanges to Interstate Highways," Vol. 124, p. 558, *Transactions,* American Society of Civil Engineers.)

shown in Fig. 5–24(e) or Fig. 5–24(f). The arrangement shown in Fig. 5–24(g) is suitable where two diamond ramps are in close proximity. Weaving movements, which would, in this case, inhibit the flows of the major route, are transferred to the parallel collector distributor roads. Fig. 5–24(h) indicates how the diamond interchange can be adapted for internal entry and exit ramps. Fig. 5–25 shows a typical rural diamond interchange, which could be expected to have minor cross-flows.

FIG. 5–25. A diamond interchange. (Courtesy Michigan Department of State Highways.)

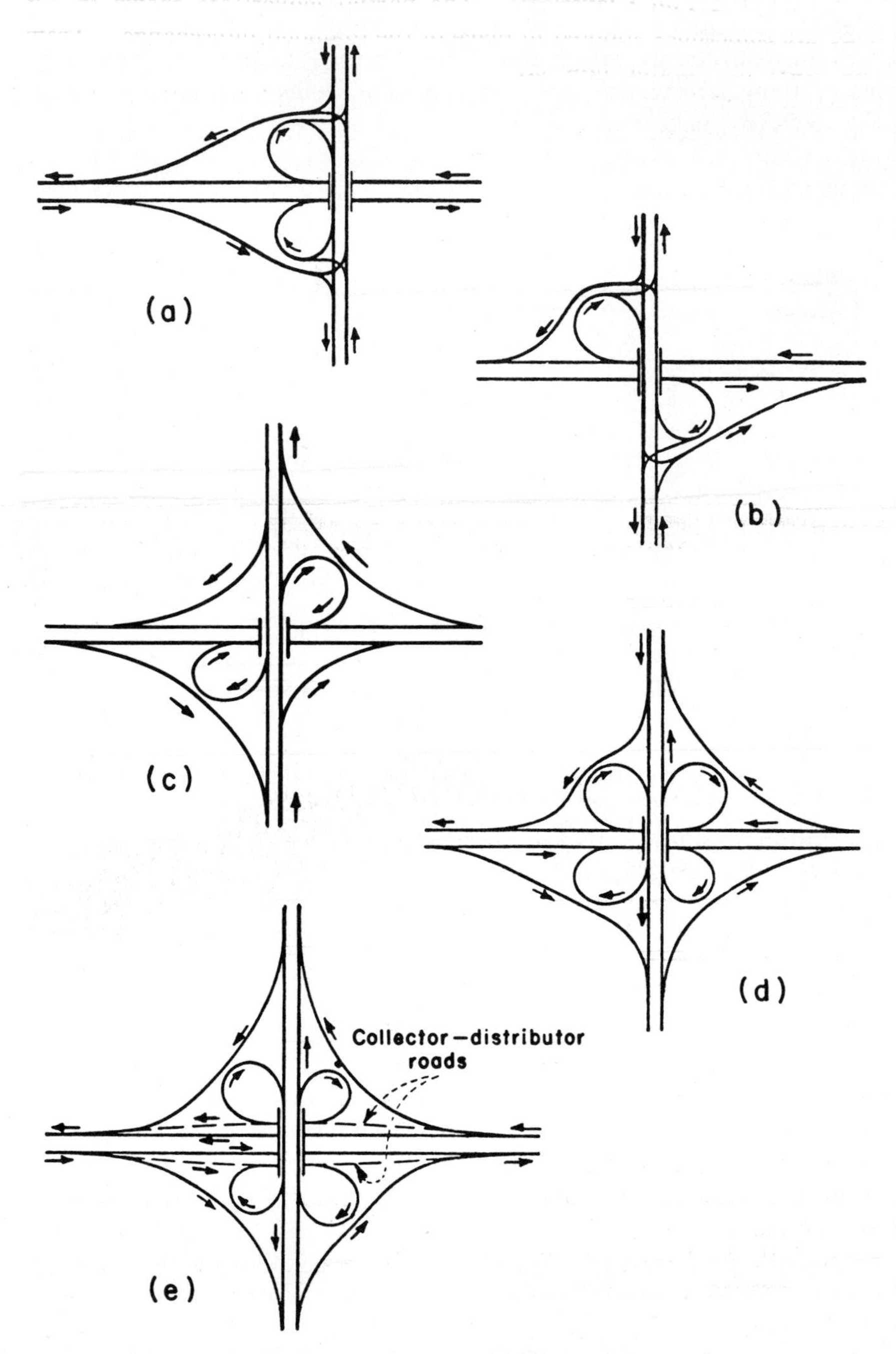

FIG. 5–26. Cloverleaf interchanges. (Adapted from *A Policy on Geometric Design of Rural Highways—1965,* AASHO.)

Partial and Full Cloverleafs. The partial cloverleafs shown in Fig. 5–26 are sometimes adopted in place of the diamond interchange. Traffic can leave the major flow either before or after the grade-separation structure, depending on the quadrant layout. The intersections at grade for the minor road are present as for the diamond interchange, but the probability of illegal turning movements can be reduced. By the provision of two on-ramps for each direction of the major route as in Fig. 5–26(c), left turn traffic on the minor route can be eliminated.

Fig. 5–27. Full cloverleaf intersection on Interstate 80 in Illinois.

The more conventional arrangement of the full cloverleaf, which can be adapted to nonorthogonal layouts, eliminates at-grade crossings of all traffic streams for both major and minor roads. The ramps may be one-way, two-way, two-way separated, or two-way unseparated roads. Although all crossing movements are eliminated, the cloverleaf design has some disadvantages: (1) the layout requires large land areas, and (2) decelerating traffic wishing to leave the through lanes must weave with accelerating traffic entering the through lanes. Fig. 5–26(e) is a layout using collector-distributor roads to overcome this second disadvantage.

Fig. 5–27 shows a typical rural full cloverleaf at the intersection of two four-lane divided highways.

Directional Interchanges. Directional interchanges are used whenever one freeway joins or intersects another freeway. The outstanding design characteristic of this type of interchange is the use of a high design speed throughout, with curved ramps and roadways of large radius. The land requirements for a directional interchange are, therefore, very large. In cases where volumes for certain turning movements are small, design speeds for these movements are reduced and the turnoff is effected within a loop. Figs. 5–28 and 5–29 show examples

Fig. 5–28. Four-level interchange in downtown Los Angeles. (Courtesy California Division of Highways.)

of interchanges with and without loops. In the highest type of design weaving sections are eliminated.

Two directional interchanges are shown in Fig. 5–30, one at either end of a bridge over the East River in New York City to join the Cross Bronx Expressway to the East River Drive and the Major Deegan Expressway. In the background, the expressway passes beneath three apartment houses and a bus terminal, then to the George Washington Bridge over the Hudson River.

5–42. Railroad Grade Intersections. In the design of a highway that intersects a railroad at grade, consideration must be given to approach

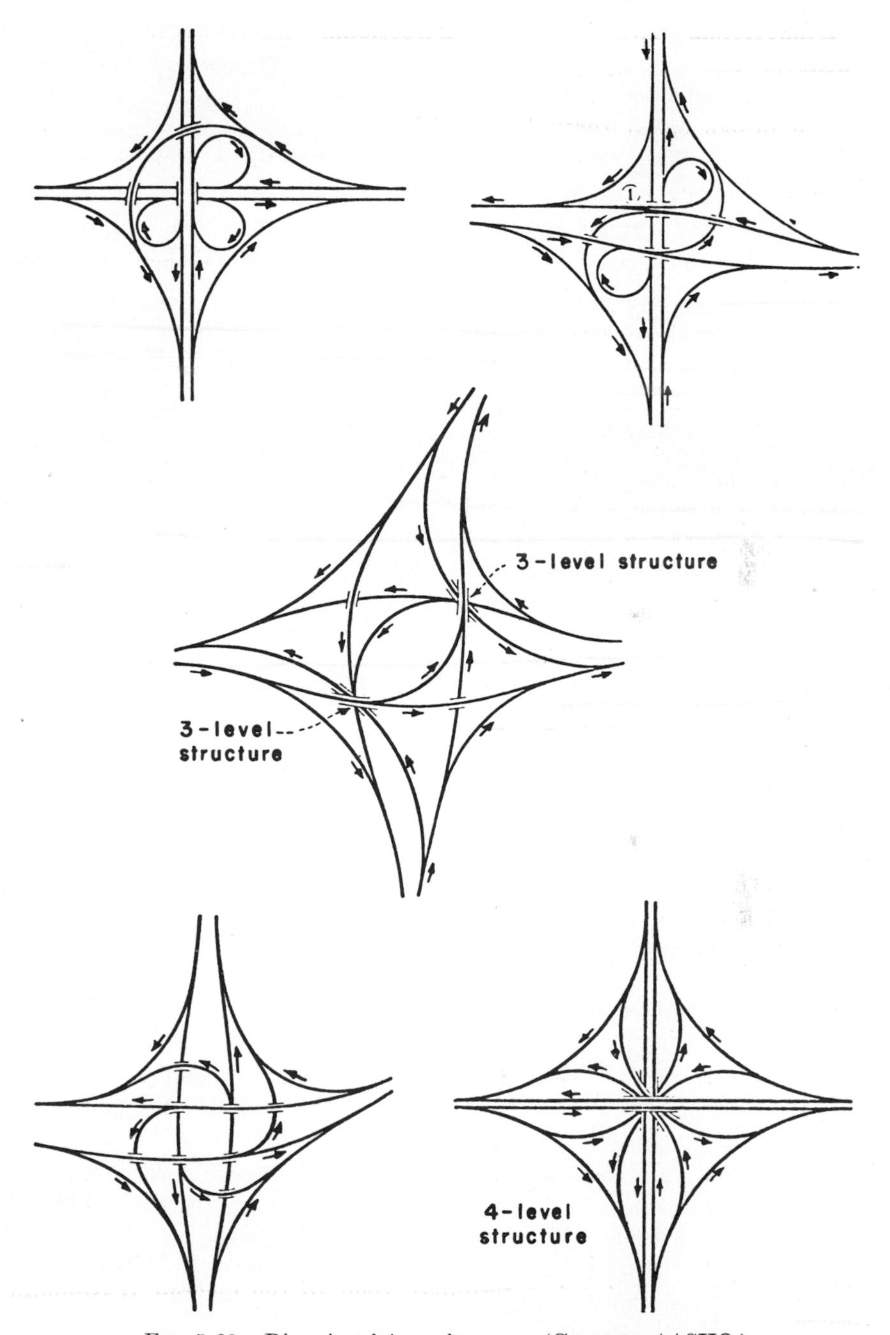

FIG. 5-29. Directional interchanges. (Courtesy AASHO.)

FIG. 5–30. An urban highway complex. (Courtesy The Port of New York Authority.)

grades, sight distance, drainage, volume of vehicular traffic, and the frequency of regular train movements at the particular intersection. The particular type of surfacing and kind of construction at railroad crossings at grade will depend upon the class of railroad and kind of roadway improvement.

All railroad intersections at grade require proper advance warning signs. At crossings on heavily traveled highways where conditions justify, automatic devices should be installed. Recommended standards for railroad-highway grade-crossing protection have been adopted by the Association of American Railroads.

The use of grade separations at railroad crossings is recommended at all main-line railroads that consist of two or more tracks and at all single-line tracks when regular train movements consist of six or more trains per day. Other considerations for separating railroad and highway traffic are the elements of delay and safety.

Railroad grade-separation structures may consist of an overpass on which the highway is carried over the railroad or an underpass which carries the highway under the railroad. The selection of the type of structure will depend in large part upon the topographical conditions and a consideration of initial cost. Drainage problems at underpasses can be serious. Pumping of surface and subsurface water may have to be carried on a large part of the time and the failure of power facilities sometimes causes flood conditions at the underpass with the resultant stoppage of traffic.

5–43. Limited-Access Highways. A limited-access highway may be defined as a highway or street, especially designed for through traffic, to which motorists and abutting property owners have only a restricted right of access. Limited or controlled-access highways may consist of (1) freeways which are open to all types of traffic and (2) parkways from which all commercial traffic is excluded. Most of our present expressway systems have been developed as freeways.

Limited-access highways may be elevated, depressed, or at grade. Many examples of the various types may be found in the United States in both rural and urban areas.

The control of access is attained by limiting the number of connections to and from the highway, facilitating the flow of traffic by separating cross traffic with overpasses or underpasses, and by eliminating or restricting direct access by abutting property owners.

The need for limited access in urban and suburban areas is apparent in planning for present and future traffic needs. High traffic densities, congestion, delays, loss of life, and property damage are factors demanding better traffic facilities. Due to traffic congestion, property values in the heart of many municipalities are diminishing and blight areas are increasing. Control of access is necessary to maintain the design capacity of the facility, reduce accidents, and protect the public investment in highways and streets. Rural highways also need control of access but for reasons slightly different from those in urban and suburban areas. While it is desirable to facilitate the flow of traffic at moderate speeds on rural routes, consideration should also be given to their use. It is important to maintain esthetic and scenic values to control roadside encroachment. Very often, when a new highway is constructed without access control, many new businesses spring up along its route. The effect of this so-called "ribbon development" in rural areas tends to make a highway obsolete long before the physical structure wears out.

Many states have adopted legislation aiding the concept of limited access and the control of highway development. The adoption of such legislation was given a tremendous boost by the Federal-Aid Highway Act of 1956, with its basic requirement of control of access on all mileage of the Interstate System. One approach to the problem, which is worthy of note, has been made by the Ohio Department of Highways. By means of reservation agreements between the Ohio Department of Highways and the owner, the state acquires specific rights in designated areas for a nominal consideration. The owner is permitted the use of the reserved areas for normal purposes, and the state is expected to prevent the erection of buildings and public utility facilities that would increase the cost of acquiring the easement. Areas thus reserved serve to control developments along the highway and make possible the easy taking of this land as needed.

The design of limited-access routes should provide for adequate width of right-of-way, adequate landscaping, prohibition of outdoor advertising on the controlled access proper, and provisions for controlling abutting service facilities such as gas stations, parking areas, and other roadside appurtenances.

In urban areas, the design of a limited-access facility is usually accompanied by the design of frontage roads, parallel to the facility, which serve local traffic and provide access to adjacent land. Such roads may be designed for either one-way or two-way operation. Reasonably convenient connections should be provided between through-traffic lanes and frontage roads.

5–44. Design Standards for the Interstate System. The National System of Interstate and Defense Highways carries more traffic per mile than any other comparable national system and includes the roads of greatest significance to the economic welfare and defense of the nation. The highways of this system must be designed with control of access to ensure safety, permanence, utility, and flexibility to provide for possible future expansion.

All interstate highways are required to meet the minimum standards established by the *Policy on Design Standards* adopted July 12, 1956, by AASHO, and later revised. The following has been abstracted from this *Policy* (*100*).

TRAFFIC BASIS. Interstate highways shall be designed to serve safely and efficiently the volumes of passenger vehicles, buses, and trucks, including tractor-trailer and semitrailer combinations and corresponding military equipment, estimated as those which will exist 20 years beyond that in which the plans, specifications, and estimates for actual construction of the section are approved, including attracted, generated, and development traffic on the basis that the entire system is completed.

The peak-hour traffic used as a basis for design shall be as high as the thirtieth highest hourly volume of the design year, DHV. Unless otherwise specified, DHV is the total, two-direction volume of mixed traffic.

CONTROL OF ACCESS. On all sections of the Interstate System, access shall be controlled by acquiring access rights outright prior to construction, by the construction of frontage roads, or both. Control of access is required for all sections of the Interstate System. Under all of the following conditions, intersections at grade may be permitted in sparsely settled rural areas which are a sufficient distance from municipalities or other traffic-generating areas to be outside their influence, and where no appreciable hazard is created thereby.

1. The interstate highway is a two-lane highway having a DHV of less than 500.
2. Each intersection at grade is with a public road or private driveway with little potential for traffic increase and on which the current average daily traffic (ADT) does not exceed 50 vehicles.
3. Such intersections do not exceed two per side of the interstate highway per mile.
4. Sufficient additional corner right-of-way at each intersection at grade is acquired to ensure that access connections on the crossroad are sufficiently removed to minimize interference with the interstate highway.
5. The right to eliminate, terminate, or re-route each such public road or private driveway is vested in the appropriate public authority at the time of initial construction.

Where a grade separation is called for under these standards, and extraordinary conditions exist under which a grade separation would not be in the public interest, an intersection at grade may be permitted through agreement between the State Highway Department and the Secretary of Commerce (now, Transportation).

Railroad Crossings. Railroad grade crossings shall be eliminated for all through-traffic lanes.

Intersections. All at-grade intersections of public highways and private driveways shall be eliminated, or the connecting road terminated, re-routed, or intercepted by frontage roads, except as otherwise provided under Control of Access.

Design Speed. The design speed of all highways on the system shall be at least 70, 60, and 50 miles per hour for flat, rolling, and mountainous topography, respectively, and depending upon the nature of terrain and development. The design speed in urban areas should be at least 50 miles per hour.

Curvature, Superelevation, and Sight Distance. These elements and allied features, such as transition curves, should be correlated with the design speed in accordance with the *Policy on Geometric Design of Rural Highways* of the AASHO (*101*).

On two-lane highways, sections with sufficient sight distance for safe passing should be frequent enough and the total length of such sections should be a sufficient percentage of the highway length to accommodate the DHV. Where it is not feasible to provide enough passing opportunities, a divided highway should be provided.

GRADIENTS. For design speeds of 70, 60, and 50 miles per hour, gradients generally shall be not steeper than 3, 4, and 5 per cent, respectively. Gradients 2 per cent steeper may be provided in rugged terrain.

WIDTH AND NUMBER OF LANES. Traffic lanes should not be less than 12 feet wide.

Where the DHV exceeds 700 or exceeds a lower two-lane design capacity applicable for the conditions on a particular section, the highway shall be a divided highway. For lower volumes, the highway shall be a two-lane highway so designed and located on the right-of-way that an additional two-lane pavement can be added in the future to form a divided highway.

Efficiency and capacity of two-lane highways may be increased by providing added climbing lanes on up-grades where critical lengths of grade are exceeded or by providing more frequent and long sections for safe passing.

MEDIANS. Medians in rural areas in flat and rolling topography shall be at least 36 feet wide. Medians in urban and mountainous areas shall be at least 16 feet wide. Narrower medians may be provided in urban areas of high right-of-way cost, on long, costly bridges, and in rugged mountainous terrain, but no median shall be less than 4 feet wide.

Curbs or other devices may be used where necessary to prevent traffic from crossing the median.

Where continuous barrier curbs are used on narrow medians, such curbs shall be offset at least 1 foot from the edge of the through-traffic lane. Where vertical elements more than 12 inches high, other than abutments, piers, or walls, are located in a median, there shall be a lateral clearance of at least 3.5 feet from the edge of a through-traffic lane to the face of such element.

SHOULDERS. Shoulders usable by all classes of vehicles in all weather shall be provided on the right of traffic. The usable width of shoulder shall be not less than 10 feet. In mountainous terrain involving high cost for additional width, the usable width of shoulder may be less but at least 6 feet. Usable width of shoulder is measured from the edge of a through-traffic lane to intersection of shoulder and fill or ditch slope except where such slope is steeper than 4:1 where it is measured to beginning of rounding.

SLOPES. Side slopes should be 4:1 or flatter where feasible and not steeper than 2:1 except in rock excavation or other special conditions.

RIGHT-OF-WAY. Fixed minimum widths of right-of-way are not given because wide widths are desirable, conditions may make narrow widths necessary, and right-of-way need not be of constant width. The following minimum widths are given as guides.

In rural areas right-of-way widths should be not less than the following, plus additional widths needed for heavy cuts and fills:

Type of Highway	Minimum Width in Feet Without Frontage Roads	Minimum Width in Feet With Frontage Roads
Two-lane	150	250
Four-lane divided	150	250
Six-lane divided	175	275
Eight-lane divided	200	300

In urban areas, right-of-way width shall be not less than that required for the necessary cross-section elements, including median, pavements, shoulders, outer separations, ramps, frontage roads, slopes, walls, border areas, and other requisite appurtenances.

BRIDGES AND OTHER STRUCTURES. The following standards apply to interstate highway bridges, overpasses, and underpasses. Standards for crossroad overpasses and underpasses are to be those for the crossroad.

1. Bridges and overpasses, preferably of deck construction, should be located to fit the over-all alignment and profile of the highway.
2. The clear height of structure shall be not less than 16 feet over the entire roadway width, including the usable width of shoulders. In urban areas, this clearance shall be applied to a single route (primarily for defense purposes). On other urban routes, the clear heights are to be not less than 14 feet. Allowance should be made for any contemplated resurfacing.
3. The width of all bridges, including grade separation structures, of a length of 250 feet or less between abutments or end supporting piers shall equal the full roadway width on the approaches, including the usable width of shoulders. Special requirements apply on longer bridges.
4. Barrier curbs on other bridges and curbs on approach highways, if used, shall be offset at least 2 feet. Offset to face of parapet or rail shall be at least 3.5 feet measured from edge of through-traffic lane and apply on right and left.
5. The lateral clearance from the edge of through-traffic lanes to the face of walls or abutments and piers at underpasses shall be the usable shoulder width but not less than 8 feet on the right and 4.5 feet on the left.
6. A safety walk shall be provided in tunnels and on long-span structures on which the full approach roadway width, including shoulders, is not continued.

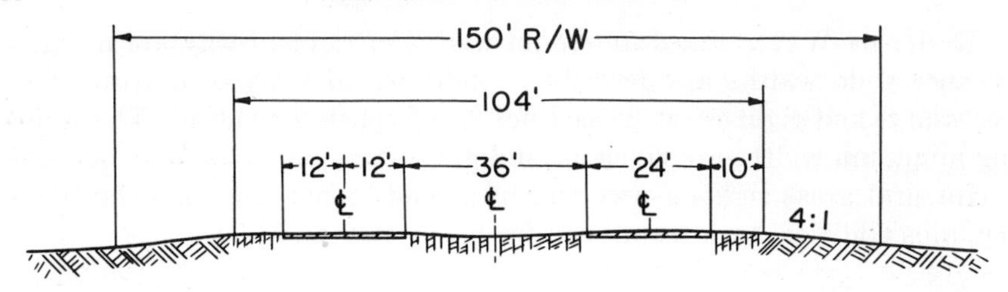

Fig. 5–31. Typical cross-section, Interstate Highway, rural type. Four lanes. Widths shown are minimum for rural flat terrain.

A typical cross-section for an interstate highway in a rural area is shown in Fig. 5–31.

5–45. Highway Beautification. The value of the highway landscape should not be underestimated, and much consideration should be given to the aesthetic details of our highway structure. Some impetus for landscaping improvement came when the Bureau of Public Roads set aside a certain percentage of funds on federal aid projects to be used for landscaping purposes. Additional federal pressure for more beautiful highways has come through the Highway Beautification Act of 1965, which allows for the withholding of certain federal funds where action is not taken to maintain an aesthetically pleasant highway. The act authorizes additional money for landscaping, and provides for control of billboards and junkyards adjacent to the highway right-of-way.

Landscaping of our highways may be considered with respect to different types of improvement: (1) the general over-all planting details for the entire length of a project, (2) roadside parks, and (3) scenic outlooks.

In the improvement of any project, plans should be checked for the care and preservation of existing trees and shrubs. Care must be exercised in roadside planting, especially in snow areas, because certain types of planting act as a snow fence, whereas other types of highway planting act as an obstruction and cause wind carrying snow to deposit its load on the highway. The creation of interest by civic groups in the planting of memorial highways is one feature that should not be overlooked.

Roadside parks and scenic outlooks, properly developed and controlled, aid the highway traveler in providing restful stopping places which are invaluable in lessening fatigue and tensions.

5–46. Roadside Lighting. Adequate lighting of highways is usually provided in urban and suburban areas, but in rural districts lighting is usually limited to certain areas on the basis of need. Grade-separation structures and bridges are usually lighted, and the use of sodium vapor lamps on these structures and at dangerous sections has been most beneficial. Significant decreases in night accident rates have been noted when highways were lighted in congested areas.

5–47. Joint Development of Urban Highways. Resistance to expressway construction in some U.S. cities in the early 1960's lead to the development of plans for joint, multiple-use development of urban highway corridors. One such proposal envisages the taking of an entire, block-wide strip, then developing this strip intensively in three dimensions. The highway would be the backbone of the development, but above and alongside the expressway would be built apartments, commercial buildings, parks and other recreational facilities. Both public and private capital might be used to finance the development. Principal advantage of the plan is the full utilization of scarce urban land.

PROBLEMS

5–1. A total crown of 3 inches is used for a 24-foot pavement. Using a modified parabolic crown, $y = dx^{3/2}/L^{3/2}$, what is the crown offset at a point 9 feet from the centerline where

y = crown offset at any point in inches
d = total crown of pavement in inches
x = horizontal distance from centerline to any point, in feet
L = half the width of the pavement in feet

5–2. A total crown of 2½ inches is used for a pavement 48 feet wide. By the use of the formula in Problem 5–1, what is the elevation of a manhole cover 15 feet from the centerline? The pavement elevation at the centerline is 500.00.

5–3. Given $\Delta = 17°\ 50'$ right, $D = 2°\ 30'$, and the P.I. station 321 + 32.00. Compute the curve data and the station of the P.T. Compute the necessary deflection angles. Use the arc definition formulas.

5–4. Given a 24-foot roadway and a curve with a 1200-foot radius, compute the theoretical superelevation required for an assumed design speed of 70 miles per hour. Compute the superelevation using a friction factor of 0.13.

5–5. The intersection of two grade tangents occurs at station 550 + 00 at an elevation of 362.50. The grade from 525 + 00 to 550 + 00 is plus 4 per cent, and the grade from 550 + 00 to 575 + 00 is plus 1.5 per cent. Calculate the centerline profile grade for each 50-foot station for a vertical curve 1000 feet in length.

5–6. A plus 4.7 per cent grade intersects a minus 3.5 per cent grade at station 365 + 85 at an elevation of 427.50. Calculate the centerline elevations for every 50-foot station for a 600-foot vertical curve.

5–7. A vertical parabolic curve is to be used under a grade-separation structure. The curve is 900 feet long and the minus grade of 5.4 per cent intersects the plus grade of 3.6 per cent at station 350 + 25. Calculate the low point of the curve if the intersection at station 350 + 25 is at elevation 542.60.

5–8. Determine by means of the chart in Fig. 5–18 the sight distance over a crest of a 1500-foot vertical curve with a plus grade of 3.6 per cent and a minus grade of 1.7 per cent.

5–9. Using Fig. 5–19, determine the horizontal sight distance in feet measured along the centerline of the pavement for a curve of 4° 20′ if the distance from the obstruction to the centerline (m) is 120 feet. Design speed is 60 mph.

5–10. For the horizontal curve of Problem 5–4, estimate the minimum length of spiral necessary for a smooth transition from tangent alignment to the circular curve.

5–11. Using the Michigan method shown in Fig. 5–12, determine the rate of superelevation for a 3° 30′ curve. Prepare a plot showing the ordinates required for the transition from a tangent section to the full superelevated section, assuming that the pavement is 24 feet wide (use Fig. 5–14 as a guide).

5–12. Using Fig. 5–16, determine the stopping sight distance over a crest if the algebraic difference in the two grades is 3.2 per cent, and the vertical curve is (a) 500 feet long, (b) 1000 feet long.

5–13. Using Fig. 5–17, determine the value of each element of the passing sight distance for a design speed of 60 mph, and the total passing sight distance.

5–14. From current literature, sketch typical cross sections for new or recently improved highway facilities of the following types: (a) a rural section on the Interstate System, (b) an urban expressway, (c) a city street.

CHAPTER 6

HIGHWAY SOIL ENGINEERING

Soil engineering is a subject of great importance to highway engineers. In highway engineering we are interested in the action of soils when used as a construction material, as in the formation of an embankment or a stabilized base course, and also when the soil serves as a foundation material in its natural condition, as it may in the case of a bridge pier founded on a natural earth deposit.

Included in this chapter are discussions of the origin and formation of soils, common soil types and terminology, basic soil properties, routine testing procedures, identification and classification of soils, soil surveying and mapping, moisture-density relationships, and frost action. Detailed explanation of several special soil problems has been reserved for later chapters. Included among these topics are soil stabilization, considerations relative to the design and construction of embankments, and the measurement of subgrade shearing strength as related to the design of flexible and rigid pavements.

Origin and Formation of Soils

Basically, the engineering definition of the word "soil" is a very broad one. Soil might be defined as all the earth material, both organic and inorganic, which blankets the rock crust of the earth. Practically all soils are products of the disintegration of the rocks of the earth's crust. This disintegration, or "weathering," has been brought about by the action of chemical and mechanical forces which have been exerted upon the parent rock formations for countless ages. Included among these forces are those of wind, running water, freezing and thawing, chemical decomposition, glacial action, and many others.

Soils may be described in terms of the principal agencies responsible for their formation and position. Thus a *residual soil* is one which, in its present situation, lies directly above the parent material from which it was derived. Residual soil deposits are characterized by varying degrees of cementation of the soil mass with depth, their generally unstable or "weathered" nature, and a gradual transition from the soil layer into bedrock. *Transported soils* are those which have been carried to their present position by some transporting agent. Transporting agents of principal importance include wind, glaciers, and water. Soils formed by the action of wind are known as *aeolian soils;* a typical example of a windblown soil deposit is seen in the very considerable deposits of loess in the Mississippi Valley. *Glacial soils* occur in many parts of the United States. An example of such a soil deposit is a glacial till, which is a deposit of tightly bonded materials containing particles ranging in size from boulders down to very finely divided mineral matter.

Soils formed through the action of water are generally termed *sedimentary soils.* These soils are of extreme importance to the engineer as many of the soils with which he deals are of this type. Typical sedimentary soils are formed by the settling of soil particles (or groups of particles) from a suspension existing in a river, lake, or ocean. Sedimentary soils may range in type from beach or river sands to highly flocculent clays of marine origin.

Soils may also be described in terms relative to the amount of organic material contained in them. Soils in which the mineral portion (soil particles) predominates are properly called *inorganic soils.* Those in which a large amount of organic matter is contained are called *organic soils.* Organic soils are usually readily identified by their dark brown to black color and distinctive odor.

One of the most important facts regarding soils and soil deposits is their normal lack of homogeneity. Due to the more or less random process of their formation, soils vary greatly in their physical and chemical composition at different locations over the surface of the earth. Although, generally speaking, soils derived from the same parent material under similar factors of geographical location, climate, and topography will be similar wherever they are found, several soil types may, and usually do, exist within a comparatively small area. Soil deposits are also characteristically varied with depth. Deposits of sedimentary soils in which layers of varying fineness were created during the process of their formation are said to be stratified, and some degree of "stratification" must be expected in any sedimentary soil deposit.

General Soil Types

Certain general soil types are of importance to the highway engineer. No attempt will be made in this section to cover all the soil types which the engineer may encounter. However, certain soil types occur generally throughout the country and will be discussed here, along with the terms which are used in describing them.

6–1. Classification by Grain Size. It is possible to classify soils into several principal groups and many subgroups on the basis of their grain size. By grain size is meant the average dimension (or diameter, in a general sense) of a soil particle contained within a soil mass. Principal divisions of any classification system based on grain size are usually gravel, sand, silt, and clay. Several different classification systems have been proposed, and one in general usage among highway engineers is shown in Fig. 6–1.

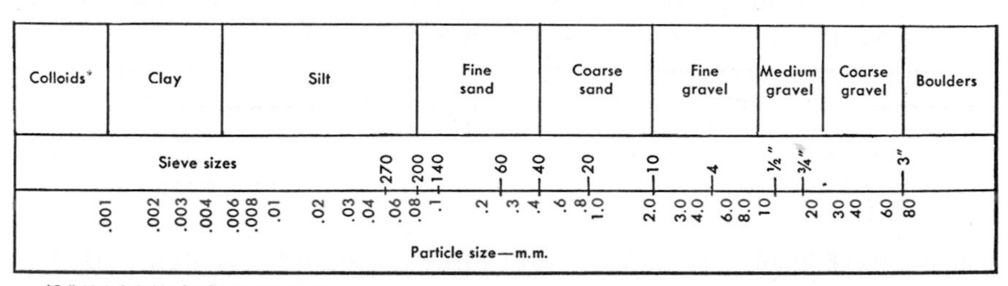

FIG. 6–1. Grain-size classification (AASHO Specification M146).

It must be emphasized that although in many cases a knowledge of the distribution of grain sizes in a soil is valuable to an understanding of soil action, this knowledge alone is not enough to give an accurate picture of the properties of a soil and its behavior under various conditions. For example, a fine silt or a rock flour may have a large percentage of particles which fall within the range of clay sizes and yet have entirely different properties from those exhibited by an inorganic clay having about the same range of particle sizes. Similarly, two sands composed of particles of about the same size may be radically different in such a property as shearing resistance if one is a beach sand composed of smooth, rounded grains and the other is composed of grains which are sharp and angular. It should therefore be recognized that the terms "clay" and "silt," at least, carry meanings which go beyond mere reference to particle size; and that for any soil, additional information must be given if the description is to prove adequate from an engineering viewpoint.

Obviously, many soil deposits will contain particles which fall into two or more of the groups described above. Thus a soil might contain percentages of sand, silt, and clay, as defined by the chart of Fig. 6–1. Terms such as "sandy clay," "silty clay," "silty sand," and many others have been employed to describe such soils. One way of systematizing the application of such names to soil mixtures is by the use of a triangular classification chart. A complete explanation of one chart of this type (that used by the U.S. Department of Agriculture) is given in reference *131*.

6–2. Common Soil Types. *Sand* and *gravel* are coarse-grained soil types possessing little or no cohesion. They are readily identified by visual inspection and are distinguished, generally speaking, by their relative stability under wheel loads when confined, by their high permeability, and by their failure to shrink or expand in deterimental amounts with change in moisture content. The term *gravel* is usually applied to natural pit, river, or bank gravels consisting largely of rounded particles; *crushed gravel,* or *crushed stone,* is the term applied to the products of crushing larger rocks into gravel sizes. The term *dirty sand* is frequently applied to sands which contain small amounts of fine material.

Silt is the term applied to fine-grained soils of low to medium plasticity, intermediate in size between sand and clay. Silts generally possess little cohesion, undergo considerable shrinkage and expansion with change in moisture content, and possess a variable amount of stability under wheel loads. If they contain large percentages of flat, scale-like particles, such as mica flakes, they are likely to be highly compressible and somewhat elastic in nature. *Organic silts* contain appreciable amounts of decomposed organic matter and are generally highly compressible and unstable.

Clays are distinguished by the occurrence of very fine grains, since many clays are principally composed of particles which are colloidal in size. Clays generally possess medium to high plasticity, have considerable strength when dry, undergo extreme changes in volume with change in moisture content, and are practically impervious to the flow of water. *Lean clay* is the term given to silty clays or clayey silts, while fine, colloidal clays of high plasticity are called *fat clays.* Clays may be further distinguished by the fact that, although they may possess considerable strength in their natural state, this strength is sharply reduced and sometimes completely destroyed when their natural structure is disturbed, i.e., when they are remolded.

Loam is an agricultural term used to describe a soil which is generally fairly well graded from coarse to fine, which is easily worked, and which is productive of plant life. This name frequently appears in engineering literature in combination with other terms. Thus a soil may be called a *sandy loam,* a *silty loam,* or a *clay loam,* depending on the size of the predominating soil fraction.

Loess is a fine-grained aeolian soil characterized by its nearly uniform grain size, predominantly silt, and by its low density. Highway cuts through loess deposits usually resemble those made in rock, in that this soil will stand on a nearly vertical slope, while it is readily eroded by rain water if flatter slopes are used.

Muck is soft silt or clay, very high in organic content, which is usually found in swampy areas and river or lake bottoms.

Peat is a soil composed principally of partially decomposed vegetable matter. Its extremely high water content, woody nature, and high compressibility make it an extremely undesirable foundation material.

Basic Soil Properties

To have an understanding of soil action, an engineer must be familiar with certain basic soil properties. We are all familiar with the basic properties of other engineering materials, such as steel, wood, and concrete. A soil engineer must have similar knowledge relative to soils. The job of the soil engineer is complicated by the fact that many soils are quite complex in nature, both physically and chemically, and that soil deposits are likey to be extremely heterogeneous in character. It must be remembered that the properties of any given soil depend not only on its general type but also upon its condition at the time at which it is being subjected to examination.

6–3. Moisture Content. Water is an extremely important constituent of soils. The moisture content is defined as the weight of water contained in a given soil mass compared with the oven-dry weight of the soil, and is usually expressed as a percentage. In the laboratory, moisture content is usually determined by selecting a small, representative sample of soil and placing it in a "tared" container, such as a pair of matched watch glasses or a moisture can. The weight of the "wet" soil sample and container is determined, and the sample is then brought to constant weight by drying in an oven at a temperature between 100 and 110°C. After constant weight has been attained, the sample is taken from the oven, cooled in a desiccator, and the oven-dry weight of the sample is determined. All weights are recorded in grams, and the following expression is used to determine the moisture content:

$$w(\%) = \frac{W_1 - W_2}{W_2 - W_c} \times 100 \tag{6–1}$$

where $w(\%)$ = moisture content, in per cent
W_c = tare weight of the container, in grams
W_1 = weight of "wet" soil and container, in grams
W_2 = weight of oven-dry soil and container, in grams

If the void spaces in a soil are completely filled with water, the soil is said to be saturated. The moisture content of a soil may then be 100 per cent or more, as might be the case in a saturated clay, muck, or peat soil.

Water in soils may be present in its normal liquid form, as when filling or partly filling the voids of a sand mass, or it may be present in the form of adsorbed water existing as films surrounding the separate soil particles or groups of particles, as in the case of the water remaining in a partially dried clay mass. The water films existing in the latter case may have properties sharply differing from those exhibited by water in its normal form. Properties of fine-grained soils are greatly dependent upon the properties and behavior of the adsorbed water films.

6–4. Specific Gravity. By the term "specific gravity" (G), as applied to soils, is meant the specific gravity of the dry soil particles or "solids." The specific gravity is frequently determined by the pycnometer method, the determination being relatively easy for a coarse-grained soil and more difficult for the finer soils. Values for the specific gravity refer to the ratio of the unit weight of soil particles to the unit weight of water at some known temperature (usually 4°C), and range numerically from 2.60 to 2.80. Within this range, the lower values for the specific gravity are typical of the coarser soils, while higher values are typical of the fine-grained soil types. Values of the specific gravity outside the range of values given may occasionally be encountered in soils derived from parent materials which contained either unusually light or unusually heavy minerals.

6–5. Unit Weight. The unit weight of a soil is the weight of the soil mass per unit of volume and is expressed in pounds per cubic foot. As commonly used in highway engineering, the term "wet unit weight" refers to the unit weight of a soil mass having a moisture content which is anything different from zero, while the "dry unit weight" refers to the unit weight of the soil mass in an oven-dry condition. The wet unit weight, dry unit weight, and moisture content are related by the following expression:

$$\text{Dry unit weight} = \frac{\text{wet unit weight}}{\dfrac{(100 + w\%)}{100}} \tag{6–2}$$

The wet unit weight of a soil may vary from 90 pounds per cubic foot or less for saturated, organic soils to 140 pounds per cubic foot or more for well-compacted granular materials.

6–6. Void Ratio. Another term that is of consequence in describing the weight–volume relationships of soils is the "void ratio" (e). The void ratio is defined as the ratio of the volume of voids contained in a soil mass to the volume of solids. Related terms are the "porosity" (n), which is defined as the ratio of the volume of voids to the total volume of the soil mass being considered, and the "degree of saturation" (S), which is defined as the ratio of the volume of water contained in the voids to the volume of voids. Both porosity and degree of saturation are normally expressed as percentages.

A simple example may serve to show the relationships existing among these latter terms, unit weight, specific gravity, and moisture content:

Given a soil mass which has a wet unit weight of 120.0 pounds per cubic foot, a moisture content of 13.5 per cent, and a specific gravity of 2.65; calculate the dry unit weight, void ratio, porosity, and degree of saturation.

The dry unit weight may be calculated by using the relationship previously given.

$$\text{Dry unit weight} = \frac{\text{wet unit weight}}{\dfrac{(100 + w\%)}{100}} = \frac{120.0}{\dfrac{(100 + 13.5)}{100}}$$

$$= \frac{120.0}{1.135} = 105.7 \text{ lb per cu ft}$$

Now consider the cubic foot of soil having this dry weight. Since the specific gravity of the soil particles is 2.65, the volume of solids contained in this volume must be:

$$V_s = \frac{105.7}{2.65(62.4)} = 0.639 \text{ cu ft}$$

Therefore the volume of voids is

$$V_v = 1.000 - 0.639 = 0.361 \text{ cu ft}$$

and

$$e = \frac{V_v}{V_s} = \frac{0.361}{0.639} = 0.56$$

also,

$$n = \frac{V_v}{V_s + V_v} = \frac{0.361}{1.000}(100) = 36.1\%$$

In this volume is contained 120.0 — 105.7 = 14.3 pounds of water. Therefore the volume of water

$$V_w = \frac{14.3}{62.4} = 0.229 \text{ cu ft}$$

and

$$S = \frac{V_w}{V_v} = \frac{0.229}{0.361}(100) = 63.4\%$$

6–7. Permeability. Permeability is the property of a soil mass which permits water to flow through it under the action of gravity or some other applied force. In soil mechanics, permeability is expressed by the term "coefficient of permeability" (k), which is frequently defined as the velocity of flow (of water) through a soil mass under the action of a unit hydraulic gradient. The coefficient of permeability of a soil is principally dependent upon its void ratio, grain size, and structure, as well as upon the density and viscosity of the water flowing through the soil. Laboratory measurements of the coefficient of permeability may be made by the use of a falling-head permeameter or a constant-head permeameter. Descriptions of this type of laboratory equipment may be found in any standard textbook on soil mechanics.

Coarse-grained soils, such as sand and gravel, have high coefficients of permeability and are said to be "pervious." Clays and other fine-grained soils have much lower coefficients of permeability and are said to be "relatively impervious" or "impervious soils."

6–8. Capillarity. By capillarity is meant that property which permits water to be drawn into the soil from a free water surface through the action of surface tension and independent of the force of gravity. Capillary flow in soils is generally associated with an upward movement of water from the water table, although in reality this type of flow can occur in any direction. Through capillary action a zone of "capillary saturation" may exist at some considerable distance above a free water surface, especially in fine-grained soils. The soil in this zone may not be completely saturated, as some air will probably remain in the void spaces and around the soil particles, but it will probably attain a high degree of saturation over a period of time. Above this zone the soil may be partially saturated by capillary action.

Capillary flow of water in soils is a complex phenomenon. However, the height to which water will be raised in a soil by capillary action is known to be primarily dependent upon the size of the void spaces and therefore upon the size of the soil particles. From the standpoint of susceptibility to capillary action and the detrimental effects accompanying the capillary flow of water in soils under certain conditions, the most critical condition is presumed to occur in a fine silt. The eventual height of capillary rise may be greater in clay soils than in silts, but the upward flow of water in clays is much slower, and the attainment of a high degree of saturation in a clay mass situated some distance above a free water surface will require a much longer time than that which would be required for the same phenomenon to occur in a silty soil. Coarse sands and gravels are not normally subject to the detrimental effects of capillary action, as the height of capillary rise in soils of this type is usually so slight as to be practically negligible. Exact values of the height of capillary rise in any soil are not easily determined for any given set of field conditions, but this height may vary from a few inches in fine sands to 9 or 10 feet in a silt or clay.

6–9. Shrinkage and Expansion. Many soils undergo very considerable reduction in volume as their moisture content is reduced from that existing when they are partially saturated or saturated. The shrinkage which occurs is greatest in clays and other fine-grained soils. Some of these soils show reduction in volume of 50 per cent or more while passing from a saturated to an oven-dry condition. The same soils may show considerable expansion or "swelling" if they have been dried and their moisture content is later increased. Sands and gravels generally show little or no change in volume with change in moisture content.

The shrinkage of the clay mass is generally supposed to be due to forces attributed to the surface tension existing in the water films created during the drying process. Simply, when the soil is saturated there exist free water surfaces, and surface tension acts at the surface of the soil mass; as evaporation proceeds, innumerable menisci are created in the pores of the soil, each of which develops forces (due to surface tension) which react upon the structure of the soil mass and force the soil particles closer together; hence the mass shrinks. As evaporation continues, these forces become greater and greater until the soil attains a certain limiting volume or else cracking occurs. The great magnitude of the internal forces thus created is also supposed to principally account for the rocklike strength of a dried clay mass.

If water is again made available to a soil mass which has undergone shrinkage, it will enter the pores of the soil and reduce or destroy the internal forces previously described. As a consequence, the soil particles move apart and a confined clay mass expands or "swells." If the mass is unconfined, it will generally disintegrate or "slake." Effects which have been described for clay soils will also generally occur in silty soils although to a lesser degree. Excessive shrinkage or swelling is a matter of great concern to the highway soils engineer, especially as related to the behavior of a subgrade soil used as a support for a flexible or rigid base or pavement.

6–10. Compressibility. By compressibility is meant the property of a soil which permits it to consolidate under the action of an applied compressive load. Deformations caused by shearing failure are excluded from this definition. The term "consolidation" is applied to the compression of a saturated clay mass under a vertical load.

In their normal condition, all soils are compressible. In a general sense the compression of a saturated soil mass is assumed to be due to a reduction in volume of the void spaces in the soil rather than to compression of the soil particles or of the water in the pores of the soil. If the soil is saturated when the load is applied, then it is apparent that some of the water occupying the original void spaces must be forced from the soil before compression can take place. The amount of compression occurring in a given soil under a given set of conditions depends upon a variety of factors, including the magnitude and method of application of the load, and the void ratio, structure, and history of the soil mass concerned. The rate of compression (or consolidation) in a saturated soil is evidently a function of the permeability.

Compression of a sand mass occurs very rapidly following the application of a load, provided that the water can escape. Compression of sands is rarely a matter of practical concern, since all the compressive deformation which is going to occur will take place during the period of load application. Deformations thus produced in sands are largely "permanent deformations" although sand masses in a compacted state may eventually attain some degree of elasticity under repeated applications of load.

The consolidation of clays and other fine-grained soils is a different matter, however, and may often be of practical concern. Such deformations occur very slowly, as the stresses created by the application of a load are slowly transferred from the water phase of the soil system to the solid phase. In thick, compressible clay deposits, consolidation may continue for many years under, for example, the applied weight of a highway embankment. The settlements resulting from consolidation are rarely uniform and serious damage may result.

6–11. Elasticity. By elasticity is meant the property of a soil which will permit it to return to its original dimensions (or nearly so) after the removal of an applied load. No soil is perfectly elastic, but some soils possess elasticity to a degree which may be detrimental when they are used in subgrade or base construction. Examples of such soils are those (usually silts and clays) which contain a sizable amount of mica, certain diatomaceous earths, and those containing a high percentage of organic colloids. "Elastic" silts and clays are quite common, while highly organic soils, such as peat, may contain organic colloids and be highly compressible and somewhat elastic.

6–12. Shearing Resistance. Failures which occur in soil masses as a result of the action of highway loads are principally shear failures. Therefore the factors which go to make up the shearing resistance of a soil are of importance. Shearing resistance within soil masses is commonly attributed to the existence of "internal friction" and "cohesion."

A simplified explanation of these properties is most easily accomplished by consideration of two extremely different types of soils; first, a cohesionless sand, and second, a highly cohesive clay in which the internal friction is assumed to be negligible. In a cohesionless sand, the force required to overcome shearing resistance on any plane is assumed to be given by the expression, $S_r = \sigma \tan \phi$, where σ is the normal force (stress) on the plane being considered, and ϕ is the angle of internal friction. The value of the angle of internal friction is assumed to include the factors of resistance to sliding (or rolling) of the soil particles over one another and any interlocking that may have to be overcome before a slip can occur. For a dry sand, ϕ is primarily dependent upon density (void ratio); the lower the void ratio the higher the value of ϕ. Grain shape is also important, as is surface texture; ϕ is higher for a rough, angular sand than for a smooth, rounded sand having the same void ratio. Gradation of a sand is also important, with ϕ being generally higher for sands which are well graded from coarse to fine. The angle of internal friction is relatively independent of the moisture content for sands; ϕ for a wet sand will be only slightly, if any, lower than ϕ for a dry sand, other conditions being the same.

In a saturated clay mass it may be assumed, for the purposes of explanation, that the angle of internal friction is equal to zero, and that the resistance to sliding on any plane is equal to the cohesion, C (usually expressed in pounds per square foot). In simple explanation, C is supposed to include both "true" cohesion, that due to intermolecular attraction, and "apparent" cohesion, that due to surface tension effects in the water contained in the clay mass. The shearing strength of most fine-grained soils decreases when their moisture content is increased and is frequently sharply reduced when their natural structure is destroyed. The interpretation of the factors influencing the shearing strength of cohesive soils is probably the most complex problem in soil mechanics, and no comprehensive explanation will be attempted here. Factors of importance include density, water content, loss of strength with remolding, drainage conditions of the clay mass subjected to stress, variation of cohesion with pressure, and variation in the angle of internal friction.

For the large majority of soils as normally encountered in the field, shearing resistance is made up of both cohesion and internal friction. For these soils the shearing resistance on any plane is frequently, although somewhat empirically, given by Coulomb's Law,

$$S_r = \sigma \tan \phi + C \tag{6-3}$$

wherein the symbols have the meanings previously indicated.

Shearing resistance may be evaluated in the laboratory by the use of the unconfined compression test, the direct shear test, or the triaxial compression test. Samples may be tested in an undisturbed condition or under conditions which are similar to those expected in the field. Direct measurements of the shearing resistance of subgrade soils may be made in the field through the use of loaded circular plates (plate bearing test). In addition, various semiempirical tests (such as the California Bearing Ratio test) have been developed to measure shearing resistance more or less directly in connection with the design of flexible and rigid pavements. Design methods in which these tests are utilized are fully described in later chapters.

Soil Classification for Highway Purposes

The objective behind the use of any soil classification system for highway purposes is to be able to predict the subgrade performance of a given soil on the basis of a few simple tests performed upon the soil in a disturbed condition. On the basis of these test results and their correlation with field experience, the soil may then be correctly identified and placed into a group of soils, all of which have similar characteristics and properties.

The student should realize, however, that placing a soil into the correct group in some classification system, although desirable in many respects, is not the final objective of the soil engineer. That is, the classifying of a soil should not be regarded as an end in itself but as a tool to further our knowledge of soil action and behavior.

The principal soil classification system in use in this country by highway engineers is the Public Roads classification system, first proposed in 1931 and later revised. Principal tests utilized by this and other classification systems are the mechanical analysis and various routine laboratory tests. The mechanical analysis is performed upon the entire sample and has as its objective the determination of the proportion of particles of various sizes in the given soil. The routine tests mentioned above are performed upon the "soil binder" or fraction passing a No. 40 sieve. Most important of the routine tests or "soil constants" are the Atterberg Limits, i.e., liquid limit, plastic limit, and shrinkage limit. The plasticity index is also of significance and is calculated from the results of the liquid limit and plastic limit determinations. Other routine tests which may be used for classification purposes include other shrinkage factors (shrinkage ratio, volumetric shrinkage, and lineal shrinkage), the field moisture equivalent, and the centrifuge moisture equivalent. The routine tests are intended to describe definite physical properties of the soil and are performed under standardized laboratory procedures. Test procedures are briefly described in the following paragraphs. For a full description, the student is referred to the appropirate publications of the American Society for Testing Materials and the American Association of State Highway Officials.

6–13. Mechanical Analysis. Separation of the soil into its fractions may be done by shaking the dry, loose material through a nest of sieves of increasing fineness, i.e., successively smaller openings. The sieve analysis may be performed directly upon soils which contain little or no fines, such as a clean sand or a soil in which the fines may be readily separated from the coarser particles. Soils which have little dry strength and can be easily crushed between the fingers would generally fall in the latter category. If the character of the fines is such that the fine material adheres to the coarser particles and is not removed by dry sieving action, the sample is *prewashed* and the fine material removed. Material which is retained on the No. 200 sieve during the washing process is then dried and subjected to sieving as before.

The practical lower limit for the use of sieves is the No. 200 sieve, which has openings which are 0.074 mm. square. Consequently, the mechanical analysis used in many soil laboratories is a combined sieve and hydrometer analysis. A representative sample of air-dry soil is selected and separated into two parts by passing it over a No. 10 sieve. The coarse material retained on the No. 10 sieve is subjected to sieve analysis and a portion of the remainder of the sample is dispersed and put into suspension in water, usually in the presence of a deflocculating agent. Changes in the specific gravity of the suspension with the passage of time are then noted by means of a hydrometer. Calculations based upon Stokes's Law can then be employed to determine the percentages of grains of various average diameters remaining in suspension (or which have settled out). In other words, for each of the time intervals specified in the test procedure, there may be determined the percentage of the dispersed sample finer than (or coarser than) a certain grain size. Following completion of the hydrometer analysis, which may require 24 hours or more if the percentage of particles of colloidal size is to be determined, the soil is washed over a No. 200 sieve. Material retained on the No. 200 sieve is oven dried and passed over a nest of sieves (frequently Nos. 20, 40, 60, 140, and 200 are used). The results of the sieve analysis and hydrometer analysis are then combined to give the complete grain-size distribution curve for the soil being tested. One method of presenting this information is shown in Fig. 6–2.

Curves of four different soils are shown. Curve "A" is that of a uniform sand; "B" is that of a poorly graded gravelly soil; "C" is a well-graded soil, with a good distribution of particles from coarse to fine; and "D" is coarse aggregate such as used in concrete.

In many cases the entire grain-size distribution curve is not required. In such circumstances, a "wet sieve analysis" may be employed, in which the soil is washed over a nest of sieves of selected sizes. One such

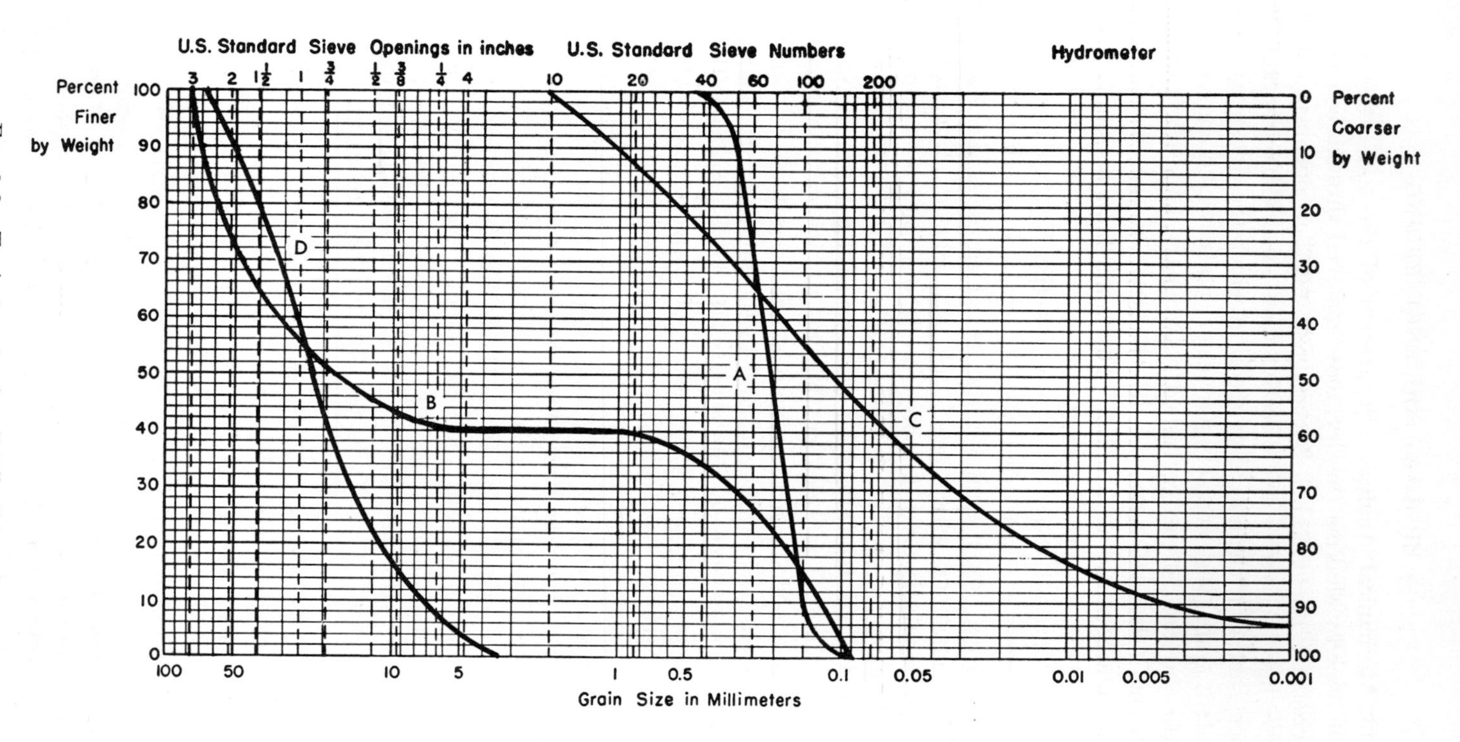

Fig. 6–2. Typical grain size distribution curves.

procedure requires the washing of the sample over the Nos. 10, 40, and 200 sieves, and a subsequent calculation of the percentage of soil retained on each of these sieves, as well as that which passes the No. 20 sieve. In this method the material passing the No. 200 sieve is designated as "combined silt and clay."

6–14. Liquid Limit. The liquid limit may be defined as the minimum moisture content at which the soil will flow under the application of a very small shearing force. At this moisture content the soil is assumed to behave practically as a liquid. The liquid limit is usually determined

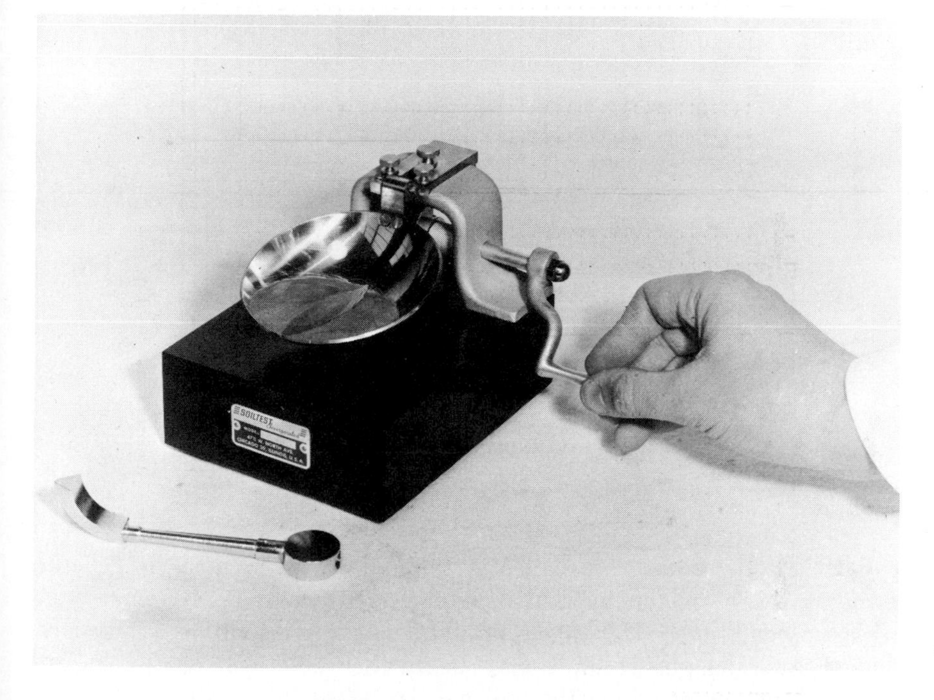

Fig. 6–3(a). Liquid limit test. (Courtesy Soiltest, Inc.)

in the laboratory by the use of the mechanical device shown in Fig. 6–3(a). When this device is used, the liquid limit may be further defined as the moisture content at which a groove of standard dimensions will just close upon application of 25 blows or "shocks," applied by raising the cup of the liquid limit device and allowing it to fall a distance of 1 centimeter. In performing the test it is usually not practicable to establish the exact moisture content corresponding to 25 blows. Consequently, a number of trials are made with the soil at moisture contents both above and below the liquid limit, and the plot of Fig. 6–3(b) is prepared. The moisture content at which the straight line (or "flow curve") crosses the line corresponding to 25 blows is then taken to be the liquid limit. Laboratories which perform a very large number of routine tests have developed various short-cut methods of determining liquid limits within permissible ranges of accuracy.

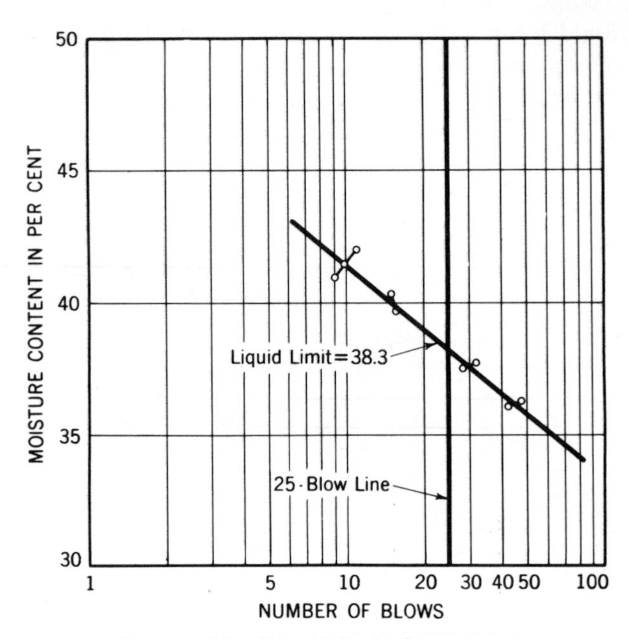

FIG. 6–3(b). Liquid limit flow curve.

6–15. Plastic Limit. The plastic limit may be defined in general terms as the minimum moisture content at which the soil remains in a plastic condition. This lower limit of plasticity is rather arbitrarily defined, and the plastic limit may be further described as the lowest moisture content at which the soil can be rolled into a thread ⅛-inch diameter without crumbling (Fig. 6–4). If a cohesive soil is wetter than the plastic limit, threads may be rolled out to diameters of less than ⅛ inch without crumbling; if drier, the soil will crumble before this diameter can be reached. When the soil has a moisture content equal to the plastic limit, a thread can be rolled out by hand pressure to ⅛-inch diameter and then will crumble or break into pieces. Certain soils, such as clean sands, are nonplastic, i.e., no plastic limit can be determined.

6–16. Plasticity Index. The plasticity index (P.I.) of a soil is defined as the numerical difference between the liquid and plastic limits. It thus indicates the range of moisture content over which the soil is in a plastic condition. Sandy soils and silts, particularly those of the rock-flour type, have characteristically low P.I.'s, while clay soils show high values of the plasticity index. Generally speaking, soils which are highly plastic, as indicated by a high value of the P.I., are also highly compressible. It is also evident that the plasticity index is a measure of cohesiveness, with a high value of the P.I. indicating a high

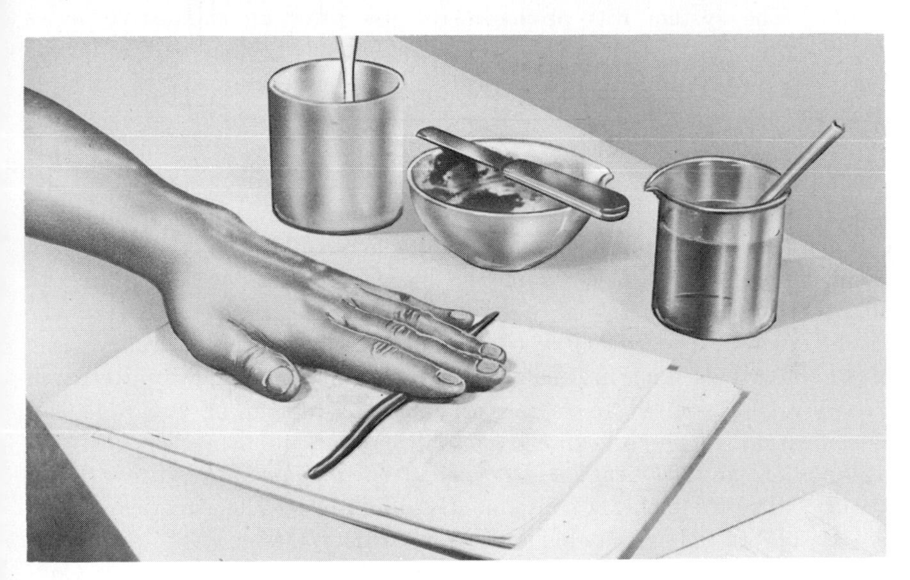

FIG. 6–4. Plastic limit test.

degree of cohesion. Soils which do not have a plastic limit, such as cohesionless sands, are reported as being nonplastic (NP).

6–17. Other Routine Tests. Shrinkage factors are determined by evaluating certain volume-weight relationships at known moisture contents. The shrinkage limit is the most important of the shrinkage factors and is defined as the maximum calculated water content at which, upon continued drying, the soil will continue to lose weight but will not decrease in volume.

The same data may be used for calculating the shrinkage ratio, volumetric shrinkage, and lineal shrinkage, where required. Some soils, such as sands, have no significant values for the shrinkage limit and other shrinkage factors. Other routine tests, now rarely used, are the centrifuge moisture equivalent and the field moisture equivalent.

6–18. Bureau of Public Roads Classification System. The Bureau of Public Roads classification system was originally presented in the June and July, 1931, issues of the magazine *Public Roads*. As originally proposed, the portion of the system pertaining to uniform highway subgrade soils consisted of eight principal groups, designated by the symbols A–1 through A–8. Each of these A–groups contained soils having similar constituents and characteristics. Classification of any particular soil into the proper group was accomplished by the use of various tables and charts based upon the mechanical analysis and subgrade soil constants. The system was proposed for use after an exhaustive series of laboratory and field tests had been completed. In the years following 1931, this system was adopted and used by a large majority of agencies in the highway field.

As time passed, it became apparent that certain of the groups established by the B.P.R. system were too broad in coverage, permitting the inclusion of dissimilar soil types under the same group symbol. Because of this fact, various highway departments and other highway agencies made such revisions in the basic system as were required to adapt it to their particular needs. In 1942 a revised version of the Public Roads classification system was published in the February issue of *Public Roads,* and in 1943 a committee of the Highway Research Board was appointed to review the entire problem of subgrade soil classification. A report of the committee was published in the *Proceedings of the 25th Annual Meeting of the Highway Research Board* (1945). The system was subsequently adopted by the American Association of State Highway Officials as AASHO Designation M145, the Classification of Soils and Soil-Aggregate Mixtures for Highway Construction Purposes (*107*).

6–19. AASHO Classification System. The subgrade soil classification system adopted by AASHO and in general use among highway engineers is a revision of the Public Roads classification system. Material given below is taken directly from the *Proceedings of the 25th Annual Meeting of the Highway Research Board,* pp. 376–382.

Objective

The objective of this report is to establish a useful classification of subgrade materials that may be made from results of the least possible number of the simple, routine tests performed by practically all highway departments.

Tests Required for Classification

Tests selected as most suitable for meeting the above-stated objective are the three most commonly made on subgrade materials, viz., sieve analysis, liquid limit, and plastic limit. The methods of test are defined in the *Standards of the American Association of State Highway Officials.*

That the simple tests here designated are useful and may supply in many instances all the information needed for the adequate and economical design of highway foundations is demonstrated by their successful use by a number of highway organizations for the test evaluation of subgrades, as well as for specific definition of suitable embankment, subbase, and base course materials.

TABLE 1

CLASSIFICATION OF HIGHWAY SUBGRADE MATERIALS

General Classification	Granular Materials (35 % or less passing No. 200)			Silt-Clay Materials (More than 35 % passing No. 200)			
Group Classification	A–1	A–3[a]	A–2	A–4	A–5	A–6	A–7
Sieve Analysis, per cent passing:							
No. 10							
No. 40	50 max.	51 min.					
No. 200	25 max.	10 max.	35 max.	36 min.	36 min.	36 min.	36 min.
Characteristics of fraction passing No. 40:							
Liquid limit				40 max.	41 min.	40 max.	41 min.
Plasticity index	6 max.	N.P.		10 max.	10 max.	11 min.	11 min.
Group Index			4 max.	8 max.	12 max.	16 max.	20 max.
General Rating as Subgrade	Excellent to Good			Fair to Poor			

Classification Procedure: With required test data available, proceed from left to right on above chart and correct group will be found by process of elimination. The first group from the left into which the test data will fit is the correct classification. (Note all limiting test values are shown as whole numbers. If fractional numbers appear on test reports, convert to nearest whole number for purposes of classification.)

[a] The placing of A–3 before A–2 is necessary in the "left to right elimination process" and does not indicate superiority of A–3 over A–2.

While the "identification" tests here designated are all that are necessary for this classification and may supply in many instances all the test information needed, the committee recognizes that there are other useful "identification" and "strength" tests as well as textural designations which should be used in some cases to supplement the classification tests for the closest possible evaluation of subgrade materials.

CLASSIFICATION PRESENTED

The subcommittee proposes the following for use in the classification of subgrade materials:

1. *Classification of Highway Subgrade Materials.* Table 1 shows the classification recommended by the subcommittee and includes test limits and group index values.

2. *Classification of Highway Subgrade Materials—with Suggested Subgroups.* Table 2 shows a suggested subdivision of groups in the classification shown in Table 1. Test limits and maximum group index values are included for each of the subgroups. The use of subgroups and group index values is recommended in instances where the main groups do not classify the soil in sufficient

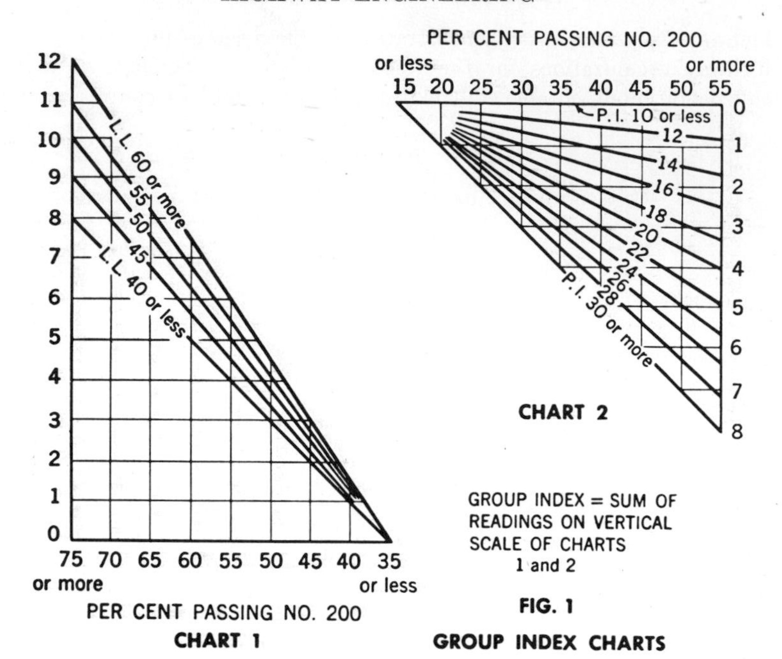
PER CENT PASSING NO. 200
or less
or more
15 20 25 30 35 40 45 50 55
P.I. 10 or less
12
14
16
18
20
22
24
26
28
P.I. 30 or more
0 1 2 3 4 5 6 7 8
CHART 2
12 11 10 9 8 7 6 5 4 3 2 1 0
L.L. 60 or more
55
50
45
L.L. 40 or less
75 70 65 60 55 50 45 40 35
or more
or less
PER CENT PASSING NO. 200
CHART 1
GROUP INDEX = SUM OF READINGS ON VERTICAL SCALE OF CHARTS 1 and 2
FIG. 1
GROUP INDEX CHARTS

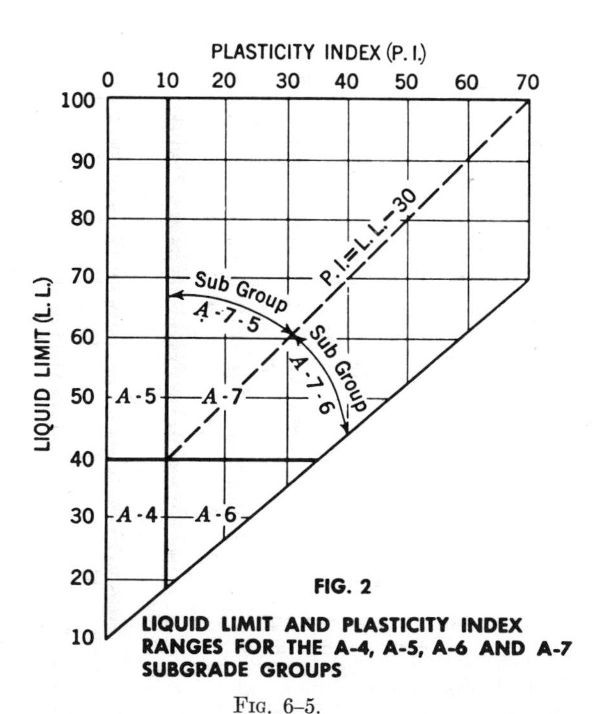
PLASTICITY INDEX (P. I.)
0 10 20 30 40 50 60 70
100 90 80 70 60 50 40 30 20 10
LIQUID LIMIT (L. L.)
P.I.=L.L.−30
Sub Group A-7-5
Sub Group A-7-6
A-5
A-7
A-4
A-6
FIG. 2
LIQUID LIMIT AND PLASTICITY INDEX RANGES FOR THE A-4, A-5, A-6 AND A-7 SUBGRADE GROUPS

Fig. 6–5.

TABLE 2

CLASSIFICATION OF HIGHWAY SUBGRADE MATERIALS

(With Suggested Subgroups)

General Classification	Granular Materials (35% or less passing No. 200)							Silt-Clay Materials (More than 35% passing No. 200)			
	A-1			A-2							A-7
Group Classification	A-1-a	A-1-b	A-3	A-2-4	A-2-5	A-2-6	A-2-7	A-4	A-5	A-6	A-7-5, A-7-6
Sieve Analysis, per cent passing:											
No. 10	50 max.										
No. 40	30 max.	50 max.	51 min.								
No. 200	15 max.	25 max.	10 max.	35 max.	35 max.	35 max.	35 max.	36 min.	36 min.	36 min.	36 min.
Characteristics of fraction passing No. 40:											
Liquid limit				40 max.	41 min.	40 max.	41 min.	40 max.	41 min.	40 max.	41 min.
Plasticity index	6 max.		N.P.	10 max.	10 max.	11 min.	11 min.	10 max.	10 max.	11 min.	11 min.[a]
Group Index[b]	0		0	0		4 max.		8 max.	12 max.	16 max.	20 max.
Usual Types of Significant Constituent Materials	Stone Fragments, Fine Gravel, and Sand			Silty or Clayey Gravel and Sand				Silty Soils		Clayey Soils	
General Rating as Subgrade	Excellent to Good							Fair to Poor			

Classification Procedure: With required test data available, proceed from left to right on above chart and correct group will be found by process of elimination. The first group from the left into which the test data will fit is the correct classification.

[a] Plasticity index of A-7-5 subgroup is equal to or less than L.L. minus 30. Plasticity index of A-7-6 subgroup is greater than L.L. minus 30 (see Figure 2, part of Fig. 6-5).

[b] See group index formula and Figure 1 (part of Fig. 6-5) for method of calculation. Group index should be shown in parentheses after group symbol as: A-2-6(3), A-4(5), A-6(12), A-7-5(17), etc.

detail and when it is desirable to differentiate between soils within the same group.

3. *Group Index.* Table 3 gives the empirical formula for obtaining the group index and shows examples of the computation. Figure 1 (part of Fig. 6–5) is a chart suitable for rapid graphical determination of the group index values.

TABLE 3

GROUP INDEX FORMULA

$$\text{Group index} = 0.2a + 0.005ac + 0.01bd$$

In which

a = That portion of percentage passing No. 200 sieve greater than 35 and not exceeding 75, expressed as a positive whole number (1 to 40).
b = That portion of percentage passing No. 200 sieve greater than 15 per cent and not exceeding 55 per cent, expressed as a positive whole number (1 to 40).
c = That portion of the numerical liquid limit greater than 40 and not exceeding 60, expressed as a positive whole number (1 to 20).
d = That portion of the numerical plasticity index greater than 10 and not exceeding 30, expressed as a positive whole number (1 to 20).

The following are examples of calculation of the group index:

(1) An A–6 material has 65 per cent passing No. 200 sieve, liquid limit of 32, and plasticity index of 13. The calculation is as follows:

$a = 65 - 35 = 30$
$b = 55 - 15 = 40$ (55 is substituted for 65, as critical range is 15 to 55.)
c = zero, since liquid limit is below 40
$d = 13 - 10 = 3$

Group index $= 0.2 \times 30 + 0.01 \times 40 \times 3 = 7.2$. (Should be recorded to nearest whole number, which is 7.)

(2) An A–7 material has 54 per cent passing No. 200 sieve, liquid limit of 62, and plasticity index of 33. The calculation is as follows:

$a = 54 - 35 = 19$
$b = 54 - 15 = 39$
$c = 60 - 40 = 20$ (60 is substituted for 62, as critical range is 40 to 60.)
$d = 30 - 10 = 20$ (30 is substituted for 33, as critical range is 10 to 30.)

$$\text{Group index} = 0.2 \times 19 + 0.005 \times 19 \times 20 + 0.01 \times 39 \times 20 = 13.5(13)$$

Charts for graphical determination of group index are shown on Figure 1 (part of Fig. 6–5).

4. *Liquid Limit and Plasticity Index Ranges.* Figure 2 (part of Fig. 6–5) shows graphically the ranges of liquid limit and plasticity index for Groups A–4, A–5, A–6, and A–7. This figure is helpful in subdividing the A–7 group into subgroups A–7–5 and A–7–6.

5. *Description of Groups.* Table 4 is a word description of materials of the various groups and subgroups.

TABLE 4

DESCRIPTION OF CLASSIFICATION GROUPS

GRANULAR MATERIALS—Containing 35 per cent or less passing the No. 200 sieve.

Group A–1. The typical material of this group is a well-graded mixture of stone fragments or gravel, coarse sand, fine sand, and a nonplastic or feebly plastic soil binder. However, this group includes also stone fragments, gravel, coarse sand, volcanic cinders, etc., without soil binder.

Subgroup A–1–a includes those materials consisting predominantly of stone fragments or gravel, either with or without a well-graded binder of fine material.

Subgroup A–1–b includes those materials consisting predominantly of coarse sand either with or without a well-graded soil binder.

Group A–3. The typical material of this group is fine beach sand or fine desert blow sand without silty or clay fines or with a very small amount of nonplastic silt. The group includes also stream-deposited mixtures of poorly graded fine sand and limited amounts of coarse sand and gravel.

Group A–2. This group includes a wide variety of "granular" materials which are borderline between the materials falling in Groups A–1 and A–3 and the silt-clay materials of Groups A–4, A–5, A–6, and A–7. It includes all materials containing 35 per cent or less passing the No. 200 sieve which cannot be classified as A–1 or A–3, due to fines content or plasticity or both, in excess of the limitations for those groups.

Subgroups A–2–4 and A–2–5 include various granular materials containing 35 per cent or less passing the No. 200 sieve and with a minus No. 40 portion having the characteristics of the A–4 and A–5 groups. These groups include such materials as gravel and coarse sand with silt contents or plasticity indexes in excess of the limitations of Group A–1, and fine sand with nonplastic silt content in excess of the limitations of Group A–3.

Subgroups A–2–6 and A–2–7 include materials similar to those described under Subgroups A–2–4 and A–2–5, except that the fine portion contains plastic clay having the characteristics of the A–6 and A–7 group. The approximate combined effects of plasticity indexes in excess of 10, and percentages passing the No. 200 sieve in excess of 15, is reflected by group index values of 0 to 4.

SILT-CLAY MATERIALS—Containing more than 35 per cent passing the No. 200 sieve.

Group A–4. The typical material of this group is a nonplastic or moderately plastic silty soil usually having 75 per cent or more passing the No. 200 sieve. The group also includes mixtures of fine silty soil and up to 64 per cent of sand and gravel retained on No. 200 sieve. The group index values range from 1 to 8, with increasing percentages of coarse material being reflected by decreasing group index values.

Group A–5. The typical material of this group is similar to that described under Group A–4, except that it is usually of diatomaceous or micaceous character and may be highly elastic, as indicated by the high liquid limit. The group index values range from 1 to 12, with increasing values indicating the combined effect of increasing liquid limits and decreasing percentages of coarse material.

Group A–6. The typical material of this group is a plastic clay soil usually having 75 per cent or more passing the No. 200 sieve. The group also includes mixtures of fine clayey soil and up to 64 per cent of sand and gravel retained on the No. 200 sieve. Materials of this group usually have high volume change between wet and dry states. The group index values range from 1 to 16, with increasing values indicating the combined effect of increasing plasticity indexes and decreasing percentages of coarse material.

TABLE 4 (*Continued*)

DESCRIPTION OF CLASSIFICATION GROUPS (*Continued*)

Group A–7. The typical material of this group is similar to that described under Group A–6, except that it has the high liquid limits characteristic of the A–5 group, and may be elastic as well as subject to high volume change. The range of group index values is 1 to 20, with increasing values indicating the combined effect of increasing liquid limits and plasticity indexes and decreasing percentages of coarse material.

Subgroup A–7–5 includes those materials with moderate plasticity indexes in relation to liquid limit and which may be highly elastic as well as subject to considerable volume change.

Subgroup A–7–6 includes those materials with high plasticity indexes in relation to liquid limit and which are subject to extremely high volume change.

DISCUSSION

Classification Group Symbols. The symbols used are A–1 to A–7 series established some sixteen years ago by publications of the Public Roads Administration. Table 2 includes suggested subdivisions and introduces the "group index" in order that the classification designation may indicate a closer evaluation of individual samples than was possible with the original Public Roads symbols alone.

Consideration was given to dropping the "A" from the symbols and also to the adoption of an entirely new set of symbols. It was determined that adjustments in the original test limitations for the Public Roads groups could be made to accomplish the objective of this report without actual radical change in the types of materials designated by the symbols. Therefore, as the use of the designations "A–3," "A–7," etc., has been well established by long usage, and as a better method of labeling was not suggested either by the committee or by the various highway engineers consulted, these symbols are retained. While some changes in test limits have been made in this revised classification, the word description of the groups as originally promulgated (see *Public Roads*, May, 1929, and June, 1931) remain generally applicable to the designated groups.

Confusion as to whether a material designated by these group symbols has been classified under the original Public Roads arrangement or under this revised arrangement can be avoided by always showing the group index value (even though zero) in parentheses after the group or subgroup symbol. For examples: A–1–a(0), A–3(0), A–2–7(3), A–4(7), A–6(12), etc.

The Grouping Arrangement and the Group Index. A major difference between this and the Public Roads classification is that the Public Roads classification applies only to the soil mortar (Minus No. 10) portion of the material, while the proposed classification applies to mixtures of coarse and fine materials as well as to materials consisting only of fine soils. As shown in Tables 1 and 2, the classifying of materials starts by first dividing into "granular materials" and "silt-clay materials" using 35 per cent passing the No. 200 sieve as the arbitrary dividing line between the two general types.

The granular materials are then divided into groups with the A–1 containing the gravels and coarse sands, either with or without limited amounts of nonplastic or feebly plastic binder, and the A–3 containing the nonplastic, binderless fine sands, leaving the A–2 as the borderline or "catch-all" group of the granular materials. The A–2–4 and A–2–5 subgroups contain gravels or coarse sands inferior to A–1, due either to excessive amounts of binder or excessive plasticity of the binder or both, and fine sands superior to A–3 due to a low-plasticity binder content which facilitates handling and compaction of such fine sands. The A–2–6 and A–2–7 subgroups contain gravels and sands with a plastic binder soil having the characteristics of the A–6 and A–7 groups. The quality of these A–2–6 and A–2–7 materials as subgrade ranges from good, where per cent passing the No. 200 is low (say less than 15 per cent), to increasingly questionable as reflected by the extent to which percentages passing the No. 200 exceed 15, and plasticity indexes exceed 10.

Thus this borderline A–2 group, as a whole, may contain materials ranging from approximately equivalent to some materials of the A–1 group to materials which are actually inferior to the best materials which can be classified in the A–6 and A–7 groups. Elimination of this condition without a tediously complicated table does not appear practicable. However, the subgroups and the group index system hereinafter discussed take care of the matter of approximate relative evaluation.

The "silt-clay materials" (those containing more than 35 per cent passing the No. 200) are divided into the four main groups A–4, A–5, A–6, and A–7 on the basis of liquid limit and plasticity index only. Inasmuch as each of these groups may contain from zero to 64 per cent material retained on the No. 200 sieve, and in view of the wide liquid limit ranges of the A–5 and A–7 groups and the wide plasticity index ranges of the A–6 and A–7 groups, it is obvious that each group may contain materials of widely different value as subgrade. In other words, the main group symbol performs the important function of indicating the general characteristics of the fine soil portion but does not evaluate the possible variations in percentages of coarse material, liquid limit, and plasticity index. This desirable within-group evaluation is taken care of by the group index system hereinafter discussed.

The empirical group index formula devised for approximate within-group evaluation of the "clayey granular materials" and the "silt-clay materials" is based on the following assumptions:

(a) Materials falling within Groups A–1–a, A–1–b, A–2–4, A–2–5, and A–3 are satisfactory as subgrade when properly drained and compacted under moderate thickness of pavement (base and/or surface course) of a type suitable for the traffic to be carried, or can be made satisfactory by the addition of small amounts of natural or artificial binders.

(b) Materials falling within the "clayey granular" Groups A–2–6 and A–2–7 and the "silt-clay" Groups A–4, A–5, A–6, and A–7 will range in quality as subgrade from the approximate equivalent of the good A–2–4 and A–2–5 subgrades to fair and poor subgrades requiring a layer of subbase material or an increased thickness of base course over that required under (a) in order to furnish adequate support for traffic loads.

(c) The assumed critical ranges of percentages passing the No. 200 are 35 to 75 neglecting plasticity, and 15 to 55 as affected by plasticity indexes greater than 10.

(d) The assumed critical range of liquid limit is 40 to 60.

(e) The assumed critical range of plasticity index is 10 to 30.

The formula will give values ranging from a fraction of 1 to 20 and is so weighted that the maximum influence of each of the three variables is in the ratio of 8 for per cent passing the No. 200 sieve, 4 for liquid limit, and 8 for plasticity index. This weighting and the adopted critical ranges represent the best judgment of the committee, based on the study of average relative evaluations placed on subgrade materials by several highway organizations which use the tests involved in this classification system.

Under average conditions of good drainage and thorough compaction, the supporting value of a material as subgrade may be assumed as an inverse ratio to its group index, that is, a group index of 0 indicates a "good" subgrade material and group index of 20 indicates a "very poor" subgrade material.

Definitions of Gravel, Sand, and Silt-Clay. The terms "gravel," "coarse sand," "fine sand," and "silt-clay," as determinable from the minimum test data required in this classification arrangement and as used in the word descriptions of this report, are defined as follows:

Gravel: Material passing sieve with 3-in.-square openings and retained on the No. 10 sieve.

Coarse Sand: Material passing the No. 10 sieve and retained on the No. 40 sieve.

Fine Sand: Material passing the No. 40 sieve and retained on the No. 200 sieve.

Combined Silt and Clay: Material passing the No. 200 sieve.

Boulders (retained on 3-in. sieve) should be excluded from the portion of the sample to which the classification is applied, but the percentage of such material, if any, in the sample should be recorded.

The term "silty" is applied to fine material having a plasticity index of 10 or less, and the term "clayey" is applied to fine material having a plasticity index of 11 or greater.

By keeping the above terms and the group tests limits in mind, it is possible, with some practice, to make fairly close approximations of the correct classifications by visual examination and handling of the materials in a damp condition.

While visual classification of subgrade materials does not take the place of laboratory tests, such visual classification by highway engineers should be encouraged.

Group A-8. The A-8 Group, included in the Public Roads classifications and used to describe peats, mucks, etc., ordinarily found in obviously unstable swampy areas, has been omitted from the classification tables included in this report. It is felt that this designation is descriptive more of the state in which the materials are found in place, viz., low density, high water content, humus content, etc., than the characteristics determinable by the laboratory tests adopted for this classification grouping.

The problem ordinarily involved with materials in this state is one of removal or consolidation in such manner as to afford a stable foundation for embankment, and description of such state and treatment belongs properly under the subject of embankment foundations rather than classification of subgrade materials. However, if desired, the A-8 designation may be used in soil reports or on soil profiles in lieu of the groups determinable from the tests used in this classification method to designate obviously unstable swampy materials which are deemed unfit either as embankment foundation or embankment material due to high humus content, etc.

6–20. Unified Soil Classification System. The Unified Soil Classification System is based upon the Airfield Classification System developed by Professor A. Casagrande of Harvard University during World War II. The system was modified slightly and became the Department of the Army Uniform Soil Classification System. In turn, this system was slightly modified and adopted by the Corps of Engineers and the Bureau of Reclamation in January, 1952.

The Unified System is based primarily upon characteristics which determine how a soil will behave when used as a construction material. Table 6–1 is the master chart for the Unified System; it contains procedures which are to be followed in identifying and classifying soils under this system.

Soils may be classified and placed into one of the 15 major soil groups in the Unified System by using either laboratory or field identification procedures. Soils are classified on the basis of (1) Percentage of gravel, sand, and fines (fraction passing the No. 200 sieve); (2) shape of the grain-size distribution curve; and (3) plasticity and compressibility characteristics.

The coefficients of uniformity (C_u) and gradation (C_g) are used to judge the shape of the grain-size distribution curve of a coarse-grained soil. The term "D_{10}" which appears in the table means the grain size (diameter) which corresponds to 10 per cent on a grain-size distribution curve of the type shown in Fig. 6–2. D_{30} and D_{60} have similar meanings.

The influence and relationship of the liquid limit and plasticity index are reflected in the plasticity chart of Table 6–1.

In general terms, clays (C) plot above the "A" line of the plasticity chart and silts (M) plot below the "A" line. The silt (M) and clay (C) groups are further divided on the basis of low (L) or high (H) liquid limit; a high liquid limit is associated with high compressibility.

Further details of the classification procedure will be revealed by close study of Table 6–1. Further information is contained in publications of the Corps of Engineers (*144,167.*)

Practical value of the Unified System is enhanced by the information contained in Table 6–2, which is a tabulation of the engineering characteristics of the various soil groups, as pertinent to roads and airfields.

TABLE

UNIFIED SOIL

<table>
<tr><th colspan="3">Major Divisions</th><th>Group Symbols</th><th>Typical Name</th><th colspan="3">Field Identification Procedures (Excluding Particles Larger Than 3 Inches and Basing Fractions on Estimated Weights)</th></tr>
<tr><td>1</td><td colspan="2">2</td><td>3</td><td>4</td><td colspan="3">5</td></tr>
<tr><td rowspan="8">Coarse-grained Soils
More than half of material is larger than No. 200 sieve size.
The No. 200 sieve size is about the smallest particle visible to the naked eye.</td><td rowspan="4">Gravels
More than half of coarse fraction is larger than No. 4 sieve size.
(For visual classification, the 1/4-inch size may be used as equivalent to the No. 4 sieve size.)</td><td rowspan="2">Clean gravels (little or no fines)</td><td>GW</td><td>Well-graded gravels, gravel-sand mixtures, little or no fines</td><td colspan="3">Wide range in grain sizes and substantial amounts of all intermediate particle sizes</td></tr>
<tr><td>GP</td><td>Poorly-graded gravels, gravel-sand mixtures, little or no fines</td><td colspan="3">Predominantly one size or a range of sizes with some intermediate sizes missing</td></tr>
<tr><td rowspan="2">Gravels with fines (appreciable amount of fines)</td><td>GM</td><td>Silty gravels, gravel-sand-silt mixtures</td><td colspan="3">Nonplastic fines or fines with low plasticity (for identification procedures see ML below)</td></tr>
<tr><td>GC</td><td>Clayey gravels, gravel-sand-clay mixtures</td><td colspan="3">Plastic fines (for identification procedures see CL below)</td></tr>
<tr><td rowspan="4">Sands
More than half of coarse fraction is smaller than No. 4 sieve size.</td><td rowspan="2">Clean sands (little or no fines)</td><td>SW</td><td>Well-graded sands, gravelly sands, little or no fines</td><td colspan="3">Wide range in grain size and substantial amounts of all intermediate particle sizes</td></tr>
<tr><td>SP</td><td>Poorly-graded sands, gravelly sands, little or no fines</td><td colspan="3">Predominantly one size or a range of sizes with some intermediate sizes missing</td></tr>
<tr><td rowspan="2">Sands with fines (appreciable amount of fines)</td><td>SM</td><td>Silty sands, sand-silt mixtures</td><td colspan="3">Nonplastic fines or fines with low plasticity (for identification procedures see ML below)</td></tr>
<tr><td>SC</td><td>Clayey sands, sand-clay mixtures</td><td colspan="3">Plastic fines (for identification procedures see CL below)</td></tr>
<tr><td rowspan="8">Fine-grained Soils
More than half of material is smaller than No. 200 sieve size.</td><td rowspan="5">Silts and Clays</td><td rowspan="5">Liquid limit less than 50</td><td rowspan="2"></td><td rowspan="2"></td><td colspan="3">Identification procedures on fraction smaller than No. 40 sieve size</td></tr>
<tr><td>Dry strength (crushing characteristics)</td><td>Dilatancy (reaction to shaking)</td><td>Toughness (consistency near PL)</td></tr>
<tr><td>ML</td><td>Inorganic silts and very fine sands, rock flour, silty or clayey fine sands with slight plasticity</td><td>None to slight</td><td>Quick to slow</td><td>None</td></tr>
<tr><td>CL</td><td>Inorganic clays of low to medium plasticity, gravelly clays, sandy clays, silty clays, lean clays</td><td>Medium to high</td><td>None to very slow</td><td>Medium</td></tr>
<tr><td>OL</td><td>Organic silts and organic silty clays of low plasticity</td><td>Slight to medium</td><td>Slow</td><td>Slight</td></tr>
<tr><td rowspan="3">Silts and Clays</td><td rowspan="3">Liquid limit greater than 50</td><td>MH</td><td>Inorganic silts, micaceous or diatomaceous fine sandy or silty soils, elastic silts</td><td>Slight to medium</td><td>Slow to none</td><td>Slight to medium</td></tr>
<tr><td>CH</td><td>Inorganic clays of high plasticity, fat clays</td><td>High to very high</td><td>None</td><td>High</td></tr>
<tr><td>OH</td><td>Organic clays of medium to high plasticity, organic silts</td><td>Medium to high</td><td>None to very slow</td><td>Slight to medium</td></tr>
<tr><td colspan="3">Highly organic soils</td><td>Pt</td><td>Peat and other highly organic soils</td><td colspan="3">Readily identified by color, odor, spongy feel and frequently by fibrous texture</td></tr>
</table>

(1) Boundary classifications: Soils possessing characteristics of two groups are designated by combinations of group symbols. For example

Adopted by Corps of Engineers and Bureau of Reclamation, January, 1952.

6–1

Classification System

Information Required for Describing Soils	Laboratory Classification Criteria			
6	7			
For undisturbed soils add information on stratification, degree of compactness, cementation, moisture conditions, and drainage characteristics Give typical name; indicate approximate percentages of sand and gravel, max. size; angularity, surface conditions, and hardness of the coarse grains; local or geologic name and other pertinent descriptive information; and symbol in parentheses Example: Silty sand, gravelly; about 20 per cent hard, angular gravel particles ½-inch maximum size; rounded and subangular sand grains coarse to fine; about 15 per cent nonplastic fines with low dry strength; well compacted and moist in place; alluvial sand; *(SM)*	Use grain-size curve in identifying the fractions as given under field identification.	Determine percentages of gravel and sand from grain-size curve. Depending on percentage of fines (fraction smaller than No. 200 sieve size) coarse-grained soils are classified as follows: Less than 5 per cent *GW, GP, SW, SP,* More than 12 per cent *GM, GC, SM, SC.* 5 per cent to 12 per cent Borderline cases requiring use of dual symbols	$C_u = \frac{D_{60}}{D_{10}}$ Greater than 4 $C_g = \frac{(D_{30})^2}{D_{10} \times D_{60}}$ Between 1 and 3	
			Not meeting all gradation requirements for *GW*	
			Atterberg limits below "*A*" line or *PI* less than 4	Above "*A*" line with *PI* between 4 and 7 are borderline cases requiring use of dual symbols
			Atterberg limits above "*A*" line with *PI* greater than 7	
			$C_u = \frac{D_{60}}{D_{10}}$ Greater than 6 $C_g = \frac{(D_{30})^2}{D_{10} \times D_{60}}$ Between 1 and 3	
			Not meeting all gradation requirements for *SW*	
			Atterberg limits below "*A*" line or *PI* less than 4	Limits plotting in hatched zone with *PI* between 4 and 7 are borderline cases requiring use of dual symbols
			Atterberg limits above "*A*" line with *PI* greater than 7	
Give typical name, indicate degree and character of plasticity, amount and maximum size of coarse grains, color in wet condition, odor if any, local or geologic name, and other pertinent descriptive information; and symbol in parenthesis For undisturbed soils add information on structure, stratification, consistency in undisturbed and remolded states, moisture and drainage conditions Example: Clayey silt, brown, slightly plastic, small percentage of fine sand, numerous vertical root holes, firm and dry in place, loess, *(ML)*		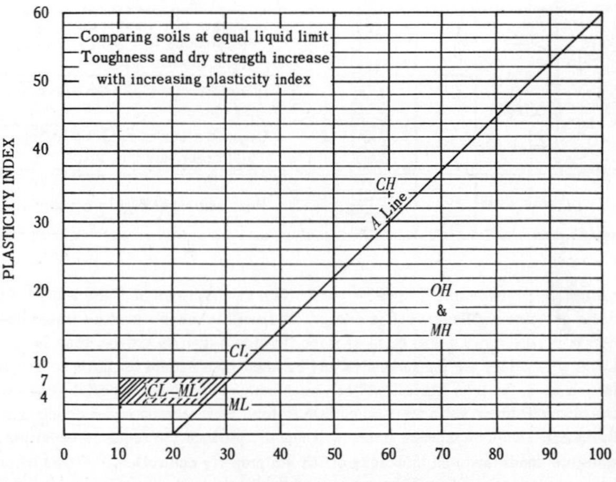 LIQUID LIMIT PLASTICITY CHART Per laboratory classification of fine-grained soils		

LIQUID LIMIT PLASTICITY CHART

Per laboratory classification of fine-grained soils

GW-OC, well-graded gravel-sand mixture with clay binder. (2) All sieve sizes on this chart are U. S. standard.

TABLE

CHARACTERISTICS PERTINENT

Major Divisions (1)	(2)	Letter (3)	Symbol Hatching (4)	Color (5)	Name (6)	Value as Foundation When Not Subject to Frost Action (7)	Value as Base Directly Under Bituminous Pavement (8)
Coarse Grained Soils	Gravel and gravelly soils	*GW*		Red	Well-graded gravels or gravel-sand mixtures, little or no fines	Excellent	Good
		GP		Red	Poorly graded gravels or gravel-sand mixtures, little or no fines	Good to excellent	Poor to fair
		GM *d*		Yellow	Silty gravels, gravel-sand-silt mixtures	Good to excellent	Fair to good
		GM *u*		Yellow		Good	Poor
		GC		Yellow	Clayey gravels, gravel-sand-clay mixtures	Good	Poor
	Sand and sandy soils	*SW*		Red	Well-graded sands or gravelly sands, little or no fines	Good	Poor
		SP		Red	Poorly graded sands or gravelly sands, little or no fines	Fair to good	Poor to not suitable
		SM *d*		Yellow	Silty sands, sand-silt mixtures	Fair to good	Poor
		SM *u*		Yellow		Fair	Not suitable
		SC		Yellow	Clayey sands, sand-clay mixtures	Poor to fair	Not suitable
Fine Grained Soils	Silts and clays LL < 50	*ML*		Green	Inorganic silts and very fine sands, rock flour, silty or clayey fine sands or clayey silts with slight plasticity	Fair to poor	Not suitable
		CL		Green	Inorganic clays of low to medium plasticity, gravelly clays, sandy clays, silty clays, lean clays	Fair to poor	Not suitable
		OL		Green	Organic silts and organic silt-clays of low plasticity	Poor	Not suitable
	Silts and clays LL > 50	*MH*		Blue	Inorganic silts, micaceous or diatomaceous fine sandy or silty soils, elastic silts	Poor	Not suitable
		CH		Blue	Inorganic clays of high plasticity, fat clays	Poor to very poor	Not suitable
		OH		Blue	Organic clays of medium to high plasticity, organic silts	Poor to very poor	Not suitable
Highly organic soils		*Pt*		Orange	Peat and other highly organic soils	Not suitable	Not suitable

Notes:

1. Column 3, Division of *GM*, and *SM* groups into subdivisions of *d* and *u* are for roads and airfields only; subdivision is on basis of Atterberg limits; suffix *d* (e.g., *GMd*) will be used when the liquid limit is 28 or less and the plasticity index is 6 or less; the suffix *u* will be used when the liquid limit is greater than 28.
2. Column 7, values are for subgrades and base courses except for base course directly under bituminous pavement.
3. In column 8, the term "excellent" has been reserved for base materials consisting of high quality processed crushed stone.
4. In column 9, these soils are susceptible to frost as indicated under conditions favorable to frost action described in the text.
5. In column 12, the equipment listed will usually produce the required densities with a reasonable number of passes when moisture conditions and thickness of lift are properly controlled. In some instances, several types of equipment are listed, because variable soil characteristics within a given soil group may require different equipment. In some instances, a combination of two types may be necessary.
   a. Processed base materials and other angular materials. Steel-wheeled rollers are recommended for hard angular materials

Potential Frost Action (9)	Compressibility and Expansion (10)	Drainage Characteristics (11)	Compaction Equipment (12)	Unit Dry Weight Lb. per Cu. Ft. (13)	Field CBR (14)	Subgrade Modulus K Lb. per Cu. In. (15)
None to very slight	Almost none	Excellent	Rubber-tired roller, steel-wheeled roller, vibratory compactor	125–140	60–80	300 or more
None to very slight	Almost none	Excellent	Rubber-tired roller, steel-wheeled roller, vibratory compactor	110–130	25–60	300 or more
Slight to medium	Very slight	Fair to poor	Rubber-tired roller, sheepsfoot roller, vibratory compactor, close control of moisture	130–145	40–80	300 or more
Slight to medium	Slight	Poor to practically impervious	Rubber-tired roller, sheepsfoot roller, vibratory compactor	120–140	20–40	200–300
Slight to medium	Slight	Poor to practically impervious	Rubber-tired roller, sheepsfoot roller, vibratory compactor	120–140	20–40	200–300
None to very slight	Almost none	Excellent	Rubber-tired roller, vibratory compactor	110–130	20–40	200–300
None to very slight	Almost none	Excellent	Rubber-tired roller, vibratory compactor	100–120	10–25	200–300
Slight to high	Very slight	Fair to poor	Rubber-tired roller, sheepsfoot roller, vibratory compactor, close control of moisture	120–135	20–40	200–300
Slight to high	Slight to medium	Poor to practically impervious	Rubber-tired roller, sheepsfoot roller, vibratory compactor	105–130	10–20	200–300
Slight to high	Slight to medium	Poor to practically impervious	Rubber-tired roller, sheepsfoot roller, vibratory compactor	105–130	10–20	200–300
Medium to very high	Slight to medium	Fair to poor	Rubber-tired roller, sheepsfoot roller, close control of moisture	100–125	5–15	100–200
Medium to high	Medium	Practically impervious	Rubber-tired roller, sheepsfoot roller	100–125	5–15	100–200
Medium to high	Medium to high	Poor	Rubber-tired roller, sheepsfoot roller	90–105	4–8	100–200
Medium to very high	High	Fair to poor	Sheepsfoot roller	80–100	4–8	100–200
Medium	High	Practically impervious	Sheepsfoot roller	90–110	3–5	50–100
Medium	High	Practically impervious	Sheepsfoot roller	80–105	3–5	50–100
Slight	Very high	Fair to poor	Compaction not practical	–	–	–

with limited fines or screenings. Rubber-tired equipment is recommended for softer materials subject to degradation.

b. Finishing. Rubber-tired equipment is recommended for rolling during final shaping operations for most soils and processed materials.

c. Equipment size. The following sizes of equipment are necessary to assure the high densities required for airfield construction:

Rubber-tired equipment – wheel load in excess of 15,000, pound wheel loads as high as 40,000 pounds may be necessary to obtain the required densities for some materials (based on contact pressure of approximately 65 to 150 psi).

Sheepsfoot roller – unit pressure (on 6- to 12-sq-in. foot) to be in excess of 250 psi and unit pressures as high as 650 psi may be necessary to obtain the required densities for some materials. The area of the feet should be at least 5 per cent of the total peripheral area of the drum, using the diameter measured to the faces of the feet.

6. Column 13, unit dry weights are for compacted soil at optimum moisture content for modified AASHO compactive effort.

6–21. Correlation Between AASHO and Unified Systems. Approximate correlations between groups in the AASHO and Unified Classification Systems are given in Table 6–3. Such relationships are approximate only but are accurate enough for purposes of comparison.

TABLE 6–3

APPROXIMATE EQUIVALENT GROUPS OF AASHO AND UNIFIED SOIL CLASSIFICATION SYSTEMS

AASHO	Unified
A–1–a	GW, GP, GM
A–1–b	SW, SM
A–2–4	GM, SM
A–2–5	GM, SM
A–2–6	GC, SC
A–2–7	GC, SC
A–3	SP
A–4	ML, OL
A–5	MH
A–6	CL
A–7–5	CL, OL
A–7–6	CH, OH

6–22. Classification of Typical Soils. In Table 6–4 are given the results of laboratory tests on four inorganic soils. Each soil may be classi-

TABLE 6–4

RESULTS OF LABORATORY TESTS ON FOUR INORGANIC SOILS

	Soil Number			
	1	2	3	4
	Mechanical Analysis			
Sieve Size	Per Cent Passing, by Weight			
3-inch	–	–	–	100.0
¾-inch	–	–	–	56.0
No. 4	–	–	–	30.0
No. 10	100.0	100.0	100.0	16.4
No. 40	85.2	97.6	85.0	7.2
No. 60	–	–	20.0	5.0
No. 200	52.1	69.8	1.2	3.5
No. 270	48.2	65.0	–	–
C_u	–	–	2.0	12.5
C_g	–	–	–	2.2
	Plasticity Characteristics			
	Per Cent, by Weight			
Liquid Limit	29.2	66.7	21.3	–
Plasticity Index	5.0	39.0	NP	–

fied under the AASHO and Unified Soil Classification systems, as follows:

Soil No. 1

AASHO System—To calculate the group index, refer to Fig. 6–5. From chart 1, read 3; from chart 2, read 0. Therefore, group index = 0 + 3 = 3. Entering Table 2, and using a left-to-right elimination process, soil cannot be in one of the *granular materials* groups, since more than 35 per cent passes a No. 200 sieve. It meets the requirements of the A–4 group. It is therefore A–4(3).

Unified—Soil is fine-grained since more than 50 per cent passes No. 200 sieve. Liquid limit is less than 50, hence must be ML or CL, since it is inorganic. On plasticity chart it falls below the "A" line, therefore ML.

Soil No. 2

AASHO System—As before, group index = 10 + 8 = 18. This soil must fall into the A–7 group, since this is the only group which will permit a group index value as high as 18. Referring to the lower chart of Fig. 6–5, it falls in the A–7–6 subgroup; hence, A–7–6(18).

Unified—Soil is fine-grained since more than 50 per cent passes No. 200 sieve. Liquid limit is more than 50, hence must be MH or CH. On plasticity chart, it falls above the "A" line, therefore CH.

Soil No. 3

AASHO—As before, group index = 0 + 0 = 0. This is one of the soils described as *granular material.* It will not meet the requirements of the A–1 group, since it contains practically no fines. It does meet the requirements of the A–3 group, hence, A–3(0).

Unified—Soil is coarse-grained, since very little passes No. 200 sieve. All passes a No. 10 sieve, hence sand. The soil contains less than 5 per cent passing No. 200, therefore it must be either an SW or SP. Value of $C_u = 2$ will not meet requirements for SW, hence SP.

Soil No. 4

AASHO—As before, group index = 0 + 0 = 0. This is obviously a granular material and meets the requirements of the A–1–a subgroup, even though the plasticity index is not known. Therefore, A–1–a (0).

Unified—Soil is coarse-grained since very little passes No. 200 sieve. Is a gravel, since more than 50 per cent of the coarse fraction is larger than No. 4 sieve. Since the soil contains less than 5 per cent passing the No. 200, it is either GW or GP. It meets the requirements relative to gradation, hence GW.

Soil Surveys for Highway Purposes

Soil surveys are made in connection with highway location, design, and construction. Many sources of information concerning soils generally are available for the area in which a highway project is to be carried out. These sources include geological and topographical maps and reports, agricultural soil maps and reports, aerial photographs, and the results of previous soil surveys in this area. Information from such sources is of importance in two general ways. First, a study of this information will aid in securing a broad understanding of soil conditions and associated engineering problems which may be encountered. Second, such information is of great value in planning, conducting, and interpreting the results of detailed soil surveys which are necessary for design and construction. Modern techniques emphasize the use of all the available information about a given area in order to minimize the amount of detailed field and laboratory work necessary for a given project.

6–23. Geological and Topographical Maps and Reports. It is apparent that there is a close relationship between geology and soil conditions. Information contained in geological maps and reports may be either of general or specific usefulness to the engineer, depending on the type of map or report available.

Geological maps and reports are available for many areas of the United States through the U.S. Geological Survey and the various state geological surveys. Certain of the maps and reports deal with surface geology, and can be of specific value.

Ordinary topographical maps may be of some value in estimating soil conditions, particularly when reviewed in conjunction with geological maps. Experience is required for such examinations, but a careful inspection of maps of known areas will provide a background for the recognition of similar features on maps of unknown areas. For example, such characteristic shapes as sand dunes, beach ridges, and alluvial fans may be detected on large-scale topographical maps, as may typical formations in glaciated areas.

6–24. Use of Pedological Classification System. Soil scientists have for many years been engaged in studying the nature and occurrence of soils important to agriculture. Loosely speaking, this branch of science has been called "Pedology," which may be defined as the study of soil in its natural state.

A great deal of information regarding soils from an agricultural viewpoint has been accumulated over the years by various agencies and in almost every state. This information, when used in connection with geological knowledge of an area or region, can readily serve as a basis for engineering soil classification and mapping. Soil maps and reports may be obtained through the U.S. Department of Agriculture, county extension agents, universities, and other groups.

Certain state highway departments have made extensive use of existing pedological information to simplify the process of preparing soil maps for engineering purposes. A system of this type has been in use for many years by the Michigan Department of State Highways (*181*).

When an engineering soils map is to be prepared for a given area in Michigan, typical soil profiles existing in the area are carefully studied. The layers of soil or "horizons" are examined in test pits, railroad or highway cuts, or other exposed places. Each of the horizons is studied from the standpoint of color, texture, structure, organic content, and consistency. In addition, such factors as drainage, topography, vegetation, agricultural land use, and geographical location are noted. On the basis of this examination each soil is classified and placed in the proper mapping unit. For mapping purposes, three principal designations may be utilized. These designations are (1) series, (2) type, and (3) phase.

The designation given as the soil series is applied to soils having the same genetic horizons, possessing similar characteristics and profiles, and derived from the same parent material. Series names follow no particular pattern but are generally taken from some geographical place name near which they were first found. The soil type refers to the texture of the upper portion of the soil profile and several soil types

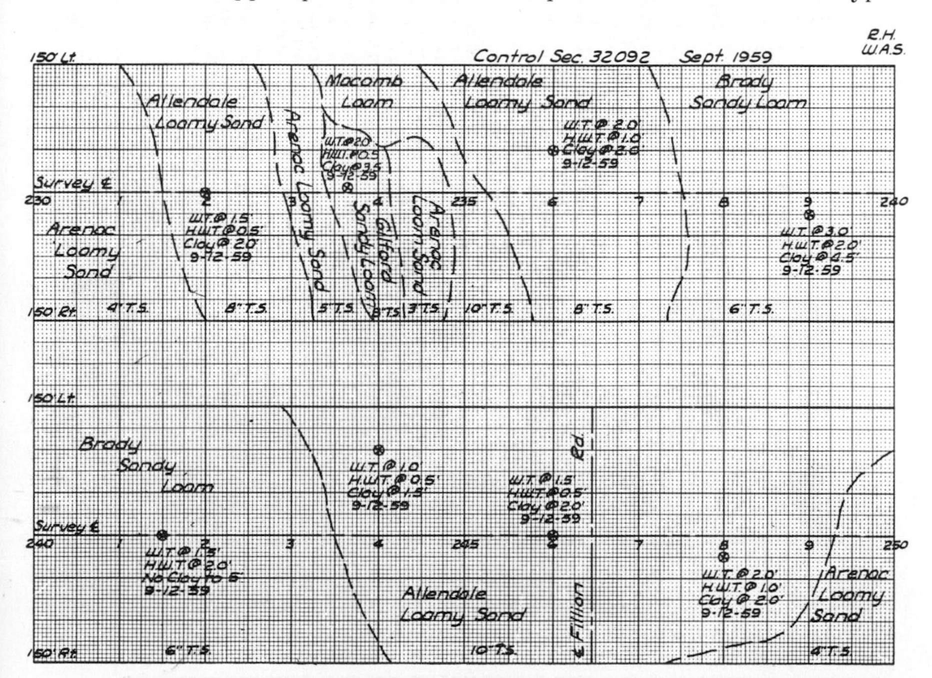

Fig. 6–6. Soil survey map. (Courtesy Michigan Department of State Highways.)

SOIL TYPE DESCRIPTION

inches: 2 4 6 8 10 12 14 16 18 20 22 24 26 28 30 32 34 36 38 40 42 44 46 48 50 52 54 56 58 60 62 64 66 68

Litter mold and humus soil.

Lavender gray loose sandy loam.

Brown sandy loam, grading with depth to a yellowish sandy loam or loamy sand.

Grayish or pale reddish sandy loam.

Reddish brown compact sandy loam.

Sandy drift with occasional layers and pockets of silt, gravelly sand and sandy clay.

Group	Series	Class
Podzol	Emmet	Sandy Loam

The texture of Emmet is usually a sandy loam.

This series occurs on well drained sandy moraines of the northern part of the Southern Peninsula and in the eastern part of the Upper Peninsula and is also found in association with the Onaway series on the till plains of Menominee County. The surface relief ranges from undulating to hilly. In places, there are sharp ridges, steep slopes and rounded depressions. Limestone, gravel, cobbles and boulders may occur on the surface and embedded in the soil.

Emmet is weathered from a sandy drift with usually a noticeable amount of limestone gravel. The drift is loose and relatively low in water holding capacity. The original forest cover was dominantly hardwood with sugar maple, birch, yellow birch and some white pine.

The Leelanau series has a reddish brown sandy loam or sandy clay loam lower "B" horizon thinner than that of the typical Emmet profile. When mapping, include Leelanau with Emmet.

Emmet is found in association with Onaway, Nester and Roselawn. The geological relation and textural profile of the Emmet is intermediate between the Onaway and Wexford. The Wexford is sandier throughout the profile. The geological relation and textural profile of Emmet is similar to Roselawn, but is distinguished from the Roselawn by the darker or stronger coloring from organic matter in the "B" horizon and a stronger limestone influence in the parent drift and by the original hardwood cover.

Emmet is similar to the Hiawatha in the Upper Peninsula in physiographical features and texture but is distinguished by the yellow color of the drift in contrast to the salmon pink color of the Hiawatha. It is also distinguished by the limestone influence in the parent material.

Typical gradation of "C" horizon:

16%—Gravel 45%—Fine Sand
12%—Coarse Sand 19%—Silt
8%—Clay

Construction Information

This material is considered excellent for grading operations during all seasons of the year. However, the loose character of the sandy material may interfere with hauling operations. Seepage may be encountered in sandy clay and silt pockets in sufficient amounts to make construction difficult in local areas.

FIG. 6–7. Soil description. (Courtesy Michigan Department of State Highways.)

may exist within a given series. Thus the designated soil types may be sand, sandy loam, silty loam, clay, or some other term relative to soil texture. By "phase" is meant some variation of the designated soil type. Such a variation is usually minor in character, although it may be of practical significance in either agriculture or engineering.

Following the field survey a map is prepared showing the nature and extent of the soils encountered in the area under consideration. This information is supplemented by facts relative to the location of the water table, depth to clay or rock, and other information of specific interest. In Fig. 6–6 is shown a typical soil map prepared from such a survey.

Engineering characteristics of the soils shown on the map are then obtained from an over-all index or "key" to soils in the State of Michigan. For example, in Fig. 6–7 is shown the tabulated information contained in the key relative to one soil type, Emmet Loamy Sand. Similar information is available for about 140 other soil types which occur in Michigan.

6–25. Use of Aerial Photographs. In recent years the technique of soil mapping has been further improved through the extensive use of aerial photographs. Soil conditions and related factors may be rapidly and accurately determined for relatively large areas by this means when details shown on the photograph are properly interpreted.

Briefly described, the use of aerial photographs for identifying soil deposits and determining subsurface soil conditions is based upon the recognition and association of various features appearing on the photograph with characteristic soil types and formations. Included among the features which provide clues to the nature and extent of soil deposits are land form, drainage patterns, presence or lack of erosion, soil color (appearing as shades of gray on the normal photograph), vegetation characteristics, and land use. In the hands of an experienced "interpreter," the photograph can yield a surprising wealth of soil information.

In addition to providing a basis for the preparation of soil maps, the photographs may aid in the location of a highway by indicating areas where unfavorable soil conditions exist; by permitting the determination of preliminary drainage requirements; by permitting ready location and identification of borrow materials, such as gravel; and they may serve a host of other purposes. Much of the United States has been photographed from the air, principally by various governmental agencies, and many of these photographs are suitable for use in highway work. Some use is being made of aerial color photographs for soil mapping and materials surveys (*10*). In Fig. 6–8(a) is shown an aerial mosaic, while Fig. 6–8(b) shows the engineering soil map prepared therefrom.

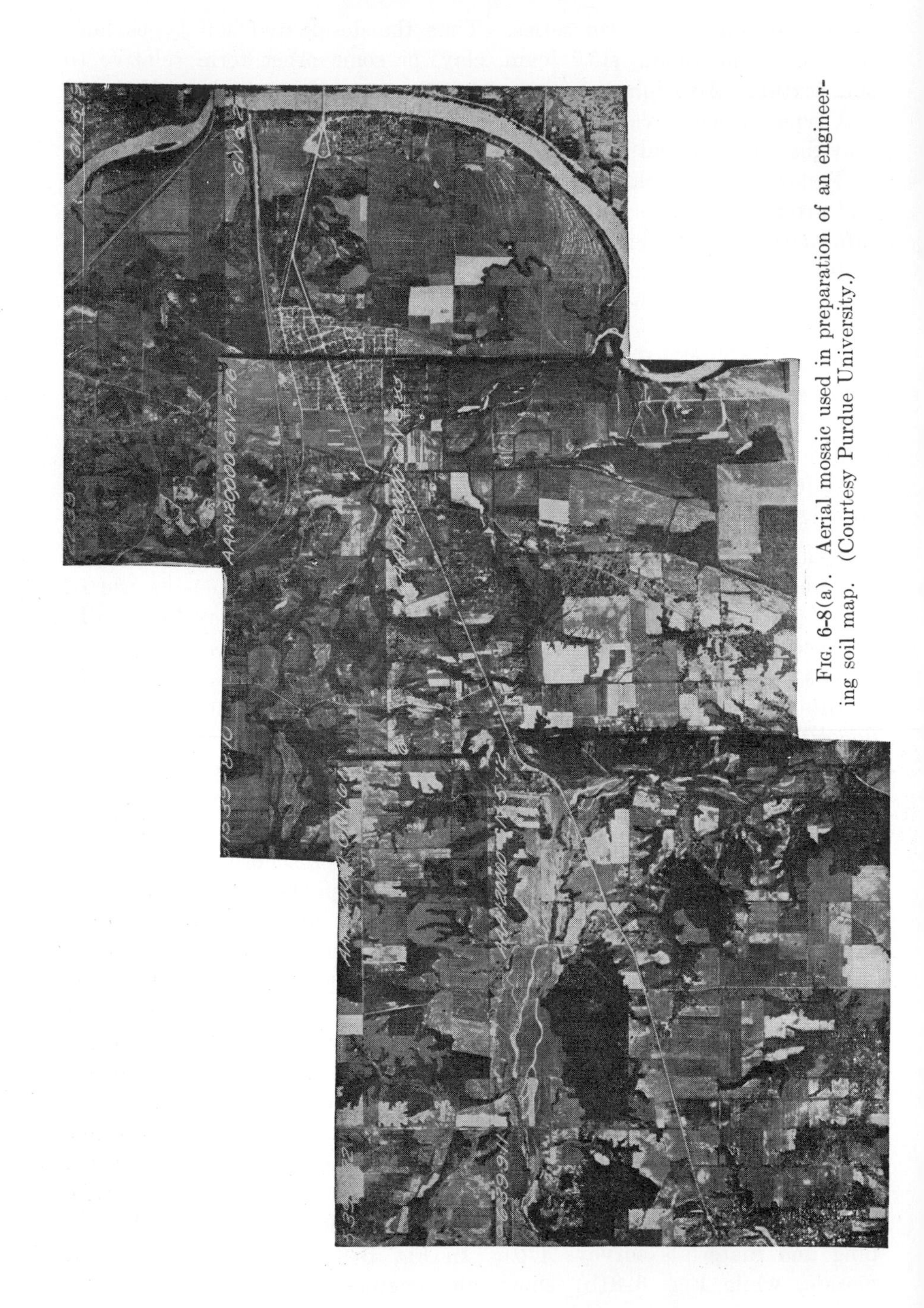

FIG. 6-8(a). Aerial mosaic used in preparation of an engineering soil map. (Courtesy Purdue University.)

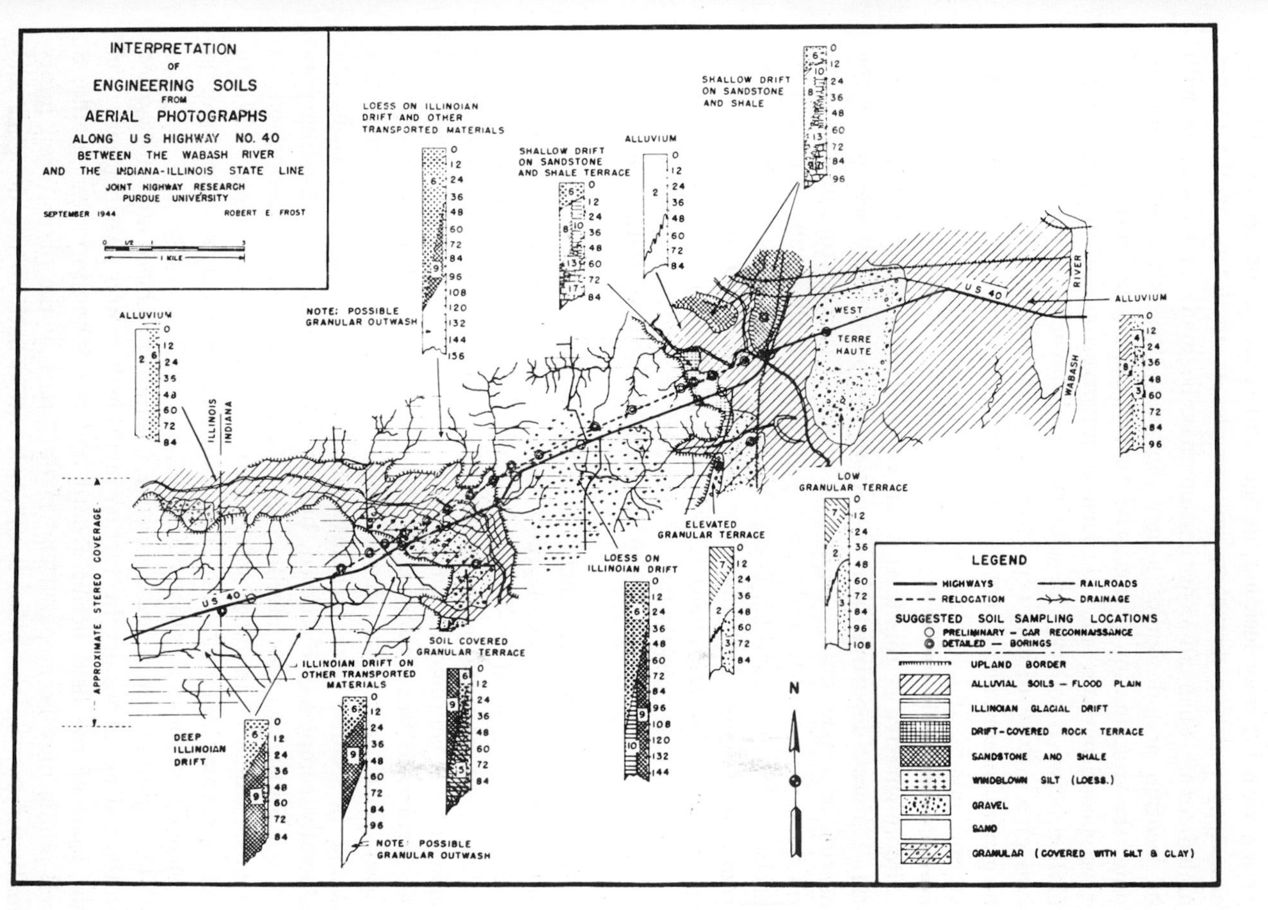

FIG. 6–8(b). Engineering soil map prepared from aerial photographs of Fig. 6–8(a). (Courtesy Purdue University.)

6–26. Detailed Soil Surveys. Detailed soil surveys required in connection with highway location, design, and construction are usually performed by securing disturbed soil samples to moderate depths over a limited area. Such an investigation may be made in connection with the location and design of a new highway, the relocation of an old route, the examination of material to be taken from borrow pits, and other special-purpose surveys, such as an investigation to establish the relation between soil type or condition and pavement behavior. Objectives of a survey of this type frequently include the determination of the general nature and extent (both horizontally and vertically) of existing soil layers, the location of the water table, and the selection of representative soil samples for classification and other testing purposes.

Representative of this kind of survey are those performed by several state highway departments as a part of the preliminary survey for the location of a new route. Equipment required for this sort of "subgrade soil survey" generally includes a soil auger, either hand operated or mechanically driven, which is capable of securing disturbed soil samples from 2 to 4 inches in diameter to depths of from 20 to 25 feet below the surface; necessary tools for the use of the sampler; cartons, jars, or sacks for the preservation of samples secured; and a notebook or set of forms for recording the information secured during the boring operations.

In general practice, borings are made at intervals not greater than 500 feet along the centerline (and at the sides of the road, as required) and to a depth of from 3 to 5 feet below the proposed grade line. In embankment sections, examination of surface soil conditions may be sufficient. A continuous record is kept of the soil types encountered from the surface to the bottom of the hole and the depth of each layer is noted. Each soil encountered is carefully described and a representative sample is secured from each layer or "stratum." Samples secured are generally completely disturbed, but are adequate for classification purposes. Pertinent groundwater and other drainage information is recorded, along with other information of special significance. Where soil conditions are quite variable, additional borings may be required at spacings of much less than 500 feet.

The samples which are secured are suitably labeled and forwarded to the testing laboratory for further identification and classification. On the basis of the field boring record and the laboratory test results, a soil profile may be prepared and made a part of the plans. Soil information shown on the plans is of great value both to the designer and the road builder. For example, the profile may show the existence of undesirable soils, such as peat and muck, which must be removed; it may show the existence of soils of low permeability which will require special drainage measures; it will indicate whether soils to be excavated will be suitable for use in embankment construction; and it may show the presence of rock layers requiring special construction methods. In Fig. 6–9 is shown a typical soil profile.

It should also be noted that this type of survey may serve as the basis for planning additional investigations which may require the securing of so-called "undisturbed samples," samples in which the soil structure, moisture content, and other important properties are carefully preserved. Samples of this type may then be obtained at selected locations for the measurement of shearing resistance, permeability, compressibility, or other important soil properties. No discussion of this type of sampling operation will be attempted here, and the student is referred to standard textbooks in soil mechanics for information regarding this subject.

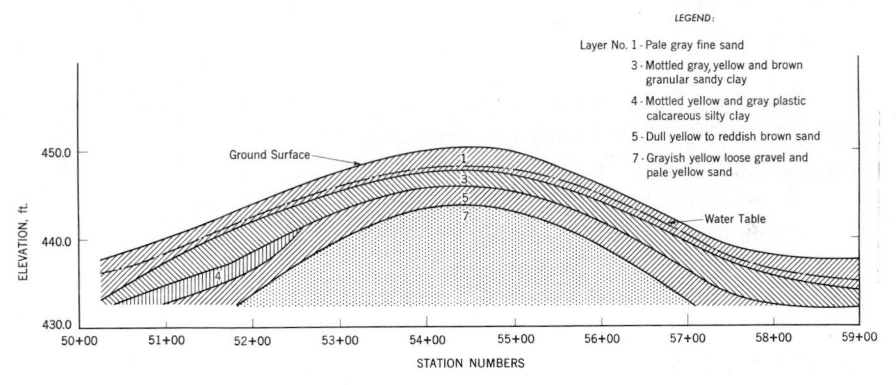

Fig. 6–9. Simplified plot of soil profile.

6–27. Geophysical Methods of Soil Exploration. Two geophysical methods of soil exploration are being used for highway purposes (31). These are the electrical resistivity and the seismic methods. Both are practical in the study of many highway construction problems. They are particularly useful in detecting and locating rock layers. Such knowledge is highly important, both from the standpoint of the designer and that of the contractor who must plan and carry out the construction work. They can be used in conjunction with borings to identify soil layers of varying characteristics, to locate the water table, and to find deposits of construction materials, such as gravel.

The seismic method of soil exploration consists of creating sound or vibration waves within the earth, usually by exploding small charges of dynamite 3 or 4 feet below the surface, and measuring the time of travel of these waves from their point of origin to each of several detectors placed at known distances from the source. The variations in mechanical energy transmitted to the detectors are converted into variations in electrical energy; in turn, these variations are used to detect light rays reflected from small mirrors that are parts of sensitive galvanometers, and these are recorded photographically on rapidly flowing film. Time data obtained from film records and the measured distances along the ground between the shot point and the detectors are plotted in the form of time-distance graphs. From these the depth and probable character of the various subsurface formations can be determined. Wave velocities range from approximately 600 feet per second in light, loose soils to approximately 18,000 to 20,000 feet per second in dense, solid rock. This wide range in wave velocities makes possible a determination of the general character of the materials encountered. By use of simple formulas, the average depth to the various substrata can be calculated. Fundamental principles of the seismic method are illustrated in Fig. 6–10.

In recent years, a simple method of seismic exploration has been developed and correlated with the "rippability" of the soil. A soil that is "rippable" is one that can be broken up enough by a ripper attached to a large crawler tractor to be loaded by scrapers (see Art. 10–14); this method of excavation often is cheaper than drilling and blasting hard soil or soft rock. In this seismic method, a blow by a hammer on a steel plate at ground level is substituted for the explosion of a small charge or blasting cap.

Experience has shown that many earth and rock materials can be identified by their reaction to the flow of a direct current of electricity. This is an action of electrolytic nature in which moisture in the soils and rocks, together with the dissolved impurities, gives to the several materials characteristic resistances to a current flow. These characteristic resistances (resistivities) may be used for locating and identifying subsurface formations. Fundamentals of the resistivity method are illustrated in Fig. 6–11. In the method commonly used, a prediction of the character of the subsurface material is attempted by measurements which indicate magnitude of resistance. Ordinary moist soils containing moderate amounts of clay or silt, with some electrolytic agent more or less active, have a comparatively low resistance. In contrast, sand, gravel, extremely dry, loose soils, and solid rock usually have high resistivity values. These classifications are too general to be useful, however, and it is necessary to calibrate the instrument with tests made over local materials which can be identified by exposed faces, test pits, core drilling records, and so on. Curves obtained later for unknown conditions may then be compared with those for known conditions. Predictions can then be made as to the materials lying below the surface.

The Michigan Department of State Highways makes extensive use of both types of geophysical equipment in soil surveys (*181*). It makes geophysical surveys on all proposed cut sections 12 feet or more in depth

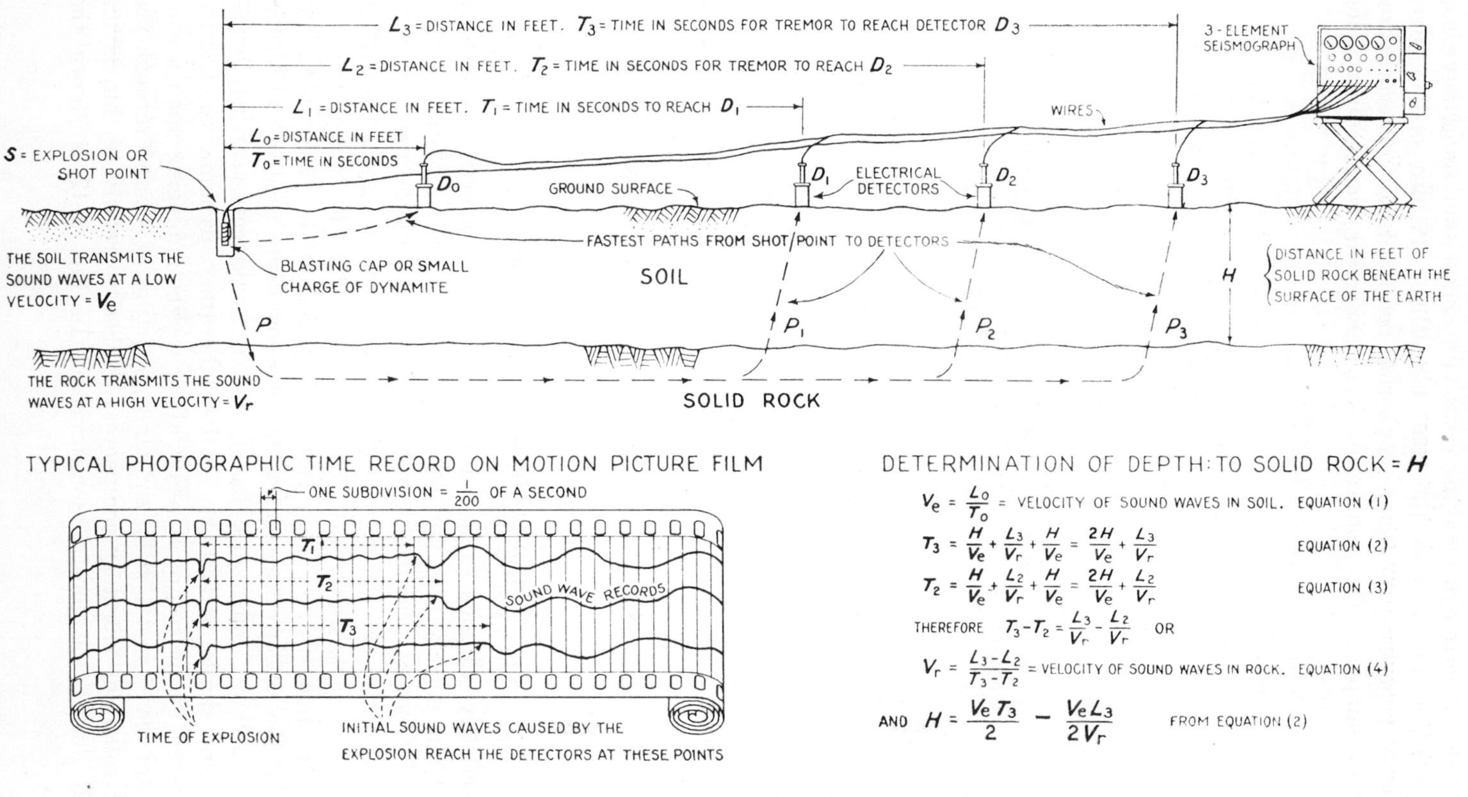

FIG. 6–10. Fundamental principles of seismic method of soil exploration.

and on many borrow pits, taking resistivity and seismic soundings along definite traverses, such as the road centerline. It also makes auger and/or wash borings, averaging one boring per 700 feet, and tests representative samples. The end result is a geological cross-section along

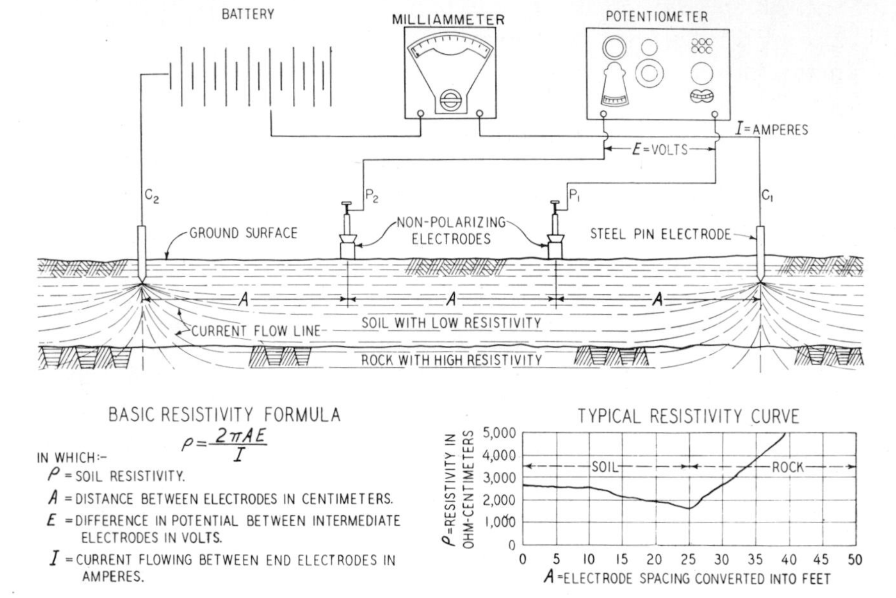

FIG. 6–11. Fundamental principles of electrical resistivity method of soil exploration.

a traverse based on the interpretation of all available geophysical, geological, pedological, boring, and laboratory test data.

Moisture-Density Relationships

Practically all soils exhibit a similar relationship between moisture content and density (dry unit weight) when subjected to dynamic compaction. That is, practically every soil has an optimum moisture content at which the soil attains maximum density under a given compactive effort. This fact, which was first stated by R. R. Proctor in a series of articles published in *Engineering News-Record* in 1933, forms the basis for the modern construction process commonly used in the formation of highway subgrades, bases, and embankments, earthen dams, levees, and similar structures. In the laboratory, dynamic compaction is achieved by the use of a freely falling weight impinging upon a confined soil mass; in the field, similar compaction is secured through the use of rollers or vibratory compactors applied to relatively thin layers of soil during the construction process. In Fig. 6–12 is shown the relationship between moisture and density for a typical soil subjected to dynamic compaction; the meaning of the terms "maximum density" and "optimum moisture" is clearly shown. This curve was obtained in the laboratory. In the field an attempt is usually made to maintain the soil at optimum moisture and bring the soil to maximum density or some specified percentage thereof.

In the laboratory, compaction is usually performed under what has come to be called the "Standard Proctor" or "Standard AASHO" method (T99). In the basic procedure used in this test (method A), the portion

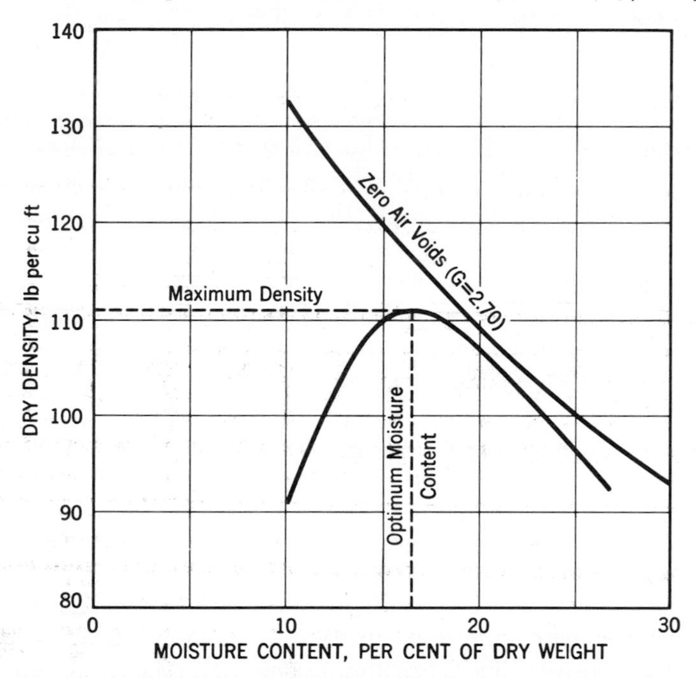

FIG. 6–12. Moisture-density relationship for a typical soil under dynamic compaction.

of the soil that passes a no. 4 sieve is placed in a mold 4 inches in diameter which has a volume of $\frac{1}{30}$ cubic foot. The soil is placed in three layers of about equal thickness, and each layer is subjected to 25 blows from a hammer weighing 5.5 pounds and having a striking face 2 inches in diameter, falling freely through a distance of 12 inches (12,375 foot-pounds per cubic foot). In recent years, heavier compaction equipment has come into widespread use, with the result that under certain conditions a greater compactive effort may be required in the laboratory. An increased compactive effort which is frequently used is known as the "Modified Proctor" or "Modified AASHO" compaction (T180), in which the soil is compacted in the Proctor mold, using 25 blows from a 10-pound hammer, dropping through a distance of 18 inches on each of five equal layers (56,250 foot-pounds per cubic foot). These are "impact" methods of compaction; some work has been done with "kneading-type" compactors, which simulate the action of a rubber-tired roller, and vibratory compaction of granular materials.

Regardless of the compactive effort used, the optimum moisture and maximum density are usually found in the laboratory by a series of determinations of wet unit weight and the corresponding moisture content. The series is started with the soil in a damp condition, at a moisture content a few per cent below optimum. The soil is compacted, the wet unit weight determined, and a sample selected from the interior of the compacted soil mass for the determination of moisture content. The soil is then broken up (or a new sample of the same soil selected); the moisture content is increased by 1 or 2 per cent, and the compaction process is repeated. This process is continued until a decrease is noted in the wet unit weight, or until the soil definitely becomes wet and shows an excess of moisture. After the moisture content determinations are completed, the dry unit weights may be calculated and plotted as shown in Fig. 6–12. The relationship between the dry unit weight, wet unit weight, and moisture content is as given in Eq. (6–2).

The behavior of the soil mass under dynamic (impact) compaction is of interest and may be explained in the following fashion. In the early stages of the compaction run, when the moisture content is considerably less than optimum, the soil does not contain sufficient moisture to flow readily under the blows of the hammer. As the moisture content is increased, the soil flows more readily under the "lubricating" effect of the additional water, and the soil particles move closer together, with a resulting increase in density. This effect is continued until the optimum moisture content is reached, at which point the soil has achieved its greatest density. The condition at optimum may be explained in terms of the degree of saturation (the ratio between the volume of water and the volume of voids); at this moisture content as large a degree of saturation has been attained as can be reached by compaction. This limiting degree of saturation is due to the existence of air entrapped in the void spaces and around the soil particles. Further increase in moisture tends to overfill the voids but does not materially decrease the air content. As a consequence, the soil particles are forced apart and the unit weight decreases. At moisture contents well beyond optimum, the soil mass tends to become somewhat elastic, the hammer "bounces," and excess moisture is generally forced out of the bottom of the mold.

Also shown on the plot of Fig. 6–12 is the "zero air voids curve" for the soil in question. Points on this curve represent the theoretical density which this soil would attain at each moisture content if all the void spaces were filled with water, i.e., if the soil were completely saturated or $S = 100$ per cent. The position of points on the curve may be calculated from the following equation if the specific gravity of the soil is known or may be assumed:

$$\text{Dry unit weight} = \frac{62.4G}{1 + \frac{wG}{100}} \text{ lb per cu ft} \qquad (6\text{–}4)$$

in which G = specific gravity of solids
w = moisture content, %

Although the densities shown on the zero air voids curve represent theoretical values which are practically unattainable, since all the air can never be exhausted from any soil system by dynamic compaction alone, the relative position of the moisture–density curve with respect to the zero air voids curve is of consequence, as the distance separating the two curves is indicative of the air voids remaining in the soil at various moisture contents. For example, for the soil shown, the sample at 17.8 per cent moisture has the same dry unit weight as it has at 15.1 per cent moisture, although the percentage of air voids at the lower moisture content (on the dry side of optimum) is much greater. Thus, if the soil is compacted at the lower moisture content and then subjected to inundation or "flooding," it might be expected that a larger percentage of air would be displaced by entering moisture and greater expansion would occur, than if the soil were compacted at the higher moisture content.

Another item of importance about the compaction process is that maximum density and optimum moisture are a function of the compactive effort. In general terms, an increase in compactive effort results in an increase in maximum density and a decrease in optimum moisture. Effect of variation in compactive effort is shown graphically in Fig. 6–13. This drawing also serves to show the similarity between laboratory and field compaction processes.

Inspection of the drawing indicates also that careful correlation must be established between the laboratory compaction procedure and the compactive effort being exerted by rollers in the field during construction if the laboratory curves are to be used as a basis for control of the construction process.

Obviously the nature of the soil itself has a great effect upon the density which is obtained under a given compactive effort. Moisture-density relationships for seven different soils are shown in Fig. 6–14. It is apparent that the different soils react somewhat differently to compaction at moisture contents which are somewhat less than optimum. Moisture content is less critical for the heavy clay than for the slightly

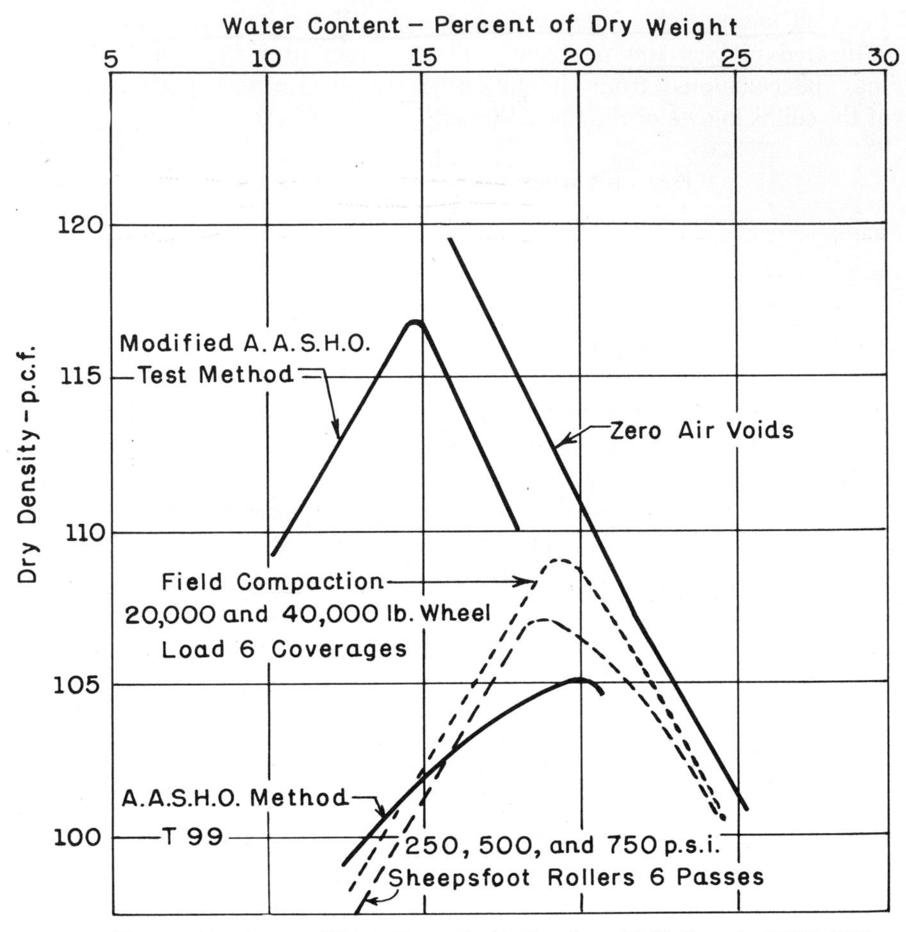

Fig. 6–13. Results of field and laboratory compaction on silty clay soil. (After A. W. Johnson.)

plastic sandy and silty soils. Heavy clays may be compacted through a relatively wide range of moisture contents below optimum with comparatively small changes in dry density. On the other hand, the granular soils which have better grading and higher densities under the same compactive effort react sharply to small changes in moisture content, with significant changes in dry density. The relatively clean, poorly graded, nonplastic sands also are relatively insensitive to changes in moisture.

Many other factors influence the results of compaction tests done in the laboratory, including the amount of coarse aggregate contained in the soil. These factors are discussed in detail in reference (*22*).

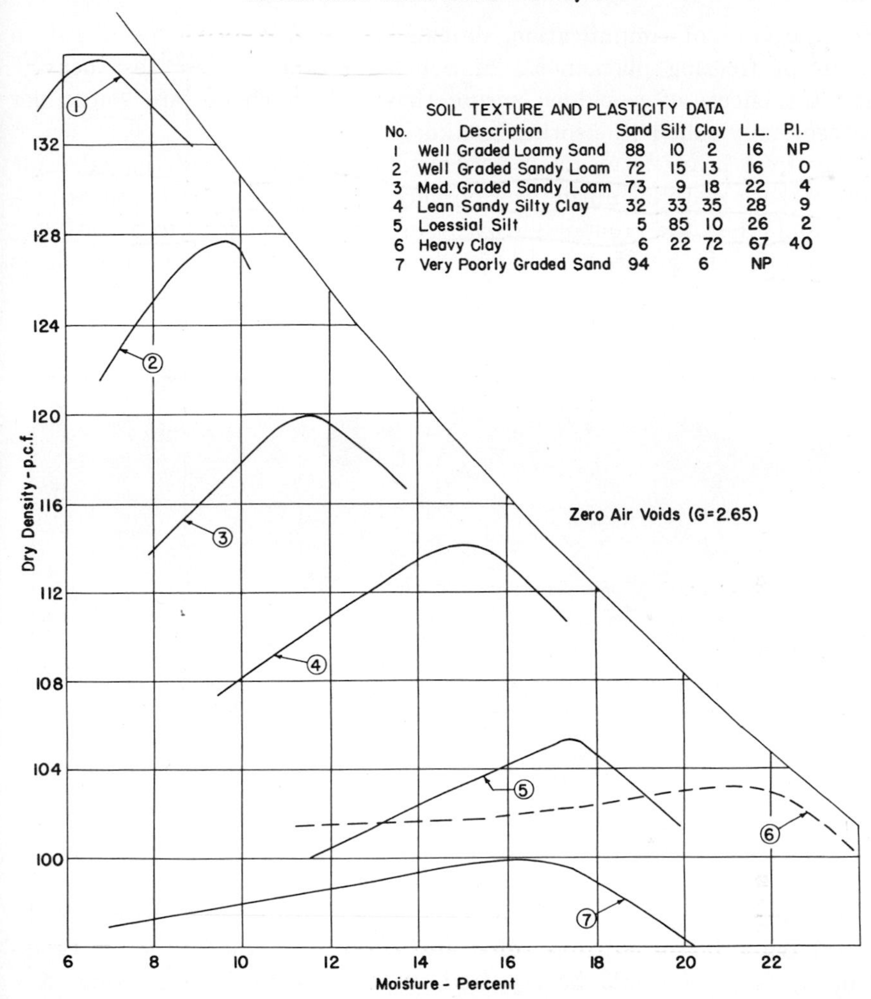

FIG. 6–14. Moisture-density relationships for seven soils compacted according to the Standard AASHO compaction procedure. (Courtesy Highway Research Board.)

Frost Action in Soils

One of the special soil problems of interest in many areas of the United States is that of damage caused to roads and streets by the freezing and thawing of subgrades and bases during the winter and spring seasons. This phenomenon is discussed here for purposes of convenience and as a background for the discussion of certain design, construction, and maintenance considerations in later chapters.

Under certain conditions, very severe damage to subgrades, bases, and pavements, both rigid and flexible, may result from frost action.

For purposes of simplification, damaging action may be visualized in terms of freezing phenomena, of which an example is "frost heave," and the effects of a sudden spring thaw, which effects are sometimes categorically called "the spring breakup."

By frost heave is meant the distortion or expansion of the subgrade soil or base material during the period when freezing temperatures prevail and when the ground is frozen (or partially frozen) to a considerable depth. In Fig. 6–15 is shown typical damage resulting from frost

Fig. 6–15. Damaging effects of frost action on a rural road in a northern state. (Courtesy Minnesota Dept. of Highways.)

action in a northern state. The phenomena of frost action are generally associated with the existence of a shallow water table and a frost-susceptible soil. It would be expected that excess moisture existing in a base or subgrade would freeze during a severe winter season. It might also be expected that the soil mass would expand, since water expands as it freezes. However, the magnitude of the severe frost heaves which occur regularly in some states cannot be explained on the basis of the freezing of existing pore water alone. Investigations have shown that, as the water in the upper soil layers freezes, ice crystals are formed, and water may be drawn from a free water surface into the zone where freezing temperatures prevail. This water then freezes, additional water may be drawn to this level, and this process continues until layers of ice or "ice lenses" of considerable thickness may be formed. The volume increase brought about by the formation of these layers of ice is the cause of frost heaving. In Fig. 6–16 are shown ice lenses which may be formed in a soil as a result of frost action.

It will be noted that a certain combination of circumstances (in addition to freezing temperatures) must usually exist before severe frost heave can occur. As previously mentioned, these include the existence of a shallow water table and a frost-susceptible soil. If the water table is not close to the surface, water is generally not available to the shallow soil layers, and the heave that can occur is limited to that caused by the expansion of freezing water in the voids of the soil. Since the change in volume from water to ice is only about 9 per cent, this expansion would generally not be sufficient to cause severe frost damage. However, it might be possible, in somewhat unusual circumstances, that sufficient free water would be available from seepage flows or some other cause

FIG. 6–16. Ice layers accumulated in a soil under freezing conditions. (Courtesy Minnesota Dept. of Highways.)

for frost heave to occur in a location where the water table is located well beneath the ground surface.

The requirement relative to capillarity seems to limit detrimental frost action to soils containing a considerable percentage of fine particles such as silt or clay. Experience has shown that frost action is generally most severe in silt soils, as the upward movement of water in clay soils is usually so slow that the amount of water made available to the freezing layer is quite small. Soils which contain less than 3 per cent by weight of material finer than 0.02 mm. in diameter are not frost-susceptible. Soils which contain more than 3 per cent of material of this size generally are susceptible to frost action, except that uniform fine sands which contain up to 10 per cent of material finer than 0.02 mm. generally are not susceptible to frost action.

Experience has also shown that frost heave is most severe when temperatures only slightly below freezing prevail for a considerable period of time so that frost penetration comes about somewhat gradually.

During the period of freezing temperatures, under the conditions described above, a considerable amount of water in the form of ice may accumulate in the shallow soil layers beneath a road surface. If a sudden rise in temperature or "thaw" occurs, as it frequently does in the spring of the year in many localities, this accumulated ice may melt rapidly. Generally speaking, it would be expected that the ice in the upper soil layers would melt first, while deeper layers would remain frozen for a longer period of time. As a consequence, excess moisture may be trapped beneath the pavement and a sharp reduction in shearing strength may result. If this happens, the pavement and base may then fail under wheel loads; this type of failure is associated with the term "spring breakup." Damage to pavements and bases incurred during this period results in severe economic losses in many states.

Certain preventive measures are commonly employed to eliminate or minimize frost damage. Probably the most common solution to the problem is to remove soils subject to frost action and replace them with suitable granular backfill to the depth of the frost line. It is not always necessary to treat unsuitable soils to the full depth of frost penetration. For example, experience in Minnesota has shown that treatments to a depth of about 3 feet are sufficient to prevent any breakup in the spring, although some slight heave may still occur. Another solution may be effected by the installation of drainage facilities in areas where frost action is likely to be detrimental, with the objective of lowering the water table or intercepting seepage flows which may cause excessive amounts of water to be accumulated in certain areas. Insulation or blanket courses may be employed as a construction measure, while some state highway departments regulate truck traffic in the spring months to prevent damage caused by heavy wheel loads on roads known to have been weakened by frost action.

PROBLEMS

6–1. The specific gravity of a soil is determined in the laboratory, following the procedure given under the standard method of test designated as No. T100 by the American Association of State Highway Officials. The following information was secured in the laboratory during the conduct of the test:

Weight of pycnometer filled with water at 25 C = 201.03 grams
Weight of oven-dried soil = 10.51 grams
Weight of pycnometer filled with water and soil at 25 C = 207.65 grams

Calculate the specific gravity of this soil.

6–2. Given a soil which has a wet unit weight of 142.4 pounds per cubic foot, dry unit weight of 124.6 pounds per cubic foot, and specific gravity of 2.65. Find the moisture content, degree of saturation, void ratio, and porosity.

6–3. A number of trials made in the laboratory, using the mechanical device of Fig. 6–3 to determine the liquid limit of a soil, have resulted in the following data. Plot the flow curve and determine the liquid limit of this soil.

Number of Blows	Moisture Content (Per Cent)
40	46.4
28	47.0
20	47.3
10	48.1

6–4. Ascertain the practice of your state highway department with regard to soil surveys and mapping, and prepare a brief report outlining the methods used in your state in this phase of highway work.

6–5. From current literature, determine and describe how seismic methods are being used to determine the "rippability" of a soil deposit. Can you find a specific example in which a contractor used this method to save money in ripping a material that he might otherwise have drilled and blasted? (NOTE: The Caterpillar Tractor Co., Peoria, Illinois, has developed a wealth of information on this subject.)

6–6. Given the following information, calculate the Group Index, and classify each of these soils into the proper subdivision of the AASHO classification system.

Sample No.	SIEVE ANALYSIS			CHARACTERISTICS OF SOIL BINDER	
	Per Cent Passing				
	#10	#40	#200	Liquid Limit	Plasticity Index
1	100.0	97.5	63.2	62.3	12.7
2	100.0	73.1	2.3	20.1	NP
1	100.0	46.2	11.9	16.9	NP
4	77.2	37.1	28.2	42.1	6.8
5	100.0	100.0	87.6	60.2	31.7
6	100.0	100.0	93.7	54.2	33.6
7	100.0	61.2	19.1	24.8	5.6
8	100.0	96.3	85.9	33.7	8.9
9	100.0	100.0	83.1	41.2	8.6
10	100.0	100.0	97.8	32.7	11.3
11	100.0	73.7	39.1	62.3	12.7
12	37.1	21.1	9.3	13.6	NP

6–7. Classify each of the soils of Problem 6–6 under the Unified Soil Classification System. The following additional data are available for certain of the samples.

Sample No. 2	$C_u = 3.5$	
Sample No. 4	Passing No. 4 sieve	92.6%
	Passing ⅜-in. sieve	100.0%
Sample No. 12	Passing No. 4 sieve	68.1%
	Passing ½-in. sieve	86.4%
	Passing 1-in. sieve	100.0%
	$C_u = 5$	
	$C_g = 2$	

6–8. Given the following information relative to a compaction test performed in the laboratory, using the Standard Proctor Compaction procedure:

Weight of Mold = 2456 grams
Volume of Mold = 1/30 cubic foot

Trial No.	Weight of Compacted Soil Plus Mold (Grams)	Moisture Content (Per Cent)
1	4136	8.5
2	4205	10.1
3	4308	12.0
4	4408	14.0
5	4398	16.4
6	4354	18.3

Draw the moisture-density curve and determine optimum moisture and maximum density for this soil.

6–9. Assuming that the soil of Problem 6–8 has a specific gravity of 2.68, make the necessary computations and plot the "zero air voids curve" on the drawing prepared in Problem 6–8.

CHAPTER 7

DRAINAGE AND DRAINAGE STRUCTURES

One of the most important aspects of the location and design of rural highways and city streets is the necessity for providing adequate drainage. Adequate and economical drainage is absolutely essential for the protection of the investment made in a highway structure and for safeguarding the lives of the persons who use it.

Highway drainage may be generally defined as the process of controlling and removing excess surface and underground water encountered within the limits of the right-of-way and adjacent territory. The flow of surface water with which the highway engineer is concerned generally results from precipitation in the form of rain or snow or melting ice. A portion of the surface water enters or "percolates" into the soil, while the remainder stays on the surface of the ground and must be carried on, beside, beneath, or away from the traveled way. Artificial drainage resulting from irrigation, street-cleaning, and similar operations may also be of consequence in some cases. In certain instances the control of underground water (ground water) may be important, as in the case of an underground flow encountered in a highway cut or in a location where the water table lies close to the surface of the ground.

Measures taken to control the flow of surface water are generally termed "surface drainage," while those dealing with ground water in its various forms are called "subsurface drainage" or, more simply, "subdrainage."

The solution of drainage problems should not be regarded as a separate element of highway or street design. Rather, considerations relative to drainage must accompany every step in location and design, so that the final design and resulting construction operations will provide for adequate drainage at reasonable cost.

As is the case with many other branches of highway engineering, knowledge relative to drainage and drainage structures has advanced rapidly in the last few years. All the specialized knowledge and experience gained by a competent drainage engineer obviously cannot be covered in a textbook of this type. In the following pages, therefore, are presented the elements of highway and street drainage. Surface drainage in essentially rural areas (as distinguished from urban areas) is discussed in considerable detail; accompanying this is a discussion of measures for the prevention of erosion of shoulders, side slopes, and side ditches. Considerable space is devoted to the location, design, and construction of culverts. Attention is briefly given to highway bridges. Material is also presented relative to subdrainage, and the chapter concludes with a brief discussion of drainage in municipal areas.

Surface Drainage

The portions of the highway structure which provide for surface drainage in rural locations include the roadway crown, shoulder and side slopes, longitudinal ditches (channels), culverts, and bridges. Divided highways in rural areas also have inlets and storm drains (underground pipes) in the median strip to handle a portion of the surface flow.

7–1. Roadway Crowns. As has been presented in an earlier chapter, roadway surfaces are normally crowned (or sloped, as in the case of a superelevated section) to facilitate the removal of surface water from the wearing surface. The amount of crown varies with the type of surface, being generally small for impervious surfaces such as portland cement concrete and large for relatively pervious surfaces such as gravel or earth. If the road is properly built and maintained to design standards, water should be quickly removed from the surface and become a part of the surface flow. A portion of the precipitation which falls on a gravel or earth surface will, of course, enter the soil. Where a bituminous or concrete pavement is used, water may also enter the base or subgrade through cracks or joints. This amount is generally quite small but it may be of considerable importance if the base is composed of relatively impervious material or has been built on an impervious subgrade.

7–2. Shoulder Slopes. Shoulders which are provided along rural roadways are sloped away from the surface. Precipitation which occurs on the shoulder area thus largely flows to the side ditches or the median swale, as does that which falls on the roadway proper. Shoulder slopes vary in magnitude from about ½ to 1 inch per foot, depending on whether the shoulder is surfaced or unsurfaced. In cases where the shoulders are of natural earth or similar material, a portion of the surface water enters the soil in this area and contributes to the underground water supply.

7–3. Side Slopes and Side Ditches. Open side ditches are generally provided in cut sections in highway locations in rural areas to provide for surface drainage. Side ditches may also be constructed along embankment sections when needed to supplement natural drainage channels. Both flat-bottomed and **V**-section ditches are used, with preference being given to the former type, with slope changes in the ditch section being rounded to improve appearance and prevent erosion. In either case, side slopes are made as flat as possible consistent with drainage requirements and limiting widths of right of way. Deep, narrow side ditches are to be avoided whenever possible because of the increased hazards presented by such construction. Where they must be used, adequate provision must be made for safeguarding traffic through the use of guardrails and similar devices. In Chapter 5, figures have been given relative to recommended side and ditch slopes, and typical ditch sections have been incorporated in drawings of typical right-of-way cross-sections.

Flow of water in side ditches, of course, occurs in a direction which is generally parallel to the roadway centerline. Grades used in open ditches may also be roughly the same as those which are used on the highway centerline; on the other hand, flat roadway grades and steeper ditch grades in the same location are very frequently used. In very flat country, ditch grades as low as 0.1 or 0.2 per cent may be used, while in rolling or mountainous terrain the maximum grade may be dictated only by the necessity for preventing erosion. The relation in elevation between the ditch and the roadway surface is important and may be determined only by a study of local conditions. Even under very favorable conditions, the vertical distance between the bottom of the ditch and the roadway surface should be at least 2 feet, and it may have to be several times that figure, in either cut or embankment sections.

Side ditches, then, provide open channels for the removal of surface water from within the limits of the highway right-of-way. In certain circumstances, areas adjacent to the right-of-way may also contribute to the flow. The water must be carried to an outlet, either in the form of a natural or an artificial drainage channel. The ditch must be hydraulically capable of handling the anticipated flow of surface water in such fashion that the roadway structure is not endangered or the safety of the motorist threatened. See Art. 7–20 for the hydraulic design of side ditches and other open channels.

7–4. Prevention of Erosion of Shoulders and Side Slopes. The flow of surface water adjacent to highways is frequently accompanied by detrimental soil erosion which may result in the destruction of productive soils, the creation of areas of unsightly appearance, and the clogging of ditches and drainage structures. Erosion may also endanger the stability of side slopes in embankment and cut sections. A number of construction measures designed to prevent or minimize erosion are discussed in the paragraphs immediately following.

Any decision which is made regarding the installation of any of the special features discussed hereafter with the objective of preventing erosion on shoulders and side slopes must be based largely on judgment and knowledge of local conditions. Many factors enter into such a decision, including such items as intensity and duration of rainfall, soil types and condition, magnitude of slopes, climatic conditions, and many others.

7–5. Lip Curbs. On rural highways constructed of portland cement concrete, lip or "roll" curbs are sometimes used to prevent the water

Fig. 7–1. Flume outlet. (Courtesy Armco Drainage & Metal Products, Inc.)

which flows from the crowned surface from spilling over and eroding shoulders and side slopes. Curbs are generally not more than 4 inches in height and are of "rolled" cross-section in order to minimize damage to vehicles which are forced up over the curb onto the shoulder. The curb is "broken" at intervals along its length—the interval between breaks being generally 300 to 500 feet. At each of these openings water is channelized into a spillway, paved ditch, or flume, and carried into the side ditch or pipe drain. Inlets are also provided at the low points of vertical curves. In Fig. 7–1 is shown a flume outlet of the type which may be used in connection with a lip curb or in other locations to channel water down a steep side slope.

Lip curbs once were widely used by some organizations. They are now seldom used on new construction. Their use presents some hazard to the driver if the shoulder is not maintained at the level of the top of the curb or if the water is not removed quickly at the outlets.

7–6. Drainage Dikes and Gutters. Drainage dikes may be used to confine surface water in a fashion similar to that of lip curbs. The dikes are low, being from 4 to 8 inches in height, and are constructed along outside shoulder lines, the shoulders generally being surfaced. Simple earth dikes are used, as are dikes made wholly or partially of a bituminous plant mixture. Shallow earth or bituminous-surfaced gutters have also been used on embankment sections in rural locations. Their function is similar to that of a drainage dike.

7–7. Turf Culture. One of the easiest and most effective ways of reducing erosion on side slopes and in side ditches is through the cultivation and development of a firm turf. Every effort is usually made in areas subject to erosion to encourage the growth of native grasses on exposed slopes. Formation of a firm turf may be accomplished by seeding, sprigging, or sodding the slope with suitable native grasses. In recent years, widespread use has been made of mats and linings of various kinds to encourage turf formation. The problems of planting and cultivating these grasses will not be discussed here. Suffice it to say that it is generally possible to establish adequate growths of grass which serve greatly to reduce the erosive effects of a flowing stream or sheet of water in areas of moderate rainfall or on moderate slopes.

7–8. Intercepting Ditches. Shallow ditches may be placed at the top of outside cut slopes to intercept surface water and thus prevent erosion. Such ditches, which are usually quite shallow and narrow, are termed "intercepting ditches." Water conveyed in such a ditch is collected at suitable intervals along the top of the slope and discharged down the slope in paved or otherwise protected spillways into the longitudinal ditch, or spread out over the adjacent land.

7–9. Slope Protection Under Severe Erosion Conditions. Highways are often subjected to very severe erosive action in which the measures thus far discussed provide only a partial solution of the problem. One example of this situation might lie in a highway location in a mountainous region subjected to heavy rain and snowfalls. In such areas, a large portion of the location will probably consist of side-hill sections, with the side slopes being made as steep as possible in the interests of economy. Slopes of 1.5 to 1, 1 to 1, or even steeper, are not unusual, and are, of course, very susceptible to erosion. The slopes may be protected, for example, by the use of riprap or hand-placed rock.

Still another situation in which a highway embankment may be subjected to very severe erosion is that existing when the location parallels a large or rapidly flowing stream. Such a location may involve a change in alignment of the stream, with resultant scour of the embankment slope. The slope may be protected by riprap, a paved revetment, steel sheet or timber piling, cribbing, or any one of a number of other protective devices.

7–10. Culverts and Bridges. Culverts and bridges are visualized as composing the "cross-drainage" system of a highway in a rural location. Their general purpose is to transmit water flowing in natural streams or collected on the high side of the right-of-way from one side of the highway to the other. Additional information relative to culverts is contained in Arts. 7–27 through 7–34; bridges, Arts. 7–35 through 7–41.

7–11. Storm Drains. Where space is restricted, in median swales, in urban areas, or where the natural slope of the ground is unsuited for drainage by open channels, storm drains are provided for the disposal of surface water. Storm drains and appurtenant structures are discussed when consideration is given to the drainage of city streets, Arts. 7–46 through 7–49.

7–12. Design of Surface Drainage Systems. The surface drainage system should be designed to remove all surface water effectively and economically from the roadway area and intercept and dispose of surface water from adjoining areas. Subdrainage systems may be used to supplement surface drainage systems, or to solve specific problems associated with the flow of underground water.

Design procedures logically may be divided into three major phases: an estimate of the quantity of water which may be expected to reach any element of the system; the hydraulic design of each element of the system; and the comparison of alternative drainage systems, alternative materials, and other variables in order to select the most economical system which can be devised. In the third phase, attention must be given to selecting the system which has the lowest annual cost, with all variables taken into consideration.

7–13. Streamflow Records. When a drainage structure is to handle the flow of an existing stream, as in the case of some culverts and most bridges, peak flow for use in hydraulic design may be based upon available records. Flood-frequency studies have been published for a number of states. Some of these studies have been made by the U.S. Geological Survey in cooperation with the various state highway departments, and studies of this type are continuing.

Flood-frequency studies of this type are based on statistical analysis of streamflow records. Usually, they indicate the mean annual flood as a function of the size of the drainage area for each hydrologic region within the state. Factors for computing floods of any frequency (within limits) in terms of the mean annual flood are given on a separate graph. The choice of flood frequency to be used for design is a matter for careful evaluation—frequencies of 10 to 25 years are common for rural highways, and frequencies as low as two years and as high as 50 years have been used. The regional flood curves usually cover drainage areas from about 100 to 2000 square miles. Separate flood curves may be given for the larger rivers, which cross regional boundaries and seldom conform to the trend given by the smaller rivers.

Basic considerations in estimating flood frequencies are contained in Reference *158*.

Many of the larger rivers are controlled by dams and other installations. Frequency and magnitude of floods in such cases depend on operation of the system and must be learned from the operating agency.

Where suitable streamflow records are not available, useful information may be gained from observations of existing structures and the natural stream. Drainage installations above and below the proposed location may be studied, and a design based upon those which have given satisfactory service on other portions of the same stream. Lacking this information, an examination of the natural channel may be made, including the evidences left by flood crests which have occurred in the past, and an estimate made of the quantity of water which has been carried by the stream during flood periods. Measurements may be made and values assigned to the slope, area, wetted perimeter, and roughness coefficient of the flood channel, and the quantity may be estimated by the Chezy formula. The Chezy formula is of the form,

$$Q = Ca\sqrt{rS} \tag{7-1}$$

where Q = quantity of flow, cubic feet per second

C = roughness coefficient, varying from 30 to 80, depending on the condition and nature of the channel

a = area of the flow cross-section, square feet

S = slope of the channel, feet per foot

r = hydraulic radius, feet, $= a/p$, where p = wetted perimeter, the length of the boundary of the cross-section of the flow channel in contact with the water

7–14. Hydrologic Criteria. In cases where streamflow records and/or examination of the existing stream are inadequate, or where the drainage facility does not carry water which flows in a presently existing stream, the estimate of quantity must be made by some other method. The quantity estimate is important, since its accuracy determines the adequacy of the drainage structure designed.

Basic assumptions which must be made in order to obtain even a reasonably accurate estimate of quantity are the following: the rainfall intensity in inches per hour for the design storm; coefficient of runoff or rate of infiltration; and the time of concentration for critical points in the drainage system.

7–15. Rainfall Intensity. Proper design of a highway drainage system is accomplished by consideration of severe storms which occur at intervals and during which the intensity of rainfall and runoff of surface water is far greater than at other periods. Rainfall intensity during the design storm is a function of the occurrence, duration, and intensity.

Selection of the frequency of occurrence of the design storm is largely a matter of experience and judgment, although departmental policy may establish the interval to be used for a given situation. As previously mentioned, frequency of occurrence is a statistical matter. For example, if the system is designed upon a frequency of occurrence of 25 years, the statistical assumption is that the system will accommodate runoff from the most severe storm to occur once in 25 years. It is apparent that selection of a frequency of 100 years, instead of 25, would mean designing for a more severe storm and, in general, a more costly system. Conversely, if the frequency is 10 years, the intensity of the design storm will be less and, in general, a less costly drainage system will result, but economic losses from use of the shorter time might offset the savings in construction costs.

On the Interstate System, the Bureau of Public Roads requires that all drainage facilities other than bridges and culverts be designed for storms with a frequency at least as great as 10 years, except that a 50-year frequency is used for underpasses or other depressed roadways where ponded water can be removed only by the storm drainage.

The intensity of rainfall for a particular frequency of occurrence varies greatly with the duration of rainfall. The average rate for a short time—such as five minutes—is much greater than for a longer period, such as an hour. In the design of many highway drainage systems, the duration chosen corresponds to the time of concentration (Art. 7–17).

An accurate estimation of the probable intensity, frequency, and duration of rainfall in a particular location can be made only if sufficient data have been collected over a period of time. If such information is available, standard curves may be developed to express rainfall-intensity relationships with an accuracy sufficient for drainage problems. A number of state highway departments have developed such curves for use in specific areas. Various types of information available from the U.S. Weather Bureau may be useful to the highway designer. One publication is a rainfall-frequency atlas (164), which contains maps of rainfall frequency for 30-minute and 1-, 2-, 3-, 6-, 12-, and 24-hour durations for periods of 1, 2, 5, 10, 25, 50, and 100 years. Another Weather Bureau publication (*165*) contains rainfall-duration-frequency curves for selected stations in the United States and Puerto Rico. Lacking such detailed relationships, recourse may be had to methods based on average relationships, such as the one explained in Art. 7–18.

7–16. Surface Runoff. When rain falls on a pervious surface, part of it passes into the soil and the remainder disappears over a period of time either by evaporation or by runoff or by both. In design of highway drainage systems, the amount of water lost by evaporation is negligible; thus, drainage must be provided for all rainfall that does not infiltrate the soil or is not stored temporarily in surface depressions within the drainage area.

The rate at which water infiltrates the soil is dependent upon the following factors: type and gradation of the soil; soil covers; moisture content of the soil; amount of organic material in the soil; temperature of the air, soil, and water; and the presence or absence of impervious layers near the surface. Rates of infiltration on bare soil are less than on turfed surfaces. Frozen soil is impervious, and rain infiltrates very little until the frozen layer thaws. The rate of infiltration is assumed to be constant during any specific design storm.

TABLE 7–1

COEFFICIENTS OF RUNOFF TO BE USED IN THE RATIONAL FORMULA

Type of Drainage Area	Coefficients of Runoff, C
Concrete and bituminous pavements	0.70–0.95
Gravel or macadam surfaces	0.40–0.70
Impervious soil	0.40–0.65
Impervious soils, with turf*	0.30–0.55
Slightly pervious soils*	0.15–0.40
Pervious soils*	0.05–0.10
Wooded areas (depending on slope and cover)	0.05–0.20

* For slopes from 1 to 2 per cent.

The rate at which runoff occurs depends upon the nature, degree of saturation, and slope of the surface. The rate of runoff is greater on smooth surfaces and initially slower where vegetation is present. On pavements and compacted surfaces runoff occurs at a high rate, which varies with the slope and character of the surface at each point. For use in connection with the design of a drainage system for a particular area, these variables are considered and a coefficient of runoff selected. Values of the coefficient of runoff for use in one method of estimating the quantity of flow (the rational method) are given in Table 7–1. If the drainage area being considered is composed of several types of surfaces, then a coefficient may be chosen for each surface, and the coefficient for the entire area computed as a weighted average of the individual areas.

7–17. Time of Concentration. After selecting the design storm frequency, computations are made to determine the duration of the rainfall

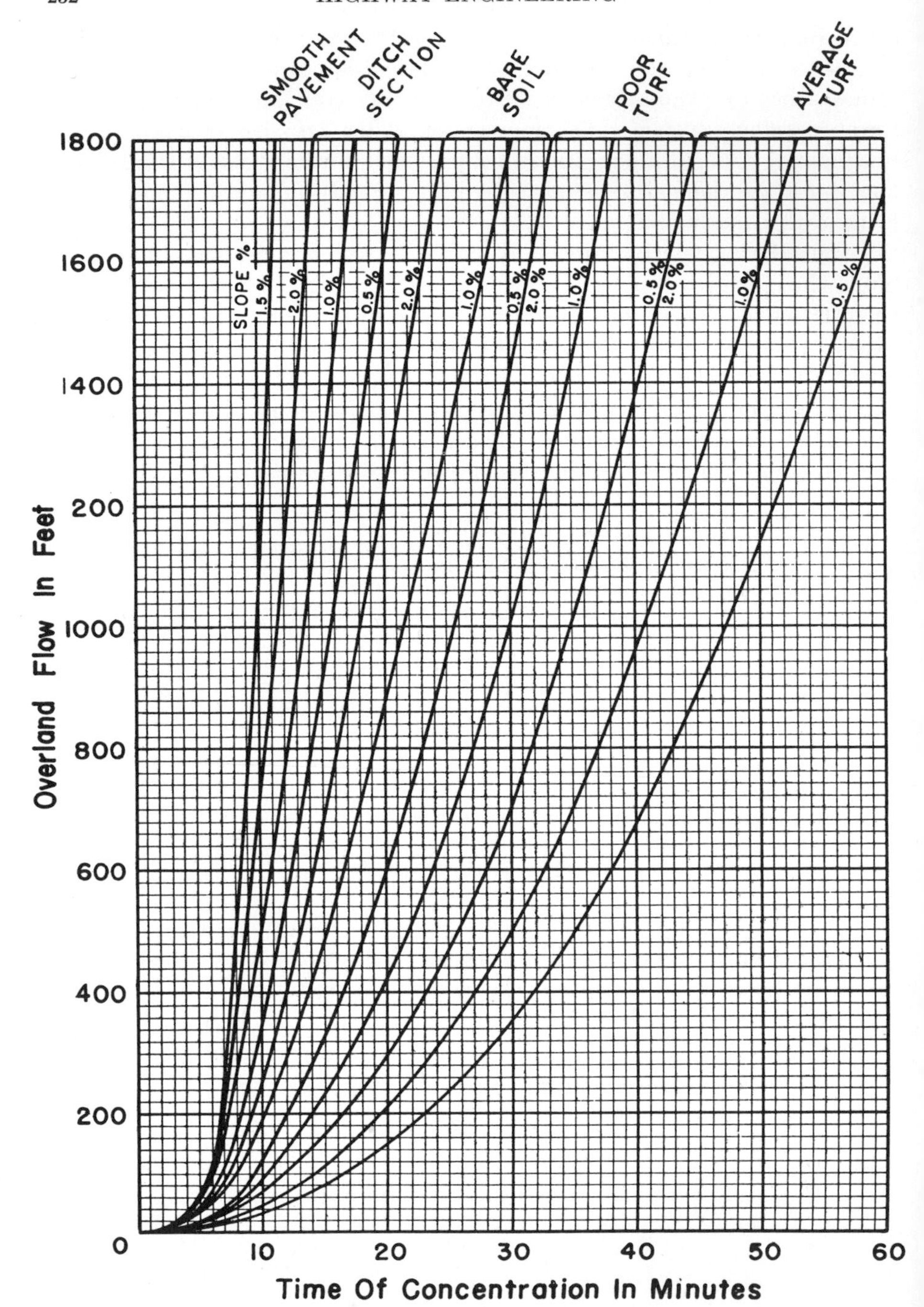

FIG. 7-2. Curves to estimate the time of overland flow.

that produces the maximum rate of runoff. The duration of rainfall required to produce the maximum rate of runoff is known as "the time of concentration." The time of concentration usually consists of the time of overland flow plus the time of flow in the drainage system. The time of overland flow is the time required for a particle of water to flow from the most remote point in any section of the drainage area being considered to the point where it enters the drainage system. To this must be added the time of flow in the drainage system, from the intake to the point being considered. The time of overland flow varies with the slope, type of surface, length, and other factors. Fig. 7–2 gives reasonable values for the time of overland flow for various conditions of cover, slope, and length. When the particular drainage area consists of several types of surfaces, the time of overland flow must be determined by adding together the respective times computed for flow over the lengths of the various surfaces from the most remote point to the inlet.

7–18. Bureau of Public Roads Chart. The Bureau of Public Roads developed Fig. 7–3 for use in determining approximate rainfall intensity-duration-frequency data (*146*). This drawing was adapted from a map in the Weather Bureau atlas and pertains to 30-minute rainfall intensity for a 2-year recurrence period.

The 2-year rainfall intensity for other durations is gotten by multiplying the 30-minute intensity for the project location by these factors: 2.22 (5 minutes); 1.71 (10 minutes); 1.44 (15 minutes); 1.25 (20 minutes); 0.8 (40 minutes); 0.6 (60 minutes); 0.5 (90 minutes); and 0.4 (120 minutes).

To convert the chart values to other intervals of recurrence, multiply by these factors: 0.75 (1 year); 1.3 (5 years); 1.6 (10 years); 1.9 (25 years); and 2.2 (50 years).

7–19. The Rational Method. One of the most common methods of estimating runoff from a drainage area is the rational method. Its popularity is due to the fact that it combines engineering judgment with calculations made from analysis, measurement, or estimation. The method is based upon the direct relationship between rainfall and runoff, and is expressed by the equation.

$$Q = CIA \tag{7–2}$$

where Q = runoff, in cubic feet per second

C = a coefficient representing the ratio of runoff to rainfall. Typical values of C are given in Table 7–1

I = intensity of rainfall, in inches per hour for the estimated time of concentration.

A = drainage area in acres. The area may be determined from field surveys, topographical maps, or aerial photographs.

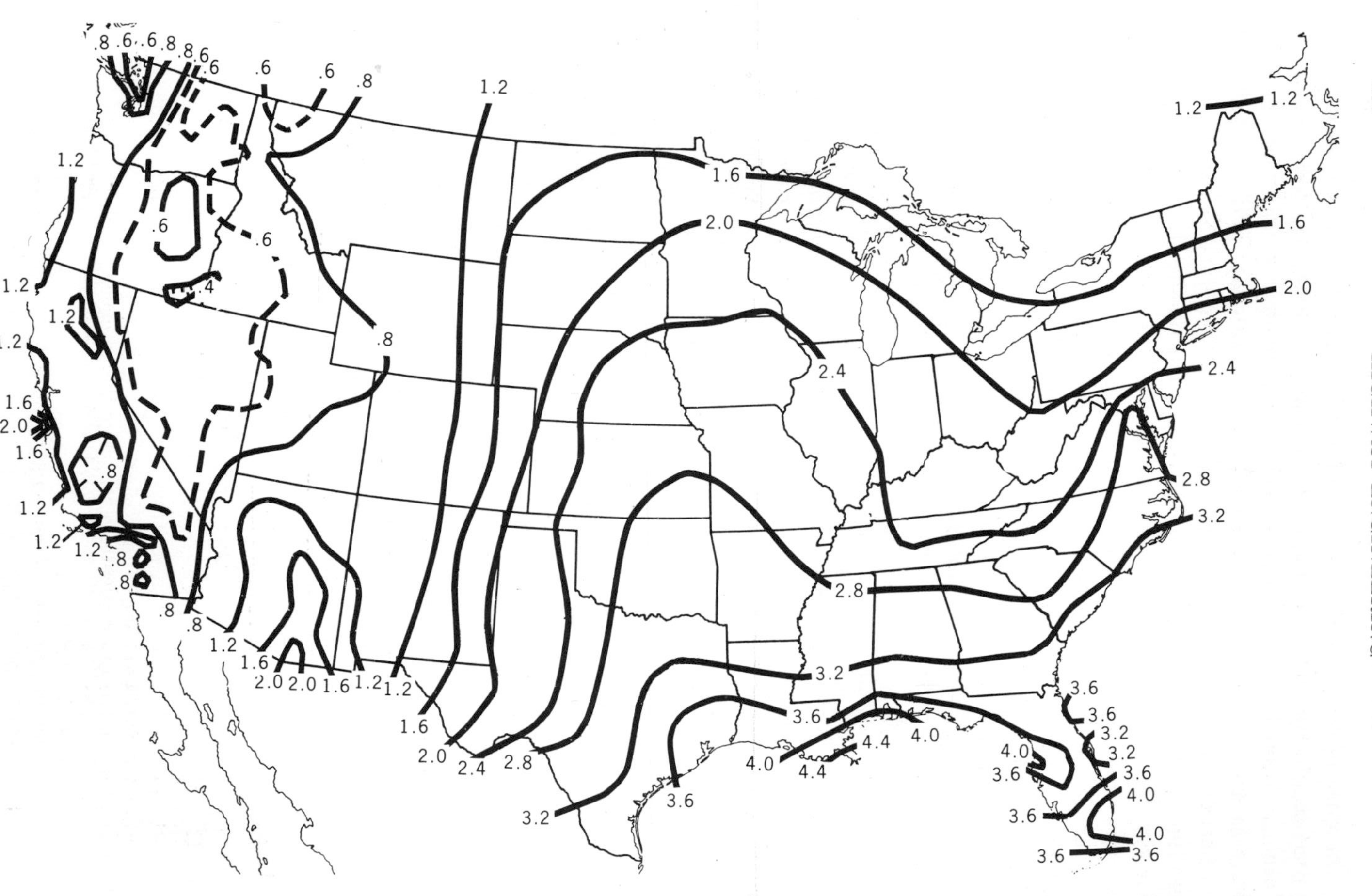

Fig. 7-3. Map of the contiguous United States, showing 2-year, 30-minute rainfall intensity. (Courtesy Bureau of Public Roads.)

In view of the preceding discussion of the many variables involved in the rainfall-runoff relationship, shortcomings of this method are apparent. Application of the method should be confined to relatively small drainage areas (up to 200 acres, according to the Bureau of Public Roads).

The Bureau of Public Roads continues to develop various publications dealing with peak rates of runoff and flood discharge. Specific data for estimating peak rates of runoff from small watersheds are available for Ohio, New Jersey, and portions of Alabama; a general discussion of this problem is contained in Reference *162*. One of the Bureau's hydraulic engineering circulars contains data for estimating flood discharges in the Piedmont Plateau (for drainage areas larger than 25 square miles).

7-20. Design of Side Ditches and Other Open Channels. With the quantity of water expected to reach any given point in the drainage system known, the design of side ditches, gutters, stream channels, and similar facilities is based upon established principles of flow in open channels. The principles also apply to flow in conduits with a free water surface. Most commonly used for design is Manning's formula, which applies to conditions of steady flow in a uniform channel and has the following form:

$$V = \frac{1.486 R^{2/3} S^{1/2}}{n} \qquad (7\text{-}3)$$

where V = mean velocity in feet per second

R = hydraulic radius in feet; this is equal to the area of the cross-section of flow (square feet) divided by the wetted perimeter (feet)

S = slope of the channel in feet per foot

n = Manning's roughness coefficient. Typical values of n are given in Table 7-2; many more are given in Reference *146*.

TABLE 7-2

VALUES OF MANNING'S ROUGHNESS COEFFICIENT
(Open Channels)

Type of Lining	Values of n
Smooth concrete	0.013
Rough concrete	0.022
Riprap	0.030
Asphalt, smooth texture	0.013
Good stand, any grass—depth of flow more than 6 inches	0.09-0.30
Good stand, any grass—depth of flow less than 6 inches	0.07-0.20
Earth, uniform section, clean	0.016
Earth, fairly uniform section, no vegetation	0.022
Channels not maintained, dense weeds	0.08

Also applicable is the equation of continuity,

$$Q = VA = \frac{1.486}{n} AR^{2/3}S^{1/2} \tag{7-4}$$

where Q = discharge, cubic feet per second
A = area of the flow cross-section

A family of charts has been developed by the Bureau of Public Roads for the solution of Manning's equation for various common channel cross-sections (*146,147*). One of these charts is shown in Fig. 7–4.

At this point it is necessary to introduce the concepts of the "theory of critical flow," as it relates to the flow of water in open channels and culverts. The theory of critical flow gives rise to the definition of "critical depth," which is the depth where the flow changes from "tranquil" to "rapid" or "shooting." For example, water flowing down a relatively flat slope in an open channel is in tranquil flow, while that tumbling down a steep slope is rapid.

In the same channel, with the same quantity of flow, the flow can be changed from tranquil to rapid by an increase in slope. In such a case, the depth of flow decreases from that existing in the section of tranquil flow to a lower value in the section of rapid flow. The decrease in depth begins at some point ahead of the crest of the steep slope and continues gradually over some distance. In the section of rapid flow, since the quantity of flow is the same, the dimensions of the channel the same, and the depth less, it follows that the velocity is greater, from the equation of continuity. Critical depth is defined as the depth corresponding to the change from tranquil to rapid flow.

Critical velocity and critical slope are the velocity and slope which correspond to uniform flow at critical depth. Critical depth is independent of channel slope and roughness, but the critical slope is a function of the slope of the channel.

Tranquil flow exists when the normal depth of water in an open channel is greater than the critical depth; conversely, when the depth is less than critical, the flow is rapid.

Rapid flow is difficult to control, because abrupt changes in alignment or cross-section produce waves that travel downstream, alternating from side to side and sometimes causing the water to overtop the sides of the channel. Changes in channel slope, shape, or roughness cannot be reflected upstream except for very short distances (this condition is called "upstream control"). This type of flow is common in steep flumes and mountain streams.

Tranquil flow is relatively easy to control. Changes in channel shape, slope, or roughness affect the stream for some distance upstream (down-

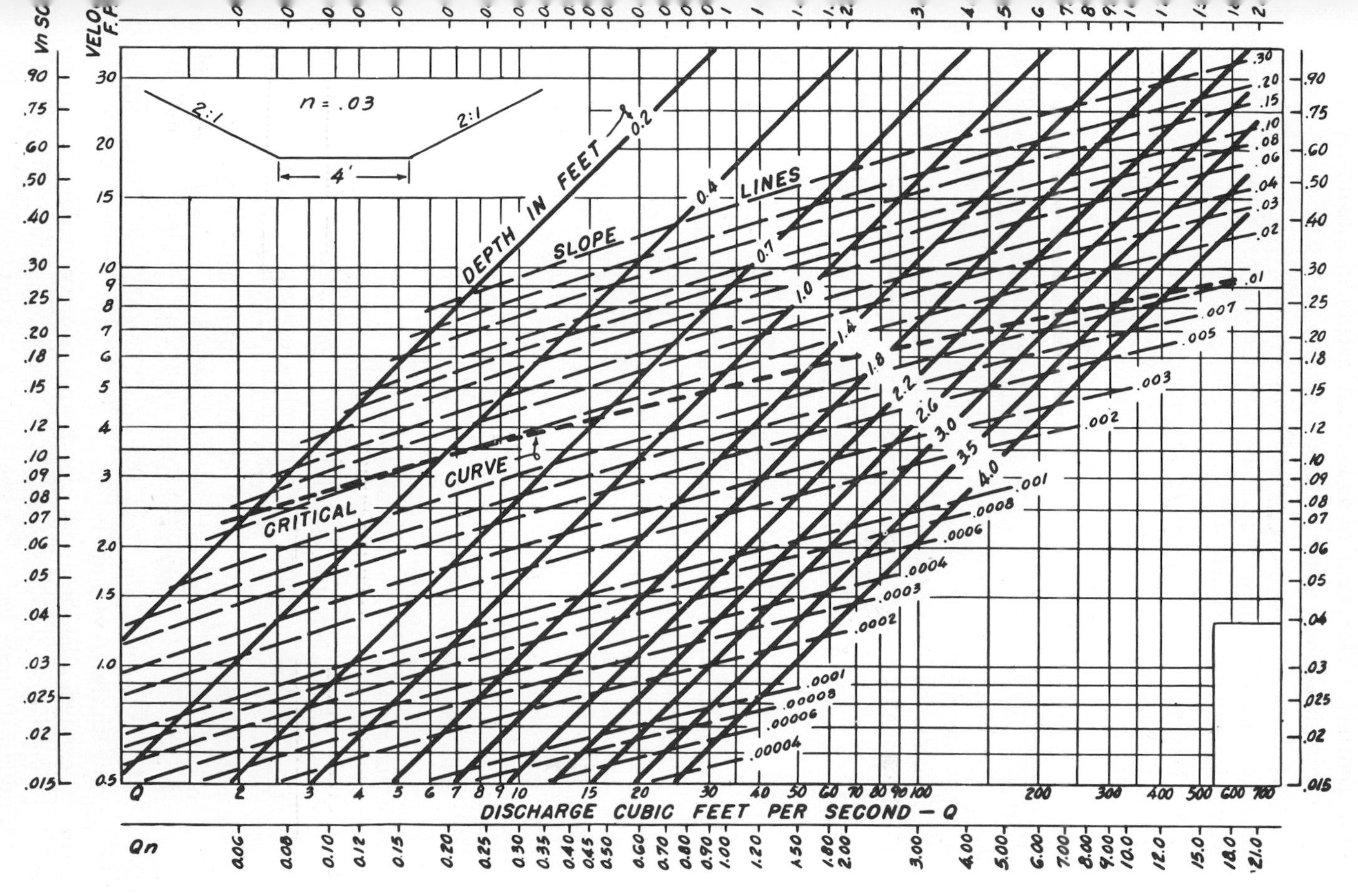

FIG. 7–4. Graphical solution of Manning's equation for one trapezoidal channel section. 2–1 side slope. Bottom width 4 ft. (Courtesy Bureau of Public Roads.)

stream control). This kind of flow often occurs in streams in plains and valley regions where slopes are relatively flat.

Critical depth is important in analysis, because it is always a hydraulic control. The flow must pass through critical depth in going from one type of flow to the other.

7–21. Design Procedure. Hydraulic design procedures are difficult to simplify, because of the wide variety of choices presented to the designer in a typical case and the various assumptions that must be made. Design is usually based upon an assumption of uniform flow, since the error involved is relatively slight in most cases.

Basically, the design of a highway drainage channel is done in two parts. The first part involves the selection of a channel section that will carry the given discharge on the available slope. The second part is the determination of the protection required (if any) to prevent erosion of the drainage channel (see Art. 7–22).

Use of charts like the one of Fig. 7–4 gives a direct solution of the Manning equation for uniform flow in trapezoidal channels with 2:1 side slopes, and with fixed bottom width (in Fig. 7–4, bottom width is 4 feet). Depths and velocities shown in the chart apply accurately only to channels in which uniform flow at normal depth has been established by sufficient length of uniform channel on a constant slope when the flow is not affected by backwater.

Depth of uniform flow for a given discharge in a given size of channel on a given slope and with $n = 0.030$ may be determined directly from the chart by entering on the Q-scale and reading normal depth at the appropriate slope line (or an interpolated slope). Normal velocity may be read on the V-scale opposite this same point. This procedure may be reversed to determine discharge at a given depth of flow.

For channel roughness other than $n = 0.030$, compute the quantity Q times n and use the $Q \cdot n$- and $V \cdot n$-scales for all readings, except those which involve values of critical depth or critical velocity. Critical depth for a given value of Q is read by interpolation from the depth lines at the point where the Q-ordinate and the critical curve intersect, regardless of channel roughness. Critical velocity is the reading on the V-scale at this same point. Where $n = 0.030$, the critical slope is read at the critical depth point. Critical slope varies with n; therefore, in order to determine the critical slope for values of n other than 0.030, it is first necessary to determine the critical depth. Critical slope is then read by interpolation from the slope lines at the intersection of this depth with the $Q \cdot n$-ordinate.

As an example, determine the depth and velocity of flow in a trapezoidal channel ($n = 0.030$) with 2:1 side slopes and 4-foot bottom width discharging 150 cubic feet per second on a 2 per cent slope ($S_0 = 0.02$).

Use Fig. 7–4. From $Q = 150$ cubic feet per second on Q-scale follow up and read normal depth, $d_n = 2.1$ feet and normal velocity, $V =$ approximately 8.5 feet per second.

To find the critical depth, velocity, and slope for these conditions, read upward from $Q = 150$ cubic feet per second to intersection with critical curve. At this point, read by interpolation, critical depth = 2.4 feet; critical velocity = 7.0 feet per second and critical slope = 0.013. The normal depth is less than the critical depth, hence the flow is rapid and not affected by backwater conditions.

If $n = 0.020$, and other conditions are the same, the procedure is similar. Use the $Q \cdot n$-scale. $Q \cdot n = 150(0.02) = 3.00$. From 3.00 on the $Q \cdot n$-scale read downward to the intersection with the $S = 0.02$ line. At this point, $d_n = 1.8$ feet and $V \cdot n = 0.23$. $V = (0.23/0.02) = 11.5$ feet per second.

7–22. Maximum Allowable Velocity in Unlined Ditches. Design of an unlined open channel is not complete unless consideration is given to the maximum allowable velocity, in order that erosion will not take place and to maintain a stable channel. The computed velocity in the channel is compared with the maximum allowable velocity, which is a function of the channel lining. If the velocity is excessive, means must be found to reduce the velocity or protect the channel. With a given discharge on a given slope, changes in width affect the velocity only slightly; changes in slope are more effective, where this is possible.

Maximum allowable velocities in unlined channels are given in Table 7–3 (*146*).

7–23. Prevention of Erosion in Side Ditches. As has been indicated, very severe erosion may occur in side ditches when water velocities are excessive or when the ditches are cut in easily eroded soils. In Fig. 7–5 is shown a condition of very severe erosion in a side ditch. In addition to the cultivation of vegetation, two principal measures are employed to minimize ditch erosion, the construction of ditch checks or paved ditches.

7–24. Ditch Checks. Ditch checks are placed at intervals along, and at right angles to, the centerline of a ditch to reduce the velocity of water flowing in the ditch and thus prevent erosion. The checks are in reality small weirs and may be constructed with notches which are triangular, rectangular, trapezoidal, or even semicircular in shape. They are built out of a variety of materials, including concrete, timber, masonry, and corrugated metal. In Fig. 7–6 are shown small corrugated-metal ditch checks.

Each ditch check creates a small reservoir or "stilling basin," with an accompanying sharp decrease in the velocity of the water. The size

TABLE 7-3

ALLOWABLE VELOCITIES FOR ERODIBLE LININGS

Uniform Flow in Continuously Wet Channels
Earth—No Vegetation*

Soil Type or Lining	Maximum Allowable Velocities: Clear Water (fps)	Water Carrying Fine Silts (Colloidal) (fps)	Water Carrying Sand and Gravel (fps)
Fine sand (noncolloidal)	1.5	2.5	1.5
Sandy loam (noncolloidal)	1.7	2.5	2.0
Silt loam (noncolloidal)	2.0	3.0	2.0
Ordinary firm loam	2.5	3.5	2.2
Volcanic ash	2.5	3.5	2.0
Fine gravel	2.5	5.0	3.7
Stiff clay (very colloidal)	3.75	5.0	3.0
Graded, loam to cobbles (noncolloidal)	3.7	5.0	5.0
Graded, silt to cobbles (colloidal)	4.0	5.5	5.0
Alluvial silts (noncol.)	2.0	3.5	2.0
Alluvial silts (colloidal)	3.7	5.0	3.0
Coarse gravel (noncol.)	4.0	6.0	6.5
Cobbles and shingles	5.0	5.5	6.5
Shales and hard pans	6.0	6.0	5.0
Soil cement		10.0	

Earth with Vegetative Cover

Type of Cover	Slope Range (%)	Maximum Allowable Velocities: Easily Eroded Soils (fps)	Erosion Resistant Soils (fps)
Bermuda grass sod	0–5	6	8
	5–10	5	7
	10+	4	6
Sod-forming grass such as Kentucky blue grass, buffalo grass, smooth brome, red top, blue grama	0–5	5	7
	5–10	4	6
	10+	3	5
Grass mixture. This is not recommended for use on slopes steeper than 10 per cent	0–5	4	5
	5–10	3	4
Bunch grasses, vines and similar open cover such as lespedeza, weeping lovegrass, ischaemum (yellow bluestem), kudzu, alfalfa, crabgrass, sudan grass, annuals (for temporary use). Not recommended for use on slopes steeper than 5 per cent	0–5	2.5	3.5

* Recommended in 1926 by Special Committee on Irrigation Research, ASCE.

FIG. 7–5. Severe erosion in a side ditch. (Courtesy Armco Drainage & Metal Products, Inc.)

and spacing of the checks are dependent upon a number of variables. The notch in each check must, of course, be sufficiently large to handle the peak flows in the ditch without overflowing the ditch banks. Generally speaking, the drop in elevation between the tops of successive checks should not be more than 3 or 4 feet. With this maximum in mind, the spacing is then dependent upon the grade of the ditch, assuming ordinary soil conditions. When the ditch grade is more than about 5 per cent, ditch checks usually become uneconomical and consideration should be given to the construction of a paved ditch.

In recent years, with the general improvement in design standards, ditch checks have become less popular in humid regions because they may be a hazard to drivers who go off the road, they are hard to maintain and unsightly, and their job can often be done well by vegetative cover. They are more popular in arid and semi-arid regions, where some channel erosion is regarded as inevitable in erodible soils where it is not economical to pave the channel.

An even simpler device has been used in some states in locations

FIG. 7-6. Corrugated metal ditch checks. (Courtesy Armco Drainage & Metal Products, Inc.)

where ditch grades are not excessive. This method consists of constructing earth dams or dikes at intervals along the ditch to retard the flow of water. The dams may be a foot or less in height and operate on the same principle as is employed in contour farming. The dams are sodded, and have proved to be very effective in combating erosion in some instances.

7-25. Paved Ditches. Erosion of side ditches is effectively prevented by paving of the ditch bottoms. A large number of erosion-resisting materials have been used to pave side ditches, including concrete, brick, masonry rubble, concrete block, broken concrete, broken stone, and various bituminous paving mixtures.

7-26. Stream Enclosures. Side ditches paralleling highway locations in rural areas are sometimes eliminated by enclosing the stream in a pipe drain. Conditions which would make this desirable include the elimination of a deep, narrow side ditch made necessary by a narrow right-of-way, the desired widening of an existing pavement built in a narrow right-of-way, and the elimination of erosion. This method has also been employed for the somewhat unusual purpose of enclosing a surface stream so that the highway could be built in the middle of a narrow valley rather than on a side-hill location.

Stream enclosures of the type that have been described are, generally speaking, storm sewers, and are designed in a fashion which is similar to that used in the design of municipal drainage systems. Appurtenant structures such as inlets, catch basins, and manholes are also normally included in the design.

Culverts

A culvert is usually, although not always, differentiated from a bridge by virtue of the fact that the top of the culvert does not form a part of the traveled roadway. More frequently, culverts are differentiated from bridges on the basis of span length. On an arbitrary basis, structures having a span of 20 feet or less will be called culverts; while those having spans of more than 20 feet will be called bridges. This line of division is by no means standard, and span lengths of from 8 to 20 feet are employed by various organizations as limiting culvert lengths. Culverts also differ from bridges in that they are usually designed to flow full under certain conditions, while bridges are designed to pass floating debris or vessels.

Culverts are to be found in three general locations: at the bottom of depressions where no natural watercourse exists; where natural streams intersect the roadway; and at locations required for passing surface drainage carried in side ditches beneath roads and driveways to adjacent property.

7–27. Principles of Culvert Location. The majority of culverts are installed in natural watercourses which cross the roadway, either at right angles or on a skew. In addition to selecting the proper location or "station number" for the culvert crossing with respect to the centerline of the road, the alignment and grade of the culvert are of importance.

The location of the centerline of the culvert on the centerline of the road may be determined by inspection of the plans or in the field. This location will generally be on the centerline of an existing watercourse or at the bottom of a depression if no natural watercourse exists.

The alignment of the culvert should generally conform to the alignment of the natural stream, and the culvert should, if possible, cross the roadway at right angles in the interests of economy. Skew culverts, located at an angle to the centerline of the road, are needed in many instances. The selection of the natural direction of the stream is somewhat difficult in some areas, where the stream bed is not in a fixed position but shifts with the passage of time. In such a case judgment must be exercised in selecting the most desirable location for the culvert, and some channel improvements may be necessary to ensure the proper functioning of the culvert after it is built. Where meandering streams are encountered, the water should be carried beneath the roadway at the earliest opportunity. Any changes which are necessary in the direction of the culvert itself should be effected gradually so that excessive head losses and consequent reduction in flow may be avoided.

Similarly, the grade of the culvert should generally conform to the existing grade of the stream. If the grade is reduced through the culvert, the velocity may be reduced, sediment carried in the water will be deposited at the mouth or in the length of the culvert, and the capacity of the structure thus further reduced. Culvert grades which are greater than those existing in the natural channel may result in higher velocities through the culvert and at the outlet end. Undesirably high velocities at the outlet will result in scour or erosion of the channel beyond the culvert and may make it necessary to install elaborate and costly protective devices. Changes in grade within the length of the culvert should also be avoided.

7–28. Hydraulics of Culverts. The following discussion is based on design concepts developed by the Bureau of Public Roads.

Wherever a constriction such as a culvert is placed in a natural open channel, there is an increase in the depth of water just upstream from the constriction. For a given size and shape of constriction the depth increases with the rate of flow or discharge. Conversely, for a given discharge the increase in depth varies with the size and shape of the constriction.

The selection or design of a culvert should be based on hydraulic principles, so that it will be of the most economical size and shape to carry a selected design discharge with a resulting headwater depth not to exceed some predetermined allowable value.

The allowable headwater depth (defined as the depth above the culvert flow line at the entrance) may be based on one of the following conditions:

1. A certain permissible freeboard below the proposed height of fill.
2. The elevation of permissible flooding upstream.
3. Where (1) and (2) will permit a large headwater depth, it may be limited by an allowable outlet velocity.
4. Some lesser headwater depth as governed by other design considerations or by departmental policy.

Hydraulic design procedures recommended by BPR are contained in two publications, *Hydraulic Engineering Circular No. 10*, "Capacity Charts for the Hydraulic Design of Highway Culverts" (*142*), and *Hy-*

draulic Engineering Circular No. 5, "Hydraulic Charts for the Selection of Highway Culverts" (*157*). Together, they provide guidance for the design of all culverts commonly employed in United States highway practice. Each is an extensive manual that can only be summarized here.

The first of the listed publications (Circular No. 10) contains a series of hydraulic capacity charts that permit the direct selection of a culvert size for a particular site without detailed computations. The charts do not replace the nomographs of Circular No. 5, which was issued earlier; the two publications complement one another. For the majority of cases, the capacity charts are adequate.

The charts are grouped by culvert shapes—square, rectangular, circular, oval, pipe arch, and barrel. Two basic types of culvert inlets are covered by the charts: headwalls with wingwalls, and projecting barrels to the toe of an embankment. The charts can be used for other entrances without appreciable error.

Each culvert capacity chart contains curves which show the discharge capacity of each of several sizes of one type of culvert for various depths of headwater. Headwater depth is given in feet above the invert of the culvert at the inlet, referenced to the first complete cross-section of the barrel. The discharge rate per barrel is given in cubic feet per second (cfs). Solid-line curves represent inlet control, and dashed-line curves outlet control.

Inlet control means that the discharge capacity is controlled at the culvert entrance by the depth of headwater (HW) and the entrance geometry, including the barrel shape, the cross-sectional area, and the type of inlet edge. Sketches of inlet-control flow for both unsubmerged and submerged projecting entrances are shown in Fig. 7–7(a) and (b). Fig. 7–7(c) shows a mitered entrance flowing submerged with inlet control.

In inlet control, the roughness and length of the culvert barrel and outlet conditions, including depth of tailwater, are *not* factors in capacity. An increase in barrel slope reduces headwater to a small degree; any correction for slope can be neglected.

Culverts flowing with outlet control can flow with the barrel full or partly full for part of the barrel length or all of it (see Fig. 7–8). The entire cross-section of the barrel is filled with water for the total length of the barrel in Fig. 7–8(a) and (b). Two other common types of outlet-control flow are shown in Fig. 7–8(c) and (d).

The inlet-control curves are plotted from model test data. The inlet edge contour and the barrel size control the depth of headwater.

The outlet-control curves were computed for culverts of various lengths with relatively flat slopes and assuming free outfall at the outlet.

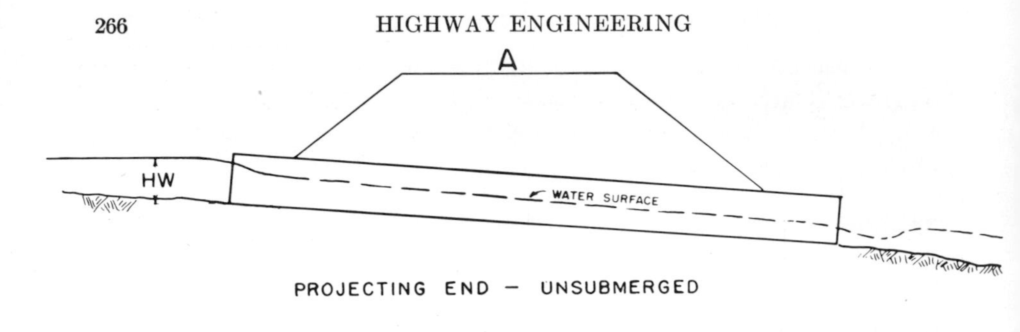

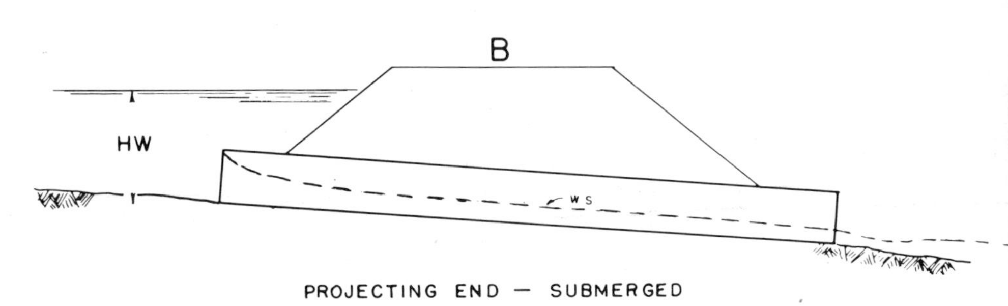

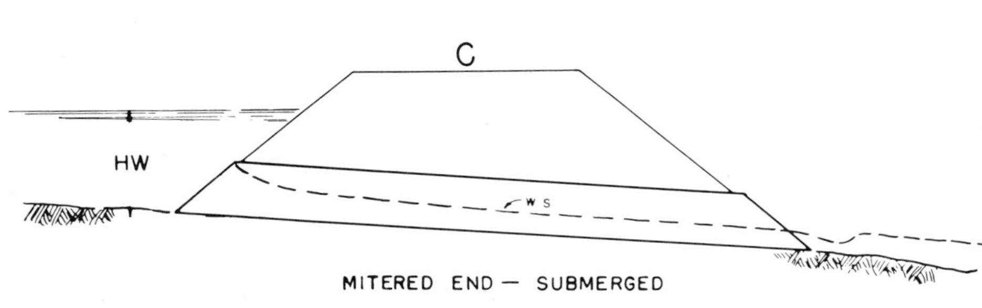

FIG. 7–7. Inlet controls for culverts. (Courtesy Bureau of Public Roads.)

For this condition, tailwater or flow in the outlet channel has no effect on culvert performance. Head losses at the entrance and resistance losses in the culvert barrel were taken into account in the computations.

Use of the charts is unrestricted for headwater depths up to about twice the barrel height (shown by a dotted line stepped across the chart curves).

The designer may use the charts to determine the headwater or the discharge of a given culvert installation or to select a size that will operate under a given set of conditions.

To illustrate the use of the charts contained in circular No. 10, sections of charts are shown in Fig. 7–9. Fig. 7–9(a) combines two separate sets of curves for a 36-inch standard corrugated metal pipe culvert with projecting entrance. Included are the inlet-control curve with $L/100S_o$ of 50 (L = length of culvert, and S_o = barrel slope) and two outlet-con-

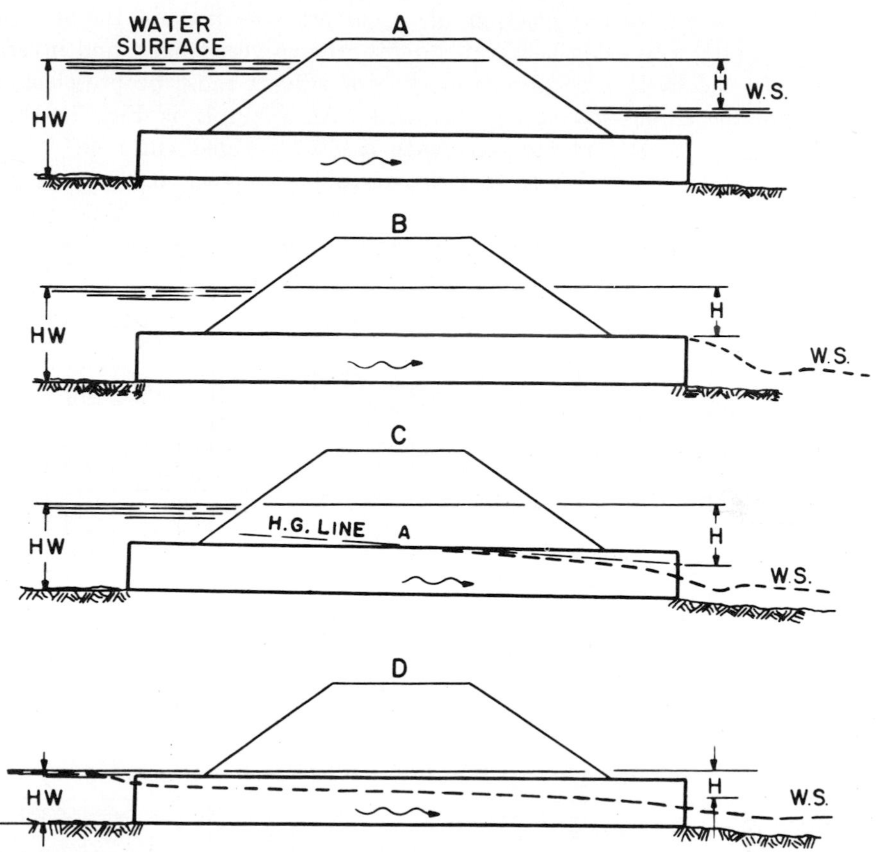

FIG. 7–8. Outlet controls for culverts. (Courtesy Bureau of Public Roads.)

trol curves with $L/100S_o$ of 250 and 450; the additional $L/100S_o$ curves were obtained by interpolation.

Headwater depth is determined by entering the chart at the design Q, then following up the chart to intersect a curve with the $L/100S_o$ value for the particular culvert under study. Headwater is read on the ordinate scale (Fig. 7–9[a]). If $L/100S_o$ is 50 or less, the culvert operates in inlet control, and headwater depth is read on the solid-line curve. No lesser depth of headwater is possible even for smaller $L/100S_o$ values. For any $L/100S_o$ value other than that shown on the curves and within the chart limit, headwater depth must be read by interpolation.

For example, consider a peak discharge rate of 50 cfs. For a 36-inch culvert 78 feet long on a slope of 0.0056 ($L/100S_o = 140$), headwater depth will be 5.2 feet. Another culvert 166 feet long at 0.5 per cent slope ($L/100S_o = 330$) will require a headwater depth of 6.3 feet for the same discharge.

Factors that govern selection of a culvert size include the design discharge rate, a limiting headwater depth, the culvert length, and invert slope. In the BPR procedure, the type of culvert must be preselected to locate the capacity chart to be used. An example is Fig. 7–9(b), which applies to a concrete pipe with a groove-edged entrance.

Assuming a culvert length of 240 feet and a slope of 0.002, a range

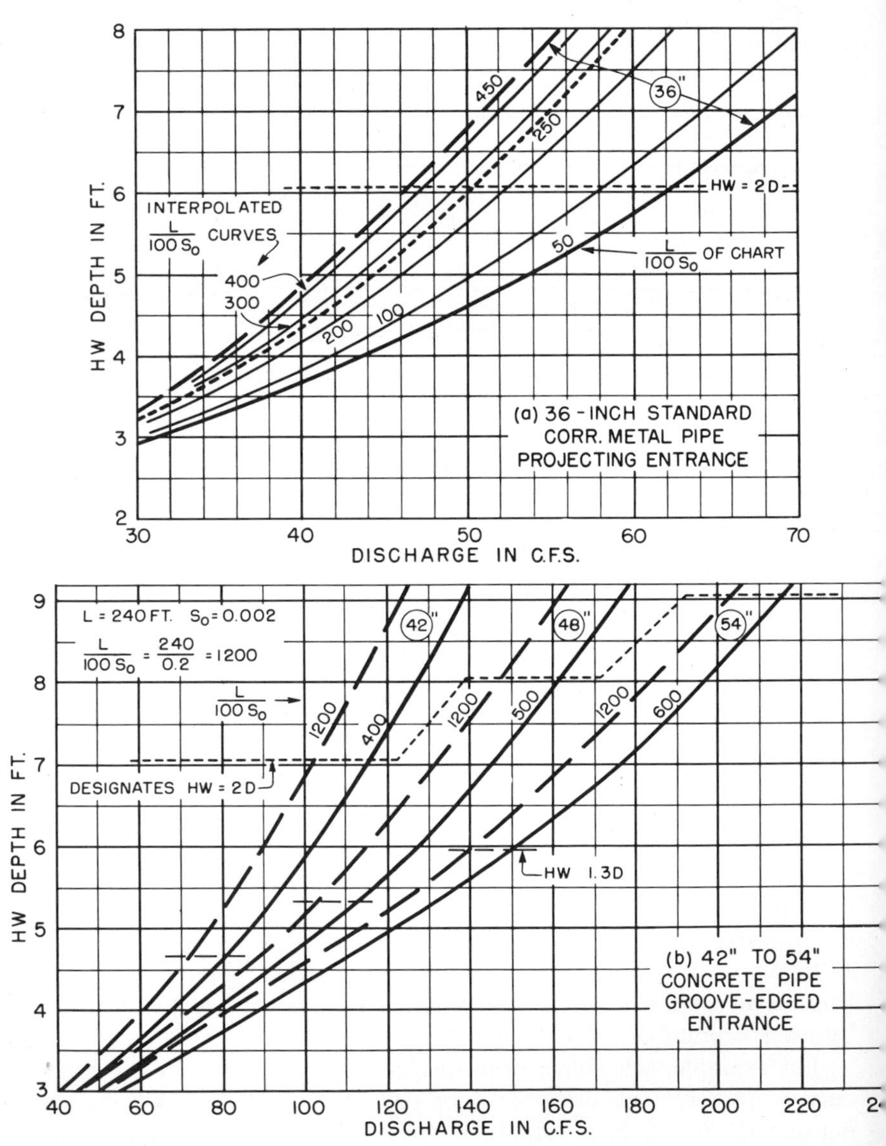

FIG. 7–9. Typical culvert capacity charts. (Courtesy Bureau of Public Roads.)

of headwaters and discharges can be studied for the three sizes of pipe shown in the chart. The $L/100S_o$ ratio is 1200. If the situation was such that the design discharge was 100 cfs for a limiting headwater depth of 6.0 feet, the designer would select a 48-inch pipe (the smaller, 42-inch, pipe gives a headwater of 6.8 feet).

To illustrate further the culvert selection process, assume a design discharge of 120 cfs and an allowable headwater depth of 6.0 feet. Inspection of Fig. 7–9(b) shows that a 42-inch pipe will not be adequate for any length-slope ratio, since the minimum headwater depth is 7.4 feet for 120 cfs. A 48-inch pipe would be a proper choice for all $L/100S_o$ ratios less than about 900. Some extreme lengths and small slope ratios would require a 54-inch pipe.

It is not possible to describe here all the limitations of, and exceptions to, the use of the charts, or the complications that may occur in selecting culverts for particular conditions. The hydraulic design of culverts is a complicated matter, much simplified by the use of the BPR charts. Suffice it to say that the two BPR publications provide tools for handling most designs with an accuracy that is compatible with that of forecasting design discharges and their frequency.

Two additional BPR charts are shown in Figs. 7–10 and 7–11. Fig. 7–10 applies to certain concrete box culverts with wingwalls flared from 30 degrees to 75 degrees with the culvert axis and a chamfered edge at the top of the entrance. Fig. 7–11 applies to certain structural plate pipe arches with inlet faces normal to their axes and projecting to or beyond the toe of the fill. They may also be used for mitered inlets and some other inlet variations. The Bureau of Public Roads recommends that such culverts be installed at a slope of 0.01 or more.

The Bureau of Public Roads recommends the following general procedure for culvert design:

1. Select the average frequency of the design flood.
2. Estimate the peak discharge of the design flood.
3. Obtain all site data. Plot a roadway cross-section at the culvert site and a stream channel profile.
4. Establish the culvert invert elevations at inlet and outlet, the culvert length. Determine the invert slope and compute $L/100S_o$.
5. Determine the allowable headwater depth (or depths).
6. Compute the depth of flow in the stream channel, including the floodplain, for the design flood, and determine the tailwater depth.
7. Select one or more appropriate culvert types (see Art. 7–29). Compute an approximate barrel area ($A_b = Q/10$) to guide selection of type and possible number and sizes of multiple barrels. Compute the discharge rate (Q) per barrel if a multiple culvert is used.
8. Select the capacity chart for the culvert and entrance type to be con-

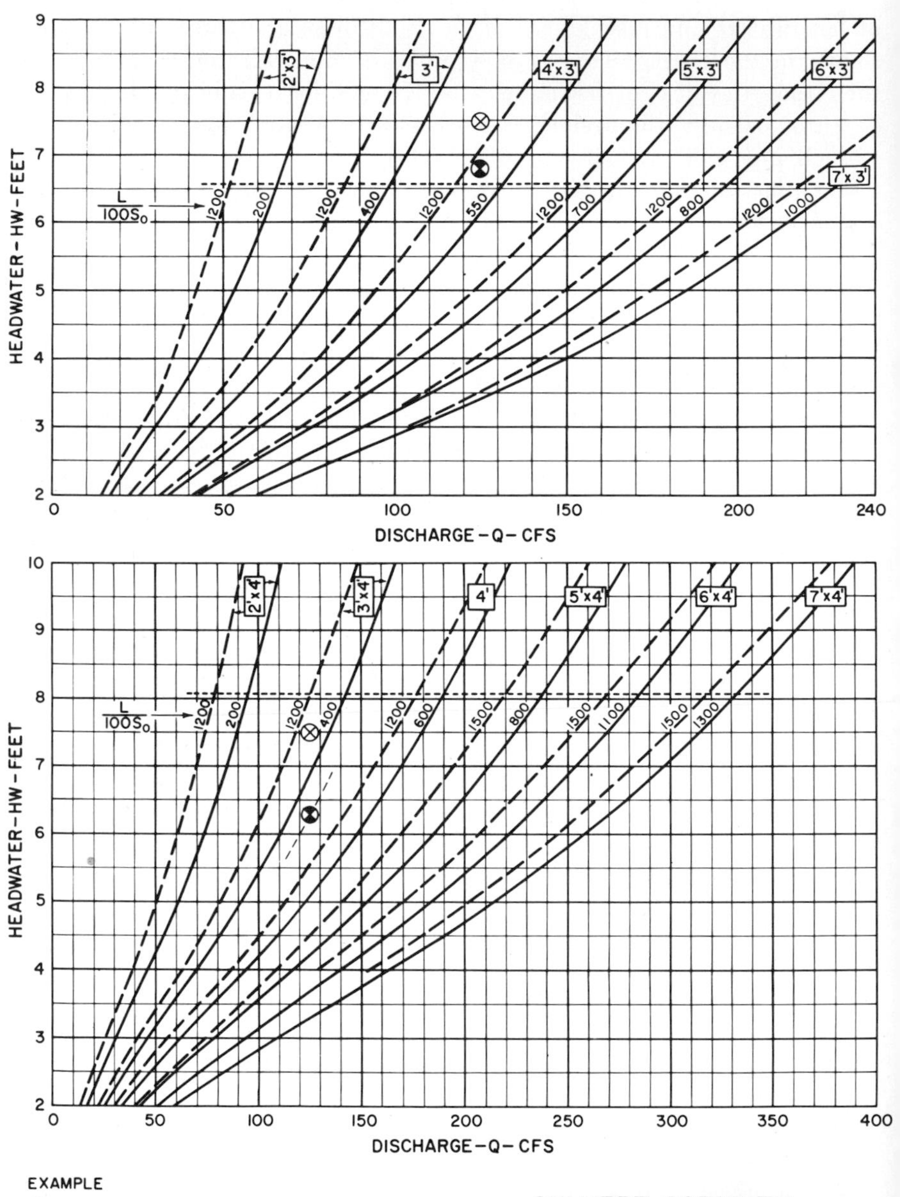

FIG. 7-10. Culvert capacity chart, rectangular concrete box culverts. (Courtesy Bureau of Public Roads.)

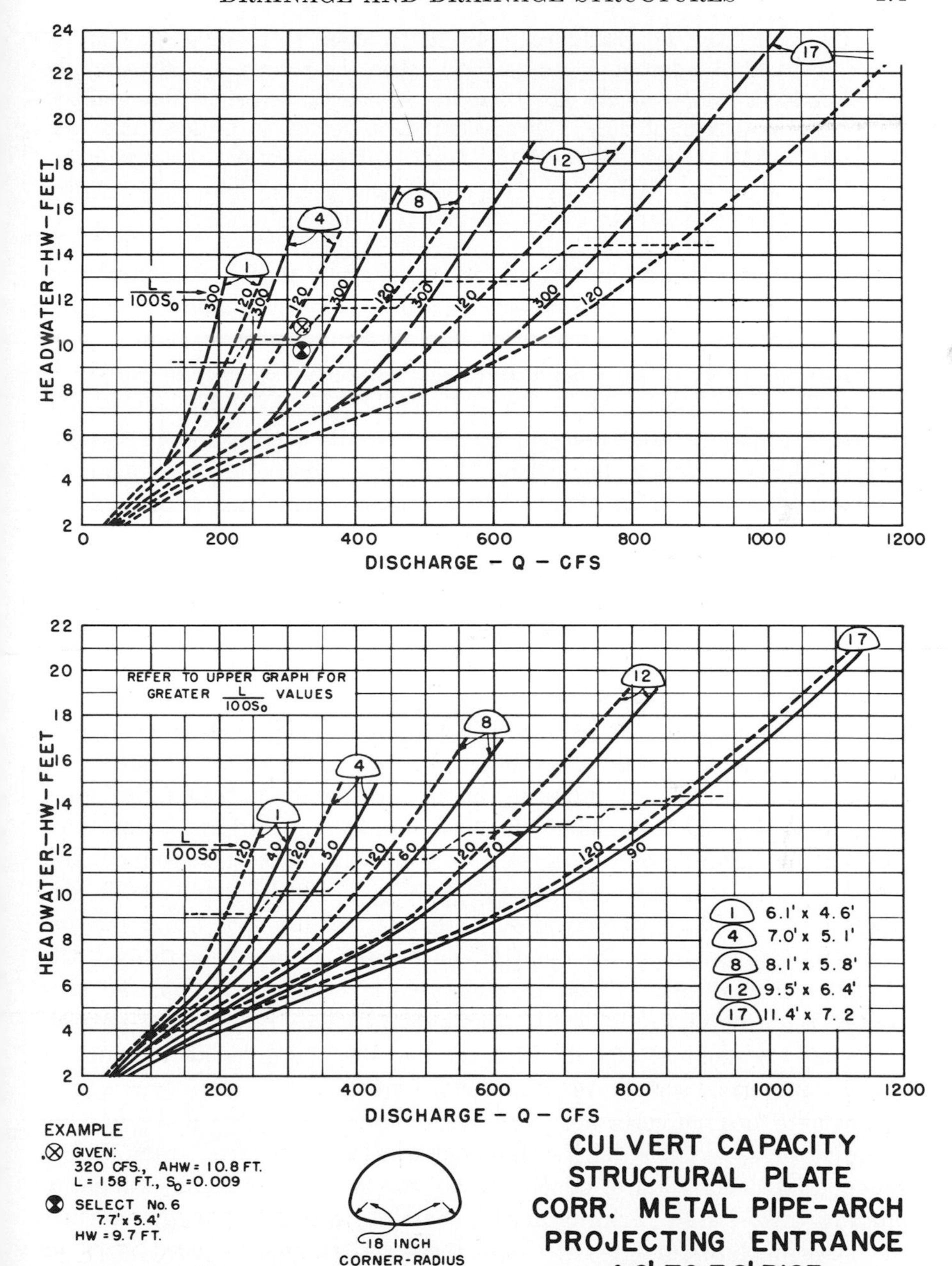

FIG. 7-11. Culvert capacity chart, structural plate corrugated metal pipe arch. (Courtesy Bureau of Public Roads.)

sidered. On the chart, locate the intersection of the Q value and the allowable headwater depth (AHW). Use the culvert $L/100S_o$ and the $L/100S_o$ values of the chart curves to determine the smallest culvert size that will produce a headwater depth equal to or less than that allowable. Check the tailwater depth. (The elevation of the tailwater in the outlet channel must not submerge critical depth at the outlet. Charts in Circular No. 5 give this information.)

7–29. Culvert Types and Materials. Materials most commonly used in the construction of culverts are reinforced concrete and corrugated metal. Less frequently, culverts are made from timber, cast-iron pipe, vitrified-clay pipe, and, occasionally, stone masonry.

Concrete Culverts. Reinforced-concrete pipe intended for use in culverts is made in diameters of from 12 to 108 inches and in various lengths, the usual length being from 4 to 8 feet. Standard specifications establish five classes of pipe with strengths increasing from class I through class V (*203*). The specifications show cross-sectional areas of reinforcing steels and concrete strengths for three series of wall thicknesses. Reinforcement may be either circular or elliptical. For special applications, reinforced-concrete culvert pipe may be manufactured with a cross-section other than circular—elliptical and "arch" shapes being in quite common use. Concrete culvert pipes have tongue-and-groove or bell-and-spigot joints; the joints are sealed during construction with portland cement mortar, rubber gaskets, or other materials. A greater or lesser amount of care is required in the preparation of the foundation upon which the pipes are to be laid. This preparation or "bedding" may vary from simple shaping of the bottom of a trench, or of the ground upon which the pipe is laid, to embedment of the pipe in a concrete cradle, depending on foundation conditions, loads on the pipe, and other factors. Pipe culverts are most frequently constructed in what is termed the "projection condition." That is, the culvert is constructed on the surface of the ground, in the open, and the fill is built around it. In such cases, only a nominal amount of attention need be given to bedding of the pipe in normal soils and normal heights of fill.

Concrete box culverts are constructed in place with square or rectangular cross-sections; single box culverts vary in size from 2 feet square to 12 feet square, depending on the required area of waterway opening. Most state highway departments use standard designs for various sizes of box culverts; perhaps the most commonly used sizes of concrete box culverts are in the range of 4 feet square to 8 feet square, including such sizes as $4' \times 4'$, $4' \times 6'$, $6' \times 6'$, $4' \times 7'$, and many others. Rectangular cross-sections are used where it is desired to reduce the height of the culvert to provide adequate cover between the top of the culvert and the roadway surface. The use of box culverts has declined in recent years, largely because of the time required for their construction.

Both concrete pipe and concrete box culverts are built with more than one opening where additional waterway area is required and when it is desired to avoid the use of excessively large single pipes or boxes. Such installations are called "multiple culverts" and may, for example, be "double" or "triple" concrete pipe or concrete box culverts.

Concrete arches are sometimes used in place of concrete box culverts, although difficulties attendant upon their proper design and construction have somewhat restricted their use. Concrete arch bridges are more frequently used.

Corrugated Metal Culverts (Steel). Corrugated steel is used in various forms in the construction of culverts for use in highway drainage.

Corrugated metal (galvanized steel) pipe is made in diameters of from 8 to 96 inches and in lengths of from 20 to 40 feet. Various thicknesses of metal are used, being generally from 16- to 8-gage. The corrugations which are formed in the sheet metal are 2⅔ inches from crest to crest, and ½ inch deep. Standard pipe is manufactured by bending the corrugated sheet metal into a circular shape and riveting the longitudinal joint. Helically corrugated pipe has a folded seam, rather than a riveted longitudinal joint. In the field, lengths of corrugated metal pipe may be joined by a pipe sleeve or by a connecting band which is several corrugations in length—angle irons are riveted to the two ends of the band and connected by bolts. The band is slipped over the ends of the pipes to be connected and the bolts drawn tight to form the connection.

The maximum desirable diameter of standard corrugated metal pipe is 8 feet. This fact led to the development of a method of construction utilizing heavier, curved, corrugated metal plates which are bolted together to form circular pipes or arches. The plates are curved, corrugated, and galvanized at the factory, shipped to the field site, and there bolted together to form the desired structure. The plates are heavier than normal corrugated metal pipe, being available in gages designated as 1, 3, 5, 7, 8, 10, and 12. They are made in various widths and in lengths of 6 and 8 feet. Pipes up to 21 feet in diameter have been fabricated by this method, while arches of almost any desired combination of rise and span are possible; maximum standard size is a rise of 13 feet, 2 inches and a span of 20 feet, 7 inches. Development of the "compression ring" method has allowed rational design of multiple-plate structures to fit any combination of circumstances.

In many culvert installations headroom is limited, and a circular pipe which has sufficient hydraulic capacity is not suitable. In such cases, corrugated metal "pipe arches" may be used. Pipe arches made of standard corrugated metal are available in sizes varying from a span of 18 inches and a rise of 11 inches, to a span of 72 inches and a rise of 44 inches. In Fig. 7–12 is shown a pipe-arch installation in a location where headroom is limited. A typical example of a location where this type of culvert may be used to advantage lies in the culvert

FIG. 7–12. Corrugated metal pipe arch. (Courtesy Armco Drainage & Metal Products, Inc.)

opening required to pass water flowing in a side ditch beneath an approach road to a rural highway.

In the interests of increased durability, corrugated metal pipe is sometimes furnished with the invert of the pipe covered with a thick bituminous mixture which completely fills the corrugations in this section of the pipe. Such pipe is called "paved-invert" pipe. The remainder of the pipe may also be coated with a mixture of bituminous material and asbestos, for very severe exposure conditions, or with bituminous material alone.

ALUMINUM CULVERTS. After extensive research and field testing (*4*), highway agencies began to use corrugated aluminum alloy for culverts in the 1960's. The principal advantage of this material is its light weight.

Corrugated aluminum culverts are very similar in shapes and dimensions to corrugated steel culverts. The material is available as standard riveted pipe, up to 8 feet in diameter, and pipe arches; helical corrugated pipe (from 6 to 48 inches); and multiple plates for field assembly (for culverts up to 16 feet or so in diameter). The pipe may be partially or wholly coated with bituminous material. Bolted corrugated couplings are used to join pipe lengths.

Fig. 7–13. Corrugated aluminum pipes, each 96 inches in diameter, form a culvert in New Mexico. (Kaiser Aluminum photo.)

Fig. 7–13 shows a multiple-barrel aluminum pipe culvert under construction in New Mexico. Each pipe is 8 feet in diameter, made of 8-gage corrugated sheet; sections are 13 feet long.

Miscellaneous Culvert Types. Other materials which may be used in the construction of culverts include vitrified-clay pipe, cast-iron pipe, and timber. In addition, masonry arch culverts are still constructed in some localities where suitable stone is cheaply available, although the use of this type of construction has declined in recent years. Each of these types of culvert has advantages and may be used in areas

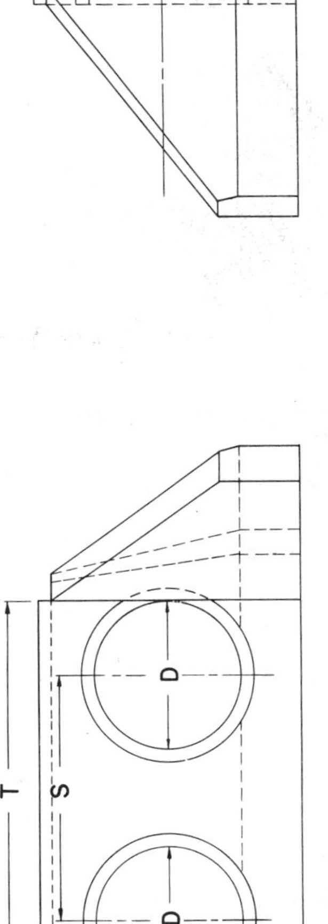

CONCRETE TO BE CLASS "A" AIR-ENTRAINED.
QUANTITIES GIVEN ARE FOR ONE ENDWALL.
ALL DIMENSIONS NOT GIVEN IN TABLE ARE SAME AS THOSE FOR SINGLE ENDWALLS FOR SAME SIZE PIPE.
Note: On shallow fills, where endwalls are 1' or less below shoulder line, the endwalls shall be constructed parallel to the line and grade of the road.

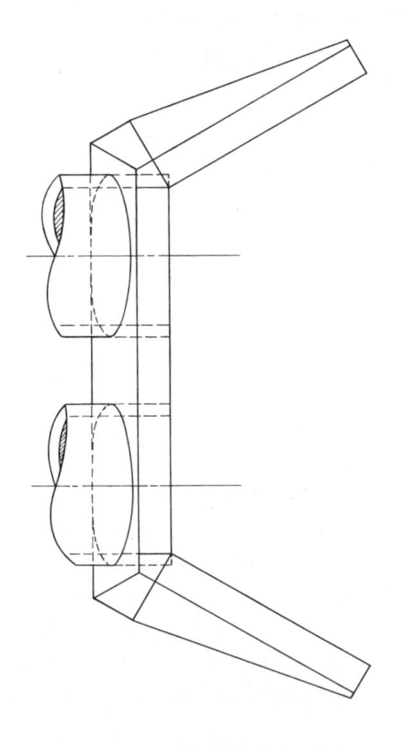

FIG. 7–14. Standard endwalls for multiple pipe culverts, 42″–84″ pipe. (Courtesy Virginia Department of Highways.)

FOR CONCRETE PIPE

DIAMETER D OF PIPE	S	T	FILL SLOPE 1½:1		FILL SLOPE 2:1	
			CU. YDS. CONC. ONE DOUBLE ENDWALL	INCREASE CU. YDS. FOR EACH ADDITIONAL PIPE	CU. YDS. CONC. ONE DOUBLE ENDWALL	INCREASE CU. YDS. FOR EACH ADDITIONAL PIPE
42"	6'-0"	9'-6"	4.829	1.271	5.493	1.255
48"	6'-10"	10'-10"	5.964	1.591	6.802	1.572
54"	7'-8"	12'-2"	7.692	2.057	8.796	2.035
60"	8'-6"	13'-6"	9.698	2.600	11.112	2.574
66"	9'-4"	14'-10"	12.016	3.240	13.811	3.209
72"	10'-2"	16'-2"	14.663	3.961	16.885	3.927
78"	11'-0"	17'-6"	17.612	4.751	20.325	4.713
84"	11'-10"	18'-10"	21.148	5.696	24.387	5.617

FOR CORRUGATED METAL PIPE

DIAMETER D OF PIPE	S	T	FILL SLOPE 1½:1		FILL SLOPE 2:1	
			CU. YDS. CONC. ONE DOUBLE ENDWALL	INCREASE CU. YDS. FOR EACH ADDITIONAL PIPE	CU. YDS. CONC. ONE DOUBLE ENDWALL	INCREASE CU. YDS. FOR EACH ADDITIONAL PIPE
42"	5'-3½"	8'-9½"	5.070	1.279	5.732	1.263
48"	6'-0½"	10'-0½"	6.296	1.616	7.132	1.596
54"	6'-9½"	11'-3½"	8.228	2.104	9.258	2.081
60"	7'-6½"	12'-6½"	10.319	2.677	11.738	2.650
66"	8'-3½"	13'-9½"	12.751	3.261	14.543	3.231
72"	9'-0½"	15'-0½"	15.673	4.068	17.889	4.033
78"	9'-9½"	16'-3½"	18.918	4.934	21.623	4.893
84"	10'-6½"	17'-6½"	22.773	5.905	25.999	5.860

FIG. 7-14. (*Continued*)

where the material is economically available and for special purpose structures.

7-30. Selection of Culvert Type. The type of culvert selected for use in a given location is dependent upon the hydraulic requirements and the strength required to sustain the weight of a fill or moving wheel loads. After these items have been established, the selection is then largely a matter of economics. Consideration must be given to durability and to the cost of the completed structure, including such items as first cost of manufactured units and costs of transportation and installation. Maintenance costs should also be considered in any over-all comparison of the cost of different culvert types. In brief, a thorough analysis should be made of the ultimate cost of all the different types of culverts which might be selected for use in a given installation. Other things being equal, the culvert selected should be the one which would be expected to show the lowest total cost over the expected life of the structure. Economic principles stated in Chapter 3 can be used in the analysis.

Information supplied by the various manufacturers' associations may be of value in making such a decision. However, selection is best made on the basis of accurate and complete records of construction and maintenance costs of similar structures. Many highway agencies keep records of this type and selection of the most economical culvert type is then made a relatively simple matter.

7–31. Culvert Appurtenances. Attention has been given thus far only to the main portion, or "barrel," of the culvert. In many locations various other features are incorporated into the design, and these features have been loosely grouped as "culvert appurtenances." Culvert appurtenances may be further subdivided into headwalls, intakes, and protective devices required at culvert outlets.

7–32. Culvert Headwalls. Headwalls are provided on culverts principally to protect the sides of the embankment against erosion. The term "endwall" is also used, being used interchangeably with "headwall" in many instances. In addition to their function in erosion control, headwalls may serve to prevent disjointing of sectional pipe culverts and to retain the fill. Materials most commonly used for culvert headwalls are concrete, masonry (stone or rubble), and metal. Of these, concrete is most widely used because of its adaptability to all types of culverts and because it lends itself to pleasing architectural treatment.

In selecting the size and type of headwall to be used in a given case, matters of economy must again be given consideration. In addition, some weight must be given to aesthetic considerations, as the headwall is the principal portion of the average culvert structure which is visible to the traveler. Headwalls are not always necessary, of course, and their use should be avoided wherever it is feasible to do so because of their cost. Headwalls are an expensive portion of the average culvert installation; the headwall should be made as small as possible consistent with adequate design. Safety of traffic must also be considered in choosing the type of headwall to be used in a given case.

Many different types of headwall are used by different highway agencies; generally, each agency has developed standard designs which are used whenever possible. No attempt will be made here to discuss all the possible variations in headwall design, as it is believed that any attempt of this sort would only be confusing to the student. Rather, several illustrations are given of typical headwall installations as currently used by various organizations.

Fig. 7–14 is a drawing that illustrates the details of standard endwalls used for multiple-pipe culverts by the Virginia Department of Highways. The flaring portions of this endwall are generally called "wingwalls." A masonry headwall used on a concrete pipe culvert is shown in Fig. 7–15, while the endwall shown in Fig. 7–12 is typical of the

Fig. 7–15. Twin 60″ reinforced concrete-pipe culverts, Grafton, N.H. (Courtesy American Concrete Pipe Assn.)

prefabricated metal end sections used in connection with small corrugated metal pipe culverts.

7–33. Culvert Intakes. In certain circumstances, e.g., in easily eroded soils when high velocities occur, unusual precautions must be taken to prevent damage to the inlet of a culvert. Such precautions may consist only of the paving of ditches which carry water to the culvert entrance. In other cases, special culvert intakes may be provided. An installation of this type appears in Fig. 7–16, which shows the details of a culvert intake for a 15-inch to 24-inch pipe culvert as used in Virginia.

As emphasized previously, consideration should also be given to the hydraulic properties of the culvert entrance. That is, the intake should

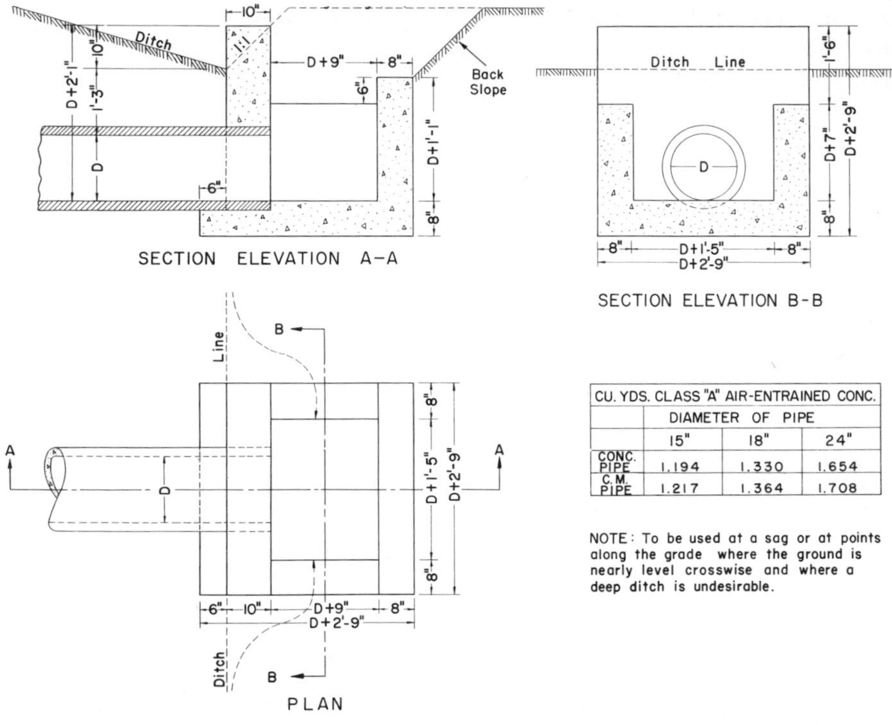

CU. YDS. CLASS "A" AIR-ENTRAINED CONC.			
	DIAMETER OF PIPE		
	15"	18"	24"
CONC. PIPE	1.194	1.330	1.654
C.M. PIPE	1.217	1.364	1.708

FIG. 7–16. Plans for intake, 15″–24″ pipe. (Courtesy Virginia Department of Highways.)

be designed to minimize losses of head due to eddies and turbulent flow. In general, sharp corners or "breaks" in the culvert entrance should be avoided. In certain installations, also, consideration must be given to prevention of the clogging of the culvert entrance by drift and rubbish. Special auxiliary devices are sometimes constructed for this purpose (*145*).

7–34. Outlet Protection. Water flowing from a culvert is frequently the cause of severe erosion or "scour" in the channel into which it discharges. If this scour is long continued, a pool of water may form below the culvert with resulting unsightly appearance, unsanitary condition, and impairment of the culvert's efficiency. This problem is most frequently overcome by constructing an apron immediately below the culvert outlet, upon which the water may impinge, and by paving, riprapping, or otherwise protecting the stream channel for some distance below the culvert outlet. In Fig. 7–17 is shown a concrete apron used in connection with a reinforced-concrete culvert pipe in Minnesota. Sills and stilling basins are used for the same purpose.

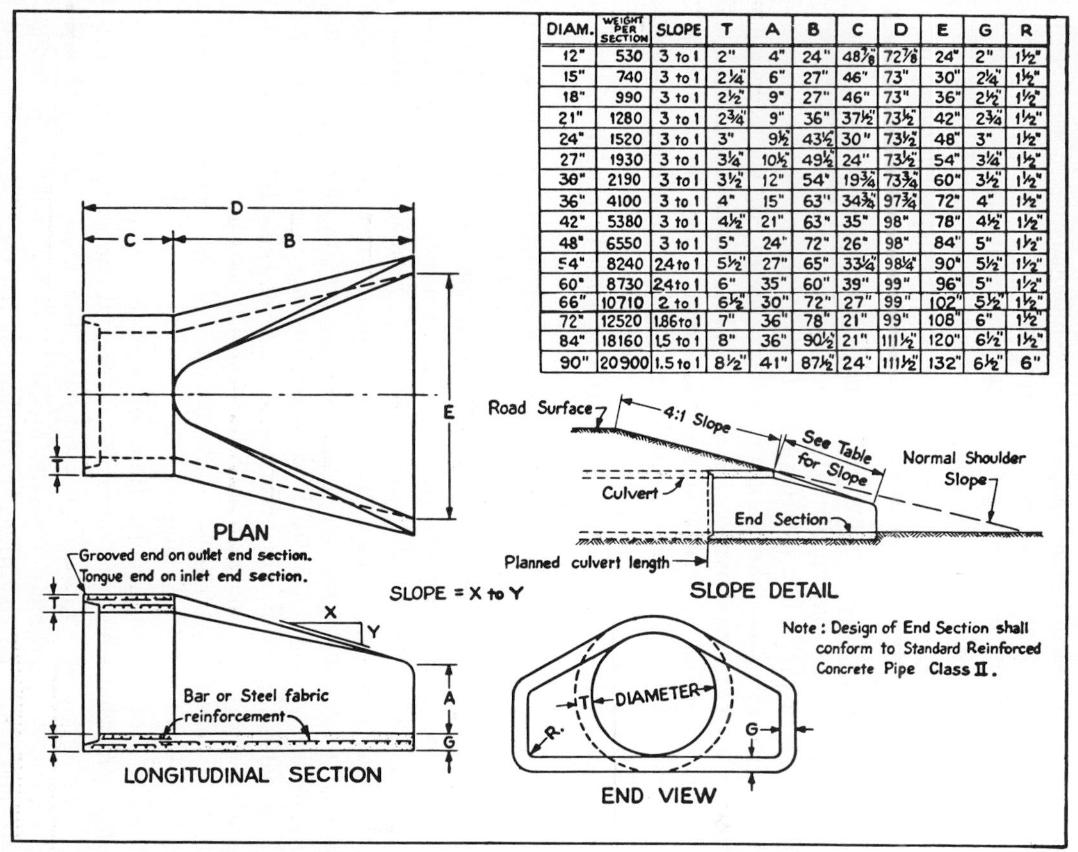

DIAM.	WEIGHT PER SECTION	SLOPE	T	A	B	C	D	E	G	R
12"	530	3 to 1	2"	4"	24"	48⅞"	72⅞"	24"	2"	1½"
15"	740	3 to 1	2¼"	6"	27"	46"	73"	30"	2¼"	1½"
18"	990	3 to 1	2½"	9"	27"	46"	73"	36"	2½"	1½"
21"	1280	3 to 1	2¾"	9"	36"	37½"	73½"	42"	2¾"	1½"
24"	1520	3 to 1	3"	9½"	43½"	30"	73½"	48"	3"	1½"
27"	1930	3 to 1	3¼"	10½"	49½"	24"	73½"	54"	3¼"	1½"
30"	2190	3 to 1	3½"	12"	54"	19¾"	73¾"	60"	3½"	1½"
36"	4100	3 to 1	4"	15"	63"	34¾"	97¾"	72"	4"	1½"
42"	5380	3 to 1	4½"	21"	63"	35"	98"	78"	4½"	1½"
48"	6550	3 to 1	5"	24"	72"	26"	98"	84"	5"	1½"
54"	8240	2.4 to 1	5½"	27"	65"	33¼"	98¼"	90"	5½"	1½"
60"	8730	2.4 to 1	6"	35"	60"	39"	99"	96"	5"	1½"
66"	10710	2 to 1	6½"	30"	72"	27"	99"	102"	5½"	1½"
72"	12520	1.86 to 1	7"	36"	78"	21"	99"	108"	6"	1½"
84"	18160	1.5 to 1	8"	36"	90½"	21"	111½"	120"	6½"	1½"
90"	20900	1.5 to 1	8½"	41"	87½"	24"	111½"	132"	6½"	6"

FIG. 7–17. Concrete apron for reinforced concrete pipe. (Courtesy Minnesota Dept. of Highways.)

Highway Bridges

As has been previously indicated, a bridge may be defined as a drainage structure which has a span of more than 20 feet. As a further distinguishing feature, bridge spans usually, although not always, rest on separate abutments, while culverts are regarded as integral structures. Although the term "bridge" is usually associated with structures which are required to carry the roadbed over an established waterway, it may also be somewhat loosely applied to grade separation structures and elevated highways in urban areas (viaducts).

7–35. Bridge Location. In modern practice, bridges of relatively short span are located to conform with the general location of the highway, which has been previously determined. That is, the tentative location for the highway is established after an analysis of all the economic and engineering factors involved, and the bridge engineer is given the problem of providing an economical and adequate bridge design to conform to the roadway location. In some cases the location of a suitable stream crossing may be the most important single factor influencing the location of the highway in a given section; such is usually the case when long bridge spans are involved.

The ideal location for a bridge crossing is, of course, one in which the crossing is made at right angles to the centerline of the stream at the narrowest point, where the alignment of the approach pavement is straight, where the approach grade is slight, and where soil conditions are adequate for the installation of the most economical foundation for the span involved. This ideal combination of circumstances is encountered all too infrequently, except in structures of short span, and many bridges have been located on skew crossings, vertical curves, or with curving alignment. In such cases, considerations relative to the general roadway location may still be regarded as controlling factors, and the required adjustments in the location of the bridge are made with these requirements in mind.

Many times, alternative locations of a proposed bridge may seem to offer somewhat similar advantages. A careful comparison must then be made of the several possible locations. The final decision should be based upon a complete analysis, including factors relative to traffic safety and operating conditions, fulfillment of the purpose of the road (e.g., the direct connection of population centers), and economics. Any complete analysis must include both the bridge and the approaches to it. A comparison of this type will generally result in the selection of one of the possible sites as the most desirable.

Once the general location is determined, the selected site must be subjected to careful scrutiny. This examination may be extremely detailed or somewhat cursory, depending largely upon the size and importance of the contemplated structure. A complete survey of the bridge site may include an examination of the channel for some distance above and below the bridge crossing, a complete topographical map of the site, and an extensive soil survey of the area, including the securing of undisturbed soil samples where required, the determination of the required waterway opening, and requirements of navigation on large streams. In Fig. 7–18 is shown a detailed bridge survey sheet prepared by the California Division of Highways. Information shown on a sheet of this type should be sufficient for the bridge designer to prepare an adequate and economical design.

Although the preceding paragraphs have dealt largely with waterway crossings, similar factors govern the design of grade separation and the bridge portions of complex interchanges. Obviously, a grade separation structure designed to carry the roadway over an existing railroad presents the location engineer and designer with similar problems, as does the design of a highway overpass. Similarly, the general location of an elevated highway in an urban area is usually determined on the basis of maximum serviceability to traffic (and availability of right-of-way) so that the engineer must prepare a design suitable to conditions in a rather limited area.

7–36. Design of Waterway Opening. In many locations the natural stream channel is somewhat constricted by the bridge structure and roadway approaches. In the interests of economy, the roadway is frequently placed on an embankment on either side of the bridge span, the distance between abutments is reduced as much as possible, and piers may be placed in the stream channel. All these things serve in many cases to reduce severely the area through which the water must pass, particularly when the stream is at flood stage. Two results may be immediately noticed—during flood stage the velocity of the water through the bridge opening may be considerably increased, with resultant danger to the bridge structure through scour at abutments and piers, and the elevation of the water on the upstream side may be increased, with the result that the area subjected to flooding above the bridge site is increased and adjacent property is subjected to overflow beyond the limits of the normal flood plain. It thus seems axiomatic that the bridge must be designed to pass the flow occurring at flood stage without excessive velocity and without damage to property located above the bridge crossing. Estimating flood flows is best done by study of stream-gaging records, but is sometimes based upon observation of high-water marks, the behavior of structures located on the same stream, and hydrological computations.

Bridge openings are also normally designed to pass floating debris

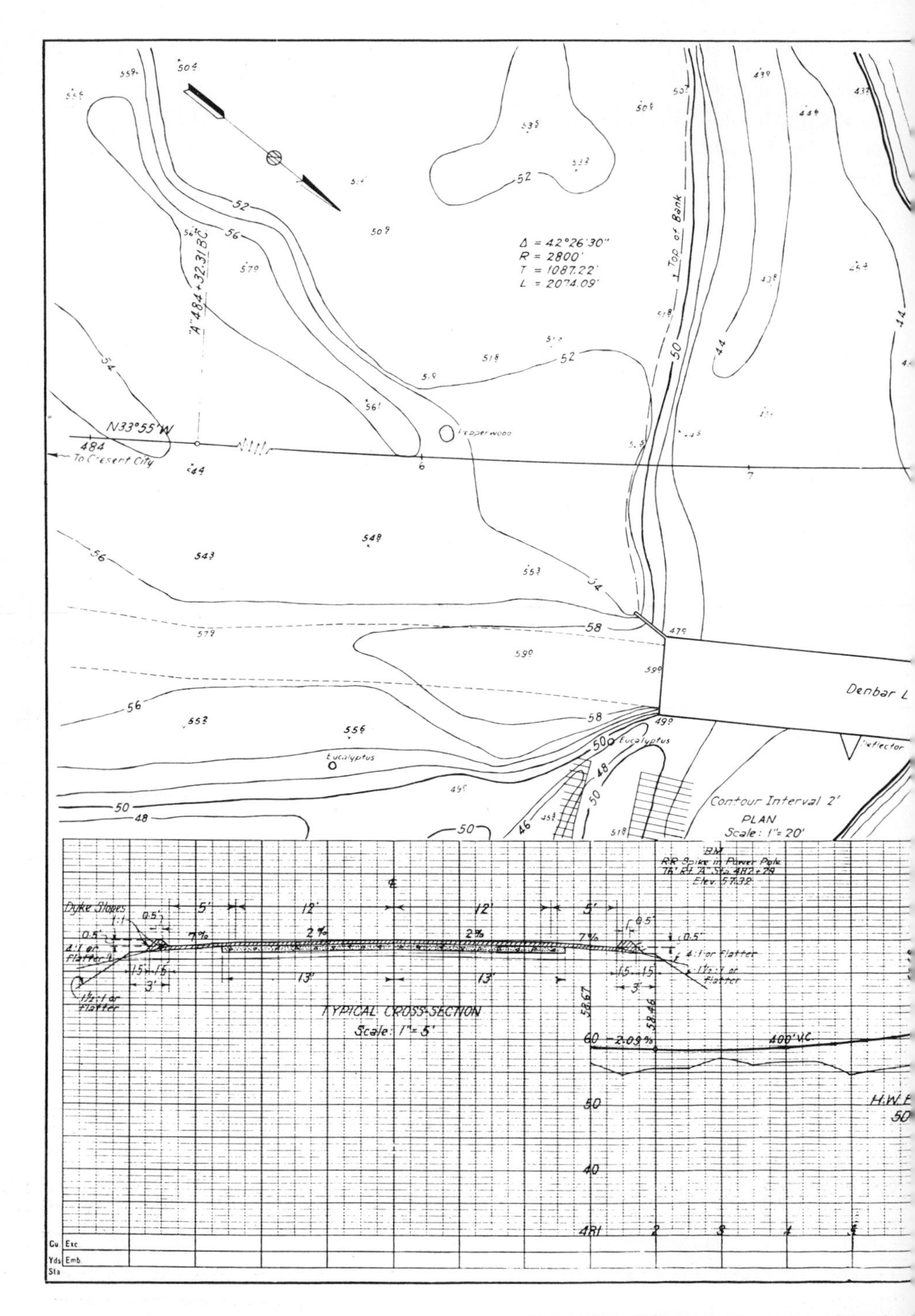

FIG. 7–18. DN-71-A site map at Rowdy

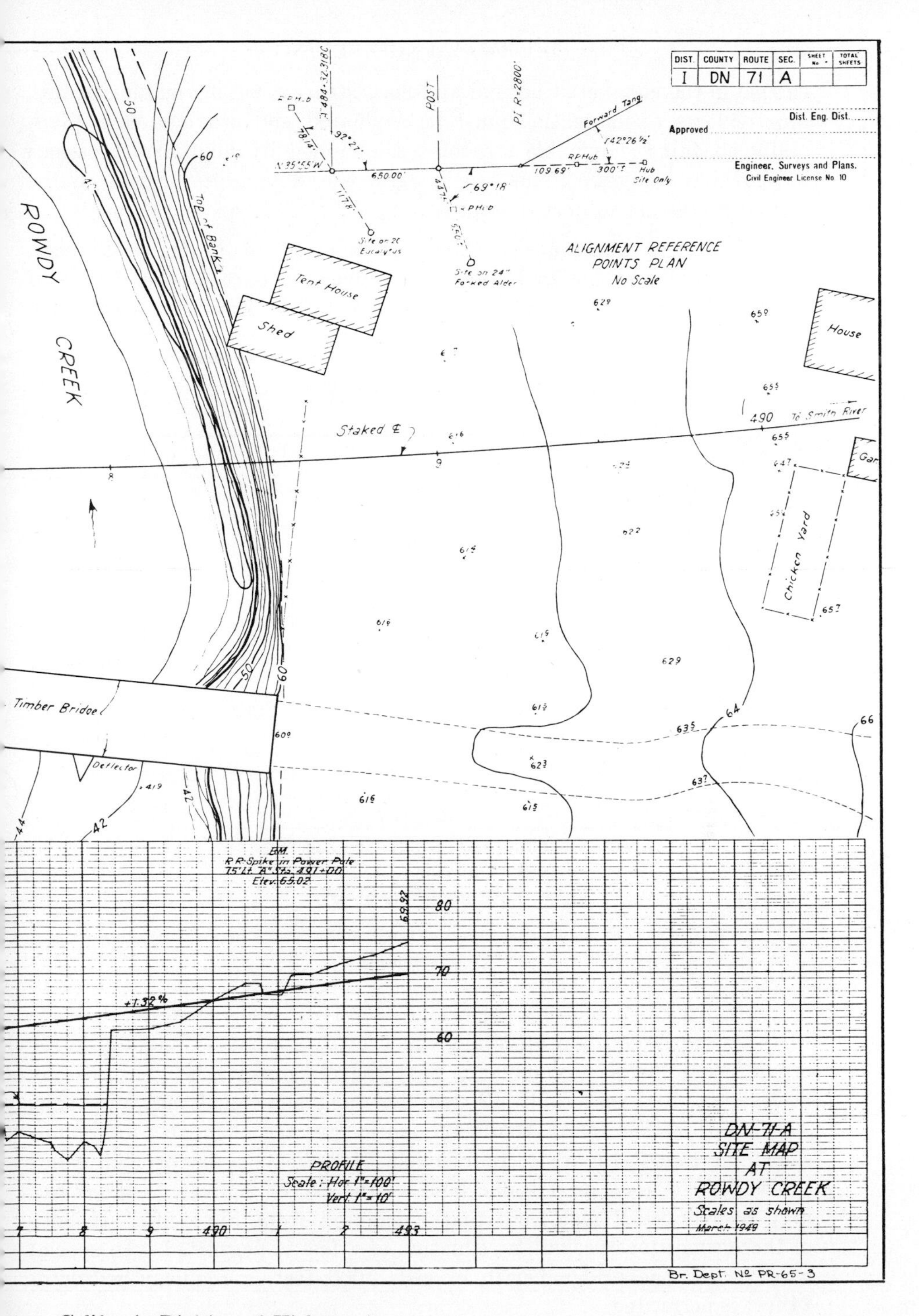

rtesy California Division of Highways.)

carried in the channel at normal and flood stages. On navigable streams, requirements of navigation must be evaluated and provided for. Generally speaking, navigable streams are spanned by high-level crossings or movable bridges. Plans for bridges over navigable streams in the United States are subject to approval by the U.S. Corps of Engineers.

7–37. **Bridge Clearances.** Standards relative to bridge clearances, both vertical and horizontal, are an important part of the design of highway bridges. Clearances recommended by the American Association

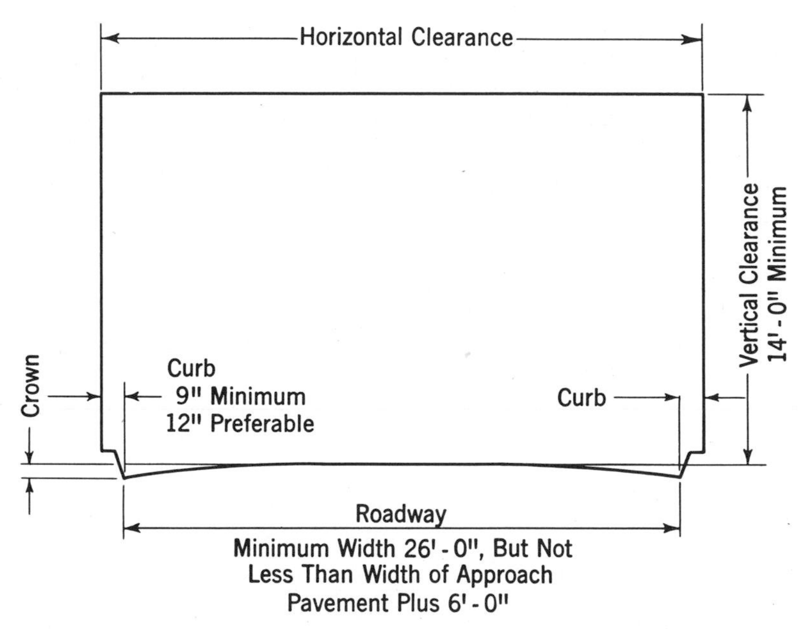

FIG. 7–19. Clearance diagram for birdges (two-way highway traffic).

of State Highway Officials for bridges (and culverts, where applicable) carrying two-way highway traffic are shown in Fig. 7–19 (*106*). This organization further recommends that the clear width be increased by at least 10, and preferably 12, feet for each additional lane of traffic. For heavy traffic roads, roadway widths greater than the minimums shown in Fig. 7–19 are recommended. If safety curbs or contiguous sidewalks are used, or if traffic-lane widths exceed 12 feet, the roadway may be reduced 2 feet from that calculated from Fig. 7–19. If both of these conditions prevail, a reduction of 4 feet is permitted. In special cases where the bridge length exceeds 1000 feet, a roadway width of 24 feet is permitted.

When the bridge is of short span or the roadway is an integral part of a culvert installation, the custom is to provide a clear width equal to the combined width of the approach pavement and shoulders, the use of curbs being optional. The provision of adequate clearances on new bridges is extremely important, as literally thousands of existing bridges which are otherwise adequate have been rendered functionally obsolete by narrow roadways. The provision of adequate width in new construction is much easier and cheaper than modifications which must be made after the design width proves to be inadequate.

7-38. Bridge Live Loads. Live loads used in the design of highway bridges are normally those established by the AASHO; two types of loadings are in use—H loadings and H-S loadings.

The H loadings consist of a two-axle truck or corresponding "lane loading," the latter being a certain uniformly distributed load and a concentrated load, which are equivalent in effect to the specified truck loading. Three H loadings are specified: H20-44, H15-44, and H10-44. The meaning of these may be explained by examination of the H15-44 loading. The gross weight of this truck is 15 tons (30,000 pounds), and this load is distributed 20 per cent (6000 pounds) on the front axle and 80 per cent (24,000 pounds) on the rear axle; the spacing between axles is 14 feet. The load has 10-foot clearance and load-lane width; the transverse spacing center-to-center of wheels is 6 feet. The suffix "44" simply indicates that this standard, as explained, was adopted in 1944. The H20 and H10 loadings are similarly specified.

The H-S loadings consist of a tractor truck with semitrailer, or the corresponding lane loading. Two H-S loadings are in common use: the H15-S12-44 and H20-S16-44. These loadings are somewhat similar to the H loadings; the spacing between axles of the truck is fixed at 14 feet, while the axle spacing of the semitrailer is from 14 to 30 feet, this distance being varied to produce maximum stresses. The total load of the H15-S12-44 loading is 27 tons (54,000 pounds), which is distributed as follows: front axle of truck, 6000 pounds; rear axle of truck, 24,000 pounds; and axle of semitrailer, 24,000 pounds. Total load of the H20-S16-44 loading is 36 tons (72,000 pounds) and it is similarly distributed.

The choice of load to be used in the design of a particular structure depends upon the traffic expected to use the bridge during its anticipated life. Heavier loadings are generally used in the design of structures on the various primary systems, and lighter loads on other components of the highway system, such as county roads. The H20-S16-44 loading is being used in the design of new bridges on the Interstate System, while many structures on the existing Federal-Aid System have been designed for H–15 loading. There is a definite trend toward the use of heavier loadings simply because the use of the heavier loads does not greatly increase the cost of the structure and seems to be consistent with the objectives of long-range planning.

7–39. Bridge Types. Under the stimulus of a greatly expanded highway construction program, engineers in the United States (and the rest of the free world) have virtually revolutionized bridge design and construction methods in the last decade. The advances apply to short-, medium-, and long-span bridges.

For permanent bridges, the most commonly used materials are steel and concrete. Bridges of many different types are built with these materials, used singly or in combination. Timber may be used for temporary above-water construction, for the elements of a structure that lie below the waterline (particularly timber piles), or for short-span bridges located on secondary roads. A few short-span aluminum bridges have been built in the United States on an experimental basis.

The principal portions of a bridge may be said to be the "substructure" and the "superstructure." This division is used here simply for convenience, since in many bridges there is no clear dividing line between the two.

Common elements of the substructure are abutments (usually at the bridge ends) and piers (between the abutments). Piers and abutments often rest on separately constructed foundations such as concrete spread footings or groups of bearing piles; these foundations are part of the substructure. Occasionally a bridge substructure comprises a series of pile bents, in which the piles extend above the waterline and are topped by a pile cap that, in turn, supports the major structural elements of the superstructure. Such bents often are used in a repetitive fashion as part of a long, low over-water crossing (Fig. 7–26).

In recent years, the dividing lines between short-, medium-, and long-span bridges have blurred somewhat. Currently, spans of 20 to 100 feet are regarded as short by many designers, who have developed many standardized designs to handle these spans economically. Medium spans range up to, perhaps, 400 feet in modern bridge practice, depending on the organization involved and the materials used. Long spans range up to 4000 feet or more, but a clear span above 1000 feet is comparatively rare.

Bridges may also be classed as "deck" or "through" types. In the deck type of bridge, the roadway is above the supporting structure; i.e., the load-carrying elements of the superstructure are below the roadway. In the through type of bridge, the roadway passes between the elements of the superstructure, as in a through steel-truss bridge. Deck structures predominate; they have a clean appearance, provide the motorist with a better view of the surrounding area, and are easier to widen if future traffic requires it.

Thousands of short-span bridges have been built in the United States in recent years under the impetus of the accelerated federal-aid highway program and the increasing use of limited access.

Steel, reinforced concrete, and prestressed concrete are used in these bridges in a variety of ways. The Bureau of Public Roads analyzed the cost of many short-span bridges and produced the comparative cost chart of Fig. 7–20.

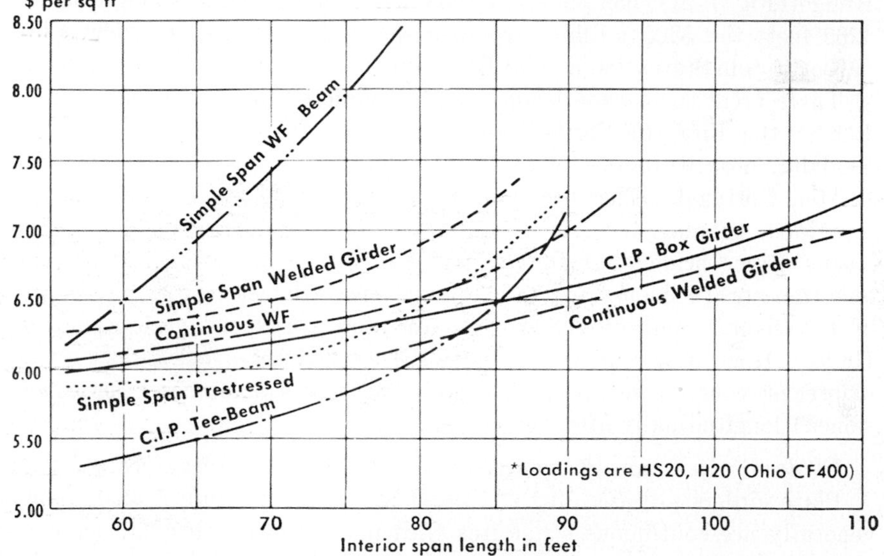

Fig. 7–20. Comparative costs of short span bridges, based on Bureau of Public Roads statistics. (Courtesy, *Engineering News-Record.*)

Four types of steel bridges and three of concrete are listed on the chart. The steel bridges are simple-span, wide-flange beam; simple-span, welded girder; continuous, wide-flange beam (the bridge is designed to act as a continuous structure over two or more supports); and continuous, welded girder. Most, if not all, of these steel bridges have reinforced-concrete decks designed for composite action with underlying steel members.

The concrete bridges are cast-in-place, reinforced-concrete **T**-beam (and slab); simple-span, prestressed (this type incorporates precast, prestressed I-girders or box girders topped by a cast-in-place deck); and cast-in-place box girder.

As the chart shows, BPR found the concrete T-beam design to be the most economical for spans of up to about 80 feet.

The designer of each medium- and long-span bridge tries to devise a structure that is best suited to the conditions encountered at that particular location. The result is an almost bewildering variety of structures that differ either in basic design principles or in design details.

General categories of steel bridges are listed below, together with examples of outstanding structures of each type.

Suspension bridges are used for very long spans, or for shorter spans where intermediate piers cannot be built. The Verrazano-Narrows Bridge (Fig. 7–21), completed in 1964, contains the world's longest span, 4260 feet; the $305 million structure spans the entrance to New York harbor to join Staten Island and Brooklyn.

Three great suspension bridges were completed in Europe in the mid-1960's: the Firth of Forth bridge, in Scotland; the Severn bridge, in England, near its border with Wales; and the Tagus River bridge, at Lisbon, Portugal. The design of the Severn bridge is unusual; it has a welded, tubular type deck supported by diagonal suspenders. The Tagus River bridge has the longest clear span in Europe, 3323 feet, and the world's deepest caisson foundation, 260 feet below the surface of the river. A short (680-foot center span) suspension bridge near Hudson Hope in a remote section of British Columbia has a deck made of precast concrete segments; the entire deck was prestressed (post-tensioned) longitudinally after fabrication.

Girder bridges come in two basic varieties—plate and box girders.

Plate girders are used in the United States for medium spans. They generally are continuous structures with maximum depth of girder over the piers and minimum depth at midspan. The plate girders generally have an "I" cross-section; they are arranged in lines that support stringers, floorbeams, and, generally, a cast-in-place concrete deck. The girders are shop fabricated by welding; field connections generally are by high strength bolts. The longest plate girder span (450 feet) in the United States is the Calcasieu River bridge near Lake Charles, Louisiana (Fig. 7–22).

Welded steel box girder structures are generally similar to plate girder spans, except for the configuration of the bridge cross-section. They have been used for spans of up to 856 feet (over the Save River in Yugoslavia). The MacDonald-Cartier bridge over the Ottawa River between Ottawa and Hull, Canada, contains box girder spans of 290—464—522—464—290 feet; the girders are 24 feet deep at the piers and

Fig. 7–21. The Verrazano-Narrows Bridge, longest clear span in the world. (Courtesy American Institute of Steel Construction.)

12 feet deep at midspan. The six-lane structure has two 40-foot roadways separated by a 4-foot median and flanked by two 6-foot sidewalks.

A few major bridges in the world are "stayed girders," meaning that long girder spans are partially supported by cables anchored in the bridge deck and passing over a tower erected over a pier.

FIG. 7–22. The Calcasieu River Bridge, near Lake Charles, La., the longest plate girder span in the United States.

Orthotropic bridges are a variation of the steel box girder design in which the upper flange of the box girder is a steel plate stiffened by longitudinal ribs; the orthotropic plate also serves as the bridge deck (Fig. 7–23). Two major orthotropic bridges have been built in the United States—the Poplar Street bridge, in St. Louis, and a portion of the San Mateo-Hayward bridge, in California. The Zoo Bridge over the Rhine, at Cologne, Germany, contains a pair of steel box girders with an orthotropic deck and a center span of 850 feet.

Fig. 7–23. An experimental orthotropic plate girder bridge under construction in California. (Courtesy U.S. Steel Corp.)

Rigid frames are used occasionally, most often for spans in the range of 75 to 100 feet and for grade-separation structures.

Arch bridges (Fig. 7–24) are used for longer spans at locations where intermediate piers cannot be used and where good rock is available to withstand the thrusts at the arch abutments. The Lewiston (New York)–Queenston (Ontario) bridge over the Niagara River is the world's longest fixed steel arch span (1000 feet). The Burro Creek bridge in Arizona, completed in 1966, is a two-hinged, trussed arch with a span of 680 feet.

A variation in the arch bridge is the tied arch, in which a horizontal tie that carries the roadway takes much of the horizontal thrust inherent in the arch form. The Port Mann bridge over the Fraser River near Vancouver, British Columbia, is a tied arch with a 1200-foot main arch span flanked by 360-foot arch side spans; it has an orthotropic plate deck.

Truss bridges are built in many forms and in many locations for medium and long spans. Both deck and through trusses are built, with cantilever and continuous trusses being the most common. Fig. 7–25 shows the Newburgh-Beacon (New York) bridge over the Hudson River. The total length of this bridge is 7786 feet. It contains three

FIG. 7–24. A steel arch, the Cold Spring Canyon Bridge near Santa Barbara, Calif. (Courtesy *Engineering News-Record.*)

types of spans; plate girder approach spans, 11 deck trusses of moderate span, and a three-span, cantilever-type through truss over the main channel. The anchor spans are 602 feet long, and the central span is 1000 feet, made up of two 250-foot cantilevers with a 500-foot suspended span between them.

One of the latest Mississippi River crossings is the Sunshine Bridge,

FIG. 7–25. A through steel truss bridge, the Newburgh-Beacon Bridge over the Hudson River in New York State.

at Donaldsonville, Louisiana. The river portion of this bridge contains a five-span, double cantilever truss and one 350-foot deck truss. The central span is 850 feet, flanked by two 800-foot spans.

Other types of steel bridges have been built in various locations. They include girders of a "delta" cross-section incorporated in a medium-span bridge in Washington State; thin-walled steel box girders strengthened by prestressing and composite with a cast-in-place concrete deck on a 330-foot-long, four-span bridge in South Africa; and short-span bridges at an expressway interchange in Detroit made of an alloy steel that never needs painting.

Concrete bridges come in nearly as great a variety as do steel bridges. Listed below are various types now in use and some examples of outstanding concrete structures.

Conventional reinforced concrete is used primarily for short spans, with a variety of innovations from the customary cast-in-place, slab-and-beam design. The Texas Highway Department pioneered the use of high-strength reinforcing steel in a 286-foot-long, two-lane, four-span continuous girder bridge completed in 1964; use of the high-strength reinforcing steel reduced the required quantity of steel by 30 per cent and the over-all cost by 10 per cent.

A bridge was built in Connecticut in 1966 by precasting concrete components and "gluing" them together with epoxies. Precast concrete piles in the structure have epoxy splices; precast abutments and piers are fixed by epoxies to the piles; precast girders for the 50-foot spans were made continuous by gluing their ends together; and the cast-in-place deck is bonded with epoxy to the girders for composite action.

Precast, prestressed concrete bridges are popular in the United States for short, simple spans; they generally comprise prestressed I-beams or box girders that support a cast-in-place deck.

One of the world's great water crossings, the 17.6-mile Chesapeake Bay Bridge-Tunnel (Fig. 7–26), contains over 12 miles of repetitive spans that incorporate prestressed concrete cylinder piles, precast pile caps, and 75-foot-long precast, prestressed deck sections. Each span contains four parallel deck sections, each with a double T shape.

A four-lane, three-span continuous bridge in Paris, France, incorporates 8-foot-long precast concrete box girders assembled by the cantilever method with a minimum of falsework (Fig. 7–27). This bridge is prestressed three ways—longitudinally, transversely (across the 93-foot-wide deck), and vertically (at the piers). Its flexible concrete piers rest on deep pile foundations and contain large precast elements. Its central span is 180 feet long.

Designers often combine cast-in-place concrete with precast, pre-

FIG. 7–26. The Chesapeake Bay Bridge-Tunnel. Concrete trestle and man-made island are in the right foreground. (Courtesy Chesapeake Bay Bridge and Tunnel District.)

FIG. 7–27. The Choisy-le-Roi Bridge over the Seine in Paris, during construction. (Courtesy *Construction Methods and Equipment*.)

stressed members. The bridge over San Sebastian Inlet, Florida, has 16 simple spans, each 73 feet long and containing five precast, prestressed **I**-beams. The three main spans comprise an anchor-cantilever unit with four longitudinal lines of girders. Each 100-foot anchor span has 65-foot-long girders of conventional concrete, extending from a pier toward the center span (this end was supported by falsework during construction). Another set of 65-foot-long girders rests on each of the channel piers, extending 35 feet back to the splice point and cantilevering 30

FIG. 7–28. The Lake Maracaibo Bridge, Venezuela. (Courtesy *Engineering News-Record.*)

feet beyond the pier. A center drop-in girder, 120 feet long, completes the 180-foot channel span. The girders and deck of this section are lightweight concrete.

One of the world's monumental concrete structures is the $5\frac{1}{2}$-mile-long bridge over Lake Maracaibo, Venezuela (Fig. 7–28). The bridge contains these elements: 79 low level, repetitive spans on four-pier bents and incorporating 153-foot-long precast, prestressed girders; 28 spans, each 279 feet long, that include 126-foot, cast-in-place, prestressed girders (on **X**-shaped trestles) that cantilever both ways to support drop-in, 153-foot prestressed girders; two 525-foot transition spans; and five 771-foot suspended spans. Each suspended span incorporates the halves of 618-foot continuous prestressed girders built on preassembled steel trusses and supported by massive tower piers; the central portion of each of these spans also incorporates the 153-foot girders. The bridge rests on hundreds of prestressed concrete pile, and the deck is prestressed transversely.

Cast-in-place prestressed concrete has been used for long-span bridges in Europe and in a few locations in the United States.

The Bendorf Bridge, over the Rhine River, at Cologne, Germany, contains the world's largest cast-in-place, prestressed box girder span

FIG. 7–29. The Bendorf Bridge over the Rhine at Cologne, Germany, during construction. (Courtesy *Engineering News-Record.*)

(682 feet) (Fig. 7–29). The bridge was built by the cantilever method, using traveling forms.

A 660-foot-long, four-span viaduct in Oakland, California, curves vertically and horizontally and is superelevated. It was prestressed (post-tensioned) with wires longitudinally, transversely, and vertically. In addition, it was stressed by a longitudinal movement of the pier bases produced by jacking after the superstructure was completed.

Arches also are built in concrete. The world's longest concrete span (1000 feet) is an arch, the Gladesville Bridge, at Sydney, Australia (Fig. 7–30). The arch is made up of 512 huge, hollow concrete boxes, each 20 feet deep. The boxes were winched to the high point of an arch form and then rolled down the form, one on one side, the next on the other, to form four voussoir ribs. The ribs are tied together by transverse stressing through diaphragms. The columns that carry the roadway over the ribs were precast and prestressed.

A fixed reinforced-concrete arch bridge in Portugal has a span of nearly 900 feet between abutments.

Fig. 7–30. The world's longest concrete span, the Gladesville Bridge at Sydney, construction. (Courtesy *Engineering News-Record.*)

A very unusual bridge in Denmark has simply supported approach spans that incorporate 111-foot-long prestressed concrete girders, and its 300-foot center span is a concrete bowstring with prestressed tie.

7–40. Movable Bridges. The three basic types of movable bridges—bascule, vertical lift, and swing—are shown in Fig. 7–31. They are used where a high-level bridge is uneconomical but provision must be made for navigation. Fig. 7–32 shows a double-leaf bascule bridge of outstanding beauty.

There is still a fourth type of movable bridge, the floating bridge. Three of these unusual structures are in the Puget Sound area of Washington State. The most recent of these incorporates a four-lane, 1.4-mile-long floating section, supported by 35 prestressed concrete pontoons, the largest of which is 360 feet long, 60 feet wide, and nearly 13 feet deep. Each pontoon is secured by cables to precast concrete anchors on the lake bottom in about 200 feet of water. A drawspan at the center provides for passage of ships.

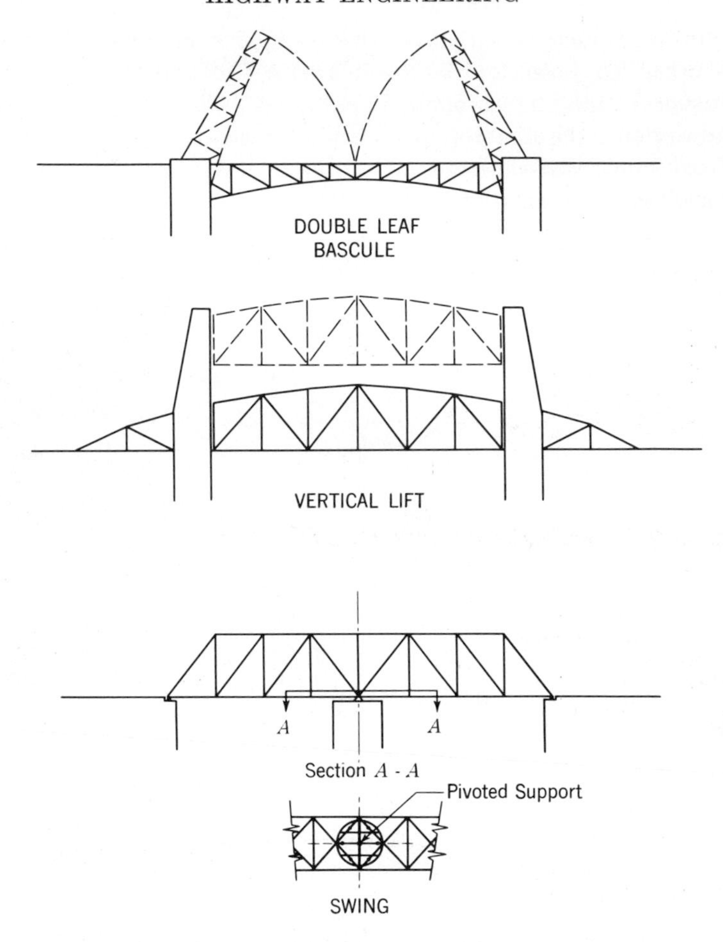

FIG. 7-31. Movable bridges.

7-41. Miscellaneous Bridge Features. Railings are a necessary part of every bridge structure and are constructed of a variety of materials. The railings must be designed to provide adequate protection for the pedestrian and motorist, and they must be of pleasing appearance. If possible, they are made low enough so that the motorist's view is not obstructed. They are placed outside the limits of the traveled way, including curbs and sidewalks, and above the top of the roadway, curb, or sidewalk.

Probably the most widely used material for bridge railings at the present time is galvanized steel. An economical, attractive design is sought; well-designed handrails enhance the appearance of the structure. They are usually of open or lattice construction in order that their weight may be reduced. Pipe sections are often used. Precast and cast-in-place concrete are still in use, and stone masonry railings are constructed in areas where stone is economically available and, for example, when a bridge is faced with stone masonry. Timber railings of different designs are used in connection with timber structures, as in park areas or on secondary roads.

As has been indicated, bridges are customarily provided with curbs, except in the case of bridges of very short span. "Safety curbs" are those designed for occasional pedestrian traffic and are not less than 18

FIG. 7–32. A double-leaf bascule, the North Dearborn Street Bridge in Chicago. (Courtesy *Engineering News-Record*.)

inches in width. The top of the curb is generally not less than 9 inches above the adjacent roadway surface, and any portion of the curb which is higher than 10 inches is sloped or stepped back so that there will be no contact between the vehicle (except the tires) and the curb. Bridge curbs are principally constructed of concrete, except on timber structures.

Sidewalks are necessary to provide for pedestrian travel on most bridges in urban areas and occasionally on rural bridges. If any considerable volume of pedestrian traffic is anticipated, the sidewalk should be separated from the traveled roadway by means of a suitable protecting curb or railing. Sidewalks are commonly constructed of concrete.

On concrete and on many steel bridges, the bridge floor is made of reinforced concrete. In some cases, as in urban areas where the bridge carries an extremely heavy volume of traffic, bituminous wearing surfaces are placed over the bridge floor. Timber plank floors are used on timber bridges that carry a small volume of traffic.

Steel-plate floors have been used on steel bridges in the past but are seldom used in modern bridge design, except on orthotropic plate bridges. Because of its smaller weight, steel grid flooring is often used on steel bridges. The grids, which are of various designs, may be filled with concrete or left open, as desired. The flooring is manufactured in sections which are of a size that can be handled conveniently, and which are generally welded together and to their supports in the field.

Highway Subdrainage

As has previously been mentioned, the term "subdrainage" relates to the control of ground water encountered in highway locations. Subdrains are a necessary part of the complete drainage system for many highways in rural areas, and they function along with adequate surface drainage facilities to prevent damage caused by water in its various forms.

A highway subdrain usually consists of a circular pipe laid at a suitable depth in a trench, which is then backfilled with porous, granular material. Materials principally used in subdrains include vitrified clay pipe, porous concrete pipe, and perforated corrugated metal pipe (both galvanized steel and aluminum). Clay pipe is usually laid with open joints, while concrete and perforated corrugated metal pipe are generally laid with sealed joints. Concrete pipe may also be laid with open joints. The size of drain is usually based upon previous experience. Hydraulic design of subdrains is difficult, but may be carried out on extensive projects (*216*). Six- and 8-inch diameter pipes are in common use although pipes for subdrains are made as large as 24 inches. The slope of the pipe should be sufficient to prevent the deposition or "settling out" of any solid material which may enter the pipe through the joints or perforations. Minimum recommended slope is from 0.15 to 0.25 foot per 100 feet.

Subdrains are installed for a number of purposes, most of which may be included in the following classifications:

1. Control of seepage in cuts or sidehill locations—these installations are generally called "intercepting drains."

2. Lowering of ground-water table, as in swampy areas.
3. Base and shallow subgrade drainage.

7–42. Intercepting Drains. In rolling or mountainous terrain, cuts made during highway construction frequently expose flowing ground water or "seepage." Seepage which occurs through the cut slope may be a source of damage to the slope and to the roadway itself. Similarly, the seepage zone itself may not be invaded by the construction of a sidehill section, but the roadway and pavement structure may be located only slightly above the zone of flowing underground water and thus be subject to the detrimental effects of capillary action. In such cases the flow of underground water is intercepted by a subdrain located on the uphill side of the section so that the water is prevented from flowing beneath the pavement. An example of such an installation is shown in Fig. 7–33. In this example, since there is an impervious soil layer

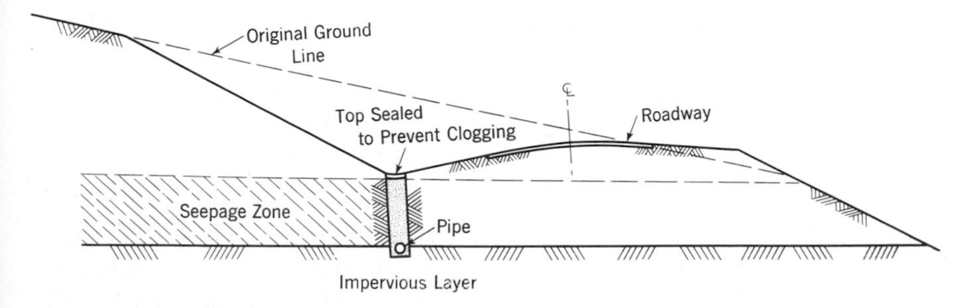

Fig. 7–33. Intercepting subdrain.

located at a relatively shallow depth, the flow of seepage water is entirely cut off by the subdrain. In cases where the seepage zone is deeper, or where no impervious layer is found, the subdrain may simply be carried to sufficient depth to eliminate the effects of capillarity on the subgrade or base. Drains of the type which have been described are normally placed parallel to the centerline and are called "longitudinal drains." In some cases, seepage flow may occur in a direction parallel to the centerline. Transverse drains beneath the pavement may then be needed to intercept this longitudinal seepage.

The location of subdrains of this type before construction begins is somewhat difficult unless a very complete study is made of subsurface conditions. Information obtained from a soil survey may permit the approximate location of subdrains. However, many organizations prefer to provide necessary subdrainage on the basis of information gained during construction. Such a procedure demands the exercise of good engineering judgment in the field, but should be more satisfactory than a decision made in the office upon the basis of incomplete information regarding ground-water conditions. In any case, water collected in the subdrains must be carried to a suitable outlet.

7–43. Lowering of the Water Table. In many locations in flat terrain, the roadway may be built on a low embankment in the interests of economy and the base may be only 2 or 3 feet above the water table. If the subgrade soil is one which is subject to capillary action, water will be drawn up into the subgrade and base with resultant loss of stability, possible frost damage, and similar detrimental effects. The answer to such a problem is simply to lower the water table a sufficient amount to prevent harmful capillary action. This is usually accomplished by the installation of parallel lines of subdrains at the edges of the roadway (or shoulders) at the proper elevation. In practice, the solution is frequently complicated by the difficulty of providing satisfactory outlets for the water collected in the subdrains. The depth required beneath the base varies with the type of soil encountered but is generally from 3 to 6 feet.

7–44. Base Drainage. Water that falls on the surface of flexible and rigid pavements may enter the base and subgrade through cracks in the surface, joints, and shoulders. If the base is relatively impervious, or if it lies above an impervious subgrade soil, this water may collect in the base and the upper portion of the subgrade. This effect may be noted even where water is not brought up into the base or subgrade from below by capillary action. Again, water thus trapped may cause weakening of the base material and it is particularly serious in areas that are subject to frost action. Subdrains of the types which have been described may be used to remove water from the base, although experience has not always shown this type of installation to be successful.

7–45. Laying and Backfilling. Elements of the construction of a pipe subdrain of the type which has been described include the excavation of a suitable trench, preparation of the trench bottom, laying of the pipe, and backfilling with pervious material.

Excavation may be accomplished by either hand or machine methods. Trenches are usually quite shallow, are excavated with vertical walls whenever possible, and lend themselves to construction through the use of a ditching machine in most localities. When the pipe is being laid in a pervious, water-bearing stratum, very little preparation of the trench bottom is necessary other than a nominal amount of shaping. In the case of an intercepting drain laid at the top of an impervious layer in order to ensure complete removal of the intercepted flow, the pipe is generally placed a short distance down into the impervious layer. If soft, unstable soils are encountered, sufficient granular material must be worked into the upper portion of the soil to ensure uniform support of the pipe.

Perforated circular pipe is laid with the perforations down, when located in pervious soils, so that there is less likelihood of clogging by the entrance of fine-grained soil into the pipe through the perforations. For the same reason, corrugated metal and concrete pipe is generally laid with sealed joints. Open joints in bell-and-spigot vitrified clay pipe are protected from silting by covering of the joints with tar paper or some similar material.

Proper backfilling around the pipe and in the trench is of extreme importance if the drain is to function properly without an excessive amount of maintenance. The backfill material must be coarse enough to allow the easy passage of water, and fine enough to prevent the intrusion of fine-grained soil into the pipe.

Criteria which have been developed (*167*), for adequate filter soils are as follows:

$$\frac{D_{15\ \text{(filter)}}}{D_{85\ \text{(protected soil)}}} \leqq 5, \text{ and}$$

$$\frac{D_{15\ \text{(filter)}}}{D_{15\ \text{(protected soil)}}} \geqq 5$$

Where D_{15} is the grain diameter which is larger than 15 per cent of the soil grains; i.e., the 15 per cent size on the grain-size distribution curve. (Similarly, for D_{85}.) For most fine-grained soils, standard concrete sand (AASHO M6 fine aggregate for portland cement concrete) forms a satisfactory filter material. When such sands contain a sufficient amount of fine gravel, and for pipes with small openings, they are safe against infiltration into the pipe. Otherwise, the relationship between the size of the filter material and the size of the perforations in the pipe is as follows:

$$\text{For slots} \quad \frac{D_{85\ \text{(filter)}}}{\text{Slot width}} > 1.2$$

$$\text{For circular openings} \quad \frac{D_{85\ \text{(filter)}}}{\text{Hole diameter}} > 1.0$$

In some states, material of this approximate gradation is used for the entire depth of the trench, with the exception of the top 8 to 12 inches, which may be filled with ordinary topsoil. In other states the upper portion of the trench is filled with a coarse material. Under certain circumstances it may be advisable to provide an impervious seal over the top of the trench to prevent washing of silt and other fine material down into the drain. Adequate compaction of the backfill is essential to prevent excessive subsidence of the ground over the subdrain, and is especially important if the drain is to be beneath the roadway (as in the case of an intercepting drain designed to prevent longitudinal seepage) or beneath a shoulder.

Drainage of City Streets

Precipitation which occurs upon city streets and adjacent areas must be rapidly and economically removed before it becomes a hazard to traffic. The removal of surface water in municipal areas is accomplished by methods similar to those employed in the drainage of rural highways, except that the surface water is commonly carried to its eventual disposal point by means of underground pipe drains or "storm drains." The storm drains may be designed to provide for the flow of ground water as well as surface water, or additional subdrains may be required in certain areas. No further mention will be made of subdrainage in this discussion, as subdrainage installations in urban areas are very similar to those required in rural areas.

The surface-water drainage system in an average city may be considered to be composed of the following basic elements: pavement crown; curb and gutter; and the storm drains themselves, including such commonly used appurtenances as inlets, catch basins, and manholes.

7–46. Pavement Crowns, Curbs, and Gutters. Water falling on the pavement surface itself is, as in the case of rural highways, removed from the surface and concentrated in the gutters by the provision of an adequate crown; crown requirements previously discussed are applicable to city streets. The crown of any street should be made as small as possible, consistent with proper drainage, in the interests of appearance and safety.

The surface channel formed by the curb and gutter fulfills about the same drainage function as a side ditch on a rural highway, and it must be designed to adequately convey the runoff from the pavement and adjacent areas to a suitable collection point. Typical curb and gutter sections have been previously illustrated in Chapter 5. The longitudinal grade of the gutter must also be sufficient to facilitate rapid removal of the water; minimum grades are generally recommended as 0.2 or 0.3 per cent. The grade of the gutter is usually the same as that of the pavement surface, although it may be different if so required for proper drainage. Hydraulically, gutters are designed as open channels of special shape.

Flow in the gutters generally continues uninterrupted to the intersection unless a low point or "sag" is reached. At such a point an inlet must be provided to carry the water into the storm-sewer system. Inlets may also be required between intersections if long blocks are used.

The paved area formed by a street intersection must also be drained, and the adjustment of the grades of city streets at intersections to provide for adequate drainage can be a very complex problem. Several analytical methods have been successfully used for determining the elevations required within the intersection to remove water from the intersection area and to adjust any differences in grades which may be present on the streets that approach and leave the intersection. The solution may also be accomplished in the field during the construction process without previous computation, especially where light street grades are involved.

7–47. Inlets. Inlets are generally provided at intersections to intercept the water flowing in the gutters before it can reach the pedestrian crosswalks. It is essential that enough inlets of sufficient size be provided to rapidly remove collected storm water. Improperly drained intersections are a constant source of annoyance and danger to both vehicular and pedestrian traffic.

The selection of the type of inlets to be used and their positioning in the area adjacent to the intersection are matters of individual choice, and considerable variation in practice occurs. As an example of the proper positioning of inlets at the intersection of two streets of normal width, the simple sketch of Fig. **7–34** has been prepared.

On lightly traveled streets it may sometimes be desirable to provide for the flow of surface water across the intersection by means of a shallow paved gutter or "dip." Such a procedure is not generally recommended, but it may be necessary in the interest of economy.

Inlets of various designs are used by different organizations. The type most generally used on new construction may be called the "combined curb-and-gutter inlet." This type of inlet basically consists of a concrete or brick box with an opening in one side, near or at the bottom, into a circular pipe forming one of the laterals of the storm-sewer system. Water enters the box through a metal grate placed over a horizontal opening in the gutter and through a relatively narrow opening in the face of the curb. This latter opening in the face of the curb may or may not be protected by a grating of some kind. Inlets of this type are either cast in place or composed of precast units. Inlets are also designed without the opening in the curb, the water flowing in only through a grated opening in the gutter. Some inlets have also been constructed with only curb openings.

While inlets of the types which have been described usually have only an outlet in the form of a circular pipe placed so that the invert of the pipe is flush (or nearly so) with the bottom of the box, they are also constructed with a pipe inlet (in addition to the curb and gutter openings) and a pipe outlet. In other words, the inlet may be a connecting link in the storm-sewer system. Still another variation in design occurs in what may be called a "drop inlet." In a drop inlet, the water entering from the gutter must drop some considerable distance to the bottom of the box or may simply be conveyed directly into the storm sewer by means of a vertical pipe.

Inlets are susceptible to clogging with debris and ice, and they must be subjected to continual inspection and maintenance if they are to function properly. The proper hydraulic design of inlets to accommodate a given amount of water under given conditions is beyond the

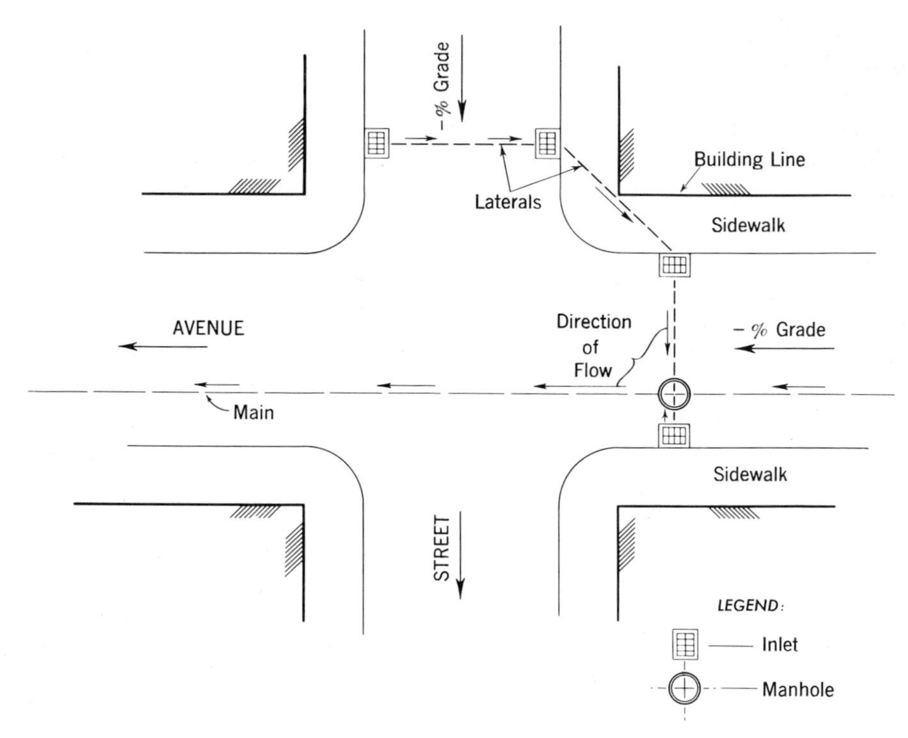

FIG. 7–34. Drainage at an urban intersection.

scope of this book. Recent research has lead to increased understanding of the hydraulic principles involved and the development of various design methods (*71,17*). Curb openings for inlets, on the average, are from 2 to 4 feet in length and from 6 to 8 inches high.

7–48. Catch Basins. Catch basins are similar to inlets both in their function and design. The principal difference between a catch basin and an inlet is that in a catch basin the outlet pipe (and inlet pipe if present) is placed some distance above the bottom of the chamber. The notion behind the installation of a catch basin is that debris flushed from the street into the drainage system is trapped in the bottom of the catch basin and so does not enter the storm sewer itself. This consideration is of less importance now than in former years because of the universal construction of permanent-type street surfaces, more efficient street-cleaning methods, and the general design of storm-sewer systems to carry water at higher velocities of flow. This latter feature has made many sewer systems practically self-cleaning. Catch basins require good maintenance; if they do silt up and the debris is not removed, they then function as inlets.

7–49. Manholes. Storm-sewer systems are subject to partial or complete clogging, and facilities must be provided for cleaning them at regular intervals. In modern sewerage practice, manholes are generally placed at points where the sewer changes grade or direction, where junctions are made, and at intermediate points, usually at intervals of from 300 to 500 feet. The opening provided must, of course, be large enough to permit a man to enter the manhole chamber and have room in which to work. They are usually about 4 feet in diameter in their main portions and are carried to a depth sufficient to perform the purpose for which they are intended. The pipes generally enter and leave the manhole at the bottom of the chamber, and various arrangements are made for carrying the flow through the manhole with a minimum loss in head. Manholes are built of concrete masonry, concrete block, or brick, and cities generally employ a standard design of manhole to be used in all installations. Manhole covers (and frames) are generally of cast iron; the covers are circular in shape and about 2 feet in diameter. When the entrance to the manhole occurs in the traveled way, as it usually does, special care must be taken in the design and placing of the cover relative to the finished street surface if traffic is not to be impeded by the presence of the manhole.

PROBLEMS

7–1. Estimate the time of concentration of water for an overland flow distance of 1000 feet, assuming that the area is poor turf with an average slope of 1 per cent. How much additional time would be required for the water above to flow a distance of 800 feet in a side ditch with a slope of 1.5 per cent?

7–2. Using the Bureau of Public Roads chart, estimate the runoff to be expected once in 25 years from a 30-minute storm on a small drainage area located near Chicago, Ill. What would be the expected rainfall intensity near Mobile, Ala.? Near Denver, Colo.? Make the same estimates for a design frequency of 10 years.

7–3. Using the rational method, determine the expected runoff from an area of 300 acres located near Chicago and subjected to the 30-minute, 25-year rainfall intensity determined in solving Problem 7–2. The area is 10 per cent bituminous pavement, 10 per cent gravel, and the remaining 80 per cent impervious soil with turf.

7–4. The drainage area which contributes runoff to a proposed culvert installation is 475 acres in extent. The rainfall intensity for which this culvert is to be designed is 3.25 inches per hour. The area is located in fairly flat farming country, with an average slope of about 2 per cent. The soil is slightly pervious in nature and the amount of vegetal cover is variable. Determine the quantity of flow for which the culvert should be designed by the rational method.

7–5. Assume that a 36-inch concrete-pipe culvert is to be built beneath a four-lane divided highway which has a total width of 96 feet from outside shoulder line to outside shoulder line. The culvert is at right angles to the centerline of the highway, and is to be placed at a slope of 0.5 per cent. The top of the culvert pipe is 12 feet below the finished surface at the centerline, and the embankment slopes are 3:1. Neglecting the shoulder slope, roadway crown, slopes in median strip, etc., compute the required total length of pipe if the culvert is to be built with the end sections shown in the drawing of Fig. 7–17.

7–6. Estimate the quantity of concrete required for the endwalls for a three-barrel, concrete pipe culvert, if the pipes are 54 inches in diameter and built in accordance with Fig. 7–14. Make the same estimate for a three-barrel corrugated metal pipe culvert of the same pipe size. Fill slope in each case is 2:1.

7–7. Determine the depth and velocity of flow in a trapezoidal channel ($n = 0.030$) with 2:1 side slopes and 4-foot bottom width discharging 90 cubic feet per second on a slope of 2.0 per cent. Determine the critical depth, velocity, and slope for these conditions. Assume that this is a side ditch which is to be cut in a clay soil. Is the velocity of flow excessive? If it is, what could be done about it, assuming that the slope of the ditch cannot be greatly changed?

7–8. Determine the depth and velocity of flow in the channel of Problem 7–7 for a value of $n = 0.013$.

7–9. (a) Given a 50-foot-long concrete pipe culvert with a groove-edge entrance to carry 120 cubic feet per second on a slope of 0.1 per cent, determine the size of pipe required if the limiting headwater depth is 6 feet. (See Fig. 7–9.)

(b) What size of concrete pipe would be required to carry 120 cfs through a long culvert (360 feet) on a slope of 0.3 per cent if the allowable headwater depth was only 5.5 feet?

(c) How many 36-inch corrugated metal pipes with projecting entrances would be required to carry 120 cfs if the pipes were to be 50 feet long and laid on a slope of 0.5 per cent? Assume that flow in a multiple-barrel culvert is divided equally among the barrels and the limiting headwater depth is 5.5 feet.

7–10. Assume that a rectangular concrete box culvert is to be 200 feet long and is designed to carry 200 cfs on a slope of 0.2 per cent with a limiting headwater depth of 7 feet. Determine the required width of the culvert if the height of the culvert is (a) 3 feet and (b) 4 feet. (See Fig. 7–10.)

7–11. A structural plate corrugated metal pipe arch with a projecting entrance is to be 100 feet long and laid on a 1 per cent slope. From Fig. 7–11 determine the size of pipe arch required to carry 400 cfs if the limiting headwater depth is 10 feet. What size is required if the slope is decreased by 0.5 per cent?

CHAPTER 8

SURVEYS AND PLANS

In relocating old highways or establishing new ones, surveys are needed to be able to prepare plans for the project and to estimate its cost. Good surveys require engineers with first-class training who are experienced in planning, design, construction, and economic analysis. The work of the highway location survey for a given project may include (1) a reconnaissance, (2) a preliminary survey, and (3) a final location survey. Each phase of the location survey is discussed in some detail in the following sections.

8–1. Available Techniques. The determination of a highway alignment may be done either by conventional ground survey methods, or by aerial photogrammetry. While the older techniques are still used on projects of medium and small size, the use of photogrammetric surveys on large works is almost universal, and is becoming more frequent on smaller projects. The adaptability of photogrammetry to computer operations has enabled substantial savings in time and money over conventional methods.

The Michigan Department of State Highways estimates that aerial surveys reduce final survey time by 30 per cent. Using conventional survey methods and a five-man crew, MSHD engineers estimate it costs about $1000 a mile for a preliminary survey; the use of aerial photogrammetry reduces the cost to $400 per mile and greatly reduces survey time.

8–2. Reconnaissance. The term "reconnaissance," as applied to highway location, may be defined as an exhaustive preliminary study of an area in which an improvement is to be made. The first step in any reconnaissance is to procure all available pertinent data. These data may be in the form of maps, aerial photographs, charts or graphs, and so on, and may require the application of a large variety of engineering and economic knowledge. All reconnaissance work is not of the same magnitude or importance. In the planning of an expressway or other high type of improvement, it can be readily seen that many factors must be considered. On secondary roads, only a few of the engineering and economic factors may warrant consideration. A large mass of engineering and economic information may be available to the location engineer, as indicated below.

8–3. Engineering Data. Engineering data help determine the feasibility of a project and engineering problems that may be encountered. The physical features, either natural or man-made; the kind, type, and number of vehicles that may be expected to use the highway; and the probable cost are items which may be included. Much information may be readily available in the form of topographical and geological maps of the particular area to be investigated. Stream and drainage basin maps, along with climatic records, may prove to be valuable in the study of drainage problems. Preliminary survey maps of previous projects in the area may also be a source of information which can be applied to the contemplated improvement. Engineering data available from planning and traffic surveys should also be incorporated in the reconnaissance. These should include maps, charts, and tables showing the type, density, and volume of traffic to be expected. Information as to practical working capacities of highways and streets is also usually available. In order to estimate the probable cost of the improvement, cost data may be readily obtained from existing tabulations of unit costs for road and bridge construction. Land usage maps and records of the physical inventory showing road surface types and widths represent other types of information which are generally available from inventory divisions of planning surveys.

8–4. Economic Data. Economic data should provide a means of evaluating the benefits to be derived from the contemplated improvement. Data relative to the possible reduction in vehicle operating costs because of time saved, distance shortened, or improvement in alignment or reduction in grade will aid in determining the dollar savings which would result from the proposed improvement. Records showing the density, distribution, and volume of population and motor car registrations are a source of information relative to benefits to be received by establishing a new or improved highway facility. Origin and destination surveys, with maps showing location and types of impediments to the free flow of traffic, should be available from planning and traffic surveys. Records of planning by other local governmental agencies should be studied so that economies may be effected by cooperative planning. Records of the various transportation media serving the community, such as buses, freight-truck lines, and so on, together with the number of scheduled trips, should also be considered. The density, distribution, and dollar volume of agriculture, manufacturing, and wholesale and retail sales will in part determine the benefits that may result from the proposed improvement.

8–5. Analysis of Data. When all the available engineering and economic data have been assembled, a detailed analysis should reveal much information pertinent to the proposed project. For example, analysis of the available information may allow the engineer to determine the advisability of selecting an entirely new location or improving the existing one. After an exhaustive study of aerial photographs, topographical maps, drainage maps, soil maps, and other data is made in the office, a series of proposed locations may then be selected for a field investigation.

8–6. Reconnaissance Survey from Aerial Photographs. On many projects, the data on available maps and photographs will be out of date or incomplete. In some cases, there are no available base maps of suitable scale and accuracy. A reconnaissance survey from aerial photographs will amplify and verify conditions determined from the preliminary study. The function of the survey is to provide sufficient information on the topography and culture of the area to enable the selection of a preliminary route location.

Control points between the two terminal points on the route are determined, and the flight lines between these control points established. It is customary for the width of coverage to be about 0.4 to 0.6 the distance between control points. From the aerial photographs, a base map is

FIG. 8–1. A stereoplotter used by the Kentucky Department of Highways.

produced, usually to a scale of 1 inch = 200 feet, with a 2- or 5-foot contour interval. A stereoscopic examination of the photographs will also permit the determination of ground cover and soil conditions (Fig. 8–1). From the base maps, and photo-interpretation of the surveyed area, likely routes can be designated.

The most likely route is that route which best satisfies the following requirements:

1. Traffic service for population and industrial areas
2. Directness of route
3. Suitability of terrain encountered
4. Adequacy and economy of crossings at water courses and at other transportation routes

The reconnaissance is an important portion of the work and should not be treated lightly. The engineer should not dismiss any route until he has carefully examined its possibilities. The use of photogrammetric base maps enables an objective study of all possible routes with greater ease than field reconnaissance.

8–7. Selection of Preliminary Centerline. Possible alternative alignments are plotted on the base map and from these alignments, with preliminary gradelines, the alternatives are compared for suitability. Such features as design capacity, safety, road-user costs, construction costs, right-of-way costs, and maintenance costs are examined. Based on the results of this preliminary analysis, a preliminary alignment is chosen. Fig. 8–2 and Fig. 8–3 show preliminary alignment locations set down on photogrammetric maps.

8–8. Mapping for Final Design. For final design and location, further photogrammetric mapping is necessary. The selected preliminary alignment is used as a guide for the strip area to be mapped. It is becoming customary to use coordinate systems in highway design, because of the increasing use of electronic computers for direction and distance calculations. Therefore, at this time, prior to taking low-level aerial photographs, a baseline is established on paper, and by monuments on the ground. The low-level aerial photographs are taken with markers on the ground to tie in the baseline. Distances between the baseline monuments often are determined by electronic measuring devices such as the Tellurometer or Geodimeter, which have excellent accuracy over long distances. The baselines, which run between monuments of calculable coordinates, can be used in the final location survey in locating the alignment in the field.

The low-level photographs are used to produce base maps along the preliminary alignment, usually at a scale of not less than 1 inch = 100

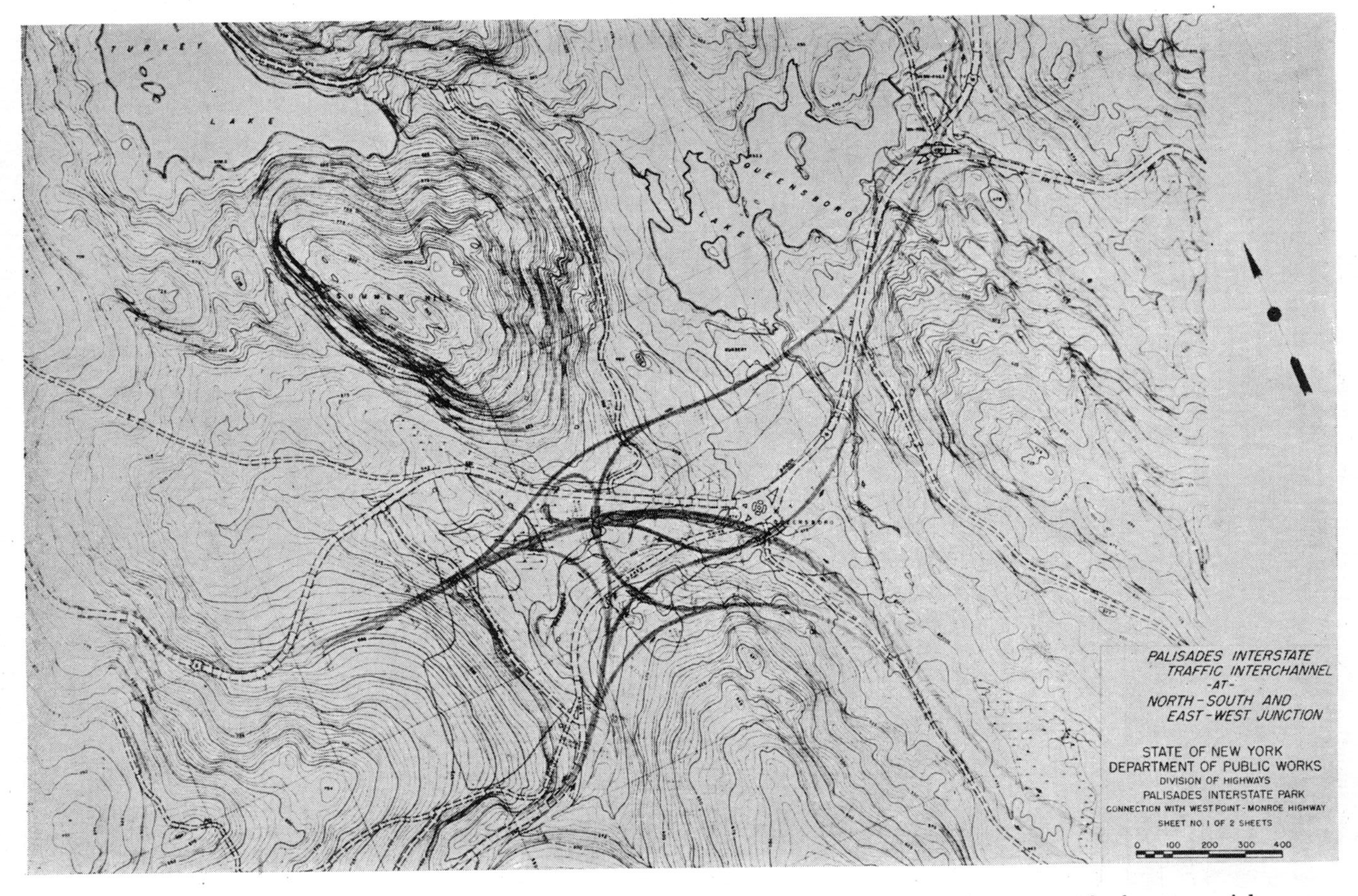

FIG. 8–2. Preliminary location of an intersection of a parkway on a photogrammetric map made by an aerial survey. (Courtesy Bureau of Public Roads.)

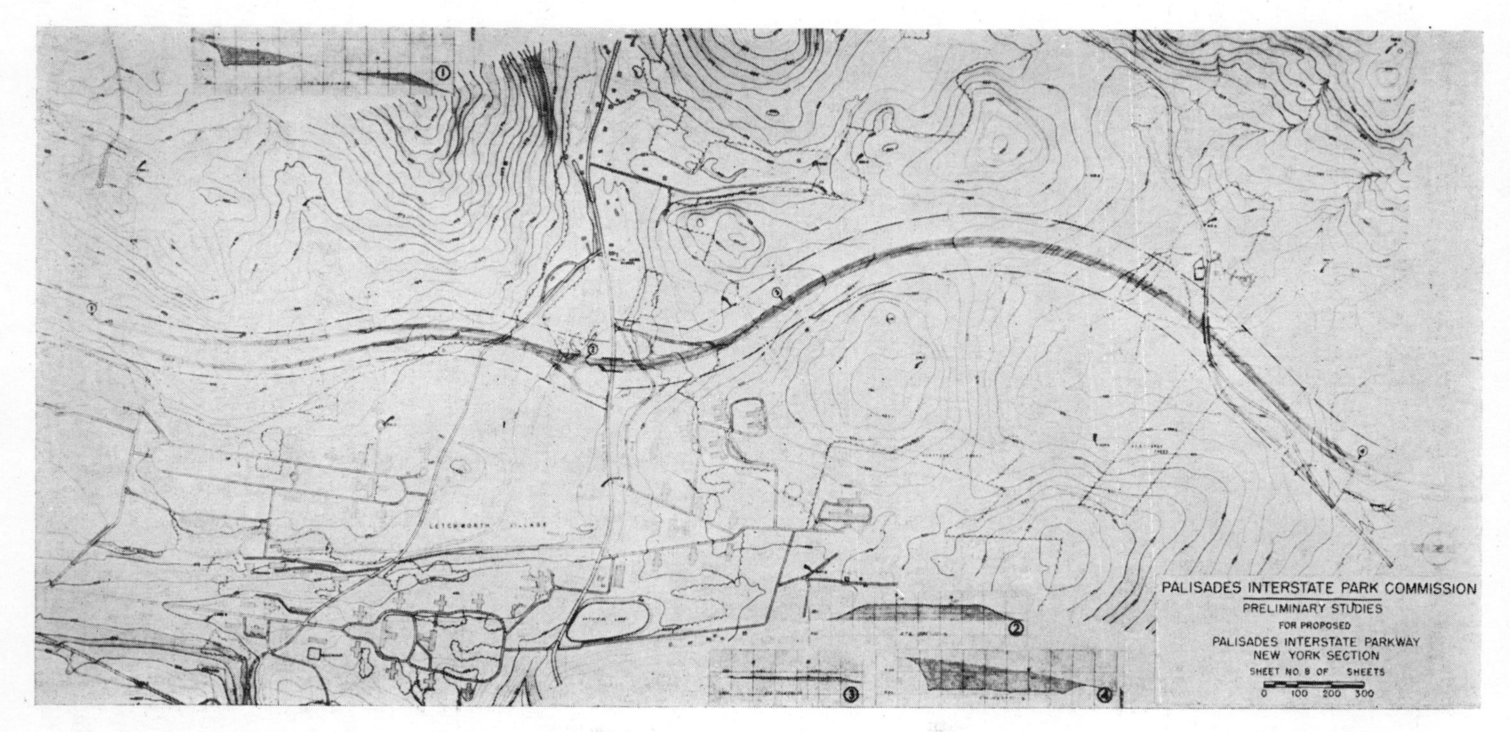

FIG. 8–3. Preliminary line laid down with a spline on a photogrammetric map made by an aerial survey. (Courtesy Bureau of Public Roads.)

feet. These base maps are more complete than those used for reconnaissance. The preliminary survey taken at a lower level can give virtually complete information on:

1. Topography, with respect to changes in elevation and drainage characteristics, and information concerning the soil conditions in the area
2. Land uses, designating type, intensity and quality
3. Transportation facilities, with respect to proximity and effect of proposed location
4. The effect of the proposed location on property within the area

For the purposes of design, however, the preliminary survey will produce topographic base maps, and information concerning property and utility location which will enable the engineer to proceed with the design of the final alignment.

8–9. Design of the Final Alignment. The process of the design of the final alignment is one of great skill and judgment. It is a trial-and-error process by which the most economic alignment is obtained. The highway is fitted by hand to the topography and land use, until the engineer is satisfied that no better fitting can be achieved. Modern practice makes great use of splines and curve templates. The spline is a flexible plastic guide, which can be bent to different positions until the most satisfactory curvilinear alignment is achieved. Curve templates can come in such simple forms as the old-fashioned highway curves, which are boxes of circular segments usually in increments of 15 minutes (degree of curve). Currently these are supplemented by transparencies, which to a standard scale display circular curves, three-center compound curves, and transition spirals.

To facilitate computation, tangent alignments can be set at even degree angles. This speeds computations by permitting the use of design tables, which can be generated within the computer. In urban areas where small changes in alignment may cause property encroachment, the use of whole-degree azimuths may not be feasible.

The method of design usually followed is to adjust the spline until a smooth, consistent alignment is achieved which is satisfactory from the viewpoint of grade, curvature, cross-sections, drainage, and stream crossings. The hand-fitted line is then converted into a defined line by fitting to it tangent lengths, transition spirals, and simple and compound curves using the visual aids of the templates. The standardized templates contain information which can be used directly to compute coordinates of the controlling points on the curvilinear alignment, and for the computation of tangents.

8–10. Final Location Survey. Conventionally, the method of establishing the final location in the field has been by direct chaining along the tangents, setting first the P.I.'s, and then closing out the horizontal curves between the tangents, using deflection angles and chaining to set the curve stations. This method is still in use, although the electronic distance-measuring devices have to a great degree replaced chaining procedures. Moreover, the use of coordinates in highway location means that the computer can be used to calculate azimuths between control points on the curvilinear alignment (e.g., P.C., P.T.) to points of known coordinates on the baseline itself. The errors along the alignment are not in this case additive. The P.C.'s and P.T.'s can be set and checked by the intersection of three azimuths instead of two. Chaining between the control points on the alignment is done to set stakes at the even 100-foot stations, and stakes on the curves can be set from the P.C. or P.T. by normal deflection angle methods.

In general, it is unnecessary for ground crews to take cross-sections along the alignment. Such cross-sections can be obtained directly from the photogrammetric maps with as much accuracy as ground surveys normally achieve.

8–11. Conventional Ground Techniques. Where photogrammetric surveying methods are unavailable, or where the size of project makes the use of aerial photographs uneconomic, it may be necessary to use the older, conventional ground surveys. Such instances are becoming less common, as the development of photogrammetry has led to the widespread availability of aerial photographic equipment. This can generally be used to economic advantage.

The steps involved in ground surveys are similar to those where photogrammetry is used:

1. Reconnaissance
2. Preliminary location survey
3. Final location survey

Reconnaissance includes the preliminary steps of evaluation of all available data as covered in Art. 8–3. The reconnaissance survey is completed by a field investigation which usually provides a means of verification of conditions as determined from the preliminary study. For example, building symbols on maps do not indicate the true value of property under consideration, and this information can usually be secured by field investigation. A study is made of the profiles and grades of all alternate routes and cost estimates made for grading, surfacing, structures, and right-of-way. A comparison of alternative routes in this fashion will aid the final selection of the most likely location.

Next, a preliminary survey is made to gather information about all the physical features which affect the tentatively accepted route. In general, the work is carried out by a regular survey party consisting of party chief, instrument men, chainmen, and rodmen. A primary traverse or baseline is established. This is an open traverse consisting of tangent distances and deflection angles following approximately the line recommended in the reconnaissance report. When the preliminary line has been established, the topographical features are recorded. The extent to the right and left of the traverse line to which the topography should be determined will vary, and is usually left to the judgment of the party chief in charge of the work. It will certainly never be less than the proposed width of right-of-way.

The amount of level work on a preliminary survey should be kept to a minimum. In flat areas no centerline profiles are needed; in rolling or mountainous terrain a profile of the traverse centerline is often sufficient. A few cross-sections are sometimes of value in exceptionally rugged areas where centerline profiles do not give proper information.

Using the preliminary survey as a basis, a preliminary survey map is drawn. The preliminary map should show all tangents with their bearings and distances, all deflection angles, section corner ties, and all witness points of the alignment. Certain topographic features are shown on the map, such as streams, water courses, lakes, hills, ravines, etc., and man-made features such as buildings, drainage structures, power lines, and other public facilities.

The design of the final alignment when ground survey methods have been used is essentially the same as the procedure indicated in Art. 8–9. The location of the centerline requires considerable skill and judgment. Trial lines are drawn on the map and are related to control points, such as road intersections, building corners, etc. The trial lines should avoid badly drained ground, rough or impassable areas, valuable property, or other obstructions. The use of splines and curve templates is recommended. Rough estimates of excavation and embankment may also influence the placing of the trial lines on the map. Essentially, the only difference encountered when conventional ground methods are used rather than aerial surveys is that it is more likely that with the latter the whole project will be tied to a coordinate grid system to facilitate the use of electronic computers in determining the distances and azimuths. On the smaller projects where ground methods are still used, the older system of chaining along the centerline, tangent distances, and deflection angles may be the essential data given for the final location survey.

The final location survey serves the dual purpose of permanently establishing the centerline and collecting the information necessary for the preparation of plans for construction. The line to be established in the field should follow as closely as is practical the line drawn on the preliminary map, conforming to the major and minor control points and the alignment which has been previously determined.

The first step in the final location survey requires the establishment of the centerline, which is used as a survey reference line, upon which property descriptions are based for the purpose of purchasing right-of-way. Modifications of the final location may also be made in the field as a result of incomplete or inaccurate preliminary information. Centerline stakes are placed at 100-foot intervals, and each stake is marked with the proper station number. Curves are staked out by deflection angles, and the error of closure is determined by an independent stake-out of the P.T. from the P.I. Particular care must be taken in the chaining, and a closure should be effected where possible on control hubs for the preliminary survey, since errors along the centerline are cumulative.

Level work is of the utmost importance, since the grade line, earthwork, and drainage are designed from the level notes. Elevations may be obtained from United States Geological Survey or other bench marks, or elevations may be assumed for a project. Bench marks are permanent reference points, and extreme care should be used when establishing benches along the route of the proposed project. The bench marks should be placed approximately 1000 feet apart in fairly level country and closer in hilly country. Benches should always be placed at least 200 feet from any proposed structure, if possible. Bench mark elevations should be established by the loop system of levels or by any other approved method where a high degree of accuracy can be obtained. The error of closure between benches should never be greater than 0.01 foot.

The importance of keeping good notebooks on alignment and level surveys cannot be overemphasized. Notes taken in the field during the conduct of the final location survey usually become a part of the permanent record of the project.

Finally, cross-section levels are taken at even 100-foot stations and at any intermediate points with abrupt slope changes. This information is necessary for accurate determination of existing grade line, design of drainage structures, computation of earthwork quantities, and, in general, estimation of quantities of work to be done. The manner in which the cross-sections are taken and the form of notes used are fairly well established. In order to obtain a fairly representative cross-section at a station, an elevation is always obtained at the centerline stake. Then elevations and distances at right angles to the centerline are recorded, first on one side and then on the other. Cross-sections should also be taken at least 500 feet beyond the limits of the point of beginning and the point of ending of a project. Centerline profile elevations should also be obtained at some distance beyond these points in order to establish proper grades for connecting projects.

8–12. Problems of Highway Location in Urban Areas. In rural areas, the location of a highway is dictated chiefly by the desired end points of travel, topography, and geology. The road should serve a desired travel pattern with an alignment that provides consistent conditions within acceptable limits of curvature and grade over terrain that is capable of supporting the proposed construction without excessive costs.

In urban areas the location process is more involved, reflecting the complexities of urban conditions. Among the conditions which control urban locations are

1. Traffic service
2. Land use
3. Off-street parking
4. Other transportation systems
5. Topography and geology
6. Sociological considerations

The traffic planner, as a result of his extensive investigations of the travel desires and existing travel patterns in an urban area, can predict the effect on travel patterns of alternative locations of proposed highways. Traffic assignment techniques can be used to analyze the probable effect of new links in the highway network. This permits the location engineer to determine how well alternative locations will fit with the existing network. Typical travel desire lines are shown in Fig. 4–8. The major lines often can show the major elements of a desirable arterial highway system in an urban area.

Land use is a major factor in the pattern of traffic generation in an urban area. Most cities have a central commercial core surrounded by areas of residential, industrial, and commercial development. Travel patterns vary with time of day, day of week, and season of year. Trips from the home to work are heavy to the commercial downtown area and the industrial areas during the morning peak hours, with a reversed pattern during the evening rush hour. During the day, shopping trips from the residential areas to the outlying commerical areas and the center core predominate. The pattern of land use affects location in other ways. Industrial and commercial areas rely heavily on truck transportation and, therefore, need service by arterial routes. Health and safety requirements may indicate that heavily traveled routes should not be located through residential areas. Aesthetic considerations may indicate that an alignment which cuts through a public open space should be avoided, if the space can be preserved by a skirting alignment. Most cities have recognized the interaction between highway location and land use and have integrated a major thoroughfare plan with the city master plan.

Parking is of major importance in urban areas. New highway locations must be considered in relation to existing or planned parking facilities for the vehicles attracted to the new facility. Ideally, the location should be close to existing and potential parking areas to minimize the amount of travel on existing streets.

Interference and interaction with other transportation systems must be considered. The aim of any new facility is to increase the over-all level of service of transportation in the urban area, and a location should not unnecessarily disrupt existing service. Over-all coordination of the various transport modes is essential in the development of the balanced transportation system so urgently needed in complex urban areas.

As in rural areas, topography and geology have a great effect on urban highway locations. Topographic barriers such as ridges, valleys, rivers, and lakes affect construction costs and determine the economy of a route. Similarly, soil and ground water conditions may render an otherwise feasible location uneconomic. The effect of topography and geology is somewhat lessened in an urban area owing to the high cost of right-of-way in certain areas. For example, the cost of a location through a heavy industrial area might greatly exceed costs involved in remedying the swampy conditions of an alternate alignment. The controls of topography remain, however, imposing grade and curvature controls on any designated alignment.

The complex of utilities which serves an urban area is usually almost totally absent from rural areas. Extensive relocation of utilities where depressed highways and underpasses are involved can add greatly to the cost of urban projects and require considerable advance planning. In some cases, as in Manhattan, the underground maze of utility lines and subways can make the cost of a depressed highway prohibitive.

The social effects of a highway project must not be forgotten in an urban area. Proposed locations should not cut up residential neighborhoods. The results could well be a drop in land values, then urban blight, and finally the creation of a slum. Open public land should be retained where possible; highways should not eat up open space desperately needed for urban recreation. Similarly, a highway can help support land values by screening industrial areas from residential communities. Among the more important social forces which affect urban location are the federal-aid programs of urban renewal, which are eradicating slums in central urban areas, creating, in many cases, ideal sites for new highways.

8–13. Completion of Location Survey. The final location survey is complete when all the necessary information is available and ready for the designer to use. This includes all information pertaining to (1) alignment, (2) topography, (3) bench mark levels, (4) cross-sections, (5) section corner ties and other land ties, (6) plats and subdivisions, (7) soil surveys, (8) drainage and utilities.

Much of the work involved in route location is repetitive and requires little exercise of the engineer's professional skill. Therefore, in the last few years, great emphasis has been placed on increasing the individual engineer's productivity by the use of automated procedures which rely extensively on electronic computers.

The photogrammetric aerial survey methods already described are readily adaptable to automation. Digital terrain models can be stored

Fig. 8–4. Wild A7 Autograph with large drawing table and coordinate printer. (Courtesy Wild Heerbrugg, Inc.)

in computer memory by linking the stereoplotter to a device which automatically records the x, y, and z coordinates of points on the stereo model. Input of a proposed alignment will result in the computation of horizontal centerline geometry, vertical profile geometry, and earthwork quantities. This allows rapid evaluation of the proposed alignment and permits the engineer to make small modifications without the tedious and expensive repetition of manual calculations.

From the stored terrain data and the input of a proposed alignment, it is possible to obtain directly cross-section data in numerical form or, by linking the computer to an automatic plotting device, to obtain cross-sections in a graphic form. The computer output can be directly utilized in the plans by the "stick-up" method described in Art. 8–27.

By the use of more sophisticated stereoplotters, the amount of ground control work required for the aerial survey can be greatly reduced and time and cost of the survey minimized. The stereoplotter shown in Fig. 8–4 is an example of a machine which uses *instrumental aerotriangulation* in a process called "bridging," which eliminates the need of ground control points for each pair of stereo photographs forming a stereo model. The computer can also be used for analytical solutions of aerotriangulation problems. Bridging is a process whereby the horizontal and, to some degree, the vertical control on one stereo model can be extended through successive models until additional ground control is reached.

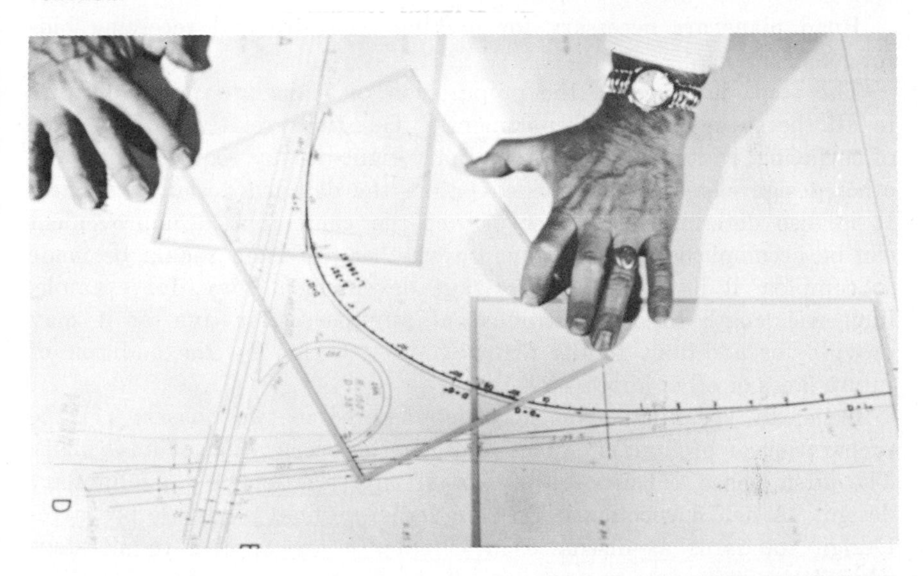

Fig. 8–5. Use of a transparent template in ramp layout. (Courtesy DeLeuw, Cather & Co.)

Most highway departments use computer programs for the adjustment of field survey traverses and for the computation of earthwork volumes from cross-section data. Modern techniques of establishing horizontal alignment rely extensively on the use of transparent templates for curve design and intersection layout. Precomputed data from these templates are compiled into design tables which can be used as input for the computation of the horizontal alignment referenced completely to a coordinate system. The use of templates and design tables relieves the designer of much computation and (possibly more important) ensures more standardization and higher design standards. Fig. 8–5 shows the use of a transparent design template in the design of an interchange ramp.

Extensive use has also been made of the computer in the layout of interchanges and bridges. The intersection or bridge is related to the coordinate system of the project. From this relation, the azimuths between any two points are readily available by machine calculation. Field layout can be done by measuring along an azimuth from a known point, or by determining the point of intersection of azimuths from at least two other points. The latter method has been found successful for setting bridge piers using three azimuths from three points, where chaining is difficult.

Preparation of Plans

Road plans are necessary for making estimates and receiving bids for construction.

The steps involved in the preparation of plans are fairly uniform in all the state highway departments. Questions relative to the type of surfacing, width of grade, preliminary right-of-way requirements, and other design elements are decided before the detailed design is started. It is also determined whether or not the contemplated improvement can be accomplished at one time, or whether provision should be made to complete it by stages. This stage development may, for example, limit the length to be constructed at any particular time or it may provide for additions to the highway cross-section by the addition of future lanes or other improvements.

Generally speaking, plans may be prepared in two phases: (1) the preparation of preliminary plans and (2) the preparation of final plans. The first phase consists of a partial investigation and preliminary design. A field inspection is then made before final plans are prepared. Design standards as discussed in Chapter 5 are applied to all steps when designing a given project.

8–14. Plan and Profile. With the information obtained from the final location survey, a plan and a profile of the proposed route are drawn on standard plan and profile sheets, as shown in Fig. 8–6. The plan of the road is drawn to some suitable scale, 1 inch equaling 100 feet being the most common. When greater detail is desired, as is often the case in urban areas, the scale may be increased to 1 inch equals 50 feet. The plan should show the proposed construction centerline with the limits of right-of-way and all important topographical features such as fences, buildings, streams, railroads, and other structures on the right-of-way. All survey information necessary for the establishment of the survey centerline should also be placed on the plan. This should include compass or computed bearings of all tangents, together with all curve data, including the point of beginning, point of ending, degree of curvature, and so on, and all witness information to the survey points. Bench marks, with their elevations, are also shown on the plan sheet. Where electronic computers have been used to determine azimuths between control points of known coordinates for the purpose of final location, the coordinates of these control points should be shown on the plan.

A notation is made on the plans of the sizes and types of all existing structures and the manner in which they are to be utilized or removed. Other details are added, such as type and depth of ditches, slopes, right-of-way and borrow requirements, etc.

The centerline profile is plotted on the lower portion of the plan and profile sheet. The horizontal scale is usually the same as that used for the plan. The vertical scale, however, is distorted, with scales of 1 inch equals 5 feet and 1 inch equals 10 feet being most commonly used. The profile usually shows the profile of the existing ground line and the profile of the proposed construction centerline, the tangents and vertical curves forming one continuous profile. All grade lines show the percentage of grade and elevation points where changes of grade occur, with the lengths of the vertical curves used at these points of intersection.

All existing drainage structures are generally shown on the plan and profile sections. These include culverts, catch basins, inlets, manholes, and so on. Any special information pertaining to the profile and affecting the design may also be added, such as curb grades, gutter grades, and sodded slopes.

8–15. Grades and Grade Control. All controls that affect the position of the grade line should be given thorough study. Ideal grades are those that have long distances between points of intersection, with long vertical curves between tangents to provide good visibility and smooth riding qualities.

One consideration in grade control is to try to fit the grade line to the natural ground surface as closely as possible. Good drainage is usually accomplished in this manner. Intersections with railroads and existing bridges are points which are generally fixed and the grade must be set to meet them. Intersections with trunk lines, other surfaced roads, unimproved roads, drives, trails, and field entrances may also be grade controls. The effect of soils on grade is also very important. In swamp or peat areas it is necessary to have fills high enough so that waste material can be disposed of at the sides of the roadway. The grade must also be above the water table, where possible.

The placing of the grade line requires much more detailed study in urban areas than in rural areas. Changes in grade affect store entrances, driveways, and sidewalks. It is often necessary to warp or tilt the pavement to meet existing curbs and to provide proper drainage.

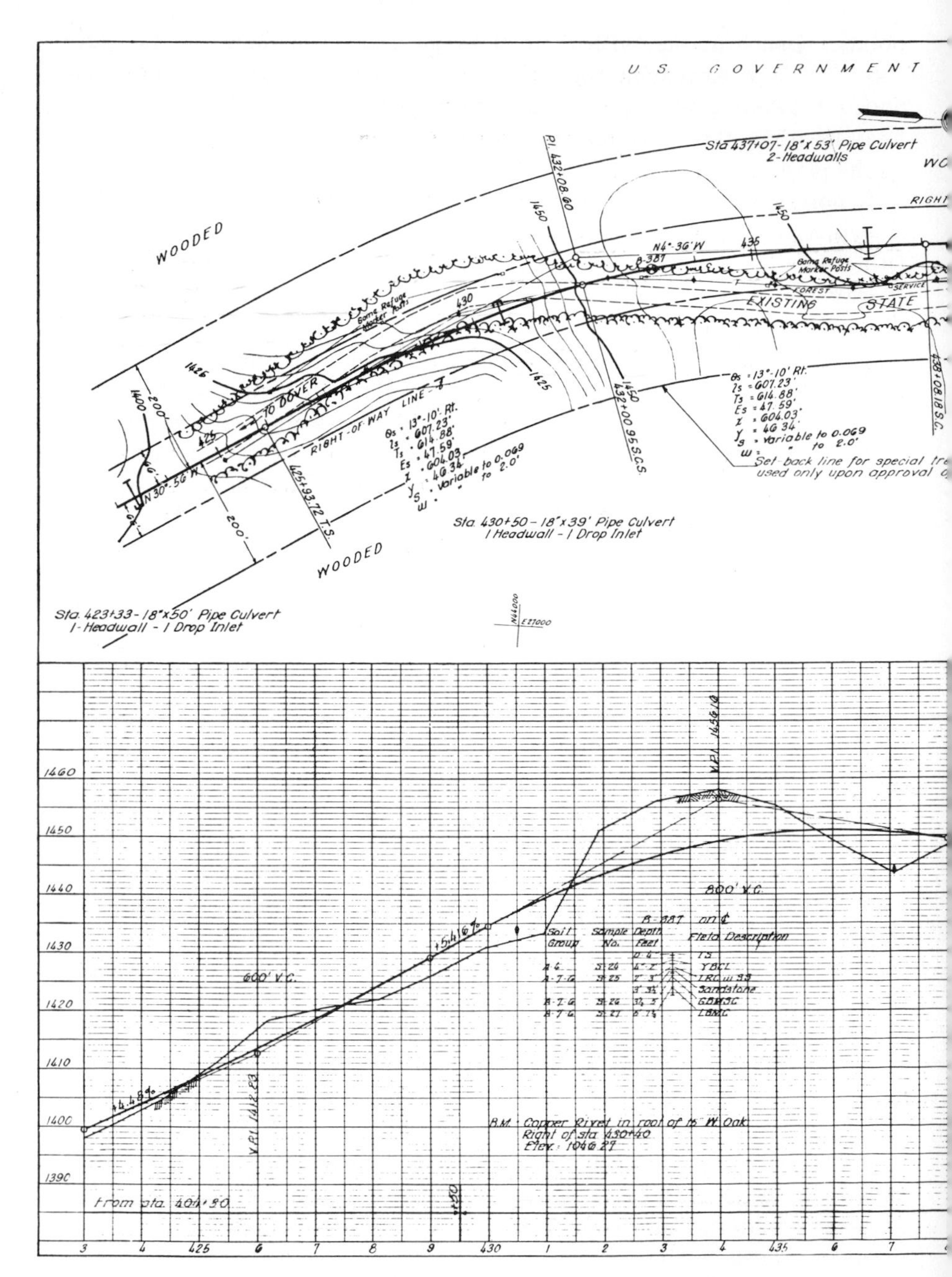

FIG. 8–6. Typical plan and profile s

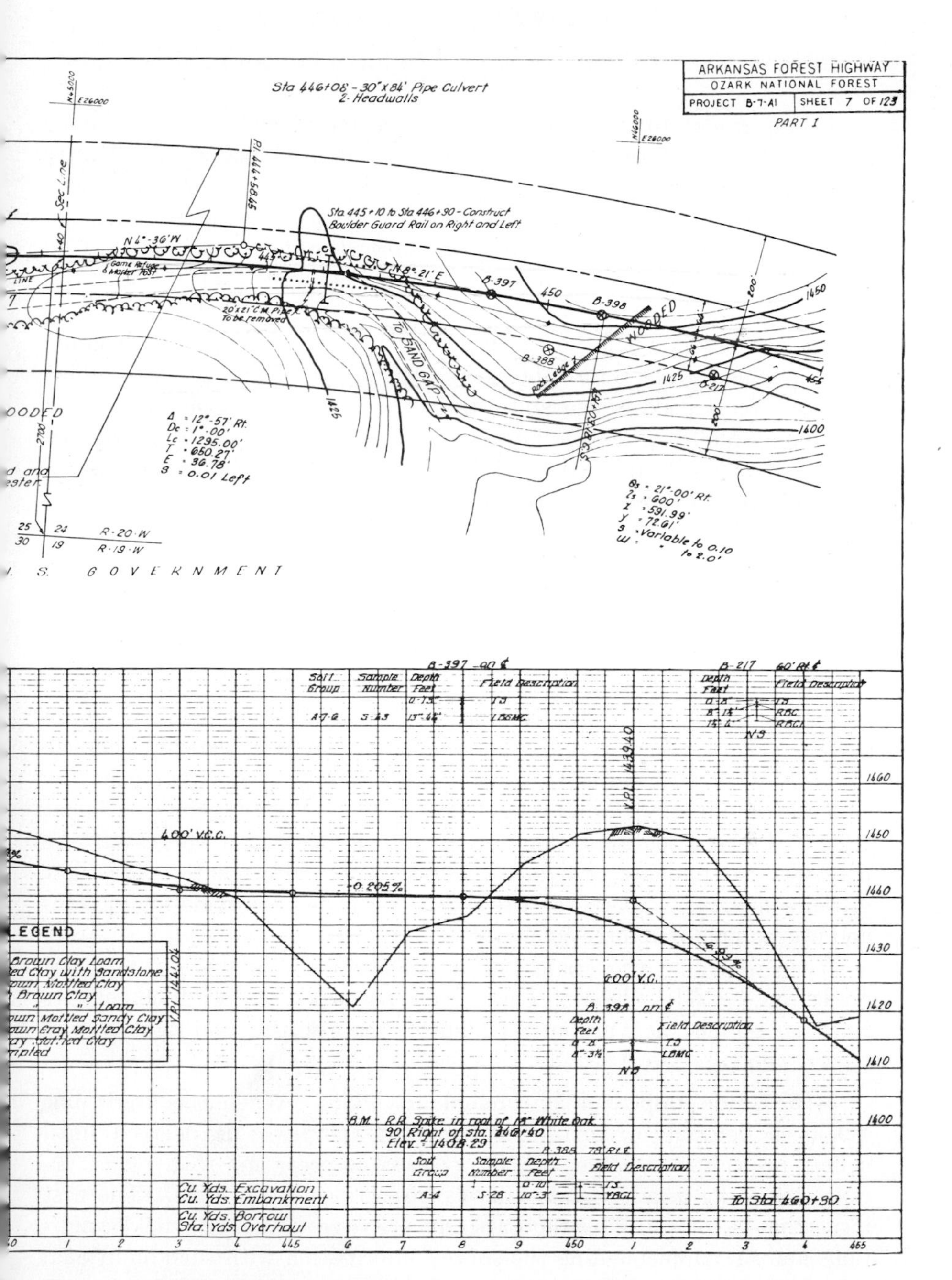

tesy Bureau of Public Roads.)

Slight distance is of fundamental importance in both rural and urban areas.

Ruling grades for the economical operation of motor vehicles have been fairly well established. These have been indicated in Chapter 5. When existing conditions indicate a new grade, care and judgment should be exercised to obtain the most economical one possible.

When establishing a grade line, an effort should be made to provide an economical balance between the quantities needed for fill sections and those required in cut sections. Changes in grade from a plus grade to a minus grade should be accomplished in cuts, and changes in grade from a minus grade to a plus grade should be placed in fills. This usually gives a good design and often avoids the appearance of building hills and producing depressions contrary to the existing slope of the land. All hauls should be downhill, if possible, and not too long.

A common method of laying the grade is to staple the various profile sheets together; with pins as points of intersections, a string or thread is drawn taut from end to end to produce a continuous profile. Shifting of the pins to various positions will give an opportunity to study the effect of different grades without drawing lines on the paper. When the grade is established, the pin points are used as control points and the elevations are noted from the profile elevations.

Horizontal control is usually well established when the location survey is made. However, certain refinements may make it necessary to change the alignment when adjusting grade controls for vertical alignment. A combination of grades and horizontal alignment may give insufficient sight distance, poor appearance, or both. Flattening a curve may be necessary to eliminate or reduce superelevation or to lengthen the sight distance in a cut. Special efforts should be made to provide long grades and proper sight distances on curves.

When designing the grade line, consideration must be given to the grades of the side ditches. The depth and width of the ditch and its side slopes affect the width of the highway cross-section and right-of-way requirements.

8–16. Grade Inspection of Preliminary Plans. When the grade line has been determined, final completion of the plans is delayed until a field inspection is made. In the process of preparing road plans, arrangements are usually made for two inspections in the field. The first inspection is for checking the partially prepared plans with field conditions and for supplementing the survey information with such additional data as are necessary for completing final plans and preparing a correct estimate of quantities. The field investigation can serve as a check for the general design and should reveal whether or not the width of right-of-way is sufficient for the proposed cross-section; how drainage structures, ditches, and slopes are affected by field conditions; and other data which may be necessary for completing the plans. Blueprints are made of the partially completed plans for field inspection purposes.

The field investigator makes his inspection by walking the entire length of the project with a set of blueprints in hand and checking the various items of the plans, giving his approval or making notations as to his recommendations for a change in design. When his inspection is completed, the blueprints that were used for his field investigation are returned to the designer. Where the surveys have been conducted by photogrammetric means, the amount of field inspection at this stage can be reduced by a restudy of the mosaics constructed from the aerial photographs.

It may be mentioned here that a final inspection of the proposed project is made when the final plans have been completed. This inspection, called the "plans, specifications, and estimates" inspection and commonly referred to as the P. S. & E. inspection, is made in order to verify the final design before the project is approved to be advertised for bids.

Preparation of Final Plans

After the field inspection is made with the partially completed plans, or "preliminary plans," as they are sometimes called, the final design steps are carried out. These include (1) plotting the cross-sections of the original ground and the placing of "templates" on original ground sections; (2) computing earthwork quantities; (3) preparing construction details for bridges, culverts, guardrails, and other items; (4) preparing summaries and estimates of quantities; and (5) preparing specifications for the materials and methods of construction.

8–17. Cross-Sections and Templates. One important phase of design is to determine the amount of earthwork necessary on a project. Earthwork includes the excavation of material and any hauling required for completing the embankment. Payment for earthwork is based on excavated quantities only and generally includes the cost of hauling the materials to the embankment. An additional item of payment called "overhaul" is often used to provide for the hauling of the excavated material beyond a certain free-haul distance. The excavated material may be obtained from within the area of the highway cross-section or from some distance outside the proposed highway. When excavation is obtained outside the proposed highway limits it is called "borrow excavation."

In order to determine earth excavation and embankment requirements by manual means, a section outline of the proposed highway, commonly referred to as a "template section," is placed on the original ground cross-section; the areas in cut and the areas in fill are determined; and the volumes between the sections are computed. Fig. 8–7 shows various conditions that may be encountered when plotting these template sections. "Cut" and "fill" are the terms that are usually used for the areas of the sections, and the terms "excavation" and "embankment" generally refer to volumes.

Cross-sections are plotted on standard cross-section paper to any convenient scale. A scale of 1 inch equals 5 feet vertically and horizontally is common practice. Each cross-section should show the location or

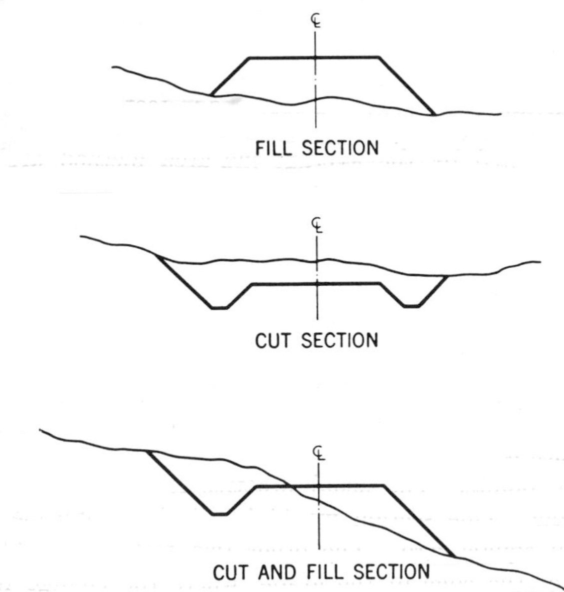

FIG. 8–7. Original ground line and template sections.

station of the original ground section and template section; the elevation of the proposed grade at that station; and the areas of cut and fill for each section. The computed volumes of excavation and embankment may also be placed on the cross-section sheet between two successive cross-sections.

To facilitate the plotting of these cross-sections, a template of the roadway which conforms to the outline of the proposed highway cross-section is made of some transparent material. The outline can be quickly drawn on the cross-section paper. When templates are placed on the cross-section paper, consideration must be given to superelevations and transitions due to horizontal curvature.

The areas of cut and fill may be measured by the use of a planimeter, a computation method using coordinates or some other suitable method.

The areas of cut and fill are recorded in square feet. These areas are then used to compute the volumes of excavation and embankment between the sections, from which a summation can be made of the total volumes of excavation and embankment for the entire project.

The average-end-area method, which is used by a majority of state highway departments, is based on the formula for the volume of a right prism whose volume is equal to the average end area multiplied by the length, or

$$V = \frac{\frac{1}{2}L(A_1 + A_2)}{27} \tag{8-1}$$

where V = volume in cubic yards
A_1 and A_2 = area of end sections in square feet
L = distance between end sections in feet

Volumes computed by the average-end-area method are slightly in excess of those computed by a prismoidal formula when the sections are not right prisms. This error is small when the sections do not change rapidly, which is usually the case when a small degree of curvature is used. On sharp curves, however, a prismoidal correction should be applied.

When a section changes from a cut section to one of fill, a point is reached when a zero cut and zero fill occurs. This usually happens along an irregular line across the roadway. In order to properly compute the volumes at these locations, it is necessary to compute volumes at intermediate points. The usual practice is to determine end areas for three sections. One section would be at the centerline, where the cut and the fill equals zero. The other two sections would be at the right and left of the edge of the grade, where the change from cut to fill or fill to cut occurs.

A typical volume sheet for computing excavation and embankment quantities is shown in Fig. 8-8. The columns 1 to 4 show end areas in square feet for cut, fill, loss, and unsuitable material. The item of loss may be explained as material in cuts which is ineffective in fills on account of leafmold, tree roots, and other vegetable matter. Likewise, where such materials occur on the original surface of fill sections, an increased amount of fill is required. This is sometimes classed as "grubbing loss." This loss becomes an item of shrinkage when it is impracticable to estimate it. The item "unsuitable material" is material considered unfit for use in an embankment.

Manual methods described above are rapidly being supplanted by methods based on the use of electronic computers. In typical cases and for earthwork quantities at the design stage, the computer input includes the cross-section data, centerline grades, and design template

32	22	18	6		96	81	8	13		102		6	+397	
33	26	22	6		89	74	7	11		92		3	+394	
34	24	20	6		93	78	8	11		97		4	+390	
35	26	24	4		93	81	8	9		98		5	+385	
36	16	20	0		78	81	8	4		93		15	+370	
37	14	18			56	70	7			77		21	+349	
38	18	22			59	74	7			81		22	+327	
39	22	24			74	85	9			94		20	+307	
40	26	30			89	100	10			110		21	+286	
41	26	28			96	107	11			118		22	+264	
42	22	24			89	96	10			106		17	+247	
43	18	20			74	81	8			89		15	+232	
44	16	32			63	96	10			106		43	+189	
45	14	48			56	148	15			163		107	+82	
46	12	94			48	263	26			289		241	−159	
47	8	138			37	430	43			473		436	−595	
48	4	172			22	574	57			631		609	−1204	
49	18	77			41	461	46			507		466	−1670	
50	16	40			63	217	22			239		176	−1846	
51	18	16			63	104	10			114		51	−1897	
52	22	34			74	93	9			102		28	−1925	
53	34	8			104	78	8			86	18		−1907	
54	14	16			89	44	4			48	41		−1866	
55	46	8			111	44	4			48	63		−1803	
56	50	12			178	37	4			41	137		−1666	
57	40	14			167	48	5			53	114		−1552	
58	46	12			159	48	5			53	106		−1446	
59	42	12			163	44	4			48	115		−1331	
60	36	16			144	52	5			57	87		−1244	
61	27	21			117	69	7			76	41		−1203	
62	24	24			94	83	8			91	3		−1200	
											724			
					7458									

447
345
35
324
58
162
5920 Cubic Yar

Balance Point

107
1200
1267
126

Balance Point

1567
1422
141
252 Cu. Yard Stations

Balance Point

2439
724
3163

FIG. 8–8. Volume

VOLUME SHEET

SHEET NO. | PROJECT NO. 10 | COUNTY | COMPUTED BY | DATE

Station	End Area, Square Feet: 1 Total Cut	2 Fill	3 Loss	4 Unsuit. Material Waste	Volume in Cubic Yards: 5 Total Cut	6 Fill	7 Shr. % 10	8 50% Loss	9 Unsuit. Material Waste	10 Fill + 7, 8 & 9	Bal. Volumes Diff. bet. 5 and 10: 11 Cut +	12 Fill −	M. O. 13
0	4	16	4										0
1	0	76	6		7	170	17	9		196		189	−189
2	0	122	8		0	367	37	13		417		417	−606
3	6	46	4		11	311	31	11		353		342	−948
4	46	12	4		96	107	11	7		125		29	−977
5	126	0	6		319	22	2	9		33	286		−691
6	120	0	6		456	0		11		11	445		−246
7	108	0	6		422	0		11		11	411		+165
8	100	0	6		385	0		11		11	374		+539
9	92	0	6		356	0		11		11	345		+884
10	68	4	6		296	7	1	11		19	277		+1161
11	56	12	4		230	30	3	9		42	188		+1349
12	42	20	0		181	59	6	4		69	112		+1461
13	28	26			130	85	9			94	36		+1497
14	32	26			111	96	10			106	5		+1502
15	20	38		0	96	119	12			131		35	+1467
16	38	42		20	107	148	15		37	200		93	+1374
17	32	46		32	130	163	16		96	275		145	+1229
18	44	52		44	141	181	18		141	340		199	+1030
19	40	32		40	156	156	16		156	328		172	+858
20	22	36		22	115	126	13		115	254		139	+719
21	12	21		0	63	106	11		41	158		95	+624
22	14	24			48	83	8			91		43	+581
23	16	22			56	85	9			94		38	+543
24	20	24			67	85	9			94		27	+516
25	18	24			70	89	9			98		28	+488
26	16	20	0		63	81	8			89		26	+462

Rock Exc.	Earth Exc.	Borrow	Fill	Shrkg. 10 %	Loss	Waste	Special Borrow	Overhaul	Borrow O'haul
	1311		977	98	142		71		

Balance Point

Cu-Yard Miles

computations.

data with provision for superelevation, curve widening, and varying side slopes. The computer produces a tabulation of the cut and fill volumes at each station and the difference between them adjusted for shrinkage and swell (see below), the cumulative volumes of cut and of fill, mass diagram ordinates, and slope stake coordinates.

8–18. Shrinkage. When earth is excavated and hauled to form an embankment, the freshly excavated material generally increases in volume. However, during the process of building the embankment, it is compacted, so that the final volume is less than when in its original condition. This difference in volume is usually defined as "shrinkage." In estimating earthwork quantities, it is necessary to make allowance for this factor. The amount of shrinkage varies with the soil type and the depth of the fill. An allowance of 10 to 15 per cent is frequently made for high fills, and 20 to 25 per cent for shallow fills. The shrinkage may be as high as 40 or 50 per cent for some soils. This generally also allows for shrinkage due to loss of material in the hauling process and loss of material at the toe of the slope. In peat or muck areas, shrinkage should not include settlement of fills due to consolidation.

When rock is excavated and placed in the embankment, the material will occupy a larger volume. This increase is called "swell" and may amount to 30 per cent or more. The amount of swell is not important when small amounts of loose rock or boulders are placed in the embankment.

8–19. The Mass Diagram. A mass diagram is a graphical representation of the amount of earth excavation and embankment involved on a project and the manner in which the earth is to be moved. It shows the location of balance points, the direction of haul, and the amount of earth taken from or hauled to any location. It is a valuable aid in the supervision of grading operations and is helpful in determining the amount of overhaul and the most economical distribution of the material.

The overhaul distance may be defined as the length of haul beyond a certain distance, known as "free haul." The free-haul distance may be as low as 500 feet and as high as 3000 feet or more. Some states do not consider any overhaul, which means that the cost of hauling the material long distances must be included in the cost of earth excavation. The overhaul distance is found from the mass diagram by determining the distance from the center of mass of the excavated material to the center of mass of the embankment. This distance may be measured in stations or in miles. Thus a "cubic-yard station" is the hauling of one cubic yard of excavation one station beyond the free-haul limits, and a "cubic-yard mile" is the hauling of 1 cubic yard a distance of 1 mile beyond the free-haul limits. The dividing line between yard stations and yard miles will vary, and has to be determined by definition in the specifications. For example, the specifications may state that excavation hauled up to ½ mile beyond the free-haul limits shall be measured in yard stations, and any excavation hauled beyond the free-haul limits over ½ mile shall be in yard miles. Thus with a free-haul distance of 1000 feet, the dividing line between yard stations and yard miles will be a distance of 3640 feet. The use of an arbitrary dividing line to determine the number of units of cubic-yard stations and cubic-yard miles provides a better means of determining the volumes hauled for short distances and those hauled for long distances. A better distribution of costs can be made than when only one distance is used between the centers of mass of the volumes being considered.

Where it is necessary to haul material long distances, it is sometimes more economical to waste material excavated from the roadway section and borrow material from a borrow pit within the free-haul distance. The length of economical haul can be determined by equating the cost of the excavation plus the cost of the overhaul, to the cost of excavation in the roadways plus the cost of excavation from the borrow pit.

If h equals the length of haul in stations beyond the free-haul distance, e equals cost of earth excavation, and o equals cost of overhaul then to move 1 cubic yard of material from cut to fill, the cost will equal $e + ho$, and the cost to excavate from cut, waste the material, borrow, and place 1 cubic yard in the fill will equal $2e$, assuming that the cost of excavation in the roadway is equal to the cost of the borrow excavation.

Then

$$e + ho = 2e$$

and

$$h = \frac{e}{o} \text{ stations}$$

For example: If the cost of earth excavation is \$0.38 per cubic yard, and if the cost of overhaul is \$0.02 per cubic-yard station, the length of the most economical haul will be $h = 38/2 = 19$ stations, or 1900 feet. Assuming a free-haul distance of 1000 feet, the limit of length of economical haul becomes 1900 feet plus 1000 feet or 2900 feet.

Columns 5 to 10 in Fig. 8–8 show volumes computed from the corresponding end areas, with an allowance for shrinkage which has been previously explained. The cut is balanced with the fill and the excess is listed in column 12; excess of cut is indicated as plus and an excess of fill is indicated as minus. The cumulative algebraic sums in the last column become the ordinates of the mass diagram.

After the mass ordinates are computed, they are plotted to form a mass diagram. The scales used are generally 1 inch equals 500 feet horizontally and 1 inch equals 500 cubic yards vertically. However, other convenient scales may be used. The mass ordinates of Fig. 8–8 were used to construct the mass diagram shown in Fig. 8–9.

Beginning at station 0 + 00 and ending at station 4 + 00, the curve of Fig. 8–9, indicates that the embankment requirements exceed the available excavation; from station 4 + 00 to 6 + 60 the available excavation is greater than the embankment requirements. This can be restated to say that the embankment from station 0 + 00 to 4 + 00 is

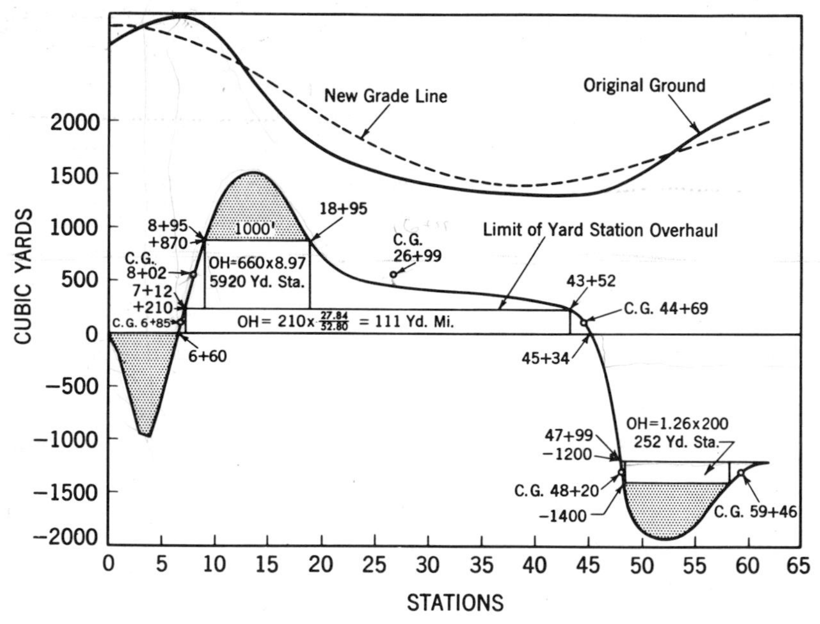

FIG. 8–9. Mass diagram.

constructed from the excavation between 4 + 00 and 6 + 60. This also means that the direction of haul has to be from station 4 + 00 and 6 + 60 to the left, to station 0 + 00 and 4 + 00. It also means that the earth quantities balance at station 6 + 60.

A balance point may then be described as a point where the volume in excavation balances the volume in the embankment. Any line drawn parallel to the base line and intersecting two points within the same curve will also indicate a balance of cut and fill quantities between these two points.

Further analysis of the curve shows other balance points at stations 45 + 34, 47 + 99, and 62 + 00; the earth excavation from 6 + 60 to 14 + 00 is used to construct the embankment between stations 14 + 00 and 45 + 34, and the direction of haul is to the right; the embankment between stations 45 + 34 and 47 + 99 is constructed from borrow; and the embankment between stations 47 + 99 and 52 + 00 is constructed from excavation between stations 52 + 00 and 62 + 00, and the direction of haul is right to left.

Where earth excavation and embankment quantities balance on the complete project, the curve of the mass diagram would end on the base line at the zero point. This condition does not always occur. In this case the mass ordinate shows a minus 1200 cubic yards which must be borrowed from some other source in order to complete the embankment. When the mass ordinate ends in a plus volume, and the last point on the curve is above the base line, there is excess material that must be wasted.

Referring to Fig. 8–9, it is seen that the quantities balance between 0 + 00 and 6 + 60, a distance of 660 feet. This is within the free-haul distance and need not be considered for overhaul. It also indicates that 977 cubic yards are to be excavated and placed in the embankment, and that no payment will be made for the hauling of this material. Payment is made, however, for 1311 cubic yards of excavation within this balance point.

The next balance occurs between stations 6 + 60 and 45 + 34. A scaled horizontal line 1000 feet long is drawn parallel to the base line and found to intersect the curve at station 8 + 95 and 18 + 95. This represents the free-haul distance, and all excavation and embankment above this line can be eliminated from further consideration as to overhaul. The length of this balance from 6 + 60 to 45 + 34 equals 3874 feet, which is over 3640 feet and indicates that there will be both yard-stations and yard-miles for overhaul. A horizontal line 3640 feet long is drawn parallel to the base line and is found to intersect the curve at station 7 + 12 and 43 + 52. The overhaul above this line will be yard-stations and below this line it will be yard-miles.

The 1000-foot free-haul line intersects the volume ordinate at +870 at station 8 + 95 and 18 + 95; the 3640-foot line intersects the volume ordinate line at 210 at station 7 + 12 and 43 + 52.

Perpendicular lines are drawn from the limits of free haul to the limits of the yard-station overhaul line at station 8 + 95 and 18 + 95; this vertical line represents the volume ordinate +210 to +870, which represents 660 cubic yards of excavation from station 7 + 12 and 8 + 95 that must be placed in the embankment between stations 18 + 95 and 43 + 52. The cubic-yard stations are determined by multiplying the excavation, which in this case is 660 cubic yards, by the distance from the center of mass of excavation to the center of mass of embankment, less 1000 feet free haul.

In order to determine the number of cubic-yard miles, perpendiculars are drawn from the limit of yard-station overhaul at station 7 + 12 and station 43 + 52; this vertical line represents the volume ordinate 0 to +210, which is 210 cubic yards from station 6 + 60 to station 7 + 12 that must be placed in the embankment between stations 43 + 52 and 45 + 34. The cubic-yard miles are determined by multiplying 210 cubic yards by the distance in feet from the center of mass of excavation to the center of mass of embankment, less 1000 feet free haul, divided by 5280.

There are three methods in common use for determining the center of mass of excavation and embankment when mass diagram calculations are done manually:

1. The graphical method
2. The method of moments
3. The planimeter method

Various computer programs have been developed to perform these calculations by machine, but using a similar approach.

8–20. The Graphical Method. The volume ordinate line from +210 to +870 represents 660 cubic yards. If a line is drawn bisecting this line at 330 and extended horizontally to the curve between stations 7 + 12 to 8 + 95 and 18 + 95 to 43 + 52, the length of this line would represent approximately the average length of haul. This method is fairly accurate when the volume curve has a fairly uniform slope such as exists between stations 7 + 12 to 8 + 95, but appreciable error is evident when finding the center of mass between stations 18 + 95 to 43 + 52. No attempt will be made to complete the analysis by the graphical method because of its limitations. The more accurate method of moments will be explained in detail below.

8–21. Method of Moments. To find the center of mass between stations 7 + 12 and 8 + 95, moments may be taken about station 7 + 12. A moment is computed for each volume that makes up the volume curve between 7 + 12 and 8 + 95. Dividing the total moments by the total volume will give the distance to the center of mass from 7 + 12. From the volumes of Fig. 8–8 and the mass diagram of Fig. 8–9, it can be determined that the volumes and distances are as indicated in the table below.

Station	Volume (cu yd)	Distance (ft)
7 + 12 to 8 + 00	329	44
8 + 00 to 8 + 95	331	135.5
Total	660	

Taking moments about station 7 + 12, we have the distance to the center of mass equal to

$$\frac{44 \times 329 + 135.5 \times 331}{660} = 90 \text{ ft}$$

The location of the center of mass equals 7 + 12 + 90 feet = station 8 + 02.

In the same manner the center of mass of the embankment between stations 18 + 95 and 43 + 52 is determined by taking moments about station 18 + 95 and using the volumes given in Fig. 8–8 between stations 18 + 95 and 43 + 52. The sum of these moments divided by the total yardage equals 804 feet. The center of mass is therefore 804 feet from 18 + 95 or at station 26 + 99.

With the center of mass of the excavation and that of the embankment being located at station 8 + 02 and 26 + 99, respectively, the distance between these points equals 1897 feet. Subtracting 1000 feet as free haul, the overhaul distance is 897 feet or 8.97 stations. The overhaul will equal 660 × 8.97 or 5920 cubic-yard stations.

Overhaul on a cubic-yard mile basis is computed in the same manner, except that the overhaul distance is expressed in miles instead of stations. In this case the centers of mass of excavation and embankment are found to be at stations 6 + 85 and 44 + 69, a distance of 3784 feet. The cubic-yard miles will then be $210\left(\frac{3784 - 1000}{5280}\right)$ or 111 cubic-yard miles.

8–22. Planimeter Method. To determine the center of mass by the use of the planimeter, the following formula is used:

$$\text{Distance to center of mass} = \frac{\text{area} \times \text{horizontal scale} \times \text{vertical scale}}{\text{volume ordinate}}$$

area = the planimetered area under the volume curve in square inches

horizontal scale = horizontal scale of the mass diagram in stations per inch

vertical scale = vertical scale of the mass diagram in cubic yards per inch

volume ordinate = the vertical ordinate in cubic yards between the base line and limits of haul

For example, to determine the center of mass of the embankment from station 18 + 95 to 43 + 52, the area under the volume curve between these stations to the line-marked limit of yard-station overhaul was planimetered and found to be 2.122 square inches. Using the above formula, the distance of the center of mass from station 18 + 95 is as follows:

$$\text{Distance to center of mass} = \frac{2.122 \times 5 \times 500}{660} = 8.04 \text{ stations}$$

Center of mass = station 18 + 95 plus 8 + 04 = 26 + 99, which is the same as was obtained by the method of moments.

8–23. Borrow Overhaul. The volume ordinate between stations 45 + 34 and stations 47 + 99 indicates 1200 cubic yards of fill for which suitable borrow must be provided. When it becomes necessary to obtain excavation from a borrow pit in order to complete the embankment, the overhaul distance is determined from the center of mass of the borrow pit to the center of mass of the embankment as measured along the haul route, less the amount of free haul.

A tabulation of earth excavation, borrow excavation, and overhaul, as determined from the mass diagram of Fig. 8–9, is given below.

	Earth Excavation	Borrow Excavation	Overhaul	
			Cu-yd Sta.	Cu-yd Mi
0 + 00 to 6 + 60	1311			
6 + 60 to 45 + 34	4473		5920	111
45 + 34 to 47 + 99	107	1200	*	*
47 + 99 to 62 + 00	1567		252	
Total	7458	1200	6172	111

* Overhaul will have to be paid for the 1200 cubic yards of borrow excavation if it is obtained beyond the free-haul limits.

8–24. Preparation of Construction Details. Earth excavation quantities, overhaul, and borrow requirements are generally the first items that are determined when drawing the plans for a project, and then the other details are prepared.

Many of the structures and miscellaneous items included in the project may be standard items for which designs have been previously approved and adopted for use. When such a design can be used, it is made a part of the plans by reference. If it cannot be used, the structure has to be designed to meet existing conditions.

8–25. Preparation of Quantities and Estimates. When the plans are completed, a summary is made of all the items of work, estimated quantities, and units of measurement. These are placed on one of the plan sheets. Some states provide regular quantity sheets for this purpose that show all items as per plan, with provision for inserting quantities as constructed. These sheets are useful for computing final quantities on the completed project. The summary of quantities may be arranged in groups such as grading items, drainage structures, surfacing items, and incidentals.

A summary of quantities is tabulated and an estimate is made of the probable cost of the project, with an allowance made for contingencies. An engineer's estimate is shown in Fig. 8–10.

8–26. Preparation of Specifications. When the plans have been completed, a proposal and specifications are then prepared for the project. When this is done, the project is ready to be advertised for bids and contracts prepared for the construction of the project according to the prepared plans. These matters are discussed in the next chapter.

8–27. Completed Final Plans. A complete set of plans will contain any number of sheets, depending upon the length and complexity of the project. Most state highway departments specify a definite order in which the plan sheets are to be arranged, and this varies considerably with the several states. A complete set of plans should contain the following:

(1) TITLE SHEET. The first sheet of a set of plans, which is known as the "title sheet," shows a map of the area in which the project is located. The map should show the point of beginning and point of ending; the main topographical features; township, county, and state lines; and possible detour routes. The name and number of the project are also indicated. The index, legend, and tabulation of standard plans used are generally included in the title sheet. Provision is also made on this sheet for the affixing of signatures approving the plans that are contained in the completed set.

(2) TYPICAL CROSS-SECTION SHEET. The typical cross-section sheet shows all earthwork and finished road sections. These should include ditch sections and any treatment needed for the subgrade section.

(3) NOTE AND SUMMARY SHEET. Notes or special provisions consist of special requirements, regulations, or directions prepared to cover the work which is not covered by the standard specifications. The summary of quantities is a tabulation of the items of work, quantities, and units of measurement for that particular project.

(4) PLAN AND PROFILE. The plan and profile sheets contain most of the details in regard to the work. These have been previously discussed. Their number depends upon the length of the project.

(5) MASS DIAGRAM. This sheet gives the contractor and engineer a comprehensive graphic representation of the source and disposition of all earth moved on the project.

(6) DRAINAGE STRUCTURES. All drainage structures are usually shown in detail. These will include detail drawings of culverts, inlets, manholes, and so on, and will show the flow lines, sizes, and locations of the

SUMMARY OF QUANTITIES AND BID UNIT PRICES

STATE AID PROJ. NO. SAP-1135-A (5).
COUNTY ATKINSON
LOCATED BETWEEN on the Pearson to Ray City Road, S.R. 64.
6.315 miles grading and bituminous surfacing

ITEM	APPROX. QUAN.	UNIT	UNIT PRICE	AMOUNT
Clearing and Grubbing Roadway	76.550	Acre	LUMP	11,200.00
Clearing and Grubbing	7	Acre	250.00	1,750.00
Unclass. Excav. & Bor. Incl. Ditches & Shldrs.	139,000	Cu. Yd.	.29	40,310.00
Overhaul on Excavation	250,000	Sta. Yd.	.005	1,250.00
Special Subgrade Compaction	6.315	Mile	200.00	1,263.00
Excavation for Culverts & Minor Structures	500	Cu. Yd.	1.50	750.00
Selected Matl. for Clvt. Found. Backfill	360	Cu. Yd.	2.00	720.00
Class "A" Concrete Culverts	132	Cu. Yd.	55.00	7,260.00
Bar Reinforcement Steel	10,600	Lb.	.14	1,484.00
18" Class "B" Concrete Pipe S.D.	980	Lin. Ft.	2.85	2,793.00
24" Class "B" Concrete Pipe S.D.	48	Lin. Ft.	4.60	220.80
18" Std. Str. Reinf. Concrete Pipe C.D.	264	Lin. Ft.	3.65	963.60
24" Std. Str. Reinf. Concrete Pipe C.D.	1,344	Lin. Ft.	5.45	7,324.80
30" Std. Str. Reinf. Concrete Pipe C.D.	96	Lin. Ft.	7.25	696.00
36" Std. Str. Reinf. Concrete Pipe C.D.	192	Lin. Ft.	10.10	1,939.20
Culvert Pipe Removed S.D. or C.D.	330	Lin. Ft.	1.25	412.50
Culvert Pipe Relaid S.D. or C.D.	50	Lin. Ft.	1.25	62.50
Sand-Cement Bag Rip Rap	460	Sq. Yd.	6.00	2,760.00
Concrete Right-of-Way Markers	48	Each	5.50	264.00
Sprigging	74,000	Sq. Yd.	.025	1,850.00
Loose Sod Rip Rap	450	Sq. Yd.	.60	270.00
Water for Grassing	75	M. Gal.	3.00	225.00
First Application Fertilizer	10	Ton	55.00	550.00
Second Application Fertilizer	2,160	Lb.	.28	604.80
Removal of Existing Wood Bridge	Lump Sum	LUMP	LUMP	200.00
Reflect. Railroad Warning Signs-Type "F"	2	Each	60.00	120.00
Reflect. Railroad Stop Signs-Type "F"	2	Each	60.00	120.00
6" Soil-Bitum. Stabilized Base Course	78,800	Sq. Yd.	.22	17,336.00
Cut-Back Asphalt - RC-3	325,200	Gal.	.145	47,154.00
Aggr. for Penetration Mac. Base-Railroad Crossing	25	Ton	7.00	175.00
Bit. Matl. for Penetration Mac. Base-Railroad Crossing	400	Gal.	.20	80.00
Asphalt Cement Surf. Treat.-Stone Size 4-Type-2	75,000	Sq. Yd.	.23	17,250.00
Asphalt Cement Bituminous Seal	75,000	Sq. Yd.	.16	12,000.00
Finishing and Dressing	55,000	Sq. Yd.	.02	1,100.00
		Total		182,458.20
		Contingency		18,241.80
				200,700.00

FIG. 8–10. Engineer's estimate.

various structures. Bridge plans are generally not included in these drawings, but in a separate set of plans.

(7) Special Details. Special details may include slab or box culverts, special intersection treatment, railroad crossings, or any special detail not covered by the standard plans or the specifications.

(8) Utility Sheet. On some projects, the number of utilities involved may be considerable, and a special sheet is devoted to their treatment.

(9) Quantity Sheets. These sheets contain all the items of work involved on a project. They have been previously discussed. They differ from the note and summary sheet by having more detail, such as stations for the location of culverts, etc.

(10) Cross-Sections. Prints are usually made of the cross-sections used to determine earth excavation quantities and are made a part of the plans. On projects of considerable length many of these sheets may be required. Many states do not attach these sheets to each set of plans but keep them at some central or division office where they are available to the contractor or others desiring this information.

When the plans are completed, a final inspection of plans, surveys, and estimates, known as the P. S. & E., is made, and when the necessary approvals are obtained the project is ready for construction.

In recent years, it has become ever more necessary to increase the productivity of engineers and draftsmen in order to keep up with the demands of the detailed design involved in current construction schedules. Time-saving devices have been introduced in the production of drawings by the use of "stick-up" plans. The "stick-up" method is a method whereby large portions of plans which are routine yet time-consuming are produced by mechanical rather than manual means. There are many applications of the technique. Quantity summaries can be typed (rather than hand-lettered), reduced, and assembled on one sheet. Standards and details do not need to be traced for each job; those which are applicable are simply assembled for reproduction. A technique has also been perfected to eliminate the time-consuming tracing of photogrammetric topography by reproducing the maps on the back sides of transparent work sheets. Fig. 8–11 is an example of a plan and profile sheet produced by these techniques. Laborious copying of computer output has been eliminated by obtaining the printout on paper which, when attached to bordered and titled clear drawing sheets, will itself produce prints. This has proved practicable in the production of quantity summaries.

The computer itself is increasingly used in the automated plotting

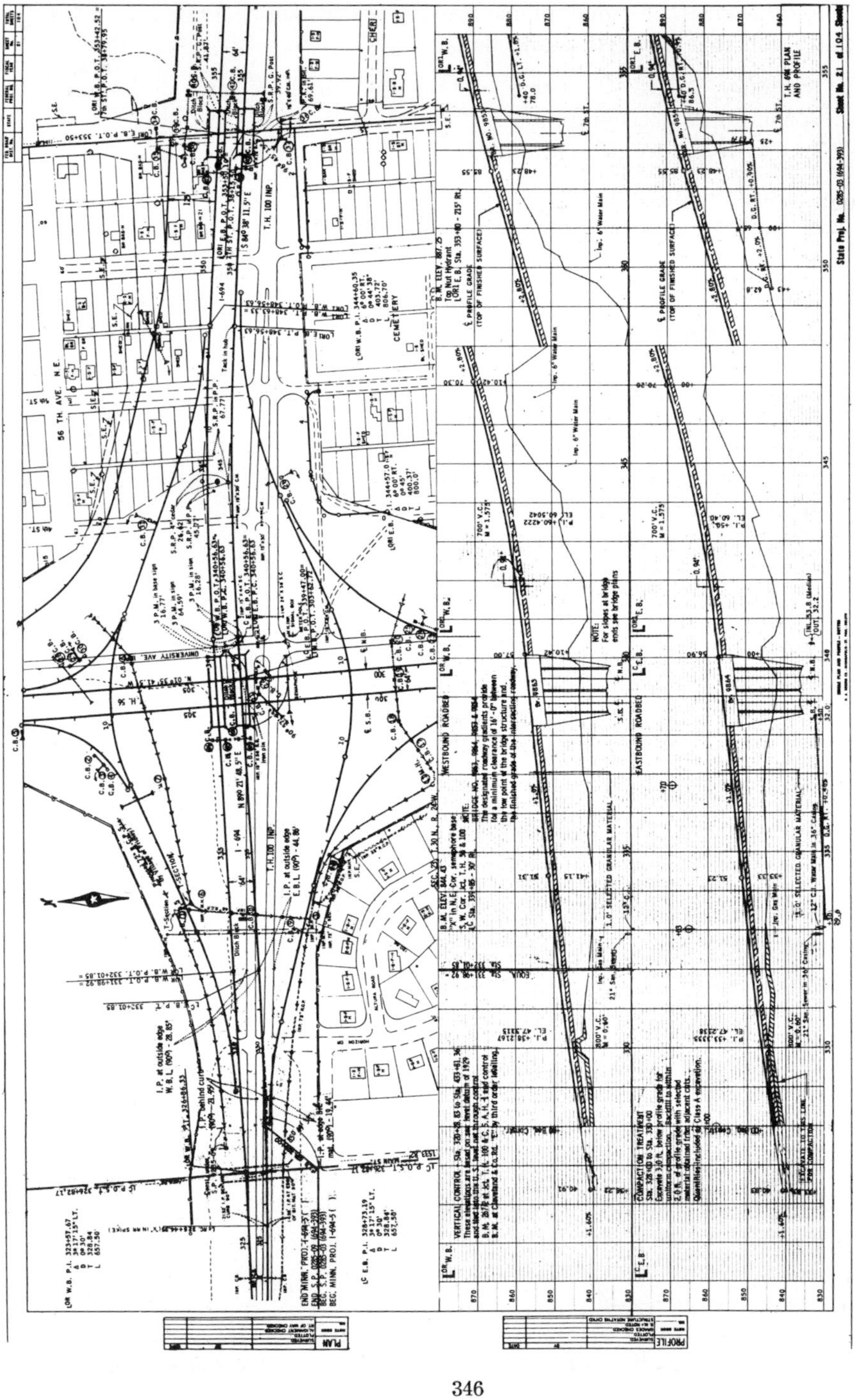

FIG. 8-11. Plan and profile sheet produced by "stick-up" method. (Courtesy *Better Roads*.)

of information which can be incorporated into final plans by using "stick-up" procedures. Programs are in use which, in addition to computing earthwork quantities from photogrammetrically derived input, give an output in the form of automatically plotted cross-sections.

Other methods of reducing drafting time have come through the use of the scribing technique. The process of scribing is one of cutting through a thin coating of opaque material which overlays a clear base material. The drawing which is produced is a negative which can be used to make a printing plate. Linework is greatly speeded by this dry process, since no time is lost in waiting for ink to dry, and errors are easily and completely removed by a touching-up solution.

PROBLEMS

8–1. The 1200 cubic yards of borrow needed in Fig. 8–9 can be secured from 800 feet right of station 77 + 40. What is the overhaul?

8–2. Assuming a freehaul distance of 1500 feet, and that the cost of excavation equals $0.49 per cubic yard, what is the limit of economical haul when the unit price of overhaul is $0.07 per cubic-yard station?

8–3. Given the following end areas in cut and fill, complete the earthwork calculations, using a shrinkage factor of 20 per cent. Prepare the mass diagram and indicate the direction of haul and whether you will have to borrow or waste any material in order to complete the project.

	End Areas			End Areas	
Station	Cut	Fill	Station	Cut	Fill
286	60	35	296	43	148
287	54	47	297	87	160
288	48	50	298	109	89
289	24	65	299	156	50
290	12	43	300	112	37
291	0	56	301	83	20
292	6	42	302	68	9
293	7	63	303	56	5
294	13	97	304	54	0
295	28	110	305	42	0

8–4. If you as a highway engineer had to lower the grade through a cut and waste a large amount of earth to obtain a passing sight distance, how would you justify the additional cost?

8–5. Given below are the ordinates of the mass diagram for a highway project. No haulage is possible past station 38 + 50, so this point may be considered to be the end of the project for this purpose. The free-haul distance is 500 feet. Draw the mass diagram, and determine (a) the volume of excavation and embankment, (b) the quantities of overhaul (station yards), waste, and borrow; and (c) the distribution of haul.

Station	Mass Ordinate, Cu. Yd.	Station	Mass Ordinate, Cu. Yd.
0	0	20	−9,122
1	+461	21	−10,437
2	+982	22	−9,333
3	+1343	23	−8,138
4	+2022	24	−8,985
5	+3589	25	−11,214
6	+5767	26	−12,482
7	+8010	27	−15,076
7 + 40	+8476	28	−17,451
8	+7982	29	−20,519
9	+7665	30	−23,048
10	+6832	31	−25,110
11	+4931	32	−26,397
12	+3214	33	−27,025
13	+865	34	−26,991
14	−437	35	−24,018
15	−2892	36	−18,212
16	−4504	37	−12,222
17	−6075	38	−7,537
18	−7111	38+ 50	−3,761
19	−7686		

8–6. Consult current literature, and write a report on the use of (a) aerial photogrammetry and associated devices, (b) electronic surveying devices, or (c) electronic computers in highway location studies by your state highway department or some other highway agency.

8–7. Consult your state or local highway agency, and write a report on its use of "stick-up" or other modern drafting techniques to simplify plan preparation.

CHAPTER 9

CONTRACTS AND SUPERVISION

Practically all highway-construction projects in the United States are public works, which are constructed with public funds. The agency authorizing this construction may be a federal, state, municipal, or county governmental unit. However, the greatest number of highway-construction projects today are authorized through the various state highway departments. More than 95 per cent of the construction done under state highway supervision is done by contract. The remaining 5 per cent is done by the state's own forces organized and equipped to do this work. The contract system has proved beneficial, and it is very unlikely that it will be changed.

Under the contract system, plans and specifications are prepared and are submitted for competitive bids. As is usually the case when public funds are used, the methods and procedures for carrying out public works projects are prescribed by law in each state and must be strictly adhered to. It is the intent to include here the procedure generally followed by most state highway departments in preparing contractual documents and in supervising construction.

9–1. Unit Price System. Nearly all highway contracts are prepared on a unit price basis. The unit price contract includes an estimate of the number of units of each type of work and the price for each unit. The unit price contract has many advantages and is well adapted to highway construction, because usually it is not possible to determine exact quantities of some items of work when plans are prepared. Sometimes, after construction has started, unusual conditions are encountered. In either event, the various quantities can be adjusted; that is, they may be increased or decreased to the quantities actually constructed. Under the unit price system, each unit includes the materials which are to be furnished, the work to be done, the method of measurement of the work completed, and the basis on which payment is to be made. These units are described in detail in the specifications.

9–2. Lump-Sum Bids and Force-Account Work. It is not always feasible to have a breakdown of all the work on a unit price basis. It may be necessary to have minor items done on a lump-sum basis, especially when some piece of work is too intangible to compute unit prices.

Some highway organizations now use lump-sum bidding to cover items which traditionally have been bid on a unit price basis. These items include clearing and grubbing, structural excavation, structural concrete, and the like, where estimated and final quantities can be expected to be essentially the same. In some cases, it is preferred to have this type of work done on a force-account basis; that is, the work may be done by the contractor, who is reimbursed on a "cost-plus" basis, i.e., the work is paid for at actual cost plus a percentage of the actual cost as the contractor's profit. By "actual cost" is meant the cost of labor and materials actually required on that particular operation. The percentage of profit allowed varies. The contractor is compensated for the use of equipment on a rental basis, generally at an agreed-on daily or hourly rate. Small tools are generally excluded in the cost of force-account work.

When large sums are involved, force-account work of this nature requires good supervision. Good practice dictates a close auditing of all work and materials. The audit should include certified copies of payrolls, original receipted bills of materials, and evidence of all other charges, such as freight bills. All force-account work should be checked and approved by both parties daily.

It is possible to have a contract in which the work to be done will include unit price items, a lump-sum bid item, and even provisions for a force-account item.

Contractual Documents

In order to put a highway construction project under contract, it is necessary to take various steps and to prepare certain documents. The contractual documents generally include (1) the advertisement, (2) instructions to bidders, (3) the proposal, (4) the agreement or contract, (5) plans, (6) the specifications, (7) bonds, and (8) all written or printed agreements and instructions pertaining to the method and manner of furnishing materials and performing the work under the contract.

9–3. The Advertisement. The advertisement or notice to contractors is a written notice to inform prospective bidders that bids are requested for certain types of work and that a contract may be entered into with the successful bidder. The advertisement is usually inserted in a newspaper or magazine which serves as a medium for supplying such information to persons interested in this particular type of construction. Advertisements or invitations for bids are also mailed to prequalified contractors. Highway departments are required by law to provide ample time between publication of the notice to contractors and the taking of bids.

This period may vary from two to several weeks, and it should allow contractors sufficient time to prepare their estimates, obtain prices for materials, and make essential arrangements necessary for filing a proper bid. On large or complex projects, arrangements often are made for a "pre-bid conference," a meeting at which details of the work are discussed with prospective bidders.

The advertisement should be brief and still include the information needed by the prospective bidder. The information usually includes the following:

1. A brief description of the type of work and the location of the project.
2. The date, time, and place where bids will be received.
3. Where plans, proposals, and specifications may be seen or obtained and the amount of fee or deposit required if any.
4. Approximate quantities of the work involved.
5. Amount of bid bond, certified check, surety, or security required and provisions for its return to the unsuccessful bidder.
6. Any restrictions relative to the qualifications of bidders.
7. Reservations for the rejection of any or all bids or waiver in bidding formalities.
8. A statement regarding minimum or prevailing wage rates, and other conditions and regulations pertaining to labor.

Fig. 9-1 shows a typical advertisement for bids (notice to contractors) used by the State of New Hampshire.

9-4. Instructions to Bidders. Some state highway departments issue additional "Instructions to bidders," a document in which more information is given about the procedure to be followed in the preparation and submission of bids, details of any formalities required in the acceptance or rejection of bids, and miscellaneous instructions relative to the work.

This particular instrument is used to expand the information given in the advertisement and it is sometimes used instead of an advertisement. When this is done, it is necessary to have all the information given in the advertisement contained in the instructions to bidders. The instructions to bidders may also include requirements for the submission of the bidder's experience record or prequalification data, the time for the starting and completion of the work, and a reference to any legal considerations in regard to the work.

It may be pointed out here that many highway departments omit the instructions to bidders as a separate document but include this information in other contractual documents such as the advertisement, the proposal, and their standard specifications which have been previously adopted and which satisfy any legal requirements included in the instructions to bidders.

THE STATE OF NEW HAMPSHIRE
DEPARTMENT OF PUBLIC WORKS AND HIGHWAYS

STATE OFFICE BUILDING CONCORD, NEW HAMPSHIRE

NOTICE TO CONTRACTORS

Sealed proposals will be received at this office until 2:00 o'clock P.M. Eastern Standard Time on the 24th day of February 1966 for the following described project:

WARNER
I 89-1(53)16, P-7407-A
1.231 mi. Type I-1 Hot Asph. Conc.Pavement
Single Span R.C. Slab on Steel Beam Br. 75± ft. long
2 Span R.C. Slab on Steel Beam Br. 240± ft. long
3 Span R.C. Slab on Steel Beam Br. 160± ft. long
CERTIFIED CHECK OR BID BOND: $10,000.00

A schedule of the minimum wages for all labor classifications as determined by the N.H. Commissioner of Labor in accordance with Chapter 280 of the Revised Laws and the Secretary of Labor pursuant to Section 115 of the Federal Aid Highway Act of 1956 is included in the proposal. All unskilled labor may be hired from lists prepared by the N.H. Dept. of Employment Security designated in the proposal.

Plans and specifications may be seen at the office of the Department of Public Works and Highways, at the office of the "Nerba", 20 Kilby St., Boston, Mass. and at the office of the Associated General Contractors of N.H. & Vt. Inc. Bridge St. F.E. Everett Turnpike, Concord, N.H. Plans, specifications and proposal forms may be obtained at the office of the Department of Public Works and Highways for six dollars ($6.00) a set, non-returnable. Check should be made payable to the "Treasurer State of New Hampshire". Cash will not be accepted.

Proposals must be completed in both words and numerals on regular proposal forms, which shall be submitted in a sealed envelope marked: "Proposal for WARNER P-7407-A", addressed and delivered to the Department of Public Works and Highways not later than the date and time mentioned above, at which time they will be publicly opened and read aloud. Proposals must be accompanied by a certified check or bid bond in the amount listed above, payable to the "Treasurer, State of New Hampshire," as security for the execution of the contract.

All individuals, firms, partnerships or corporations intending to bid, before obtaining plans, specifications and proposal forms, must file with the Department of Public Works and Highways on forms prepared for that purpose at least eight (8) days prior to opening of bids, a statement showing their qualifications.

The successful bidder will be required to furnish a contract bond in the amount of one hundred (100) per cent of his bid on forms furnished by the Department.

The successful bidder will be required to execute a sworn statement pursuant to Section 112(c) of Title 23 USC, certifying that he has not, either directly or indirectly, entered into any agreement, participated in any collusion or otherwise taken any action in restraint of free competitive bidding in connection with such contract.

The right is reserved to waive any informalities in or to reject any or all proposals.

February 2, 1966 Deputy Commissioner

FIG. 9-1. Typical "notice to contractors" (advertisement) for a highway construction project. A more detailed description of the location and nature of the project, including estimated quantities, is given in an accompanying "Information Report" not shown here.

9–5. The Proposal. The form on which the bidder submits his bid is called the "proposal" and is prepared by the highway department. The proposal is arranged so that all the items of work and the quantities are tabulated, with spaces for the bidder to insert his prices and extensions for each item and a space for the total amount for all items of work. The proposal also contains the statement that the bidder has carefully examined the plans and specifications, and that he is fully informed as to the nature of the work and the conditions relating to its performance. The proposal also states that he agrees to do the work in the time proposed, and that he agrees to forfeit his "proposal guarantee" in case his bid is accepted and he fails to enter into a contract. He also agrees to furnish the necessary bonds or security for carrying out the work.

The amount of the proposal guarantee (certified check, bid bond, or other security) required for submission with the proposal, and the manner in which it is to be returned, are also specified in the proposal. Space also may be provided for the listing of any subcontractors the bidder may require, for the proper signatures of the bidders, and for the address of their place of business.

A proposal generally includes the specifications and plans. As previously mentioned, all state highway departments have prepared and adopted standard specifications for road and bridge construction. The standard specifications are generally in book form and include general requirements and covenants, items of work, materials, methods of measurement, and payment of quantities. The standard specifications are always referred to in the other contract documents and in this case become a part of the proposal by reference. Due to certain conditions in the design or construction of a particular project, it is often necessary to supplement existing specifications as well as to add certain special provisions necessary to the project. It is quite convenient to add the supplemental specifications and any special provisions to the proposal. This gives the bidder the information he needs in order to prepare his bid.

9–6. The Agreement or Contract. The agreement is the formal part of the contract documents, and from a legal point of view is the strongest of them all. The advertisement, instructions to bidders, and the proposal are generally classed as contract documents; but they are not, strictly speaking, parts of a contract but are preliminary to it. If any statements of contractual importance are included in any of these documents, they should be made a part of the agreement or contract by reference or they must be repeated in the formal contract.

The contract must be in accord with the form prescribed by law and must contain the following:

1. The declaration of the agreement, names of the parties, their legal residence, and the date of execution of the agreement.
2. The consideration for work to be done, the items of work, quantities and unit prices with reference to the plans, specifications, and proposal.
3. The time for starting and the date for completion of the work, and provisions for damages, if any.
4. Signatures and witnesses.

9–7. Plans. Plans are approved drawings or reproductions of drawings pertaining to the work covered by the contract. They are usually complete in all details before construction is started. The plans, together with the specifications, are an essential part of the contract documents.

9–8. Specifications. The specifications are the written instructions which accompany or supplement the plans and form a guide for standards required in the prosecution of the work. These standard specifications, which are the result of experience and knowledge acquired over a period of years, cover the quality of materials, workmanship, and other technical data. Standard specifications are often revised from time to time in order to keep up with improved processes and the use of different materials. It is sometimes necessary to deviate from the standard specifications, and this is generally done by adding supplemental specifications either to the plans or to the proposal.

9–9. Bonds. All contracts require some form of security as a guaranty of faithful performance of the work. This security may be in the form of collateral deposited with the contracting agency by the contractor, or it may be in the form of surety bonds.

Private individuals may serve as sureties, but as a general rule surety bonds are issued by bonding companies. Bonds issued in connection with highway construction projects are of three common types: (1) bid bonds, (2) performance bonds, (3) lien bonds.

The bid bond, previously mentioned in connection with the advertisement and proposal, is one form of the proposal guaranty required when the contractor submits his bid. This is the contractor's guaranty that he will enter into a contract if his bid is accepted. In the event that the contractor fails to enter into a contract if his bid is accepted, the surety or bonding company is required to pay the owner damages in the amount specified in the bond. Bid bonds generally range from 5 to 10 per cent of the bid amount, or 5 to 10 per cent of the estimated cost of the project, based on the engineer's estimate prepared for the owner.

Performance bonds, as the name implies, guarantee that the work will be completed as required by the state.

Lien bonds, or "payment bonds," are used by some organizations and guarantee that the contractor will pay all obligations for labor and materials incurred in the performance of the contract.

The amount of bonds required is generally 100 per cent of the bid or contract price. Some states require performance bonds only, in the amount of 100 per cent of the bid price, with provisions in the bond for the payment of all bills and protection against all liens. Some states require bonds in the amount of 100 per cent of the bid or contract price, but make a distribution of 50 per cent for performance bonds and 50 per cent for lien or payment bonds. Some states require both, each in the amount of 100 per cent of the contract price.

Surety bonds are generally furnished before or at the time the contract is executed. When bonds are executed, they require the signatures of the three parties: (1) the owner, (2) the contractor, and (3) the surety.

Surety bonds are contracts between the owner and the surety. Since most highway construction contracts provide for changes and modifications in the contract, it is essential that the surety be informed of any modification of the contract between the owner and the contractor. In case the contractor fails to fulfill the terms of the contract, it becomes the responsibility of the bonding company to do so. This is usually done by letting the contractor complete the work with his own forces under the direction of an engineer supplied by the surety. This method has been found to be of maximum benefit to the owner and to the contractor in that the work is completed with a minimum of delay and expense to the owner and at the same time lessens the loss entailed by the contractor. If this procedure is not followed, the bonding company may let a new contract for the remainder of the work or may elect to pay the penalty and let the owner complete the work, provided that the owner wishes to do so.

9–10. Supplemental Agreements. After the contract has been executed and construction work has been started, it is often necessary to make changes in the contract. Conditions commonly arising which require supplemental agreements are the time necessary to complete the work, changes in the quantities of work involved, and the addition of extra work not provided for in the original contract. The standard specifications usually provide for these conditions by granting an extension of time and authorizing any changes or extra work necessary to complete the project. All supplemental agreements must conform to the same basic requirements as the original contract as to competent parties, monetary consideration, legality of subject matter, and mutual agreement. All supplemental agreements should be in writing. In addition, unforeseen conditions may force an adjustment in the contract to accommodate these "changed conditions" (see Art. 9–14).

9–11. Extension of Time. The contract usually designates the time when work is to begin and when it shall be completed. The time to begin work may be designated as a certain number of days after the contract has been executed. The completion date may be a specific calendar date or it may be a specific number of calendar days in which the contractor is to complete the work. In most states it is understood and agreed that time is of the essence of the contract, and failure to complete the contract on time may make the contractor liable for damages. These damages may be a definite amount per day for each day beyond the time required for completion of the work. These are known as "liquidated" damages. In lieu of liquidated damages, the owner may charge actual damages incurred, such as for engineering costs and supervision, maintenance of necessary detours, or any other direct charges caused by the delay. The American Association of State Highway Officials recommends a schedule of liquidated damages ranging from $30 to $420 per day, depending on the size of the contract (*94*).

From a legal point of view, no work can be done beyond the date of completion, as listed in the contract. It is necessary to make a new agreement or extend the original agreement by requesting an extension of time to complete the contract. This should be done within a reasonable time before the completion date of the contract will have arrived. If the contractor requests an extension of time due to extra work or other extenuating circumstances beyond his control, such time should be extended without penalty. The agreement extending the time is also a formal document and requires all the necessary elements of a legal contract. When properly executed, it is as binding as the original contract.

9–12. Authorizations for Changes in the Work. It is impossible to determine exact quantities for some items of work (e.g., cubic yards of excavation) when designing a project. When a contractor bids on the various unit price items, as set up in the proposal, he is reasonably certain that the final quantities may be increased or decreased from the bid amount. To make the necessary adjustment from the quantities as stated in the contract to the quantities as actually constructed, written authorizations are made in accordance with methods of procedure set forth in the specifications.

On major items where large quantities are involved, some specifications provide for an increase of the unit price if the quantities are drastically reduced, and, conversely, if the quantities are materially increased, the unit price may be decreased. Some states require an adjustment of the unit price when the quantities vary more than 20 or 25 per cent from estimated quantities and when the value of the work exceeds a certain sum.

9-13. Authorization for Extra Work. Some states make a distinction between a change authorization and an extra authorization. A "change authorization" may be defined as one in which there is a change in the quantities at the unit price as listed in the contract, while an "extra authorization" is for extra work for which no unit price is listed. The unit price for extra work is usually agreed upon at the time the extra work is discussed, and it should always be approved before any extra work is started.

9-14. Changed Conditions. The federal government and a few of the state highway departments incorporate in their contracts a "changed conditions clause," one that provides for adjustment in the contract if conditions at the site are substantially different from those anticipated. Use of such a clause by the states is growing, but the idea has been slow to gain acceptance, as many organizations regard it as unnecessary.

AASHO suggests (94) the following changed conditions clause: "Should the contractor encounter or the department discover during the progress of the work subsurface or latent physical conditions at the site differing materially from those indicated in this contract, or unknown physical conditions at the site of an unusual nature, differing materially from those ordinarily encountered and generally recognized as inhering in work of the character provided for in the contract, the engineer shall be promptly notified in writing of such conditions before they are disturbed. The engineer will thereupon promptly investigate the conditions and if he finds they do so materially differ and cause an increase or decrease in the cost of, or the time required for performance of the contract, an equitable adjustment will be made and the contract modified in writing accordingly."

9-15. Subcontract. The use of a subcontract in highway work is often necessary. A subcontract may be defined as a contract for the performance of a part or all the work previously contracted for. When this is done, the subcontract should have all the elements of a formal contract. The scope of the work to be performed, the compensation to be received, and the terms and conditions of the original contract with which the subcontractor must comply should be clearly stated. The owner holds the original contractor directly responsible for the work of the subcontractor. The subcontractor, however, has the legal right to hold the owner responsible for any default of payment due him for labor or supplies by the contractor. The owner usually protects himself by having the contractor furnish the necessary bonds that guarantee payment of all labor and materials.

On state highway contracts the amount of work which a contractor may sublet is usually limited; in some cases, up to 50 per cent of the value of the work contracted for. The state highway departments usually reserve the right to approve or disapprove a subcontract, and they may require information in regard to the qualifications of the subcontractor and his ability to do the work. Copies of all subcontracts, including prices, are usually required before a subcontract may be approved.

9–16. Prequalification of Contractors. On most public works, contracts are awarded to the lowest responsible bidder and it is the problem of the state agency to determine the responsibility of the lowest bidder after the bids are taken. This is necessary because contractors with limited financial resources and limited experience may be competing with reliable contractors for an award of a contract. Such an award to an unqualified contractor often results in unnecessary delay, unsatisfactory work, and sometimes in failure to complete the work, with the corresponding extra expense to the state agency requesting bids. In order to avoid or reduce these dangers, many states have adopted prequalification laws and regulations. The fundamental purpose of the prequalification requirements is to determine beforehand whether or not the contractor is responsible and competent to bid on a project, and to permit only those properly qualified to do so. A typical procedure followed in prequalifying contractors is described below.

To comply with prequalification requirements, the prospective bidder is requested to submit a comprehensive experience and financial statement. From the information thus supplied, the prospective bidder is given a numerical rating which indicates the maximum amount of work in dollar volume he is permitted to have at any one time, and is further classified for the various types of work he is permitted to bid on. The major rating factors for determining a prospective bidder's qualifications are his experience record, the amount of equipment he owns, and his financial resources at the time of filing his statement.

The numerical rating and classification required for each project may be given in the advertisement. Prequalification of contractors is highly desirable in that there is more time to investigate the bidders' qualifications beforehand, and in that it saves the time and expense of contractors who are not qualified to do the work. The legality of prequalification has been established by the rulings of the various courts. Opponents of prequalification object to the lengthy questionnaires and the time and expense of preparing statements, and they state that it restricts competition to those with large resources. Additional objection to prequalification based on equipment owned is raised when suitable equipment may often be rented.

9–17. State-and-Federal Agreements. On state projects in which the federal government is to participate in the cost, certain conditions and agreements are necessary to qualify for these funds. In the apportionment of these funds for highway purposes, each state is required to submit for approval a program of projects to be constructed on a predetermined system of federal-aid highways, the total cost of which must not exceed that state's apportionment. Plans and specifications are then prepared, and a detailed estimate based on current prices is submitted to the Bureau of Public Roads for approval. Approval of this agency is required before advertising the project and before the award is made to the low bidder. After the award is made by the state to the contractor, a detailed estimate is made using the contractor's unit bid prices; and it is upon this estimate that the "project agreement" is to be based. The contractor's bid, plus a contingency not to exceed 10 per cent of the bid, is usually the amount of the agreement. This project agreement is the formal document or contract between the state and the federal government. There is one other formal agreement that may be mentioned here and that is the "maintenance agreement." In this document the state agency agrees to the maintenance and upkeep of the project after it is constructed. The maintenance agreement is usually made after plans are completed for the particular project and before federal approval is given to advertise the project. Engineers of the federal agency inspect the work as it progresses, and their approval is required for acceptance of the project when all work is completed.

9–18. Mechanics of Bidding Procedure. When plans, specifications, and estimates are completed, and when the necessary approvals are obtained, the project is ready for advertising. It is always preferable to have several projects ready for advertising at the same time. The advertisements are written with the required information previously discussed, and the date is set in accordance with statute requirements. Plans and proposals are then requested by those desiring to submit bids on the projects advertised. Bids may be mailed or submitted in person at the designated time and place. At large lettings, when the number of projects and bidders are numerous, much detailed work is necessary in order to check the many details that arise when bids are received. The sealed bids are opened, the bid security is checked, and the proposed details are scrutinized for irregularities. The total amount of the bid is then read aloud. Any irregularities found before the reading of the bid are usually a cause for rejection. For example, when bid bonds are requested to accompany the proposal, the bidder may neglect to submit his bid bond or it may not be in the amount requested. The omission of a unit price may also be a cause for rejection. Rejection at this point is desirable in that the bid security can be returned without undue delay.

After all bids are read aloud and the apparent low bidders are determined, all bids are checked for errors. The bids are then tabulated in the order of first, second, third, etc. After the low bidder is determined on each project, the detailed unit prices of the successful bidder are then read aloud. Electronic computers are being used successfully in this phase of the work of the state highway departments, also, with great savings in time and manpower. The contract is then prepared, bonds are furnished by the successful bidder, and an award is made within a reasonable time. On federal-aid projects concurrence in the award of a contract to a contractor must also be received from the federal agency. When the contractor furnishes the necessary bond, arrangements are made for the return of the other bidders' security. A majority of states return the bid security of all bidders except the lowest two or three immediately after a letting. Up to 30 days is sometimes required after the low bidder has been determined in order to make an award of a contract.

9–19. The Unbalanced Bid. On unit price contracts it is expected that each bid item will carry its proportionate share of the cost as well as the profit. It is possible for a contractor to raise certain unit prices on some items and lower unit prices on other items without changing the total of his bid. The result is usually an unbalanced bid. Generally speaking, unbalanced bids are undesirable, and when detected are usually a cause for rejection. An example of an unbalanced bid may be shown in the bidding of the following items:

	Balanced		Unbalanced	
	Unit Price	Amount	Unit Price	Amount
12,000 cu yd earth excavation	$1.00	$12,000	$ 0.50	$ 6,000
500 cu yd rock excavation	6.00	3,000	18.00	9,000
Total Cost		$15,000		$15,000

Assuming that when final quantities were determined the amount of earth excavation was 10,000 cubic yards, and the rock excavation was 2500 cubic yards, the final cost would be as follows:

	Balanced		Unbalanced	
	Unit Price	Amount	Unit Price	Amount
10,000 cu yd earth excavation	$1.00	$10,000	$ 0.50	$ 5,000
2,500 cu yd rock excavation	6.00	15,000	18.00	45,000
Total Cost		$25,000		$50,000

9–20. Right-of-Way Agreements. Contracts between a state highway agency and a property owner are often necessary in order to obtain the right-of-way for highway construction purposes. These contracts may be for the outright purchase of the land needed or as an easement for necessary grading operations. Most of these agreements are negotiated agreements between the property owner and a representative of the state highway department. When land is vitally needed for a project and no agreement can be reached by negotiation for its purchase, the land may be taken under what is known as the "right of eminent domain." The process of acquiring the necessary property for public use, which is referred to as "condemnation proceedings," follows a definite legal procedure. Resort to the courts should be avoided if possible, due to the extra cost and delay involved. However, some states prefer to resort to court procedure in the procurement of right-of-way. It is claimed that the benefits of such procedure include a sense of fairness to all parties selling property, since court-appointed appraisers determine all values, and a less vulnerable position for the state, since they cannot be criticized for paying exorbitant prices in order to speed acquisition.

Detailed right-of-way procurement procedures are beyond the scope of this book, but have become very important in recent years, with the great increase in land-taking brought about by the Interstate program.

Supervision of Construction

After a contract is awarded, arrangements are made by the contractor to move his equipment and personnel to the job site in order to start construction operations. The time it takes to do this will depend upon the distance involved. However, he must comply with contract requirements which usually state that construction work must start within so many days of the award of the contract. The manner in which supervision of construction is carried out is described below and is considered representative of the general practice of many highway departments.

9–21. Project Supervision. The work of the project conducted under the supervision of a state highway department is usually under the administration of the project engineer. It is his duty to see that all phases of the work are carried out in accordance with plans and specifications. He receives his instructions from, and is usually responsible to, a district or division engineer. The project engineer is assisted by a number of trained persons, including a survey crew, inspectors, office engineer, and others. The number of people obviously depends upon the length, type, and complexity of the project.

On highway work where federal funds are used, engineers from the

Bureau of Public Roads make periodic inspection trips, usually accompanied by the district or division engineer of the state highway department.

9–22. Duties of the Project Engineer. It is necessary that the project engineer secure a high standard of work, and this can be done only by observing whether or not the contract requirements and specifications are being complied with. This function is usually delegated to the various inspectors on the project who are responsible to the project engineer.

In recent years, congressional investigations and other criticisms of the highway program have focused attention upon the fact that absolute compliance with typical plans and specifications is neither desirable nor necessary in this type of work. From a legal point of view, the concept has presented some problems. For example, if the plans call for a pavement to be 6 inches thick, engineers know that it may be 5¾ or 6¼ inches without creating any problems; but the lawyers don't see it that way. They want it to be 6 inches thick, no more, no less, despite the fact that the field construction process is hardly a precise one.

For this reason, many state highway agencies are including in their specifications the "substantial compliance" clause recommended by AASHO (*94*). This clause reads as follows:

"All work performed and all materials furnished shall be in reasonably close conformity with the lines, grades, cross sections, dimensions and material requirements, including tolerances, shown on the plans or indicated in the specifications.

"In the event the engineer finds the materials or the finished product in which materials are used not within reasonably close conformity with the plans and specifications but that reasonably acceptable work has been produced, he shall then make a determination if the work shall be accepted and remain in place. In this event, the engineer will document the basis of acceptance by contract modification which will provide for an appropriate adjustment in the contract price for such work or materials as he deems necessary to conform to his determination based on engineering judgment.

"In the event the engineer finds the materials or the finished product in which the materials are used or the work performed are not in reasonably close conformity with the plans and specifications and have resulted in an inferior or unsatisfactory product, the work or materials shall be removed and replaced or otherwise corrected by and at the expense of the contractor."

It is the project engineer's duty to see that all engineering work, such as the setting of stakes for alignment and grade, is performed well in advance of the contractor's work so that there will be no delay. It may be necessary to effect a change in design during the progress of the work, and each state defines the procedure to be followed in such matters.

The necessary reports required by the state highway department and the Bureau of Public Roads are the responsibility of the project engineer. These reports include the preparation of estimates for partial payments and also for final payment.

As a representative of the highway department on a project, the project engineer must keep others properly informed as to the progress of the work, such as municipalities and utility companies who must frequently be contacted on most construction projects. The proper authority should be informed of any additional borrow requirements or on matters involving disputes in regard to the right-of-way. Good administration of a project usually results in a well-constructed project.

9–23. Project Engineer-Contractor Relationship. The highest standards of work are developed when there is complete cooperation and accord between the contractor's employees and those of the state highway department. It must be remembered that the contractor is entitled to a fair profit and is due every consideration consistent with the specification requirements and quality of work demanded of him.

In case of a claim by the contractor, a careful record should be made of all work in question and the information made available to those responsible for the adjustment of claims.

The contractor should make available at all times any information that may be required, such as transcripts of payrolls and receipted invoices, etc. The project engineer should see that the contractor complies with all regulations regarding labor, equipment, and so on.

9–24. Construction and Inspection Details. All materials used on a project must be inspected for compliance with applicable specifications. These materials may be inspected at the site of the project, or if quantities are sufficient to warrant it, they may be tested and approved before shipment is made to the project. When this is done, the testing is frequently done by an approved commercial testing laboratory. This does not mean that materials may not be rejected if during shipment they become damaged and cannot comply with specifications or if a defect was not detected in the inspection at the source.

Most highway departments maintain well-equipped testing laboratories of their own. The contractor is required to submit samples of the materials to be used, for testing purposes, and to name the sources of the materials. For concrete and bituminous paving mixtures, proportioning of the various ingredients is usually under laboratory control and the various details are carried out by inspectors assigned to the project.

The number and kind of inspectors needed for each project will depend upon the kind and type of work to be done. The project may require the services of a grade inspector, a culvert inspector, a gravel plant inspector, a plant inspector for the preparation of various types of plant mixtures, and others. The inspector not only inspects the materials but also exercises control over the manner in which the work is done. The importance of the work of the inspector should never be underrated.

9–25. Quality Control of Materials and Construction. Quality control of materials by tests on small samples has been the traditional practice in highway construction. However, interpreting the results of such tests has usually involved subjective judgment because little information was known regarding randomized sampling techniques, over-all sample size, and the probability risks involved in the acceptance of material based on the usual sampling methods. Furthermore, little information was available regarding the variable nature of materials and test results, although many engineers were well aware that test results vary.

Literal interpretation of many requirements of existing specifications does not permit realistic variations in the quality of the material. In actual construction operations, such inflexible enforcement of specifications is often impossible, highly impractical from the standpoint of economics, and not worth the extra cost from the standpoint of the functioning and serviceability of the finished product.

Development of clear and definite procedures for using statistical techniques for quality control and acceptability with reasonable tolerances of major components of highway construction is under way. Model specifications are expected to result that will reflect the variable nature of particular product or process and the ranges of acceptable tolerances for quality control. These model requirements will also reflect realistic risks in the acceptance of the work performed under them.

9–26. Construction Surveys. The project engineer and his party are usually assigned to a construction project at about the same time that the plans and specifications are completed and when it is reasonably certain that bids will be received within a short time. This permits the construction party to do necessary engineering work before the contractor arrives.

A good plan to follow when the project engineer arrives at the new job location is to make an inspection of the whole project with his entire party, on foot and with plans in hand. This permits each member of the party to become familiar with the various physical features of the area and their relation to the proposed design. When this is done, the construction surveys can be made. "Construction surveys" comprise all the engineering surveys necessary for the prosecution of the work to its final completion.

The first problem in construction surveys is to check the alignment to see if it is in accord with the final plans. In many cases this means reestablishing the centerline of the entire project. It is also necessary to check all witness points of alignment; and if there is a possibility that these witness points may be in the way of construction, they should be reestablished beyond construction limits. Consideration should also be given to the establishment of additional reference points on the road centerline to facilitate the instrument work when setting stakes for alignment and grade.

All bench marks should also be checked for location and elevation. Any benches that have been damaged or destroyed should be reestablished. All benches should be outside the limits of construction.

9–27. Staking Requirements. After the road centerline and the bench marks have been checked or reestablished, the major portion of the survey work consists of setting stakes for the guidance and control of construction operations. All stakes that are set by the engineer should be marked in such a manner that they can be easily interpreted.

The first stakes that are set are usually clearing and grubbing stakes. The distances out from the road centerline are generally determined by using the plans and scaling the distances to the limits of construction. The stakes may consist of small saplings cut to 4- or 5-foot lengths, with a blaze mark facing the road centerline. These stakes are usually placed at even 100-foot intervals.

9–28. Cross-Sections and Slope Stakes. The construction survey party is responsible for all earthwork calculations. It is therefore important that they take cross-sections over the entire project in many cases. Such a procedure may not be necessary in cases where cross-sections (and slope stake positions) are determined from aerial photographs and by the use of computers, as discussed in Chapter 8. The original cross-sections provide a record of original ground surface elevations which can never be obtained again after construction has started. The cross-sections taken on the project which served as a basis for the final design may not be suitable. The period of time between the taking of the cross-sections for design purposes and letting a contract for construction may vary from several weeks to two years or more. Taking cross-sections just before construction operations begin eliminates one of the many possibilities of disputes over earthwork quantities.

A slope stake is set at the point where the proposed fill slope or cut slope intersects the original ground. The stakes may be set when cross-sections of the original ground are taken. Slope stakes for grading are usually set on both sides of the road at 50-foot intervals at the station and half-station points. All information necessary for grading opertions should be shown on the stake. It is sometimes necessary to offset slope stakes. When this is done, the amount of offset, which is the distance from the offset location of the stake to its intended location, is noted on the stake. The cut or fill reading is taken from the ground location of the stake, as set. Fig. 9–2 shows typical cross-sections and positions of slope stakes.

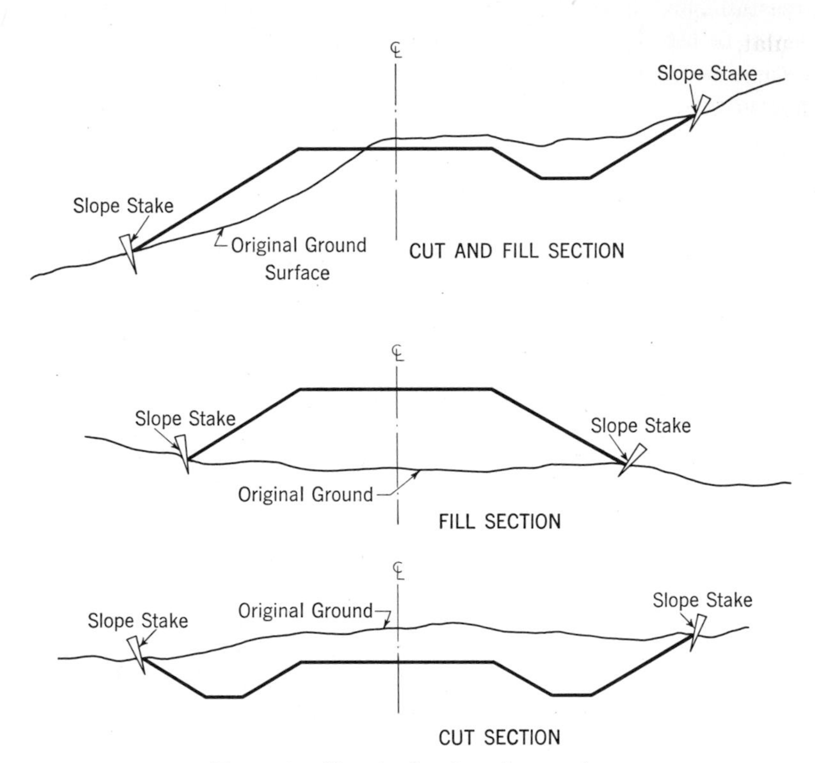

Fig. 9–2. Sketch showing slope stakes.

When setting slope stakes, the cut or fill is usually marked on the centerline stakes to serve as an aid to grading operations. Many states set slope stakes in heavy cuts only, and control grading operations in other areas by the use of earth-grade stakes.

9–29. Earth-Grade Stakes. When the rough grading operations have been completed, it is generally necessary to set finishing grade stakes to control the finishing of the earth grade. These stakes are generally set on the shoulder line at every station and half station. Some operators require that the finish grade stakes be set on the road centerline instead of the shoulders. The grade stakes may be set in such a manner that the top of the stake is at grade. The stake may be left high and the amount of cut or fill noted from a keel mark on the stake.

9–30. Final Cross-Sections. When grading operations are complete, it is necessary to take final cross-sections of the completed work in order to determine the final quantities of excavation and embankment completed on the project. The final cross-sections should be taken at the same points as the original sections. Final cross-sections may be taken as soon as grading operations are completed. This permits the calculation of earth quantities as the work progresses and generally shortens the time necessary for the preparation of the final estimate when the project is completed.

9–31. Borrow-Pit Measurements. When additional earth for embankment sections must be obtained from borrow pits, it is necessary to

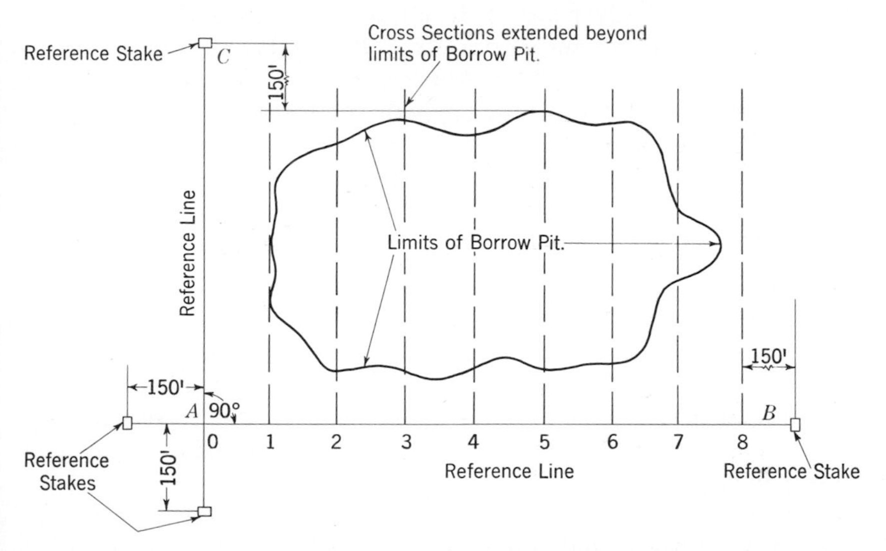

FIG. 9–3. A method of staking and cross-sectioning a borrow pit.

take cross-sections before and after the removal of earth in order to determine the quantities used. The area from which the earth is to be excavated should be staked with one or more baselines, and cross-sections should be taken at 25-foot intervals. The base lines should be well referenced so that they may be relocated for the final cross-sections. Sketches of borrow-pit layout and reference lines should always be made. After the borrow earth has been removed, cross-sections are again taken on the original lines. Both original and final cross-sections are plotted, and the volume of earth removed is determined by the average-end area method or some other acceptable method. Fig. 9–3 shows a typical method of staking a borrow pit for cross-sectioning. Quantities of borrow material may also be determined by weighing loaded hauling units.

9–32. Side Borrow. Most borrow pits are located at some distance from the roadway. Sometimes it may be feasible to obtain additional earth from the side of the roadway in large cuts by extending the side slopes. In order to obtain an accurate measurement of the earth, sections should be taken at 25-foot intervals, just as in borrow pits and the method of computing volumes may be the same.

9–33. Grade and Line Stakes for Pavements. The first stakes required for a surface paving project are generally for trenching or for the establishment of a uniform grade in accordance with the plan centerline profile. These stakes may be placed on the side of the road and may be used later for form stakes when placing concrete pavement. The form stakes are usually placed 2 feet from the edge of the slab unless otherwise requested by the contractor. The stakes should be set 50 feet apart on tangents, and 25 feet apart on horizontal and vertical curves. Form stakes are placed on both sides of the road. On many projects, final finishing of the subgrade, base, or subbase is done by a machine with automatic controls operating on a taut wire stretched between stakes or pins; the field survey party must establish and check the position of the wire.

9–34. Culvert Stakes. The plans give the size, location, and lengths of all culverts on the project. They also give the elevations of the flow line and the angle of crossing of the road centerline. Field conditions should always be checked to see whether or not the proposed installation will be satisfactory. Fig. 9–4 shows two typical culvert installations, one at right angles to the centerline and the other at an angle of less than 90 degrees.

The stakes required for a culvert installation consist of the centerline stakes of the culvert, and reference stakes which usually indicate the length of the culvert. The flowline elevations are generally given on the centerline reference stakes, as indicated in Fig. 9–4. All stakes should be offset so they will not be disturbed during construction. On large box culverts the same method of staking is used.

9–35. Miscellaneous Engineering Work. The engineer may be required to furnish stakes for the location of manholes, catch basins, and inlets, with the necessary elevations for flow lines and tops of covers. Curb-and-gutter stakes and sidewalk stakes showing grade and alignment may also be required.

9–36. The Preparation of Estimates. As the work progresses, the contractor is entitled to partial payment for the work completed. These payments are generally shown on forms called "estimates" and are prepared by the project engineer. Estimates may be made monthly or bi-monthly. On lump-sum projects, the amount of the estimate will be a percentage of the whole. On unit price contracts, the individual items of work completed are listed and the value of the work is computed.

It is sometimes possible to pay for materials delivered to the site of the project or furnished for later use, such as gravel (stockpiled), reinforcing steel, etc. However, such procedure should be specified in the proposal or specifications, or definite means for providing authorizations should be made.

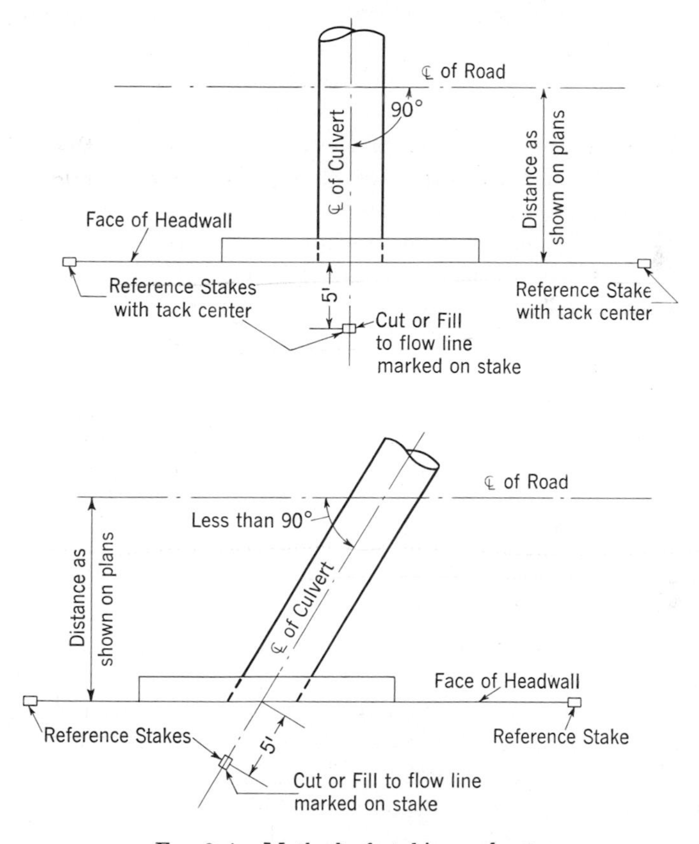

FIG. 9–4. Method of staking culverts.

On partial payments the full amount is not always paid to the contractor, but a percentage, usually 10 per cent, is retained. This protects the owner in case of an underrun of estimated quantities, as a guaranty for faulty work, or for minor claims for materials. On large contracts, 10 per cent sometimes results in a large amount of capital being tied up for reserve purposes. Some highway departments arrange a reduction of this reserve to 5 per cent when the work has been completed and accepted, and pending the preparation of the final estimate. When the final estimate is complete (and certain delays in office procedure delay its processing), this reserve may be reduced to a small lump sum.

In the preparation of the final estimate, all items of work are measured as constructed in the field. If these quantities differ from those in the contract as authorized and explained under changes in the work, it is necessary to prepare authorizations that decrease or increase the quantities, as the case may be.

9–37. Final Acceptance. When all work has been completed, inspection of the whole project is made by the authorized party for the highway department. If the project is one in which federal funds were used, the final inspection is made by an engineer of the Bureau of Public Roads. When all parties indicate their acceptance, the final estimate, which is usually in the process of completion, is also approved, and recommendation for final payment is made.

During construction, it is the responsibility of the project engineer to supervise all phases of the work, approve or disapprove all authorizations for changes or extra work, and see that the work follows the schedule of operations as set up in the contract. As no work can be done beyond the completion date, he must notify the contractor in due time and prepare to extend the contract by supplemental agreements, such as an extension of time. When all the work is completed, the recommendations of the project engineer as to the contractor's performance, quality of work, and cooperation are necessary for his rating factor, as mentioned under the prequalification of contractors. The preparation of weekly and monthly reports, a history of the project, and administrative functions are always incidental to the supervision of a construction project.

CHAPTER 10

EARTHWORK OPERATIONS AND EQUIPMENT

Nearly all highway construction jobs, especially those in new locations, involve a considerable amount of earthwork. In general terms, earthwork operations are those construction processes which involve the soil or earth in its natural form and which precede the building of the pavement structure itself.

Basic earthwork operations may be classified as follows: clearing and grubbing; roadway, borrow, structural, and several special forms of excavation; the formation of embankments; and finishing operations. Any or all of these construction processes may be performed on a given highway project and they may overlap to a certain extent.

By "clearing" is meant the removal of trees, brush, shrubs, etc., from within the limits of a designated area, while "grubbing" usually refers to the removal of roots, stumps, and similar obstacles to a nominal depth below the existing ground surface. Frequently clearing and grubbing comprise a single contract item and include the removal of topsoil to a shallow depth. The term "excavation" refers to the removal of earth in its various forms from its natural resting place and for various purposes. Excavated material must generally be transported to a different place, and these earthmoving operations are generally included as a part of excavation. Embankments required in highway construction are usually formed in relatively thin layers or "lifts" of soil compacted to a high degree of density. Such embankments are called "rolled-earth embankments" or "rolled-earth fills." Hydraulic fills are also occasionally required in highway construction. Finishing operations include such items as the trimming and finishing of slopes and the fine-grading operations required to bring the subgrade to the final desired elevation. Broadly speaking, earthwork operations may be said to include virtually all the operations, exclusive of those relative to structures, involved in bringing the highway structure to the stage where, in the normal sequence of events, construction of the base (or subbase) is about to begin, unless the subgrade requires special treatment.

Equipment Used in Earthwork Operations

In modern highway practice in the United States, earthwork operations are largely accomplished by the use of a large number of highly efficient and versatile machines. The development and application of such machines in the last five decades has been an outstanding phase of highway progress. Machines have been developed which are capable of efficiently and economically performing every form of earthwork. Some machines are tailored to perform a specific operation or series of operations, while others have a wide variety of uses. The proper selection and efficient use of these machines are an important part of the highway scene.

To the beginner in this subject, the large number and almost infinite variety of machines used in earthwork operations are undoubtedly bewildering. Not only are there many different basic types of machines and modifications of them, but there are many manufacturers engaged in the production of machines of similar characteristics and uses. The discussion which follows has been deliberately simplified in the interests of clarity and brevity.

10–1. Tractors. One of the basic tools used in highway construction is the tractor. Crawler and rubber-tired tractors provide the power for an almost unbelievable number of construction operations, as indicated in the following paragraphs. The tractor serves to push or pull construction equipment. Various kinds of equipment may be mounted on the tractor, or the tractor may serve as a power take-off unit. There are many types, sizes, and models of tractors now available; they range from extremely small utility units up to the 700-horsepower giants, They may be grouped into three broad classes—crawlers, large rubber-tired units, and smaller utility-type tractors.

Crawler tractors are rugged work-horses on any large earthmoving project. They are used to push-load scrapers, spread material at the fill, maintain haul roads, rip hard ground or soft rock (with a hydraulic ripper mounted on the back of the machine), and perform many other tasks. Their greatest usefulness is on rough ground, on rock, on slopes, and on relatively short hauls. Crawler tractors range in weight from a few thousand to more than 60,000 pounds. Diesel engines are used in all except the smallest models.

Primarily because of the demand for greater speed in earthmoving operations, developments in recent years have led to the greatly increased use of rubber-tired tractors. Some of these machines are very large and powerful. For example, one unit of this type has a shipping weight of over 70 tons. A turbo-charged 12-cylinder diesel engine powers the unit at speeds up to 28 miles per hour, both forward and reverse. Lighter units in this category have even higher over-the-road speeds.

In recent years, light agricultural-type tractors have emerged as full-blown construction tools. The principal reason for this is the hydraulically operated attachments which have been made available for these units, particularly front-end loaders and backhoes. With one unit, an operator can load trucks, stockpile material, dig trenches, lay pipe, excavate for structures, backfill excavations, and so on. The front-end loader is described in more detail in Art. 10–4.

Several manufacturers now produce large rubber-tired tractors that are hinged in the middle (articulated). Advantages of this design include better maneuverability, shorter turning radius, and, at the same time, a longer wheelbase.

10–2. 'Dozer Units. One group of earth-moving tools which utilize the basic tractor includes the bulldozer and related tools of the 'dozer

FIG. 10–1. This crawler tractor carries a hydraulically operated bulldozer blade. (Courtesy Allis-Chalmers Mfg. Co.)

family. The bulldozer basically consists of a heavy curved blade which is mounted on the front of the tractor. The blade is fastened to the tractor by means of two long beams, one on either side of the tractor unit. The bulldozer blade may be raised and lowered by a cable arrangement or hydraulically. The blade of a bulldozer is usually only slightly wider than the tractor and is of extremely rugged construction. A typical application of a heavy-duty bulldozer is shown in Fig. 10–1.

One modification of the bulldozer is given various trade names but may be referred to simply as a "bulldozer with angling blade." On this type of machine, in addition to its vertical travel, the blade may be angled from a horizontal position so that one corner of it may be dug into the ground. In certain machines the top of the blade may also be tilted slightly forward. The blade used on this type of machine is generally somewhat wider and slightly lighter than a bulldozer blade that is designed for use on the same tractor.

Many other modifications have been made of the basic bulldozer unit for special job applications. An example of this lies in the replacement of the 'dozer blade with a large, heavy circular steel plate, which is often cushioned by heavy springs or hydraulic cylinders, for use in the pusher-loading of scraper units. Another example is the "tree-dozer" which has an extra-heavy beam which may be placed high against a tree in order to push it over. In still other machines the blade is replaced by a giant toothed rake which is useful in land-clearing operations.

10–3. Scraper Units. Another group of earthmoving tools of fundamental importance includes tractor-drawn and self-propelled scrapers. These tools will dig, load, carry, dump, and spread earth in an independent, self-sufficient operation under conditions which are suited to their use. Basic elements of a scraper unit include a cutting edge which may be lowered into the ground to make a shallow cut; the apron in front of the bowl, which opens and closes in order to regulate the flow of dirt into and out of the bowl; the bowl itself, in which the dirt

Fig. 10–2. Twin-bowl scraper has three engines, carries 28 cubic yards of earth, and is being push-loaded by a crawler tractor. (Courtesy Goodyear Tire & Rubber Co.)

is carried; and a tail gate or other ejection mechanism which is shoved from the rear of the bowl forward in order to push the dirt out of the bowl during the unloading operation. During the "haul" portion of the cycle, the bowl is held up off the ground and the weight is supported on huge pneumatic tires.

There are three basic types of prime movers employed with modern scrapers: crawlers, four-wheeled rubber-tired tractors, and two-wheeled "overhung" (self-propelled) rubber-tired tractors. In the last type, the scraper section has only two wheels, as does the tractor unit. To gain more power, manufacturers developed the "twin power" scraper, which has tractors fore and aft, driving on wheels in front of and behind the scraper unit. The tandem unit in Fig. 10–2, which has two bowls and three engines (front, center, and rear), is shown being push-loaded by a crawler tractor. Capacities of scrapers range from about 7 to 60 cubic yards.

Present use of crawler-powered scrapers is limited to very short hauls, or very bad ground. The high-speed, four-wheeled units seem to be ideal for long, open hauls. The highly maneuverable, two-wheeled units seem to be best for shorter hauls, in tighter quarters, and in materials offering poorer traction. Control systems on all types of scrapers may be cable, hydraulic, or electrical. Many of these units are also used in conjunction with loaders or shovels as wagons. The larger units of this type require pusher help to attain their maximum load capacity.

A variation of the two-wheel scraper is the "elevating" scraper, which can load itself without a pusher in many soils (Fig. 10–3). In this type of machine, a hydraulically operated elevator is mounted at the front of the scraper bowl; material that is cut and loosened by the bowl cutting edge is carried upward by the flights of the elevator and

Fig. 10–3. Soil loosened by the cutting edge is lifted by the elevator flights into the bowl of this elevating scraper. (Courtesy Caterpillar Tractor Co.)

dumped into the scraper bowl. Capacities of these machines range up to about 30 cubic yards.

10–4. Front-End Loaders. A tractor may be equipped with a front-end bucket and thus serve as a digging, hauling, and loading unit. These units are called "front-end loaders" and are not considered as basic earthmoving tools in most highway work; but they perform very useful functions in miscellaneous applications associated with earthwork, such as backfilling trenches or around structures, excavation in limited areas, cleaning up around aggregate stockpiles, and they may even be used in borrow-pit excavation under unusual circumstances. The buckets used in these applications vary in capacity up to about 6 cubic yards. The bucket unit generally is raised, lowered, and tipped back hydraulically. Both rubber-tired and crawler tractors are used for these units.

FIG. 10–4. This rubber-tired tractor carries a front end bucket and a rear-mounted backhoe: both attachments are hydraulically operated. (Courtesy International Harvester Co.)

Smaller, rubber-tired units of this type often carry a hydraulic backhoe on the rear end (Fig. 10–4). Either the front bucket or the hoe may be replaced with one of a large variety of other hydraulic attachments.

10–5. Tractor-Drawn Belt Loaders. Under certain conditions, particularly where long, flat grades can be maintained and when the excavation is performed in average soils, a tremendous volume of earth can be moved by a tractor-drawn belt loader (elevating grader). This unit, which has its own power for operating a belt conveyor, is towed by a tractor. As the tractor moves along, a cutting edge is forced into the ground and the dirt thus dug is forced onto a conveyor belt. The dirt is then carried up and to one side where it falls off the belt into a truck or wagon.

10–6. Shovel-Crane Units. Another group of machines of importance in this phase of highway construction may be designated as "shovel-crane units." The five basic units which are included in this classification are shown by the sketches of Fig. 10–5. These units are the shovel, crane, dragline, clamshell, and backhoe. The crane unit, as such, is

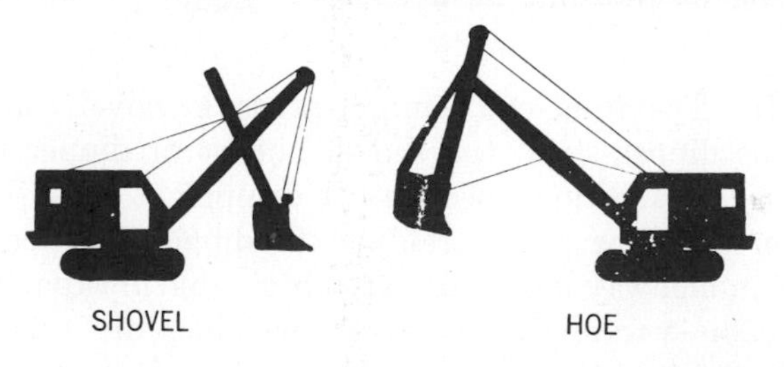

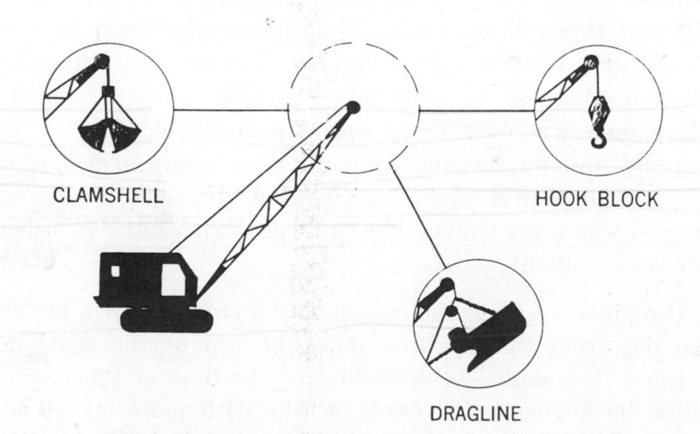

Fig. 10–5. Shovel-crane units.

not of importance so far as earthwork is concerned but is included in the group because it is the basic unit upon which the dragline and clamshell attachments are superimposed. All these units have the ability to perform several basic operations. These include (1) the ability to travel from place to place, (2) the lifting or hoisting of a load, (3) the swinging of the load to a new location (all commercial units of this type can revolve 360 degrees), and (4), in the case of a shovel, the ability to thrust or "crowd" the shovel bucket into a bank or other mass in order to accomplish the digging operation.

The description of these units may be further simplified by dividing each machine into its basic elements—the mounting, the revolving superstructure, and the front-end attachments. Units of this type may be mounted on crawler treads, rubber tires, or on rubber-tired truck units. The revolving superstructure is common to all these machines and is basically the platform upon which are mounted the engine and necessary mechanisms for performing the hoist, swing, and travel functions. Modern shovel-crane units may be powered by gasoline, diesel, or electric motors. The front-end attachments include all the equipment necessary for the operation of the unit as a shovel, dragline, clamshell, and so on.

10–7. Shovels. The front-end equipment on a shovel consists of the shovel boom, the dipper stick, the shovel bucket or dipper proper, and the dipper trip. In addition, mechanical equipment must be provided on the boom for crowding and retracting the dipper. Dippers provided on shovels used in highway operations are obtainable in capacities ranging from 3/8 to 5 cubic yards. Much larger machines are used in certain industrial applications. An average size shovel used on normal highway operations would be one having a capacity of 2½ to 4 yards.

The shovel is a very versatile tool which may be used in handling many different types of material. Conditions which are best suited to shovel operations include excavation in firm or hard material where the loading or disposal of materials can take place at or above the ground level and where sufficient space is available for the shovel to work properly and for hauling units to have ready access to the machine. A typical shovel operation involves borrow-pit excavation in hard material where the shovel can work against a bank or other nearly vertical face (Fig. 10–6).

10–8. Draglines. A dragline is basically a scoop or open bucket which is thrown out from the machine onto the top of the material to be excavated and then pulled back in toward the base of the machine. As the dragline bucket is pulled back, it fills with material. The bucket may then be lifted and swung, and the material dumped into hauling units or wasted. Dragline buckets, as used in normal highway operations, are available in approximately the same range of sizes as are shovel buckets.

The dragline is best suited for excavation in soft materials and at or below the existing ground level. Draglines are particularly adapted to such operations as ditch or canal excavation in swampy, wet ground, particularly where conditions are such that the dragline can remain on

FIG. 10–6. A 2½-cubic-yard shovel loads material into a 30-ton rear dump truck. (Courtesy Thew-Lorain Division, The Koehring Co.)

higher firm ground. These units also have the ability to cast or dump material at a comparatively long distance from the base of the machine.

10–9. Clamshells. The clamshell bucket is composed of two halves or "shells" which are hinged together at the top so that the bucket may be opened or so that the two halves may be brought together to form a closed bowl. In operation, the open clamshell bucket is first placed on top of the material to be dug; as the bucket is pulled up the two halves come together, digging into the material and filling the bowl as it closes. The capacities of clamshell buckets are in the same range as was previously mentioned for shovels and draglines.

Probably the most common application of the clamshell in highway work is in the placing and handling of aggregates at a stockpile or at a central mixing plant. They are also useful for structural excavation in soft to medium materials and at or below the existing ground surface.

10–10. Backhoes. The backhoe (or hoe) can best be described as a cross between a shovel and a dragline, in that it combines the digging action of a shovel with the pulling action of a dragline. A backhoe is illustrated in Fig. 10–7. In using this machine the boom, with the dipper stick extended, is lowered into the excavation; the dipper is then pulled in toward the machine, filling with material in the process; the

FIG. 10–7. Crawler-mounted backhoe. (Courtesy Link-Belt Speeder Co.)

whole assembly is hoisted, swung, and then dumped by extending the dipper stick.

The principal use of the hoe is in digging trenches required for drainage. It functions best when the excavation is below existing ground level, when the material is firm, and where the material may be disposed of within a relatively short distance.

10–11. Cranes. No further discussion of the basic crane unit will be given in this chapter. Many different attachments are utilized on cranes for particular construction operations. Included among these are concrete buckets, various hooks, magnets, "skull-crackers" (pavement breakers), and pile-driving attachments.

10–12. Special Purpose Units. Several variations of the conventional crane-shovel units have been developed in recent years. One of these is a fully hydraulic unit, like the one shown in Fig. 10–8, which has an action very closely resembling that of the human arm and wrist. Truck- or crawler-mounted, the boom can telescope, swing, raise or lower, rotate, and mount a variety of attachments. The unit can dress slopes,

FIG. 10–8. Truck-mounted hydraulic excavator. (Courtesy Warner & Swasey Co.)

clean ditches, dig trenches, backfill, place pipe, and so on. Many smaller, hydraulic crane units are used, primarily in materials handling.

10–13. Trucks and Wagons. The shovel-crane units which have been described are not, of course, hauling units. If the excavated earth is to be transported to some location beyond the range of the shovel or crane boom, this must be accomplished by a separate hauling unit.

Motor trucks with dump bodies and dirt or rock "wagons" are widely used for this purpose. Commonly used dump trucks vary in size from 1.5 to 15 cubic yards, are of rugged construction, and may be either rear- or side-dump models; off-highway dump trucks carry up to 65 tons (even more in mining applications). The size of the truck used on a particular job should be keyed to the size of the shovel unit, the larger models being used with larger excavating units. An International Harvester "Payhauler" is shown in Fig. 10–9.

FIG. 10–9. Rear-dump, off-highway hauling unit has a capacity of 45 tons. (Courtesy International Harvester Co.)

In recent years large dirt or rock wagons have gained steadily in favor, particularly where large shovel-crane units are used for excavation and for long hauls. These units are generally rubber-tired semitrailers drawn by rubber-tired tractor units. They may be rear-, side-, or bottom-dump machines. Wagon trailers somewhat resembling scrapers are also in use; prime movers may be interchangeable with scraper units. They are suitable for hauling in rough terrain, and they develop speeds equivalent to those of regular dump trucks when traveling on well-maintained haul roads.

Some very large hauling units use electric wheel drive. In these machines, a diesel engine drives an electric generator that, in turn, powers individual electric motors in each wheel. One big advantage of this arrangement is that the maintenance problems associated with a conventional power train (transmission, drive line, axle, etc.) are eliminated. Future hauling units of this type may well use gas turbines for power.

10–14. Rippers. A hydraulic ripper, integrally mounted on the back of a crawler tractor (Fig. 10–10), is an important tool on many earth-

Fig. 10–10. Integral hydraulic ripper with single shank rips soft rock. (Courtesy Caterpillar Tractor Co.)

moving projects. The ripper's function is to loosen and break up hard soil or soft rock so that the material can be loaded by scrapers.

Recent advancements in rippers and ripping techniques have made it possible to handle with scrapers many materials that formerly were drilled, blasted, and loaded by power shovels. Rippers used in earthmoving may have one, two, or three teeth of various designs. One or more hydraulic cylinders put down pressure on the teeth.

In general terms, a ripper-scraper operation, where feasible, is more economical than drilling and blasting. (Blasting is more economical now than in previous eras because of the widespread use of bulk ammonium nitrate as an explosive, rather than dynamite.)

Developments in shallow seismic exploration methods (Art. 6–27) also have extended the use of rippers. Manufacturers, particularly the Caterpillar Tractor Co., have correlated the wave velocities obtained in seismic studies with ripping feasibility and cost. Thus, a contractor can determine beforehand with reasonable accuracy whether he can rip a deposit or must drill and blast it.

10–15. Rollers. Three principal types of rollers are used in highway construction: sheep's-foot or "tamping" rollers, pneumatic-tired rollers, and steel-wheel rollers. The "smooth" rollers of the last type are used to some extent in earthwork operations, but their principal use is in the construction of bases and wearing surfaces. They will be described in a later chapter.

10–16. Sheep's-Foot Rollers. A sheep's-foot roller consists of a large (typically 40 inches in diameter and 4 or 5 feet long) steel drum on which are fastened a number of steel projections, or "feet." A roller of this type will generally weigh from 3000 to 10,000 pounds empty, depending on the size and number of tamping feet, and it is arranged so that it may be filled with water or sand in order to gain the increased weight desired for the compaction of certain soil types. A single roller may be used, or a single towed unit may consist of two or three rollers operating from the same pull-bar. The mechanism which connects the roller to the towing tractor is usually arranged so that the roller may oscillate freely during travel. Frequently, two or more roller units are pulled by the same tractor. Large, self-propelled sheep's-foot rollers (Fig. 10–11) used on large projects have from one to four drums. A few vibrating rollers of this type have been used in special situations.

The sheep's-foot roller is the basic tool employed in the compaction of many soils used in the formation of rolled-earth fills. Sheep's-foot rollers are manufactured with various combinations of size (area), shape, and spacing of feet. If desired, the proper combination of these variables, together with weight of the roller, can be secured for compacting the soil types encountered on a particular project. Some soils, such as clean sands of uniform size, cannot be successfully compacted by a sheep's-foot roller.

A roller which is similar in function to the sheep's-foot roller and which is of interest is the "grid" roller. This roller consists of a pair of steel drums approximately 5 feet in diameter arranged in parallel. The face of each drum is composed of an open grid of 1.5-inch diameter

Fig. 10–11. Self-propelled, dual-drum sheep's-foot roller. Courtesy *Engineering News-Record.*)

alloy-steel bars with net openings 3.5 inches square. The roller may be equipped with a cleaning device and a counterweight box to permit increases in the weight of the roller. A somewhat similar tool is the segmented steel-wheel roller, which is popular for use in certain non-granular soils because of the tamping action of the feet and because it is self-propelled.

10–17. Pneumatic-Tired Rollers. A compaction unit which has been gaining in popularity is the pneumatic-tired "wobble-wheel" roller. Towed or self-propelled, a wobble-wheel roller consists basically of a ballast box mounted on two lines of oscillating axles carrying small, smooth-faced pneumatic tires. Tires on the rear axle are placed so that they ride over areas left between each set of front tires. Gross weights of these units go up to 30 tons or so; tire pressures can be varied "on the run" to give the most efficient compaction for the soil involved.

Another compaction tool is a combination steel-wheel and pneumatic-tired rolling unit. Pulled by a rubber-tired tractor, the machine consists of a semitrailer ballast box mounted over a single axle of wobble-wheel tires. Directly in front of the tires is a retractable steel roller; after initial compaction, the steel roller is lowered to iron out the ridges left by the pneumatic tires.

Fig. 10–12. Towed "super-compactor" (large pneumatic-tired roller). (Courtesy Southwest Welding & Manufacturing Co.)

Use is also made of giant rubber-tired rollers. On these units, straight (nonoscillating) axles are used. These are semitrailer units, consisting of a very large ballast box mounted on a single axle (Fig. 10–12). Standard weight of a roller of this type is 50 tons, although they have been ballasted up to 200 tons, usually for test purposes.

10–18. Vibratory Compactors and Pneumatic Tampers. Since World War II, the use of vibratory compactors has become widespread, particularly for the compaction of granular materials. Primarily designed for the compaction of subgrades and base courses, these units are also used in embankment construction. Two principal types of vibratory units are available—the towed vibratory roller (Fig. 10–13) and the self-propelled, multiple-shoe vibratory compactor.

In a typical vibratory roller, vibration of a steel drum is caused by the fall of eccentric weights rotating on a shaft driven by a gasoline engine mounted on the unit. The frequency of vibration varies, but is usually in the range of 110 to 2000 vibrations per minute.

Vibration is obtained either mechanically or electrically in the self-propelled, multiple-shoe vibratory compactors, of which there are several

Fig. 10–13. Towed vibratory roller is effective on granular materials. (Courtesy Buffalo-Springfield Co.)

models. One compactor of this type is made up of from four to six compacting units mounted at the front of a pneumatic-tired tractor. The individual units can be arranged in a straight line (six units, working width, 13 feet 6 inches) or in various tandem or staggered patterns. Units can also be towed at the side for use on road widening projects. Each of the individual units is 26 inches in width and delivers up to four thousand five hundred 2400-pound blows per minute. Working speed ranges from 20 to 90 feet per minute.

Small plate-type vibrators and pneumatic tampers are useful in compacting granular or sandy soils around structures and in backfilled trenches. A tamper of this type may be fashioned by attaching a small plate to a regular air-hammer or similar pneumatic tool. Specially designed ramming and vibrating units are also available. All these units are of small size and are consequently useful only in rather limited areas. They have proved to be very effective in increasing the density of granular materials.

10–19. Motor Graders. Still another basic tool used in earthwork is the motor grader. This device is used in a whole series of operations, including the shaping of subgrades, shoulders, and ditches, back sloping, light grading operations, and the spreading of dumped materials. The self-propelled grader is also widely used as a maintenance unit, for mixing materials in road construction, and for snow removal. The smaller motor graders usually have a blade which is about 10 feet long and are sometimes called "patrol graders" or "motor patrols." Patrol graders are primarily intended for use in maintenance operations and will not be further discussed here.

Large motor graders are generally powered by diesel engines and are equipped with 12- to 14-foot blades. They are powerful, fast-moving units which are specially designed to perform a wide variety of operations. A grader of this type is shown in Fig. 10–14. The circle on which the blade is mounted generally can revolve a full 360 degrees so that the blade may be placed at any desired angle with the line of travel of the machines. The blade can generally be extended several feet beyond the outside line of the grader wheels, and it can be placed at varying angles with the horizontal so that back slopes up to 90 degrees may be readily cut. Power is supplied to rear tandem driving wheels, and the front wheels may be "leaned" to facilitate steering. Some models are capable of travel speeds up to 30 miles per hour. Various attachments are available for motor graders, including scarifiers, bulldozer blades, snow plows, vibratory compactors, and rollers. A grader may be equipped with an automatic control which maintains any desired slope of the blade, regardless of roughness of the terrain. A dial is set to the desired slope and an electronic device controls finished slopes to a high degree of accuracy. Although the operator can raise or lower the blade to control the depth of cut, the slope will remain constant.

10–20. Other Tools. Beyond the scope of this text, but of great importance in rock cuts and the production of aggregate for highway construction, are the equipment and materials associated with drilling and blasting. These items include truck and track-mounted drilling rigs, air compressors, drill bits and accessories, explosives, and detonators. Another type of "earthmoving" tool not covered here is the trenching machine, which is primarily used in the installation of subsurface drains on highway projects.

Having discussed the principal tools which are used in earthwork operations, we are now ready to consider the various procedures included in this phase of highway construction and the application of this equipment to these operations. It should be remembered that the discussion which follows will deal largely with average practice, and that the specific procedures followed by any one organization may differ considerably

Fig. 10–14. Heavy duty motor grader cutting a side ditch. (Courtesy Caterpillar Tractor Co.)

from those given here. It should also be noted that the construction specifications which are discussed are intended to be typical of those used by highway agencies in the direction and control of contract work.

Clearing and Grubbing

Clearing and grubbing are usually included as a single contract item by highway agencies, although some organizations pay separately for these two operations. This is generally the first operation to be undertaken on any project involving earthwork, and thus it precedes any excavation.

Clearing and grubbing may be defined as the removal of trees, stumps, roots, snags, down timber, rotten wood, rubbish, and other objectionable material from an area marked on the plans or otherwise designated by the engineer. The area designated will usually fall within the limits of proposed excavation and embankment, although it might be said more generally that clearing and grubbing is performed in all areas where the obstructions which are to be removed would interfere with the conduct of the proposed work.

In cut areas, in addition to the removal of vegetation and other obstructions from the surface of the ground, specifications generally require that all stumps and roots be removed to a depth of not less than 1 foot below the proposed grade. This requirement also holds in embankment areas where the height of the fill is less than 2 or 3 feet. In embankment sections, when the embankment height is to be more than 3 feet, trees and stumps may be left in place and cut off at ground level or at a height of from 6 to 12 inches above the existing ground surface. Some agencies require that all large trees (18 inches or more in diameter) be removed, regardless of the height of the fill. It is usually provided that excavations made for the purpose of removing stumps and similar obstructions be filled to a level consistent with the surrounding ground before construction proceeds.

Large trees usually receive special consideration and may be left in place at the discretion of the engineer. In such cases trees which overhang the roadway are usually required to be properly trimmed to furnish a clearance of from 14 to 16 feet. It might be noted that very frequently areas within the limits of the right-of-way but beyond the limits of excavation and embankment are subjected to "partial" or "selective" clearing in which certain trees and shrubs are left standing. The debris removed from the designated areas is usually disposed of by burning, although in most cases provisions are made for the storing and salvage of marketable timber. It should also be noted that certain organizations include the removal and disposal of buildings, fences, and so on in clearing and grubbing, while others include a special contract item when obstacles of this nature are to be removed.

Measurements of this item for the purpose of payment are generally made in acres; that is, the bid unit is 1 acre. Where only clearing operations are involved, special payment may be made for large trees, i.e., so much per tree. Clearing and grubbing operations are generally performed largely by machines, with the bulldozer and its various attachments being the basic devices, or by companion tools such as rock rakes, rippers, and tree-'dozers. A considerable amount of hand labor may also be necessary, and liberal use is made of power saws and similar tools.

Excavation

In the interests of simplicity the following discussion will deal with three types of excavation: roadway and drainage excavation, excavation for structures, and borrow excavation. Before beginning the discussion of each of these types, it is desirable to briefly discuss the subject of "classification of excavation."

10–21. Classification of Excavation. It seems fairly obvious that the difficulties encountered in the excavation of materials which require blasting for removal, such as solid rock, will generally be much greater than those encountered in the removal of other earth materials. This fact generally increases the cost of excavation in rock, and for this reason some highway agencies find it desirable to divide excavation into two separate classes. These classes are generally termed "rock (or solid rock) excavation" and "common (or earth) excavation." These terms may be defined as follows:

ROCK EXCAVATION. The excavation, removal, and disposal of all boulders which are 0.5 cubic yards or more in volume, and all hard rock which, in the opinion of the engineer, can only be removed by blasting. In some cases the requirement relative to removal by blasting is not specified, and a phrase such as "solid, well-defined ledges of rock" is substituted in the definition. Even though a contractor may choose to rip some materials, this operation may still be called rock excavation by some agencies.

COMMON EXCAVATION. This consists of all excavation not included in the definition of rock excavation.

The method of classification presented above has certain disadvantages, as borderline cases frequently arise in which there may be some dispute as to the proper classification of a portion of the work. Many organizations obviate this difficulty by using the term "unclassified excavation" to describe the excavation of all materials, regardless of their nature. This latter procedure is now preferred by a majority of highway agencies.

10–22. Roadway and Drainage Excavation. Roadway and drainage excavation, as defined here, means the excavating and grading of the roadway and ditches, including the removal and disposal of all excavated material and all work needed for the construction and completion of the cuts, embankments, slopes, ditches, approaches, intersections, and similar portions of the work.

All excavated materials which are suitable are used in constructing the elements of the roadway structure, including embankments, shoulders, subgrade, slopes, backfill when required for structures, and so on. Provision is made for the disposal of unsuitable or surplus excavated material. It must again be emphasized that, under the definitions which we have given here and in accordance with normal practice in this country, payment made for excavation items is not just payment for excavating the earth. On the contrary, it frequently includes nearly all the work needed in many jobs for forming the earthwork portions of the highway structure. This, again, would normally include such items as the formation of embankments and the finishing operations that are to be discussed later. However, practice is not uniform in this respect, and some agencies pay separately for compaction, addition of water during compaction, fine grading, and other items.

Many miscellaneous provisions are generally included in specifications for roadway and drainage excavation. Provision is usually made for the removal of vegetation if the area has not previously been cleared and grubbed, and also for the salvage of portions of the excavated materials which are to be used in miscellaneous sections of the roadway. This latter item, for example, might provide for the use of suitable subgrade materials, the saving of topsoil for use on side slopes, and work of a similar nature. When unsuitable materials are encountered in cut or embankment sections, provision is generally made for their removal and replacement with satisfactory materials. Muck excavation may be a separate contract item.

The specifications may also include instructions for grading the roadbed, intersections, etc., in accordance with the required alignment, grade, and cross-section. In connection with this portion of the work, the roadway is usually required to be well drained at all times during the course of construction operations. Provision is also made for the grading of shoulders, slopes, ditches, etc., in accordance with the plans, sometimes including requirements relative to the flattening, rounding, and warping of side and ditch slopes where applicable. When rock is encountered in cut sections, it is generally required that it be excavated to a depth of 12 inches or more below the grade line, and that suitable backfill material then be used in bringing the cut section up to the proposed grade. Another provision frequently included is that of maintenance of the graded roadbed until final acceptance of the work or until construction of the base or surface course begins. Provisions under this item might also include the scarification and compaction of cut sections, although they would not normally include subgrade soil stabilization.

Measurement of excavated material is usually based upon a determination of the volume, in cubic yards in original position, of material actually excavated and disposed of as directed. Volumes are usually determined by the average-end area method, and payment is made for the number of cubic yards of roadway and drainage excavation and for overhaul.

Construction methods used in performing the operations required in this type of excavation may involve nearly every one of the types of equipment previously described. For example, tractor-scraper units are normally used in the basic operation of excavating, hauling, and spreading ordinary soils; if hard, compacted materials are encountered, rippers may be required for breaking up the material so that the scrapers may load. Scrapers with crawler tractors are used for relatively short hauls or in rough terrain; rubber-tired units are used for longer hauls where suitable haul roads can be established. In the case of rock excavation, drilling and blasting will be necessary, and a shovel may be needed to load the blasted rock into trucks or rock wagons for transportation to its disposal point. A dragline is frequently used for excavation in swampy land, and a backhoe for trench excavation. In excavation of this type 'dozer units may be extremely useful in performing excavations involving very short hauls, for spreading dumped materials, and for similar operations. A motor grader may be used for shaping ditches, trimming slopes, and maintaining haul roads, as well as for a host of other functions.

10–23. Excavation for Structures. Structural excavation refers to the excavating of material in order to permit the construction of pipe culverts, concrete box culverts, foundations for bridges, retaining walls, and practically all other structures which may be required in a particular job. Payment is usually made separately for this item, except in the case of structures like manholes and catch basins, where the payment made for completion of the structure includes payment for excavation. Some organizations also include in this item the satisfactory placing of backfill around the completed structure.

Specifications of the various state highway departments covering this item provide for the satisfactory removal of all material, regardless of its nature, which is encountered in structural excavation. In certain cases, particularly where large excavations are required, as in the case of the construction of piers and abutments for large bridges, it may be necessary to employ caissons or cofferdams to facilitate the excavation process. The construction and subsequent removal of these items is also generally provided for in the specifications. Structural excavation is measured and paid for by the cubic yard; the volume upon which payment is based is usually the actual volume of material excavated in order to place the given foundation structure. The pay quantity is generally limited, however, to a volume computed by the imagined placing of planes parallel to the lines of the foundation but located from 1 to 2 feet outside these lines, as shown on the plans. Provision is also frequently made for an increase in depth of the foundation, with consequent increased payment to the contractor.

Suitable materials taken from excavations for structures are used either in backfilling around the completed structure or in other portions of the roadway structure. Unsatisfactory or surplus excavated materials are, of course, wasted. Requirements relative to the placing of suitable backfill are usually quite rigid, particularly with regard to the production of a satisfactory density in the backfill. This density may be obtained by rolling or vibrating the material.

Both machines and hand methods are used in the performance of structural excavation, with more hand labor being required in this operation than in the other types of excavation described in this chapter. When working in average soils, the clamshell is particularly adapted to this type of work because of its ability to work to vertical lines and in close proximity to forms and protective devices such as the bracing and sheathing required in deep excavations. Smaller units of excavating machinery, such as the tractor shovel, also find application in this work. The bulldozer, which is present on nearly every earthwork operation, can also be used to good advantage in excavating and backfilling for structures. The vibratory and tamping tools previously discussed are very useful for compacting backfills, especially in restricted spaces.

10–24. Borrow Excavation. When sufficient material for the formation of embankments and other elements of the roadway structure is not available from excavation performed within the limits of the right-of-way, additional suitable material is generally taken from borrow pits. The contracting agency may indicate and furnish the contractor with suitable borrow pits, or the contractor may obtain specified and suitable materials from locations of his own choosing. Specifications relating to borrow excavation are generally about the same as those relating to roadway excavation; that is, they provide for the satisfactory excavation of material from an indicated area, and for the incorporation of that material into the roadway structure. Additional requirements usually pertain to the condition of the borrow pit during and after construction. For example, it may be stipulated that when construction operations are complete, borrow pits are to be left in such a condition that water will not collect or stand in them. Payment for borrow excavation is generally by volume in cubic yards of borrowed material, measured in place (original volume) and actually incorporated into the designated sections of the work.

The opening and working of borrow pits is frequently a large-scale operation which requires the best in mass-production methods of earth moving. If the bulk of the material is to be incorporated into embankment, and if it is soil, as distinguished from rock or rocklike materials, there is little doubt but that borrow-pit excavation in most cases may be most efficiently effected by the use of scraper units. This is, of course, because of the inherent nature of the scraper unit which, with one operator and one power unit, will dig, load, haul, and spread material. Either crawler or rubber-tired tractor units may be used; the latter units are generally preferred where long hauls (above about 1200 feet for large units) are to be made. Where rock or cemented soils are to be excavated, the shovel is usually the best tool unless the material can be ripped economically; the shovel loads into trucks or wagons which haul the material to the fill where the load is dumped and spread by a 'dozer or a motor grader. Shovel-truck groups are capable of very large production in suitable circumstances, but the interdependence of the units involved makes their comparison with scraper units generally unfavorable for this operation, except when the excavation is being made in materials or under conditions unsuited to scraper operations.

10–25. Nuclear Excavation. In the early 1960's, attention was focused on the possibilities of using nuclear explosives for large-scale excavation. Experimental blasts showed the method to be feasible, safe, and economical.

Among the projects being studied is Project Carryall, which would involve the nuclear excavation of an estimated 68 million cubic yards of earth and rock in a cut 340 feet deep in a thinly populated section of the southern California desert (*48*). Purpose of the planned excavation is to provide a favorable location for a transcontinental railroad and an Interstate Highway. Nuclear excavation costs were projected to be much less than conventional methods. However, international political considerations have delayed the project.

Embankments

Embankments are used in highway construction when it is required that the grade line of the roadway be raised some distance above the level of the existing ground surface in order to maintain design standards or prevent damage to the highway through the action of surface or ground water. Both rolled-earth and hydraulic-fill embankments are built, with the vast majority of highway embankments being of the former type. From the standpoint of average practice it is also likely that the large majority of fills used in highway construction are 15 feet or less in height. Many fills are 2, 3, or 4 feet in height, and fills as high as 300 feet have been reported.

Before discussing construction procedures relative to embankments, it seems desirable to briefly discuss some of the basic factors involved in the design of such structures. These design elements may be classed as height, stability of slopes, stability of foundations, and the selection of embankment materials.

10–26. Height. As has been indicated, the height of an embankment is generally fixed by considerations relative to the general location of the highway in the area. Thus an embankment may be needed to maintain the grade to some fixed point, as in the approaches to a bridge or a large culvert structure, or to maintain desirable grades in rolling or mountainous country. Similar considerations relative to alignment may also force the use of embankments rather than the possible use of side-hill sections. In rolling country it is frequently possible to reduce embankment construction to a minimum on secondary roads by the simple expedient of following closely the shape of the existing ground. Such a procedure is usually not desirable on primary routes because of sight distance requirements.

In low-lying areas where the water table is at or close to the surface of the ground, the minimum height of embankment is frequently established by the desirability of preventing the intrusion of ground water into the subgrade and base. In such cases the elevation of the top of the subgrade is generally required to be at least 2 feet above the water table, and it may be considerably more when soils subject to capillarity are used in the construction of the embankment or in areas subject to frost. If free water is expected in the area crossed by the embankment, the minimum distance above the expected water level may be further increased in the interests of protecting the embankment and pavement structure.

10–27. Embankment Slopes. In simple essence, the cross-section of a highway embankment consists of a flat, horizontal top section of variable width and generally symmetrical uniform side slopes on either side, which begin at the top and continue until they intersect the natural ground surface. The top section would have a minimum width of about 40 feet in our simplified view of the embankment cross-section. Minimum-side slopes vary from about 1.5:1 to 3:1, depending on a number of variables, including the height of the fill, the type of soil used in its formation, and its probable condition during its anticipated life. Recommendations relative to side slopes in relation to type of soil, height, and exposure condition are given in Table 10–1 (*12*).

All major fills should be subject to stability analyses and designed in accordance with established principles of soil mechanics.

10–28. Embankment Foundations. As has been indicated, a portion of the analysis required for the design of a highway embankment consists

TABLE 10-1

RECOMMENDED MINIMUM REQUIREMENTS FOR COMPACTION OF EMBANKMENTS
Condition of Exposure

Class of Soil (AASHO M 145-49)	CONDITION 1 (Not Subject to Inundation)			CONDITION 2 (Subject to Periods of Inundation)		
	Height of Fill (feet)	Slope	Compaction (Per Cent of AASHO Maximum Density)	Height of Fill (feet)	Slope	Compaction (Per Cent of AASHO Maximum Density)
A-1	Not critical	1.5 to 1	95+	Not critical	2 to 1	95
A-3	Not critical	1.5 to 1	100+	Not critical	2 to 1	100+
A-2-4	Less than 50	2 to 1	95+	Less than 10	3 to 1	95
A-2-5	Less than 50			10 to 50		95 to 100
A-4	Less than 50	2 to 1	95+	Less than 50	3 to 1	95 to 100
A-5	Less than 50					
A-6	Less than 50	2 to 1	90 to 95*	Less than 50	3 to 1	95 to 100
A-7						

Note—Recommendations for Condition 2 depend upon height of fills. Higher fills of the order of 35 to 50 feet should be compacted to 100 per cent, at least for part of fills subject to periods of inundation. Unusual soils which have low resistance to shear deformation should be analyzed by soil-mechanics methods to determine permissible slopes and minimum compacted densities.

* The lower values of minimum requirements will hold only for low fills of the order of 10 to 15 feet or less and for roads not subject to inundation nor carrying large volumes of very heavy loads.

of checking the stability of the side slopes or the body of the embankment itself. The design should also include an examination of the soil beneath the embankment proper, or the embankment "foundation." An embankment may fail because the stresses imposed upon the underlying soil layer due to the weight of the fill are greater than the shearing resistance of the foundation soils. In such a case, the underlying soil layer generally would flow laterally, with resulting subsidence of the fill. The embankment upon which a pavement surface has been placed may also fail in its supporting function by continued settlement due to consolidation of the underlying soil layers. Either of these situations is likely to occur when a high fill is founded on cohesive soil, or where the foundation material is a soft, compressible, fine-grained soil, as an organic silt or clay, peat or muck. Where any doubt exists as to the stability of the embankment foundation, the shearing stresses which will be created in the foundation should be compared with the available shearing resistance of the soils involved, the consolidation characteristics of the layer determined, and a settlement analysis made. The measurement of the shearing resistance and the consolidation characteristics must both be based upon the laboratory testing of undisturbed samples taken from the foundation soil.

Construction procedures required for the improvement of the structural properties of the foundation are generally involved and costly. Relocation of the route should be considered in every possible case where extremely weak soils are encountered to considerable depth. In event that relocation is not feasible, several methods have been evolved for the treatment or displacement of the underlying soil layer. These methods are discussed in detail in later portions of this chapter.

10–29. Selection of Embankment Materials. Many different soils may be satisfactorily used in the construction of rolled-earth highway embankments. In terms of the AASHO classification system outlined in a previous chapter, and in general statement, soils of the A–1, A–2, and A–3 groups can be used successfully in embankment construction, as can certain soils falling in the A–4 group. Soils of the A–5, A–6, and A–7 groups are generally regarded as being less desirable for embankments, while soils of the A–8 group are unsuitable for this purpose.

The A–1 soils are highly desirable for inclusion in embankments since they may be compacted to a high degree of stability and density by the use of normal compaction equipment. A–2 soils are also entirely suitable for this purpose, although they require a little more careful control of the compaction process during construction. The A–3 group consists of cohesionless sands which can generally be satisfactorily used in embankment construction, although they cannot be satisfactorily compacted by the use of sheep's-foot rollers. Adequate compaction may be obtained by the use of pneumatic-tired rollers, construction equipment, or vibratory compactors.

With regard to groups A–4 (and A–6), the sandy and silty soils included in this group, as well as some of the inorganic clays, may be satisfactory for inclusion in embankments under certain conditions. These conditions might be said to include low height of fill, careful control of the compaction process, and utilization in areas where the moisture content of the soil might be expected to remain the same or less during service than that which was utilized in the construction process. These soils are more difficult to compact than those of the first three groups because of their high moisture-retaining characteristics. The moisture content of these soils during construction must generally be maintained within relatively narrow limits in order to secure adequate density and stability. The elastic soils included in the A–5 and A–7 groups are generally regarded as unsatisfactory for embankment construction. In addition, the clay soils of the A–7 groups are subject to very high volume change with change in moisture content. Despite some of the faults exhibited by several of the soils included in the A–4, A–5, A–6, and A–7 groups, conditions are sometimes such that these are the only soils available within a reasonable haul distance and their use is therefore dictated by considerations of economy. These soils must all be carefully handled during the construction process.

Construction of Rolled-Earth Embankments

10–30. Formation. Rolled-earth embankments are constructed in relatively thin layers of loose soil. Each layer is rolled to a satisfactory degree of density before the next layer is placed, and the fill is thus built up to the desired height by the formation of successive layers or "lifts." The majority of state highway departments at the present time require layers to be from 6 to 12 inches thick before compaction begins, when normal soils are involved. Specifications may permit an increase in layer thickness where large rocks are being used in the lower portion of a fill, up to a maximum thickness of 24 inches.

The layers are required to be formed by spreading the soil to approximately uniform thickness over the entire width and length of the embankment section at the level concerned. "End-dumping" from trucks without spreading is usually specifically prohibited. An exception to this latter requirement may be permitted when the embankment foundation is of such a character that it cannot support the weight of spreading and compacting equipment. In such cases the lower portion of the fill may be placed by end-dumping until sufficient thickness is developed to permit the passage of equipment. Where unstable soils comprise the foundation, the special construction measures described later in the chapter may be required.

Two general construction methods are used in embankment formation. The first of these is the direct dumping and spreading of the soil by scraper units in one operation. This, of course, is a function for which these units are ideally suited. A second method involves the dumping of the material in the proper location by trucks or wagons. The dumped material must then be spread to the required uniform thickness by 'dozer units or occasionally by motor graders. In addition, on very short fills, 'dozer units might conceivably form the embankment, working alone by moving material short distances from cut into embankment sections.

10–31. Compaction. Requirements of the various highway agencies relative to the compaction of soils in embankments (and subgrades) have undergone considerable change since World War II. So, also, have various elements of the understanding of the behavior of soils undergoing compaction, and the end results of such compaction. Considerable variation in requirements for compaction exists among the various agencies.

However, the general nature and purpose of compaction are well understood. The general notion is that soils should be compacted by rolling, at or near optimum moisture content, and to some percentage of the maximum density established through the use of a known laboratory compactive effort. It has been established that certain advantages may be expected to accrue from the use of this compaction process. Included among the more important of these advantages are the increased shearing resistance or "stability" of the soil, its decreased permeability, and the minimizing of future settlement of the embankment itself (but not, necessarily, settlement due to consolidation or shearing failure of the embankment foundation).

Present specifications of the various state highway departments relative to compaction are not uniform (*23*). A large majority of the states require compaction to a certain minimum percentage of the maximum density, as determined by the standard AASHO laboratory compaction procedure (Method T-99). (See Chapter 6.) The minimum requirement most often is in the range of from 90 to 100 per cent of standard AASHO maximum density. In some states, maximum compaction requirements are stated in terms of the maximum laboratory density; e.g., for a soil with a maximum density of 99.9 pounds per cubic foot or less, the required density in the field must be 100 per cent of the laboratory value, for 100 to 119.9 pounds per cubic foot, 95 per cent, and so on. A few states still specify the number of passes to be made by specified equipment. Many specifications contain requirements relative to the size and nature of compaction equipment. In recent years, the tendency has been toward the incorporation of "end-result" specifications for compaction, with desired end results specified but leaving the choice of equipment up to the contractor. Moisture content during compaction is controlled, sometimes by general specification statements, sometimes by numerical values (e.g., ± 2 per cent of optimum moisture).

The establishment of compaction requirements by the design engineer is a matter which requires careful study of all the factors involved (see reference (*12*) in the bibliography for a detailed discussion of this point). State highway department specifications represent a compromise of the many variables and reflect requirements which are reasonable for the conditions which exist in each state. On large or unusual projects, special provisions may be written and used to control the work. General recommendations for density requirements for highway embankments are contained in Table 10–1.

10–32. Control of Compaction. Generally speaking, the field compaction process is controlled by making relatively frequent checks of the density and moisture content of the soil which is undergoing compaction. The measured density is the wet unit weight; the dry unit weight is calculated on the basis of this figure and the measured moisture content. The dry unit weight may then be compared with the compaction curve for the soil and compactive effort involved to see if the density being obtained in the field meets the requirement established in the laboratory. If different soils are being placed in different sections of the fill, it is understood that the curve (or figure) which is being used to check the rolling process is the proper one for the soil concerned.

Three methods have been used widely by highway agencies for determining the density of the soil. These may be designated as the sand, balloon, and heavy oil methods. In each of these the density determination is begun by carefully excavating a cylindrical hole in the soil layer; the hole is usually about four inches in diameter and the full depth of the layer. All the material taken from the hole is carefully saved, placed in a sealed container, and weighed as quickly as possible. The weight determination may be made immediately on a field balance if desired. The volume of the hole may then be determined by the use of sand, a balloon apparatus, or heavy oil. Some use has also been made of small undisturbed samples taken with a cylindrical drive sampler; in such cases, wet unit weights may be used for the comparison.

The apparatus required for the sand density determination is shown in Fig. 10–15. In using this apparatus, the area around the hole is carefully leveled, the jar—with funnel attached and filled with "standard" sand—inverted over the hole, and the valve opened. The sand flows into and fills the hole. When the hole is filled the valve is closed, and the jar and the remaining sand weighed. The weight of sand required to fill the hole may then be determined. Since the same procedure is carefully followed in each determination, and since the same dry sand is used, a previously established relationship between a given weight of sand and the volume occupied by this amount of sand may be used to ascertain the volume of the hole. Since the weight of soil taken from the hole and its volume are then known, the wet unit weight may be calculated. A balloon filled with water may also be used to determine the volume of the hole, and the volume of the hole measured directly. This volume may also be obtained by filling the hole with a heavy oil of known specific gravity. This latter procedure is more widely used for checking the densities of bases than embankments.

If the soil is known to be at or near optimum moisture content, the density check may be made by comparing the measured wet unit weight with the wet unit weight established by laboratory procedure. Generally, however, the procedure which is used is to make an additional moisture determination so that this factor may be checked and the dry unit weight determined. If a moisture content determination is deemed necessary, then speed is of the essence. The determination may be made by rapid drying over a field stove. Drying is also frequently accomplished by adding alcohol, or some other volatile solvent, to the soil and then removing the water by igniting the solvent. Time is seldom available for the use of more accurate methods because the decision as to whether or not a satisfactory density has been obtained must be quickly made so that construction operations will not be delayed.

FIG. 10–15. Density determination—sand method. (Courtesy Virginia Department of Highways.)

Many highway agencies use nuclear devices to measure in-place density and moisture content. Unit weight is measured by directing gamma rays of known intensity into the soil and measuring the intensity of gamma rays reflected back (*49*). As the unit weight increases, the intensity of reflected gamma rays increases.

To determine moisture content, "fast" neutrons of known intensity are directed into the soil and the intensity of "slow" neutrons reflected back measured. "Fast" neutrons are slowed principally by collisions with hydrogen atoms. Thus, the number of reflected "slow" neutrons becomes a measure of the moisture content.

A commonly used source of gamma rays is cesium-137. A radium-beryllium mixture may be used as a "fast" neutron source.

There are two basic types of nuclear gages; the probe, which is lowered into the ground through an access tube driven into the soil, and the surface gage. One type of surface gage is shown in Fig. 10–16.

FIG. 10–16. Surface type of nuclear gage used to measure soil density and moisture content. (Courtesy *Engineering News-Record.*)

The radioactive source is at the left; on the right is the electronic counter (scaler), on which moisture and density values are shown. Lane-Wells' "Road Logger" is a specially equipped truck-and-trailer that contains nuclear devices and travels at speeds up to 100 fpm to produce a continuous record of moisture and density as it moves over a strip of compacted material.

Principal advantage of the nuclear gages is speed (as little as one-fifth the time required for conventional tests). They also do not disturb the soil, can be used on a wide range of materials, and require little operator training. There is little danger from radioactivity, but the devices must be calibrated carefully.

If the density determined as outlined above is equal to or greater than that required, the rolling may generally be judged to be satisfactory and construction of another lift may proceed. If the density is lower than that specified, then additional rolling may be required or the moisture content may have to be adjusted. If these measures fail, then it is reasonable to assume that the rolling equipment being used is not capable of producing the required density or that the moisture–density relationship which is being used is not the one for the soil which is actually being compacted.

The procedure which has been outlined for the control of compaction may seem unduly cumbersome and complicated. Actually, after compaction of a given section has begun and the initial adjustments have been made, the situation is usually considerably simplified. The inspector will, for example, quickly learn to judge the moisture content of the soil by appearance and feel. He may also quickly decide that, with the proper moisture content being maintained, a certain number of passes of the roller—say six or eight—will accomplish the required densification. His job may then be simplified into a count of the roller trips, with occasional density checks being made as judgment dictates.

The number and frequency of density and moisture checks which are required for adequate control are, of course, functions of job conditions. On an average basis, perhaps one check will be required every hour on a fill of moderate length. Some other standard may be applied, such as a minimum of one test per 1000 cubic yards of compacted material. Some soils are more sensitive to moisture change than are others. For example, sands usually have a moisture–density curve which shows a sharp peak, with the optimum moisture content being clearly defined, while clay soils usually have a flatter curve so that a small variation in moisture content does not greatly affect the density. It is impossible, generally speaking, to maintain the moisture content exactly at optimum for any substantial length of time, so that some tolerance must be permitted in the moisture content.

10–33. Compaction Equipment. Principal factors involved in the selection of compaction equipment for use on a given project include the specifications under which the work is to be carried out and the choice of equipment available to the contractor. More than one type of equipment and compaction procedure can often be used successfully for a given set of circumstances; however, there usually is one combination which will give the most economical result. General factors to be considered in the selection of compaction equipment are given in Table 10–2 (*210*). The term "super-compactors" refers to the very large pneumatic-tired rollers. In addition, vibratory compactors are very efficient tools for the compaction of granular soils.

A detailed discussion of the most efficient and economical use of a given type of equipment for a given project is beyond the scope of this text; refer to references *12*, *210*, and *23* in the bibliography for

TABLE 10–2

CHARACTERISTICS OF STANDARD ROLLERS

Roller Type	Range of Total Weight (tons)	Range of Pressure	Recommended Lift Height (loose-in.) for 8 passes* Heavier Units	Lighter Units	Operating Speed (mph)	Most Suitable Soil Type for Use of Highway Embankment Construction
Three wheel	5–20	60–500 lbs./in.	6–8	4–6	1–5	Granular, particularly where crushing is desirable
Tandem-two axle	3–16	60–500 lbs./in.	6–8	4–6	1–5	Granular, particularly where crushing is desirable
Tandem-three axle	12–20	60–500 lbs./in.	6–8	4–6	1–5	Granular, particularly where crushing is desirable
Pneumatic	3–12	20–30 p.s.i.	6–8	4–6	1–15	Sandy, sand-clay, and silts
Super-compactors	20–50	50–90 p.s.i.	12–24	12–18	5–10	All types
Sheep's-foot	2–20	100–800 p.s.i.	8–12	6–9	5–10	Clays and silty-clays

* A pass is defined as a single application of a given roller.

a complete discussion of the subject. Pertinent factors which must be considered include thickness of lift, total weights and unit pressures exerted on the soil, required number of passes, operating speeds, soil type and moisture content, and others. On large projects, relationships among the variables involved may often be resolved by the construction of test fills prior to the start of actual construction, or by trial during early phases of construction. For economical operation, there must be a proper balance between rolling equipment and hauling equipment on the job.

Additional water required for compaction is generally added by ordinary pressure distributors. Rotary speed mixers may be used to mix the water and soil in some instances. Aeration of soils which are too wet for compaction may be accomplished by the use of agricultural tools, such as harrows and discs, by turning the soil with blade graders, or by rotary speed mixers with the hood up. The problem also may be handled by "sandwiching" alternate layers of wet and dry materials. In many cases aeration is accomplished by simply allowing the soil to dry of its own accord.

Hydraulic Fills

Hydraulic fills, which are occasionally used in highway construction, are made by dredging or pumping material from rivers, channels, swamps, and similar places and placing it by hydraulic means to form an embankment of the desired height. Generally speaking, hydraulic fills are objectionable for highways, principally because of the fact that this type of structure is subject to continued settlement over a considerable period of time. The placing of a permanent pavement surface is then not feasible for some time following construction of the fill.

Special Treatment of Embankment Foundations

The foundations of embankments which are constructed in swampy areas—particularly where peat or other highly organic soils are encountered—frequently require special treatment if failures of the embankment or pavement structure due to consolidation or displacement of the underlying soil layer are to be avoided. The methods in general use may be classified as follows:

1. Gravity subsidence
2. Partial or total excavation
3. Blasting
4. Jetting
5. Vertical sand drains

Various combinations of these methods may also be used in specific locations.

10–34. Gravity Subsidence. In some instances, a fill may simply be placed on the surface of an unsatisfactory foundation soil and allowed to settle as it will, with no special treatment of the underlying soil. Where soils of high compressibility and low shearing resistance are present in the embankment foundation, this method is obviously unsuitable for any except secondary roads which carry extremely low volumes of traffic. The fill may settle rapidly during or shortly after the construction period, due to shearing failure and displacement of the foundation soil. Settlement due to consolidation may continue for many years. In either case the satisfactory placing of any wearing surface is virtually impossible and a long period of continuous maintenance must be anticipated, as more and more soil must be added to the fill as time passes, until the embankment eventually comes to a stable condition.

In some cases a surcharge of extra height of fill has been used to speed up settlement; the surcharge is later removed. A variation of this method was used by the Louisiana Department of Highways to build a section of Interstate Highway near New Orleans. They used a "rolling" surcharge over a 10-to-20-ft-thick deposit of peat and muck. In this process 4 feet of sand surcharge (of a total of 5½ feet) was placed at one end of the fill and allowed to remain in place for a specified time; the sand then was removed and placed at the other end of the fill. Thus, the fill "rolled" slowly ahead in a series of steps.

10–35. Partial or Total Excavation. Where the undesirable soil is of shallow depth (up to from 8 to 12 feet) and is underlain by a soil of satisfactory character or by rock, many highway agencies excavate the entire depth of unsatifactory soil. In New York state, total excavation has been carried out to a depth of 35 feet. The excavated area is then backfilled with suitable granular material. This method is expensive, of course, but has the advantage of permitting the immediate placing of a pavement on the completed fill. When such treatment is used, the backfill is composed of sand and gravel of suitable gradation.

A procedure sometimes used on secondary roads where deep swamp deposits are encountered consists of partial excavation of the foundation soil. That is, the soil is excavated to a depth of several feet and backfill is placed as before. In this operation it is essential that the backfill be placed as excavation proceeds. This method is open to the same criticisms which are applicable to the method which we have called "gravity subsidence." Displacement of the deep underlying soil may still occur under the weight of the backfill and embankment, while settlements due to consolidation are also to be expected, although they should be somewhat smaller than if the compressible soil had not been partially excavated. It is again usually impossible to place any permanent type of wearing surface without the expectation of very high maintenance costs.

10–36. Blasting. Explosives have been widely used to aid in the placing of embankments in swampy areas. When this method is used, the embankment is generally constructed to a level considerably above the final grade and the displacement of the underlying soil is accelerated by the detonation of deep charges of explosives. The blasts serve a dual purpose in that some of the material may be displaced by the force of the explosions, and the remainder of the unstable soil is liquefied to a greater or lesser extent and can thus be displaced more readily by the weight of the overlying fill. As the underlying soil is displaced, the fill subsides; and it may then be finished to the required elevation and cross-section. This method requires careful control but should result in the formation of an embankment which will be stable in a relatively short period of time.

Blasting was used to build the embankment sections of the Rainy Lake Causeway in Ontario (*15*). The explosions remolded a soft to firm layer of varved clay up to 50 feet deep and allowed its displacement by rock fill.

10–37. Jetting. The process of "jetting" involves the pumping of water into the underlying soil layer in order to liquefy it and thus aid in the displacement of the layer by the weight of the embankment. Jetting is carried out in conjunction with the placing of the embankment material. In typical installations, a single line of jets is placed along one side of the embankment. The jets are spaced from 10 to 25 feet apart, and they extend about two-thirds of the way into the underlying layer. The fill settles or "subsides," as before, and must be brought to the desired cross-section after a stable condition has been attained. Trenches may be excavated along one or both sides of the embankment to assist in the displacement of the weak soil layer. In some cases the embankment material, rather than the underlying soil, may be jetted; this results in an increase in the weight of the fill.

10–38. Vertical Sand Drains. In recent years a type of construction utilizing vertical sand drains has been increasingly used in the treatment of embankment foundations. Vertical sand drains generally consist of circular holes or "shafts" from 18 to 24 inches in diameter, which are spaced from 6 to 20 feet apart on centers beneath the embankment section, and which are carried completely through the layer of compressible soil. The holes are fashioned by rotary drilling or by the use of a hollow or plugged-end mandrel. The holes are backfilled with suitable granular material. A "sand blanket" from 3 to 5 feet in thickness is generally placed at the top of the drains (and the existing soil) extending across the entire width of the embankment section. The embankment is then constructed by normal methods on top of the sand blanket.

The basic purpose served by the installation of vertical sand drains is the acceleration of consolidation of the compressible soil layer. Consolidation is, of course, due to the forcing of water from the voids of the soil and the rate of consolidation is dependent, among other things, upon the distance which the water has to travel to escape. Without sand drains the water may escape only at the top of the soil layer if there is an underlying layer of impervious material, or at the top and bottom of the layer. If the layer is of considerable thickness, settlements due to consolidation may occur over a period of many years. Vertical sand drains speed up the consolidation process by providing many shortened paths through which the water may escape. Vertical drainage at the top (or top and bottom) continues, and in addition radial horizontal drainage takes place. Water which is thus forced into the sand drains moves up the drains and out through the sand blanket. In addition to the fact that radial drainage is provided this method takes advantage of the fact that many fine-grained sedimentary soils have a permeability which is several times greater in a horizontal direction than in a vertical direction. This fact also contributes to the accel-

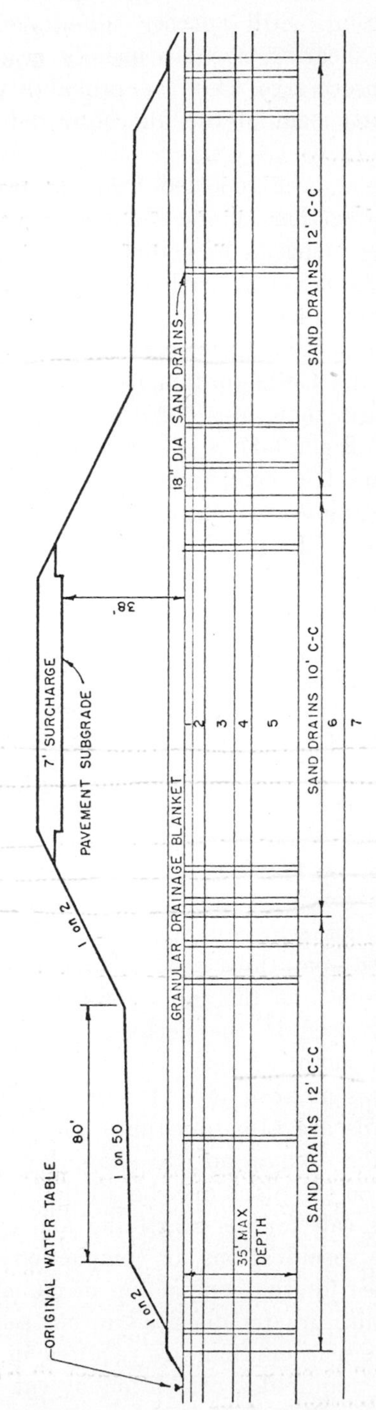

Fig. 10-17. Typical section of sand drain installation near Syracuse, N.Y. (After Lyndon H. Moore, Courtesy New York State Department of Public Works.)

erated rate of consolidation. Still another advantage lies in the fact that a rapid increase in shearing strength usually accompanies the accelerated consolidation process so that the probability of lateral flow is diminished and the settlement of the fill confined to that caused by consolidation.

Phenomenal results have been reported from the use of this method on many jobs in different sections of the country. In many cases, even where high fills and deep deposits of compressible soil were involved, 90 per cent or more of the ultimate settlement has occurred during the construction period or very shortly thereafter. This behavior makes possible the placing of a permanent-type surface within a few months after the fill is completed. Costs of this method have been generally reported as somewhat less than most of the "conventional" methods previously described. In Fig. 10–17 is shown a profile of an installation of vertical sand drains used in connection with the construction of an embankment across a swampy area encountered in the construction by the New York Department of Public Works of a highway near Syracuse, N.Y. (*194*). Fills up to 45 feet high were built on surface organic soils underlain by soft clays to a depth of 35 feet.

Finishing Operations

Finishing operations refer to the final series of operations which are required to complete the earthwork operations involved in a typical highway project. These operations are here taken to include such things as trimming of shoulders, side, and ditch slopes; the fine-grading operations required to bring the earthwork sections to their final grade and cross-section; the compaction of cut sections; and similar items. Actually, most of these are not separate operations which are performed after all other earthwork operations are completed, but are generally carried along and performed as the job approaches completion.

The tool most widely used in the performance of the majority of finishing operations is the motor grader. Its long wheel base and ability to work to close tolerances make it ideal for this application. 'Dozers, scraper units, and gradalls are also sometimes used in finishing operations. In cut sections, the excavation and shaping of the cross-section to the dimensions shown on the plans may be all that is required. However, many specifications call for the scarifying and compaction of the soil in cut sections. The specifications for compaction, generally speaking, are the same as those for the compaction of embankments. There is some tendency to require greater densities in cut sections and in the top layer of embankment sections than are required in normal embankment construction. The required scarification of cut sections may be accomplished by rippers, discs and harrows of various types, or by scarifier attachments on motor graders. The required depth of scarification and compaction in these sections is generally from 8 to 12 inches. The compaction operation, as before, may be accomplished through the use of various types of rollers, with the sheep's-foot roller the predominating tool. The section would generally be shaped to final dimensions through the use of a motor grader.

It should be noted that the finishing operations and other earthwork items which have been described are in many cases practically all the work involved in construction prior to the placing of the base and surface. In other cases, additional finishing operations may be required before this phase of construction can begin. For example, the subgrade soil may require stabilization. The placing of high-type pavements and bases, such as portland cement concrete, usually involve additional subgrade preparation beyond that which has been described in this section. These items and others of a similar nature are discussed in later chapters.

Planning and Execution of Construction Projects

Since the majority of highway construction is performed by contract, the principles stated in the following sections are primarily intended to apply to the planning and execution of contract work. Most of the items which are discussed are also applicable to work which is performed by force-account. Many contractors prepare tentative job plans as a part of their bid computations, and on large projects it may be required that the job plan be submitted with the bid. In any event, a comprehensive job plan should be prepared by the successful bidder immediately after the contract is awarded and before construction operations begin. Space does not permit lengthy exposition of the many details which enter into the successful planning and conduct of the earthwork operations involved in a typical highway project, but some of the steps which are necessary are presented here.

The first step in the preparation of a schedule of construction operations is a detailed analysis of the amount of work of different types involved in the job. By this is meant such items as the number of acres to be cleared and grubbed, the number of cubic yards of roadway excavation and the haul distances, and so on. The quantities used are generally those of the engineer's estimate. The plans are carefully examined, and the location, extent, and conditions affecting the execution of each step of the work determined. A detailed, step-by-step analysis is then generally made of the equipment, labor, and material requirements for each phase of the project.

In establishing the equipment estimate sheet, for example, it would be necessary to tabulate each operation and the quantity of work involved in that operation, to select the units of equipment to be assigned to this portion of the job, to estimate the hourly production rate for each of the machines to be used, and to calculate the total number of equipment hours required for completion of the operation. If the time to be consumed in this operation was limited and if sufficient equipment was available, a sufficient number of machines would simply be assigned to the operation to complete it in the given time. The labor required for each operation would probably be estimated at the same time, while material estimates would probably be made separately.

10–39. Job Scheduling. On the basis of the information tabulated in the work estimates previously referred to, a detailed construction schedule may be prepared. This schedule may be presented in several different forms, and one method of presentation is shown in Fig. 10–18. As

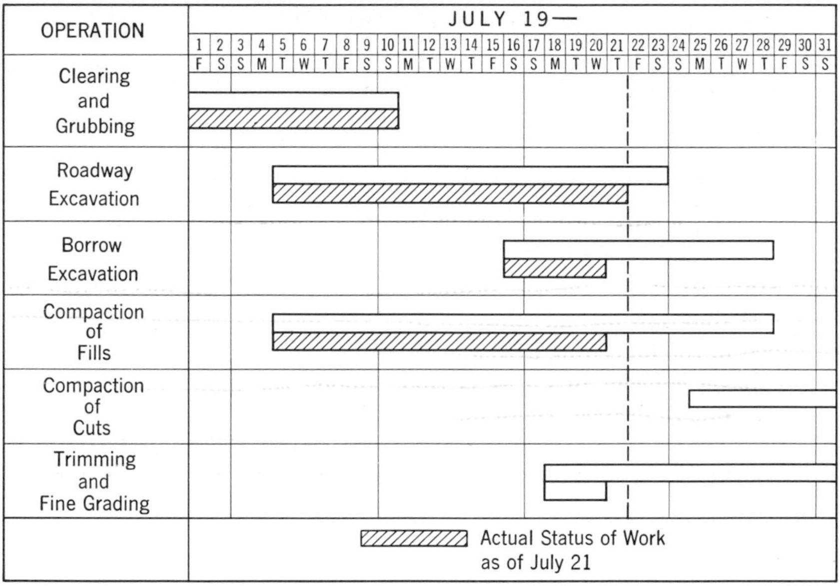

OPERATION
JULY 19—
1 2 3 4 5 6 7 8 9 10 11 12 13 14 15 16 17 18 19 20 21 22 23 24 25 26 27 28 29 30 31
F S S M T W T F S S M T W T F S S M T W T F S S M T W T F S S
Clearing and Grubbing
Roadway Excavation
Borrow Excavation
Compaction of Fills
Compaction of Cuts
Trimming and Fine Grading
Actual Status of Work as of July 21

Fig. 10–18. Schedule of construction operations.

will be noted, each of the general operations involved in the project is tabulated, and the time assigned for completion of each operation is shown by the plain bar. The sequence of these operations is also shown. As indicated by the chart, the progress which has been made on the project at any particular time is also shown by the cross-hatched bars. Using this sheet, the contractor may then quickly determine the status of the work.

It will be noted that certain operations overlap, as would be expected. As a matter of fact, the separate operations must be carefully coordinated and balanced so that they will be completed in the proper sequence, and so that equipment on the job will be utilized to the fullest possible extent. Although the schedule which is illustrated here is for earthwork operations only, a similar schedule would be prepared for every phase of the complete project.

In addition to the construction schedule, contractors on many jobs prepare detailed sheets which show equipment, labor, and material requirements for every day of the project. Preparation of these sheets enables the contractor to further plan his operations so as to make use of the available equipment and labor in the best possible way, and they help to maintain a smooth flow of materials, fuel, lubricants, and so on to the job.

10–40. Critical Path Method. Many highway contractors now use the critical path method, developed in the mid-1950's by engineers from DuPont and the Univac Division of Sperry Rand, for construction project scheduling and control. State highway departments also use the method for planning and programming, scheduling of design activities, and other activities.

CPM is an analysis and management tool that is useful in planning and scheduling any project made up of various subunits of work (as are most highway projects), some of which can be done sequentially from start to finish of the project (*237*). Many engineers feel that the CPM diagram has advantages over the more conventional bar chart described in the preceding article.

The basic concept of CPM is that, among the many operations that form a complete project, there must be one series of sequential operations that takes longer than any other possible series. The shortest possible time in which the project can be completed is the time needed for the longest series. This critical series is the "critical path." Determination of the critical path is in itself an exercise in logic that promotes good management of the project.

Determination of the critical path is aided by the construction of an "arrow diagram," such as the one shown in Fig. 10–19. Detailed instructions for the arrow diagram and its application to highway problems are contained in reference *256*.

In the planning phase of CPM, the first step in drawing the arrow diagram is to break the project into its subunits (activities). On paper, an activity is represented by an arrow, drawn from left to right. At each end of the arrow is placed a circle or node, to represent an "event." The head end (right end) of the arrow merely indicates the flow of work; the arrow is not to scale.

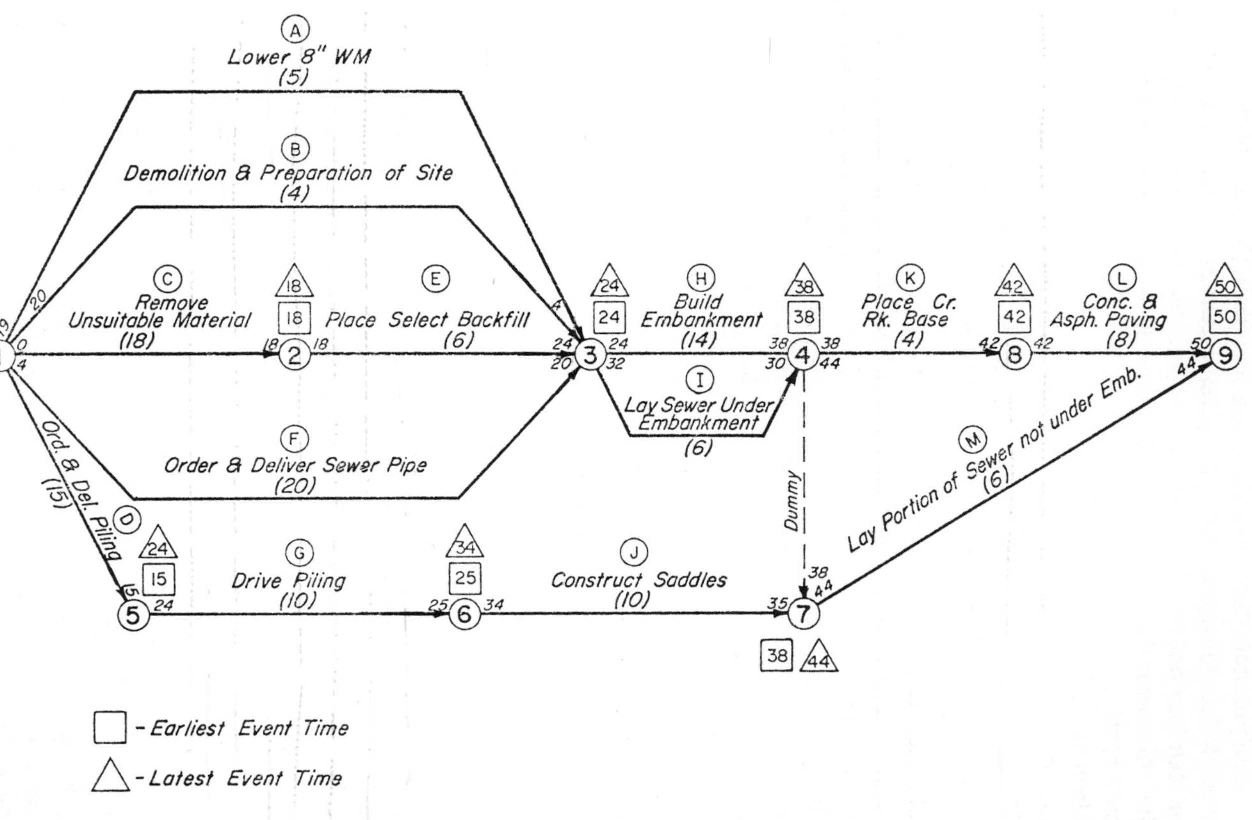

Fig. 10-19. Typical critical path diagram for a simple highway project. (After Douglas L. Jonas.)

In CPM terminology, the event at the tail of the arrow is the "i" event; it represents the instant of time at which this event may start and the instant at which the preceding event must end. The event at the head of the arrow is the "j" event; it represents the instant at which the activity indicated by the arrow ends and subsequent activities may begin.

A logical analysis of a project may produce the general form of the diagram of Fig. 10–19. In this diagram, activities A, B, and F may proceed simultaneously with C and E. However, note that E cannot begin until C is complete, and H and I, although they can be done concurrently, cannot begin until A, B, E, and F are completed.

Note also that activities D, G, and J must follow one another in that order. The dotted line between nodes 4 and 7 is a "dummy" activity put there only to show a logic restraint. In this case, it means that K cannot proceed until J is complete (as well as H). It is made necessary on the diagram because of the fact that the arrow for H and I could not be drawn to nodes 4 and 7 at the same time. (In the approach used in reference *256,* activities are designated by connected nodes, e.g., 1–3; in that case, additional dummy activities must be used to keep the tabulations straight.)

In Fig. 10–19, the remaining steps, K and L, must follow one another, while M can be done concurrently.

In the usual approach, the drawing of a generalized diagram and a check of its logic, forward and backward, complete the "planning" phase of the analysis. The "scheduling" phase may then begin, working with the basic diagram.

The first step in the scheduling phase is to estimate the time required for each activity. The usual procedure is for persons experienced in the work to estimate the normal time required for each operation. At this stage, no attention is paid to other than normal resources of men, time, machinery, and money. Later, adjustments can be made if desired to obtain the most efficient utilization of one or more of these resources. The normal time, in days, for each of the operations of Fig. 10–19 is shown in parentheses along each arrow, e.g., "(20)" days for activity F.

Study of the diagram will reveal that, since activity C takes 18 days, activity E cannot begin before that time has elapsed. It follows that, since activity E takes 6 days, activity H cannot begin until 24 days after the start of activity C. In CPM terminology, the earliest start time for activity E would be the end of the eighteenth day; this is called the "earliest event time" and is shown in a square on the diagram. The "earliest finish time" for E would be 24 days.

Note also that each of the activities concurrent with C and E (i.e., A, B, and F) requires less than 24 days to complete; respectively, they take 5, 4, and 20 days to do. Putting it another way, free time, called "float," is available for each of the activities.

The same sort of analysis can be followed from start to finish of the project to find the earliest finish time for the entire project (in this case, 50 days). Examination also will show that one sequence of activities has no float time; this sequence is, of course, the critical path. In Fig. 10–19, the critical path is the horizontal line connecting nodes 1, 2, 3, 4, 8, and 9 (activities C, E, H, K, and L).

One further step can be taken at this stage. That is, one can start at the right end of the diagram, with the earliest finish time, and work backward through the diagram to determine the latest time that each activity could start or finish without affecting the project completion date.

Basic construction of the CPM diagram for a simple project thus is a relatively simple matter; calculations may be done manually. On a complex project with hundreds of subitems, the CPM diagram can be highly complex; an electronic computer often is used to speed calculations.

If the time of 50 days is satisfactory, the CPM diagram of Fig. 10–19 may be used to control the project. Detailed equipment, labor, and material schedules may be prepared on a day-to-day basis. If job conditions change, the CPM diagram may have to be adjusted to fit the changed conditions. If the time of completion is to be advanced ("crashed"), it is obvious that it is most fruitful to crash the activities on the critical path; crashing noncritical activities will not advance the completion date.

As mentioned previously, one of the greatest values of the CPM approach is in resources planning. Detailed resources planning is beyond the scope of this text. However, its importance can not be overemphasized. Obvious applications are to make the most efficient use of equipment, to utilize skilled labor in the most productive way, and to determine the optimum job schedule from the standpoint of over-all cost. The contractor or owner can, if desired, determine the cost of crashing a project schedule and compare the cost with anticipated benefits. In all these approaches, a computer is an essential tool for a complex project.

CHAPTER 11

BITUMINOUS MATERIALS

Bituminous materials, which are very widely used in the United States for road purposes, may be generally defined as substances consisting primarily of bitumen or containing a large percentage of bitumen. Bitumen, in turn, may be said to be a complex organic material. This organic material may occur naturally or it may be artificially created during a production process. Bitumen is, by definition, a material which is soluble in carbon disulfide. Bituminous materials are frequently divided on the basis of their consistency into liquid, semisolid, and solid materials.

The definitions given above are somewhat unsatisfactory from the viewpoint of common experience, and these materials may be further defined as "sticky, adhesive substances which are dark brown to black in color, frequently associated with sharp, characteristic odors, and which are usually liquid at the time of their application to mineral aggregates or similar materials in road construction." This latter description is hardly adequate from any scientific point of view, but it may serve to identify these substances from the standpoint of the casual observer who may not associate solubility in carbon disulfide with his everyday experience.

Bituminous materials are principally of value to the highway engineer because of their binding or cementing power and their waterproofing properties. They are now used in the construction of a large number of different types of wearing surfaces and bases, ranging in application from low-cost, light-traffic roads to high-type bituminous pavements intended to carry extremely heavy volumes of traffic. A large majority of the surfaced highways and streets existing in the United States today have been built by some process involving the use of bituminous materials.

For the purpose of simplification, bituminous materials used in road construction in this country may be divided as follows:

1. Asphalts
   a. Petroleum
   b. Native
2. Tars

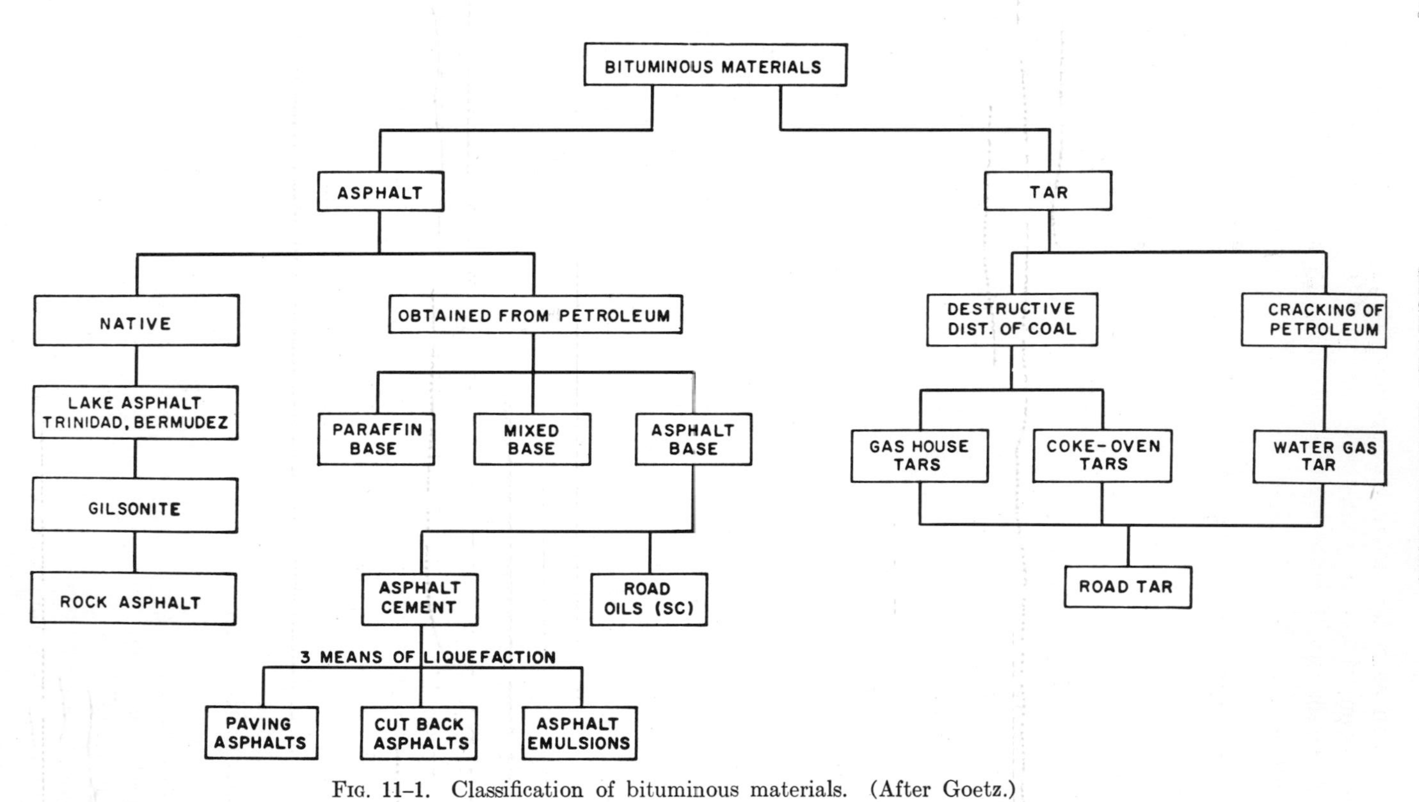

Fig. 11-1. Classification of bituminous materials. (After Goetz.)

Although asphalts and tars are somewhat similar in appearance and general composition, they are produced by greatly dissimilar processes and differ sharply in certain properties which are of importance in highway work. Generally speaking, tars are more susceptible to temperature change than asphalts of similar grade, they are toxic in nature, they possess more free carbon than do asphalts, and they harden more rapidly when exposed to the air after being incorporated into a wearing surface or pavement. These and other differences in properties and behavior have led to the selective use of tars and asphalts in highway construction.

The production and refining of the broad classes of materials given above are briefly discussed in the following paragraphs. Additional terms relative to tars and asphalts are also introduced and defined. Fig. 11–1 shows the broad classification of bituminous materials used in road construction.

Production of Petroleum Asphalts

As the term implies, petroleum or "oil" asphalts are asphalts produced from petroleum oil. A wide variety of refining processes may be used to produce petroleum asphalts suitable for use in highway construction. However, it may be stated that petroleum asphalts are, in general, obtained by distillation and by blowing with air at high temperatures.

Crude petroleums obtained from different parts of the world show great differences in composition and may be generally divided into three classes, i.e., asphaltic, mixed, and paraffin base petroleums. The majority of asphaltic products used for highway purposes are derived from asphaltic base petroleums, although it is possible to derive asphaltic materials from other types of petroleum.

11–1. Dehydration of Petroleum. Almost all crude petroleums contain appreciable amounts of water which must be removed before the crude oil can be subjected to additional refining processes. Frequently, the crude oil is held in storage for a long enough period for the water and other inorganic impurities, such as sand, to be removed by a simple process of sedimentation. If excessive amounts of water remain at the end of the storage period, any one of several methods may be employed to separate the water from the oil.

One of the most common dehydrating processes consists of passing the oil through a tube or pipe still under pressure and at a temperature slightly above the boiling point of water. The vapors thus produced are then condensed and the water separated out mechanically. In this process a certain amount of volatile material other than water may be removed from the crude petroleum. If such is the case, the crude oil is said to have been "topped" or "skimmed," and may be referred to as a "topped crude."

11–2. Distillation of Petroleum Oils. As previously stated, petroleum asphalts may be produced by distillation. Two basic types of distillation may be briefly explained at this point.

(1) *Fractional distillation,* in which removal of the various volatile materials contained in the crude oil is accomplished by more or less direct or "mechanical" separation. That is, the volatile constituents are driven off at successively higher temperatures without substantial chemical change. In the distillation of asphaltic petroleums, the successive fractions obtained would be those which yield gasoline, naphtha, kerosene, diesel oil, and lubricating oil. The residue obtained would be petroleum asphalt. Fractional distillation may be facilitated through the use of steam or a vacuum.

(2) *Destructive distillation,* in which the material undergoes chemical change under application of extreme heat and pressure. This process is generally employed when greater yields of the lighter fractions are desired. Destructive distillation is usually associated with the manufacture of tars, which will be described later in this chapter. However, a refining process known as "cracking" is a form of destructive distillation, and the "cracked residues" or "pressure tars" which are produced may be used in the manufacture of asphaltic paving materials.

11–3. Steam Distillation. Both batch and tube (or pipe) stills may be employed in the steam distillation of petroleums to produce petroleum asphalts. Tube or pipe stills are generally preferred in modern refineries because of their greater efficiency and flexibility of operation. However, the process of steam distillation of an asphaltic petroleum in a batch still will be described here because of its greater simplicity (see Fig. 11–2).

In this process the crude oil is placed in a large cylindrical still and its temperature raised by the application of heat at the bottom of the vessel. Dry steam is introduced into the still to aid in the vaporization of the more volatile constituents of the petroleum and to minimize decomposition of the distillates and residue. The volatile constituents are collected, condensed, and the various fractions stored for further refining if desired.

The distillation process may be continued until the residue attains the desired consistency. The "stopping point" may be determined by sampling the residue, noting the character of the distillate, or by measuring the temperature of the residue. The longer the process of distillation is continued, the harder will be the residue which is produced. The

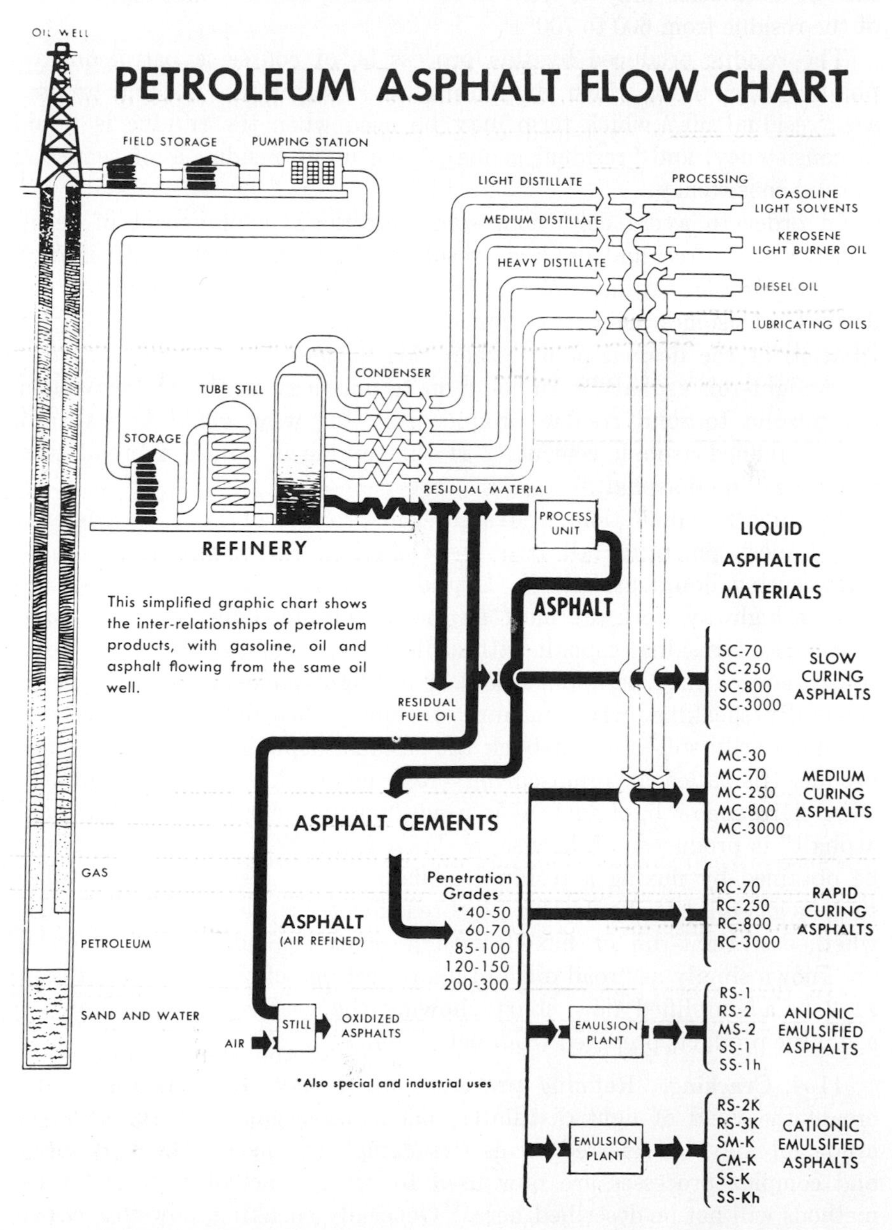

FIG. 11-2. Simplified flow chart of recovery and refining of petroleum asphalts. (Courtesy the Asphalt Institute.)

time of distillation may be from 12 to 36 hours, and the final temperature of the residue from 600 to 700°F.

The residue produced by this process is, of course, a petroleum asphalt. Other terms which may be used to describe this residual product are "residual oil," which term may be used when the residue is liquid in consistency, and "residual asphalt," when the residue is of semisolid to solid consistency.

In order to avoid confusion, some additional explanation of terms relative to the residual products resulting from this type of distillation seems desirable. It must again be emphasized that a residue of any desired consistency may be obtained in the process by stopping the distillation at the desired point; when this is done the residue is termed a "straight-run asphalt." In common terminology in the highway field, a semisolid to solid residue produced in this way would be called a "steam-refined asphalt cement." If the residue is liquid, a broad term would be "liquid asphalt."

In further explanation, a straight-run product of liquid consistency would most generally fall into the classification commonly known as "slow-curing liquid asphalts." Liquid asphalts (other than emulsions) used in highway work are most frequently obtained by fluxing or "cutting-back" a residual asphalt with a distillate. The "cut-back asphalts" thus produced may be applied cold or at slightly elevated temperatures. Generally speaking, when gasoline or some other light solvent is used to temporarily soften or cut-back the residual asphalt, the liquid asphalt produced is called a "rapid-curing (RC) cut-back asphalt." Similarly, when kerosene or light fuel oil is used, "medium-curing (MC) cut-back asphalt" is produced. Likewise, a slow-curing (SC) liquid asphalt may be obtained by fluxing a residual asphalt with a less volatile distillate such as gas-oil. It should also be noted that these slow-curing materials, whether straight-run or fluxed with a relatively nonvolatile material, are known simply as "road oils" in many sections of the country. Figure 11–2 is a simplified flow chart showing the recovery and refining of asphaltic products prepared from petroleum.

11–4. Cracking. Refining processes which have been devised to increase the yield of light distillates, particularly gasoline, from a given crude oil may be referred to as "cracking" methods. Many different and complex processes are now used to "crack" petroleums, and these methods will not be described here. Generally speaking, however, cracking is accomplished by heating the crude oil to a high temperature under pressure. Temperatures as high as 1100°F and pressures in excess of 50 atmospheres are used in some cracking processes. The residual which is produced is called "pressure tar" or "cracked residue," and is of some consequence in highway usage.

The cracked residues derived from these processes, if liquid, may be used directly for road purposes in areas where they are plentiful and cheap, as, for example, in the "oiling" of natural earth roads in oil fields or refineries. They may also be mixed in suitable proportions with residual asphalts obtained by the regular steam distillation process, or steam-distilled or air-blown to produce asphalts of the desired grade. In general, asphalts prepared from cracked products are regarded as being less weather-resistant and durable than those prepared from uncracked materials.

11-5. Air Blowing. Air blowing is a refining process used in the production of asphalts from residual oils. Advantages claimed for this type of refining process include greater yields of asphalt, better control of the consistency or "grade" of asphalt which is produced, and the production of a material which is less susceptible to temperature change than comparable residual asphalts produced by steam distillation. It is also claimed that blown asphalts are more weather-resistant than residual asphalts. It should be noted that carelessly blown asphalts are apt to be lacking in ductility and may contain relatively large amounts of material which is insoluble in carbon disulfide. Both reduction in temperature susceptibility and loss of ductility are a function of the amount of blowing. Since high ductility is generally associated with desirable cementing power in the past blown asphalts were not widely used as binding agents but were extensively used as crack and joint fillers. It must be emphasized that modern refining methods permit the production of slightly blown asphalts which possess all the desirable binding properties associated with steam-refined asphalt cements and which, in addition, are more weather-resistant than comparable residual asphalts. This fact has led to the increasing use of blown asphalts as binding agents in the construction of bituminous pavements.

As the term indicates, air-blown ("oxidized") asphalts are produced by blowing air through residual oil at moderately high temperatures. The residual oil is placed in a still and the temperature raised to from 525 to 575°F. Air is introduced into the bottom of the still and blown through the heated mass under pressure or by applying a partial vacuum at the top of the still. This process is continued for several hours or until the residue has attained the consistency desired. As the air is blown through the heated residual oil, chemical changes occur, some of the volatile materials are removed, and water and carbon dioxide are given off.

11-6. Emulsification. In order that bituminous materials may be used with unheated aggregate, they must be rendered liquid. One method is by fluxing, as was previously described for cut-back asphalts.

A second method of making asphalts liquid is emulsification. Petroleum asphalts are usually employed in this process, and an emulsified asphalt may be defined as a dispersion of asphalt particles, ranging around 3 microns in size, in water in the presence of an emulsifying agent. The emulsifying agents commonly used are soaps, used alone or in combination with other substances. This type of asphalt emulsion is further generally described as an "oil-in-water" emulsion.

The principal objective of the liquefaction of asphaltic materials by this method is to allow their application at normal temperatures without the use of heat. The emulsion breaks down when sprayed or mixed with mineral aggregate in a field construction process; the water is removed, and the asphalt remains as a film on the surface of the aggregate. The water simply serves as a temporary transporting medium and may be compared to the volatile constituents of a cut-back asphalt in that both largely disappear during the mixing process, leaving the semisolid asphalt as a binding material. Three principal classes of asphalt emulsions are used for highway purposes. These are called "rapid-setting," "medium-setting," or "slow-setting" emulsions, depending upon the speed at which the emulsion breaks down when placed in contact with mineral aggregates.

Emulsified asphalts may be of the anionic or cationic types, depending upon the emulsifying agent. The anionic type contains electronegatively charged asphalt globules and the cationic type contains electropositively charged asphalt globules. It is only within the last few years that the latter type has been produced commercially.

Asphaltic emulsions of the types described are usually manufactured by use of a high-speed disintegrator or "colloid mill." In this process the asphaltic material, which may be a cold or heated liquid or even a comminuted solid, and the mixture of water and emulsifying agent are fed simultaneously into the disintegrator. Disintegration is accomplished by passage of the material between revolving or fixed blades, or between smooth or slightly roughened plate or conical surfaces, one of which rotates while the other is stationary, or which rotate in opposite directions. The contiguous surfaces of the latter type of machine are very closely spaced, the clearance being in the range of 0.002 to 0.003 inch, and the surfaces rotate at very high relative velocities. The asphaltic material is reduced to elements which are in the range of colloidal sizes, and is dispersed in the water. The emulsions thus produced are quite stable and may be transported or stored for long periods of time. In Fig. 11–3 is shown a simplified sketch of a colloid mill employing smooth surfaces.

Asphaltic emulsions offer certain advantages in construction, particularly when used in connection with moist aggregates or in wet weather. Their principal disadvantage lies in the fact that the water occupies

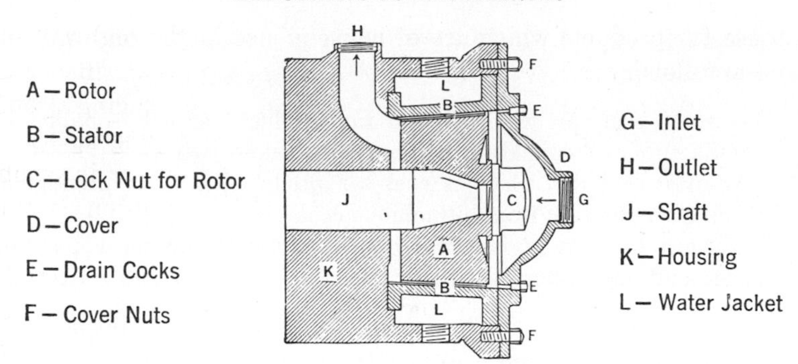

FIG. 11-3. The Charlotte colloid mill. (Courtesy Chemicolloid Laboratories, Inc.)

space and has weight. The user, then, pays for a large percentage of water for which he has no use save as a transporting medium for the asphalt. This factor may be of special significance in areas where distances are great and freight rates are high.

Production of Native Asphalts

Native asphalt may be defined as asphalt which occurs in a pure or nearly pure state in nature. Native asphalts which are associated with a large proportion of mineral matter are generally termed "rock asphalts." Several extensive deposits of rock asphalt occur in the United States and their production and use will be considered in a later chapter.

Native asphalts were extensively used in highway construction in this country in the period from about 1880 to the end of the First World War. From that time onward the use of petroleum asphalts has steadily increased until native asphalts (not including rock asphalts) play a distinctly minor role in highway construction at the present time. Deposits of native asphalt occur in various parts of the world, although few of the deposits are of commercial importance.

Probably the most important native asphalt, from the standpoint of past and present usage in the United States, is Trinidad Lake Asphalt. Other native asphalts which have been used in highway construction in this country at various times include Bermudez (Venezuela) and Cuban native asphalts, and Gilsonite, an asphaltite.

Production of Road Tars

Tar may be generally defined as a substance obtained by the condensation of distillates resulting from the destructive distillation of organic material. Many organic substances may be treated to yield tar. However, those tar products which are of consequence in the highway industry are the following:

1. Gas-works coal tar, which is coal tar produced as a by-product in the manufacture of illuminating gas from bituminous coal.
2. Coke-oven coal tar, which is coal tar produced as a by-product in the manufacture of coke from bituminous coal.
3. Water-gas tar, which is tar produced by cracking oil vapors in the manufacture of carburetted water gas.

These various tars are further refined and may then be used singly or in combination with one another to produce the various grades of road tar which range in consistency from semisolid or very viscous liquid materials to very thin, highly volatile liquids. The production of each of the above types of tars is briefly described in the following paragraphs.

11–7. Gas-Works Coal Tar. In the production of illuminating gas from bituminous coal, heating takes place in ovens which are referred to as horizontal, inclined, or vertical retorts, depending on the position in which they are supported. Coal in the retorts is heated with water gas obtained by passing air and steam through incandescent coke. As heating continues, vapors are driven off at temperatures varying from 1300 to 1800°F. These volatile materials are collected in a closed trough or "hydraulic main," which is partially filled with water. The temperature of the vapors is reduced to about 150°F in the hydraulic main, which cooling causes a large percentage of the tar to condense in the water, from which it is later removed. The remaining vapors are then passed through a sequence of operations utilizing various types of equipment, including condensers, tar extractors, scrubbers, and purifiers. No further explanation of this process will be given here, but the interested student is referred to item (*198*) of the Bibliography for complete descriptions of the several units mentioned.

11–8. Coke-Oven Coal Tar. The basic process involved in the manufacture of coke from bituminous coal is one of destructive distillation. Distillation takes place in retorts in which the coal is heated to temperatures as high as 2200°F. Various gases are produced as a consequence of this heating, some of which are burned to supply additional heat. Much excess gas is produced, and these vapors are collected and passed through various forms of condensers and scrubbers. Coal tar is thus manufactured as a by-product in the production of coke.

The coal tars produced as described above are further refined before they are suitable for use in road construction. They must normally be dehydrated, a process which may be carried on by settling, centrifuging, or through the use of various stills or heaters. Further refining usually consists of fractional distillation, which may be accomplished in any one of a number of different types of still. The distillates which are produced have many industrial uses. The residue which is obtained is termed "coal-tar pitch" or "refined coal tar," depending on its physical characteristics.

11–9. Water-Gas Tar. As has been stated, water gas is produced by allowing steam to pass through anthracite coal which has been heated to a point of incandescence. Since water gas is not luminous, it is "enriched" by the addition of a petroleum distillate (gas oil). The gas oil is sprayed into a closed chamber (carburetor) maintained at the correct temperature, where it is vaporized and combined with the gas produced in the heating chamber or "generator." From the carburetor, the volatiles pass into a superheater maintained at temperatures between 1200 and 1300°F, and are there "cracked" into permanent gases. As an incidental part of this cracking process, tarry substances are produced. The carburetted water gas is then passed through a series of operations involving the use of a wash box (corresponding to a hydraulic main), scrubber, condenser, and tar extractor to remove the last traces of tar before the refined gas is ready for consumption. The water-gas tar produced is dehydrated by processes similar to those used in the manufacture of coal tars; the refined product may then be used in the production of various grades of road tar.

Shipment and Sampling of Bituminous Materials

Shipment of bituminous materials for use in highway construction takes place in both small or "packaged" quantities and large or "bulk" quantities. Relatively small quantities of liquid or semisolid material may be shipped in light steel drums. Drums used in the shipment of liquids are usually provided with screw-type openings through which the material may be poured, and which are closed by plugs during transit. Drums containing semisolid substances have friction tops.

Packaged shipment such as described above is economical only when the required amount of bituminous material is quite small. In most highway construction operations, large quantities of bituminous materials are required and shipments are made in bulk quantities. Bulk shipment is usually made in railroad tank cars equipped with steam-heating coils. When the car is to be unloaded, steam is passed through the heating coils and the material is brought to the temperature desired for unloading. The fluid may then be pumped into storage tanks or directly into the distributors being used in construction operations. The steam for heating the tank cars described may be supplied by an ordinary steam boiler, but more often portable tank-car heaters are used. On shorter hauls bulk quantities are most frequently moved by motor trucks.

Payment for bituminous materials is made on the basis of either weight or volume, with the ton or gallon being common units of measurement. The volume at 60°F is frequently taken as standard for volume measurements. Since all bituminous materials expand when heated, volume measurements which are made at temperatures other than standard must be corrected. Tables of temperature-volume relationships at various temperatures and for materials of different specific gravities are available in various publications of the Asphalt Institute. The following simple formula may be used for approximate calculations:

$$V_t = V_{60}[1 + c(t - 60)] \tag{11-1}$$

where V_t = volume at t°F
V_{60} = volume at 60°F
c = 0.00035 for asphalts of specific gravity greater than 0.966
c = 0.00040 for asphalts of specific gravity less than 0.966

Bituminous materials purchased for use in highway construction must, of course, meet certain specifications. Specifications employed to obtain bituminous substances suitable to the use for which they are intended are presented in later portions of this chapter. However, it can be stated here that acceptance of the material is based upon the results of certain tests which are performed upon representative samples. Detailed instructions for sampling, employed by many highway agencies, are the "Standard Methods of Sampling Bituminous Materials," AASHO Designation T40 and ASTM Designation D140.

Laboratory Tests

A large number of different laboratory tests are performed upon bituminous materials for the purpose of checking compliance with the specifications which are being used. Most of the tests which are performed in the laboratories of the various highway agencies have come to be more or less routine in nature and are conducted in accordance with methods of test established by the American Association of State Highway Officials and the American Society for Testing Materials. Some of the tests are intended to measure specific properties of the material, while others are used primarily as identification tests or in checking uniformity of the material.

In order to permit the student to obtain a better concept of the number of tests in common use, Table 11-1 has been prepared. This lists the names of the most commonly used tests, the corresponding AASHO and ASTM designations for each method of test, and indicates whether the test is normally performed upon asphaltic materials only, upon tar products only, or upon both. In the interests of further simplification, the test methods have been placed in five groups: general, solubility, consistency, ductility, and volatility. The same system of grouping has been used in the discussion of the various test procedures. The test

TABLE 11–1

LABORATORY TESTS OF BITUMINOUS MATERIALS USED IN ROAD CONSTRUCTION

Name of Test	AASHO Designation	ASTM Designation	Applicability*
General			
Specific gravity	T43	D70–D71	3
Flash point (Open Cup)	T48	D92	1
Flash point ("Tag" Open Cup)	T79	–	1
Water in petroleum products, etc.	T55	D95	3
Solubility			
Determination of bitumen (soluble in CS_2)	T44	D4	3
Determination of bitumen (soluble in CCl_4)	T45	D165	1
Spot test	T102	–	1
Sulfonation index	T108	D872	2
Consistency			
Specific viscosity (Engler)	T54	D1665	2
Viscosity by means of the Saybolt viscosimeter	T72	D88	1
Kinematic viscosity	T201	D2170	3
Float test	T50	D139	3
Penetration	T49	D5	1
Softening point (Ring-and-Ball Method)	T53	D36	3
Ductility			
Ductility	T51	D113	1
Volatility			
Distillation:			
Cut-back asphaltic products	T78	D402	1
Tar products	T52	D20	2
Loss on heating	T47	D6	1
Thin film oven test	T179	D1754	1
Residue of specified penetration	T56	D243	1

* 1—Asphaltic materials only.
2—Tar products only.
3—Both asphalts and tars.

methods are only briefly outlined in the following pages, and the student is referred to the designated methods of test for more complete descriptions. In the following discussion an attempt is made to give some idea of the significance of the various tests. This is complicated by the fact that many of the tests seem to have little significance when compared with the desired properties of bituminous materials as used in highway construction. Tests performed upon emulsified asphalts are mentioned separately, and no attempt has been made to discuss the testing of various miscellaneous substances derived from bituminous materials and of use in a minor way in highway work.

General Tests

11–10. Specific Gravity. The specific gravity of a bituminous material is defined as the ratio of the weight of a given volume of the material at 25°C to that of an equal volume of water at the same temperature. The specific gravity may be determined by the use of a hydrometer, by displacement, or by the use of a pycnometer. The hydrometer method has been used in the evaluation of thin fluid materials. In this method, the fluid is brought to the desired temperature (25°C), and the specific gravity is measured directly by immersing a suitably calibrated and marked hydrometer in the fluid mass. The displacement method may be used to determine the specific gravity of bituminous materials which are sufficiently hard and solid to be handled in fragments. The method involves the determination of the weight in air of an irregular piece or a ½-inch cube of the material, its complete immersion in distilled water maintained at the standard temperature, and the determination of the corresponding "weight in water." A simple calculation may then be employed to determine the specific gravity. Both the hydrometer and displacement methods are seldom used, the majority of laboratory determinations being made by the pycnometer method.

Suitable pycnometers used in this test procedure are made of glass; two types are illustrated in Fig. 11–4. The clean and dry pycnometer, with stopper, is placed on an analytical balance and weight W_0 determined. The pycnometer is then completely filled with distilled water at the standard temperature, the stopper inserted, and weight W_w determined. In the case of liquid materials which are sufficiently fluid to flow readily through the hole in the glass stopper, the water is removed and the pycnometer completely filled with the liquid. This weight may be called W_1. The specific gravity is then

$$G_b = \frac{W_1 - W_0}{W_w - W_0} \tag{11-2}$$

where all weights are in grams.

In the case of semisolid or very viscous materials, a small portion of the substance is heated gently, poured into the pycnometer, and then allowed to cool. This weight of pycnometer and contents is called W_2.

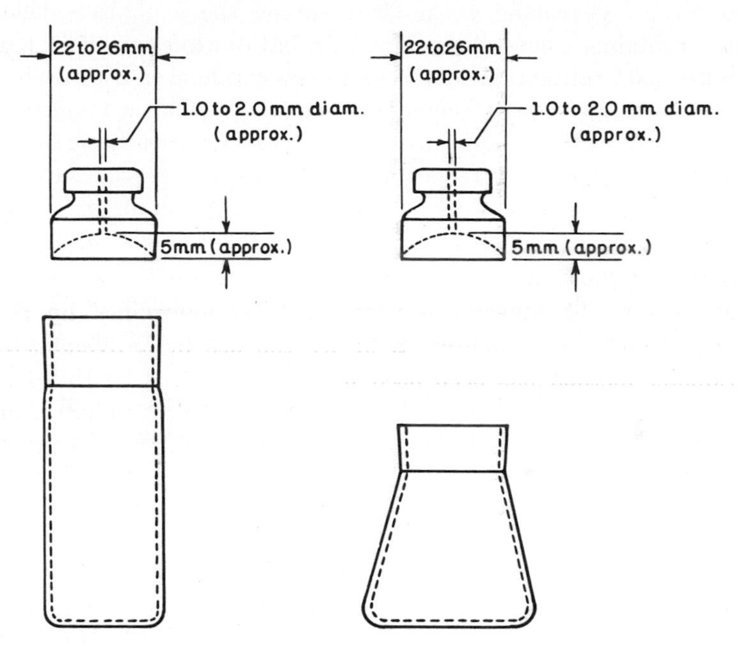

Fig. 11–4. Pycnometers used for specific gravity determinations.

The pycnometer is then completely filled with water, the stopper inserted as before, and weighed again. This weight may be called W_3. The specific gravity is then given by the expression,

$$G_b = \frac{W_2 - W_0}{(W_w - W_0) - (W_3 - W_2)} \tag{11–3}$$

all weights again being in grams.

The specific gravity of asphaltic products prepared from petroleum varies from about 0.92 for thin, watery liquids to about 1.06 for semisolid materials. Refined tars (road tars) used in highway construction vary in specific gravity from about 1.08 to 1.24. The specific gravity of a bituminous material is of some value for identification purposes and for checking uniformity. Its principal use, however, is in converting from volume to weight measurements, and vice versa.

11–11. Flash Point. Two methods are in common use for the determination of the flash point of native and petroleum asphalts used in highway work. The flash point of cut-back asphalts (RC and MC) is generally determined by the use of a Tagliabue Open Cup apparatus, while the Cleveland Open Cup is used for flash-point determinations on other asphaltic materials. The two methods are essentially the same. However, in the Cleveland Open Cup method the asphalt is heated in a metal container suspended in an air bath, while in the "Tag" Open Cup method heating takes place in a glass cup held in a water bath.

In both methods, the temperature of the asphaltic material is gradually increased at a uniform rate. As the temperature rises, a small, open flame is passed at specified intervals across the surface of the heated material. Volatile constituents of the asphalt are driven off as the temperature is increased. As the test proceeds, a point will be reached where sufficient volatiles will collect at the surface of the heated asphalt to briefly ignite, and cause a distinct flicker or "flash" across the surface of the material. The minimum temperature at which this flash occurs is called the "flash point."

The flash point is an indirect measurement of the quantity and kind of volatiles present in the asphalt which is being tested. Generally speaking, rapid-cure cut-back asphalts will have very low flash points (100°F or less); medium-cure cut-backs will have slightly higher flash points; slow-curing liquid asphalts higher; and semisolid materials, such as asphalt cements, will have flash points of 450°F or more. In addition to giving some information as to the amount of volatile constituents which are present in the material, the flash point is indicative of the temperature to which the asphalt can be safely heated under working conditions, especially where heating is being accomplished in an open vessel over an open flame.

11–12. Determination of Water. Specifications for bituminous materials to be used in highway construction usually state that the material "shall be free from water," or the amount of water which may be present is limited to a small percentage of the volume of the material. This requirement tends to aid in insuring careful manufacture, storage, and shipment, and has the definite purpose of preventing "boiling" or "foaming" of the substance when it is heated. Excessive foaming may readily occur when bituminous materials containing any sizable amount of water are heated to normal working temperatures.

The amount of water present in a given bituminous substance may be determined by heating a sample of the given material which has been combined with an equal amount of suitable petroleum (or coal tar) distillate in a metal still or glass flask. As the temperature is increased, the volatiles, including the water, are driven off and pass through a condenser-and-trap arrangement. The process is continued until all the water has been removed from the sample. The volume of water may be measured in the trap and the per cent, by volume, of water in the sample calculated.

Solubility Tests

11–13. Solubility in Carbon Disulfide (Determination of Bitumen). As previously noted, bitumen is, by definition, a substance which is soluble in carbon disulfide. Petroleum asphalts are practically 100 per cent soluble in carbon disulfide, while most native asphalts and tars contain considerable percentages of nonbituminous materials, or "impurities," which may be either organic or inorganic in nature. Obviously then, this test has value in identifying the nature and source of a given bituminous material.

In the laboratory this test is conducted by dissolving a known weight of the material in a given volume of carbon disulfide and filtering the solution through an asbestos mat formed in the bottom of a Gooch crucible. Insoluble material retained in the crucible is dried and cooled, and the increase in weight of the crucible determined. If only a small portion of the material is retained, the test may be stopped at this point. Further steps which may be taken include the ignition of the residue in the crucible and the evaporation and ignition of the filtrate. Results are reported simply as "per cent bitumen" or "per cent soluble in carbon disulfide."

11–14. Solubility in Carbon Tetrachloride. Certain substances, designated as "carbenes," are soluble in carbon disulfide but are not soluble in carbon tetrachloride. Since the presence of these insoluble materials is generally supposed to be an evidence of careless refining, overblowing, or "burning" of the material, this test is sometimes used as a check on the refining processes accompanying the production of petroleum asphalts. Considerable disagreement exists as to the real significance of this test. The test itself is performed in a manner similar to that employed in the determination of bitumen, and the results are reported as "per cent *bitumen* soluble in carbon tetrachloride."

11–15. Spot Test. More complex tests are sometimes employed by various highway agencies to further check the processes used in refining petroleum asphalts for use in highway construction. One such test is the "spot test," which is designed to determine the solubility of the material in a standard petroleum naphtha. The procedure calls for the dissolving of a specified amount of the original material (or the residue from a distillation test) in the stated solvent, under very carefully controlled conditions. The degree of solution is judged by placing a drop of the resulting solution on a piece of filter paper. If a uniform brown stain is produced in original and repeat tests, the test is reported as "negative" and the material is judged to have passed the test satisfactorily. On the other hand, if a nonhomogeneous spot (usually darker in the middle than at the edges) is formed and confirmed, the result is said to be "positive," and rejection of the material may result from this evidence of the presence of certain compounds assumed to result from careless manufacture.

Several variations of the standard method are now in use. One widely used variation with positive spot materials is to repeat the test using a mixture of the standard naphtha and varying percentages of a spot-retarding agent, usually xylene. The xylene is normally added in 5 per cent increments to determine the percentage required to change the spot from positive to negative. The specification in this case sets an upper limit on the amount of xylene required to suppress a positive spot. Several western states prefer to use normal heptane in place of the standard solvent naphtha; other variations in the solvent are used.

11–16. Sulfonation Index of Road Tars. A control test which is used by some agencies in specifications for road tars is the sulfonation index, which is defined as the number of milliliters of unsulfonated residue per 100 grams of tar when determined in accordance with the standard method. The test is performed upon the distillate obtained by performance of the standard distillation test on road tars. A carefully controlled procedure involves the addition of a specified quantity of sulfuric acid to the distillate, with the acid being added in increments followed by vigorous shakings. After addition of the acid, the mixture is centrifuged to remove material in solution. The material which remains is unsulfonated residue; the sulfonation index expresses the relationship between the volume of unsulfonated residue, the percentage of distillate in the tar, and the weight in grams of the sample of distillate.

Consistency Tests

Several consistency tests are commonly used in connection with the laboratory examination of bituminous materials. Two types of viscosity measurements have been employed to measure the consistency of liquid substances. These are the Engler Specific Viscosity, commonly used in connection with the testing of road tars, and the Viscosity by Means of the Saybolt Viscosimeter, which is used in the evaluation of liquid asphalts. The trend is to supplant the latter test with kinematic viscosity determinations. The Float Test is used to measure the consistency of materials which are too viscous for testing in a viscosimeter and too fluid for the Penetration Test, and also in the testing of solid and semisolid tar products. The Penetration Test is used in evaluating the consistency of solid and semisolid asphalts. The determination of the softening point by the Ring-and-Ball Method is also classed as a consistency test, as it is visualized as an indication of the consistency of the material at elevated temperatures.

11–17. Engler Specific Viscosity. In simple explanation, the Engler Viscosimeter consists of a cylindrical brass vessel, approximately 4 inches in diameter and 3 inches high, with a conical bottom and a fitted cover. A larger, surrounding cylinder, which is filled with water or oil, aids in securing uniform temperature in the sample which is being tested. The brass vessel has an orifice of standard dimensions in the bottom, and the entire assembly is mounted in a metal frame. The container is filled to a specified level with the bituminous material being subjected to test, a uniform temperature is established, and the stopper is removed from the orifice. The time required for exactly 50 milliliters of the substance at the specified temperature to flow from the orifice is determined. This time is then compared with that required for water at 25°C to flow from the orifice under similar conditions. The ratio of these times is the "Engler Specific Viscosity." A temperature of either 40 or 50°C is generally specified in the testing of road tars.

11–18. Saybolt Viscosity. Fig. 11–5 is a sketch of the "oil tube" which forms the integral part of a Saybolt Viscosimeter. Over-all dimensions of this tube are approximately 5 inches long and slightly more than 1 inch in diameter. Outlet tubes having two different sizes of orifice are provided. One, which is 0.1765 cm. inside diameter, is termed a

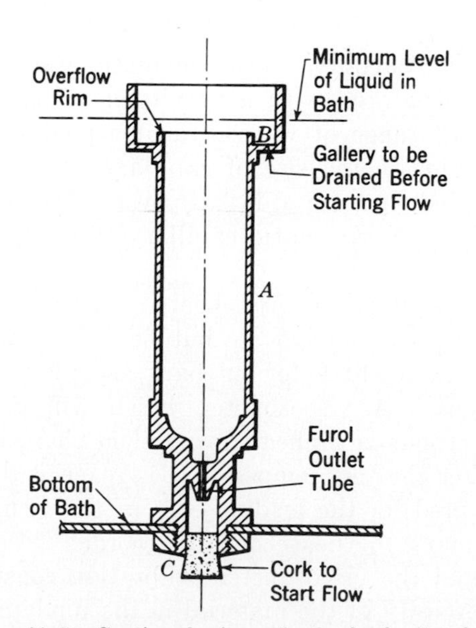

FIG. 11–5. Sectional view of standard oil tube.

"Universal" outlet, while another, which is 0.315 cm. inside diameter, is termed a "Furol" outlet. The Saybolt Furol Viscosimeter is normally used in the testing of liquid asphalts. The outlet tubes are precisely machined to meet definite specifications.

The outlet tube is encased in a much larger water or oil bath which is usually equipped with an electrical heater and a stirring device. The material to be tested is placed in the oil tube and the temperature of the bath and material is brought to the desired level. Temperatures of 77, 122, 140, and 180°F are used in testing highway materials. As soon as the desired temperature has been maintained in both the bath and the tube for a stated length of time, the stopper closing the outlet tube is removed, a stop-watch is started, and the time required for exactly 60 milliliters to flow into a standard glass container is noted. This time is then reported as the "Saybolt Furol Viscosity" at the specified temperature.

The principal use of the viscosity determination is, of course, for securing the consistency desired in the material under the temperature specified. Liquid asphalts, as will be shown later, are specified in various grades, and one of the principal differences in these grades lies in the variation in the Furol viscosity. Similarly, the grade of a liquid road tar is based upon the Engler Specific Viscosity. The temperatures used in conducting the test may be further selected to supply information about the fluidity of the material at the anticipated temperature of application. A tentative procedure for determining the viscosity of asphalts at elevated temperatures has been adopted by ASTM (E102).

11–19. Kinematic Viscosity. The kinematic viscosity of an asphalt cement is usually measured with a gravity flow capillary viscosimeter. Because of the wide range of viscosities of asphalts, several calibrated viscosimeters differing in the size of capillary are necessary. The viscosimeters are placed in a suitable bath where the temperature can be carefully controlled. A kinematic capillary viscosity test apparatus is shown in Fig. 11–6.

The sample to be tested is first heated (for asphalt cements only) and stirred carefully to remove air bubbles. The bath is maintained at its test temperature, 140°F for cut-back asphalt and road oils, 275°F for asphalt cements. A viscosimeter which will give an efflux time greater than 60 seconds is preheated and then charged with the sample to be tested. After the test temperature is reached the valve is opened and the time required for the leading edge of the meniscus to pass from the first timing mark to the second is recorded. Using the measured time in seconds and the viscosimeter calibration constant, it is possible to compute the viscosity of the material in the fundamental units, stokes and centistokes.

The relationship of centistokes (kinematic viscosity) to seconds (Saybolt Furol viscosity, as previously described) is approximately 2:1. Factors for converting kinematic viscosities to Saybolt Furol seconds for a limited range of temperatures are given in ASTM D 2161. This comparison for old and new grades of liquid asphalts is shown in Fig. 11–7. The complete test procedure for kinematic viscosity is described in AASHO T 201 and ASTM D 1270.

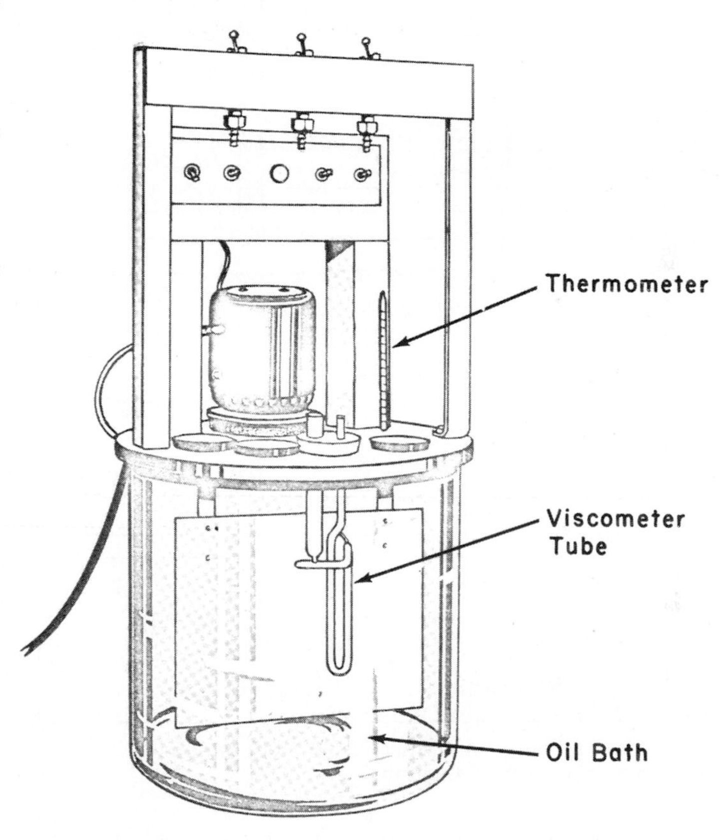

FIG. 11–6. Apparatus for kinematic capillary viscosity test. (Courtesy the Asphalt Institute.)

Another test which can be used to measure viscosity of asphalt cements at low temperatures is described in AASHO T 202, and ASTM D 2171. This method gives the absolute viscosity in poises.

A method using a sliding plate microviscosimeter has been developed by the Shell Oil Company, and not only measures the absolute viscosity of asphalts, but extends the temperature range at which viscosities can be measured reliably.

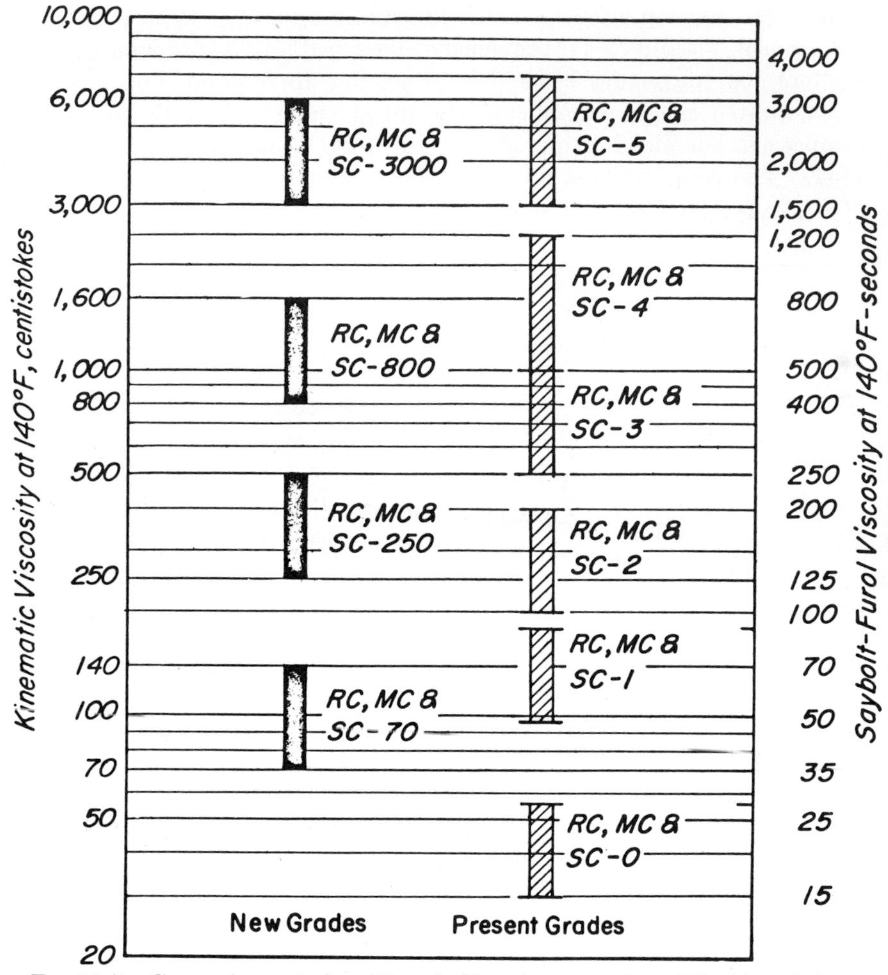

Fig. 11–7. Comparison of viscosities of old and new grades of liquid asphalts at 140°F. (Courtesy the Asphalt Institute.)

Essentially, the sliding plate microviscosimeter (Fig. 11–8) is a device capable of measuring the shear rate of a sample at different temperature levels. A sample of the asphalt is placed between two glass plates in a specified manner. One plate is held fixed, and weights are applied to the other until a weight sufficient to move the plate is observed. The viscosity in poises is computed from the shear stress. This test promises to be one of the more widely used asphalt consistency tests in the future.

11–20. Float Test. As has been previously indicated, the Float Test is used for determining the consistency of semisolid to solid road tars. Specified grades of heavier road tars are based upon this test, performed at a temperature of either 32 or 50°C. The test is also useful in evaluating the consistency of the residues obtained in the distillation of slow-curing liquid asphalts.

This test, which is quite arbitrary in nature, involves the use of the apparatus shown in Fig. 11–9. In the conduct of the test, the brass collar, *B*, shown in Fig. 11–9 is filled with the material to be tested and brought to a temperature of 5°C by immersing it in ice water. A

Fig. 11–8. A sliding plate microviscosimeter. (Courtesy the Asphalt Institute.)

water bath is prepared, and the temperature of the water in the bath is brought to the desired level (32 or 50°C). The brass collar containing the chilled bituminous material is then screwed into the bottom of the float and the assembly is floated in the water bath maintained at the desired temperature. As time passes, the material in the plug softens, and, after the passage of some time, the water breaks through the plug of bituminous material into the upper part of the float. The time, in seconds, which elapses between the instant when the assembly is floated in the water and that when the water breaks through is the reported value. This time is held to be a measure of consistency.

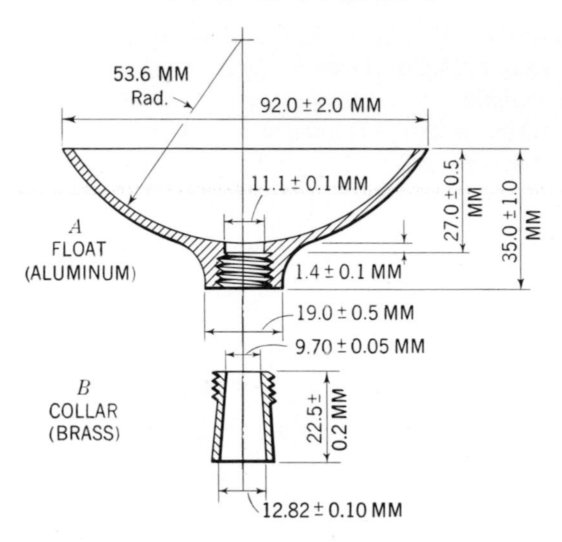

FIG. 11–9. Apparatus for float test.

11–21. Penetration. The penetration of a bituminous substance may be defined as the distance (in hundredths of a centimeter) which a standard needle penetrates the material under known conditions of time, loading, and temperature. This test is used for evaluating the consistency of asphaltic materials, and is not regarded as suitable for use in connection with the testing of road tars because of the high surface tension exhibited by these materials, and the fact that they contain relatively large amounts of free carbon.

The standard penetration test procedure involves the use of the standard needle under a load of 100 grams, for five seconds, and at a temperature of 25°C. In Fig. 11–10 is shown a so-called "precision-type" penetrometer which may be used for laboratory measurements of penetration. Simpler and more rugged types of penetrometers may also be used. The grade of semisolid and solid asphaltic materials is usually designated by the penetration. For example, penetration ranges such as 30–40, 40–50, 50–60, 60–70, 70–85, 85–100, etc., may be used in specifying the desired grades of asphalt cements prepared from petroleum. The test may be performed under other conditions of load, time, and temperature. This is frequently done, for instance, when it is desired to study the effect of changes of temperature upon the consistency of a given material.

At the beginning of this chapter it was mentioned that bituminous materials could be classed as liquid, semisolid, or solid on the basis of their consistency. This classification may be arbitrarily based upon the penetration test, as shown in Table 11–2.

FIG. 11–10. Penetrometer.

TABLE 11–2

PENETRATION LIMITS OF CONSISTENCY

Classification	Penetration	Load (grams)	Time (seconds)	Temperature (°C)
Liquid	351 or more	50	1	25
Semisolid	350 or less	50	1	25
	11 or more	100	5	25
Solid	10 or less	100	5	25

11–22. Softening Point (Ring-and-Ball Method). Since the softening of a bituminous material does not take place at any definite temperature, but rather involves a gradual change in consistency with increasing temperature, any procedure which is adopted for determining the softening point must be of a somewhat arbitrary nature. The procedure in common use in highway materials laboratories is known as the "ring-and-ball method" and may be applied to semisolid and solid materials.

In this test procedure, the sample is melted and poured into a brass ring which is of ⅝-inch inside diameter and is ¼ inch high. After cooling, the ring is suspended in a water bath maintained at a temperature of 5°C, in such a fashion that the bottom of the filled ring is exactly one inch above the bottom of the bath. A steel ball ⅜ inch in diameter is placed on the surface of the bituminous material contained in the ring, and the temperature of the water is elevated at a standard rate. As the temperature is increased, the bituminous material softens and the ball sinks through the ring, carrying a portion of the material with it. The behavior of the ball is observed, and the softening point is taken to be the temperature at which the bituminous material touches the bottom of the container.

Since this test is so arbitrary in nature, it is difficult to see its exact significance when applied to bituminous materials used in highway work. It is useful for identification purposes and gives some indication of the behavior of the material at elevated temperatures. For the same penetration range, a bituminous material which has a high softening point may be less susceptible to temperature change than one having a lower value. The softening point may be run on materials differing widely in consistency, and it furnishes a convenient means for a direct comparison of these materials. For example:

Material	Softening Point (Approximate)
SC–3	50 F
200 Pen. Asphalt	100 F
50 Pen. Asphalt	120 F
Crack Filler Asphalt	130 F
Brick Filler Asphalt	170 F
Roofing Asphalt	180 F and higher

Ductility

By ductility is generally meant that property of a material which permits it to undergo great deformation (elongation) without breaking. In regard to bituminous materials, ductility may be further defined as the distance, in centimeters, to which a standard sample or "briquette" of the material may be elongated without breaking. The test is applicable only to semisolid (or solid) asphaltic material which is melted by gentle application of heat and poured into a standard mold. Dimensions of the mold are such that the minimum cross-section of the briquette thus formed is exactly 1 centimeter square. The mold is then immersed in a water bath maintained at the desired temperature of the test, which is usually 77°F.

After the material has attained the desired temperature, the sample is placed in a ductility machine. The machine is so arranged that one end of the mold is held in a fixed position, while the other end is pulled horizontally at a standard rate. The behavior of the "thread" of material is noted, and elongation is continued until the thread breaks. The distance (in centimeters) which the machine has traveled is the "ductility" of the material. Many asphalt cements have a ductility of 100 or more.

In interpreting the results of this test, it is generally assumed that ductility is a measure of the cementing power of the asphaltic material. Since high cementing qualities are desirable in most applications, it is held that an asphalt which is to be used as a binder should be ductile. However, it should be noted that the exact value of the ductility is not so important as the mere presence or lack of ductility. That is, an asphalt which has a ductility of 60 may be just as good a binder as one which has a ductility of 100 or more. However, if the ductility is 0 or 4, or some other low figure, experience has, in general, shown that the material will not have satisfactory binding qualities.

It is further believed that asphalts which are highly ductile are also highly susceptible to temperature change. Although the test is generally performed at 77°F, nevertheless it may be conducted at lower (or higher) temperatures to evaluate the effect of temperature change upon the properties of the material undergoing examination.

Volatility Tests

11–23. Distillation. The objective of a distillation test performed on a bituminous material is simply to separate the volatile from the nonvolatile substances. By fractional distillation the relative proportions of volatile materials driven off at various specified temperatures may be determined, and the residue may be tested to ascertain the properties of the residual substance contained in the original fluid material.

In testing cut-back asphaltic products, including slow-curing liquid asphalts, the apparatus shown in Fig. 11–11 is used. As the distillation proceeds, the volatiles are condensed and collected in the graduated cylinder of Fig. 11–11. The original volume of the material contained in the flask is 200 milliliters, and the temperature of the material is increased at a gradual rate. It may be desirable to know the proportion of distillate driven off at various selected temperatures. For example, the volume of distillate may be read at temperatures of 374, 437, 500, and 600°F in evaluating rapid-curing cut-back asphalts. The end temperature usually employed is 680°F. The percentage of volatiles collected to any given temperature may be determined on a volumetric basis. The residue is cooled and set aside for additional testing.

Tar products intended for use in road construction are evaluated by means of a distillation test which is very similar to that which has previously been described. The apparatus used is somewhat simpler but the general conduct of the test is practically the same. Temperatures used in separating the various fractions of the distillate are generally 170, 270, and 300°C.

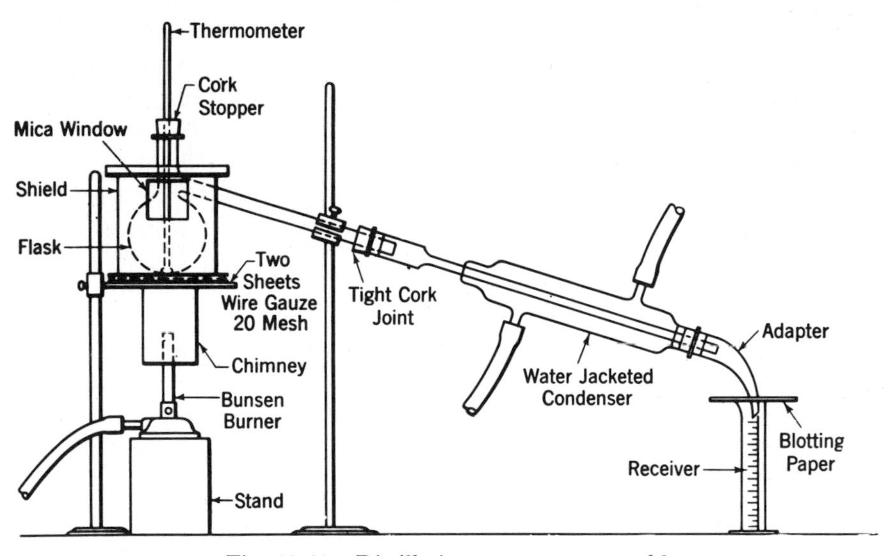

Fig. 11–11. Distillation apparatus assembly.

The distillation test supplies a very considerable amount of information, as it permits the testing engineer to ascertain the kind and amount of volatile material which has been used in fluxing or "cutting-back" a semisolid to produce a liquid substance. Furthermore, by removal of the volatile substances, the engineer creates in the laboratory a material resembling that which will exist in the field, i.e., the volatiles are removed, leaving the residual which will go to make up the binder in a mixture of the bituminous material and mineral aggregate. The residue may then be tested to determine whether or not it has the qualities which the engineer deems desirable and necessary.

11–24. Loss on Heating. In determining the percentage of volatile material which will be removed from the asphaltic substance being examined under the imposed conditions of the loss on heating or "volatilization" test, 50 grams of the selected material is carefully weighed into a standard, flat cylindrical container. The sample is then placed in a specially constructed oven maintained at 163°C, and allowed to remain there for a period of five hours. At the end of this period the sample is removed from the oven, cooled, and weighed. The loss in weight may then be determined, and the percentage of loss on heating, based on the weight of the original sample, may then be calculated.

This test is regarded as an "accelerated" test, the concept being that the losses which occur on heating under the test conditions somewhat resemble the losses which occur in the construction process and during the service life of the pavement. However, the conditions prevailing during this test are in no way comparable to the hardening of asphalt in thin films in asphalt aggregate mixtures. As commonly specified, the test is unable to distinguish between asphalts that differ widely in resistance to weathering, and failure to meet specifications in this test is rare. Quality tests may be performed upon the residue, and the test may be useful for identification and control purposes. This test is being superseded by the Thin Film Oven Test described in the next article.

11–25. Thin Film Oven Test. Relatively high temperatures are used in the plant mixing of asphalt cements and aggregates. Excessively high temperatures, however, are detrimental, hardening the mixture and reducing pavement life.

In measuring the amount of hardening of an asphalt cement which may be expected to occur during plant mixing, the standard penetration test is made before and after the Thin Film Oven Test. Hardening is recorded as the percentage of penetration of the sample before the test. A representative sample, 50 cc in volume, is poured into a pan of specified dimensions such that a film approximately ⅛ inch thick is tested. This sample is then placed on an aluminum rotating shelf in a ventilated oven at 325°F, and rotated at a specified rate for 5 hours. It is then poured into a standard container and its penetration determined. The percentage of the penetration before the Thin Film Oven Test is calculated and reported. Specifications usually prescribe minimum values for the per cent of retained penetration for the various grades of asphalt cement.

11–26. Residue of Specified Penetration. The objective of this test to determine the percentage of residue of given penetration (usually 100 under "standard" penetration conditions) contained in a liquid asphaltic material. It is normally applicable only to slow-curing liquid asphalts. In order to accomplish this purpose, a specified weight of a given material is heated under standard conditions. Heating is continued for some indefinite time, the duration of the heating period being left to the operator's judgment. The heating is then stopped, the material is cooled, and the penetration is determined. If the desired penetration has not been attained, heating is resumed and continued until the operator decides that a second trial is desirable. This process is continued until enough volatile material has been removed to bring the residue to the desired penetration. When this point has been reached, the residue is weighed and the percentage of the original sample is determined.

This test has been regarded as indicating the percentage of "asphalt" contained in a liquid asphaltic substance. However, since it is known that asphaltic substances may be created during the heating process involved in this test, interpretation of the test results is somewhat obscure. The test does, of course, give some indication of the relative amount of volatile material contained in the original sample. Use of this test has decreased in recent years and it is to be expected that this trend will continue. Several organizations have substituted a vacuum-distillation procedure for the test described above and ASTM has proposed a tentative method of test (D1189).

Testing of Emulsified Asphalts

Emulsified asphalts are specified on the basis of the results of a complete series of tests, some of which are of a specialized nature. Standard methods of testing emulsified asphalts are designated as AASHO T59 and ASTM D244. Tests of the following categories are included: composition, consistency, and stability. Composition tests include water content, in which a 50- or 100-gram sample is mixed with an equal volume of water-immiscible solvent, distilled, and the quantity of water determined by volume; residue by distillation, in which the percentage by weight of residue is obtained by distillation of a 200-gram sample in an iron still to 500°F; and residue by evaporation, which is less frequently used. Consistency is measured by use of the standard Saybolt viscosimeter with the sample at 77 or 122°F.

Stability tests include demulsibility, in which a 100-gram sample is mixed with a dilute solution of calcium chloride, and the percentage by weight of asphalt which fails to pass a No. 14 wire cloth determined; settlement, in which the difference in asphalt content of top and bottom samples is determined on a 500-milliliter sample after allowing it to stand for five days; cement mixing, in which the sample is diluted to 55 per cent residue with water, mixed with cement, and the percentage by weight of coagulated material which fails to pass a No. 14 sieve determined; a sieve test, in which the percentage by weight which fails to pass the No. 20 sieve is determined; a coating test, in which the ability of a sample to coat a specified stone is judged visually after mixing; and miscibility with water, in which approximately 50 milliliters of sample is mixed with 150 milliliters of water and the visible coagulation after two hours observed. Tests are also performed upon the residue, including specific gravity, ash content, solubility in CS_2, penetration, and ductility.

A particle charge test is made to identify cationic emulsions and a pH test is performed to determine their acidity.

Specifications for Bituminous Materials

A large number of different specifications for bituminous materials are in use by the various highway agencies in the United States. Some organizations have developed their own specifications on the basis of local experience and to permit the use of local materials. Others have adopted the specifications of the AASHO, ASTM, the Asphalt Institute, or some similar body. In any case, the objective of the specifications is the same—to ensure that materials will be secured which will be satisfactory for the use for which they are intended. In the following pages examples of specifications for various types of bituminous materials are presented. These specifications are intended as examples only, and there is no intention of trying to present specifications covering all the bituminous substances used in highway construction and maintenance operations. The following specifications are included (all except Table 11–9 are Asphalt Institute specifications):

Table	Material
11–3	Asphalt cement (prepared from petroleum)
11–4	Cut-back asphalt—rapid curing type
11–5	Cut-back asphalt—medium curing type
11–6	Slow-curing liquid asphaltic materials
11–7	Anionic emulsified asphalts
11–8	Cationic emulsified asphalts
11–9	Tars for use in road construction

Uses of Bituminous Materials

The bituminous materials which have been described in the preceding portions of this chapter find many and varied uses in highway construc-

TABLE 11-3

Specifications for Asphalt Cements
(The Asphalt Institute)

Characteristics	AASHO Test Method	ASTM Test Method	Grade				
Penetration, 77°F, 100 g., 5 sec.	T49	D5	40–50[2]	60–70	85–100	120–150	200–300
Viscosity at 275°F							
Kinematic, centistokes	T201	D2170	240+	200+	170+	140+	100+
Saybolt Furol, SSF	–	E102	120+	100+	85+	70+	50+
Flash point (Cleveland Open Cup), °F	T48	D92	450+	450+	450+	425+	350+
Thin Film Oven Test	T179	D1754	–	–	–	–	–
Penetration after test, 77°F, 100 g., 5 sec., % of original	T49	D5	55+	52+	47+	42+	37+
Ductility							
At 77°F, cm.	T51	D113	100+	100+	100+	60+	
At 60°F, cm.			–	–	–	–	60+
Solubility in carbon tetrachloride, %	T44[1]	D4[1]	99.5+	99.5+	99.5+	99.5+	99.5+
General requirements	The asphalt shall be prepared by the refining of petroleum. It shall be uniform in character and shall not foam when heated to 350°F.						

[1] Except that carbon tetrachloride is used instead of carbon disulphide as solvent, Method No. 1 in AASHO Method T44 or Procedure No. 1 in ASTM Method D4.

[2] Also special and industrial uses.

TABLE 11-4

Specifications for Rapid-Curing (RC) Liquid Asphalts
(The Asphalt Institute)

Characteristics	AASHO Test Method	ASTM Test Method	Grades			
			RC-70	RC-250	RC-800	RC-3000
Kinematic viscosity at 140°F, cs[1]	T201	D2170	70–140	250–500	800–1600	3000-6000
Flash point (open tag.), °F	T79	D1310	–	80+	80+	80+
Distillation						
Distillate (per cent of total distillate to 680°F):						
To 374°F			10+	–	–	–
To 437°F	T78	D402	50+	35+	15+	–
To 500°F			70+	60+	45+	25+
To 600°F			85+	80+	75+	70+
Residue from distillation to 680°F, per cent by volume			55+	65+	75+	80+
Tests on residue from distillation:						
Penetration, 77°F, 100 g., 5 sec.	T49	D5	80–120	80–120	80–120	80–120
Ductility, 77°F, cm.	T51	D113	100+	100+	100+	100+
Solubility in carbon tetrachloride, %	T44[2]	D4[2]	99.5+	99.5+	99.5+	99.5+
Water, %	T55	D95	0.2–	0.2–	0.2–	0.2–
General requirement	The material shall not foam when heated to application temperature recommended by The Asphalt Institute.					

Note: When the Heptane-Xylene Equivalent Test is specified by the consumer, a negative test with 35 per cent xylene after 1 hour will be required, AASHO Method T102.

[1] As an alternate, Saybolt Furol Viscosities may be specified.

[2] Except that carbon tetrachloride or trichloroethylene is used instead of carbon disulphide as solvent, Method No. 1 in AASHO Method T44 or Procedure No. 1 in ASTM Method D4.

tion and maintenance in this country. No attempt will be made here to tabulate the uses for which the various materials are suited, principally because of the large number of different bituminous materials in use in different sections of the country and the wide variation in conditions under which they are used. In later chapters, in which various bituminous wearing surfaces and bases are considered, mention will be made of the particular types of bituminous material which are best suited for use in the various types of construction described. Recommended spraying and mixing temperatures are given in Table 11–10 (p. 453).

TABLE 11–5

SPECIFICATIONS FOR MEDIUM-CURING (MC) LIQUID ASPHALTS
(The Asphalt Institute)

Characteristics	AASHO Test Method	ASTM Test Method	Grades				
			MC-30	MC-70	MC-250	MC-800	MC-3000
Kinematic viscosity at 140°F, cs[2]	T201	D2170	30–60	70–140	250–500	800–1600	3000–6000
Flash point (open tag.), °F[1]	T79	D1310	100+	100+	150+	150+	150+
Distillation							
Distillate (per cent of total distillate to 680°F):							
To 437°F			25−	20−	0–10	-	-
To 500°F	T78	D402	40–70	20–60	15–55	35−	15−
To 600°F			75–93	65–90	60–87	45–80	15–75
Residue from distillation to 680°F, per cent by volume			50+	55+	67+	75+	80+
Tests on residue from distillation:							
Penetration, 77°F, 100 g., 5 sec.	T49	D5	120–250	120–250	120–250	120–250	120–250
Ductility, 77°F, cm[3]	T51	D113	100+	100+	100+	100+	100+
Solubility in carbon tetrachloride, %	T44[4]	D4[4]	99.5+	99.5+	99.5+	99.5+	99.5+
Water, %	T55	D95	0.2−	0.2−	0.2−	0.2−	0.2−
General requirement	The material shall not foam when heated to application temperature recommended by The Asphalt Institute.						

NOTE: When the Heptane-Xylene Equivalent Test is specified by the consumer, a negative test with 35 per cent xylene after 1 hour will be required, AASHO Method T102.

[1] Flash point by Cleveland Open Cup may be used for products having a flash point greater than 175°F.

[2] As an alternate, Saybolt Furol Viscosities may be specified.

[3] If penetration of residue is more than 200 and its ductility at 77°F is less than 100, the material will be acceptable if its ductility at 60°F is 100+.

[4] Except that carbon tetrachloride or trichloroethylene is used instead of carbon disulphide as solvent, Method No. 1 in AASHO Method T44 or Procedure No. 1 in ASTM Method D4.

TABLE 11-6

SPECIFICATIONS FOR SLOW-CURING (SC) LIQUID ASPHALTS
(The Asphalt Institute)

Characteristics	AASHO Test Method	ASTM Test Method	Grades			
			SC-70	SC-250	SC-800	SC-3000
Kinematic viscosity at 140°F, cs[1]	T201	D2170	70–140	250–500	800–1600	3000–6000
Flash point (Cleveland Open Cup), °F	T48	D92	150+	175+	200+	225+
Distillation						
Total distillate to 680°F, % by volume	T78	D402	10–30	4–20	2–12	5−
Float Test on distillation residue at 122°F, sec.	T50	D139	20–100	25–110	50–140	75–200
Asphalt Residue of 100 Penetration, %	T56	D243	50+	60+	70+	80+
Ductility of 100 Penetration Asphalt Residue at 77°F, cm.	T51	D113	100+	100+	100+	100+
Solubility in carbon tetrachloride, %	T44[2]	D4[2]	99.5+	99.5+	99.5+	99.5+
Water, %	T55	D95	0.5−	0.5−	0.5−	0.5−
General requirement	The material shall not foam when heated to application temperature recommended by The Asphalt Institute.					

NOTE: When the Heptane-Xylene Equivalent Test is specified by the consumer, a negative test with 35 per cent xylene after 1 hour will be required, AASHO Method T102.

[1] As an alternate, Saybolt Furol Viscosities may be specified.

[2] Except that carbon tetrachloride or trichloroethylene is used instead of carbon disulphide as solvent, Method No. 1 in AASHO Method T44 or Procedure No. 1 in ASTM Method D4.

TABLE 11-7

SPECIFICATIONS FOR ANIONIC EMULSIFIED ASPHALTS
(The Asphalt Institute)

Characteristics	AASHO Test Method	ASTM Test Method	Grades				
			Rapid Setting		Medium Setting	Slow Setting	
			RS-1	RS-2	MS-2	SS-1	SS-1h
Tests on emulsion							
Furol Viscosity at 77°F, sec.			20–100	–	100+	20–100	20–100
Furol Viscosity at 122°F, sec.			–	75–400	–	–	–
Residue from distillation, % by weight			57+	62+	62+	57+	57+
Settlement, 5 days, % difference	T59	D244	3−	3−	3−	3−	3−
Demulsibility:							
35 ml. of 0.02 N $CaCl_2$, %			60+	50+	–	–	–
50 ml. of 0.10 N $CaCl_2$, %			–	–	30−	–	–
Sieve Test (retained on No. 20), %			0.10−	0.10−	0.10−	0.10−	0.10−
Cement-Mixing Test, %			–	–	–	2.0−	2.0−
Tests on residue							
Penetration, 77°F, 100 g., 5 sec.	T49	D5	100–200	100–200	100–200	100–200	40–90
Solubility in carbon tetrachloride, %	T44[1]	D4[1]	97.5+	97.5+	97.5+	97.5+	97.5+
Ductility, 77°F, cms.	T51	D113	40+	40+	40+	40+	40+

[1] Except that carbon tetrachloride is used instead of carbon disulphide as solvent. Method No. 1 in AASHO Method T44 or Procedure No. 1 in ASTM Method D4.

TABLE 11-8

SPECIFICATIONS FOR CATIONIC EMULSIFIED ASPHALTS
(The Asphalt Institute)

Characteristics	AASHO Test Method	ASTM Test Method	Grades					
			Rapid Setting		Medium Setting		Slow Setting	
			RS-2K	RS-3K	SM-K	CM-K	SS-K	SS-Kh
Tests on emulsion								
Furol Viscosity at 77°F, sec.	T59	D244	–	–	–	–	20–100	20–100
Furol Viscosity at 122°F, sec.	T59	D244	20–100	100–400	50–500	50–500	–	–
Residue from Distillation								
Residue, % by weight	T59	D244	60+	65+	60+	65+	57+	57+
Oil Distillate, % by Volume of Emulsion	T59	D244	5–	5–	20–	12–	–	–
Settlement, 7 days, % difference	T59	D244	3–	3–	3–	3–	3–	3–
Sieve Test (Retained on No. 20), %	T59[1]	D244[1]	0.10–	0.10–	0.10–	0.10–	0.10–	0.10–
Aggregate Coating—Water Resistance Test	–	D244						
Dry Aggregate (Job), % Coated			–	–	80+	80+	–	–
Wet Aggregate (Job), % Coated			–	–	60+	60+	–	–
Cement Mixing Test, %	T59	D244	–	–	–	–	2–	2–
Particle Charge Test	T59A	D244	Positive	Positive	Positive	Positive	–	–
pH	T200	E70	–	–	–	–	6.7–	6.7–
Tests on residue								
Penetration, 77°F, 100 g., 5 sec.	T49	D5	100–250	100–250	100–250	100–250	100–200	40–90
Solubility in Carbon Tetrachloride, %	T44[2]	D4[2]	97.0+	97.0+	97.0+	97.0+	97.0+	97.0+
Ductility, 77°F, cm.	T51	D113	40+	40+	40+	40+	40+	40+

[1] Except that distilled water is used instead of sodium oleate solution.

[2] Except that carbon tetrachloride is used instead of carbon disulphide as solvent, Method No. 1 in AASHO Method T44 or Procedure No. 1 in ASTM Method D4.

Note: a) "K" in grade designations signifies cationic type

b) In Medium Setting Grades—
"SM" indicates sand mixing grade
"CM" indicates coarse aggregate mixing grade

PROBLEMS

11–1. If 12,300 gallons of cut-back asphalt (specific gravity 0.94) are pumped from a railroad tank car at an average temperature of 160°F, determine the quantity upon which payment will be based.

11–2. The unit price of the liquid asphalt of Problem 11–1 is $0.22 per gallon, delivered. Calculate the total price of this quantity of bituminous material. Also calculate the unit price of this material per ton.

TABLE 11–9

Specifications for Tars for Use in Road Construction
(American Association of State Highway Officials, Designation M52)

Grades	RT-1	RT-2	RT-3	RT-4	RT-5	RT-6	RT-7	RT-8	RT-9	RT-10	RT-11	RT-12	RTCB-5	RTCB-6
Consistency:														
Engler Sp. Visc. at 40°C	5–8	8–13	13–22	22–35		–	–	–	–	–	–	–	–	–
Engler Sp. Visc. at 50°C	–	–	–	–	17–26	26–40	–	–	–	–	–	–	17–26	26–40
Float Test at 32°C	–	–	–	–	–	–	50–80	80–120	102–200	–	–	–	–	–
Float Test at 50°C	–	–	–	–	–	–	–	–	–	75–100	100–150	150–220	–	–
Sp. Gr. at 25°C/25°C	1.08+	1.08+	1.09+	1.09+	1.10+	1.10+	1.12+	1.14+	1.14+	1.15+	1.16+	1.16+	1.09+	1.09+
Total Bitumen, % by wt.	88+	88+	88+	88+	83+	83+	78+	78+	78+	75+	75+	75+	80+	80+
Water, % by volume	2.0–	2.0–	2.0–	2.0–	1.5–	1.5–	1.0–	0	0	0	0	0	1.0–	1.0–
Distillation, % by wt.														
To 170°C	7.0–	7.0–	7.0–	5.0–	5.0–	5.0–	3.0–	1.0–	1.0–	1.0–	1.0–	1.0–	2.0–8.0	2.0–8.0
To 200°C	–	–	–	–	–	–	–	–	–	–	–	–	5.0+	5.0+
To 235°C	–	–	–	–	–	–	–	–	–	–	–	–	8.0–18.0	8.0–18.0
To 270°C	35.0–	35.0–	30.0–	30.0–	25.0–	25.0–	20.0–	15.0–	15.0–	10.0–	10.0–	10.0–	–	–
To 300°C	45.0–	45.0–	40.0–	40.0–	35.0–	35.0–	30.0–	25.0–	25.0–	20.0–	20.0–	20.0–	35.0–	35.0–
Softening point of distillation residue, °C	30–60	30–60	35–65	35–65	35–70	35–70	35–70	35–70	35–70	40–70	40–70	40–70	40–70	40–70
Sulfonation index (when specified) on distillate														
To 300°C	8–	7–	6–	6–	5–	5–								
300°C to 500°C	1.5–	1.5–	1.5–	1.5–	1.5–	1.5–								

\+ Sign indicates that value shown is the minimum allowable.
– Sign indicates that value shown is the maximum allowable.

TABLE 11-10

SPRAYING AND MIXING TEMPERATURES

(The Asphalt Institute)

Type	Spraying Temperature, °F		Mixing Temperature, °F	
	Min.	Max.	Min.	Max.
Liquid Asphalts				
SC—MC—RC 70	105	175	90	155
SC—MC—RC 250	140	225	125	200
SC—MC—RC 800	175	255	160	225
SC—MC—RC 3000	215	290	200	260
Penetration asphalts				
60–70, 85–100			250	325
120–150, 200–300			250	275
Road tars				
RT 1, 2	60	125		
RT 3, 4	80	150		
RT 5, 6	80	160	80	160
RT 7, 8, 9	150	225	150	225
RT 10, 11, 12	175	250	175	250
RTCB 5, 6	60	120	60	170

11–3. Determine the average unit price of the following materials, delivered in your city in bulk quantities:

Cut-back asphalt
Asphalt cement
Emulsified asphalt
Road tar

Note: There will be some difference in price with the "grade" of each of these materials, but average figures are desired.

11–4. The following weights are recorded during the determination of the specific gravity of a bituminous material by the pycnometer method:

Weight of pycnometer, empty	34.316 grams
Weight of pycnometer, filled with water	60.000 grams
Weight of pycnometer, filled with bituminous material	57.942 grams
Temperature (all determinations)	25°C

Calculate the specific gravity of this substance. From the material given in the test identify this substance as to its general class of bituminous material.

11–5. The following are the results of laboratory tests performed upon a sample of medium-curing liquid asphalt (MC800) for the purpose of checking compliance with the specifications:

Flash point	108°F
Kinematic viscosity at 140°F	850 centistokes
Distillation	
To 437°F	15.0%
To 500°F	40.5%
To 600°F	85.2%
Residue from distillation to 680°F, by volume	75%
Tests on residue from distillation	
Penetration (standard conditions)	105
Ductility, 77°F	100 cm.
Solubility in CCl_4	98.5%

Does this material meet Asphalt Institute specifications? If your answer is "no," explain. What do you think happened to this material?

11-6. The following results were obtained from an emulsified asphalt. Determine the grade.

Tests on emulsion	
Furol viscosity at 122°F	125 sec.
Residue from distillation, by weight	68%
Settlement 5 days, by difference	2%
Demulsibility	
35 ml. of 0.02 N $CaCl_2$	55%
Tests on residue	
Penetration, 77°F, 100 g., 5 sec.	125
Solubility in carbon tetrachloride	98.5%
Ductility, 77°F	55 cm.

11-7. Given the following results of laboratory tests performed upon a representative sample of a bituminous material:

Semisolid consistency	
Specific gravity	1.165
Solubility in CS_2	78%
Float test at 32°C	95 sec.

Determine the general nature and grade of this bituminous material.

11-8. Consult the "Standard Specifications" in use by the highway department in your state and list the bituminous materials covered by these specifications. Determine the types of road surfaces in which each of these bituminous materials is used.

CHAPTER 12

FLEXIBLE PAVEMENT DESIGN

Generally speaking, pavements (and bases) may be divided into two broad classifications or types, i.e., rigid and flexible. As commonly used in the United States, the term "rigid pavement" is applied to wearing surfaces constructed of portland cement concrete. A pavement constructed of concrete is assumed to possess considerable flexural strength which will permit it to act as a beam and allow it to bridge over minor irregularities which may occur in the base or subgrade upon which it rests; hence, the term "rigid." Similarly, a concrete base which supports a brick or block pavement might be described as "rigid." The structural design of this type of pavement deserves special consideration and is discussed in detail in Chapter 18.

12–1. General Discussion of Flexible Pavements. All other types of pavements and bases are classed as "flexible." In Fig. 12–1 is shown

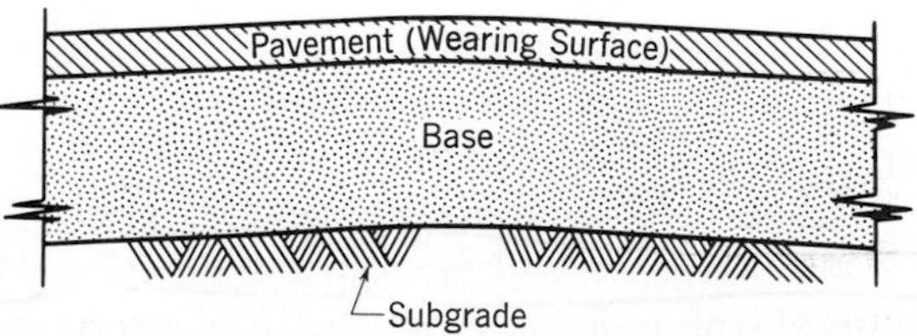

Fig. 12–1. Simplified flexible pavement structure.

the usual arrangement of the structural elements of a flexible pavement. The principal elements of this structure are shown to be the pavement or "wearing surface," base, and subgrade. Both the wearing surface and the base often comprise two or more layers that are somewhat different in composition and that are put down in separate construction operations. In addition on many heavy-duty pavements there is one more layer, generally of clean, granular material, between the base and subgrade; this layer is called the "subbase." It is assumed that any distortion or displacement of the lower layers will eventually be reflected in the wearing surface. Thus distortion occurring in the subgrade will be reflected in the base, and distortions in the base will be reflected in the wearing surface. Hence, the term "flexible" is used to denote the tendency of all layers in this type of structure to conform to the same shape under traffic.

In a typical case, the uppermost layer of this type of roadway is some kind of bituminous wearing surface and is comparatively thin. The wearing surface may range in thickness from less than 1 inch, in the case of a bituminous surface treatment used for a low-cost, light-traffic road, to the 3 inches or more of bituminous concrete used for heavily traveled routes. The wearing surface must be capable of withstanding the wear and abrasive effects of moving vehicles and must possess sufficient stability to prevent it from displacing under traffic loads. In addition, it serves a useful purpose in preventing the entrance of excessive quantities of surface water into the base and subgrade from directly above.

The base is a layer (or layers) of very high stability and density. Its principal purpose is to distribute or "spread" the stresses created by wheel loads acting on the wearing surface so that the stresses transmitted to the subgrade will not be sufficiently great to result in excessive deformation or the displacement of that foundation layer. The base must also be of such character that it is not damaged by capillary water, thus reducing the possibility of damaging frost action or other detrimental effects accompanying the flow of capillary water. Locally available materials are extensively used for base construction, and materials preferred for this type of construction vary widely in different sections of the country. For example, the base may be composed of gravel or crushed rock; it may be a granular stabilized soil mixture, a bituminous mixture, or any one of a large number of suitable base materials.

A subbase may be used in areas where frost action is severe or in locations where the subgrade soil is extremely weak. It may also be used in the interests of economy, in locations where suitable subbase materials are cheaper than base materials of higher quality.

The subgrade is the foundation layer, the structure which must eventually support all the loads which come onto the pavement. In some cases this layer will simply be the natural earth surface. In other and more usual instances it will be the compacted soil existing in a cut section or the upper layer of an embankment section. In the fundamental concept of the action of flexible pavements, the combined thickness of subbase (if used), base, and wearing surface must be great enough to reduce the stresses occurring in the subgrade to values which are not sufficiently great to cause excessive distortion or displacement of the subgrade soil layer.

12–2. Status of Thickness Design. Before World War II, the determination of the combined thickness of flexible pavement and base required in any set of circumstances was largely a matter of judgment and experience. This is attributed (*156*) to the fact that (1) this method appeared to be giving satisfactory results, (2) much of the basic knowledge required for a more scientific approach to the problem was not available, and (3) the methods of stage construction then in wide use did not seem to require the evolution of more scientific design methods. However, in the period immediately prior to the war and during its early stages, the necessity for economically designing and constructing a truly enormous mileage of airport runways and access roads to military installations focused attention upon this design problem. Efforts directed toward the solution of the problem of economical design of flexible pavements and bases were further emphasized by the increased cost of road building during and after the war years.

As a consequence, many governmental agencies entered upon comprehensive investigations that led to development of a large number of different design methods. There was (and is) no universal agreement among practicing engineers as to which of these is the best.

The start of construction of the Interstate System in 1956 greatly accelerated efforts to improve pavement design methods. The AASHO road test (Art. 12–7) was the most extensive field test in roadbuilding history. One major phase of the road test, which was completed in 1961, concerned itself with flexible pavements. Since that time, the results of the road test have been distributed widely; they have had a profound influence on flexible pavement design procedures. A great deal of work is still under way to extend the findings of the test to other subgrade conditions, pavement materials, traffic loads, and climatic conditions.

For many years, some engineers have wanted to design flexible pavements on an analytical basis, i.e., by applying the theory of elasticity or similar concepts to the behavior of the layered surface that is a flexible pavement. They have been hampered by the complexity of the problem, and by the lack of certain parameters to fit into the design equations. At present, aided by the accumulation of vast amounts of experimental data, engineers in various parts of the world seem to be hovering on the edge of a breakthrough in this approach to pavement design. Analytical methods of design are discussed briefly in Art. 12–19.

12–3. Elements of Thickness Design. From a somewhat simplified viewpoint, the following are the principal factors entering into the problem of the thickness design of flexible pavements:

1. Magnitude, method of application, and number of wheel loads.
2. Function of pavement and base in transmitting the load to the subgrade.
3. Measurement of subgrade supporting power.

A number of variables influence each of the principal factors that are listed above, and the student should realize that the procedure of evaluating these variables is by no means a simple task. In fact, it seems unlikely that any extremely simple method of approach will prove entirely successful.

12–4. Considerations Relative to Wheel Loads. In evaluating the magnitude of the wheel load to be selected as a design criterion, consideration must be given to maximum legal load limits prevailing in the state concerned, to any overloads which may be expected to occur, and to the results of traffic surveys made on routes similar to the one for which the design is being prepared. The maximum legal wheel load in many states is 9000 pounds, and the results of traffic surveys and loadometer studies should show the magnitude and occurrence of heavy wheel loads to be considered in the design procedure.

The wheel load does not act as a point load; rather, it is distributed in some fashion over the area of contact between the tire and the pavement. The load is frequently assumed to be spread uniformly over a circular or elliptical area of contact. Tire contact pressures at the surface of the pavement may be calculated from information supplied by tire manufacturers or may be estimated from actual observations. The heavier wheel loads with which the designer is concerned are often carried on dual tires, and consideration must be given to this fact in evaluating the effect of wheel loads on the pavement structure.

As indicated above, it seems likely that the designer could select the maximum wheel load to which the pavement would be subjected, and make some reasonable assumption as to the area over which this load was distributed. This procedure should prove entirely satisfactory if a single static load is chosen for design purposes. However, there is a growing trend on the part of designers to recognize the fact that they are in reality dealing with a large number of repeated load applications of variable magnitude. With this in mind, most modern design methods include an estimate of the number and type of wheel loads which may be expected to occur on a given pavement section during its service life.

12–5. Load-Transmitting Function of Pavement and Base. As has been previously mentioned, the chief function of the base (and subbase) is to distribute the wheel load to the subgrade. The pavement, or wearing surface, serves the same function. If it is quite thin its load-distributing qualities often are not considered separately but are considered to be similar to those exhibited by the base material. The surface, base, and subbase must be stable enough to resist stresses imposed upon them.

The matter of distribution of stress or "pressure" through the various layers to the subgrade has been a subject of considerable debate among pavement designers for many years. This portion of the design procedure may be approached by recognizing that the load is transmitted in such fashion that the unit pressure on any plane beneath the area of contact of the load is less than the contact pressure. Also, the unit pressure on any horizontal plane decreases with an increase in depth of the plane under consideration. Thus the distribution of pressure is assumed to be due to an increase in loaded area with depth.

In early attempts to develop a simplified design procedure, it was assumed that the stressed area increased uniformly with depth. However this concept does not agree with theoretical pressure distributions developed by application of the theory of elasticity. Certainly it seems to be an oversimplification of the situation to assume that the pressure distribution is the same in all bases, regardless of their type or condition. Factors which may affect the distribution of pressure include density, moisture content, gradation, and the shape of particles.

More recent investigations have led some designers to advocate various theories of curved-line distribution of pressure with depth. Some research studies have shown that all base materials do not function the same way in transmitting load, and lead to the proposition that this property of bases might well be evaluated for each individual type of base proposed for use in a given case.

12–6. Measurement of Subgrade Supporting Power. The final and most important element in the simplified design concepts which have been discussed thus far lies in the evaluation of the subgrade supporting power. More properly, this step in the procedure involves the measurement of the shearing resistance of the subgrade soil.

Design methods presently employed by different agencies vary widely in their approach to the problem of evaluating this property of the subgrade soil. No general agreement exists among investigators in this field as to how this quantity should be determined. Generally speaking, however, the approach most frequently used is to measure the shearing resistance of the subgrade soil by employing some type of shear test performed on the subgrade soil in its natural condition or in the condition in which it is expected to exist in the field after completion of the construction process. This concept is further expanded by certain investigators who agree that the criterion to be applied in evaluating subgrade soils for this purpose should involve the measurement of shearing resistance at small deformations.

Various methods are in current use to accomplish this purpose. One method which has been used involves the use of large circular plates loaded to produce certain critical values of the deformation in the subgrade soil. This procedure is called the "plate-bearing test" and is termed a "direct" measurement of subgrade supporting power. Other design methods place reliance upon various types of shear tests performed upon the soil in place, or in the laboratory, to obtain "indirect" measurements of subgrade supporting power. Tests most often used are the California Bearing Ratio test, triaxial test, and determinations by the use of the Hveem stabilometer. Still other methods do not require the performance of any shear test but attempt to correlate the results of routine laboratory tests with pavement performance to estimate the required thickness of pavement and base. Some design methods rely almost completely upon properties of the subgrade to determine thickness, while others extend this concept to the other layers of the pavement structure, evaluating the shearing resistance of intermediate layers which may exist between the wearing surface and the foundation soil.

It must be emphasized that any test performed for the purpose of evaluating the shearing resistance of a subgrade soil must be made with the soil in a condition similar to that which it is anticipated will exist in the field. Requirements relative to density and moisture content seem to be of special importance. Some design methods involve testing of the soil in the most unfavorable condition imaginable, which usually involves saturation of the soil concerned. Still others do not require saturation, but do require that the density and moisture content of the soil be adjusted to fit service conditions.

12–7. The AASHO Road Test. The AASHO road test was a $27 million project undertaken cooperatively by 49 of the states, the District of Columbia, Puerto Rico, the Bureau of Public Roads, and various industry groups; the test was administered by the Highway Research Board.

Construction began in April, 1956, and test traffic began in October, 1958. Except for some special tests, traffic ceased on the test road in November, 1960.

The test road was located near Ottawa, Illinois, about 80 miles southwest of Chicago. Major portions of the test dealt with flexible pavements, rigid pavements, and short-span bridges. Only one subgrade soil, an A–6, was used in the pavement tests.

The test sections comprised four major loops and two smaller ones. Each loop was a segment of a four-lane divided highway with the parallel roadways connected by a turnaround at each end of the loop; tangent sections of the large loops were 6,800 feet long.

Each tangent was built as a succession of pavement sections arranged in such a way that pavement design could be varied from one section to another. The minimum length of any section in the main loops was 100 feet.

In the principal flexible pavement test sections, the surface course was bituminous concrete; the base course, a well-graded crushed limestone; and the subbase, a uniformly graded sand-gravel mixture.

Major design factors in the principal experiments were surface, base, and subbase thicknesses. In the main factorial experiments, three levels of surface thickness existed in combination with three levels of base thickness; each of these nine combinations existed in combination with three levels of subbase thickness. Over the test road as a whole, surface thickness ranged from 1 to 6 inches; base thickness, from 0 to 9 inches; and subbase thickness, from 0 to 16 inches.

Special experiments on the flexible pavements included three other design variables—base type, bituminous surface treatment, and shoulder paving. Besides the crushed limestone, bases included a well-graded uncrushed gravel, a bituminous plant mixture, and a cement-treated aggregate.

Test traffic included both single- and tandem-axle vehicles, with ten different axle arrangement–axle load combinations. Single-axle loads ranged from 2000 to 30,000 pounds; tandem-axle loads, from 24,000 to 48,000 pounds. Each pavement section was tested with one of the ten combinations; each section was subjected to thousands of load repetitions before it was taken out of the test.

Hundreds of thousands of "bits" of data were gathered by investigators during the road test. These included observations and measurements of pavement condition (cracking and patching done to keep the section in service), longitudinal profile, and transverse profile (to determine rutting and other transverse distortions). Other measurements included surface deflection under loaded vehicles moving at creep speed, deflections at different levels in the pavement structure under vehicles operated at various speeds, curvature of the pavement under vehicles operated at various speeds, pressures transmitted to the surface of the subgrade soil, and temperature distribution in the pavement layers.

Data gathered in the test were subjected to exhaustive analysis by the road test staff and by other interested organizations.

12–8. The Pavement Serviceability Concept. One of the products of the AASHO road test was the "pavement serviceability concept." In essence, this involves the measurement, in numerical terms, of the behavior of the pavement under traffic, its ability to serve traffic at some instant during its life (*2*).

Such an evaluation can be made on the basis of a systematic, but subjective, rating of the riding surface by individuals who travel over it. Or pavement serviceability can be evaluated by means of certain measurements made on the surface, as was done at the test road.

For flexible pavements, researchers at the test road established that the serviceability index (p) at any time is a function of slope variance in the two wheelpaths, the extent and type of cracking (and patching) of the pavement, and the rutting of the surface. The slope variance is an expression of variations in the longitudinal profile, or longitudinal roughness.

At the start of the test, researchers at the test road determined the average initial serviceability index to be 4.2 for the flexible pavements. In general, the value of the index declined gradually under traffic. When the serviceability index (as measured every two weeks) dropped to 1.5 for any section, that section was taken out of test. A value of p of,

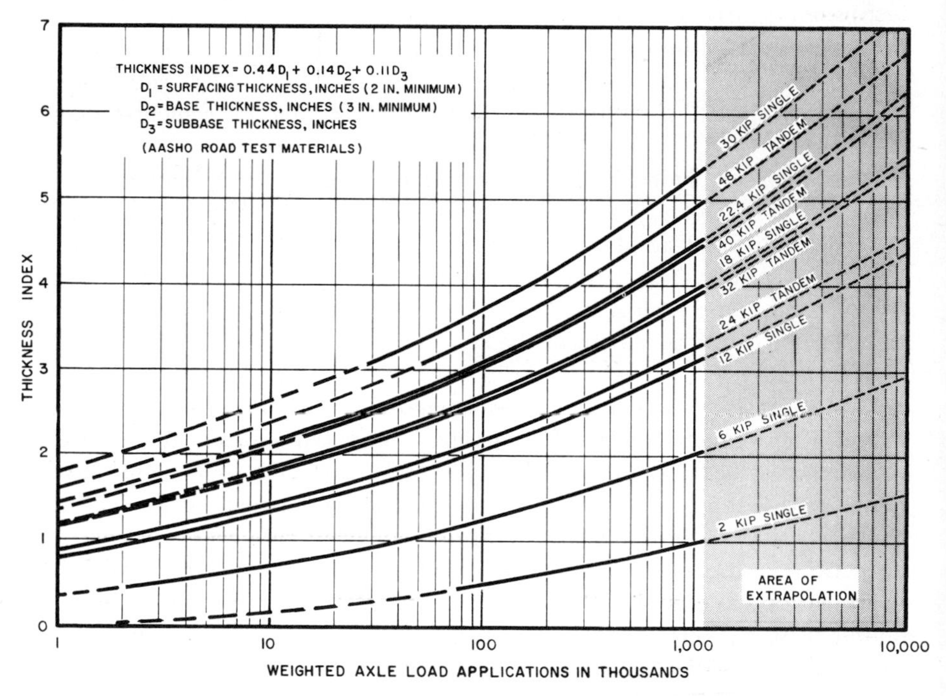

FIG. 12–2. AASHO road test relationship between thickness index and axle loads at $p = 2.5$. (Courtesy Highway Research Board.)

say, 2.5 is an intermediate one between initial construction and failure to render adequate traffic service.

Analysis of relationships among the flexible pavement variables considered in the test resulted in a series of equations, none of which are given here. Fig. 12–2 is one way of showing the relationships among the thickness index, axle load, and number of load applications. The thickness index expresses the relative effectiveness of each layer in the pavement structure (see Fig. 12–2).

As an example of the use of this chart, assume that it is desired to determine a pavement structure that would survive 1 million applications of a 22,400-pound single-axle load before its serviceability index dropped to 2.5. From Fig. 12–2, the thickness index is about 4.5. Many combinations of asphaltic concrete surface, base, and subbase will meet the conditions of the thickness index equation. One such combination is 4 inches of asphaltic concrete, 10 inches of base, and 12 inches of subbase. Fig. 12–2 is not intended for use in design, but it can serve as a basis for the development of design procedures such as the one explained in Art. 12–9.

Other important conclusions about flexible pavements from the road test were

1. One inch of surfacing was about three times as effective as an inch of granular base and about four times as effective as an inch of subbase in improving pavement performance in the range studied.
2. Four types of granular base were studied—crushed stone, gravel, cement-treated gravel, and bituminous-treated gravel. Researchers found the performance of the treated gravel base to be "definitely superior" to that of the untreated crushed stone. Most of the untreated gravel bases failed early in the test.
3. Asphaltic concrete overlays were very effective in extending the service life of flexible pavement sections.

12–9. AASHO Design Chart. Coincident with the release of results of the road test, the AASHO Operating Committee on Design released recommendations for design methods for both flexible and rigid pavements (*248*).

The AASHO Committee accepted the pavement serviceability concept (Art. 12–8) and the basic equations developed at the test road without modification. To make the procedure practically useful, however, it had to extend the equations to cover mixed traffic and locations where soils, materials, and climate are different from those at the test site.

One result of the Committee's work was the design chart of Fig. 12–3, which applies to primary highways with an acceptable value of the pavement serviceability index (p) of 2.5 after 20 years of service. The chart incorporates scales for regional factors (R), equivalent daily 18-kip single-axle load applications, soil support value (S), structural number ($\overline{SN}$), and weighted structural number (SN). Important facts relative to each of these quantities are summarized below.

1. *Regional factor.* The regional factor recognizes that the load-carrying ability of a pavement is a function of its environment; e.g., greater damage may be inflicted by traffic during the spring break-up, when the subgrade soil is saturated, than at other times. The Committee suggested factors for weighting traffic applications at different seasons

of the year to take into account general variations of this type in the strength of subgrade soils.

In the plot of Fig. 12-3, a regional factor of 1.0 would have no effect on the structural number. A value above 1.0 would increase the structural number, i.e., call for a thicker pavement for the same values of soil support and traffic applications. A regional factor of less than 1.0

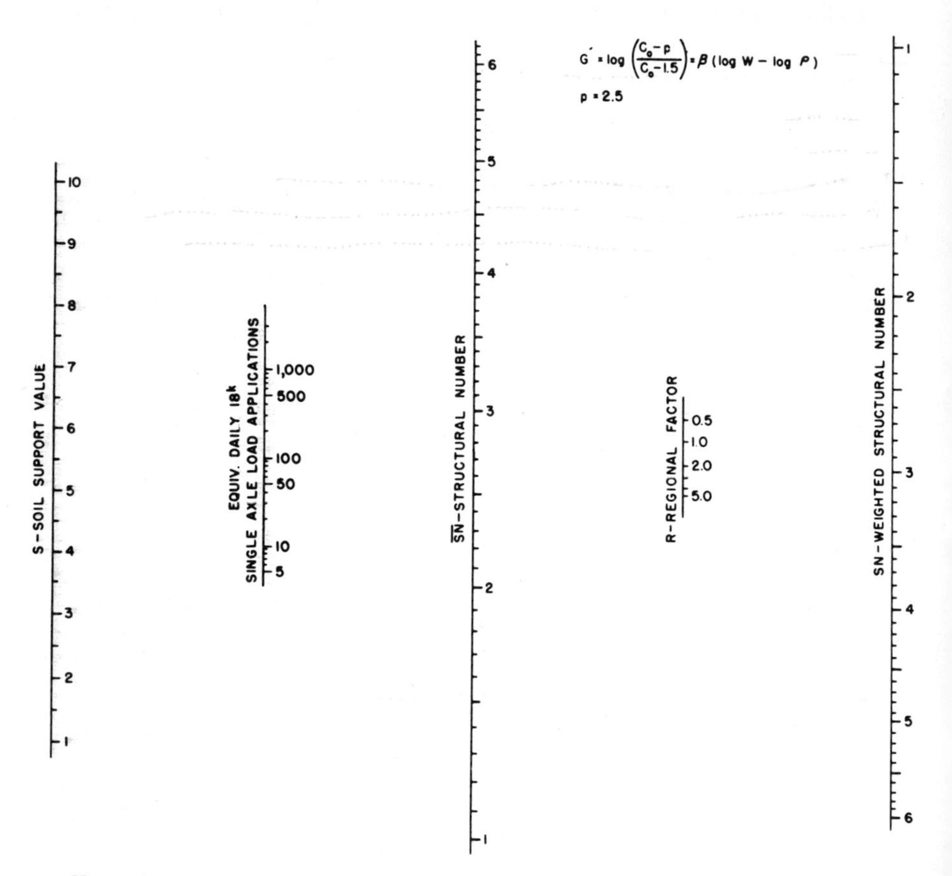

FIG. 12-3. AASHO flexible pavement design chart (20-year traffic analysis). (After W. J. Liddle, Courtesy University of Michigan and International Conference on the Structural Design of Asphalt Pavement.)

is a favorable condition, requiring a thinner pavement structure. The Committee suggested that values could range from 0.5 to 3.0 for the continental United States.

The regional factor can also take into account such unfavorable conditions as a high water table, steep grades, and areas of concentrated turning or stopping movements. Adjustment of the factor for these conditions is a matter of judgement.

2. *Load applications.* The Committee handled the problem of mixed traffic by first adopting an 18,000-pound single axle load as a standard and then developing a series of "equivalence factors" for each axle weight group. If the estimated traffic to be used in design can be broken down into axle load groupings, the number of load applications in each group can be multiplied by the equivalence factor to determine the number of 18,000-pound axle loads that would have an equivalent effect on the pavement structure (if desired, some axle load group other than 18,000 pounds could be chosen as a standard).

The traffic scale in Fig. 12–3 is expressed as daily 18-kip single-axle load applications. The Committee tabulated equivalence factors for a range of axle loads, structural numbers, and p values.

3. *Soil support value.* The soil support value scale shown in Fig. 12–3 was established by experience and two numerical values established at the test road–3.0 for the subgrade there and 10.0 for the crushed stone base course.

If one knows the soil support value, the equivalent traffic, and the regional factor, one can find a structural number from Fig. 12–3. The procedure recommended by the Committee for determining the thicknesses of the various layers in a flexible pavement from the structural number is the same as that explained in Art. 12–8, using the equation shown in Fig. 12–2; the "structural number" is the "thickness index" of Fig. 12–2.

12–10. The Asphalt Institute Method. Engineers of the Asphalt Institute observed the AASHO road test and served in an advisory capacity, as did engineers of the Portland Cement Association on the rigid pavement sections. After the test was over, Asphalt Institute engineers and mathematicians did an independent analysis of the data secured in the test.

Their analysis produced a revised method of design for asphalt pavements, which was made readily available to practicing engineers in 1963 (*123*). The published method is intended for use under a wide range of conditions, hence it is somewhat less precise than some methods developed by local agencies for use in their own particular situations.

The Asphalt Institute method relies heavily on the AASHO road test, earlier road tests, and accumulated experience of many years. Basic elements of the design are traffic analysis and evaluation of subgrade, subbase, and base materials. An economic comparison of alternatives should be a part of pavement design; methods of economic analysis given in Chapter 3 should be used, although, in practice, the choice often is made on a first-cost basis.

The Asphalt Institute method employs a "design traffic number," which is the number of equivalent 18,000 pound single-axle load applications expected for the design lane during the design period. An equivalent 18,000-pound single-axle load is defined by equating the effect on pavement performance of one or more axle loads of any magnitude to the number of 18,000-pound single-axle loads required to produce the same effect. The design lane is either lane of a two-lane road, generally the outside lane of a multilane highway.

The design period is the number of years until the first major resurfacing is anticipated. This period often is taken as 20 years, but procedures are available for using other design periods. The end of the design period is the estimated time when the serviceability index (p) will reach 2.5.

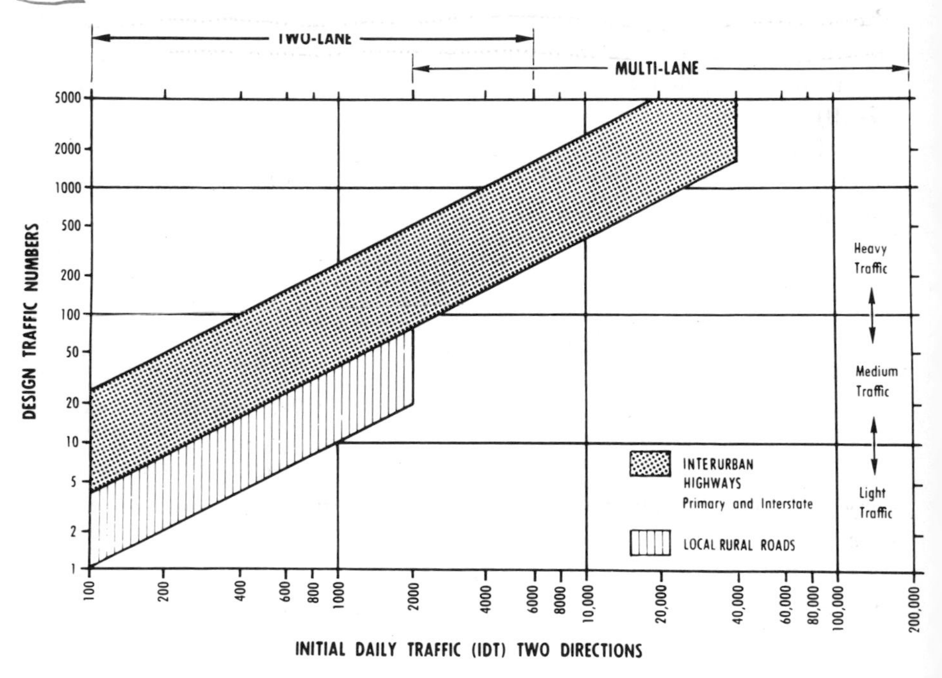

Fig. 12–4. Traffic analysis chart for flexible pavement design. (Courtesy the Asphalt Institute.)

The Asphalt Institute design manual contains procedures for determining the design traffic number from loadometer surveys, traffic forecasts, and load equivalence factors. However, the design traffic number may be approximated from Fig. 12–4, which is based on a design period of 20 years and an anticipated uniform growth of traffic during the design period of about 80 per cent of the initial daily traffic. If the design traffic number is less than 10, traffic is termed "light"; if between 10 and 100, "medium"; if above 100, "heavy."

With the estimated initial daily traffic known, including all vehicles, enter the chart of Fig. 12–4 at that point on the horizontal scale, then move vertically to the band representing the class of highway involved. Judgment is necessary in deciding what portion of the diagonal band to use; the primary consideration is the relative proportion of heavy trucks. After this decision is made, the selected point is projected horizontally to the vertical scale and the design traffic number is read.

For medium and heavy traffic conditions, the Asphalt Institute design method is based upon a measurement of the strength of the subgrade soil in the laboratory or in the field, with the material in a suitable condition as regards compaction, water content, and saturation. Any three measures of soil strength may be used—the California Bearing Ratio (Art. 12–11); the Bearing Value (psi) as determined by a plate bearing test with a 12-inch-diameter plate, 0.2-inch deflection, and 10 repetitions of load (Art. 12–16); or the Resistance Value (see Art. 12–19).

If feasible, the same tests should be used to design rural roads and city streets that carry light traffic. If such a procedure is not feasible, the approximate relationships among groups of the AASHO soil classification system (see Art. 6–19) or the Unified Soil Classification System (see Art. 6–20) and the California Bearing Ratio given in Fig. 12–5 may be used.

Fig. 12–6 is one of two basic design charts contained in the Asphalt Institute manual (the other is based on the Resistance Value). The chart gives required thickness (T_A) of a pavement structure composed entirely of asphaltic concrete.

Alternate designs incorporating granular base and subbase courses may be derived from the chart by using appropriate "substitution factors." Recommended substitution ratios are 2:1 for granular base (2 inches of granular base for 1 inch of asphaltic concrete), 2.7:1 for granular subbase and asphaltic concrete, and 1.35:1 for granular subbase and granular base.

The "A"-line on the chart establishes the minimum thickness of asphaltic concrete surface and base courses. The "B"-line determines if subbase can be used, and it also is useful in determining required thicknesses of base and subbase.

Minimum recommended thickness of asphaltic concrete surface for light traffic is 1 inch; for medium traffic, 1½ inches; and, for heavy traffic, 2 inches. From a construction standpoint, the minimum practical thickness of a granular base or subbase is 3 inches.

To determine T_A from Fig. 12–6, locate the soil strength on the horizontal scale. Draw a line vertically from this point to the diagonal line representing the design traffic number, interpolating as necessary.

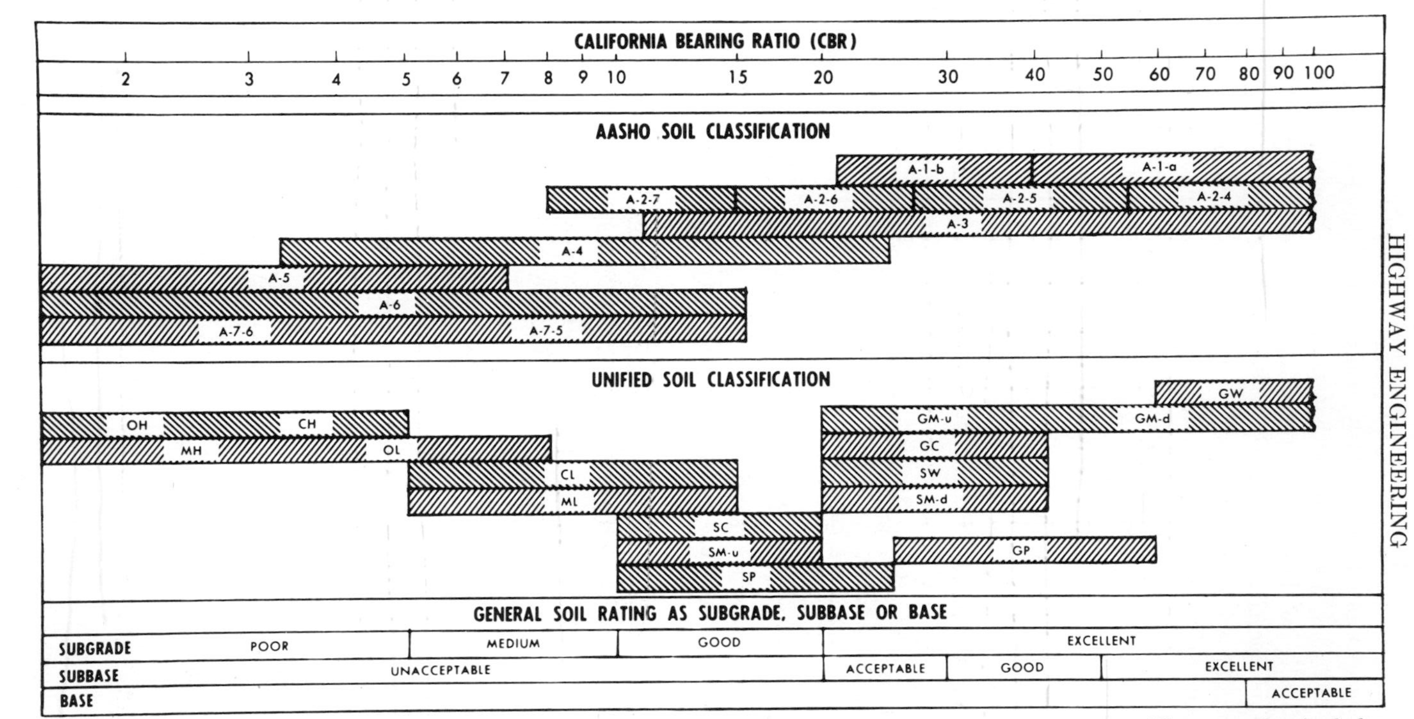

FIG. 12–5. Approximate correlation of soil ratings for use with Fig. 12–6 in design of light pavements. (Courtesy the Asphalt Institute.)

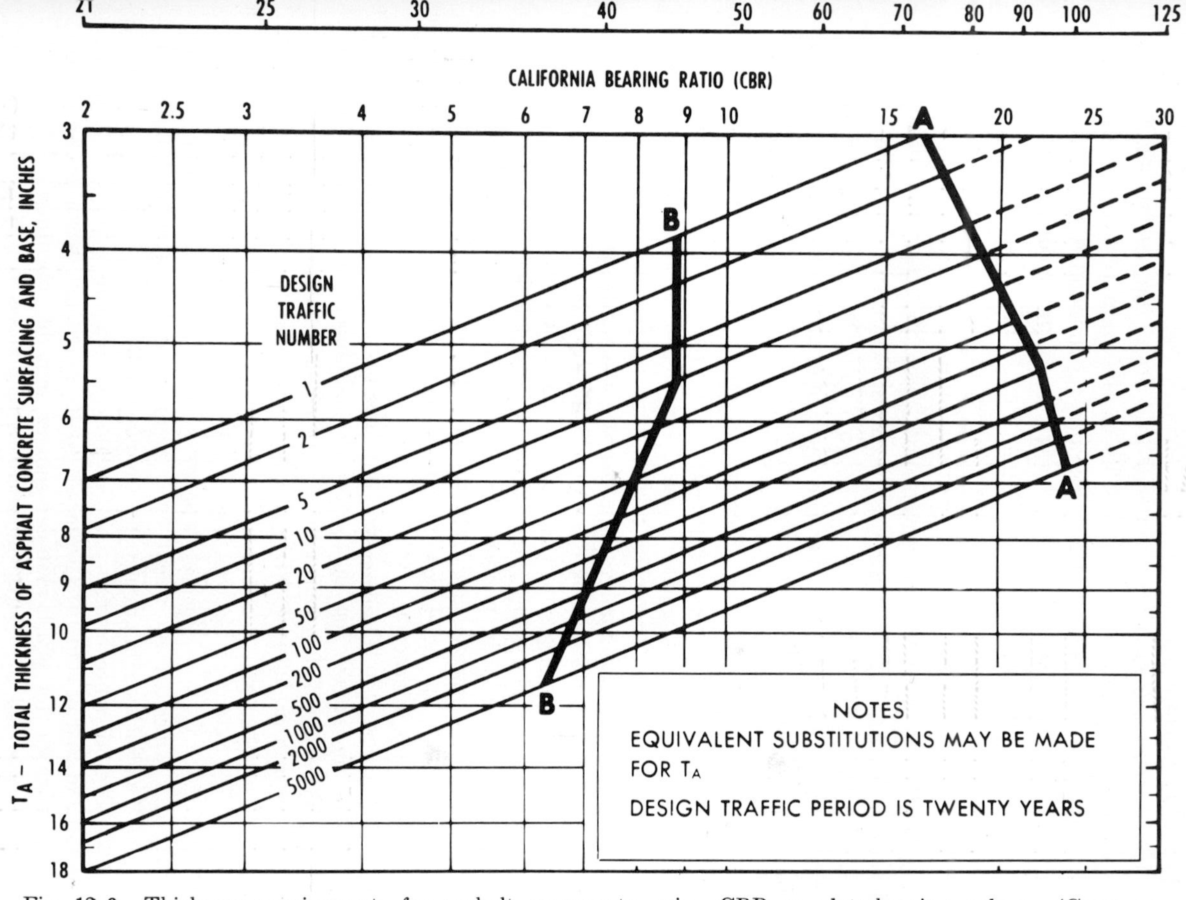

Fig. 12–6. Thickness requirements for asphalt pavements, using CBR or plate-bearing values. (Courtesy the Asphalt Institute.)

The next step is to draw a horizontal line from this point to the T_A scale, reading the required total thickness of asphaltic concrete to the next highest half-inch.

If the intersection of the vertical line from the soil strength value with the traffic number diagonal falls to the right of the "A"-line, the thickness will be less than the minimum recommended by the Asphalt Institute. In this case, the value of T_A to be used corresponds to the intersection of the traffic number diagonal with the "A"-line.

For many pavements, some combination of asphaltic concrete surface and base with granular base and granular subbase will provide a satisfactory pavement structure at lower cost than an all-asphaltic concrete design. The Asphalt Institute design manual covers in detail the various possible general combinations. The following is the design procedure, with a numerical example, for the most complicated case. Successive steps in the design are

(1) Assume a section of interurban highway has an estimated initial traffic of 8,000 vehicles per day. The proportion of truck traffic in relation to total traffic volume is expected to be about normal for this class of road. From Fig. 12–4, reading the central portion of the interurban highway band, the approximate value of the traffic number is 500.
(2) The California Bearing Ratio of the subgrade soil is 4.
(3) Using the values of (1) and (2), the required T_A, is found (Fig. 12–6) to be between 11 and 12 inches. Select 12 inches.
(4) To determine the minimum permissible thickness of asphaltic concrete, draw a horizontal line from the intersection of the 500-traffic-number line and the "A"-line to the T_A scale at the left of the chart. This value is 6 inches.
(5) The next step is to determine the value on the T_A scale that corresponds to the intersection of the 500-traffic-number line and the "B"-line. This value is 9.5 inches. It represents the minimum permissible thickness of asphaltic concrete if full depth asphaltic concrete surface and base are used
(6) The difference between the values obtained in steps (4) and (5) [9.5 − 6.0 = 3.5 inches] represents the thickness of asphaltic concrete for which an equivalent layer of granular base may be substituted. The thickness of granular base required is 2(3.5) = 7 inches.
(7) The difference between the thickness obtained in step (3) [12 inches] and that of step (5) [9.5 inches] is the maximum thickness of asphaltic concrete for which granular subbase may be substituted. This value is 2.5, and the equivalent thickness of subbase is 2.7(2.5) = 7 inches.
(8) Thus, the recommended pavement structure has a total depth of 20 inches: 2 inches of asphaltic concrete surface, 4 inches of asphaltic concrete base, 7 inches of granular base, and 7 inches of granular subbase. All layers are assumed to meet Asphalt Institute requirements for composition and placement.

Other layer thicknesses may be used, provided recommendations relative to minimum thickness of asphaltic concrete and maximum thickness of subbase are met.

12–11. California Bearing Ratio Test. The basic testing procedure employed in the determination of the California Bearing Ratio was developed by the California Division of Highways before World War II and was used by that agency in the design of flexible pavements. The basic procedures of this test were adopted by the Corps of Engineers of the United States Army during the early stages of the war, and served as a basis for the development of design curves which were used for determining the required thickness of flexible pavements for airport runways and taxiways. Certain modifications were made in the test procedure which had been used in California. The modified method adopted by the Corps of Engineers has come to be regarded as the standard method of determining the California Bearing Ratio. It will be briefly described. The description here given will be further limited to that pertaining to the testing of disturbed samples in the laboratory.

The selected sample of subgrade soil is compacted in a mold which is 6 inches in diameter and from 6 to 7 inches high. The moisture content, density, and compactive effort used in molding the sample are selected to correspond to expected field conditions. After the sample has been compacted, a surcharge weight equivalent to the estimated weight of pavement and base is placed on the sample, and the entire assembly is immersed in water for four days. At the completion of this soaking period the sample is removed from the water and allowed to drain for a period of 15 minutes. The sample, with the same surcharge imposed upon it, is immediately subjected to penetration by a piston 1.95 inches in diameter moving at a speed of 0.05 inch per minute. The total loads corresponding to penetrations of 0.1, 0.2, 0.3, 0.4, and 0.5 inch are recorded.

A load-penetration curve is then drawn, any necessary corrections made, and the corrected value of the unit load corresponding to 0.1 inch penetration determined. This value is then compared with a value of 1000 pounds per square inch required to effect the same penetration in standard crushed rock. The California Bearing Ratio (C.B.R.) may then be calculated by the expression:

$$\text{C.B.R. (\%)} = \frac{\text{unit load at 0.1-inch penetration}}{1000}(100) \qquad (12\text{–}1)$$

In Fig. 12–7 is shown the apparatus required for the conduct of the California Bearing Ratio test. It should be noted that this penetration shear test may also be performed in the field or upon "undisturbed"

Fig. 12–7. Apparatus for California Bearing Ratio test. (Courtesy Soiltest, Inc.)

samples. In certain cases the bearing ratio is computed at 0.2-inch penetration rather than at the standard 0.1 inch. The value for standard crushed rock at 0.2-inch penetration is 1500 pounds per square inch. Various modifications of the basic test procedure have been adopted by highway agencies which use the C.B.R. method of design.

12–12. California Bearing Ratio Method (Virginia). The Virginia Department of Highways has used a flexible pavement design method in which the strength of the subgrade soil is measured by a modified California Bearing Ratio test. Tables 12–1 and 12–2 are the basic charts used in the design process (*26*).

Virginia uses the AASHO method for determining anticipated daily 18,000-pound equivalent axle loads. The traffic analysis is based on a

TABLE 12–1

FLEXIBLE PAVEMENT DESIGN CHART, PRIMARY AND INTERSTATE SYSTEM, VIRGINIA DEPT. OF HIGHWAYS

Traffic Category	Daily Equivalent 18-Kip Axle Loads, 1 Direction-Design Lane (a)	*Subgrade Stabilization (b)	*Subbase Thickness	Base Thickness
I	0–9	6″	None	4–6″(c)
IA	10–19	6″	None	6–8″(c)
II	20–149	6″	4–8″	3″B-3(d)
IIA	150–269	6″	4–8″	3″B-3
III	270–399	6″	4–8″	5½″B-3
IV	400–519	6″	6–10″	5½″B-3
V	520–799	6–12″	6–10″	7½″B-3
VI	800 and over	6–12″	8–12″	7½″B-3

(a) Estimated daily number of equivalent 18-kip axle loads anticipated in one direction of travel in the heaviest traveled lane (each lane of single lane or outside lane of dual lane road) during the mean year. Traffic analysis based on a 20-year analysis period with a 10-year mean period.

(b) Where subgrade stabilization is not feasible, an equal or greater thickness of select material is required.

(c) Aggregate Base Course Material.

(d) B-3 designates an untreated crushed stone base.

* Minimum depth of subgrade stabilization and subbase will depend on soil conditions and design C.B.R. value.

20-year period following completion of the pavement, with a 10-year mean traffic figure used for design.

Table 12–1 is used to select the depths of the various layers of the pavement; fixed values of base, binder and surface depths, and variable depths of subbase and subgrade stabilization, depending upon the design C.B.R. of the average soil on the project. Depths of binder and surface courses are not shown in Table 12–1; they range from 0 (thin surface treatments) to 1½ inches, depending on the traffic and the section of the state in which the road is located. For heavy traffic, binder course thickness is 1½ inches and surface course, 1 inch.

Table 12–2, which was compiled from standard wheel load–C.B.R.

curves, is used to select total pavement depth for each traffic category and corresponding C.B.R.

To determine the minimum thickness of pavement required, the designer must use both charts. Table 12–2 gives the total depth of pavement required, using a "design C.B.R.," which is ⅔ of the average C.B.R. of the average soil on the job, and the traffic category. From Table 12–1, the designer can then choose a design that will correspond. Since the depth of base, binder course and surface course are fixed for each

TABLE 12–2

FLEXIBLE PAVEMENT DESIGN CHART, VIRGINIA DEPARTMENT OF HIGHWAYS, TOTAL DEPTH IN INCHES

C.B.R. Value of Soil	Traffic Category					
	I	1A-II	IIA-III	IV	V	VI
1	36	40	44	52	60	70
2	25	28	31	37	43	50
3	20	22	25	28	32	36
4	17	19	21	24	27	30
5	15	17	19	21	23	25
6	14	16	17	19	21	23
7	13	15	16	17	19	21
8	12	14	15	16	18	20
9	11	13	14	15	17	19
10	10	12	13	14	16	18
12	9	11	12	13	14	16
14	8	10	11	12	13	14
16	7	9	10	11	12	13
18	7	8	9	10	11	12
20	6	7	8	9	10	11
25	6	7	8	9	10	11
30	4	6	7	8	9	10
35	4	6	7	8	9	10
40	4	6	7	8	9	10
45	4	6	7	8	9	10
50	3	6	7	8	9	10

traffic category and section of the state, he must choose a depth of subgrade stabilization and a subbase thickness that will give the total minimum required thickness determined from Table 12–2. The specified depths of the various layers are based primarily on experience.

Equivalences of bituminous concrete base versus untreated crushed stone (1.5 to 1) and cement-treated base versus crushed stone (1.7 to 1) are used with engineering judgment.

12–13. General Observations Relative to the C.B.R. Method. It will be seen that the design procedure utilizing the C.B.R. test method is an entirely empirical one. Aside from the fact that the C.B.R. determination is a type of shear test and does give the designer a measure of the shearing resistance of a soil mass at a low value of deformation, this design method bears no easily discernible relationship to the basic elements of the problem as outlined in the early pages of this chapter. However, such a procedure should prove to be perfectly satisfactory if the variables affecting the stability of the various layers can be properly evaluated and if the test results are correlated with pavement performance. The principal advantage of the C.B.R. method lies in the fact that the test can be performed in the laboratory upon comparatively small samples.

12–14. Plate-Bearing Test Method. During 1945 and 1946 the Department of Transport of Canada undertook a very comprehensive investigation of runways at a number of the principal airports in Canada. This work was reported by Norman W. McLeod (*3*) and is abstracted in the following paragraphs in order to illustrate this method of approach to the problem. Later, McLeod expanded the results of this study and related it to pavement design methods based on the theory of elasticity (*25,26*).

Included among the principal objectives of the Canadian investigation was that of obtaining test data which could be employed in designing flexible pavements on the basis of plate-bearing tests (repetitive) performed on the subgrade, base, and pavement. It should be noted that the investigation was extremely wide in scope and was directed toward several additional objectives. However, only that portion of the report dealing with certain aspects of the use of plate-bearing tests for design purposes will be presented here.

12–15. Apparatus Used for Plate-Bearing Tests. The schematic arrangement of equipment used in performing the repetitive load tests is shown in Fig. 12–8. Various sizes of steel plates were used; most of the tests were conducted through the use of a circular plate 1 inch in thickness and 30 inches in diameter. Load was applied to the plate by means of a hydraulic jack having a capacity of 100 tons and equipped with pressure gages graduated in increments of 1000 or 2000 pounds. Reaction to the load applied by the jack was supplied in each test by a loaded tractor-trailer unit capable of applying loads from 70,000 to over 150,000 pounds.

Deflections of the bearing plate were measured to the nearest 0.0001 inch by means of two Ames dials set on a diameter of the plate, as shown in Fig. 12–8. During a portion of the investigation, deflections of the adjacent surface of existing airport pavements were measured by the additional Ames dials shown.

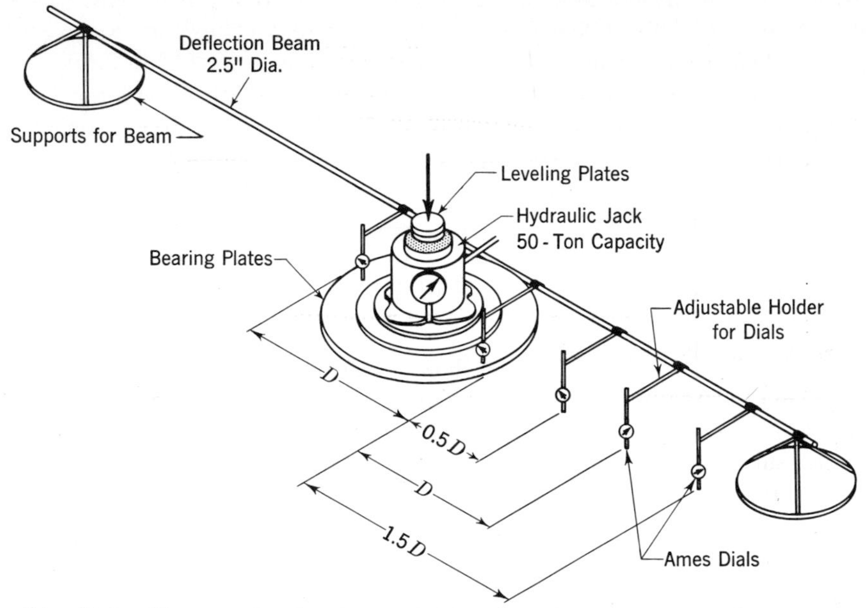

Fig. 12–8. Diagram showing arrangement of equipment for plate-bearing test. (After N. W. McLeod.)

12–16. Procedure of Plate-Bearing Test. Load tests were performed on the surface of existing pavements, on the surface of base courses, and on the surface of subgrades. In testing a base course with a 30-inch plate, the pavement was removed over an area having a diameter of 12 feet, and the load test made in the center of this area. Similar excavations were made in conducting the load tests on the various subgrades. This procedure was followed so that the subgrade load tests would be completely unconfined. The apparatus was then assembled, placed in position as indicated in Fig. 12–8, and loading was begun.

As the first step in the loading procedure, a relatively light load (4000 pounds in the case of a 30-inch plate) was applied and quickly released. This initial load served the purpose of "seating" the loaded plate on the area being tested. A load of 2000 pounds was then applied to the plate, and the deflection gages brought to zero reading. A load giving a deflection of from 0.03 to 0.04 inch was then applied and maintained until the increase in deflection was 0.001 inch or less per minute for each of three successive minutes. The load was then completely released and deflection again observed until such time as the rate of recovery became 0.001 inch or less per minute for each of three successive minutes. This procedure of loading and unloading (with the same magnitude of load) was repeated six times. It is this cyclic loading procedure which gives rise to the term "repetitive" plate-bearing test.

At the conclusion of this series of load applications and releases, the load on the plate was increased to an amount required to produce a deflection of 0.2 inch. The same loading and unloading process was applied from four to six times. In each case the load was not applied or released until the rate of deflection had attained the desired value of 0.001 inch per minute for each of three successive minutes. The load was finally increased to that required to produce a deflection of from 0.35 to 0.40 inch, and the same repetitive system employed to alternately apply and release the load.

12–17. Development of Design Method. Many load tests were performed at a number of Canadian airports, and the results of the tests were subjected to exhaustive scrutiny. The analysis employed is far too detailed to be presented here. However, certain significant factors important to the development of a design method based on plate-loading tests may be mentioned. For example, it was demonstrated that the cyclic load procedure used, which involved only a comparatively few repetitions of load, gave results which could be extrapolated to 100, 1000, and 10,000 load repetitions. Thus it seemed likely that the results could be applied to the design of runways (or highways) which would be subjected to many thousands of repetitions of load during their service lives.

A deflection of 0.5 inch was regarded as critical for flexible pavement design. It was demonstrated that relationships exist which will permit determination of load-deflection relationships over the entire range of deflection from 0.0 to 0.7 inch if the load corresponding to 0.2 inch deflection of a 30-inch plate is known. This factor is of practical significance in that a deflection of 0.2 inch of a 30-inch plate will occur in many subgrade soils at a load within the limits of equipment such as has been described, while the attainment of a 0.5-inch deflection on a similar plate might require much heavier equipment. Another relationship which was developed demonstrated that if the load required to produce 0.2-inch deflection of a 30-inch plate were known, the load for any other plate over the range of 12 to 42 inches in diameter, and for a range of deflections from 0.0 to 0.7 inch, could be calculated. Analysis of the test results produced additional facts which will not be considered here.

However, results of the investigation led to the conclusion, among others, that the load which can be supported by a given thickness of base varies directly with the strength of the subgrade upon which it is placed. This fact led to the development of a design equation which is given as follows:

$$T = K \log \left(\frac{P}{S}\right) \tag{12–2}$$

where T = required thickness of granular base, in inches
P = applied load, in kips at 0.2-inch deflection
S = total subgrade support, in kips at 0.2-inch deflection, for the same contact area as P
K = constant depending upon the base course and wheel load.

Later work performed in this investigation and reported by McLeod (*3*) in 1948 verified the general form of Eq. (12–2) but somewhat altered

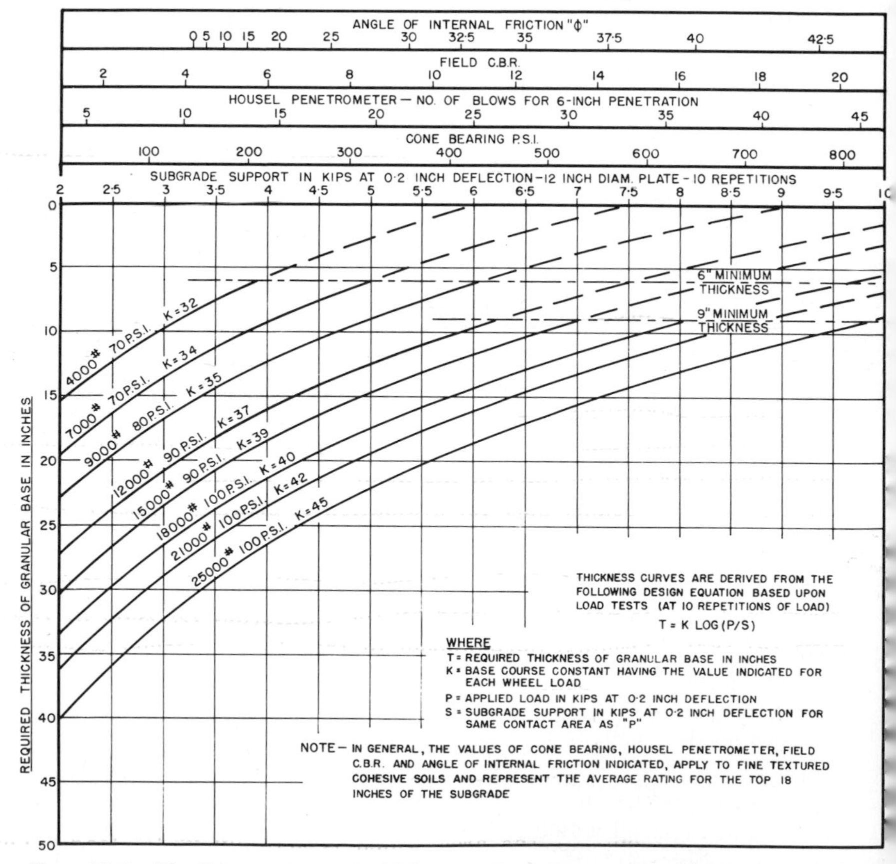

Fig. 12–9. Flexible pavement thickness requirements for highways carrying maximum traffic volume (full load on single tire). (After N. W. McLeod.)

the concept of the base course constant, K, which is an inverse measure of the supporting value of the base course per unit thickness.

Although the work undertaken in this investigation was intended primarily for application to the design of airport runways and taxiways, the design concepts developed were extended to cover the design of highways as well. In Fig. 12–9 are shown design curves for highway loadings based on Eq. (12–2).

12–18. General Observations Relative to Plate-Bearing Tests. From the standpoint of theory and practicality, design methods based upon repetitive plate-bearing tests seem to have many virtues. One of the principal drawbacks accompanying the use of this method lies in the fact that very heavy equipment is required when large plates are used, and that the tests themselves are laborious and time-consuming. Another disadvantage indicated in the Canadian investigation is that the tests must be conducted in the field after the subgrade has been constructed. This would seem to indicate that the required thickness of base and pavement would not be known until after a part of the construction had been completed. This seems to present difficulties on any highway job except very large projects. Then, too, if soil conditions were encountered which were substantially different from those for which a design equation similar to Eq. (12–2) had been developed, this would presumably call for the construction and evaluation of a number of test sections, together with the evaluation of existing pavements, in order to verify the design equation.

For a time, a design method incorporating the results of plate-bearing tests was used in North Carolina where the bearing capacity of the subgrade soil was determined on the basis of small-scale load tests conducted in the laboratory. The tests were performed upon base, subbase, and subgrade materials which had been brought to conditions of moisture and density expected to exist in the field. The method was later modified to provide a broader base for evaluation of bearing capacity, including the performance of C.B.R. tests.

12–19. California Method (Hveem). Still another method of thickness design is now being used by the California Division of Highways (*24,182*). Original development of this design method was based upon the following considerations.

Rather than assuming that the wheel load is distributed or "spread" over some area which varies with depth of the pavement, base, or subbase, this method assumes that the surface and base layers must be designed to resist the potential upward thrust of the subgrade at points adjacent to, but outside, the area which is actually under load. This approach is based upon the observed tendency for particles in any layer to be displaced along a curved path and thus exert an upward thrust against the bottom of the base or surface. The principle involved in this concept is shown in Fig. 12–10 for a failure involving displacement of the subgrade soil. The concept is that the effect of applying a load over a limited area on a bed of granular material is to force a cone-shaped mass of the material downward, displacing the adjacent material laterally and upward. In this analysis, the pressure on some plane beneath the surface may be greater than at the surface. Present California practice stems from empirical relationships developed from test track and other experiments, plus observation of roads in service.

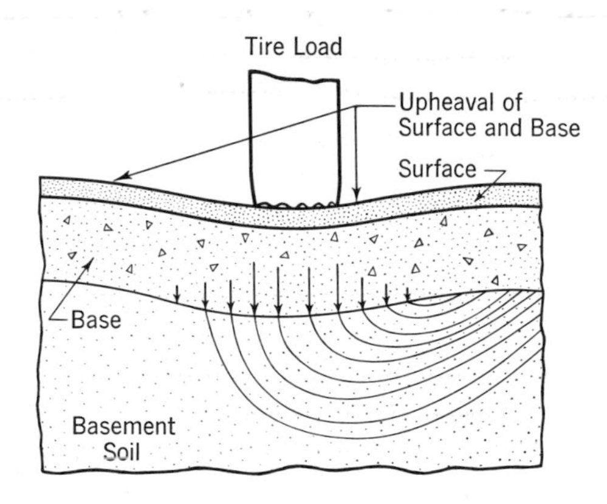

FIG. 12–10. Basement soil failure, California method. (Hveem.)

Four factors are considered to be of primary importance in evolving a design method for the thickness of a flexible pavement structure based on the concept outlined above (*193*). These factors are the unit resistance, R, of the soil material over which the pavement is to be placed; second, the accumulative destructive or deforming effect of all the wheel loads which will pass over a given portion of the pavement; third, the tensile strength of the pavement and base, which strength, in combination with the thickness (weight) of overlying material, prevents the underlying soil from being displaced by the action of traffic loads; and fourth, economic factors, including both first cost and future maintenance costs.

The general equation used in the California method is as follows:

$$GE = 0.032(TI)(100 - R) \tag{12–3}$$

where TI = traffic index
R = resistance value of the underlying soil or layer
GE = thickness of the pavement in terms of the gravel equivalent, in feet

N /

The resistance value, R, of the soil is determined by the stabilometer test method. The stabilometer, in simple explanation, is a cell which is capable of measuring the lateral pressure produced by application of a vertical load to a confined soil mass. The test specimen is 2.5 inches high and 4 inches in diameter. Materials are rated on a scale of 0 to 100 in proportion to their ability to sustain load without transmitting pressure to the side walls of the instrument. A liquid, therefore, would have an R value of 100 in the range of loads applied.

Traffic data required for design include traffic counts that represent present and estimated future average daily two-way truck volumes. Trucks are classified by number of axles, and buses are considered as trucks. These data are used to estimate the number of equivalent 5000-pound wheel loads in one direction (*EWL*) to be expected during the 20-year period following construction (or some other period). The relationship between traffic index and *EWL* is:

$$TI = 6.7(EWL/10^6)^{0.119} \tag{12–4}$$

EWL values can be corrected to *TI* values by use of a chart (not shown).

Calculations of *EWL* involve the multiplication of certain fixed constants which convert average daily traffic to yearly traffic in one direction for each axle group. Applicable constants are as follows:

Number of axles	*EWL* Constant
2	110
3	920
4	1520
5	3390
6	2230

To calculate the *EWL*, the designer multiplies the present average daily traffic volume in both directions for each truck class first by an appropriate expansion factor (to 10 years after construction) and then by the applicable *EWL* constant. The annual average *EWL*'s for all groups are added together and then multiplied by 20 to determine the 20-year *EWL* (the numbers, of course, are in the millions). He then uses Eq. 12–4 to determine the *TI*. Design of two- and four-lane highways is based on the *TI* value determined; lesser values are used in designing the inside lane of six- and eight-lane highways.

The design procedure being described results in a theoretical total thickness expressed in terms of gravel equivalence (*GE*). There must be sufficient depth to protect the material against displacement due to traffic loads. The material must also weigh enough to prevent its expansion beyond the moisture-density condition on which the design is based.

The theoretical thickness of the individual layers must be converted to actual thickness by the use of gravel equivalence factors, which are given in Table 12–3.

Two cases are encountered in California design practice. The first case is that of an expansive subgrade material. The design procedure involves the determination of various R values for the expansive material, which will vary with moisture content and assumed structural sections. These designs are worked out in the laboratory by a procedure that is too detailed to cover here. Reference *182* gives details of the procedure.

The second, and more usual, case is one in which R values of base and subbase are specified and the minimum R values of the basement soil can be determined because, under the weight of any structural section based on traffic load, the basement soil will not be expansive. In such a case, it is not necessary to assume a structural section to perform the laboratory calculations for R value. Thus, various designs can be determined readily and the most economical section can be determined.

An example will illustrate the design procedure for the second case (*193*). Assume a minimum R value of the basement soil of 10, a four-lane divided highway, and a lane TI of 8.0.

Equation 12–3 is used to determine the gravel equivalent: $GE = 2.30$ feet.

For the first alternate design, assume an aggregate base course with a specified R value of 78. Then determine the gravel equivalent of an asphaltic concrete surfacing by the use of Eq. 12–3. This value is 0.56 foot. From Table 12–3, under TI 7.5 and 8.0, determine a gravel equivalent factor of 2.01. Dividing 0.56 by 2.01 gives an actual thickness of 0.28 foot, which is rounded up to 0.30.

The next step is to choose a subbase for this design from whatever materials are economically available; assume a subbase with an R value of 50. When one uses the same equation, the gravel equivalent of the combined surfacing and base would be 1.28 feet over this subbase. Subtracting the gravel equivalent for the 0.30 foot of asphaltic concrete (from Table 12–3, this is 0.60) from 1.28 gives a gravel equivalent of 0.68 for the base. In the "Aggregate Base" column of Table 12–3, the next higher value listed is 0.72, a value that corresponds to an actual thickness of 0.65 foot, which should be used.

To determine the thickness of subbase, the gravel equivalents of the asphaltic concrete and the aggregate base are subtracted from the total required gravel equivalent previously determined. Thus, the required gravel equivalent of subbase is $2.30 - 0.60 - 0.72 = 0.98$ foot. Since

TABLE 12–3

Gravel Equivalents of Structural Layers in Feet, California Method

Actual Thickness of Layer Feet	Asphalt Concrete										RMS BTB and LTB	Cement-Treated Base			Aggregate Base	Aggregate sub-Base
	Traffic Index (TI)											Class				
	5 and below	5.5 6.0	6.5 7.0	7.5 8.0	8.5 9.0	9.5 10.0	10.5 11.0	11.5 12.0	12.5 13.0	13.5 14.0		A	B	C		
	Gravel Equivalent Factor (G_f)										G_f		G_f		G_f	G_f
	2.50	2.32	2.14	2.01	1.89	1.79	1.71	1.64	1.57	1.52	1.2	1.7	1.5	1.2	1.1	1.0
0.10	0.25	0.23	0.21	0.20	0.19	0.18	0.17	0.16	0.16	0.15	0.12					
0.15	0.38	0.35	0.32	0.30	0.28	0.27	0.26	0.25	0.24	0.23	0.18					
0.20	0.50	0.46	0.43	0.40	0.38	0.36	0.34	0.33	0.31	0.30	0.24					
0.25	0.63	0.58	0.54	0.50	0.47	0.45	0.43	0.41	0.39	0.38	0.30					
0.30	0.75	0.70	0.64	0.60	0.57	0.54	0.51	0.49	0.47	0.46	0.36					
0.35	0.88	0.81	0.75	0.70	0.66	0.63	0.60	0.57	0.55	0.53	0.42				0.39	0.35
0.40	1.00	0.93	0.86	0.80	0.76	0.72	0.68	0.66	0.63	0.61	0.48				0.44	0.40
0.45		1.04	0.96	0.90	0.85	0.81	0.77	0.74	0.71	0.68	0.54	0.77	0.68	0.54	0.50	0.45
0.50		1.16	1.07	1.01	0.95	0.90	0.86	0.82	0.79	0.76	0.60	0.85	0.75	0.60	0.55	0.50
0.55			1.18	1.11	1.04	0.98	0.94	0.90	0.86	0.84	0.66	0.94	0.83	0.66	0.61	0.55
0.60				1.21	1.13	1.07	1.03	1.98	0.94	0.91	0.72	1.02	0.90	0.72	0.66	0.60
0.65				1.31	1.23	1.16	1.11	1.07	1.02	0.99	0.78	1.11	0.98	0.78	0.72	0.65
0.70					1.32	1.25	1.20	1.15	1.10	1.06	0.84	1.19	1.05	0.84	0.77	0.70
0.75						1.34	1.28	1.23	1.18	1.14	0.90	1.28	1.13	0.90	0.83	0.75
0.80						1.43	1.37	1.31	1.26	1.22	0.96	1.36	1.20	0.96	0.88	0.80
0.85						1.52	1.45	1.39	1.33	1.29	1.02	1.45	1.28	1.02	0.94	0.85
0.90							1.54	1.48	1.41	1.37	1.08	1.53	1.35	1.08	0.99	0.90
0.95								1.56	1.49	1.44	1.14	1.62	1.43	1.14	1.05	0.95
1.00								1.64	1.57	1.52	1.20	1.70	1.50	1.20	1.10	1.00
1.05									1.65	1.60	1.26	1.79	1.58	1.26	1.16	1.05

Notes: RMS is road-mixed asphalt surfacing.
BTB is bituminous-treated base.
LTB is lime-treated base.

aggregate subbase has a gravel equivalent factor of 1.0, the required thickness of subbase is 1.00 ft.

The completed design for the example then involves 0.30 foot of asphaltic concrete over 0.65 foot of aggregate base over 1.00 foot of subbase. Alternate designs can be prepared, using other available materials; costs can be compared; and a final design can then be selected.

12–20. Rational Design Methods. So-called rational (theoretical) methods of designing flexible pavements fall into two groups (*76*)—those based on considerations of ultimate strength of the pavement components and those based on considerations relative to pavement and subgrade stresses and deflections in the range of working loads.

The *ultimate strength methods* are concerned with pavement behavior at failure. The basic concept is that the pavement must be designed with a factor of safety against shear failure of the materials. This approach has not been widely used in the United States but seems to hold promise for further development.

The second approach, the *elasticity method,* considers the behavior of the pavement under working conditions; deflections are assumed to be proportional to applied loads. Stresses and strains in the pavement structure are determined by the theory of elasticity.

Many individuals and groups are using these methods to approach the problem of pavement design, and some such methods have been widely used in practice. All of them consider the pavement structure to be a layered system in which the individual layers are homogeneous, isotropic, and elastic. The behavior of the layers under load is taken to be a function of their moduli of elasticity and Poisson's ratios.

One method of this type that has been used for many years was developed by the Kansas Highway Department (*76*). Stresses and displacements are evaluated by the Boussinesq solution for a homogeneous solid with a "stiffness factor" derived by consideration of the slab action of the upper layers. Triaxial tests are used to determine the deformation moduli of the pavement materials and of the subgrade. The design limits the theoretical deflection of the pavement surface under load to 0.1 inch.

In recent years, attention has focused on the fact that stress-strain relationships for the materials used in flexible pavements are time-dependent. This has led to heightened interest in the possible application of viscoelastic concepts to pavement behavior; such work is in its preliminary stages. Investigators also are studying (and restudying) rapid methods of testing, to more nearly simulate field loading conditions.

CHAPTER 13

SOIL-STABILIZED ROADS

Having assimilated the information which has been given in previous chapters, the student is now ready to devote his attention to the design and construction of the elements of the pavement structure itself, including various wearing surfaces, bases, and subbases. This chapter is devoted to soil-aggregate roads and soil stabilization in its various forms and is followed by chapters devoted to macadam roads, the several types of bituminous wearing surfaces, high-type bituminous pavements, and portland cement concrete pavements and bases. The sequence of these chapters might be generally said to follow a line of increasing costs so far as the pavement structure itself is concerned. Any classification of this sort will be open to question and subject to exceptions, but this chapter is primarily devoted to "low-cost" construction. Chapters 14, 15, and 16 deal principally with "intermediate" or "medium-cost" treatments, while the pavements discussed in Chapters 17, 18, and 19 may be definitely described as "high-cost" construction. The "cost" which is referred to is the initial cost of the treatment. Considerations which have been previously presented in Chapter 6, "Highway Soil Engineering," and Chapter 12, "Flexible Pavement Design," will be of particular value in the study of this chapter.

13–1. Definitions. Soil-aggregate roads may be defined as those which consist of a substantial layer of a properly proportioned and blended mixture of soil and aggregate compacted to form a road which is capable of supporting traffic in all weather conditions. The AASHO definition of soil-aggregate (dense-graded aggregate) is "natural or prepared mixtures consisting predominantly of stone, gravel or sand and containing silt-clay" (*107*). In addition to serving as wearing courses, generally for light traffic, soil-aggregate mixtures are very widely used as bases and subbases.

A wide variety of materials are used in soil-aggregate mixtures, including sand-clay, gravel, and stone or slag screenings; sand, crushed stone, or slag combined with soil mortar; and various combinations of these materials. Where two or more materials are blended, the result may be termed a "granular soil stabilized mixture." Included are "gravel" roads, which are those constructed from "natural" gravel, such as pit-run or bank-run gravel, crushed gravel, crushed stone, slag, and so on. Many miles of gravel roads still exist in many sections of the country, particularly on the various secondary systems, including so-called "farm-to-market" roads. The total mileage of this type of road in the United States has been estimated at about 1.3 million miles.

Many times, a wearing surface, base, or subbase of this type is built without incorporating the existing subgrade soil into the mixture. Regardless of whether or not the existing soil is used, the end result is the same.

In a broad sense, soil stabilization may be defined as the combination and manipulation of soils, with or without admixtures, to produce a firm mass which is capable of supporting traffic in all weather conditions. An even simpler definition, which is attributed to the late Roy W. Crum, former Director of the Highway Research Board, is that "a stabilized fill, subgrade, road surface, or road base, is one that will stay put, and stabilization is the process by which it has been made that way." If a stabilized soil is to be truly "stable," then it must have sufficient shearing strength to withstand the stresses imposed upon it by traffic loads in all kinds of weather without excessive deformation. In addition, if the stabilized soil mixture is to be used as a wearing surface, then it must be capable of withstanding the abrasive effects of modern traffic.

In this branch of highway design and construction the emphasis is definitely placed upon the effective utilization of local materials, with a view toward decreased construction costs. In some areas the natural soils are of an unfavorable character and require modification through the use of suitable mineral constituents such as gravel or crushed stone or clay binder. In still other areas, admixtures such as bituminous materials, portland cement, salt, or lime are used for effective stabilization. The type and degree of stabilization required in any given instance is largely a function of the availability and cost of the required materials, as well as the use which is to be made of the stabilized soil mixture.

13–2. Uses of Stabilized Soil Mixtures. The primary use of stabilized soil mixtures at the present time is in base and subbase construction. A stabilized soil base or subbase may provide the support for a relatively thin wearing surface which will be subjected to light or moderate amounts of traffic, or it may function as a base for a high-type pavement which will be subjected to very heavy volumes of traffic. Certain stabilized soil mixtures, including granular soil stabilized mixtures, serve as wearing surfaces on light-traffic roads.

The principles of soil stabilization have found increasing application by highway agencies of all types in the last few years. This has been primarily due to two factors: first, in many areas the supply of economically available materials suitable for these uses without substantial modification has been depleted; second, the rising costs of highway construction have forced the use of locally available materials in increasing quantities for the construction of all elements of highway and street systems. Perhaps the greatest use of stabilized soil mixtures is in the improvement of roads which carry light to moderate amounts of traffic. Similarly, many residential streets in municipal areas are being built from stabilized soil mixtures.

Stabilized soil mixtures also lend themselves readily to the process of "stage construction," which involves the gradual improvement of the individual units of a highway system as the demands of traffic increase. Thus a properly designed stabilized soil mixture might function briefly as a wearing surface, receive a thin bituminous surface treatment as traffic increased, and eventually serve as a support for a high-type bituminous pavement which will serve a very heavy volume of traffic.

13–3. Types of Stabilization. Since the emphasis in this type of construction is upon the use of locally available materials, it should be noted that many different substances and processes have been successfully used in soil stabilization. As has been indicated, certain naturally occurring soils may require only compaction and drainage for stabilization. Other soils require treatment in various fashions and with various materials in order to satisfactorily perform their intended function. This chapter is primarily devoted to these latter combinations of materials. In our presentation of the subject, soil-aggregate roads and "granular" stabilization are discussed first. "Bituminous soil stabilization," in which bituminous materials are applied to the soil to supply cohesion or for waterproofing purposes, is then discussed; the oiling of natural earth roads is included in this section. The design and construction of soil-cement roads is also considered in some detail, and the use of salt, lime, and various other chemicals for soil stabilization is discussed briefly.

Soil-Aggregate Roads and Granular Stabilization

In this discussion the term "aggregate" refers to the portion of a granular mixture or natural soil which is retained on a No. 200 sieve. By AASHO definition, "stone" and "gravel" are particles that will pass a 3-inch sieve, but be retained on a No. 10 sieve; "sand" will pass a No. 10 sieve, be retained on a No. 200. Practically all materials of a granular nature can function satisfactorily as aggregates in this type of mixture, although the best aggregates are those which are composed of hard, durable particles. The function of this portion of a stable soil-aggregate mixture is primarily to contribute internal friction. The gradation of this portion of the mixture is important, as the most suitable aggregates, generally speaking, are those which are well graded from coarse to fine. One approach to the problem of gradation is to establish limits for the various sizes of particles which will result in a combination which gives maximum density; high density, in turn, is associated with high stability. It is not wise to lay too much stress upon gradation, however, as many aggregates have been used which are not well graded in the usual sense. Specifications concerning gradation for this purpose are usually quite tolerant, with the idea, again, of making the best possible use of locally available materials.

The "fine-soil fraction" is that portion of the mixture which passes a No. 200 sieve. It will be noticed that, under the definition, this fraction might be composed of either silt or clay, or both, since the opening of a No. 200 sieve is 0.074 millimeters. The function of this portion of the soil-aggregate mixture is to act as a filler for the remainder of the mixture and to supply cohesion which will aid in the retention of stability during dry weather, while the swelling of the clay contained in this mixture serves to retard the penetration of water in wet weather. The nature and amount of this fine material must be carefully controlled if the mixture is to function properly, since an excessive amount of clay and silt may result in excessive volume change with change in moisture content. The properties of the fine-soil fraction are frequently determined by measurement of the plasticity index, or P.I. As will be recalled, the P.I. is the difference between the liquid limit and the plastic limit. Both of these latter tests are customarily performed upon the portion of the soil which passes a No. 40 sieve; this is frequently called the "soil binder."

It must again be noted that, for satisfactory results, the soil-aggregate mixture must be properly proportioned. In some cases natural soils may have satisfactory properties and thus may require only compaction and drainage for stabilization. Far more frequently, however, the existing soil will be missing in some of the important soil elements. Thus a granular soil or "aggregate" may require the addition of fine material for binder, or a fine-grained soil may require the addition of aggregate for stability and to prevent excessive volume change. In certain cases the soil existing in the roadbed may only require the addition of a small amount of material which may be brought in, distributed along the roadway in the proper amounts, and blended with the existing soil to produce a satisfactory soil-aggregate mixture. In other cases the desired soil-aggregate mixture may be compounded in the proper proportions at a central mixing plant and then deposited on top of the existing roadway soil. In any case the properties of the final mixture are generally judged and controlled by the mechanical analysis, the liquid limit, and the plasticity index. In addition, since the mixture is generally required to be compacted to a high degree of density, the moisture–density relationships for the mixture must be determined for a given compactive effort so that the field compaction process may be properly controlled.

Before presenting detailed recommendations as to the composition of satisfactory soil-aggregate combinations, it seems desirable to discuss briefly the difference in requirements for the composition of mixtures intended for use as bases and those for use as wearing surfaces. If the mixture is to withstand the abrasive effects of modern traffic it should have more binder than is required in a base course, so that it will be more impervious to precipitation which falls upon the surface of the road and can, to a certain extent, replace by capillarity moisture which is lost by evaporation. Bases do not require as high a proportion of binder soil, since they may generally be required to have high stability and low capillarity so that they do not tend to soften because of the accumulation of capillary moisture. This consideration becomes of special importance when stage construction is being employed, where the mixture may serve as a wearing surface for a time and then as a base, so that the mixture must be capable of performing both functions satisfactorily during its service life. Where frost action is a factor, the percentage of material passing the No. 200 sieve may be reduced to prevent damage to a base or subbase.

13–4. Specifications for Granular Soil Stabilized Mixtures. Many different specifications are used for these soil-aggregate mixtures by

TABLE 13–1

GRADATION LIMITS FOR SOIL-AGGREGATE SUBBASE, BASE, AND SURFACE COURSES (*107*)

Sieve Designation Grading	A	B	C	D	E	F
2-inch	100	100	–	–	–	–
1-inch	–	75–95	100	100	100	100
⅜-inch	30–65	40–75	50–85	60–100	–	–
No. 4	25–55	30–60	35–65	50–85	55–100	70–100
No. 10	15–40	20–45	25–50	40–70	40–100	55–100
No. 40	8–20	15–30	15–30	25–45	20–50	30–70
No. 200	2–8	5–20	5–15	10–25	6–20	8–25

different highway agencies. Indicative of the requirements are those given in Table 13–1, which is contained in AASHO Designation M147, Materials for Soil-Aggregate Subbase, Base, and Surface Courses (*107*).

It is recommended that the fraction passing the No. 200 sieve be less than two-thirds of the fraction passing the No. 40 sieve. For bases, subbases, and temporary wearing surfaces, the plasticity index should not exceed 6 and the liquid limit should not exceed 25. Maximum size of aggregate for wearing surfaces is recommended as one inch. In other words, gradings A and B of Table 13–1 are not recommended for wearing surfaces. Coarse aggregate (retained on No. 10 sieve) should have maximum percentage of wear, by the Los Angeles abrasion test (see Art. 14–2) of 50, according to the AASHO specification.

When a mixture of this type is to be used as a wearing surface for several years, the minimum amount of material passing the No. 200 sieve should be 8 per cent, the maximum liquid limit is 35, and the P.I. should be from 4 to 9.

13–5. Use of Calcium Chloride. Calcium chloride is very widely used in connection with the construction of soil-aggregate mixes of the type which have been described. Calcium magnesium chloride and sodium

Fig. 13–1. Spreading calcium chloride on a gravel road. (Courtesy Calcium Chloride Institute.)

chloride have also been used in some areas. Calcium chloride is a hygroscopic material; that is, it possesses the ability to attract and retain moisture from the air. Calcium chloride is used in this type of construction in order to expedite the compaction process by slowing the rate of evaporation of moisture from the mixture during compaction, and to aid in the retention of moisture by the soil-aggregate mixture during its service life. This is especially important for wearing surfaces as the presence of calcium chloride will aid in maintaining the moisture content at that desired for maximum stability, prevent raveling of the surface, and reduce dust. Calcium chloride may be incorporated into the mix during the construction process and (on wearing surfaces) by subsequent applications, as desired.

The incorporation of small amounts of calcium chloride will, in many instances, result in increased density for a given compactive effort. It has also been indicated that calcium chloride may be useful in reducing the effects of frost action in certain soil types during the winter and spring seasons. The amount of calcium chloride required in any case is not large; a frequent initial application in base construction is 0.5 pound per square yard per inch of thickness (road-mix construction). In plant mixtures, the recommended quantity is 10 pounds of calcium chloride per ton of mixture.

Figure 13–1 shows the application of calcium chloride to a gravel road.

13–6. Construction Procedures. Three basic procedures are used in the construction of soil-aggregate roads. These methods, which are equally applicable to the production of bases and wearing surfaces, as well as to the less frequent purposes of subbase construction and subgrade improvement, are as follows:

1. Road-mix construction
2. Traveling plant construction
3. Central plant construction

In all these methods the basic processes are the same. That is, the soil elements must be properly selected and proportioned; the soils pulverized and uniformly blended; water and chemicals, if desired, added and uniformly mixed with the soil; the blended materials spread in a thin layer of uniform thickness, properly compacted and finished. The several steps may be performed separately, or certain of the steps may be combined in one operation. Before beginning a discussion of the actual construction processes listed above, some mention should be made of the design procedure of selecting and proportioning the components of the stabilized soil mixture.

13–7. Selection and Proportioning of Soil Elements. It is, of course, frequently desired to use the existing roadbed soil in the stabilized mixture in the interests of economy. Information regarding this material, both in cut and embankment sections, will generally be available from the detailed soil survey for the project under consideration. The results of the mechanical analysis and routine soil tests will then be available for each of the soils encountered in the section concerned before this phase of construction begins. The availability and location of granular and binder materials suitable for use in stabilization will also generally be known. Many state highway departments keep a detailed inventory of gravel pits, sand and clay deposits, and so on so that the information is readily available to the engineers and contractors involved in a particular job.

Assuming that the nature and location of suitable materials are known, the problem is to determine the proportion of "borrow" material which must be combined with the existing soil to produce a granular soil stabilized mixture which will meet the specifications of the organization concerned. It should also be noted that in many cases the entire soil-aggregate mixture may be imported and placed on top of the existing subgrade or previously constructed subbase; the material may come from a single source or may be a blend of materials from two or more sources. Trial combinations of materials are usually made in the laboratory on the basis of the mechanical analyses of the materials concerned. In other words, calculations are made in order to determine the sieve analysis of the combined materials, and the proportions of the various components are adjusted to fall within the limits of the specified gradation.

Numerical proportioning is a simple process for two materials. For example, assume that the materials of Table 13–2 are available, and

TABLE 13–2

EXAMPLE OF NUMERICAL PROPORTIONING

Per Cent Passing (by Weight)

	Subgrade Soil (A)	Borrow Material (B)	1A, 1B	1A, 3B
	Mechanical Analysis			
1-inch	–	100	100	100
¾-inch	–	90	95	93
⅜-inch	100	80	90	85
No. 4	98	62	80	71
No. 10	96	45	70	58
No. 40	63	13	38	25
No. 200	38	1	20	10
	Plasticity Characteristics			
Liquid Limit, %	35	20	31	25
Plasticity Index, %	10	2	9	7

are to be blended to meet the requirements of Table 13–1 (Grading D).

By simple trial, it can be established that any blend between the ranges of one part A and one part B, to one part A and three parts B will be satisfactory; i.e., will meet the grading requirements of Table 13–1. Using a greater amount of material A will abrogate the requirement relative to the maximum passing the No. 10 sieve. At the other end of the scale, use of an excessive amount of borrow material will abrogate the maximum passing the No. 40 sieve. Presumably, the borrow material is more expensive than the existing soil; hence, it would be desirable to use as little of this material as possible. In other words, use the 1:1 mix.

The plasticity index of the proposed mixture is then determined in the laboratory and compared with the specified limits. If this value is satisfactory, the blend may then be assumed to be satisfactory. If the plasticity characteristics of the first mixture are not within the specified limits, then additional trials must be made until a satisfactory blend is obtained. The proportions thus secured may then be used in the field-construction process.

In certain cases, where the agency is familiar with the soils concerned, materials of known P.I.'s may be combined in trial proportions established by a design "formula" in order to produce a mixture of given plasticity characteristics. An example of such an equation is as follows:

$$\text{P I (mixture)} = \frac{\left\{ \begin{array}{l} (\%\text{A})(\%\text{ Passing No. 40, A})(\text{P I, A}) \\ \qquad + (\%\text{B})(\%\text{ Passing No. 40, B})(\text{P.I.,B}) \end{array} \right\}}{(\%\text{A})(\%\text{ Passing No. 40, A}) + (\%\text{B})(\%\text{ Passing No. 40, B})} \qquad (13\text{–}1)$$

Such proportioning will furnish an approximation of the P. I. which will serve until an actual test is made. For the 1:1 mixture of Table 13–2, the approximate value of the

$$\text{P.I.} = \frac{50\ (63)\ 10 + 50\ (13)\ 2}{50\ (63) + 50\ (13)} = 8.6.$$

A similar equation may be used to determine an approximate value of the liquid limit; in this case, 31. In any event, a mixture is generally sought which will be well graded from coarse to fine and which will have the desired plasticity. The laboratory examination is generally completed by the determination of the compaction characteristics of the stabilized mixture so that the compaction process may be satisfactorily controlled in the field.

Graphical methods are also available, and are particularly useful in the proportioning of more than two materials (*260*). A method of blending natural earth deposits was developed by Ritter and Shaffer; the procedure requires an electronic computer and uses linear programming to determine materials and sources for least cost (*261*).

13-8. Road-Mix Construction. In the basic process of road mixing, the proper proportions of soil elements which are to form the base or wearing surface are mixed directly on the surface of the subgrade or subbase. Any combination of circumstances may occur, but it may be assumed that the binder soil is largely available from near-by sources, such as the roadbed, shoulders, side slopes, and so on, while the aggregate material is obtained from a borrow pit. In such a case the existing soil would probably be cut by a motor grader and spread on the surface to dry. When this soil fraction had attained a satisfactory moisture condition it would be thoroughly pulverized by the use of suitable farm equipment such as discs, plows, and toothed harrows or by the use of a high-speed rotary mixer. When this fine soil material had been adequately pulverized it would then generally be bladed to the side of the road. The granular material would then be hauled onto the road in trucks and spread in the proper amounts. The previously windrowed soil would next be bladed back onto the road and combined with the imported material, again in the proper proportions. Calcium chloride may also be added at this stage. It is absolutely essential that the materials be blended to a high degree of uniformity if the best results are to be secured. In fact, this difficulty of securing uniformly mixed materials is one of the chief drawbacks of the use of farm tools and motor graders for mixing in the road-mix method. One of the machines which has been developed to overcome the difficulties of blending in this process is shown in Fig. 13-2. This machine is a high-speed rotary mixer and is often successfully used to pulverize and blend materials in the process of soil stabilization. It may be used to incorporate water in the mixture or to combine calcium chloride, portland cement, or bituminous material with the soil, as well as for a host of other activities. Where large coarse particles are present, available in-place pulverizers have proved to be very effective.

After the soil elements have been properly blended, the mix must then be spread in a uniform layer for compaction. If the moisture content of the mixture is less than the optimum moisture content, water must be added and blended with the soil to form a uniform mixture. After the water has been added, the soil is spread in a uniform layer and compacted. Thickness of layer is quite variable. For example, in the Midwest common practice is to spread the material in thin lifts (4 inches maximum compacted thickness) by a motor grader and begin rolling immediately; each layer is compacted before the next layer is spread. Where large rollers or vibratory compactors are used, material may be spread in much thicker layers, up to about 8 inches loose thickness. Initial compaction of a single layer may be accomplished by the use of sheep's-foot rollers. More commonly, pneumatic-tired rollers are used. Vibratory compactors are very effective on the more granular soil-aggregate mixtures. Following initial compaction, the surface is generally shaped to the proper cross-section by a blade grader and final rolling done by pneumatic-tired or smooth-wheel rollers. It is absolutely essential that adequate compaction be secured in this type

FIG. 13–2. A high-speed rotary mixer processes material in place. (Courtesy National Lime Association.)

of construction, and rolling should be continued until satisfactory density is achieved.

13–9. Traveling Plant Construction. When a traveling plant is used in this type of construction, the construction process is not greatly different from the road-mix method previously described. However, the operations of mixing the various soil elements, calcium chloride, and water are accomplished by the travel plant, usually in a single pass of the machine.

A windrow-type travel plant may be used in this type of construction. In using a typical machine of this type, the granular and pulverized fine-soil fractions are combined in proper proportions in a windrow of known size on the subgrade or subbase. The material is then picked up by the travel plant by means of a bucket or similar type of loader and mixed with the proper quantity of water in a mixing unit, which is generally of the pugmill type. Calcium chloride may also be added at the mixer. The uniformly mixed material is fed out of the rear of the machine, spread to uniform thickness, and compacted. The mixture is generally close to optimum moisture as it leaves the plant. If some time elapses before rolling takes place, additional water may be added. Final finishing and rolling are the same as previously discussed.

Large flat-type traveling plants are also available. Two machines of this type are shown in Fig. 13–3. This machine will excavate and pulverize the existing soil to a limiting depth of about 12 inches, mix the dry materials thoroughly, including calcium chloride or similar materials which are spread on the soil in front of the machine, add water

FIG. 13–3. A pair of P & H single-pass soil stabilizers process materials for a granular base. (Courtesy The Koehring Co.)

(and bituminous material, when used), thoroughly mix the dry and wet materials, and spread the mixture in a uniform layer behind the machine.

The more elaborate models of high-speed rotary mixers, such as the one shown in Fig. 13–2, are traveling plants; multiple passes are generally required to complete the processing of the soil.

13–10. Central Plant Method. Many organizations which build a large mileage of granular soil stabilized roads have found it desirable to use central mixing plants for the production of stabilized soil mixtures. Usually mixtures produced in this fashion cost slightly more than those formed by road-mixing methods, but certain advantages accrue

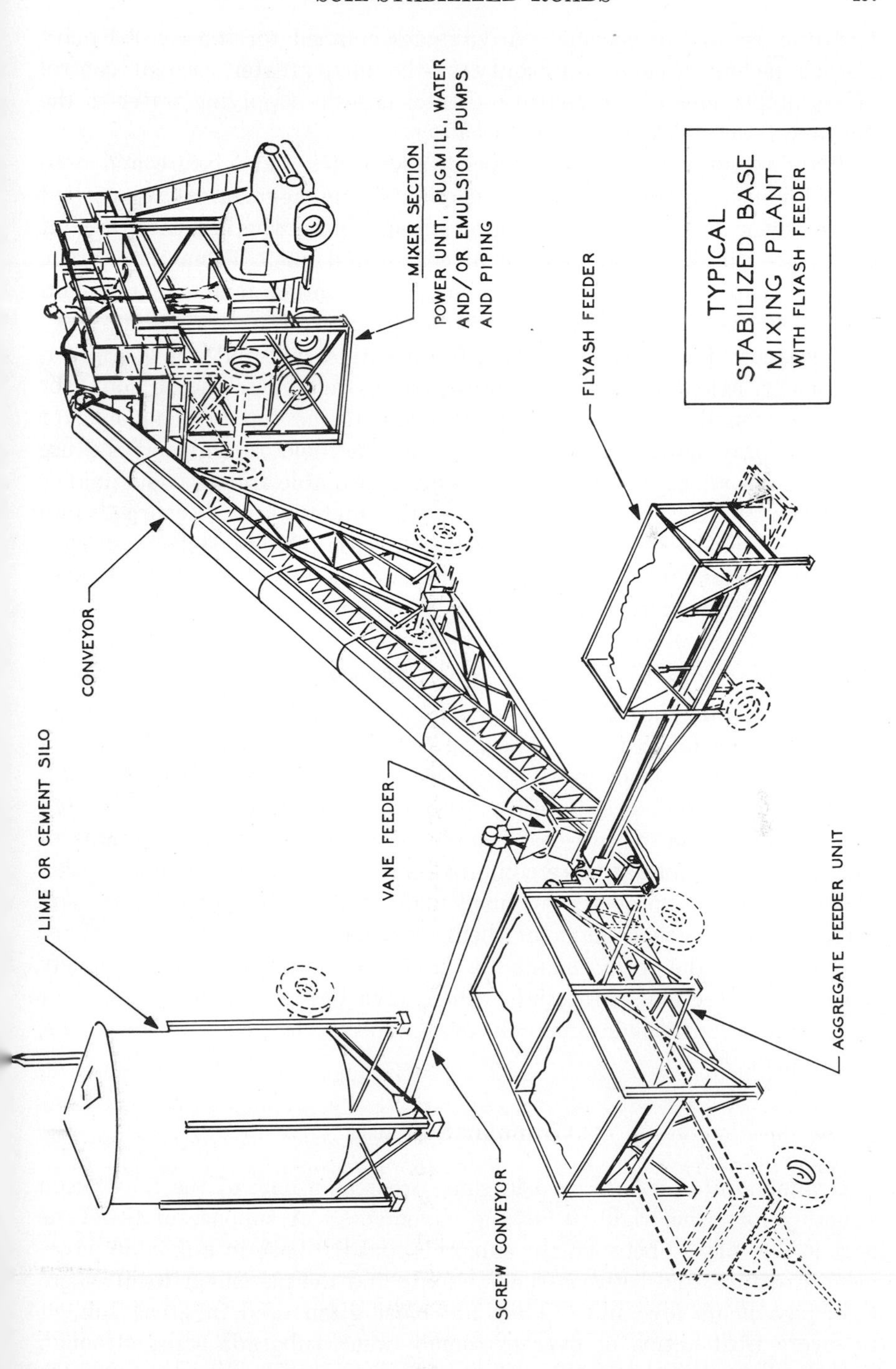
TYPICAL
STABILIZED BASE
MIXING PLANT
WITH FLYASH FEEDER
MIXER SECTION
POWER UNIT, PUGMILL, WATER
AND/OR EMULSION PUMPS
AND PIPING
FLYASH FEEDER
CONVEYOR
LIME OR CEMENT SILO
VANE FEEDER
AGGREGATE FEEDER UNIT
SCREW CONVEYOR

from the use of this scheme. Advantages claimed for the central plant method include greater uniformity of the mix, greater ease of control of the proportions of the mixture, greater ease in supplying water to the mix, and fewer delays due to bad weather.

Plant setups vary with the type of aggregates and mixture and with the equipment available. In most plants, the aggregates are carried by conveyor belt into one or more storage bins, whether they are pit or crusher run, or a combination of coarse and fine. From the storage bin, the material is conveyed to a pugmill or other mixing equipment. Where clay is to be added to coarser aggregates, a separate setup may be necessary for pulverizing and feeding this material to the mixer.

Usually, calcium chloride is spread on the aggregate in the conveyor through a small hopper. Water is added at the top of the conveyor through spray bars. Most authorities in this field recommend the use of a pugmill mixer. Suitable pugmills are available from various manufacturers. When mixing is complete, the material is discharged into trucks for transportation to the job site. At the job site, steps of spreading, compaction, finishing, and curing are generally the same as those described for in-place construction. However, spreading often is done with a self-propelled aggregate spreader or with a bituminous paver (Art. 16–17). Fig. 13–4 shows a typical plant setup.

13–11. Local Variations in Granular Soil Stabilization. It must always be remembered in considering this subject that the most important factor is the economical and proper use of local materials. Because of this fact many types of soil-aggregate combinations have been used in the construction of low-cost roads and bases in different sections of this country. Many of these combinations of materials and construction methods might be properly included under the heading of granular soil stabilization, although many of them, as used by various organizations, may not meet the requirements for gradation and plasticity previously cited in this section. Materials which have been successfully used for soil stabilization include limerock, marl, topsoil, slag, chat, oyster shells, and cinders.

Granular Subbases

Granular subbases are used in some areas as a part of the foundation structure for either rigid or flexible pavements. A subbase of this type is generally placed between the subgrade and the base in a flexible pavement structure, and between the subgrade and the pavement itself where rigid pavements are built. They are most often used in areas subject to severe frost action or over extremely weak subgrade soils, although they may also be used to prevent pumping in concrete pavements, which action will be described in a later chapter. Subbase thicknesses vary from 4 to 24 inches, while some have been built with thicknesses as great as 36 inches. The required thickness is generally determined by local experience and judgment and is dependent upon such factors as the amount and kind of traffic, climatic conditions, subgrade soil type and condition, and the position of the water table.

Materials used for subbase construction should have high stability and should not contain enough fine material to possess detrimental capillarity. Many different materials are used for this purpose, including gravel, gravel-sand mixtures, crushed gravel, crushed stone, and slag. The maximum size of aggregate permitted is generally from 3 to 3.5 inches. The requirements for gradation are usually quite liberal or "open" in order to permit the widest possible use of local materials. Natural or artificial soil-aggregate mixtures of the type which have been discussed may be used. They are selected on the basis of the desired stability of the compacted material in the presence of moisture.

Subbases of this type are usually constructed in the same way as base courses, but may be built by the "trench" method. That is, the undesirable subgrade soil is excavated in accordance with the plans, and the selected granular material is then used to replace the excavated soil. The material is generally hauled in, dumped, and spread in uniform layers without segregation. The thickness of each layer or "course" is frequently limited to 6 inches, although much greater thicknesses are permitted in some states. Each layer is required to be compacted to an adequate density before the next layer is placed. Compaction may be achieved by vibratory compactors or by pneumatic-tired or steel-wheel rollers. When the required total thickness has been placed, the subbase must be shaped to the final section shown on the plans.

Bituminous Soil Stabilized Roads

In many cases a soil-aggregate mixture which will perform satisfactorily is not economically available in a given area. In such cases economical stabilization may be effected through the use of any one of a number of admixtures. The following sections are devoted to the use of bituminous materials for this purpose.

Bituminous materials are used in conjunction with soils (and soil-aggregate mixtures) for two general purposes. In one application the bituminous material may supply cohesion to the stabilized soil mixture, as in the case of the stabilization of sands or very sandy soils. On the other hand, bituminous material may be incorporated into a natural or artificial soil mixture for the purpose of "waterproofing" it. That is, enough bituminous material is added to reduce the detrimental effects of water which may enter the soil during its service life. In some cases the bituminous material may perform the dual functions of supplying cohesion and necessary waterproofing. Bituminous soil stabilized mixtures are most frequently used for base construction.

An excellent summary of knowledge and practice relative to bituminous soil stabilization after World War II is contained in a publication of the Highway Research Board (*63*).

One of the most confusing aspects of the subject of bituminous soil stabilization was the lack of an adequate terminology for describing the types of soil mixtures utilized under this heading. This lack of a uniform terminology had impeded the ready exchange of information relative to the subject, since many of the terms which were used had primarily local significance. The Committee on Soil-Bituminous Roads, in the above-mentioned publication, proposed the use of four terms—soil-bitumen, sand-bitumen, waterproofed mechanical stabilization, and oiled earth—to describe the various types of bituminous soil stabilization in common use. It was also assumed that terminology relative to specific mixtures or projects could be contained within the framework of these terms. For example, such descriptive terms as "asphalt emulsion soil stabilization," "tar soil stabilized roads," and so on, were (and still are) used. It should further be noted that the inclusion of the type of construction known as "oiled earth" represented a deviation of some importance from classifications which had been used in the past.

Definitions for the four terms listed above are dependent both upon the physical characteristics of the soil which is being processed and the function of the bituminous material used in its treatment. The definitions given were as follows:

Soil-bitumen: A waterproofed, cohesive soil system.

Sand-bitumen: A system in which loose beach, dune, pit, or river sand is cemented together by bituminous material.

Waterproofed mechanical stabilization: A system in which a soil material possessing good gradation of particles from coarse to fine, and having high potential density, is waterproofed by the uniform distribution of very small amounts of bitumen.

Oiled earth: An earth road surface made water- and abrasion-resistant by the application of slow- or medium-curing road oils.

In the years following the committee report, use of the terms defined above has been modified somewhat. The term "waterproofed mechanical stabilization," and the type of construction which it represents, have not come into widespread use, being generally included within the framework of "soil-bitumen." The other three terms are widely used, and form the basic framework of design and construction in this field.

Some authorities have used the term "bituminous soil stabilization" in a very broad sense, even applying it to mixtures using wholly manufactured aggregates and incorporating semisolid bituminous materials (e.g., asphalt cement). That practice is not followed in this book, coverage in this chapter being limited to mixtures of liquid bituminous materials with soils and soil-aggregate materials. Mixtures using semisolid bituminous materials are described in Chapter 16; some of these mixtures resemble closely the best of the mixtures described in this chapter.

Soil-Bitumen

13–12. Material Requirements. The soil portion of a soil-bitumen mixture will generally be well graded, with particles ranging from coarse to fine and including an amount of material passing the No. 200 sieve. Generally speaking, the more fine material included in the mixture, the greater will be the amount of bitumen required for satisfactory waterproofing. Very fine, plastic soils generally cannot be stabilized economically with bituminous materials because of inherent difficulties of pulverization and mixing. Well-graded, coarse materials in the mixture, such as gravel, sand, and crushed stone, will lessen the amount of bituminous material required and contribute to over-all stability. According to the committee, best results were secured with soils which have the following approximate limits of gradation:

Maximum size: Not greater than approximately one-third the compacted thickness, or of the same size as the thickness of a compacted lift, if the latter is a fraction of the thickness of the base.
Passing No. 4 Sieve: More than 50 per cent.
Passing No. 40 Sieve: 35 to 100 per cent.
Passing No. 200 Sieve: Not more than 50 per cent nor less than 10 per cent.

In addition, the portion of the soil which passes a No. 40 sieve (soil binder) should have a liquid limit which is less than 40 and a plasticity index less than 18. Soils which do not meet the requirements given above have also been successfully stabilized in some instances.

The ARBA Committee on Soil Asphalt Stabilization (*90*) made similar recommendations. This group recommended that the percentage passing the No. 40 sieve be from 50 to 100; that passing the No. 200 sieve be not more than 35; that the liquid limit be less than 30 per cent; and that the plasticity index be less than 10 per cent.

More recent specifications are generally similar. The Asphalt Institute recommends (*110*) for either "sand or soil"; not more than 25 per cent passing a No. 200 sieve, a "sand equivalent" of not less than

25 or a plasticity index of not more than 6. The sand equivalent is a value determined by an accelerated grain-size test (AASHO T176, Art. 14–10).

In addition to the soil itself, the components of a soil-bitumen mixture usually are water and bituminous material. Water is used to facilitate compaction of the mixture and to aid in the dispersion of the bituminous material uniformly throughout the mix. In general, the amount of bituminous material which has been used has varied from 4 to 7 per cent. The sum of the percentages of water and bituminous material used should not exceed that which will fill the voids of the compacted mixture. Many different types of bituminous materials have been used in this type of stabilization, including rapid-curing, medium-curing, and slow-curing liquid asphalts. Tars have also been used, with grades RT-3 through RT-6 most widely recommended. Asphalt emulsions have also been used in some areas.

The choice of a bituminous material for use in a given project is primarily dependent upon local experience and comparative costs.

Of the cut-backs, rapid-curing grades are recommended for use in extremely sandy soils, or those containing a minimum of silt and clay particles. As the percentage of silt and clay increases, a medium-curing cut-back is recommended, since it will mix better and give a more homogeneous mixture. Slow-curing liquid asphalts have been used successfully with soils which contain from 30 to 40 per cent of silt and clay, particularly in arid regions. As to grade of cut-back asphalt, the general rule is "use the heaviest asphalt that can be readily worked into the soil." Asphalt emulsion should be of mixing grade.

The proportions of soil, water, and bituminous material are generally determined in the laboratory. Laboratory practice in this respect varies from organization to organization. ASTM has adopted a tentative standard method of test for evaluating mixtures of this type, ASTM D915, Tentative Method of Testing Soil Bituminous Mixtures. In this method, the required bitumen content is based upon two determinations. First, the shearing resistance or "stability" is determined by an "extrusion" shear test which is a modified Hubbard-Field Stability Test which is described in Chapter 17. Second, the absorption characteristics of the mixture being considered are determined by observation of changes in volume and weight of a sample exposed to partial immersion. In a typical series of tests a number of specimens would be prepared with different bitumen contents, and the stability and absorption characteristics corresponding to each bitumen content measured. A standardized procedure is used in the preparation of the test specimens, water may be included in the mixtures, and the mixtures may be cured before test specimens are formed, as deemed desirable. The bitumen content which would be selected would be the minimum amount giving satisfactory stability and waterproofing.

Suggested standards (*137*) for the evaluation of cut-back asphaltic mixtures tested in accordance with ASTM D915 are as follows, based upon uncured specimens:

Extrusion value before the absorption test—1000 pounds minimum.
Extrusion value after the absorption test—400 pounds minimum.
Expansion during the absorption test—5 per cent maximum.
Per cent water absorption during the absorption test—7 per cent maximum.

Specimens containing emulsified asphalt are usually tested in a cured condition.

Tests used by the Asphalt Institute in evaluating mixtures of cut-back asphalt with a silt, a clay, and a loess soil included ASTM D915, California Stabilometer Measurements (Art. 12–19), and the Marshall Stability Test (Art. 17–7) (*62*).

The optimum moisture content corresponding to the compactive effort to be used in the field may also be determined for the mix finally selected. Many construction agencies work from the optimum moisture content of the raw soil, considering that the bituminous material has from one-half to equal the wetting effect of water.

13–13. Construction Methods. The thickness of soil-bitumen bases varies from as little as 3 to more than 9 inches. In any case of flexible pavement design, the base thickness required should be based upon a procedure similar to those described in Chapter 12. In the case of soil-bitumen bases, the soil material may be the existing subgrade soil, the existing soil used in combination with imported materials, or wholly imported materials. The pulverized soil mixture may be combined with the necessary bituminous material by road-mixing, by the use of a traveling plant, or by central mixing methods. Methods of mixing which are used are generally the same as those which have been previously described for granular soil stabilized mixtures. An additional step, of course, involves the application of the necessary bituminous material and its uniform distribution through the mixture. In the case of road-mixing, the bituminous material, which sometimes has been heated moderately to the specified temperature for application, is applied in measured amounts by a pressure distributor, along with water applied separately, and the entire mix thoroughly blended by motor graders, disc harrows, rotary speed mixers, and so on. Where a traveling or central mixing plant is used, the measured amount of bituminous material is incorporated in the plant. In Fig. 13–5 is shown a traveling plant engaged in bituminous soil stabilization. The mixture must be thoroughly aerated before compaction, and most of the moisture and volatiles removed.

When aeration is complete, the material is spread in a layer of uniform thickness ready for compaction. The thickness of the layer may vary from 2 to 6 inches, depending on job conditions. A "layer" method of construction is preferred in some areas, with the required thickness of the base being built up by the construction of successive layers, approximately 2 inches in compacted thickness. Compaction usually is

Fig. 13–5. A Barber-Greene traveling plant in action on a county road project in Chippewa County, Wisconsin. (Courtesy Barber-Greene Co.)

done with pneumatic-tired rollers. Compaction is continued until a specified density (expressed in terms of laboratory density) is achieved. After compaction is complete, it may be required that some "curing" time elapse before the next layer is placed so that the water content after construction may be maintained at the desired figure. It should also be noted, at this stage, that the mixture may be compacted without the addition of water, as such, especially when asphalt emulsions are being used. Rolling of the final layer may be accomplished by the use of steel-wheel or pneumatic-tired rollers, and the base must be finished to the final section shown on the plans.

It should be emphasized that this type of soil stabilized mixture is generally only satisfactory for base construction. Placing of a wearing surface would then generally follow shortly after completion of the base and before any considerable amount of traffic would be permitted to use the road.

Sand-Bitumen

The type of construction which is covered under this heading refers to construction of a base course of variable thickness—usually from 3 to 10 inches—using sand, which may or may not be combined with a mineral admixture, in combination with various bituminous materials. The sand may be that existing in the roadway, or river, pit, beach, or dune sand, all of which should be free from vegetable matter and deleterious clay.

13–14. Material Requirements. The Highway Research Board committee, in the publication previously referred to, presented definite recommendations relative to the sand which should be used in combination with various bituminous materials. Recommendations relative to sand were presented in two parts, a requirement relative to the size and nature of the sand used, regardless of the bituminous material, and a stability requirement for the sand when used in combination with one of the three general types of bituminous materials.

The following are the recommendations of the committee relative to the size and nature of the sand, which are quoted directly, and those for stability, which are summarized.

13–15. Size and Nature of Sand.

> The sand or sand-mineral admixture blend should contain not more than 12 per cent of material passing the No. 200 sieve. When wind blown or dune sand is used, up to 25 per cent passing the No. 200 sieve may be allowed, provided that the portion of the sand or sand-mineral admixture blend passing the No. 40 sieve meets the following requirements:
>
> Field moisture equivalent, not more than 20 per cent.
> Lineal shrinkage, not more than 5 per cent.

13–16. Stability of Sand for Cut-back Asphalts. The sand used should be of such a nature that a mixture of the sand and cut-back asphalt which is to be used, with the cut-back asphalt in the amount of 4.5 per cent by weight of the dry sand, after curing in an oven at 100°F in a loose layer for 16 hours; molded, compacted, and tested at 77°F; will develop a Hubbard-Field Stability of 1200 pounds or more. If the sand does not meet this requirement, it may be judged satisfactory if a combination of the sand and mineral admixture (crushed stone, gravel, slag, etc.) produces a mixture which does meet this requirement and which also meets the requirements given above under "Size and Nature," Section 13–15.

13–17. Stability of Sands for Tars. The sand used should have a Florida Bearing Value of 25 pounds per square inch or more. The Florida Bearing Value is a punching shear test involving the loading of a circular plate, 1 square inch in area, placed on the surface of a compacted sample of relatively dry sand. Florida Bearing Values range from 8 or 10 pounds per square inch for rounded sands of nearly uniform size ("ball-bearing sand") to 220 or more for crushed rock. Somewhat similar requirements were given by the committee for the sands in mixes using emulsified asphalt.

13–18. Sands for Emulsified Asphalts. The American Bitumuls and Asphalt Company has presented recommendations (*199*) relative to gradation and bearing value for emulsified asphalt-sand mixtures for base courses. Recommended gradation is as follows:

	Per Cent Passing, by Weight
1-inch sieve	80–100
No. 4 sieve	60–100
No. 10 sieve	50–100
No. 200 sieve	0–25

Untreated sand for use in mixtures of this type should have a minimum bearing value of 30 pounds per square inch.

13–19. Bituminous Materials. The following bituminous materials are in general use:

Cut-back asphalts: RC–70, 250 and 800; MC–250 and 800.
Road tars: RT–6 through RT–10.
Emulsified asphalts: slow-setting or "slow-breaking" types.

13–20. Amount of Bituminous Material. The amount of bituminous material to be used in this type of construction is dependent largely upon local knowledge and field experience. Water may also be added to facilitate mixing and compaction. The amount of bitumen required may be based upon the results of a shearing test, such as the Hubbard-Field Stability Test or some other, which has been correlated with field experience. ASTM D915 is also applicable to this type of bituminous mixture. The amount of water to be used will generally be based upon moisture–density relationships.

13–21. Construction Methods. Methods of construction used in the placing of sand-bitumen bases are generally the same as those which were described for soil-bitumen roads. A road-mix process (Fig. 13–6) may be used or travel or central plant facilities may be employed. If the base is to exceed five inches in compacted thickness, then it should be constructed in two layers. In the road-mixing process it is recommended that the bituminous material be applied in successive applications of 0.5 gallon per square yard until the desired amount of bituminous

Fig. 13–6. A high-speed rotary mixer processes a sand-asphalt base course mixture in Oklahoma. (Courtesy Buffalo-Springfield Division, The Koehring Co.)

material has been incorporated into the mix (0.5 gallon per square yard would be approximately one per cent by weight for a 5-inch compacted thickness). When emulsified asphalts are used, very small quantities of hydrated lime or portland cement may be added last in the mixing process to facilitate drying and setting. After mixing is complete, the material must be aerated or "cured" until the liquid content has been judged satisfactory for compaction. This curing period may involve several days or weeks, depending on job conditions, and frequently it is continued until nearly all the volatile materials are removed. Aeration is generally accomplished by spreading the material in thin layers, followed by frequent manipulation or "turning" of the mix with blade graders or multiple-blade drags.

Compaction is generally achieved through the use of pneumatic rollers, the final rolling being performed with tandem rollers. Blading and final rolling are continued until the specified cross-section is secured. A seal coat consisting of an application of 0.2 to 0.3 gallon per square yard of the same bituminous material used as a binder, covered with coarse sand or fine stone chips, may be placed two or three weeks after final rolling. A higher type of bituminous wearing surface may, of course, be employed if traffic conditions make it necessary.

Oiled Earth Surfaces

This type of treatment is quite different from the other types of bituminous soil stabilization which have been discussed. No mechanical mixing is involved in this construction process, as the liquid bituminous material is simply applied to a properly prepared natural soil surface. The material penetrates a short distance into the soil layer, thus preventing dust and forming a thin "stabilized" surface which will be capable of supporting a limited number of light vehicles. Over a period of time, with proper annual retreatment, a stabilized surface may be built up to a thickness of several inches. Its principal use has been in the construction of very lightly traveled local or "service" roads. This type of construction also has been widely used in areas where cracked or uncracked residual oils are cheaply available.

13–22. Material Requirements. The soil, of course, will be that naturally existing in the section concerned, and the method is principally applicable to silt and clay soils. The bituminous material which is used should generally be a product of low viscosity which contains a wide range of volatile materials or "light fractions." The material which has probably been most widely used is slow-curing liquid asphalt, generally grade SC–70 or SC–250. Medium-curing cut-backs, MC–70 and MC–250, have also been successfully used. It should be noted here that a large number of cracked residuals have also been used, as in the construction of access roads in oil fields. Tars have been used to some extent, with grade RT–2 being generally preferred. The amount of bituminous material or "road oil" which has been used in initial treatment during the first year is generally about 1 gallon per square yard. Applications during successive years are somewhat less.

13–23. Construction Methods. The natural soil surface must be adequately prepared by crowning for drainage and by compaction. The surface to which the oil is applied should be free from dust and should be slightly moist in order to facilitate penetration of the oil. Generally speaking, treatment should follow immediately after the final shaping (grading) operation. The oil is usually applied in two or three applications, the total amount applied being as previously indicated. Some time should be allowed to elapse between applications, and traffic should not be allowed on the road until the oil has completely penetrated the surface. After each application, the surface should be compacted under traffic or by the use of pneumatic-tired rollers.

Soil-Cement Roads

A type of soil stabilization which has enjoyed increasing popularity in recent years involves the incorporation of portland cement, in amounts generally varying from 7 to 14 per cent by volume of the compacted mixture, with naturally occurring or artificially created soils or soil-aggregate mixtures. This type of construction is generally employed in the formation of base courses, usually with thicknesses varying from 4 to 6 inches. A soil-cement mixture may serve as a base for a thin wearing surface which will be subjected to light or medium traffic, or as a support for a high-type flexible or rigid pavement. Bases of this type have been successfully used in the construction of city streets, and soil-cement is also suitable for use in driveways, shoulders, parking areas, some airport runways, and so on. In some areas, particularly California, this general type of construction is termed "cement-treated base."

Modern use of soil-cement began in South Carolina in 1933 with the construction of experimental sections, and the quantity of soil-cement used in highway construction has steadily increased since that time. Many millions of square yards of soil-cement have been constructed in the United States.

13–24. Material Requirements. Nearly all subgrade soils, with the exception of those which contain high percentages of organic material, may be stabilized through the use of portland cement. Soils which contain high amounts of fine material, such as silts and clays, generally require large percentages of cement for successful stabilization. Gradation requirements for soils to be used in soil-cement mixtures are nearly nonexistent, since practically any soil may be utilized. Sandy and gravelly soils with from 10 to 35 per cent silt and clay combined have the most favorable characteristics. Glacial and water-deposited sands and gravels, crusher-run limestone, *caliche,* limerock, and almost all granular materials are good, if they contain 55 per cent or more material passing the No. 4 sieve. Exceptionally well-graded materials may contain up to 65 per cent material retained on the No. 4 sieve, and still have sufficient fine material for adequate binding. The maximum size of aggregate in soil-cement mixtures should not be more than 3 inches. Materials in old gravel or crushed stone roads or bases make excellent soil-cement. An important factor in determining the suitability of a given soil for soil-cement construction is the ease with which the soil may be pulverized. Soils which contain large amounts of clay will be more difficult to pulverize, will require more cement, and will demand more careful control during the construction process.

Obviously, with requirements relative to gradation being as loose as they are, the subgrade soil naturally existing in the roadbed will usually be suitable for stabilization. However, it must be noted that in many cases, especially where the existing soil contains large amounts of silt or clay, economies may be effected by the use of imported soil materials to improve the gradation and reduce the plasticity of the soil which is to be stabilized. A procedure of this sort might well result in a reduction in the cement content required for stabilization. The relative economy of using imported soil materials in an effort to improve the characteristics of the soil to be stabilized, as compared with the use of increased quantities of cement, can only be determined by laboratory study coupled with knowledge of local materials and cost factors.

The cement used is generally standard portland cement, Types 1 and 1A. The third ingredient of soil-cement mixtures is water, which is necessary to aid in the compaction of the loose mixture and for the hydration of the cement in the mix. In this respect, soil-cement mixtures are somewhat similar to concrete in that water is a necessary ingredient, and the loss of excessive amounts of water during the curing period must be prevented if the mixture is to harden properly. Practically any normal source of water may be used, although it should be quite clean and free from excessive amounts of organic matter, acids, or alkalies. Fresh water is, of course, normally used, although sea water has been used in a few instances. The proper amounts of cement and water to be used on a given project are determined by a series of laboratory tests which are briefly described in the following paragraphs. Information secured in the laboratory will also be used in the control of construction, particularly in the control of the compaction of the mixture.

13–25. Laboratory Tests. Detailed laboratory testing procedures for soil-cement have been developed by the Portland Cement Association (*130*) and adopted by AASHO and ASTM. Three general procedures are in existence, depending on the time available for testing and the size of the project.

On major projects, when time is available for testing, and with all soils, a detailed procedure to establish the satisfactory minimum cement content is recommended. As a first step in the laboratory procedure, the subgrade soil encountered in the project would be subjected to routine classification and placed in one of the groups of the AASHO classification system. With this information as a background, cement contents to be used in the preparation of trial mixtures would then be established on the basis of past experience. For example, it might be decided that mixes would be prepared using 8, 10, and 12 per cent of cement by volume. It should be emphasized that the percentage of cement used here is expressed in terms of the volume of the compacted mixture rather than by weight; that is, the per cent of a sack of cement per cubic foot of compacted soil cement. Thus, 10 per cent by volume means that there is 10 per cent of 94 pounds, or 9.4 pounds, of cement in 1 cubic foot of finished, compacted roadway.

Moisture-density relationships for soil-cement mixtures containing the selected amounts of cement are then determined, using a procedure entitled "Methods of Test for Moisture-Density Relations of Soil-Cement Mixtures," ASTM, Designation D558. The procedure used is very similar to that employed in the standard AASHO compaction procedure. The percentage of cement required for satisfactory stabilization is based upon two series of durability tests, and the moisture-density tests supply information necessary for the formation of the durability test specimens containing the desired amounts of cement. The durability tests used involve the subjecting of the various soil-cement mixtures to cycles of freezing and thawing, and of wetting and drying. The test methods used are "Methods of Wetting-and-Drying Test of Compacted Soil-Cement Mixtures," ASTM, Designation D559, and "Methods of Freezing-and-Thawing Test of Compacted Soil-Cement Mixtures," ASTM, Designation D560.

The detailed laboratory procedures involved in the durability tests are much too lengthy to be presented here. However, they may be generally described. For each of the durability tests and for each of the selected cement contents, two specimens are prepared by compaction at optimum moisture to maximum density. Both samples are cured for seven days before testing is begun. After curing, both samples are subjected to 12 cycles of alternate wetting and drying. At the end of each cycle, one specimen is vigorously brushed, using a standard procedure. Moisture and density determinations are made on the other, unbrushed specimen during each test cycle. At the end of the test, the specimens are oven-dried to constant weight, and the weight losses determined. A similar procedure of alternate freezing and thawing is used to test the two specimens prepared for that test. Although the cement requirements are principally established on the basis of the results of the described durability tests, in some cases the compressive strengths of the test mixtures also are determined. The following are the recommendations of the Portland Cement Association relative to the behavior of the test specimens (*130*).

1. Losses during 12 cycles of either the wet-dry or freeze-thaw test shall conform to the following standards:
   Soil groups A–1–a, A–1–b, A–3, A–2–4, and A–2–5, not over 14 per cent.
   Soil groups A–2–6, A–2–7, A–4, and A–5, not over 10 per cent.
   Soil groups A–6, A–7–5, and A–7–6, not over 7 per cent.
2. Compressive strengths of soil-cement test specimens should increase with age, and with increases in cement content in the ranges of cement content producing results meeting the requirements given above.

A short-cut test method has been developed (*130*) by the Portland Cement Association for determining cement factors for sandy soils. The procedure is designed for use on smaller projects, particularly those where testing facilities and manpower are limited; it results in safe, but not necessarily minimum, cement contents for these soils. The only tests required are grain size analysis, moisture-density tests, and compressive strength tests. Results of the tests are used in conjunction with charts based upon past experience involving thousands of test specimens to establish safe cement content.

Further simplification of the testing procedure is possible for very small projects or under emergency conditions. One such method is to use available information regarding the soil series (pedological soil classification system) and cement content necessary for stabilization on previous projects involving the same soil series. Another rapid method involves the molding of three specimens at optimum moisture and maximum density, at cement contents of 10, 14, and 18 per cent. After a day or two of hardening, during which they are kept moist, and after a three-hour soaking, the specimens are inspected by "picking" with a sharp-pointed instrument and by sharply "clicking" each specimen against a hard object to determine its relative hardness when wet. Selection of the cement content is based upon the results of the "pick" and "click" tests.

13–26. Construction Methods. The basic steps in the construction of a soil-cement base, assuming that the subgrade beneath the base requires no special treatment or has previously been brought to the desired condition, may be listed as follows:

1. Pulverizing of the soil which is to be processed.
2. Spreading of the required amount of cement and mixing with the soil.
3. Addition of the required amount of water and its incorporation with the soil-cement mixture.
4. Thorough compaction, including final rolling and finishing.
5. Curing of the completed soil-cement base.

Before discussing these operations in some detail, it must be emphasized that the operations must be carefully coordinated and controlled if best results are to be secured. On many jobs the processing operations are carried on in a planned sequence, with little time elapsing between separate steps. Equipment is frequently operated in complete units, each containing several pieces, which are often termed "trains." The "train" might include spreaders and mixers, rollers, and finishing tools. Usually one lane is processed at a time. Thus, pulverization, "dry" mixing, the addition of water and "wet" mixing, and compaction would follow step by step with little lost time or motion. Often the "dry" mixing step is omitted, with all the water added at one time, or the soil may be prewetted before mixing begins. For effective results, careful control must be exercised to see that a uniform mixture of the proper proportions of soil, cement, and water is secured and that the mixture is adequately compacted. Prompt mixing and laying of the mixture are necessary in order to ensure final compaction before appreciable setting of the cement occurs.

Soil-cement mixtures may be "processed" by road-mixing equipment, by traveling plants, or by central mixing plants. In the large majority of cases, traveling plants are used, including those of the flat type, the windrow type, and rotary speed mixers. The operation of traveling and central mixing plants in other types of soil stabilization has been discussed earlier in this chapter. Their use in this type of construction is not greatly different from that which has been previously described.

13–27. Pulverization. With soils which are difficult to process, the soil which is to be used must be thoroughly pulverized before cement is added. If the existing soil is to be used it sometimes must be scarified to the desired depth, by the use of a ripper or a scarifier attachment on a motor grader. If imported material is being used, it must be hauled in and spread on the existing soil to the required depth. The soil may then be pulverized by the use of offset disc harrows, gang plows, or rotary speed mixers. It is generally required that the soil, with the exception of gravel or stone, be pulverized until at the time of compaction 100 per cent of the soil-cement mixture will pass a 1-inch sieve and at least 80 per cent will pass a No. 4 sieve. It is especially important that the moisture content of fine-grained soils be within rather narrow limits during pulverization. For some soils the correct moisture content for pulverization will be at or near the optimum moisture content. Aeration of a wet soil or the addition of water to a dry soil may be necessary, and in some cases steps may be taken to protect the pulverized soil from moisture change before the next step in the construction process is begun. As indicated previously, the ease with which a soil may be pulverized is frequently an important factor when soils are being selected for soil-cement construction. With easily pulverized soils, this preliminary step is not necessary.

13–28. Incorporation of Cement. The proper amount of cement may be spread upon the surface of the soil either by hand or by mechanical means. On small jobs, bags of cement are "spotted" by hand along the surface in rows of predetermined spacing. The bags are then opened and the cement spread in uniform transverse rows, also by hand. Spreading is completed by the use of a spike-toothed harrow. On most projects, bulk cement is used and a mechanical spreader is utilized (Fig. 13–7). When spreading of the cement has been properly completed, dry mixing is ready to begin.

FIG. 13–7. Spreading bulk cement on a road stabilization project in Wisconsin. (Courtesy *Engineering News-Record*.)

13–29. Mixing. In the road-mix process, blending of the soil and cement may be done by the use of such tools as gang plows, disc harrows, spring-tooth cultivators, and rotary tillers (Fig. 13–8), but usually is done with a traveling plant. When a traveling plant is used, mixing, of course, is carried out within the plant; exact processing procedure will vary with the type of equipment used. With a central plant, mixing is carried out in the plant and the mixture hauled to the job site in trucks; such a procedure is used on projects involving borrow soils. Careful control is necessary in a road-mix operation in order that the mixing be carried to the proper depth and that a uniform mix be secured. The equipment available for this purpose will generally be operated in a train, with additional units being added to the train as construction proceeds, to supply water and complete the wet mixing operation. In many cases dry mixing and wet mixing are not essentially separate operations but are carried along together in an integrated series of operations. The proper amount of water is added, and the soil, cement, and

FIG. 13-8. A pulverizer-mixer blends cement and sand to form a base for a concrete pavement. (Courtesy *Engineering News-Record.*)

water are thoroughly blended to a uniform mixture, ready for compaction. The amount of water added is generally enough to increase the moisture content to 1 or 2 per cent more than the optimum desired for compaction. This is done because of the loss of water which may occur during road mixing operations, as compared with laboratory mixing. In other words, if a slight excess of water is incorporated during mixing, the soil may then generally be expected to be very close to optimum by the time compaction is ready to begin.

13-30. Compaction and Finishing. Initial compaction of a soil-cement base is generally achieved by the use of sheep's-foot rollers. After the base has been compacted to within 2 or 3 inches of the surface, the layer is brought to shape with a motor grader. Rolling then generally continues until satisfactory density has been secured to within about 1 inch of the surface. In very sandy soils, of course, compaction may not be possible with sheep's-foot rollers, and densification may then be accomplished by the use of pneumatic-tired rollers. Steel-wheel rollers have been used with granular soils, and newer types of compaction equipment, including vibratory compactors, grid, and segmented rollers are finding increasing use. After the soil-cement mixture has been compacted as described, the rollers are taken off and the surface brought

FIG. 13–9. A pneumatic-tired roller compacts a granular base. (Courtesy The Galion Iron Works & Mfg. Co.)

to final shape. The final shaping is done in a variety of ways, including steps to remove the compaction planes left in the surface by the rollers. Final compaction is secured by rolling with pneumatic-tired rollers alone or in combination with steel rollers weighing from 3 to 12 tons, depending on the soil involved. Rolling is continued until a tightly packed surface is secured. In Fig. 13–9 is shown a pneumatic-tired roller engaged in rolling of a section of granular base; the operation is the same on soil-cement. During all the rolling operations which have been described, it is essential that the moisture content be maintained at optimum. Frequent checks on the density and moisture content are, of course, necessary.

13–31. Curing. The water contained in the soil-cement mixture is necessary to the hardening of the cement, and steps must be taken to prevent loss of moisture from the completed base by evaporation. Curing commonly is done by the application of a light coat of bituminous material in most cases; commonly used bituminous materials are RC–250, MC–800, RT–5, and asphalt emulsions. Rate of application varies from 0.15 to 0.30 gallon per square yard. A bituminous wearing surface should be placed on the completed soil-cement base as soon as practical.

Miscellaneous Materials Used in Soil Stabilization

Primarily because of the dictates of economy, many materials other than those which have been discussed here have been used for soil stabilization at various times and in various places. A few of the more important of these materials are discussed in the following paragraphs. The discussion is not intended to be complete but will serve only to indicate the wide range of materials which may be used in construction of this sort.

13–32. Lime. The incorporation of small amounts of hydrated lime has proved to be effective in the improvement of certain plastic clay soils. Lime has been used—principally in states along the Gulf Coast—to reduce the plasticity, shrinkage, and swell of clay soils, at the same time somewhat increasing their bearing capacity. In essence, its use has permitted the upgrading of certain marginal and sub-marginal soils into satisfactory base and subbase materials.

The use of lime also has certain advantages from the standpoint of the construction process by improving the workability of plastic soils (making them easier to pulverize). It tends to waterproof the soil to some extent and allow it to dry out more quickly when saturated, thus speeding construction.

The quantity of lime used in subgrade treatment is generally from 3 to 6 per cent. Quantities as low as 1 per cent have been used, while amounts much above 6 per cent are not economical. Depth of treatment is usually 6 inches. The subgrade soil is scarified and pulverized, followed by spreading of lime, usually with a mechanical spreader or a bulk hauler (Fig. 13–10). Enough water is added to bring the moisture content to 5 per cent or more above optimum, and the water distributed by a rotary speed mixer. The lime-soil mixture is allowed to cure for periods from 1 to 7 days. Mixing and pulverizing then continue until all the material will pass a 1-inch sieve, and at least 60 per cent will pass a No. 4 sieve. Compaction is done by pneumatic rollers or vibratory compactors, and the compacted layer is allowed to cure for from 3 to 7 days before the next layer is placed. On some projects, the lime is added to the soil in the form of a slurry.

Construction procedure is quite similar to that used for soil-cement.

FIG. 13-10. Spreading lime on an airfield stabilization project. (Courtesy National Lime Association.)

A considerable amount of research has also been devoted to the use of a combination of lime and flyash for soil stabilization.

13-33. Salt. Rock salt has been successfully used in the stabilization of soils for bases and wearing surfaces subjected to very light traffic. The procedure is very similar to that described for mixtures in which calcium chloride is used. Recommended quantity of salt is generally from 1 to 2 pounds per square yard per inch of loose thickness.

13-34. Vinsol Resin. Vinsol Resin is a powdered substance which is obtained by the steam distillation of pine stumps. It is a water-repellent material, and when mixed with certain soils it serves to improve their quality by its waterproofing action. In the construction process, the Vinsol Resin is incorporated in small amounts as determined by laboratory tests and mixed "dry" with the soil to be stabilized. An alkaline solution is then added, and the materials wet-mixed. The stabilized mixture is then compacted and allowed to cure before a wearing surface is placed on the base. This material was used to some extent during World War II, when more conventional materials were in short supply.

13–35. Other Materials. The use of lignin binder, which is produced as a by-product in the manufacture of paper from wood by the use of the sulfite process, has been reported experimentally, and a limited mileage of roads has been constructed by the use of this material. Similarly, various silicates of soda have been used experimentally to improve certain clay soils.

A number of resinous waterproofing materials, including Stabinol and Vinsol, have been used experimentally; none is in widespread use. A considerable amount of study has been devoted to a variety of resinous bonding materials for soil stabilization. Most successful of these have been a combination of aniline and furfural, and calcium acrylate; both offer considerable promise for the future, but are currently too expensive for ordinary civilian use.

13–36. Foamed Plastic. Although it is not soil stabilization in the sense that the term has been used in this chapter, considerable interest has been shown in the use of sheets of foamed plastic to prevent frost damage to pavements.

Experiments of this nature have been conducted in Minnesota, Iowa, Ontario, and Manitoba. The plastic is foamed polystyrene. The sheets are ¾ to 1¾ inches thick, have been laid in a double layer in some cases, and are placed directly atop the subgrade beneath the base course of a flexible pavement or directly beneath a concrete pavement.

PROBLEMS

13–1. It is desired to combine Material A, which is the soil existing in the roadbed, with Material B, which may be obtained from near-by borrow sources at low cost, to form a stabilized soil-aggregate surface course conforming to the specification recommended by the AASHO, Table 13–1, Grading B. The sieve analyses of Materials A and B are as follows:

	Per Cent Passing (by weight)	
Sieve Designation	Material A	Material B
2-inch	100	–
1-inch	89	100
⅜-inch	68	45
No. 4	42	27
No. 10	33	15
No. 40	28	5
No. 200	10	1

Determine approximately the limiting proportions of A and B which should be used to produce a stabilized mixture of the desired gradation. That is, determine the greatest proportion of A that can be used in a mixture and still meet the specifications, then the smallest proportion of A that will meet the same objective. Plot the gradation curves of the two blends and select the mixture which you think best fits the specifications. What other requirements would normally be included in the specification for this material?

13–2. For the materials of Problem 13–1, the following are the plasticity characteristics. Material A: liquid limit, 30; plasticity index, 8. Material B: liquid limit, 12; plasticity index, 2. Estimate the liquid limit and plasticity index of the aggregate blend selected in Problem 13–1.

13–3. Assuming that the compacted mixture of Problem 13–1 has a dry unit weight of 121 pounds per cubic foot, determine how many tons of Material B will be required per mile of road, assuming that the surface is to be 24 feet wide and will have a compacted thickness of 8 inches. Estimate the number of cubic yards (loose measure) of this material that will be required per mile, assuming that the loose material weighs 107 pounds per cubic foot.

13–4. How many tons of calcium chloride will be required per mile of the surface of Problem 13–1 if the initial application of this material is made at the rate of 0.5 pound per square yard per inch of thickness? What will be the estimated cost of this calcium chloride in your area?

13–5. The optimum moisture content of the specified surface material of Problem 13–1 is 16 per cent. If the moisture content of the combined surface material, after mixing, is 9 per cent, how many gallons of water will have to be added for each mile of road to obtain optimum moisture? Ignore losses by evaporation.

13–6. A soil-bitumen mixture has a compacted dry unit weight of 112 pounds per cubic foot. How many gallons of cut-back asphalt will be required per mile of base if 6 per cent (by weight) of bituminous material is used, and the base is to be 24 feet wide, with a compacted depth of 6 inches? Estimate the percentage of asphalt binder (cement) which will remain in the soil after all the volatiles have been removed by evaporation if the material originally added to the soil is MC–800 (Table 11–5).

13–7. How many sacks of cement will be required per 100-foot station for a soil-cement mixture containing 11 per cent cement by volume if the base is 24 feet wide and 6 inches thick?

13–8. Consult current literature (for example, the technical bulletins of the American Road Builders Association) and prepare a report on a highway or airport project on which lime stabilization was used successfully.

13–9. What types of soil stabilization are used in your area? Select and briefly describe a project in your area which involved soil stabilization with locally available materials. Give a clear idea of the construction methods used.

13–10. From current literature, prepare a brief report describing a laboratory investigation or field project in which a soil was stabilized by the use of some material other than a granular material, calcium chloride, lime, cement, bituminous material, or salt.

CHAPTER 14

MACADAM ROADS

The term "macadam," as applied to roads in the United States at the present time, has come to mean road surfaces and bases which are constructed of crushed or "broken" stone, gravel or slag fragments cemented together by the action of traffic, rolling, and water, portland cement or bituminous materials.

Crushed-stone roads first constructed in Europe during the latter part of the eighteenth and the early part of the nineteenth centuries were the forerunners of modern macadam roads. Construction of this type was introduced in France by Tresaguet about 1765, and by Telford in England shortly after 1800. Both of these early engineers used large pieces of broken stone in the lower course and placed a layer of finer crushed stone as a wearing surface. Telford advocated the use of large, flat pieces of stone, which were hand placed to form the base course. McAdam, whose name was later to be immortalized among highway engineers by its application to this type of construction, insisted upon the use of smaller stone—about 1.5-inch maximum size—for the entire thickness of the "pavement." The first roads of this type were built by him in England early in the nineteenth century. Both Tresaguet and McAdam laid emphasis upon the grading and drainage of the subgrade and the crowning of the crushed-rock surface to aid in the removal of rainfall from the surface of the road. The technique of macadam construction improved rapidly over the years, and by the end of the nineteenth century many miles of this type of surface had been built and were in service as rural roads and city streets in the United States. With the advent of the motor vehicle, the basic macadam or broken-stone surface became less suited to the demands of traffic, and this type of pavement assumed less importance in highway construction. Roads of the fundamental macadam type were (and are) basically unsuited for use as wearing surfaces carrying large numbers of rapidly moving pneumatic-tired vehicles as they became quite rough and dusty under this type of traffic.

The types of macadam roads which are of greater or lesser importance in modern highway construction are four in number and may be defined as follows:

Traffic-bound macadam is a wearing surface composed of broken stone which is consolidated by the action of traffic, intermittent blading or dragging, and rain. This type of surface is generally built up gradually by the successive application of two or more layers, each of which is one or two inches in compacted thickness. Construction methods for this type of macadam are quite similar to those of the crushed-rock variety of soil-aggregate roads described in the preceding chapter. For this reason, no further information relative to them will be presented in this chapter.

Water-bound macadam is a layer composed of broken-stone (or crushed gravel or slag) fragments which are bound together by stone dust and water applied during construction, in connection with consolidation of the layer by a heavy roller or a vibratory compactor. This is the type of modern macadam road which most closely resembles those so widely used in the early days of road building, although it is now principally used only in base construction.

Bituminous macadam is a crushed-stone or crushed slag base or wearing surface in which the fragments are bound together by bituminous material; the aggregate layer is compacted, and bituminous material is applied to the surface of the layer. The bituminous material then penetrates into the voids of the compacted layer and serves to bind the fragments together. This type is frequently termed simply "penetration macadam."

Cement-bound macadam is a type of pavement which resembles the bituminous macadam defined above, except that the bituminous material is replaced by portland cement mortar or "grout," which is forced into the voids of the compacted stone layer. This type of macadam is rarely used in highway construction at the present time, and will not be discussed in detail here.

Two of the above types are of considerable prominence in modern highway practice. These types are the water-bound macadam, which is generally used in the construction of bases, and bituminous macadam, which is used both in base and surface construction. The distribution of macadam construction of all types is not uniform over the country, geographically speaking, and is generally limited to areas in which suitable aggregates are cheaply available.

As has been emphasized several times in previous sections of this book, the preparation and condition of the subgrade are vital factors in the satisfactory performance of any base or wearing surface. Macadam roads are flexible or semirigid in nature, and failures and deformations which occur in the subgrade would certainly be expected to show up in the base and wearing surface after the passage of some time. For this reason and, in addition, because of the fact that surface irregularities which do occur in this type of pavement are much more difficult to correct than in some other types of construction, the condition of the subgrade before construction of a macadam layer begins is of great importance. Weak spots in the subgrade must be eliminated before the base or surface is placed, and the subgrade soil brought to a high degree of uniformity, density, and stability.

Water-Bound Macadam

As has been indicated, water-bound macadams are road bases (or surfaces) which are constructed from crushed stone or crushed slag, cemented together by stone or slag dust and water. Limestone or dolomite are generally preferred when crushed stone is used because the dust produced by the crushing of these materials has some natural cementing properties. Many other types of stone have been used successfully however, including granite, traprock, and similar hard stones, and such relatively soft materials as limerock. Crushed and uncrushed gravels have also been used in this type of construction. Water-bound macadams are constructed in thicknesses ranging from about 3 to 12 inches, depending on the purpose for which they are intended. In almost all cases water-bound macadam layers are covered with a bituminous wearing surface of some sort soon after their construction, and the following discussion is primarily based on the assumption that the water-bound macadam will serve as a base course.

14–1. Material Requirements. Material requirements for this type of construction are well established, since roads of this kind have been built for many years in this country. Materials involved in this process are the crushed stone or slag and the screenings which are used to fill the surface voids in the compacted layer. Several different sizes of crushed stone are used by various organizations, depending on the material which is locally available and previous experience in this type of construction.

Grading requirements for aggregates to be used in water-bound macadams are given in Table 14–1 (ASTM D694–62). For a given base course, one of the three sizes of stone given in the table would be selected and used in combination with the screenings. In addition, crushed stone has a maximum permissible loss in the Los Angeles Abrasion Test of 50 per cent, and a maximum permissible loss of 20 per cent in the sodium sulfate soundness test (five cycles) or 30 per cent when magnesium sulfate is used. Crushed slag is subject to the same requirements relative to loss in the Los Angeles Test and must have a unit weight in excess

TABLE 14–1

GRADING REQUIREMENTS FOR AGGREGATES FOR USE IN WATER-BOUND MACADAM BASE COURSES (*135*)

Coarse Aggregate	Per Cent Pasing (by Weight)
	2–1-Inch Size
2½-inch	100
2-inch	90–100
1½-inch	35–70
1-inch	0–15
½-inch	0–5
	2½–1½-Inch Size
3-inch	100
2½-inch	90–100
2-inch	35–70
1½-inch	0–15
¾-inch	0–5
	3½–1½-Inch Size
4-inch	100
3½-inch	90–100
2½-inch	25–60
1½-inch	0–15
¾-inch	0–5
Screenings	
	No. 4 to 0 Size
⅜-inch	100
No. 4	85–100
No. 100	10–30

of 60 pounds per cubic foot. The portion of the screenings passing a No. 40 sieve must have a liquid limit of no more than 30 and a plasticity index not greater than 6.

14–2. Los Angeles Abrasion Test. The Los Angeles Abrasion Test is a quality test for coarse aggregates which are used for various purposes in highway construction, including those used in concrete, bituminous wearing surfaces of various types, and macadam roads. The standard method used in the conduct of this test is that entitled "Abrasion of Coarse Aggregate by Use of the Los Angeles Machine," AASHO, Designation T96. The principal element of the machine used in this test is a hollow steel cylinder which is 28 inches in inside diameter and 20 inches long. A shelf (usually a steel angle) three and a half inches wide and running the full length of the cylinder is mounted on the interior surface. In the conduct of the test, a sample of 5000 grams of aggregate is used. The sample is made to conform to one of four standard gradings, depending on the gradation of the coarse aggregate which is being subjected to the test. The prepared sample is put into the cylinder, along with a number of cast-iron or steel spheres, each of which is approximately 1⅞ inches in diameter. The number of spheres used with a given sample is again dependent upon the gradation, and varies from 6 to 12.

After the proper number of spheres has been placed in the cylinder, the cover is fixed in position and the cylinder caused to rotate at a speed between 30 and 33 revolutions per minute. Rotation is generally continued until 500 revolutions have been made, although various organizations specify a total number of revolutions varying from 100 to 500 for various purposes. After the requisite number of revolutions has been accomplished, the sample is taken from the machine and washed over a No. 12 sieve. The material retained on the No. 12 sieve is then dried and weighed. This weight is recorded, in grams, as the final weight of the test sample. The difference between the initial weight of the test sample and the final weight is divided by the initial weight and expressed as the "percentage of wear" in the Los Angeles Abrasion Test.

Specifications utilizing this test usually state the *maximum* percentage of wear which will be permitted in aggregates to be used for a given purpose. Thus the allowable per cent of wear might be some figure like 25 or 40, depending on the circumstances. With particular regard to macadam construction, this test serves to indicate the ability of the aggregate to resist breakdown under traffic compaction or rolling.

14–3. Construction Methods. As has been indicated, careful preparation of the subgrade is a necessity for any type of macadam construction. The subgrade is brought to the desired elevation and cross-section before any stone is laid, and this underlying layer must be uniform, dense, and well-drained. Some states require the placing of a 1- or 2-inch course of fine stone, or stone and sand, directly over the prepared subgrade before the placing of a water-bound macadam base. The purpose of this layer, sometimes called an "inverted choke," is to prevent the intrusion of fine soil into the lower portion of the macadam under service conditions, and it is especially important in areas where fine-grained soils, such as silt and clay, are encountered in the subgrade. This "insulation" course is carefully compacted and shaped to the desired section before the next layer is placed.

At the present time, two somewhat different construction methods are being used for water-bound macadam. The older method (described first) depends upon heavy steel-wheel rollers for compaction; the newer method, which is being used more and more, relies largely upon vibratory compactors. When vibrators are not used, and if the compacted thickness of the water-bound layer is not more than about 5 inches, it may generally be placed in one course, and some states permit one-course construction up to a compacted thickness of 6 inches. Two-course construction is used up to a total thickness of about 9 inches, while bases thicker than this would generally be placed in three lifts. String lines usually are set along the sides of the pavement at the elevation of the top of the loose layer. A side berm or the road shoulder generally is built up to confine the loose material. It is essential that the stone be laid in a uniform layer before rolling begins, as irregularities are much easier to correct in the loose layer than later. The compacted thickness will, of course, be less than that of the loose layer. The layer

Fig. 14–1. A self-propelled aggregate spreader laying base course material. (Courtesy Blaw-Knox Company.)

may generally be expected to compress to from 75 to 80 per cent of the loose thickness. For example, a 6-inch loose layer might become 4.5 inches.

Stone for use in the base course may be available from near-by quarries or crushing plants or it may be shipped to a convenient point by rail. In any event, the stone is usually brought to the job site in trucks and then must be spread in a uniform layer. The stone is spread by spreader boxes towed by trucks or large self-propelled aggregate spreaders, like the one shown in Fig. 14–1. Before rolling begins, all surface irregularities must be eliminated, and liberal use is made of templates and long straightedges in checking the uniformity of the spread material.

Rolling is done with a three-wheel power roller. Rollers used in this type of work are normally required to have a total weight of not less than 10 tons, although 8-ton rollers are used in some instances. Rolling is usually begun at the sides and progresses toward the center, except on superelevated curves where rolling begins on the low side and progresses toward the high side. Passes of the roller are made parallel to the centerline and are overlapped in such a fashion that the entire surface will be covered by the rear wheels as rolling proceeds. Rolling is generally continued until the material does not creep or wave ahead of the roller. Irregularities and weaknesses in the coarse-stone layer are corrected as rolling proceeds, with additional stone being added as required. The objective of this rolling is to thoroughly key the coarse material.

The next step is the application and keying of the stone screenings into the coarse layer. The stone screenings are usually spread by the use of mechanical spreaders. Spreading, brooming, and rolling operations are usually carried on at the same time and in conjunction with one another. Enough screenings are applied to fill the surface voids of the coarse layer and the screenings are broomed into the voids. The brooming may be accomplished by means of fiber or wire brooms attached to the roller units, by wire-broom drags, or by hand street brooms. Rolling is conducted as before and continued in conjunction with the spreading and brooming until the surface is firm and thoroughly compacted. Additional screenings may be added as needed, and any excess fine material is removed before the next step begins.

Following the application of stone screenings and the "dry-rolling" described above, the surface of the layer is sprinkled with water and rolled again. The sprinkling and rolling are continued until all the voids are filled and a wave of grout flushes ahead of the roller. Water is applied by a sprinkler mounted on pneumatic tires and hand brooms may be used to distribute the wet stone screenings evenly over the surface. The amount of water and screenings applied is dependent upon a number of factors, including the size and nature of the coarse stone, the properties of the screenings, and the type of surface desired on the lift or completed base course. The amount of each of these items actually used is largely dependent on the judgment of the engineer in the field.

If the completed layer is to be the bottom or intermediate course of a multiple-course water-bound macadam, it is allowed to dry thoroughly before the next course is placed. After the course has cured satisfactorily the next layer would be placed as before. The course which is to be the top layer of a water-bound macadam base or wearing surface may contain more stone screenings than the lower course in the interest of producing a more tightly bonded and smoother surface. At the time the top course is completed the surface is finally checked for irregularities and deviations of more than one-quarter (or one-half) of an inch, as measured by a 10-foot straightedge, from the specified cross-section and longitudinal grade corrected. The completed macadam sometimes is allowed to cure under traffic for a period from two weeks to 30 days before a bituminous surface is placed or before final acceptance if the layer is to function as a surface course. During this period additional sprinkling and rolling may be required if needed to satisfactorily bond the surface. In some cases calcium chloride is applied to the completed surface during this curing period.

Using vibratory compactors, recent experience has shown that courses of double the thickness formerly constructed can be satisfactorily keyed and filled. Several procedures are used to compact the layer of coarse stone. One method is as follows. To build an 8-inch macadam base, crushed stone is spread to a thickness of 11 inches, loose. The layer is vibrated first and then rolled; rolling and/or vibration is continued until the material is keyed thoroughly.

Vibrators are very effective in placing the screenings. An 8-inch course can be filled with only three applications of stone screenings. For the first application, one-half of the screenings required to fill the 8-inch course is spread evenly with a mechanical spreader over the keyed and compacted stone. One or two passes (depending on the size of coarse stone used) are then made with the vibrator. Then one-fourth of the screenings are applied and subjected to one pass of the vibrator; this is followed by spreading of the remaining screenings and an additional vibrator pass. This is followed by rolling, addition of fines where needed, and the application of water.

The vibrators (and rollers) are equipped with broom drags which are used to uniformly distribute the screenings. Filling the voids in the layer of coarse stone with screenings should follow closely behind the initial operations, so as not to leave long sections of the layer unfilled. Otherwise, rain may penetrate the layer and soften the subgrade. A multiple-shoe vibratory compactor is shown in Fig. 14–2.

In Fig. 14–3 is shown a completed water-bound macadam base course which has been swept preparatory to the placing of a primer and a bituminous wearing surface.

It should be emphasized that in many sections of the country bases (and surfaces) are constructed by the use of methods which are similar to those which have been described and by utilizing local materials which will not, in general, meet the material requirements given above. The wide use of soft limestone (limerock) for this type of construction in Florida is an example.

FIG. 14–2. A multiple-shoe vibrator compacting a granular base. (Courtesy Baldwin-Lima-Hamilton Corp.)

Aggregates for Bituminous Wearing Surfaces and Bases

It seems appropriate at this stage in the discussion, before embarking on the presentation of considerations relative to the design and construction of various types of wearing surfaces and bases involving the use of bituminous materials, to review briefly the more important facts relative to the quality and nature of mineral aggregates intended for use in this type of construction.

The characteristics of an ideal aggregate for bituminous construction have been listed (*156*) as follows:

1. Strength and toughness.
2. Good crushing characteristics which would result in the production of a high percentage of chunky particles and a minimum of flakes and slivers or unduly thin and elongated pieces or dust.
3. Low porosity. (Porosity should not, however, be completely lacking.)
4. Hydrophobic characteristics.
5. Particle size and gradation appropriate to the type of construction.

The listed qualities are desirable in all aggregates, regardless of whether they are naturally occurring materials, as those which are obtained from river beds, pits, and so on, or are produced "artificially" by crushing and screening stone, slag, or gravel. It must be emphasized that aggre-

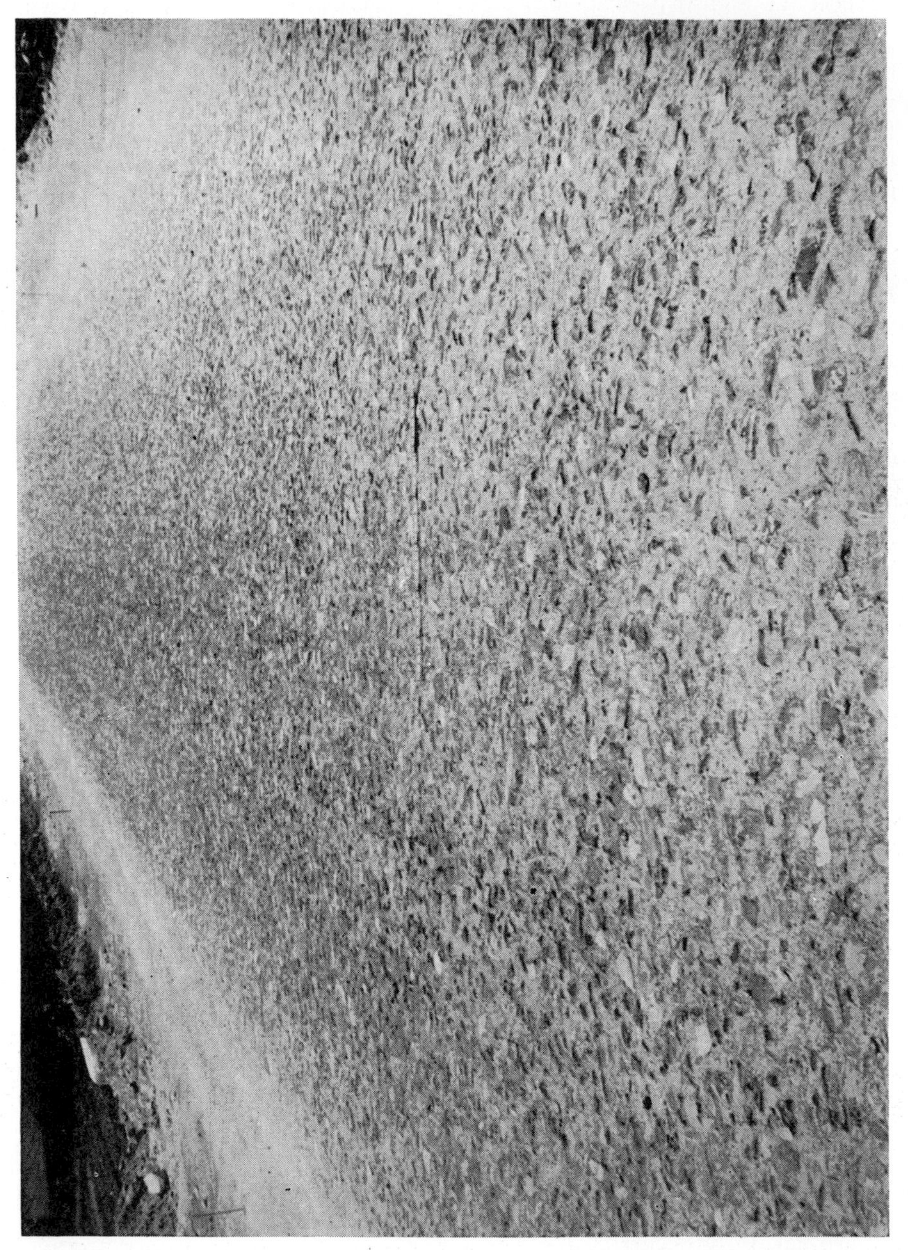

Fig. 14-3. Completed water-bound macadam base course. (Courtesy Pennsylvania Department of Highways.)

gates having all these desirable qualities may not be economically available in many areas. The locally available aggregates, must, of course, be used in highway construction, and construction methods have been devised which will permit the satisfactory use of these materials. Generally speaking, extra costs entailed by the enforcement of rigid specifications relative to certain of the properties listed above will be justified for high-type pavements carrying large amounts of traffic and involving high initial costs of construction, while less expensive aggregates may be used for other purposes.

With regard to the first two of the properties listed above, little discussion need be presented here. The percentage of wear in the Los Angeles Abrasion Test is most widely accepted as a measure of strength and toughness, and values may range from about 10 to nearly 100 for some aggregates. Detailed requirements relative to this important characteristic will generally be presented when each type of construction is discussed in detail in subsequent sections; desirable maximum values of the per cent of wear in the indicated test may generally range from 25 to 65, depending on the type of construction. The test often is applied to crushed slag, but has no significance, according to the National Slag Association. Generally speaking, again, this quality is probably most important in its relation to the prevention of excessive breakdown or "degradation" of the aggregate in processes involving the use of heavy rollers. The California Division of Highways uses a "washing" test that is quite similar to the sand equivalent test (Art. 14–10) to measure the "durability factor," a value indicating the relative resistance of an aggregate to producing detrimental, claylike fines when degraded. It is also noted in the second item listed above that certain otherwise highly desirable aggregates have a tendency, upon crushing, to break into flaky or slivery particles. This tendency may be controlled to some extent by controlling the crushing operations. In many types of bituminous construction, the presence of these particles may be more of an annoyance than totally undesirable, with the amount of such particles and other deleterious substances controlled by applicable portions of the specifications. The presence of particles such as have been mentioned is most detrimental in bituminous macadams and surface treatments. The amount of dust may generally also be controlled by screening and washing operations.

The remaining three items listed above will be discussed in more detail in the following paragraphs.

14–4. Porosity. The porosity of aggregates used in highway construction is generally measured by their absorption. That is, aggregates which have high absorption of moisture, as measured by standard test procedures, are taken to have high porosity, and conversely. A slightly porous aggregate is deemed desirable in this type of construction because a slight penetration of the bituminous material into the pores of the material improves the adhesion of the bituminous material to the aggregate. On the other hand, a high degree of porosity is undesirable, since more bituminous material will then be required in the mixture. The desirable amount of porosity may be said to be from ½ to 1 per cent (as measured by the weight of water taken up by the aggregate in the standard procedure of the absorption test). If the absorption value is above about 1 per cent, the amount of bituminous material required will generally have to be increased for satisfactory results. Such considerations are of importance when relatively porous aggregates, such as slag or limerock, are used.

14–5. Hydrophobic Characteristics. An aggregate which is "hydrophobic" in nature may be said to be one which exhibits a high degree of resistance to film stripping in the presence of water. This concept may be further explained as follows. The bituminous substance in a bituminous mixture may be generally assumed to be present in the form of thin films surrounding the aggregate particles, and filling, or partially filling, the void spaces between adjacent particles. These thin films of bituminous material adhere to the surface of normal aggregates and contribute to the shearing resistance of the mixture, this effect being generally considered as a part of the "cohesion" of the mix. Upon continued exposure to water, either in the laboratory or in the field, bituminous mixtures containing certain aggregates show a definite tendency to lose shearing resistance or "strength" because of a decrease in cohesion due primarily to the replacement of the bituminous films surrounding the aggregate particles with similar films of water. Aggregates which exhibit this tendency to a marked, detrimental degree are termed "hydrophilic" aggregates, hydrophilic meaning "water-loving." Conversely, aggregates which show little or no decrease in strength due to film stripping are called "hydrophobic" or "water-hating."

The exact reasons for the behavior of hydrophobic or hydrophilic aggregates are not completely understood. These phenomena are sometimes explained in terms of physical chemistry, and such things as surface energy, surface texture, and porosity of the aggregate are frequently mentioned as contributing factors. The highway engineer, however, is not so much interested in the exact explanation of why resistance to film stripping is developed in some aggregates and not in others as he is in being able to detect this undesirable characteristic, possessed by certain aggregates, principally in order to avoid their use under conditions which are known to be conducive to film stripping. Certain aggregates, such as limestone, traprock, and limerock, are generally strongly hydrophobic in nature, while quartzite, some granites, and silica dust are examples of strongly hydrophilic materials. Other aggregates, and even some of those listed when obtained from certain sources, show intermediate qualities in this respect. Aggregates which are markedly hydrophilic should be avoided whenever possible. When these aggregates must be used, it may be possible to blend them with other materials to improve their resistance to film stripping. Special precautions may also be taken with these aggregates, including such measures as thorough drying of the aggregate before it is combined with the bituminous material, or "precoating" of the aggregate with a small amount of bituminous material. Thorough drying of the aggregates is a routine procedure in the preparation of high-type bituminous mixtures and is especially important when hydrophilic aggregates are encountered. Treatment of aggregates with certain chemical compounds or "nonstripping agents" has also been effective in some instances.

In judging the relative resistance to film stripping of aggregates in the laboratory, several different procedures have been used. The general procedures which have been used most often are three in number and are briefly described in the following paragraphs. Many other procedures have been employed in various laboratories, and research directed toward finding a completely satisfactory method continues. Standard procedures that may be used are ASTM D1664–64T and AASHO T182–57.

14–6. Stripping Test. A type of test which has been used to evaluate the film stripping occurring when a given aggregate-bituminous mix is exposed to water may be categorically termed the "stripping test." Various procedures have been used in this test by different laboratories. A general procedure, however, would call for the mixing of the aggregate with the bituminous material which will be used in the final mixture. The prepared mixture would then generally be placed in a container of water and set aside for some time, say 18 hours. At the end of this period the mixture would be examined for evidence of film stripping. Aggregates which showed extensive film stripping (above 5 per cent, measured visually) at this stage would probably be classed as strongly hydrophilic. Generally speaking, an aggregate which sustained the immersion periods with little or no stripping of the bituminous film from the aggregate would be judged to be satisfactory. The test has been most widely used in evaluating coarse aggregates and is subject to several criticisms, the principal one being the personal factor involved in judging the extent of stripping in a given sample.

14–7. Swell Test. Another general type of test which has been used for this purpose is the "swell test" (AASHO T101–60). In this procedure a cylindrical sample is prepared by a standard procedure from the total aggregate and the bituminous material which will be used in the given mixture. The compacted specimen is then immersed in water for 24 hours and the change in height or "swell" is measured with an Ames dial. The swell which occurs is due to the absorption of water by the specimen and thus is a measure of the degree to which the bitumen film has been supplanted and replaced by water. The results may be correlated with field experience, and a critical value established.

14–8. Immersion-Compression Test. The immersion-compression test, developed by the Bureau of Public Roads (*152,173*), has been used by highway agencies and seems to provide a logical approach to the problems of evaluating the resistance of aggregates to film stripping. This test involves a comparison of the compressive strengths of cylindrical specimens of a bituminous mixture—prepared, molded, and tested in a standard fashion—with those of duplicate specimens which have been subjected to immersion in water for a definite and standard time. The specimens are molded from samples containing the proportions of aggregate and bituminous material which are to be used in the field-construction process. A standard method of test involving this procedure is AASHO T165–55.

The cylindrical specimen subjected to compression is not supported laterally during the strength determination. The compressive strength of a bituminous mixture under these conditions is largely dependent upon the cohesion supplied by the bituminous material. Any stripping or loss of adhesion to the aggregate will result in a decrease in the cohesive strength of the mixture, and this loss will be reflected by a loss in compressive strength. The comparison between the compressive strengths obtained for samples which have been immersed in water, and the strengths which have been obtained for samples not subjected to immersion, is then taken to be an accurate indication of the effects of water action on the mixture undergoing examination. Aggregates which are hydrophilic in nature may then be rejected or improved by the addition of more suitable materials.

14–9. Gradation. Concepts presented in this section are primarily applicable to bituminous mixtures rather than to bituminous macadams. They are presented here for the sake of completeness and clarity. Gradation of the aggregate used in a given bituminous mixture is obviously an important factor, being most closely related to workability and density. Recommended aggregate gradings for each specific type of bituminous mix will be presented in detail in the appropriate sections. However, some general remarks relative to gradation may be appropriate at this point. For certain types of mixtures, such as bituminous surface treatments, aggregate of essentially uniform size is desired, with the thickness of the surface course being practically equal to the maximum size of particle used. On the other hand, aggregates for dense mixtures, such as bituminous concrete, are more or less uniformly graded from coarse material down to dust. The distribution of particle sizes is important to density, and an aggregate is usually sought which is well graded from coarse to fine, although many dense mixtures have been produced when using aggregates which were not well graded in the usual sense. In certain cases, as when the pavement is to be subjected to very heavy loads or a large amount of traffic, the high cost in producing and controlling the use of aggregates of rigidly specified gradation will be justified. In other cases, as with "intermediate-type" bituminous mixtures, grading requirements may not be so rigid, thus permitting the use of more readily available materials which cost less.

Aggregates used in high-type bituminous paving mixtures are normally made up by blending aggregates from two or more sources. Such a procedure is necessary frequently because no single aggregate can be produced from locally available sources which will meet the rigid grading requirements. Proper blending of available aggregates is highly important to the production of mixtures which are both satisfactory and economical. A detailed discussion of aggregate blending is beyond the scope of this text, and the interested student is referred to current literature for additional information.

In relation to bituminous mixtures, the term "coarse aggregate" frequently refers to that material which is retained on a No. 10 sieve (the No. 8 sieve is also used as the dividing line, as are other sieve sizes). "Fine aggregate" is the material passing a No. 10 sieve; this portion of the aggregate may be divided into two or more general fractions, as desired. Fine material, most of which passes the No. 200 sieve, is generally referred to as "filler." The amount and maximum size of the coarse aggregate used in a mixture have a pronounced effect upon its workability. Mixtures which contain particles more than one inch in diameter are subject to segregation which may result in a poor surface finish, or in a surface which is inadequately sealed unless the mixture is heavily sanded. Mixtures containing a disproportionate amount of coarse material are likely to be harsh and difficult to finish. Mixtures of this general type frequently have very high stability, however, and a relatively thin seal coat may be used to smooth out surface irregularities caused by large amounts of coarse aggregate. The filler most generally used is limestone dust, although portland cement, slag dust, slate dust, and other materials have been successfully used. The filler should, in general, be hydrophobic in nature and well graded to a certain extent. The amount of filler used in a given case is also limited by practical considerations, since mixtures which contain excessive amounts of filler are likely to be gummy and difficult to work and will require excessive amounts of bituminous material.

A rapid field test to show the presence or absence of detrimental fines or clay-like materials in soils or mineral aggregates has been developed in California (*61*); it is called the "sand-equivalent test" (ASTM D2419–65T). The test provides a quick means of separating the finer clay-like particles or sand sizes; the relative proportions are compared on an arbitrary volume basis by a simple procedure which tends to magnify or expand the volume of clay somewhat in proportion to its detrimental effects. The test is applied to the fraction of material passing the No. 4 sieve. The test results indicate whether the volume of "sand," as compared with the volume of the very fine materials, is either high or low; acceptable standards have been established for aggregates to be used for various purposes.

Bituminous Macadam

As has been indicated, the term "bituminous macadam" refers to a type of macadam "pavement" in which the aggregates are bonded together by bituminous material. Both wearing surfaces and bases are constructed by this method and, at least as far as primary highway systems in the United States are concerned, this is the most important and extensively used of the macadams. Use of this method of construction is quite widespread, although practice relative to bituminous macadams has been most widely developed in areas in which suitable aggregates are readily available. There is some variation in practice in different sections, and for the purpose of simplification, bituminous macadams are here divided into two somewhat different types which will be designated as "penetration macadam" and "modified penetration macadam." Construction methods used in connection with the two types are essentially the same, the principal differences between the two being in the aggregates used and the type of bituminous material.

14–10. Penetration Macadam. This is the basic type of bituminous macadam in which "hard" stone is used and the bituminous binder is asphalt cement or one of the heavier grades of road tars. In forming a layer of penetration macadam, the coarse aggregate, which is of essentially uniform size, is placed on the prepared subgrade or base and thoroughly rolled. After this rolling the hot bituminous material is applied to the surface of the compacted layer. The bituminous material penetrates the layer of coarse stone and an intermediate size of aggregate, which is frequently called "key stone" or "choke stone," is applied and the layer again rolled. If the layer is to be a wearing surface, a seal coat is then applied consisting of another application of bituminous material, followed by the spreading and rolling of still smaller aggregate. If the layer is to function as a base, the seal coat may be omitted.

Wearing surfaces constructed in this fashion are generally about 3 inches in compacted thickness, although thicknesses from 2 to 6 inches have been reported. They are generally placed upon a macadam or crushed-stone base. Base courses of this type are more frequently constructed and usually serve as a support for a bituminous concrete wearing surface. In either case the crown would be about 1/4 inch per foot.

14–11. Modified Penetration Macadam. The method of construction which is here called "modified penetration macadam" has been developed in areas in which the larger sizes of coarse stone (or other suitable aggregates) were not readily available, and it also differs from the type described above in that liquid bituminous materials, such as cut-back asphalt or emulsified asphalt, are used as the binder. This type of macadam is used as a wearing surface and generally has a thickness of from 1 to 2 inches. The thinner surfaces closely resemble the double surface treatments described in the next chapter. Either two or three sizes of stone may be used, with the thickness of the layer being essentially the same as the maximum size of coarse aggregate. The construction methods used are very similar to those outlined for penetration macadam.

Material Requirements

As before, detailed requirements for materials used in bituminous macadams vary considerably from state to state and the following sections are intended only to indicate requirements which are typical of those in current use. Requirements relative to the two types previously described will be presented separately.

14–12. Penetration Macadam—Aggregates. The aggregate which is used is crushed stone or crushed slag. Very hard stone, such as traprock, is used in many areas, while crushed limestone and sandstone are also used. The aggregate is required to be quite uniform in character and free from deleterious amounts of dust, flat or elongated pieces, and other objectionable elements. It may be subject to requirements for soundness. Another typical requirement is that the maximum loss in the Los Angeles Abrasion Test be 40 per cent for wearing surfaces and 50 per cent for base courses. In states where unusually hard stone is abundant, the maximum loss permitted may be as low as 25 per cent. Some few states still base this requirement upon the results of the Deval Abrasion Test. Crushed slag is generally required to weigh not less than 70 pounds per cubic foot.

As has been indicated, three sizes of aggregate are normally used.

Typical requirements relative to gradation of these aggregates are those used for base courses in Maine and shown in Table 14–2 (*186*).

TABLE 14–2

GRADING REQUIREMENTS (PENETRATION MACADAM—MAINE)

Sieve Designation	Per Cent Passing (By Weight)
Coarse Aggregate:	
2½-inch	100
2-inch	95–100
1½-inch	35–70
1-inch	0–15
½-inch	0–5
Intermediate Aggregate (Key Stone):	
1-inch	100
¾-inch	95–100
½-inch	20–55
⅜-inch	0–15
No. 4	0–5

14–13. Penetration Macadam—Bituminous Materials. The bituminous materials used in this type of macadam are heated to make them liquid at the time of application to the compacted aggregates. Asphalt cements of penetration grades 85–100 and 120–150 are used, and they are heated to temperatures which generally do not exceed 350°F and applied at temperatures between 275 and 350°F. Some organizations allow use of the heavier grades of cutback asphalts in this type of pavement when finer aggregates are used. Road tars are also used, with grades RT–10, 11, and 12 being preferred. These latter materials are heated to about 250°F and applied at temperatures from about 175 to 250°F.

14–14. Modified Penetration Macadam—Aggregates. General requirements relative to the type of aggregate used in what we have called "modified penetration macadam" are the same as those given above. Crushed stone and slag are used, and crushed gravel is permitted in some locations. The distinguishing characteristics of the aggregates used here, as compared with those used in penetration macadams, are that smaller sizes of coarse aggregate are used and greater proportionate amounts of the smaller sizes. This latter characteristic is desirable because of the necessity for creating a layer which will have small void spaces, since the liquid bituminous material must be prevented from penetrating entirely through the layer. Practice varies so much in the construction of this type of wearing surface that it is impossible to give average requirements relative to the gradation of the aggregates used.

14–15. Modified Penetration Macadam—Bituminous Materials. The emulsified asphalts used are grades RS–1, RS–2, RS–2K, or RS–3K. Rapid-curing cut-back asphalts are also used, with grade RC–250 being generally preferred. Road tars have also been used to some extent in this work, although they are not in general use.

Construction Methods—Penetration Macadam

The discussion which follows relative to the construction of penetration macadams is primarily concerned with wearing surfaces, although most of the principles given are also applicable to base construction. It is further assumed that the macadam layer is to be constructed in one course. Before actual construction of the macadam may begin, the foundation layer (base, subbase, or subgrade) must be brought to the proper line and section and thoroughly cleaned of all loose material. The construction of a penetration macadam may be generally broken down into the following basic steps.

1. Spreading and rolling of coarse aggregate.
2. Initial application of bituminous material.
3. Spreading and rolling of key aggregate.
4. Application of seal coat.

These basic steps are discussed in sequence in the following sections.

14–16. Spreading and Rolling of Coarse Aggregate. Coarse aggregate generally is placed between well-compacted shoulders with vertical faces, using mechanical spreaders. The material must be placed in a uniform loose layer of the required thickness. Precautions are taken to avoid segregation of the aggregate and to prevent the material from becoming mixed with dirt or other foreign material during and after the spreading operation. Before rolling begins, the layer is subjected to careful visual inspection and excessively large and/or flat or elongated pieces are removed. Areas which show an excess of fine material are also subject to correction, and the layer is generally brought to a uniform condition.

When spreading of the coarse aggregate has been satisfactorily completed, the layer is rolled with a three-wheel roller weighing not less than 10 tons in a fashion similar to that which has been previously described for water-bound macadams. Rolling is continued until the surface is thoroughly compacted and the stone keyed together. Material which crushes under the roller in such a way as to interfere with the proper penetration of the bituminous material is removed, and weak spots are corrected and irregularities removed as rolling continues. The finished surface of the coarse layer must be true to the cross-section shown on the plans and at the desired elevation. An excessive amount of rolling is not desirable, since enough surface voids must remain to permit ready penetration of the bituminous binder. In Fig. 14–4 is shown a close-up view of a properly compacted layer of coarse slag.

14–17. Initial Application of Bituminous Material. The initial application of bituminous material is then made by means of a pressure distributor, with the material at the correct temperature and in the amount prescribed by the engineer. The quantity recommended by the Asphalt Institute is 1.75 to 2.25 gallons per square yard for a compacted thickness of 3½–4 inches; this would correspond to a first application of 350–400 pounds of aggregate (specific gravity, 2.55 to 2.75) and a

Fig. 14–4. Air-cooled blast furnace slag shown compacted on the left and bonded with an application of hot asphalt by the penetration method on the right. (Courtesy Macadam Pavements, Inc.)

second application of 35–50 pounds. It is important that the bituminous material be spread uniformly over the layer in the correct amounts. Precautions are usually taken to prevent application of excessive amounts of bitumen at the points where successive applications overlap. This may be accomplished by spreading tar paper over the end of the preceding spread of bituminous material so that the sprays may be opened over the paper and uniform distribution over the untreated section assured.

14–18. Pressure Distributors. The distributor is the key machine on many types of bituminous paving work, including macadam, surface treatment, and road-mix construction. The machine must be capable of applying an accurately measured amount of bituminous material over the distance required to empty the load, regardless of changes which may occur in grade or direction. Distributors are available in several sizes, with capacities from 800 to 5500 gallons.

A typical distributor of this type consists of a rubber-tired truck, truck-drawn trailer, or semitrailer, on which is mounted an insulated tank or one which is equipped with heating facilities. Quite frequently an oil burner is used to supply heat, with the burner being located at the bottom and rear of the tank. Heat from the oil burner passes through the interior of the tank, close to the bottom, and out at the top of the tank through a number of heating flues. The distributor is also supplied with an engine-driven pump, so designed that it may handle products ranging from light "oils" which are fluid at air temperature to semisolid materials which are fluid at temperatures of 350°F or more.

The bituminous material is applied to the surface of any desired area through a system of nozzles mounted on a spray bar at the rear of the tank. The fluid material is pumped under pressure through this spray bar, which may be of varying width, with the minimum width being about 6 feet and the maximum width being 24 feet. Most modern distributors have a circulating system that keeps the material flowing when it is not being sprayed. The distributor is further equipped with a thermometer installed in the tank, and with a hose connection to which may be attached a single nozzle in order that the material may be readily applied to a limited area or at a specific point.

When application is to be made through the spray bar and nozzles, rather simple calculations may be made to relate the variables involved in the problem of applying the fluid material at a uniform, specified rate. Assuming that the width of the spray bar, the quantity (gallons) handled by each revolution of the pump, and the speed of the pump are known, the speed at which the truck must be driven to secure a given rate of application may be readily obtained. The rate of application is controlled by a "bitumeter," which comprises a small wheel beneath the truck, connected by a cable to a dial in the cab; the dial usually shows the speed in feet per minute, but may show gallons per square yard directly. A schematic drawing of a typical pressure distributor is shown in Fig. 14–5. A bituminous distributor in action is shown in Fig. 14–6.

The description of the computation of the application rate of a bituminous distributor assumes that all the material pumped goes through the nozzles. Many distributors have circulating spray bars so that part of the material pumped returns to the distributor. For a given pump speed or pressure, a given bituminous material, and a given temperature, a given flow through the nozzles will result. Charts or tables showing

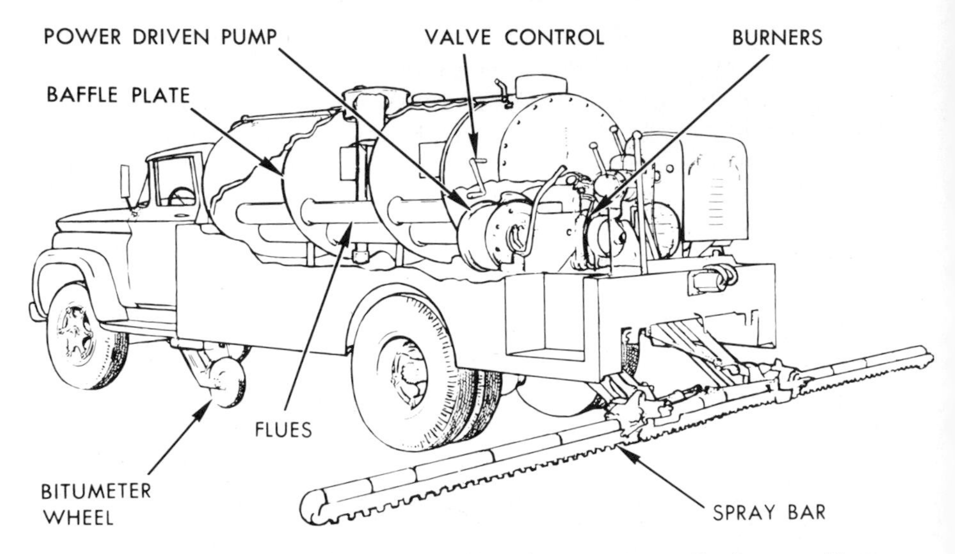

FIG. 14-5. Schematic drawing of a bituminous pressure distributor. (Courtesy the Asphalt Institute.)

FIG. 14-6. A bituminous pressure distributor in action. (Courtesy E. D. Etnyre & Co.)

these data assist the inspector or driver in the selection of vehicle speed. The application rate is then verified by the distance traveled.

14-19. Spreading and Rolling of Key Aggregate. Immediately after the initial application of bituminous material is made, the surface is lightly covered with key aggregate in sufficient quantity to prevent the roller wheels from sticking. The layer is then rolled until thoroughly compacted. The key aggregate is then spread over the surface in the desired quantity so that the surface voids of the layer will be completely filled, but there is no excess aggregate remaining on the surface. The surface is then rolled with the same equipment as before. Simultaneously with the rolling, the key aggregate may be lightly broomed to effect an even distribution of the material, and additional small amounts may be added as needed. Vibratory compactors are used in some states, in conjunction with rollers. This process is continued until the key aggregate is firmly embedded in the bituminous material and the surface is firmly compacted. In Fig. 14–7 is shown a properly bonded bituminous macadam after the application and rolling of the key aggregate. The layer is then ready for the application of the seal

Fig. 14–7. Close-up of bituminous penetration macadam showing the texture after the first and second penetration applications of bituminous material and choke stone have been applied. (Courtesy Macadam Pavements, Inc.)

coat if it is to function as a wearing surface; if it is to be a base, the seal coat usually is omitted.

14–20. Application of Seal Coat. Before placing of the seal coat is begun, the surface is swept, usually by means of a broom drag, for the purpose of either removing all loose material or of uniformly distributing any loose key aggregate which may remain after the previous rolling. Bituminous material is then applied to the surface in the amount determined by the engineer, and is immediately covered with the fine aggregate (stone chips). Rolling and brooming then is begun again and continued until the cover stone is firmly bonded and completely fills the voids in the finished surface. The required amount of rolling cannot be definitely stated, and this process is continued until the surface is completed to the satisfaction of the supervising engineer.

Pneumatic rollers may be used on this final layer in some states. The completed surface must, of course, be correct to grade and cross-section. The finished surface is tested by templates and straightedges with deviations up to one-quarter of an inch in 10 feet being permitted in the majority of states performing this work. The surface may be opened to traffic soon after the final rolling is complete.

Construction Methods—Modified Penetration Macadam

It is not feasible to give a comprehensive statement of "average" or "typical" practice with regard to this type of surface, since practice varies so widely. The process is, in general, similar to that used in the construction of penetration macadams. The following general deviations from the procedure presented above may be listed as being applicable to this type of construction:

1. The surface of the base is primed by an application of liquid bituminous material before construction of the wearing surface begins. The purpose in this application of bituminous material is to bond the wearing surface to the base.
2. The surface is built up by means of a larger number of separate applications of aggregate and bituminous material than are used in penetration macadam construction.
3. A lighter roller may be used in the compaction of each layer, or the rolling may be less extensive.
4. The first layer of coarse aggregate may be "choked" by the application of finer aggregate before any bituminous material is applied.
5. Some time may be allowed to elapse between successive applications of aggregate and bituminous material in order that some "curing" may take place. This is especially important when asphalt emulsions are used, and the time between applications may be from 12 to 24 hours. A considerable time may be permitted to elapse between the completion of the principal thickness of the macadam layer and the application of the seal coat.

PROBLEMS

14–1. A sample of aggregate which is being contemplated for use in a water-bound macadam base course is subjected to the Los Angeles Abrasion Test. After completion of the prescribed test procedure, the final weight of the test specimen is 2783 grams. Determine the per cent of wear for this sample. On the basis of this single test result, would this material generally be regarded as satisfactory for this type of construction?

14–2. An 8-inch compacted thickness of water-bound macadam is constructed from Ocala limerock. Assume that the bid price of this item is $2.10 per square yard, complete and in place, and that the compacted material has a dry unit weight of 108 pounds per cubic foot. Determine the equivalent bid price for this base on the basis of tons of dry aggregate.

14–3. Assume that a penetration macadam wearing surface is being constructed in accordance with applicable specifications. Based upon a specific gravity of 2.65, the required quantity of aggregate is 390 pounds per square yard. The actual weight of aggregate is directly proportional to the specific gravity of the aggregate used. The bulk specific gravity of the aggregate being used is 2.55, and the amount of bituminous material required is 2.25 gallons per square yard. Assume that the aggregate costs $4.00 per ton and the bituminous material $0.23 per gallon. Determine the cost of this pavement per square yard.

14–4. Assume that a pressure distributor has a spray bar which is twelve feet long. If the pump discharge is at the rate of 250 gallons per minute, with the pump operating at the proper speed for the material which is being used, determine the rate of application of the bituminous material, in gallons per square yard, when the distributor is traveling at the rate of 800 feet per minute. Also determine the reading which must be maintained on the truck tachometer to maintain this rate of application if the tachometer is calibrated to read 100 times the truck speed in miles per hour. Assume that all the pumped material passes out through the spray bars and is applied to the road surface.

14–5. From current literature or contacts with equipment manufacturers, prepare a report on one of the following:

(a) Available types and sizes of vibratory compactors used in the construction of granular base courses.

(b) Specific construction procedure used on a base construction project on which vibratory compactors were used. Describe the equipment used, number of passes made, densities obtained, etc.

(c) Results of a research study concerning the effectiveness of vibratory compaction.

14–6. Consult engineering agencies or material suppliers in your area to determine the use of anti-stripping agents in asphaltic mixtures. What commercially available products are most commonly used?

CHAPTER 15

BITUMINOUS SURFACE TREATMENTS

For the purpose of clarification, it seems necessary at this point to introduce a classification of bituminous wearing surfaces. Such surfaces may be classified in many different ways, and any listing of this type is subject to some criticism and will contain certain discrepancies. The problem is made more difficult by the lack of uniform terminology relative to the subject and by widespread variations in design and construction procedures in different sections of the country. Given below is a simplified classification of bituminous wearing surfaces. Those types which have not been discussed in a previous chapter will be discussed in detail in this and other sections of this book.

A. Surface treatments
   1. Prime and tack coats
   2. Single- and multiple-surface treatments
   3. Seal coats
B. Rock asphalt
C. Bituminous macadam
D. Road-mix surfaces, including those constructed by the use of traveling plants
   1. Open graded
   2. Dense graded
   3. Mixed-in-place sand asphalt
E. Intermediate-type bituminous plant mixes (liquid binders)
   1. Open graded
   2. Dense graded
F. High-type bituminous plant mixes
   1. Hot-mixed, hot-laid types
      (a) Bituminous concrete
      (b) Sheet asphalt
      (c) Sand asphalt
   2. Cold-mixed, cold-laid types

Several comments may be made relative to the listing given above. In the first place, the surfaces listed are arranged approximately in the order of increasing cost. This statement may be regarded as being at least generally true, although cost is, among other things, a function of the nature and availability of aggregate materials suitable to each of the listed types of construction. For example, road-mix surfaces may be much less expensive in some sections than is bituminous macadam because of a scarcity of aggregates suitable to the latter type of construction. Similarly, rock asphalt surfaces are normally economical only in areas where this material is readily available, and their cost would be very high in other sections. Secondly, the listing may be said to be arranged approximately in the order of increasing complexity of the construction process, both in its conduct and control. Following this line of reasoning, surface treatments are certainly the most simple of the types listed, road-mix surfaces are a little more complex, and hot-mixed pavements are the most difficult type to construct if they are to properly perform their intended function. No erroneous conclusions should be drawn from these latter statements, however, as careful conduct and control of the construction processes are absolutely necessary in all types of bituminous construction if satisfactory results are to be achieved.

Generally speaking, it may also be said that the surfaces listed above are arranged in order of increasing ability to withstand large volumes of traffic, heavy wheel loads, and severe service conditions. This statement is made with respect to the wearing surfaces alone, since the structural integrity of any bituminous surface is absolutely dependent upon the adequate support of the foundation beneath it. If the listing is considered from this viewpoint, bituminous macadam pavements and rock asphalt surfaces may be somewhat out of place when considered with regard to certain road-mix types and some low-cost bituminous plant mixes. Generalizing further, it may be possible to classify bituminous wearing surfaces somewhat as follows: Surface treatments and some road-mix surfaces as "low-type" or "low-cost" surfaces; rock asphalt, bituminous macadam, certain road-mix types, and some of the cold bituminous plant mixtures as "intermediate-type" surfaces; other cold-plant mixes and hot-plant mixtures as "high-type" surfaces. In any event, and regardless of any classification system which may be adopted, the proper selection of the type of bituminous wearing surface to be used in a given set of circumstances must be based upon the proper and economical use of locally available materials in a fashion which is best suited to the purposes of the road involved.

The surface types which are included under headings A and B in the listing previously given are discussed in this chapter. For convenience, the first portion of the chapter is devoted to those given under item A, while rock asphalt surfaces are discussed in subsequent sections.

Definitions

15–1. Prime Coat. When a bituminous wearing surface is to be placed upon a previously untreated foundation layer, such as earth, gravel, stabilized soil, or water-bound macadam base courses, the base is generally "primed" by a single application of liquid bituminous material. This application is called a "prime coat." In certain cases a small amount of fine aggregate or sand may be applied immediately after the application of bituminous material; this very thin treatment may also be called a prime coat. Neither prime nor tack coats, which are defined in the next paragraph, are properly called wearing surfaces. They are included here because of their distinctive nature and their function in the construction of bituminous wearing surfaces.

15–2. Tack Coat. A tack coat may be defined as a single initial application of bituminous material to an existing bituminous, portland cement concrete, brick, or block surface or base. The purpose of a tack coat is simply to insure adhesion between the existing surface and the new bituminous wearing surface.

15–3. Surface Treatment. A bituminous surface treatment is a type of wearing surface which is generally constructed by making an application of bituminous material, immediately followed by the application of covering mineral aggregate. The surface is then rolled and a wearing surface of appreciable thickness is produced. If a single application of bituminous material is made, followed by a single application of cover aggregate, it may be called a "single surface treatment." In this case the thickness of the surface treatment would be practically the same as the maximum size of the aggregate used. The thickness of this type of surface may be increased by the application of additional bituminous material and additional spreads of successively finer sizes of cover aggregate. Two, three, or four successive applications may be made as desired, and this type of surface may be generally called a "multiple surface treatment." The terms "double" and "triple" surface treatment are also frequently used. Wearing surfaces constructed by this latter method are sometimes locally known as "armor coats" or "multiple lift" surfaces. The type of construction which is involved is quite frequently termed an "inverted penetration" method for reasons which are fairly obvious.

The term "surface treatment" is also applied to thin bituminous surfaces which are constructed by road-mixing methods, generally through the use of a multiple-blade drag or blade grader. The construction methods used in this type of surface treatment closely resemble those used in the construction of road-mix surfaces using open-graded aggregates, which are described in the next chapter; for that reason they will not be discussed further here. They are called surface treatments because they are quite thin. This type of construction is also termed a "drag treatment" by some organizations.

15–4. Seal Coat. A seal coat may be defined as a very thin surface treatment which is applied as a final step in the construction of certain bituminous wearing surfaces. The purpose in applying a seal coat is generally to "waterproof" the wearing surface and in some cases to provide a more desirable surface texture. It is seen that a seal coat could not ordinarily be considered as a wearing surface in itself, but rather it is a necessary part of many bituminous wearing surfaces.

A seal coat may also be employed on an existing surface in order to enliven or "rejuvenate" a dry or badly weathered surface, to improve night visibility, to reduce slipperiness, or to build up the pavement structure to a limited degree.

Prime Coats

As noted, a prime coat is an application of liquid bituminous material (and sand in some cases) to a previously untreated base or wearing surface. The bituminous material penetrates the surface and is generally completely absorbed. Prime coats serve several definite purposes, as follows. First, the prime coat serves to promote adhesion or "bond" between the base and wearing surface; this is probably its most important function. Second, it serves to consolidate the surface upon which the new treatment is to be placed, and this may aid in maintaining the integrity of this underlying layer during the construction of a surface treatment or some other type of wearing surface. Third, it may function to a certain extent as a deterrent to the rise of capillary moisture into the wearing surface.

15–5. Materials. The material used in the prime coat should be a liquid asphalt or road tar of low viscosity. It is important that the material used have high penetrating qualities and it should be of such a nature that it will leave a high-viscosity residue in the void spaces of the upper part of the surface which is being primed. Of the asphaltic materials, grades MC–30 and MC–70 are probably most widely used. Grade MC–30 is generally most useful in the priming of dense tightly bonded surfaces, while MC–70 would be used in the treatment of surfaces having a more open texture. MC–250 is recommended in some cases, and slow-curing liquid asphalts, grades SC–70 and SC–250, have also been used. In priming very open surfaces, as, for example, very sandy soil bases, grade RC–70 has some application but is not generally used because of its rapid increase in viscosity after application.

Road tars are very widely used as prime coats in areas where these materials are economically available. The first three grades of road tar—RT-2, 3, and 4—are all in current use, with the lighter grades being used in the priming of dense surfaces and the heavier in the treatment of more open surfaces. Probably grades RT–2 and 3 are the most widely used of those listed. It is quite common practice in many states to use a tar prime, even though the wearing surface is to be constructed by using an asphaltic material.

The selection of the type and particular grade of the bituminous material to be used as a prime coat for a given project is largely dependent upon previous experience and engineering judgment.

15–6. Materials—Amount Required. The amount of bituminous material which is used varies over a considerable range, depending upon the material selected and job conditions. If a general criterion may be established, it may be said that the amount of bituminous material generally required for priming is from 0.20 to 0.50 gallon per square yard of surface. The lower limit applies to tight, fine-grained surfaces; the upper to loose, open-textured surfaces. Another criterion which is sometimes advanced as to the amount of primer to be used is that the prime material should be completely absorbed within 24 hours. The amount of prime required in a given case will again be largely dependent upon judgment and previous experience.

In certain cases the prime coat may not completely penetrate the surface within a reasonable length of time (24 hours) and an excess of material will remain on the surface. When this occurs it is customary to apply a very light coating of sand to absorb or "blot up" the excess priming material. The sand used for this purpose should be dry and clean. The Asphalt Institute recommends (*114*) that 100 per cent of the sand pass a No. 4 sieve and not more than 2 per cent pass a No. 200 sieve. Any excess sand which may remain after this operation is removed before construction of the bituminous wearing surface begins.

15–7. Construction Methods. Before the prime coat may be applied, the surface which is to be treated must be brought to a satisfactory condition. This may involve the correction of depressions or holes in the surface which require patching, or the removal of minor surface irregularities by the scarifying and recompacting of the base course. Probably the most important single item in the preparation of a surface which is to be primed is the removal of loose material, excess amounts of dust, and other foreign matter. This is usually accomplished by the use of power sweepers and/or blowers, and the complete removal of all loose material is absolutely essential if the prime coat is to penetrate properly and provide adequate bond between the foundation course and the new wearing surface. In Fig. 15–1 is shown a power sweeper engaged in the preparation of a granular base course prior to the application of a prime coat. On occasion, the engineer may order a light application of water to the surface immediately before the bituminous material is applied. As with all bituminous construction, best results are obtained in hot, dry weather; some specifications prohibit prime coating when the air temperature is less than 50°F.

Immediately after the surface has been cleaned, the bituminous material should be applied at the uniform, specified rate by means of a pressure distributor such as was described in the preceding chapter. The material should be applied at the correct temperature for the grade which is being used. See Table 11–10 for correct spraying temperatures. Precautions must be taken to insure a uniform spread of bituminous

FIG. 15–1. Power sweeper. (Courtesy W. E. Grace Mfg. Co.)

material, and areas in which successive applications overlap should be protected against the application of excessive amounts of fluid material. The surface is generally allowed to cure at least 24 hours before construction of the wearing surface begins. If excessive amounts of primer remain on the surface at the end of this time, a blotter coat of sand should be applied, as previously indicated. Traffic may be allowed on the surface within a fairly short time after the application of the prime coat if necessary. Traffic should not be permitted to use the road during the period when sufficient fluid material remains on the surface to be picked up by the wheels of moving vehicles. Before the wearing surface is placed, the primed area is carefully examined, and areas of unsatisfactory or deficient priming are corrected. It may be further noted that in certain cases a prime coat may be applied to a new course several days or weeks ahead of the construction of the wearing surface. In such cases a tack coat is generally applied to the primed surface before construction of the wearing surface begins.

Tack Coats

A tack coat is simply a single application of bituminous material to a previously prepared or treated surface as a first step in the placing of a bituminous wearing surface. A tack coat would then be needed as a part of the construction of a new wearing surface upon a new bituminous base course, a previously primed granular-type base, an old bituminous wearing surface which is being resurfaced, a new or old concrete base, or an old concrete, brick, or block pavement which is to be covered with a bituminous wearing surface. The purpose behind the application of a tack coat is to provide adhesion between the previously prepared foundation layer and the new wearing surface. A tack coat would not usually be required if the wearing surface is to be a single or multiple surface treatment of the type described in this chapter.

15–8. Materials—Nature and Quantity. A considerable number of different bituminous materials are used as tack coats by different organizations and in the treatment of different types of foundation layers. The type of bituminous material to be used is dependent largely upon the experience of the individual organization, and several alternate materials may be permitted to be used at the option of the engineer or contractor. Liquid asphalts of grades RC–70 through RC–250 are frequently used as tack coats, while medium-curing liquid asphalts of grade MC–250 has also been used. Road tars are used to some extent for this purpose, with grade RT–2 being used by some organizations, and the heavier grades, such as RTCB–5 and RTCB–6, by others. Emulsified asphalts diluted with water are specified by some agencies. Recommended spraying temperatures are given in Table 11–10.

The quantity of bituminous material which is normally required for a tack coat is quite small, with the stipulated amount generally ranging from 0.05 to 0.10 gallon per square yard.

15–9. Construction Methods. Tack coats, as has been indicated, are very frequently applied to old surfaces of various types as a first step in the construction of a new bituminous wearing surface. In many cases the old surface will require extensive correction and remedial treatment before construction of the new surface may begin because of the rough or inadequate condition of the old surface. Remedial measures which may have to be applied to an old surface before it may adequately serve as a foundation course for the new surface are discussed in Art. 20–11. It should be emphasized that after the necessary steps have been completed and the surface adequately prepared to fulfill its new function, a tack coat may or may not be applied, depending on the type of bituminous wearing surface which is to be constructed.

Assuming that the base has been brought to the desired condition, the construction of the tack coat requires only a single application of fluid bituminous material. As in the case of a prime coat, it is absolutely essential that the surface to which the material is applied be clean and dry. In most cases the surface is swept clean of all loose material just before the application is made. If asphalt emulsion is used, the surface may be slightly damp without harmful effect. The bituminous material is then applied at the specified rate with a pressure distributor, with the material being maintained at the proper temperature for application. The treated surface is then protected from traffic and allowed to dry until it reaches the proper stage of stickiness or "tackiness" for the application of the bituminous wearing surface. Any breaks which may occur in the tack coat may be corrected by the application of additional bituminous material at these points.

Single- and Multiple-Lift Surface Treatments

This type of surface treatment is very widely used in many sections of the country for the surfacing of primary and secondary roads which carry light to moderate amounts of traffic. The thickness of a single surface treatment is usually from one-half to three-quarters of an inch, and may be as much as 2.5 inches when multiple-lift treatments are used. A surface constructed by this method is frequently quite "open" in character and has excellent nonskid and visibility characteristics. These surfaces are generally constructed with a crown of about one-quarter of an inch per foot. The type of construction which is described in this section is primarily that which would normally be used in the surface treatment of a granular base such as water-bound macadam. This type of surface is generally so thin that it has little load-supporting value in itself and must depend upon an entirely adequate base and subgrade to function satisfactorily.

15–10. Material Requirements. Materials which are used in this type of surfacing are extremely varied, with a large number of different aggregates and bituminous materials being successfully used in different areas. In the interests of somewhat simplifying the coverage of this subject, it has been decided to present the detailed material requirements of only one organization, the Bureau of Public Roads, in the belief that these will be typical of those in current use in this country. With this viewpoint in mind, the following sections have been abstracted from Section 314, Bituminous Surface Treatments, of the specifications of that organization (*169*). This specification is quite complex in nature and it provides for the construction of six different surfaces, utilizing seven different aggregate gradings and a variety of bituminous materials, and ranging from a single surface treatment to a four-lift treatment of the general type being considered. These surfaces are to be constructed upon a previously prepared and bitumized base course. In other words, if the surface were to be placed upon a new granular base, it would previously have been primed. Requirements of any single state would not be expected to be as complex in nature as those indicated in this specification, since a much more limited range of materials would normally be economically available in a given area.

15–11. Aggregate—General Requirements (B.P.R.). The aggregates which are used in this type of surface are crushed stone, crushed gravel, or crushed slag. The aggregates are required to be clean, hard, tough, and durable in nature, free from an excess amount of flat or elongated pieces, dust, clay films, clay balls, and other objectionable material. They are also required to be hydrophobic in nature, as measured by a stripping test (AASHO T 182). The material must have a swell of not more than 1.5 per cent (AASHO T 101, Method A). The percentage of wear in the Los Angeles Abrasion Test must not exceed 40, and the aggregate must have a weighted loss of not more than 12 per cent after five cycles of the sodium sulfate soundness test (AASHO T 104). Crushed slag is required to be air-cooled blast furnace slag which is uniform in nature and which weighs not less than 70 pounds per cubic foot. Crushed gravel for use in this type of treatment should be such that not less than 90 per cent by weight is made up of particles having at least one fractured face.

15–12. Aggregate—Gradation (B.P.R.). Seven aggregate gradations are included in the specification. They are designated Gradings A through G, and are given in Table 15–1.

15–13. Bituminous Materials (B.P.R.). The bituminous materials which may be used under this specification include rapid-curing cut-back asphalt (grades RC–2, 3, 4, and 5); 150–200 and 200–300 penetration grade asphalt cement; emulsified asphalt (S–1 and RS–2); and road tar (RT–5 through RT–12). See Fig. 11–7 for approximate equivalent new grades of cut-back asphalts. Included in the specification are recommendations relative to the application temperature of each of these products.

It might be generally noted at this point that these products are typical of those in use in various sections of the country under other specifications, although additional grades of liquid asphalts, as well as some additional grades of asphalt cement, are permitted by some organizations. In general there seems to be a trend toward the use of more viscous bituminous materials by some organizations engaged in this type of work. Specific aggregate gradation requirements would also be expected to be different with different organizations.

TABLE 15–1

AGGREGATE GRADATIONS

Sieve Designation	Percentage Passing (by Weight)						
	A	B	C	D	E	F	G
1½-inch	100	–	–	–	–	–	–
1-inch	90–100	100	–	–	–	–	–
¾-inch	–	90–100	100	–	–	–	–
½-inch	0–15	20–55	90–100	100	100	–	–
⅜-inch	–	0–15	40–75	90–100	90–100	100	100
No. 4	–	–	0–15	0–20	10–30	75–100	85–100
No. 8	–	–	0–5	0–5	0–8	0–10	60–100
No. 200	0–2	0–2	0–2	0–2	0–2	0–2	0–10

15–14. Composition of Surface Treatments (B.P.R.). The specification which is being analyzed contains quantities of materials and indicated sequences of operations for six different wearing surfaces utilizing the listed bituminous materials, excluding emulsified asphalts. Surfaces constructed by the use of emulsified asphalt are also included in the specification, but are listed separately and will not be included here. The surfaces which are constructed by the use of the other indicated bituminous materials are designated as AT–25, AT–35, AT–50, AT–60, AT–70, and AT–110. Indicated quantities and sequences of operations are as presented in Table 15–2. In that table, the indicated quantities of aggregate are based upon a bulk specific gravity of 2.65 and must be corrected if the specific gravity of the aggregate used is more than 2.75 or less than 2.55. The total quantity of aggregate used in each of the surfaces given in the table, corrected for specific gravity and except for some stock-piling if desired, is fixed while the amount of bituminous material actually used may be varied somewhat to meet job conditions.

TABLE 15–2

QUANTITIES AND SEQUENCES OF OPERATIONS, USING CUT-BACK ASPHALT, ASPHALT CEMENT, OR TAR

Operations (as indicated in sequence)	Quantities of Materials per Square Yard (Bituminous material, gallons; aggregate, lb) Surface Designation					
	AT–25	AT–35	AT–50	AT–60	AT–70	AT–110
First Layer						
Apply Bituminous Material	0.30	0.22	0.25	0.15	0.30	0.20
Spread Aggregate						
Gradation E	25					
Gradation D		25				
Gradation C			35			
Gradation B				40	50	
Gradation A						70
Second Layer						
Apply Bituminous Material	–	0.13	0.25	0.30	0.35	0.40
Spread Aggregate						
Gradation F		10	15			
Gradation D				12		
Gradation E					20	
Gradation C						20
Third Layer						
Apply Bituminous Material	–	–	–	0.15	–	0.20
Spread Aggregate						
Gradation F				8		12
Fourth Layer						
Apply Bituminous Material	–	–	–	–	–	0.20
Spread Aggregate						
Gradation G						8
Totals						
Bituminous Material	0.30	0.35	0.50	0.60	0.65	1.00
Aggregate	25	35	50	60	70	110

15–15. Composition of Surface Treatments (Texas Practice). Fred J. Benson (*215*) has given an authoritative discussion of surface treatments, as used in Texas and other southern states. He suggests that the quantity of a given aggregate for single surface treatments and seal coats is best determined by direct measurement. To carry out this measurement, an area 1 yard square is laid off on any reasonably smooth surface, e.g., a piece of plywood. A representative sample of the aggregate to be used is then spread over the area so that it is completely covered with a layer one stone thick. The aggregate is then recovered and weighed. The quantity determined in the test is increased by 10 per cent in order to compensate for spreading inaccuracies and the failure of all particles to adhere to the bituminous material. Values obtained can be converted to a volume basis, by determining the loose unit weight.

The average thickness of the aggregate in the layer can be computed from the spread quantity and the loose unit weight. The calculation is based upon the assumption that the particles have the same arrangement in the one-stone spread as they do in the container when the loose unit weight is being determined. For example, assume that the unit loose weight of aggregate is 96 pounds per cubic foot and the amount required to cover 1 square yard is 18.5 pounds. For a layer one inch thick, the required weight of aggregate would be $3(3)96(1/12) = 0.75(96) = 72$ pounds. Average mat thickness then is $(18.5/72) = 0.26$ inch.

J. P. Kearby of the Texas Highway Department developed a procedure for determining the quantity of bituminous material on the basis of proper embedment of the aggregate; his work was later verified by research workers at Texas A & M. Figure 15–2 shows the required percentage of

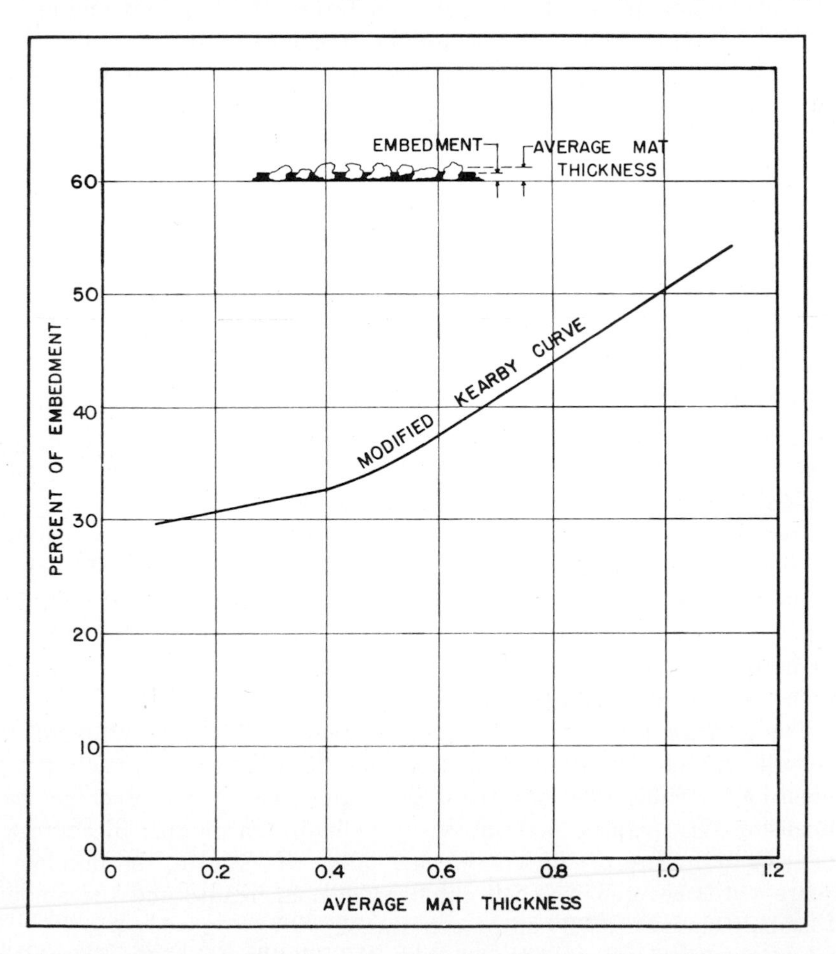

FIG. 15–2. Relationship between per cent embedment and average mat thickness, bituminous surface treatments. (After J. P. Kearby.)

embedment for aggregate of various sizes on the basis of average mat thickness. In order to compute the quantity of residual bituminous material required to fill the area not occupied by the stone it is necessary to know the average mat thickness, loose unit weight, and the specific gravity of the stone. For example, the aggregate previously considered has an average mat thickness of 0.26 inch, unit weight of 96 pounds per cubic foot, and specific gravity of 2.65. From Fig. 15–2, the required percentage of embedment is 32. Thus, the depth of residual bituminous material required is 0.26(0.32) = 0.083 inch. The volume of bituminous material per square yard is $3(3)(1/12)(0.083)\left[1 - \frac{96}{2.65(62.5)}\right] = 0.026$ cubic foot. The term 96/62.5(2.65) is the absolute volume of the aggregate, and the term in brackets is the proportion of total volume to be filled with bituminous material. In common terms, the required amount of residual bituminous material would be 0.026(7.5) = 0.195 gallons per square yard. Since this is residual material, allowance must be made for the volatile material in a cut-back asphalt or the water in an asphalt emulsion, when computing the amount to be applied by the distributor in the field.

Later research by John L. Saner and Moreland Herrin, of the University of Illinois, appeared to establish that the volume of voids does not vary linearly with depth within the aggregate layer (*72*). They found that the relationship is curvilinear and varies with different aggregates, and they suggested further research, particularly to relate the volume of voids to the shape of the aggregate.

15–16. Design of Surface Treatments (Asphalt Institute). The Asphalt Institute has developed design methods for single and multiple surface treatments (*114*) based on work done by F. M. Hanson, of New Zealand, and later modifications of his work by other engineers.

The method is based upon these concepts, for aggregates of essentially one size: When the aggregate particles are spread on a film of asphalt, the void spaces between the particles are about 50 per cent. Rolling shifts the particles, reducing the voids to 30 per cent. Under traffic, the particles are repositioned to lie on their flatest sides and the void spaces become about 20 per cent.

The average thickness of a surface treatment is a function of the over-all average smallest dimension of the aggregate particles; this dimension is called the "average least dimension." The average least dimension (H_L) can be determined by calipering a number of aggregate particles or by use of slotted screens. With H_L known, the number of square yards covered by each cubic yard of aggregate and the amount of asphalt binder needed can be determined.

For good performance, the asphalt should fill about 70 per cent of the 20 per cent void space if the traffic volume is low. The asphalt should fill not more than 60 per cent of the 20 per cent void space if the traffic volume is high.

The Asphalt Institute recommends that the maximum size of aggregate used in a surface treatment should be ½ inch and that the ratio of maximum to minimum size should be 2:1, with a reasonable tolerance for undersized and oversized particles to allow economical production.

The following equations are applicable to single surface treatments and seal coats:

$$S = 34.7G_m H_L E \tag{15-1}$$

$$A = 1.22TH_L + V \tag{15-2}$$

where S = aggregate spread, pounds per square yard
G_m = bulk specific gravity of the aggregate
H_L = average least dimension of the aggregate
A = asphalt spread, gallons per square yard
T = traffic factor, from 0.85 to 0.60, with increasing traffic volume
E = aggregate wastage factor, ranging from 1.01 to 1.15, depending on the estimated waste
V = variable factor, in gallons per square yard, to allow for absorption by the surface being treated, from 0.00 for a nonporous surface to 0.09 for a very porous one

The Asphalt Institute publication (*114*) referred to above also gives design procedures for graded aggregates and double surface treatments.

15–17. Construction Methods—Single- and Multiple-Lift Surface Treatments. It will be assumed in this discussion that the surface is to be applied to a granular type base which has been previously primed so that the application of the prime coat is not included in the procedure presented here. The base is assumed to be in a satisfactory condition to receive the surface treatment, either because of the application of suitable reconditioning measures or because of its recent completion. It is particularly important that no depressions exist in the base course into which the fluid bituminous material may flow when applied to the surface and collect at these points in excess quantities. The surface should be free from dust and dry, unless emulsified asphalt is used, when the surface may be slightly damp.

The construction of surface treatments of this type is quite simple in basic procedure, with the surface being built up in lifts through the successive application of bituminous material and progressively smaller aggregate spread in the correct amounts. The layer formed by each application of bitumen and aggregate is rolled, and this basic procedure is continued until the specified number of lifts has been completed.

The initial application of bituminous material to the prepared base is made by a pressure distributor in the specified amount and at the proper temperature of application. It is essential that this application be uniform over the desired width and length of treatment. The desired amount of aggregate is then immediately spread over the bituminous material by means of approved mechanical spreading devices, including self-propelled spreaders. Normally, trucks are used to transport the aggregate to the point of application, and specifications covering this type of work usually require that the trucks be brought into position by traveling over previously spread aggregate. They are not allowed

Fig. 15–3. Self-propelled aggregate spreader. (Courtesy Highway Equipment Company.)

to pass over the portion of the surface on which no cover aggregate has been placed. Spreading of the aggregate is illustrated in Fig. 15–3.

Rolling begins immediately after the aggregate has been spread. Both pneumatic-tired and steel-wheel rollers are used in this type of work, either alone or in combination with one another. The Asphalt Institute recommends pneumatic rollers, because of the more uniform pressure they exert. Where pneumatic-tired rollers are used, they are operated at comparatively low tire pressures in order to avoid displacement and crushing of the aggregate.

The weight of steel-wheel rollers which is most generally used in this type of work is 5 tons, and the weight may vary from 3 to 8 tons under different specifications. Fig. 15–4 shows a tandem roller engaged in this type of work. Rolling of the aggregate is continued until the aggregate particles are firmly embedded in the bituminous material and sufficiently well bonded so that they will not pick up under traffic, but discontinued after the asphalt has taken its initial set. Excessive rolling, which results in the crushing of aggregate particles or the breaking of the asphalt bond, is, of course, undesirable.

If a single surface treatment is being constructed, this is the end of the construction process and the surface may be opened to traffic. If the wearing surface is to be a multiple-surface treatment, construction

Fig. 15–4. A tandem roller compacts a surface-treated shoulder. The light-colored aggregate provides a contrast with the asphaltic surface of the travel lanes. (Courtesy The Galion Iron Works & Mfg. Co.)

continues by the application of another "shot" of bituminous material which is immediately covered by another spread of aggregate, and the layer shaped and rolled as before. This process is simply continued until the desired number of lifts is completed. Generally speaking, very little time need elapse between construction of successive lifts except when emulsified asphalts are used, and then the time between applications of bituminous material may be required to be at least 24 hours. Similar waiting periods may be enforced by some agencies when other materials are used, particularly in the top layer of three- or four-lift construction. Surface treatment work is most successfully done in warm weather, and should not be attempted when air temperatures are low (say 50°F or less). Under cold weather conditions, surface treatments have been constructed by preheating the aggregate; costs are considerably increased when this is done.

Some states, particularly Texas, have experimented with the use of precoated aggregates in surface treatments and seal coats. In this method, the aggregate and bituminous material are premixed, generally in a central mixing plan. The coated aggregate is then hauled to the job site and spread on the prepared base by the use of mechanical spreaders.

The use of precoated aggregate has certain advantages from a construction standpoint. Advantages include elimination of dust, reduction in loss of covering aggregate through spread of excess material, and somewhat less rolling.

Seal Coats

In line with the concepts which have been advanced in this chapter, a seal coat is a very thin single surface treatment which is usually less than one-half inch in thickness. Seal coats are applied as a final step in the construction of many types of bituminous wearing surfaces, their primary purpose in this application being to waterproof or "seal" the surface. They are also used to rejuvenate or revitalize old bituminous wearing surfaces, to "nonskid" slippery surfaces, and to improve night visibility. A seal coat usually consists of a single application of bituminous material which is covered by a light spreading of fine aggregate or sand. It may also be only a single application of bituminous material in some cases. The discussion which follows is primarily devoted to the case in which the seal coat consists of a single application of bituminous material, upon which a light cover of fine aggregate or sand is placed. The selection of the bituminous material and cover aggregate to be used in a given case is usually dependent upon the type of wearing surface which is being sealed, as well as on the nature of locally available materials. Specifications for seal coats are usually not given separately but are made a part of the specification for each particular type of wearing surface. A large number of different bituminous materials are used in this work, and aggregate gradings vary widely in different sections of the country.

15–18. Material Requirements. Very frequently, practice indicates that the bituminous material used as a binder in the construction of a given type of wearing surface is also used in the application of the seal coat when one is required. Thus, rapid-curing, medium-curing, and slow-curing liquid asphalts have all been used in the construction of seal coats. If any general preference may be detected in the use of these materials, it seems to be in the use of rapid-curing cut-back asphalts of the lighter grades. Emulsified asphalts are also used, particularly when they have been used in the construction of the wearing surface itself. The bituminous material in a "fog seal" (no aggregate) is an asphalt emulsion of mixing grade diluted with an equal amount of water. Paving grades of asphalt cement are widely used, as are the heavier grades of road tar.

Certain fundamental facts should be kept in mind in the selection of a bituminous material to be used in a seal coat. The material must have sufficient cohesion to hold the screenings in place, and this cohesion should be developed rapidly in order to prevent the loss of the aggregates from the surface under traffic. Thus the general preference is for rapid-curing cut-back asphalts and similar materials. The use of asphalt cement on dense surfaces is sometimes troublesome because of the tendency to get too heavy an application in proportion to the amount of cover material which can be retained, with the possibility of subsequent bleeding. The material chosen must, of course, be weather-resistant and durable, develop good adhesion to the cover aggregate and existing surface, and be fluid enough to permit ready application.

The requirements for aggregates to be used in seal coats also vary widely, and gradation requirements are usually quite "open" in nature. Sand is widely used for this purpose and is usually simply required to be of a clean and durable nature, free from an excess of dust and vegetable matter. Fine aggregate may consist of crushed stone, crushed slag, or crushed gravel, and requirements relative to the nature of these materials are generally the same as those which are applicable when they are to be used in other surface treatments.

Specifications (*169*) of the Bureau of Public Roads are indicative of the grading of aggregates used for seal coats, and are given in Table 15–3. Cover aggregate for Type 2 seal coat is sand or fine screenings,

TABLE 15–3

Requirements for Grading of Aggregate for Seal Coats (*169*)

Sieve Designation	Type 2 Cover Aggregate	Per Cent Passing (by Weight) Type 3 Cover Aggregate		
		Grading A	Grading B	Grading C
½-inch	–	–	–	100
⅜-inch	100	100	100	90–100
No. 4	85–100	85–100	60–100	10–30
No. 8	–	0–25	0–10	0–8
No. 50	0–20	–	–	–
No. 200	0–5	0–2	0–2	0–2

and for Type 3 is screenings of crushed stone, crushed gravel, or crushed slag.

15–19. Construction Methods. The construction methods used in the placing of a seal coat are generally the same as those which are used in the placing of a single surface treatment. Quantities of aggregate and bituminous material are quite small, with typical applications ranging from 0.10 to 0.33 gallon of bituminous material and from 10 to 25 pounds of aggregate per square yard, depending upon the nature of the materials used and the thickness desired. On tight surfaces with sand cover, not over 0.10 gallon per square yard would be used in order to avoid immersion of the sand. With 3/8-inch aggregate, the 0.33-gallon rate might be used. A general rule calls for the use of enough asphalt to seal the surface and fill from 50 to 70 per cent of the voids in the compacted cover aggregate.

The Wisconsin State Highway Commission has resurfaced many miles of highways with a seal coat containing a blend of rapid-setting asphalt emulsion and a synthetic latex compound; the rubber compound comprises only 2 per cent of the binder. The material is applied at a rate of about 0.30 gallon per square yard, covered with 20 to 24 pounds per square yard of stone chips and compacted with a rubber-tired roller.

New York City resurfaced several steep streets with a two-part epoxy compound and "ceramic slag," particles of 1/8-inch maximum size produced from the residue of oil furnaces. Workmen spread the epoxy compound with handbrushes, then covered it with the aggregate cast on by hand. Various epoxy-aggregate seals also have been used to resurface slippery bridge decks.

15–20. Colored Paving. Colored pavements have been placed experimentally in several cities to aid traffic control and distribution; colors include red, green, yellow, tan, and white.

The mixtures are bituminous hot mixtures prepared in a pug mill (see Art. 16–15) using a light-colored synthetic resin instead of asphalt and a pigment to give the desired color to the mixture. Light-colored aggregates are used, and in some cases aggregates have been selected to have about the same color as the final surface. Aggregates are fine and well graded, with at least 95 per cent passing the No. 8 sieve.

A tack coat of cutback synthetic resin is applied if necessary and the mixture placed with a conventional asphalt paver or by hand. Thickness of this type of surface generally is 1/2 to 1 inch.

Slurry Seal Coats

The slurry seal coat, pioneered in California and now used extensively there and in the Southwest, is used principally for the resealing of old bituminous wearing surfaces. The seal coat is quite thin, up to a maximum of ¼ inch.

The mixture for the slurry consists of a fine sand or crusher dust, or a blend of the two, a mixing grade asphalt emulsion, and water. California specifications (*192*) call for all of the aggregate to pass a No. 4 sieve, 95 to 100 per cent of it to pass a No. 8 sieve, and 8 to 15 per cent to pass a No. 200 sieve. At least half the material must be a crushed product that meets requirements of the sand equivalent test and a film stripping test. Typical proportions are 1550 pounds of aggregate (dry weight), 36 gallons of asphalt emulsion, and 20 gallons of water, including moisture in the aggregates.

One district of the Texas Highway Department uses limestone rock asphalt aggregate in a slurry seal coat. The mixture includes about 35 gallons of emulsion and 50 gallons of water per ton of dry aggregate. It is applied at a rate of about 6 pounds per square yard.

The surface to be treated is swept and cleaned prior to application of the seal coat. The work may be done only in warm, dry weather. Before application of the slurry, the surface is sprinkled with water. In some cases, a dilute asphalt emulsion prime coat has been used.

Standard batching equipment can be used to proportion the materials. Batched materials are loaded directly into concrete transit mixers. Mixing takes place enroute to the job site.

Fig. 15–5. Self-contained machine applies a slurry seal coat in one pass. (Courtesy Rex Chainbelt Co.)

Methods of applying the slurry vary somewhat. One method used successfully is as follows. The slurry is poured from the mixer into a rectangular spreader box, 10 to 12 feet wide and approximately 8 feet long. Near the rear of the spreader box there is a neoprene screed, which can be adjusted to control the thickness of the application. The spreader box is pulled forward by the mixer truck and the screed squeegees the slurry into the cracks in the old pavement. Thickness of treatment in one pass varies from ⅛ to ½ inch. A thicker coat should be placed in two passes, with ample curing time between applications.

In recent years, self-contained slurry seal machines have been developed that incorporate a pugmill mixer and do all the mixing and placing operations in one pass. Such a machine is shown in Fig. 15–5.

Traffic should be kept off the new surface for a period of from two to four hours, depending on the weather, thickness of the seal coat, and other factors. Traffic control should be exercised for an additional 4 to 20 hours.

Rock Asphalt Surfaces

"Rock asphalt," as the term is applied in highway work in the United States, refers to sandstone or limestone which is naturally impregnated with bitumen. This material should not be confused with native asphalts like Trinidad Lake Asphalt, which are primarily naturally occurring bitumens containing a large percentage of mineral matter. Rock asphalt is primarily mineral matter containing a small amount of natural bituminous binder. The amount of bitumen contained in rock asphalts which have been used in highway construction in this country has generally varied from about 3 to 12 per cent.

Rock asphalt occurs in a number of states and has been produced on a commercial basis and in sizable quantity in Kentucky, Texas, Oklahoma, and Alabama. Material of this nature produced in these states generally requires some modification for use in highway work. Both hot and cold paving mixtures have been produced through the use of these rock asphalts. These paving mixtures are used as wearing surfaces on new or old bases, in the reconstruction of old pavements, and for maintenance operations such as patching. Their use at the present time is largely confined to the states in which they are found or in neighboring areas. Construction methods used for rock asphalt pavements can be compared to those used in low-cost plant mixes, while in riding quality and durability these surfaces compare favorably with other immediate-type bituminous surfaces.

No attempt will be made to discuss the nature, occurrence, and use of all the different types of rock asphalt which have been or are produced and used in this country. Rather, the discussion will be confined to the use of two types of rock asphalt: Kentucky rock asphalt and Uvalde (Texas) rock asphalt. The rock asphalt found in Kentucky is a sandstone-asphalt combination, while that produced in Texas is limestone rock asphalt.

Kentucky Rock Asphalt Surfaces

Sandstone rock asphalt deposits are found in great abundance in western central Kentucky. The material lies under a considerable overburden in many areas, and may be obtained either from surface quarries or by tunnel-mining operations. The deposits are somewhat nonuniform in character, and the natural material is carefully selected, crushed, and blended in order to obtain a commercial product which is uniform in character. The sandstone itself is high in silica content, with about 95 per cent of the sand being SiO_2. The commercial product which is used in highway work contains from 6 to 11 per cent of bitumen.

At the job site, construction is accomplished by heating the material to make it easier to handle, spreading it in a thin, uniform layer, and rolling until a satisfactory density is secured. Wearing surfaces constructed in this fashion are usually from ½ to 1 inch thick, and are frequently placed upon a binder or leveling course which may be from 1 to 3 inches in thickness.

Specifications that have been used in Kentucky (*188*) require the sand contained in the natural sandstone rock asphalt to contain not less than 90 per cent silica and to be of a sharp and angular nature. The rock asphalt itself should be free from dirt, vegetable and other foreign matter, and any appreciable amount of uncoated sand. The gradation of the crushed material should be within the following limits:

Sieve Designation	Per Cent Passing (by weight)
¾-inch	Not less than 100
½-inch	Not less than 99
No. 4	Not less than 80

The amount of bitumen in the rock asphalt should be from 7.2 to 11.5 per cent if an ignition method is used for the determination, and from 6.2 to 10.5 per cent if the amount of bitumen is determined by extraction. Specifications also sometimes require that no natural material which contains less than 6 per cent of bitumen be used in blending the rock asphalt. Specifications may provide for surfaces of both natural and "processed" rock asphalt; one processed mixture that has been used contains 1 per cent powdered asphalt and another 2.5 per cent of asphalt cement.

The material is prepared for use by either dry or steam heating. If dry heating is performed, the material is passed through a drier which is continuously agitated and the temperature of the material is carefully controlled. Material is generally transported to the site of the work in tarpaulin covered trucks. Temperature of the material must be 170°F or above when delivered at the job site.

At the job site the material is spread in a uniform layer by a mechanical spreader or finishing machine. The spread material is then rolled with steel-wheel rollers, either of the three-wheel or tandem type, weighing from 7 to 10 tons. Pneumatic-tired rollers may be used for compaction, also. It is estimated that the rock asphalt will compact about one-third of its depth under rolling, and that the compacted material will weigh about 100 pounds per square yard per inch of compacted thickness.

Texas Rock Asphalt Surfaces

The rock asphalt produced in Uvalde County, Texas, and known commercially as "Uvalde Rock Asphalt," is a porous shell limestone impregnated with bitumen by natural forces. This material is found in horizontal strata which have a total thickness of approximately 120 feet, and lie under an overburden from 20 to 28 feet thick. It is mined by surface methods. As quarried, the material contains from 3 to 12 per cent bitumen and is selected and blended to produce a mixture of the desired characteristics at the crushing, screening, and mixing plant. The bitumen contained in Uvalde rock asphalt is very hard and must be softened by the addition of a suitable flux oil before it can be used in paving work.

Information supplied by the Uvalde Rock Asphalt Company of San Antonio, Texas, indicates that four types of rock asphalt paving mixtures have been produced. One of the most frequently used of these mixtures is *Duraco,* which is a patented paving mixture consisting of the limestone rock asphalt, flux oil, and coarse aggregate. This material closely resembles asphaltic concrete and may be used as a wearing surface which is placed on top of a new base or an existing surface.

Texas Highway Department specifications (*189*) permit the use of four different sizes of "cold mix limestone rock asphalt pavement." The rock asphalt used in these paving mixtures is required to be uniform and well graded and to contain from 5 to 9 per cent of asphalt. The specifications contain other requirements relative to the cleanliness, nature, and composition of the rock asphalt.

One flux "oil" which has been used is practically free from water and other foreign material and meets the following specifications:

Furol viscosity, 122°F (seconds)	50–100
Flash point (open cup)	250°F or more
Loss on heating (5 hours at 325°F)	5% or less
Asphalt content of 85 to 115 penetration by vacuum distillation, by weight	25% or more

The paving mixture is manufactured by combining the rock asphalt, flux oil, and water (a maximum of 4 per cent if needed to prevent the mixture from setting up during shipment). Grading limits for one of the mixtures (Type A) are as follows:

Sieve Size, Retained on	Per cent by weight
1-inch	0
7/8-inch	0– 2
5/8-inch	5–15
3/8-inch	25–35
No. 4	50–60
Passing No. 10	25–35

The amount of flux oil is not controlled directly by the specifications, but is established by the engineer for the particular mixture involved. The mixture, however, is subject to a requirement for stability.

15–21. Construction Methods. Assuming that the base or old surface upon which the rock asphalt mixture is to be placed has been brought

FIG. 15–6. Completed Duraco pavement. (Courtesy Uvalde Rock Asphalt Co.)

to a satisfactory condition, it is finally prepared by sweeping and the application of a tack coat of a suitable bituminous material. The "cold" process is usually used, and the prepared mixture is simply delivered to the job site in trucks without prior heating. The material is then spread in a uniform layer of the loose thickness required to give the compacted depth shown in the plans. Spreading and aeration of the mix may be accomplished by motor graders or by the use of mechanical spreading and finishing machines.

When spreading has been satisfactorily completed, rolling is begun. Rollers used may be of the three-wheel, tandem, or pneumatic type, and are generally required to weigh from 8 to 10 tons. Rolling is continued until no further compression of the layer can be achieved. The surface of the pavement is then tested for smoothness and compliance to grade and cross-section, and corrections are made as needed. A view of a completed Duraco pavement is shown in Fig. 15–6.

PROBLEMS

15–1. A new granular type base is to be covered with a double-surface treatment, meeting the requirements of Public Roads Type AT–60, as outlined in the text. Assuming that the base is to be 24 feet wide, compute the quantities of bituminous material and crushed stone which are required for a mile of pavement if the specific gravity of the stone is 2.70. Estimate the cost of a double-surface treatment of this type in your area.

15–2. Refer to Art. 15–15 for a discussion of one method of determining the amount of asphalt required for a single surface treatment. Applicable data are loose unit weight of aggregate, 98 pounds; specific gravity, 2.68. The weight of stone required to cover 1 square yard (one-stone thick) is 19 pounds. Estimate the residual amount of asphalt required for proper embedment of the aggregate.

15–3. A single surface treatment using aggregate of essentially one size is to be designed in accordance with the Asphalt Institute method described in Art. 15–16. The bulk specific gravity of the aggregate is 2.68, and its average least dimension is 0.27 inch. The aggregate wastage factor is estimated at 1.06, and the traffic factor at 0.70. Allow 0.03 gallon per square yard for absorption of the asphalt by the surface to which the treatment is being applied. Determine the aggregate spread in pounds per square yard and the asphalt spread in gallons per square yard.

15–4. An old bituminous wearing surface is to be resurfaced with Kentucky rock asphalt. If the rock asphalt layer will have an average compacted thickness of 1.25 inches, and the new surface is to be 24 feet wide, determine the tons of rock asphalt required per mile of road. What will be the average loose thickness of the spread layer of rock asphalt?

CHAPTER 16

ROAD-MIX AND PLANT-MIX SURFACES

The bituminous wearing surfaces discussed in this chapter would generally be classified as "intermediate-type" surfaces. They are utilized on highways subjected to from relatively light to moderately heavy traffic, and they may be used on old or new bases and, for the most part, in the resurfacing of old pavements of various sorts.

The term "road-mix" refers to a type of construction in which the mineral aggregates and bituminous material are intimately blended by mixing on top of the existing surface or base. The upper portion of the existing layer may or may not be included as a part of the road-mix surface, as circumstances warrant. The surfaces produced by this type of construction are also sometimes designated as "mixed-in-place" surfaces. The term "plant-mix" surface refers to a type of surface in which the mineral aggregates and bituminous material are proportioned and mixed at a central mixing plant. The mixture is then hauled to the job site and the surface is prepared by spreading, compacting, and finishing the prepared mixture. The designation "intermediate-type bituminous plant-mix" is somewhat ambiguous, but it is employed here to distinguish these surfaces from the "high-type bituminous pavements" described in the next chapter.

Road-Mix Surfaces

Road-mix bituminous wearing surfaces are widely used in this country for the surfacing of both primary and secondary roads, and they have been used to some extent in the surfacing of city streets. The initial development of this type of surface was probably primarily due to the need for the construction of a large mileage of surfaced roads at low cost, using local materials in areas of relatively sparse population, as in certain of the western states. Their use is now nationwide, although there is considerable variation in methods of constructing them in different sections of the country. Three general types of road-mix surfaces will be discussed here. These types are designated as follows:

1. Road-mix surface, open-graded aggregate type
2. Road-mix surface, dense-graded aggregate type
3. Mixed-in-place sand asphalt

The *open-graded aggregate* type of road-mix surface is also sometimes designated as the "macadam aggregate" type. This type of road-mix surface is distinguished by the use of aggregate of essentially uniform size, the aggregate being generally somewhat similar to that used in surface treatments. These surfaces are built in thicknesses varying from about 1 inch to 3 inches, with the majority of such surfaces being from 2 to 2.5 inches in thickness.

The *dense-graded aggregate* type of road-mix surface is marked by the use of aggregates which are locally available from such locations as river banks or pits or from the roadway itself. The aggregate is usually more or less uniformly graded from a maximum size of 1.5 inches or so down to dust. The majority of roads constructed by this method are from 2 to 3 inches in thickness but thicknesses as great as 12 inches have been reported.

The term *mixed-in-place sand asphalt* refers to a type of construction in which naturally occurring sand is mixed with liquid asphalt to form the wearing surface. Thickness of this type of treatment would generally vary from 3 to 6 inches. No separate base is employed in this type of construction, with the wearing surface resting directly upon the natural or stabilized sand subgrade. This type of surface has been widely employed in certain of the states along the Gulf Coast.

Road-Mix Surfaces, Open-Graded Aggregate Type

The macadam aggregate type of road-mix surface has been extensively used in the resurfacing of old pavements in areas where suitable crushed aggregates are readily available. This type of surface is known in some areas as a "retread," since it is so widely used for resurfacing, although the same term is also applied to other road-mix surfaces and in some cases to surface treatments.

16–1. Material Requirements—Aggregates. The aggregates used in this type of work include crushed stone, crushed gravel, and crushed slag. They are generally required to be clean, tough, and durable in nature, free from an excess of flat or elongated pieces. Both crushed stone and crushed gravel are required to have a per cent of wear of not more than 40 in the Los Angeles Abrasion Test, while crushed slag may in some instances be subjected to this same requirement. More often, however, crushed slag is simply required to be air-cooled blast furnace slag which is reasonably uniform in quality and density and which weighs not less than 70 pounds per cubic foot.

In the usual open-graded road-mix surface, only one size of coarse aggregate is used to form the principal thickness of wearing surface. This surface is then usually finished by the application of a seal coat of bituminous material and fine aggregate which serves to make the surface less open in texture. In some cases, however, a second size of intermediate or choke stone is used to somewhat seal the surface of the first layer and before the seal coat itself is placed. In this latter case, three sizes of aggregate would then be included in the specification. Given in Table 16–1 are aggregate gradings given in AASHO M-63, which are typical of this kind of construction (*107*).

TABLE 16–1

Aggregate Gradings, Open-Graded Bituminous Road-Mix Surface Course

Sieve Designation	Per Cent Passing (by Weight)				
	Road-Mix Aggregate				Seal Coat
	¾-inch—No. 4	1-inch—⅜-inch	1-inch—½-inch	1½-inch—No. 4	⅜-inch—No. 8
2-inch	–	–	–	100	–
1½-inch	–	100	100	95–100	–
1-inch	100	90–100	90–100	–	–
¾-inch	90–100	40–75	20–55	35–70	–
½-inch	–	15–35	0–10	–	100
⅜-inch	20–55	0–15	0–5	10–30	85–100
No. 4	0–10	0–5	–	0–5	10–30
No. 8	0–5	–	–	–	0–10
No. 16	–	–	–	–	0–5

16–2. Material Requirements—Bituminous Materials. A number of different bituminous materials have been used in the construction of this type of surface, including cut-back asphalts, road tars, and asphalt emulsions. So far as cut-back asphalts are concerned, both rapid-curing (grades RC–250 and 800) and medium-curing (grades MC–250 and 800) types have been used. Grades RT–6, 7, 8, and 9 of the road tars are used, with the heavier grades being used during the normal summer construction months and the lighter grades recommended for use at other times of the year. Asphalt emulsions which are used should be of the medium- or slow-setting type, particularly when mixing is done with blades or drags, so that the emulsion will not break down too quickly during the mixing process. As has been mentioned before in relation to other types of construction, the exact grade of bituminous material selected for a given project will probably depend largely upon previous experience gained on similar projects.

16–3. Quantities of Materials. No analytical method is generally used to determine the exact quantities of aggregate and bituminous material which will be required to construct a wearing surface of a given thickness, using the materials indicated above. The amount used is again largely based upon judgment and experience.

The quantity of bituminous material may be estimated at 0.6 to 0.8 gallon per square yard, for 2-inch compacted thickness and conventional aggregates. The required amount of coarse aggregate may be generally based upon a weight for the compacted aggregate, in place, of 100 pounds per square yard per inch of compacted thickness. This figure should be used with caution, as this weight may vary from 80 to 120 pounds per square yard per inch, depending on the material and the method of construction involved. A typical seal coat may take 15 to 20 pounds of aggregate per square yard, and from 0.15 to 0.20 gallons of bituminous material.

The base or old surface upon which this type of wearing surface is to be placed must be adequately prepared through the correction of irregularities, depressions, or other faults, and, generally speaking, by the application of a suitable prime or tack coat.

16–4. Road Mixing (Blading and Dragging). In this process of road mixing, the coarse aggregate usually is spread in the quantity desired in a layer of uniform thickness by the use of spreader boxes or other mechanical spreading devices. It may also be spread in a sized windrow. The aggregate must be reasonably dry (less than 3 per cent, according to Asphalt Institute Specification RM–1) before any bituminous material may be applied to it, and some preliminary aeration may be necessary to bring the aggregate to the desired condition. This kind of surface is generally placed upon a firm, dense base or old surface, and it is essential that the underlying layer not be cut into or otherwise disturbed during mixing operations.

The bituminous material which is to be used as a binder is then applied to the aggregate in the required amount, usually from 4 to 6 per cent by weight. The bituminous material is applied by a pressure distributor, and the entire amount may be applied in one "pass." More generally, however, the bituminous material is incorporated in two equal applications of up to 0.5 gallon per square yard each. If such is the case, after the initial application of bituminous material has been made, the aggregate and bituminous material are thoroughly mixed, generally through the use of motor graders, multiple-blade drags, or other tools which will accomplish the same objective. Rotary speed mixers may also be used to advantage in this phase of the construction process. Manipulation, which may involve the moving of the aggregate-bitumen mixture across the road several times, is continued until a uniform mixture is obtained. The second application of bituminous material is then made, and mixing is continued. The road-mixing process is carried on until a uniform final mixture is secured and until sufficient of the volatile materials contained in the bituminous material have been removed to permit proper spreading and compaction. The mixture is then spread in a uniform layer of the desired thickness and is ready for compaction and finishing.

16–5. Road Mixing (Travel Plant). Many of the travel plants of the types which are used in the construction of road-mix surfaces are arranged so that they will pick up the aggregate from a windrow, apply and mix the required amount of bituminous material with the aggregate in a pugmill mixer, and spread the mixture in a windrow or uniform layer behind the machine. One machine of this type is shown in Fig. 16–1. Other plants process the material in place. Two distinct advan-

FIG. 16–1. Wood Roadmixer engaged in processing a bituminous mixture. (Courtesy Pettibone-Wood Mfg. Co.)

tages may be expected to accrue from the use of a travel plant in road-mix construction: First, it is possible to more accurately control the proportions of the mineral aggregate and bituminous material; second, the mixing is more thorough and uniform than can be readily and easily achieved by blading and dragging. It would also be expected that less time would be required for the completion of a road-mix surface when a travel plant is used. However, the use of travel plants is being supplanted in many sections of the country by the central plants described later in this chapter.

16–6. Compaction and Finishing Operations. After the material has been satisfactorily spread, it is then rolled by pneumatic-tired or steel-wheel rollers weighing from 7 to 10 tons until it is uniformly and thoroughly compacted. At this stage the surface is tested for smoothness. Irregularities greater than one quarter of an inch (as measured by a

10-foot straightedge) are removed by means of a long-base planer or motor grader. Within a period of a week or so after the completion of this rolling, fine aggregate may be uniformly spread upon the surface in an amount which is sufficient to fill the surface voids. This material is uniformly distributed and keyed into the surface by use of brooms or broom drags. Another light application of bituminous material is then made and the surface is subjected to additional brooming and rolling. In certain cases the surface is completed by the application of a seal coat on top of this layer, or directly upon the road-mixed surface course.

Road-Mix Surfaces, Dense-Graded Aggregate Type

This variety of road-mix surface differs from the macadam aggregate type principally by the use of a wider range of aggregate types, including naturally occurring gravels, which are graded from relatively coarse to fine sizes. This difference in aggregate materials necessitates some difference in the two road-mix types from the standpoint of the kind and amount of bituminous material used, and the employment of somewhat different construction procedures. It may also help the reader to distinguish between the two types if it is noted that the dense-graded aggregate type of bituminous wearing surface is frequently constructed when it is desired to "hard-surface" an existing gravel or crushed rock surface. Thus the existing surface may supply all or part of the aggregate for the new wearing surface. The macadam aggregate type of surface would generally be built entirely of "imported" aggregates. Requirements for materials to be used in this type of mixture vary widely with the different organizations engaged in this type of work. Recommendations relative to materials as given in the following sections are intended only to be representative of those in current use.

16–7. Material Requirements—Aggregates. The total aggregate for the mixture is generally required to be well graded from coarse to fine. Both crushed and uncrushed aggregates are used in this work, and a crusher-run material may contain all the required sizes, as may a natural gravel aggregate. Provision is also frequently made in the specification for the blending of the coarse aggregate, which may be any one of a large number of materials including crushed stone, slag, and crushed or uncrushed gravel; fine aggregate, which might generally be stone screenings or sand; and filler, which might be limestone dust or some locally available fine material of suitable characteristics. These materials would be combined in the proportions required to meet the grading requirements established in a particular specification.

General requirements relative to aggregates for this purpose are similar to those previously given. They are generally required to be clean, hard, tough, and durable in nature, free from an excess of deleterious materials. Specifications relative to resistance to wear are a little less severe than in some other types of bituminous surfacing, with values of the maximum loss in the Los Angeles Abrasion Test frequently being in the range from 45 to 60 per cent, with 50 per cent probably being a typical figure. Mineral fillers which may be added are required to meet standards established for the same type of material when it is to be used in a high-type bituminous paving mixture, as presented in the next chapter. Since the total aggregate in this type of mix contains a considerable amount of material passing a No. 200 sieve, specifications frequently include a reference to the desired hydrophobic nature of the aggregate, and may include a definite requirement relative to the results of a stripping, swell, or immersion-compression test. The specification may also include a requirement relative to the plasticity index of the material passing a No. 40 sieve and/or the sand equivalent.

The requirements of two different state highway departments relative to the grading of aggregates for this type of mixture are listed in Tables 16–2 and 16–3. They are taken to be typical of those now being used.

TABLE 16–2

SPECIFICATION FOR AGGREGATE GRADING FOR ROAD-MIXED ASPHALT SURFACING (*192*)

Sieve Designation	Per Cent Passing (by Weight)
1-inch	100
¾-inch	95–100
⅜-inch	75–85
No. 4	50–65
No. 8	37–52
No. 30	18–30
No. 200	4–10

16–8. Material Requirements—Bituminous Material. The bituminous materials which have been most widely used in this type of work are medium-curing (grades MC–250 and 800) and slow-curing liquid asphalts (grades SC–250 and 800). The slow-curing liquid asphalts have been extensively used in the arid and semiarid regions of the western states and with aggregates containing relatively large amounts of material passing a No. 200 sieve. Rapid-curing cut-back asphalts have been used in some cases where little or no material passing the No. 200 sieve was contained in the aggregate, and emulsified asphalts have also been used. Road tars which are suitable for this work include grades RT–5, 6, 7, and 8.

TABLE 16-3

SPECIFICATION FOR BITUMINOUS BASE AND BITUMINOUS MAT SURFACE COURSE (*187*)

Sieve Designation	Per Cent Retained On (by Weight)			
	A	B	C	D
¾-inch	0–5	0–5	0–5	0–5
⅜-inch	0–35	–	–	–
No. 4	20–50	–	–	–
No. 10	45–65	30–60	–	–
No. 20	55–75	45–70	–	–
No. 50	70–85	60–80	40 Min.	–
No. 100	75–96	70–90	54 Min.	25 Min.
Passing No. 200	4–8	10–24	12–24	12–30

16–9. Quantity of Bituminous Material. The required amount of bituminous binder in a dense-graded road-mix surface generally varies from about 3 to 7 per cent. The quantity of material required on a given project may be based upon experience and judgment. However, a number of states determine the required amount by the use of a "proportioning formula," of which several are in existence. Each of these formulas has demonstrated its value to the organizations which use it, and each has been developed as a result of intensive laboratory and field investigation.

An example of the type of formula which may be used in proportioning the materials for a dense-graded road-mix surface is seen in those which are used in Nebraska. The formulas, which are given below, are a part of the specification previously referred to and have the following form:

$$P = 0.02a + 0.06b + 0.1c + Sd \qquad (16\text{-}1)$$

where P = amount of bitumen required in the mixture prior to laying, per cent by weight

a = aggregate retained on the No. 50 sieve, per cent by weight

b = aggregate passing the No. 50 sieve and retained on the No. 100 sieve, per cent by weight

c = aggregate passing the No. 100 sieve and retained on the No. 200 sieve, per cent by weight

d = combined aggregate passing the No. 200 sieve, per cent by weight

S = a factor depending upon the fineness and absorptive quality of the effective filler material and determined by a special laboratory procedure. S is frequently taken to vary from practically zero up to a maximum of 0.20.

The percentage of bituminous material (oil) to be applied in the field is determined by a second equation, as follows:

$$\text{Per cent of oil to be applied} = \frac{F(P)K(100)}{100 - P} \qquad (16\text{-}2)$$

where F = the oil application factor to be determined by the laboratory.

In general F will be in the range of 0.70 to 0.90 for Gradings A, B, and C (Table 16–3) and from 0.50 to 0.75 for Grading D.

$$K = 1.00 + 0.0075 \text{ (per cent distillate to 680°F)}$$

A method for determining the amount of bituminous material required for dense-graded bituminous mixtures has been evolved in California over a period of years and as a result of intensive field and laboratory experience. A complete explanation of the method is given in Art. 17–11.

The formulas and methods which have been cited in the preceding paragraphs give the approximate amount of bituminous material which may be required with a given aggregate. Some slight modification of this quantity may be necessary in the field as construction proceeds. Such modifications, if required, are simply made on the basis of experience gained in the construction of similar surfaces.

16–10. Construction Methods. The basic construction procedure used in the formation of dense-graded road-mix surfaces is quite similar to that used in the construction of road-mix surfaces in which macadam-type aggregates are used. Certain differences in the procedure arise, however, principally because of the difference in aggregates used in the two types and the slightly different bituminous materials. The procedure which is discussed here is more or less an average one, and the detailed process used by a particular agency may differ slightly from that presented below.

Quite frequently the material existing in an old gravel or crushed-rock road is utilized in the construction of the new road-mix surface. If such is the case, this existing layer would be scarified to the desired depth and the loosened material bladed into a windrow at the side of the road. Accurately sized windrows are essential in order to permit accurate determination of the quantities of aggregate involved in the mixing process. A windrow sizer is shown in Fig. 16–2. The underlying subgrade may be watered and compacted as necessary. In the normal case, this aggregate material would be sampled and subjected to laboratory examination before road-mixing is started. If this material met the aggregate requirements of the particular organization involved, the mixing process could then begin. Probably in most instances, however, the grading of this material would be such as to require some modification, usually by the addition of relatively small amounts of suitable coarse aggregate, sand, or filler. The procedure then would generally be to windrow the material taken from the existing surface and add the required additional aggregate materials to the windrow. From that stage on the procedure differs slightly, depending on whether the surface is to be blade-mixed or prepared in a traveling plant.

In some instances, of course, the road-mix surface is to be constructed without utilizing any material from the existing layer. This would generally be the situation, for example, when the surface is to be built

Fig. 16–2. Windrow sizer. (Courtesy Pettibone-Wood Mfg. Co.)

upon a new base or an old bituminous wearing surface. Before the new surface can be constructed, the existing base or surface must be brought to a satisfactory condition, including the elimination of weak spots, holes, depressions, or other irregularities. After this underlying layer has been properly prepared, and primed if required, the aggregate is then usually brought in and placed in a windrow of known dimensions and weight, immediately prior to the beginning of road-mixing operations.

If the mixing is to be accomplished by blading or dragging, or by any similar means, the loose aggregate materials would then be thoroughly and uniformly blended by the use of a motor grader. Even if blending of the aggregates is not necessary, manipulation may be necessary to reduce the moisture content of the aggregates. Before any bituminous material may be added, it is necessary that the aggregate material be quite dry, with from 3 to 5 per cent of moisture generally being the maximum permitted. The material would then be bladed back into a single windrow for final measurement, testing, and adjustment if required.

When the dry and blended aggregates have been finally approved, the mixing process is begun by spreading the aggregate in a uniform layer of convenient width. The bituminous material is then applied to the aggregate by means of a pressure distributor. The bituminous material is generally applied in several applications or "shots," with the average amount per application being approximately from 0.25 to 0.50 gallon per square yard. From two to as many as six or eight separate applications may be made in varying quantities, depending on the individual practice of the organization concerned. The bituminous material must be applied at a temperature which is suited to the particular grade which is being used. Specifications covering this type of work usually include a provision that no bituminous material is to be applied when the air temperature is below a certain figure, typically 60°F.

Immediately after each application of bituminous material has been made, the binder is thoroughly blended with the mineral aggregate, usually by the use of disk harrows. Harrowing is continued after each increment of binder material is added until the bituminous material has been absorbed. As soon as a section of the surface has received the entire amount of bituminous material which is to be applied and has been thoroughly disked, the mixing process is continued by the use of blade graders, multiple-blade drags, and similar tools. If a blade grader is used, the material would generally be moved into a windrow at one side of the road by several passes of the machine. That is, the grader would make a light cut and move the material to the side; another light cut would then be made, and more material moved to the side; and so on. This process would continue until the entire width and depth of the treated material had been moved into a windrow at one side of the road, care being taken not to mix in material from the underlying layer. The material in the windrow would then be moved back over the road in a series of thin layers until the entire mixture was again spread out on the surface. This process is repeated until the desired mixing has been accomplished. Figure 16-3 shows a blade grader engaged in the mixing of materials for a dense-graded road-mix surface. If the material becomes wet during mixing, the excess moisture must be removed. This is done by continuing the manipulation of the mixture until it has dried enough for construction to continue. Regardless of the individual tools and methods which are used in the process, mixing must be continued until the aggregate and bituminous material have been completely and uniformly blended. Volatiles in the mixture, and its moisture content, must be reduced to a certain level before compaction can begin. For example, an organization may require that the mixture not contain more than 50 per cent of the original volatiles and that the moisture content be less than 2 per cent (up to 5 per cent

Fig. 16–3. Motor grader being used in the construction of a dense-graded bituminous mixture. (Courtesy Huber-Warco Co.)

for emulsified asphalts). When the mixture is judged to be satisfactory, it is then generally bladed back into a single central windrow.

If a traveling plant is used, the blending of the aggregate fractions and the bituminous material would generally take place in the plant. The mixture is checked for compliance with requirements relative to aggregate gradation and bitumen content. If the mixture coming from the plant is nonuniform in character, additional processing may be required through the use of blade graders, drags, and so on. Even after mixing is completed, some additional manipulation may be required to permit aeration and securing of the proper degree of tackiness for proper laying. This is particularly true for MC materials.

At this stage it will be assumed that the satisfactory mixture has been placed in a windrow in the center of the road, regardless of the construction procedure used to accomplish the mixing operation. In the normal sequence of events, the mixture would then be spread out over the road in thin layers (not more than 2 or 3 inches compacted thickness). Each layer is rolled after spreading. Pneumatic-tired rollers are preferred for initial rolling. Steel-wheel or pneumatic-tired rollers may be used for final finishing. Requirements relative to the

FIG. 16–4. A properly constructed road-mix surface in which dense-graded aggregates were used. (Courtesy Dept. of Roads and Irrigation, State of Nebraska.)

smoothness of the finished surface are generally the same as those for open-graded road-mix surfaces.

Surfaces of this type frequently require a fairly long curing period before a stable condition is secured, and may require considerable maintenance during the first few weeks (or months) of service, particularly when slow-curing liquid asphalts are used. The placing of a seal coat is frequently delayed until the surface is hard and thoroughly cured. Fig. 16–4 shows a close-up view of a properly constructed dense-graded road-mix surface.

16–11. Mixed-in-Place Sand Asphalt. The construction methods used in the building of mixed-in-place sand asphalt pavements, which might more generally be called sand-bituminous road-mix surfaces, differ little from those which have been described for other types of road-mix surfaces. This type of construction has been of particular importance in certain states along the Gulf Coast and the Atlantic seaboard where naturally occurring sands are the only aggregates which are economically available for the construction of bituminous wearing surfaces on roads which carry light to moderate amounts of traffic.

Little difference exists between this type of construction and the sand-bituminous stabilization described in an earlier chapter. In fact, a mixed-in-place sand asphalt surface might be regarded as a specific type of sand-bituminous stabilization, with the further observation that these mixtures are primarily intended to serve as wearing surfaces, while in the previous discussion of sand-bituminous stabilization the emphasis was placed upon the construction of base courses. Sand-bituminous road-mix surfaces are usually constructed in thicknesses ranging from 3 to 6 inches, and the finished surface generally has a crown of about one-quarter of an inch per foot. Wearing surfaces of this type make excellent bases for high-type bituminous pavements if increases in traffic make the application of an additional thickness of wearing surface necessary at a later date.

Intermediate-Type Bituminous Plant-Mix Surfaces

The bituminous wearing surfaces described in this section are constructed from mixtures produced at a central mixing plant. In many "low-cost" bituminous plant mixtures, liquid bituminous materials—including liquid asphaltic products, asphalt emulsions, and road tars—are used as the binder. From the standpoint of classification, this type of surface occupies an intermediate position between the road-mix surfaces which were described in earlier portions of this chapter and the high-type bituminous pavements which are discussed in the next. The term "intermediate mixes" has frequently been applied to the mixtures discussed here.

The development of these intermediate surfaces was a natural outgrowth of certain difficulties encountered in the road-mix process. Preparation of the paving mixture at a central plant offers such advantages as more careful proportioning of the ingredients, more uniform and thorough mixing with consequent production of more uniform mixtures, less dependence upon favorable weather conditions, and the use of more viscous bituminous materials. Mixtures which are being included here under the classification of intermediate-type plant-mixes range in nature from those which differ only very slightly from the road-mix surfaces, which have previously been described, to those which very closely resemble the high-type mixtures discussed in the next chapter. They also cover a fairly wide range in cost, being generally slightly more expensive than comparable road-mix surfaces and somewhat less expensive than high-type mixtures. The more expensive of the intermediate mixes described here may be said to differ from high-type mixtures in the following respects:

1. Softer bituminous materials which require only a moderate amount of heating (or no heating, as in the case of asphalt emulsions) are used in the intermediate types, while the majority of the high-type mixtures utilize semisolid binders and are mixed and laid at elevated temperatures.
2. The nature and grading of the aggregate components of a high-type mixture, and the composition of the final mixture, are more carefully controlled than for the intermediate type.
3. Requirements relative to the density and stability of the compacted mixture are generally much more rigid for the high-type pavement than for the low-cost plant-mix surface.

Specifications relating to the composition and control of the low-cost bituminous plant mixtures show extreme variation, with all sorts of mixtures which would fall into this classification being utilized by different highway organizations in line with their particular experience and to fit their particular needs. Certain types of patented paving mixtures might also be placed in this classification, although none of these is discussed here. No attempt will be made here to describe all these mixtures. Rather, a number of them will be discussed which are typical of those in present use.

16–12. Open-Graded Aggregate Mixtures. Plant mixtures composed of open-graded aggregate and liquid bituminous materials very closely resemble those used in road-mix construction of the same type. Requirements relative to the grading and nature of the aggregates, and the composition of the mixture, are usually the same as those used by an organization for a similar road-mix surface. The chief difference between a plant-mix and road-mix surface of this particular type, so far as the mixture itself is concerned, is that slightly more viscous bituminous materials are generally used in the plant mixture. For example, if an MC–250 were used in the road-mix surface, then an MC–800 might be used in the plant mixture. The use of more viscous bituminous materials may be said to be an objective in the use of plant mixtures and it is accomplished largely by the lessened time of mixing in the plant as compared with mixing on the road, with a consequent smaller amount of volatiles lost in the mixing process, together with the heating of the aggregate.

16–13. Dense-Graded Aggregate Mixtures. Mixtures of this type vary widely, some of them being practically identical with similar road mixes, while others closely resemble the high-type mixtures. In California, for example, the requirements given in Table 16–2 also apply if the material is mixed in a central plant. Generally speaking, plant mixtures of this kind vary from similar road-mix surfaces in that the grading of the aggregate is more closely controlled and more viscous bituminous materials are used. Very soft asphalt cements are sometimes used in this work, in which case they very closely approximate the high-type mixtures. Closer control over the grading of the aggregate is sometimes secured by requiring that the aggregate be divided into two fractions which are recombined in the mixing process.

Two examples of the requirements relative to gradation of the aggregates used in this type of mixture are presented in Tables 16–5 and

TABLE 16–4

Grading of Aggregate for Use in Dense-Graded Bituminous Road and Plant-Mix Surface Courses, AASHO M62 (*107*)

Sieve Designation	Per Cent Passing, by Weight
1-inch	100
¾-inch	85–100
No. 4	45–65
No. 10	30–50
No. 200	5–10

16–6. The inclusion of these gradings is not intended as a representation of typical or "average" requirements in this respect, but only to indicate the type of specification which may be used.

Table 16–4 shows the grading requirements for aggregates given in AASHO Specification M62, which covers crushed stone, crushed slag, and crushed gravel for dense-graded bituminous road and plant-mix surface courses.

In addition to the usual stipulations, this specification contains a requirement relative to the uniformity of the aggregates. During any run, aggregates passing the No. 4 sieve should not vary by more than 5 per cent (±) from the average; passing the No. 10 sieve, ±5 per cent; and passing the No. 200 sieve, ±2 per cent.

Table 16–5 gives the grading requirements for aggregates used in Bitumuls concrete, densely graded type 4 (*225*). The binder in this mix is mixing grade asphalt emulsion.

TABLE 16–5

GRADING OF AGGREGATE FOR USE IN BITUMULS CONCRETE, DENSELY GRADED TYPE 4 (*225*)

Sieve Designation	Per Cent Passing, by Weight
¾-inch	100
½-inch	90–100
No. 4	60–80
No. 10	40–55
No. 40	15–30
No. 80	8–18
No. 200	3–8

The quantity of emulsion to use in this plant mixture is determined by the following formula:

$$P = 0.05A + 0.10B + 0.50C \tag{16–3}$$

where P = asphalt emulsion, pounds per cubic foot of loose, dry aggregate
A = aggregate retained on the No. 10 sieve, per cent
B = aggregate passing the No. 10 sieve and retained on the No. 200 sieve, per cent
C = aggregate passing the No. 200 sieve, per cent

The quantity given in Eq. 16–3 must be increased if unusually absorptive aggregates are used.

16–14. Construction Methods—Intermediate-Type Bituminous Plant-Mix Surfaces. Certain of the steps involved in the construction of plant-mix surfaces such as have been described above are essentially the same as those which are employed in the construction of similar road-mix surfaces. For example, the preparation of the base course is essentially the same for the two types of surfaces—a prime or tack coat may be required before the placing of the new surface. Likewise, the rolling and final finishing steps are practically the same, with three-wheel, tandem, and pneumatic-tired rollers of various weights being used in the placing of low-cost plant-mix surfaces by various organizations. Since the procedures used in the steps mentioned are so similar for the two types of surfaces, no additional space will be devoted to the discussion of these steps in the construction process.

The plant-mix and road-mix methods are markedly different in two major steps—the preparation of the mixture and the spreading (and preliminary finishing) operations. These two steps will be discussed in detail in the following sections as they apply to low-cost plant mixtures. A large part of the discussion will be devoted to a description of the equipment which is used in these two steps in the construction of plant-mix surfaces. The same (or similar) equipment is employed in the preparation of the high-type surfaces described in the next chapter, and this fact will be emphasized at various points in the following discussion.

16–15. Central Mixing Plants. The term "central mixing plant" refers to the plant or "factory" at which the bituminous paving mixture is produced, in a process beginning with the aggregates and bituminous materials and ending with the discharge of the mixture into hauling units for transportation to the job site. The paving plant may be small and very simple in nature or it may be large and complex, depending largely on the type and quantity of bituminous mixture which it is designed to produce.

The range in size and complexity of paving plants may be indicated by mention of two extreme installations of this type. The smallest type of machine or "plant" applicable to this work might consist of a simple concrete mixer of the drum type in which the aggregate and a cold bituminous material, such as asphalt emulsion, are proportioned by shovels and buckets. This type of machine might be used in the production of bituminous mixtures to be used in patching operations conducted on a small scale, with the capacity of the "plant" perhaps being in the neighborhood of five cubic feet of mixture per minute. On the other end of the scale are the tremendously large and complex stationary plants, such as might be installed by a city or county agency, which are capable of producing several thousand tons per day of "hot mixes," such as bituminous concrete. In between these two extremes may be placed a large number of plants of different manufacture, design, and capacity.

Plants of intermediate size and capacity are generally employed in the production of the plant mixtures with which we are concerned in this chapter. Typical plants of this type will be described in the following sections. No attempt will be made to describe all the plants of this sort which are available today, as such an undertaking is clearly beyond the purpose of this text. Certain basic units may be regarded as being common to most of the paving plants normally employed in the production of sizable quantities of plant mixtures which utilize liquid bituminous materials as the binder, and these basic units will be described in some detail. Typical combinations of these fundamental units for the purpose of producing intermediate-type plant mixtures will also be described. Some of the basic units described in this chapter are also used in plants designed to produce the high-type mixtures described in the next chapter. The description of these is intended to suffice for both purposes, but additional units required for the production of the high-type mixes and their relation to the basic units described here will be discussed at a later time.

Central mixing plants for the production of bituminous paving mixtures are frequently described as being portable, semiportable or stationary in nature. The term "portable" is applied to relatively small units which are self-contained and wheel-mounted, and it is also applied to larger mixing plants in which the separate units are themselves easily moved from one place to another. The term "semiportable" is reserved for those plants in which the separate units must be taken down, transported on trailers, trucks, or railroad cars to a new location, and then reassembled, which process may require only a few hours or several days, depending on the plant involved. "Stationary plants" are those which are permanently constructed in one location and are not designed to be moved from place to place. Portable and semiportable plants are much more numerous than stationary plants and are widely used in the construction of rural highways. The capacities of these two types of plants range up to about 400 tons of mixture per hour.

Central mixing plants may also be distinguished as "batch" and "continuous flow" plants. In the batch type of plant, the correct amounts of aggregate and bituminous material, determined by weight, are fed into the mixing unit of the plant; the "batch" is then mixed and discharged from the mixer before additional materials are introduced. The size of batch which may be handled by a given plant varies from 1000 to 12,000 pounds, except for small portable units which are primarily intended for use in maintenance. In the continuous-flow type of plant, the aggregate and bituminous material, in the correct amounts as governed by volumetric controls, are continuously fed into the mixer and the completed mixture is discharged in a continuous flow of material. Both types of plant are generally designed to be automatic, or nearly so, in their operation. Automatic control (and recording) systems are described in Art. 17–16.

One of the basic units contained in many bituminous paving plants is the drier. This unit is a necessary part of all "hot-mix" plants, and is also generally used in the preparation of low-cost plant mixtures. The drier may not be needed for some mixes of the latter type, as, for example, those in which asphalt emulsion is used as the binder. The drier consists primarily of a large (from 3 to 10 feet in diameter and from 20 to 40 feet long) rotating cylinder which is equipped with a heating unit at one end, usually a low-pressure air atomization system utilizing fuel oil. The drier is mounted at an angle with the horizontal, with the heating element being located at the lower end. Hot gases from the burner pass from the lower end up the cylinder and out at the upper end. The so-called cold aggregate is fed into the upper end of the drier, picked up by steel angles or blades set on the inside face of the cylinder, and dropped in "veils" through the burner flame and hot gases, and it moves down the cylinder because of the rotating action and the force of gravity. The hot aggregate then discharges from the lower end of the drier, generally onto an open conveyor or enclosed "hot elevator" which transports it to the screens and storage bins mounted in conjunction with the mixing unit. A drier of this conventional type is shown in Fig. 16–5.

FIG. 16–5. Drier used in a central mixing plant. (Courtesy Iowa Mfg. Co.)

In the preparation of the hot mixtures described in the next chapter, the temperature of the aggregates may be raised to 325°F or more, and practically all the moisture in them removed. In the preparation of the "cold" mixtures which we are discussing here, such complete drying may be neither necessary nor particularly desirable. The moisture content may be reduced to about 1 per cent in many cases, and excessively high temperatures are not permissible in the dried aggregates when they are to be combined with, for example, cut-back asphalt. The dried aggregates may be cooled in various ways, including movement on an open conveyor to the next step in the process, the use of another drier (no heat) as a cooling unit, longer retention time in the storage bins, and so on. The drier is the key machine in many plant-mix operations, with the amount of mixture which can be produced being largely dependent upon the capacity of the drier

As indicated above, the dried aggregates are generally transported from the drier into a unit which contains screens, storage bins, and aggregate proportioning devices. This assembly is another of the basic units which frequently are included in a bituminous paving plant. In this unit the dried aggregates pass over vibrating screens which separate them into the desired number of fractions, usually from two to four. Each fraction is then passed into a storage bin or hopper ready to be used in the mixture. The capacity of the storage bins varies with the size and type of plant being used. Where the aggregates are to be recombined by weight, each storage bin is equipped with a discharge gate which is located directly over the so-called "weigh box." In proportioning the batch in this type of plant, the required weight of each size of aggregate is placed into the weigh box. When the desired aggregate batch has been obtained, the contents of the weigh box are then discharged into the mixer. When the continuous flow type of plant is being used, the aggregates are continuously fed into the mixer through controlled mechanical apron feeders. It should be noted at this point that plants which are intended exclusively to handle some types of low-cost plant mixtures frequently do not contain the unit described immediately above, the aggregate being used directly from a single stock pile without gradation control in the plant. If such is the case, the screens and separate compartments would not be needed, although the storage hopper and weighing devices would be necessary in a batch-type plant.

Facilities must also be provided for the storage, heating, and proportioning of the bituminous material which is to go into the mixture. When heating is necessary, the bituminous material is heated in a storage tank or asphalt kettle, generally by steam, hot oil or electrical heaters. Devices are provided for the measurement and control of the temperature of the bituminous material and for agitating the heated mass. The heated material is fed into an "asphalt bucket" in the batch-type plant, with facilities being provided for securing the correct weight of material for each batch and for emptying the bucket into the mixing unit. If the materials are being proportioned on a volumetric basis, as in a continuous-flow plant, the bituminous material is continuously fed into the mixer by a metered pump.

The final basic unit which will be described here is the mixer itself. The mixer which is most commonly used is a twin pugmill of the type shown in Fig. 16–6. In a batch-type plant this unit is mounted directly beneath the weigh box and asphalt bucket, and high enough so that it may discharge the mixture into a truck or other hauling unit. The

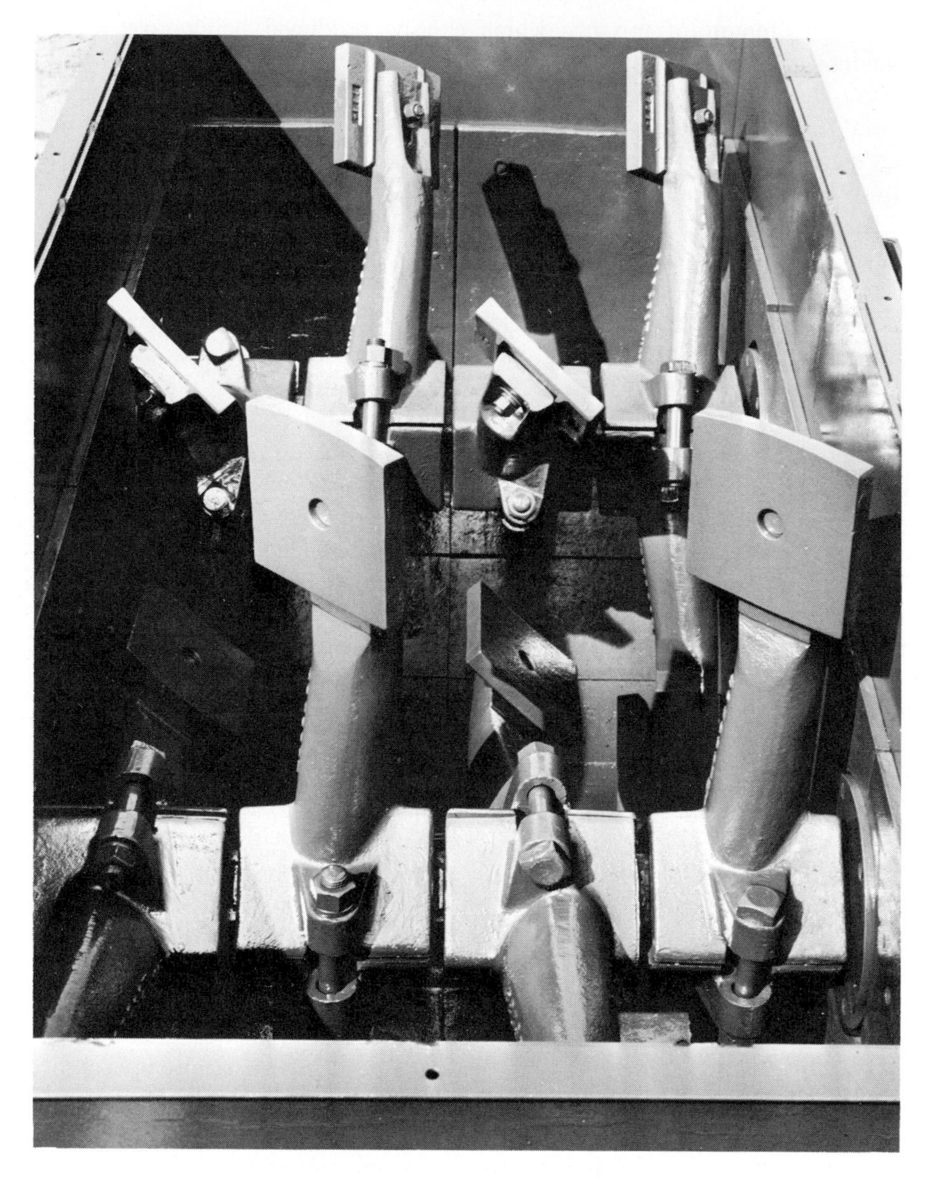

FIG. 16–6. Twin pugmill mixer. (Courtesy Barber-Greene Company.)

aggregates and bituminous material are dumped into the mixer through the top and the unit is then closed. The mixing is accomplished by the mixing blades mounted on the two shafts, which rotate in opposite directions. The space between blades is generally about three-quarters of an inch. Mixing is continued for a fixed length of time and the mixture is then discharged into the waiting truck. In a continuous-flow plant the same type of mixing unit is employed, except that the aggregate and bituminous material enter one end of the mixer and the mixture is continuously discharged at the other.

A method recently introduced into the United States, after extensive use in Europe, is "impact mixing." In this method, the bituminous material is introduced into the pugmill in the form of vapor under pressures of 250 to 300 pounds per square inch with high-speed mixing which tosses rather than kneads the mixture.

This concludes the discussion which will be given here relative to the basic units which may be included in a central mixing plant. Additional units will be described in the next chapter. In the paragraphs below are described two central mixing plants which are typical of those used in the preparation of low-cost bituminous plant mixtures.

16–16. Typical Central Plants (Intermediate Mixtures). Fig. 16–7 shows the layout of a central mixing plant intended for the preparation

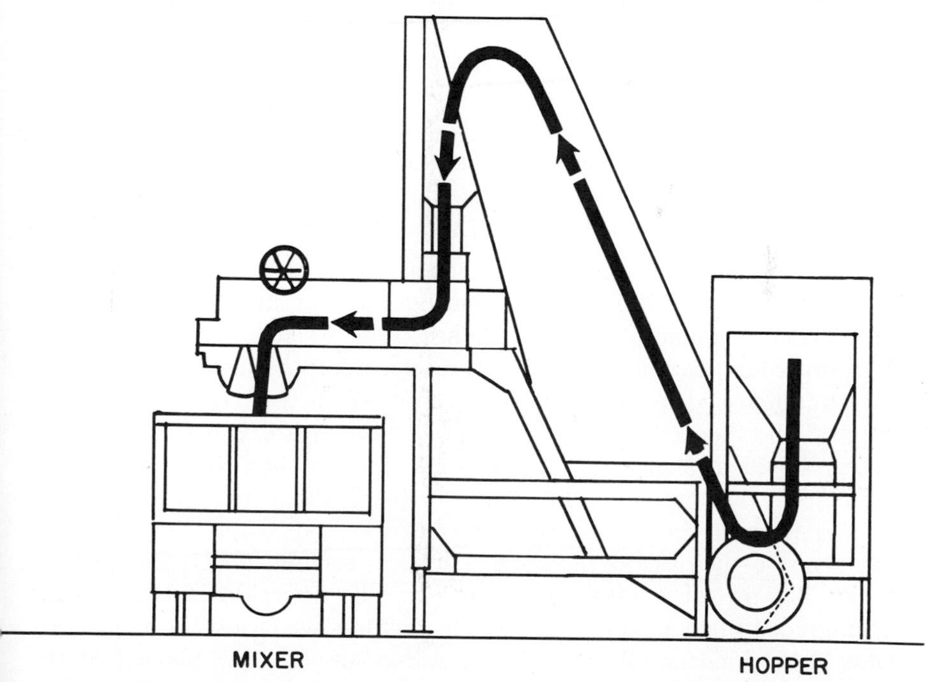

FIG. 16–7. Central mixing plant for intermediate (single-aggregate) mixtures. (Courtesy Barber-Greene Company.)

of intermediate mixtures and utilizing equipment manufactured by the Barber-Greene Company of Aurora, Illinois. The diagram is largely self-explanatory as the sequence of operations is clearly shown. It will

FIG. 16–8. Portable plant for single-aggregate mixtures. (Courtesy Barber-Greene Company.)

be noted that no "gradation control" is exercised in this plant, the aggregate being fed directly from a storage bin to the hot elevator and thence to the mixer. A plant of this type is shown in Fig. 16–8.

16–17. Spreaders and Finishers. Some low-cost plant mixtures may be spread by the use of long-base blade graders or multiple-blade drags in a fashion similar to that which has been described for road mixes. This procedure may be desirable when some aeration of the mix is necessary to remove some of the volatiles contained in the bituminous binder, as might be the case when a dense-graded aggregate is used with slow-curing liquid asphalt. In the majority of cases, however, plant mixes are placed by the use of a "bituminous paver" which spreads the mixture in a uniform layer of the desired thickness and shapes, or "finishes" the layer to the desired elevation and cross-section, ready for compaction. The use of a machine of this type is desirable in the case of open-graded plant mixtures and other mixtures which utilize highly viscous liquid binders. They are also widely used in the placing of the hot mixes described in the following chapter, which must be placed and finished rapidly so that they may be compacted while still hot. Bituminous pavers have not been previously described and will be briefly discussed in the following paragraphs.

The typical modern bituminous paver combines the functions of spreading and finishing in one machine. A machine of this type operates without the use of side forms, being supported on crawler treads or wheels. The machine is of sufficiently long wheel-base to eliminate the necessity for forms and to minimize irregularities occurring in the existing base or surface. These machines will process pavements from a fraction of 1 inch up to 10 inches in thickness over a width of from 6 to 14 feet, with working speeds generally ranging from 10 to 70 feet per minute; one model has working speeds up to 102 feet per minute, although most construction specifications do not permit this high a lay-down speed.

The mixture which is to be placed by the paver is dumped into a receiving hopper from trucks or other hauling units. Material is fed from the hopper toward the finishing section of the machine and is spread and agitated by means of screws or other agitator-distributors which serve the purposes of ensuring uniformity of the spread mixture, spread it uniformly the full width of processing, and loosen or "fluff up" the material. The spread material is then struck off at the desired elevation and cross-section by one or more screeds. In some cases a single cutter bar may be used to strike off and compact the mixture to some extent. In other machines, oscillating or vibratory screeds are used to strike off and initially compact the mixture. Several machines of this type employ a tamping mechanism in conjunction with the screed to provide additional compaction. Cutter bars and screeds are usually provided with heating units to prevent pick-up of the material during the spreading and finishing operations. All machines of the type which has been described are fully adjustable to ensure a uniform flow of material through the machine and to produce a smooth, even layer of the desired thickness and cross-section.

Most bituminous pavers are now fitted with electronic screed control systems, which were introduced in 1961. The various systems differ in detail but perform the same general functions. A sensor operates on a reference profile, senses changes in the position of the floating screed element of the paver or of the reference profile, and then applies any necessary correction to the angle of the screed so that the surface being laid will be parallel to the reference profile. Usually, the reference profile controls the longitudinal profile of the surface at one side of the machine and a slope sensor controls transverse slope, hence the pavement cross-section. However, both sides of the paver can be controlled by reference profiles, if desired.

Any of several reference profiles can be used, depending on the smoothness of the base or other surface upon which the pavement is being laid, surface smoothness and thickness requirements, and so on. Reference profiles can be fixed or mobile. A typical fixed reference profile is a taut wire carefully stretched between stakes or pins set at

FIG. 16–9. Bituminous paver operating from a taut wire. (Courtesy Barber-Greene Company.)

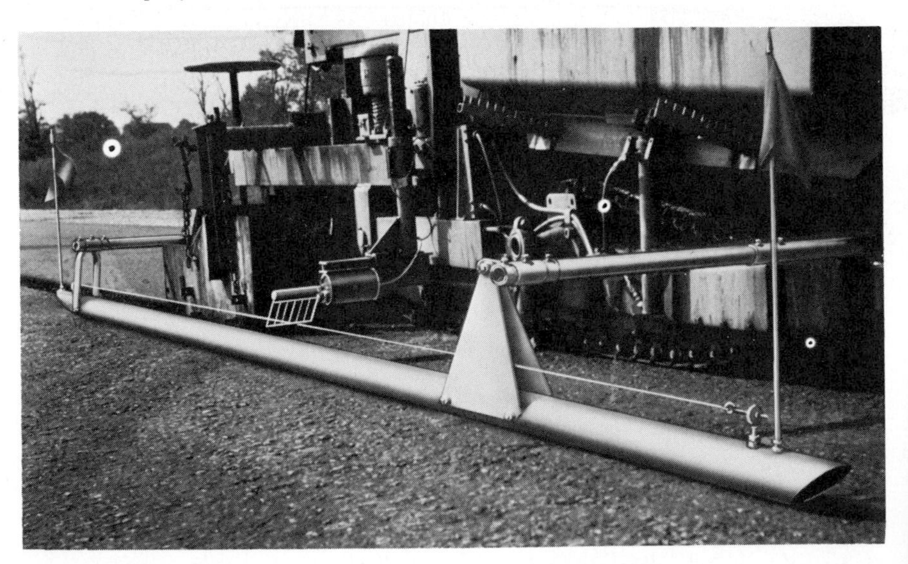

FIG. 16–10. Sensor operates from 20-ft-long ski to guide bituminous paver. (Courtesy Iowa Manufacturing Company.)

close spacings, usually 25 feet or less (see Fig. 16–9). A moving reference profile is attached to the side of the paver; it may be a long (up to 40 feet) rigid or articulated ski that incorporates a taut wire on which the sensor rides, a rolling stringline (see Fig. 16–10), or a short shoe that rides, for example, on the new surface of an adjacent lane.

Some pavers also have automatic control of material feed to the screed; sensing elements at the ends of the augers interlock with material conveyors to control the flow.

Fig. 16–11. Rubber-tired bituminous paver. (Courtesy Blaw-Knox Company.)

The paver shown in Fig. 16–11 is mounted on rubber tires, with top paving speed of approximately 60 feet per minute.

16–18. Summary of Construction Methods (Intermediate Mixes). As indicated earlier in this chapter, many of the operations involved in the construction of intermediate-type plant-mixed bituminous wearing surfaces are virtually the same as those employed for similar road-mix surfaces. Construction specifications covering this type of work generally include additional sections relative to the nature, operation, and control of central mixing plants, the transportation of the mixture from the plant to the job site, the placing of the mixture by means of mechanical pavers, and the treatment of joints formed by successive trips of the paving machine or the temporary cessation of paving operations. These elements of the construction specifications and procedure will not be discussed at this point but will be described in some detail in the next chapter, particularly as applied to the construction of high-type pavements.

PROBLEMS

16–1. Estimate the quantity of coarse aggregate, fine aggregate, and bituminous material which will be required in the construction of a mile of road-mix surface which is to be built in accordance with specification RM–1 of the Asphalt Institute. The surface will have a compacted thickness of 2.0 inches and will be 24 feet wide. Crushed stone and stone screenings of average characteristics will be used. Refer to Art. 16–3.

16–2. A bituminous road-mix wearing surface is to be constructed from an aggregate which has the following characteristics:

GRADATION

Sieve Designation	Per Cent Passing (by Weight)
¾-inch	97
No. 10	50
No. 20	40
No. 50	27
No. 100	18
No. 200	13

Also, $S = 0.14$
$F = 0.78$
Per cent distilled to 680°F = 10

Determine the quantity of slow-curing liquid asphalt required in this mixture on the basis of the formulas used by the Nebraska Highway Department.

16–3. Determine the quantity of bituminous material (in gallons per square yard) which is required for a mixed-in-place sand asphalt pavement which contains 6.5 per cent of bituminous material. The percentage of bituminous material is in this case based upon the dry weight of sand (not the total mix). The specific gravity of the cut-back asphalt to be used may be taken as 1.0 for this purpose, and the pavement is to be 6 inches in compacted thickness. Estimate the approximate quantities of material per mile, if the pavement is to be 24 feet wide; assume the compacted mixture weighs 132 pounds per cubic foot.

16–4. For a mixture of one of the types described in this chapter and in use in your area, determine the types and gradings of the aggregates used, the type and amount of bituminous material, and proportions of the final mixture. Explain the procedure used to establish the correct quantity of bituminous material to be used in the mixture.

16–5. Visit a bituminous paving plant in your area and write a report detailing the layout and equipment being used. Where do the aggregates come from? How are they fed to the plant? How is the mixture transported to the place where it is to be used?

16–6. Does the plant of Problem 16–5 have an automatic control system? If it does, describe each element of the system, what it does and how it works. What are the advantages of automatic controls? If it does not, refer to current literature and describe a typical control system installed at some other plant.

16–7. Consult a highway construction agency in your area, and determine the smoothness requirements being applied to new asphalt pavements. Are contractors using electronic screed controls to meet these requirements? How do the controls work?

CHAPTER 17

HIGH-TYPE BITUMINOUS PAVEMENTS

The bituminous wearing surfaces discussed in this chapter represent the highest types of bituminous pavements which are in current use in this country. They are widely used both on rural highways and city streets which are subjected to large volumes of traffic and severe service conditions. Properly designed and constructed surfaces of this general type are capable of carrying almost unlimited volumes of passenger, mixed, or truck traffic, provided only that they are supported by adequate foundation structures. The majority of these surfaces might be expected to have an economic life of 20 years or more.

All of the high-type bituminous mixtures discussed in this chapter are prepared in central mixing plants. The thickness of these surfaces may vary from slightly less than 1 inch to several inches, depending on the type of surface and its purpose. These mixtures are, generally speaking, marked by the use of carefully selected and graded aggregates in conjunction with semisolid bituminous binders. The composition of the mixture is more rigidly specified and controlled than for other types of bituminous surfaces. Similarly, the preparation, placing, and finishing of the mixture is performed under more rigid specifications and control, with emphasis being generally placed upon the securing of very densely compacted stable mixtures. Surfaces of this type are generally more expensive than those which have been described in previous chapters, and less emphasis is placed upon the use of local materials in the preparation of most of the mixes which are described here.

It seems desirable at this point, in the interests of clarification, to tabulate the mixtures which will be discussed in subsequent sections. These mixtures will be discussed in the order indicated in the tabulation. The high-type surfaces discussed in this chapter, then, are as follows:

I. Hot-mix, hot-laid pavements
   A. Asphaltic types
      1. Asphaltic concrete
         (a) Open graded
         (b) Coarse graded
         (c) Dense graded
         (d) Fine graded
      2. Stone-filled sheet asphalt

3. Sheet asphalt
4. Sand asphalt

B. Tar concrete

II. Cold-mix, cold-laid pavements

A. Asphaltic concrete
B. Tar concrete

III. Patented mixtures

IV. Miscellaneous mixtures

A large bulk of the detailed discussion which follows will be devoted to hot-mix, hot-laid asphaltic concrete, which is by far the most commonly used type. Before beginning the discussion of the specific types listed, space will be devoted to the consideration of certain factors relative to the nature and design of high-type bituminous paving mixtures.

Design of High-Type Bituminous Paving Mixtures

It is the purpose of this section to set forth some of the underlying principles upon which the design of a high-type bituminous paving mixture is based. Statements made in this section are intended to apply specifically to the design of the hot-mixed, hot-laid bituminous concrete mixtures described in this chapter, although they may also generally apply to those mixtures which do not contain coarse aggregate, such as sheet asphalt and sand asphalt, and to the cold-mixed, cold-laid types. The subject of the design of the high-type mixtures being considered here is not a simple one, and authorities are not entirely in accord with one another either as to certain of the fundamental principles involved or as to the detailed procedure to be used in the design of a particular mixture. The subject has been deliberately simplified in some respects in order to reduce the number and complexity of the factors which enter into the problem.

The two fundamental properties of a bituminous paving mixture which are held to be of the utmost importance are stability and durability. By "stability" is meant that property of the compacted mixture which enables it to withstand the stresses imposed upon it by moving wheel loads without sustaining substantial permanent deformation. Lack of stability in a compacted mixture may be evidenced by displacement of the completed pavement, as by rutting or shoving. By "durability" is meant that property of the compacted mixture which permits it to withstand the detrimental effects of air, water, and temperature change. For successful results the pavement must be both durable and stable during its entire service life.

Both stability and durability are intimately related to the density of the mixture. The density of a compacted mixture may be expressed in several different ways, as will be shown in later sections, but the fundamental importance of this factor cannot be overemphasized. The density of a given mixture is frequently expressed in this work in terms of the "voids" in the mixture, meaning, in general, the amount of space in the compacted mixture which is not filled with aggregate or bituminous material, i.e., is filled with air. Thus a dense mixture would have a low percentage of voids, while a loose mixture would have a high percentage of voids, all other things being the same. Many modern designers stress what might be called the "stability-voids" theory in proportioning bituminous paving mixtures. The fundamental concept underlying this approach to the problem is that mixtures which are high in density (low in voids) will be high in stability, and that, in turn, properly designed mixtures which have high stability will also be durable. There are certain limitations to this concept, as will be pointed out in succeeding paragraphs, but this basic principle is generally subscribed to in the discussions which are presented in this chapter.

In addition to stability and durability, a well-designed mixture must also be skid resistant. Lack of skid resistance is most often associated with an excess of bituminous material, although it may also be caused by aggregates which become excessively smooth or polished under the action of traffic.

The desirable qualities of a bituminous paving mixture are dependent to a considerable degree upon the nature of the aggregates used. *Coarse aggregate* may be defined as that portion of the aggregate material contained in a bituminous paving mixture which is retained on a No. 10 sieve. *Fine aggregate* is that portion of the aggregate material which passes a No. 10 and is retained on a No. 200 sieve. It should be noted that the Asphalt Institute (*122*) uses the No. 8 sieve as the dividing line between coarse aggregate and fine aggregate. *Filler* is that mineral material which passes a No. 200 sieve. The function of the coarse aggregate in contributing to the stability of a bituminous paving mixture is largely due to interlocking and the frictional resistance of adjacent aggregate particles to displacement. Thus crushed aggregates which contain a preponderance of rough, angular particles are most generally used, including crushed stone, crushed gravel, and crushed slag. Similarly, the fine aggregate or "sand" contributes to stability through interlocking and internal friction, so that particles which are sharp and angular in nature are preferred. Sands are also held to contribute to stability by their function in filling the voids in coarse aggregate. The filler itself is largely visualized as a void-filling agent, with limestone dust being most widely used for this purpose. Portland cement, slag dust, local sands, and other materials have also been successfully used as fillers.

Most commercial materials used as fillers contain a small proportion that is retained on the No. 200 sieve.

Aggregates which are well graded from coarse to fine are generally sought in high-type bituminous paving mixtures. The basic concept behind this statement is that well-graded materials produce the most dense mixtures and therefore the most durable, requiring minimum bitumen content for satisfactory results. This statement is regarded as being generally true, even though some very dense, durable, and stable mixtures have been produced by the use of aggregates which were not well graded in the usual sense. In most specifications, each of the separate aggregates—coarse, fine, and filler—is required to be well graded from coarse to fine as is the combination of these materials. The concept simply is that in a well-graded aggregate each smaller size or "fraction" of aggregate serves to fill the voids in the next larger one, with the result that a very dense aggregate combination may be secured.

Mixtures which contain more than about 70 per cent of coarse aggregate are likely to be harsh and difficult to place and finish, being subject to segregation. The amount of filler used in bituminous concrete generally varies from about 3 to 10 per cent, and up to about 16 per cent for sheet asphalt. Excessive amounts of filler may result in a gummy mixture which is difficult to finish and which may become brittle and subject to cracking in cold weather. The filler itself should be well graded to a certain extent. It should be kept in mind that the filler is all the material which passes the No. 200 sieve and comes from all coarse and fine aggregate sizes and the commercial filler added to make the total.

The kind and nature of the bituminous material used obviously have some effect upon the properties of the bituminous paving mixture. The majority of the mixtures described in this chapter use semisolid asphalt cement as the binder material. In mixtures in which this material is used as the binder, it is rendered temporarily liquid during mixing by the application of heat or by fluxing. The nature and quality of the asphalt cement used are controlled by specifications such as were illustrated in the chapter on bituminous materials. The penetration of the asphalt cements which have been used in the various mixtures ranges all the way from 40 to 150 or more. At the present time, asphalt cements having a penetration range of 85–100 are most widely used in mixtures of the types which are being discussed here.

Another basic factor which influences the behavior of a completed bituminous pavement is the degree of compaction of the mixture. A high degree of compaction is regarded as desirable, being associated with high density and stability. Mixtures of the kind being considered are generally compacted with heavy rollers, with rolling being continued until a high degree of density is achieved. However, rolling should not be continued until the aggregate breaks down or "degrades" under the action of the roller. Field rolling is controlled by requiring the attainment of a certain percentage of the density established in the laboratory by some standard compaction procedure.

At this stage in our outline of the factors which are of importance in the design of a bituminous paving mixture, it will be assumed that the aggregates which are to be used have been selected and their gradation adjusted to comply with the grading requirements for the mixture; the kind of bituminous material to be used has been established; and the desired compactive effort determined. The remaining design element to be determined, then, is the optimum bitumen content (usually asphalt). This is probably the most important single factor in the satisfactory design of a bituminous paving mixture. Detailed procedures to be followed in the determination of the optimum asphalt content for a given case will be given later, but certain general principles can be stated here. In order to explain these principles, consider the following line of reasoning. Consider first the mineral aggregates alone, without any bituminous material, assuming that the mineral aggregates have been compacted to a density which is equivalent to that which will be attained in the field process. This dense aggregate alone will possess a certain density and a relatively low stability. As small quantities of bituminous material are added to the mixture, both the density and stability will increase. At this stage the bituminous material would generally be present in the form of thin films surrounding, or partially surrounding, the aggregate particles. If more bituminous material is added, a point will be reached at which the aggregate particles are completely covered with bituminous material; additional bituminous material would serve to fill the void spaces between the aggregate particles. For many mixtures, both stability and density would increase until the voids in the mixture had been completely filled. The addition of any more bituminous material would result in a decrease in both density and stability, as the aggregates would generally be forced apart by the excess bituminous material.

It would thus appear that the most desirable bitumen content for many mixtures would then be that which would just fill the voids in the compacted mixture. This concept is, however, modified by certain practical considerations, as follows. In the first place, any bituminous material expands with an increase in temperature. Thus if the voids in the mixture were completely filled at the time of placing, any increase in temperature subsequent to that time would result in overfilling of the voids in the mixture with consequent "bleeding" of the pavement and loss of stability. Similarly, the density of the compacted mixture may increase under the action of traffic until it exceeds that used in the design. If this happened, an excess of bituminous material might again be present. As a result of these considerations and others, a compromise is frequently made in the selection of the optimum bitumen content. Assuming that all mixtures must possess adequate stability, the amount of bituminous material used would be such as to produce a mixture which would have, on the one hand, more than a certain *minimum* void content and, on the other, less than a certain *maximum* void content. The selection of the maximum void content which could be permitted would generally be based upon durability considerations; air and water permeability also may be considered. For example, the percentage of bituminous material chosen for a given mix might be such that the compacted mixture would have a percentage of voids (calculated as explained later) which was not less than two nor more than five. A final fundamental requirement for the design of a mixture is that it be economical.

Hot-Mix, Hot-Laid Asphaltic Concrete

The first of the high-type bituminous paving mixtures which will be discussed is hot-mix, hot-laid asphaltic concrete. The discussion of this type of paving mixture will be quite comprehensive and complete, with many of the statements made also being applicable to the other mixes discussed in this chapter.

Asphaltic concrete may be generally defined as an intimate mixture of coarse aggregate, fine aggregate, mineral filler, and asphalt cement. The Asphalt Institute (*122*) has codified hot asphalt paving mixtures, as shown in Fig. 17–1. The mixtures which come under our definition of asphaltic concrete are those designated *open graded, coarse graded, dense graded,* and *fine graded.* Leveling mixes are those used in an intermediate layer, often to eliminate irregularities in contour of an existing surface, prior to construction of a new layer.

In Fig. 17–1, each mix is designated by specific limits for aggregates retained on and passing the No. 8 sieve. Similarly, the chart also indicates the maximum size of aggregate normally used for a given mix designation, and the limiting amounts of mineral dust (filler) generally found satisfactory for base, leveling, or surface courses. Primary consideration is given to the proportions of fine aggregate and coarse aggregate. Also of consequence is the proportion of filler; normal limits for this proportion are indicated by a shaded band. The band is used to donate flexible rather than fixed numerical limits. As a general rule, base course mixtures will fall on the left side of the shaded

PAVING MIX DESIGNATION		MAXIMUM SIZE AGGREGATE NORMALLY USED	
Type	DESCRIPTION	SURFACE AND LEVELING MIXES	BASE AND LEVELING MIXES
I	MACADAM		$2\frac{1}{2}$"
II	OPEN GRADED	$\frac{3}{8}$" — $\frac{3}{4}$"	$\frac{3}{4}$" — $1\frac{1}{2}$"
III	COARSE GRADED	$\frac{1}{2}$" — $\frac{3}{4}$"	$\frac{3}{4}$" — $1\frac{1}{2}$"
IV	DENSE GRADED	$\frac{1}{2}$" — 1"	1" — $1\frac{1}{2}$"
V	FINE GRADED	$\frac{1}{2}$" — $\frac{3}{4}$"	$\frac{3}{4}$"
VI	STONE SHEET	$\frac{1}{2}$" — $\frac{3}{4}$"	$\frac{3}{4}$"
VII	SAND SHEET	$\frac{3}{8}$"	
VIII	FINE SHEET	No.4	No.4

1 CRITICAL ZONE→ DUST CONTENTS IN THIS REGION SHOULD NOT BE USED WITHOUT A SUBSTANTIAL BACKGROUND OF EXPERIENCE WITH SUCH MIXES AND/OR SUITABLE JUSTIFICATION BY LABORATORY DESIGN TESTS.

2 INTERMEDIATE ZONE→ DUST CONTENTS IN THIS REGION SOMETIMES USED IN SURFACE AND LEVELING MIXES AS WELL AS IN BASE AND BINDER MIXES.

FIG. 17–1. The Asphalt Institute classification of hot-

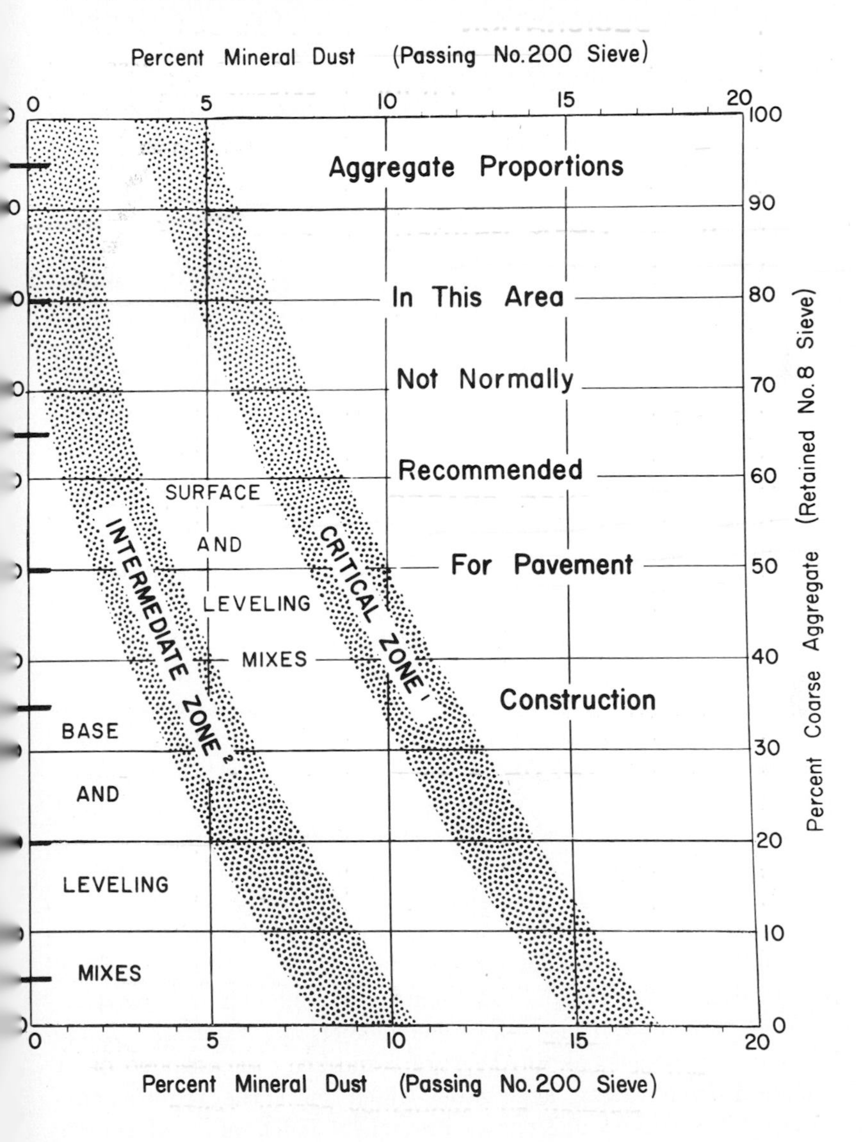

mix asphalt paving. (Courtesy the Asphalt Institute.)

band of Fig. 17–1, with surface course mixtures on the right. Leveling course mixtures may fall in either area.

Terms and limits shown in Fig. 17–1 are in general conformance with construction practices over the country as a whole, although there is still some variation in local practices. Such variations are of little significance in our discussion of the subject.

Typical cross-sections of hot-mix asphaltic pavements are shown in Fig. 17–21.

17–1. General Requirements for Aggregates. The coarse aggregates most frequently used in asphaltic concrete paving mixtures are crushed stone, crushed gravel, and crushed slag. Uncrushed gravel, which may also be used in some cases, is required to consist largely of rough-textured particles. The coarse aggregates are, as has been the case for most of the aggregates discussed in previous chapters, required to be hard, tough, clean, and durable in nature, free from an excess of flat, or elongated pieces, clay material, and other deleterious substances. Aggregates known to polish under traffic are not used in surface courses.

Crushed stone and gravel are generally permitted to have percentages of wear in the Los Angeles Abrasion Test of not more than 40, although this figure is increased to 50 in some instances. They are also sometimes subjected to requirements for soundness (AASHO M76); requirements in this respect frequently call for the weight loss incurred when the material is subjected to five cycles of the sodium (or magnesium) sulfate soundness test (AASHO Method T104) not to exceed 12 or 15 per cent. Crushed slag is usually subject to the same requirements relative to abrasive resistance as stone and gravel; and, in addition, it is generally required to weigh not less than 70 pounds per cubic foot. Certain additional requirements relative to crushed and uncrushed gravels are found in some specifications. General requirements relative to coarse aggregates to be used in asphaltic concrete base courses may be somewhat less severe than those indicated above.

Requirements relative to sands to be used in asphaltic concrete mixtures are usually quite general in nature. Sands are generally required to be composed of tough, clean, rough-surfaced, and angular particles, free from excess amounts of clay and other deleterious materials. Stone screenings may also be used as fine aggregate and are sometimes required to be produced from coarse aggregates which meet the requirements given above. The crushed aggregates previously listed may, of course, contain a considerable amount of material which passes a No. 10 sieve and which is classed as fine aggregate. The fine aggregate for a given mixture may consist of sand alone, stone or slag screenings alone, or a blend of sand and stone or slag screenings.

A variety of materials are used for mineral filler, when required, and specifications in use by a given agency generally list the specific materials which may be used for this purpose. Commonly used materials include limestone dust, dolomite dust, portland cement, slag dust, and other similar inert mineral substances. The mineral filler is generally required to be dry and free from lumps. It is desirable that this material be definitely hydrophobic in nature.

17–2. General Requirements for Bituminous Materials. The bituminous material used in the preparation of hot-mix, hot-laid asphaltic concrete mixtures is semisolid asphalt cement. The nature and quality of the asphalt cement are controlled by specifications like those given in Chapter 11. The Asphalt Institute recommends the use of 60–70 penetration asphalt for highways which carry heavy traffic in hot climates, with 85–100 penetration used for other climatic conditions; 85–100 penetration is recommended for medium to light traffic, except in cold climates where 120–150 penetration grade is suggested.

17–3. Requirements Relative to Aggregate Gradation and Composition of Mixtures. Separate grading requirements are usually given for the separate aggregate components of the mixture—coarse aggregate, fine aggregate, and filler—and then an over-all specification is given pertaining to the composition of the paving mixture itself, including the bituminous material. The specifications of two highway agencies have been abstracted and are presented in Tables 17–1 and 17–2. These specifications are taken to be typical of those in present use.

The various specifications cited are intended to represent the requirements of typical specifications relative to aggregate gradings and mix compositions. The limits placed upon the percentages of the various sizes of aggregates and the amount of bituminous material to be used in a given mixture vary to some extent within the limits of each specification. Before the preparation of the mixture can actually begin, a single figure must be definitely established for each of the aggregate fractions contained in the specification and for the bituminous material. In other words, the composition of the mix must be definitely established. The process of doing this, which is frequently called the "design" of the mixture, results in the establishment of a definite "job-mix formula" which will then be applicable to the mixture being used on a particular project. Considerations relative to the determination of the job-mix formula will be given in the following sections.

Once the job-mix formula has been established, most specifications allow small variations from it during production of the mix. The purpose of these tolerances is to obtain a uniform mixture in reasonable

TABLE 17–1

AGGREGATE GRADING REQUIREMENTS FOR ASPHALTIC CONCRETE BASE AND SURFACE COURSES (*192*)

Sieve Designation	Combined Aggregate Per Cent Passing (by weight)		
	Base Course	¾-in. Max. Surface Course	½-in. Max. Surface Course
1¼-inch	100	–	–
1-inch	95–100	100	–
¾-inch	80–95	95–100	100
½-inch	–	–	95–100
⅜-inch	50–65	65–80	80–95
No. 4	35–50	45–60	55–72
No. 8	–	30–45	38–55
No. 30	12–25	15–25	18–33
No. 200	2–7	3–7	4–8

Note 1: The asphalt cement is to have a penetration within one of the following ranges: 60–70, 85–100, or 120–150.

Note 2: The combined aggregates are subjected to various tests including Los Angeles Abrasion, Stabilometer, Swell, Determination of *K*-factor (Centrifuge Kerosene Equivalent), and Sand Equivalent.

TABLE 17–2

COMPOSITION OF BITUMINOUS PLANT MIXTURES, TYPE 1A (*184*)

Grading Requirements

Sieve Designation	Per Cent Passing (by weight)	
	Base Course	Top Course
2-inch	100	–
1½-inch	75–100	–
1-inch	55–80	100
½-inch	23–42	95–100
¼-inch	5–20	65–85
⅛-inch	2–15	32–65
No. 20	–	15–39
No. 40	–	7–25
No. 80	–	3–12
No. 200	–	2–6

Note: The binder in these mixtures is asphalt cement, 85–100 penetration grade. Quantity of asphalt cement is 2.5 to 5.0% for the base course, 5.8 to 7.0% for the top course.

compliance with the specifications, since it is not possible to produce the exact job mix every time. A representative job-mix tolerance specification appears in Table 17–3.

TABLE 17–3

JOB-MIX FORMULA TOLERANCE
(Specification for Asphaltic Concrete) (*122*)

Sieve Size	Permissible Variation, % by Weight of Total Mixture
No. 4 and larger	5.0
No. 8	4.0
No. 30	3.0
No. 200	1.0
Asphalt	0.3

17–4. Determination of the Job-Mix Formula. The procedure which is presented here for the determination of the job-mix formula is intended to indicate the procedure which might be followed by an imaginary highway agency in determining the composition of a typical dense-graded asphaltic concrete surface mixture. Since detailed procedures in the determination of the job-mix formula vary so much, it is not deemed wise to try to present typical or "average" practice in this respect. The procedure indicated is generally similar to that followed by some agencies and has been somewhat simplified in the interests of clarity and brevity. Two general steps are involved in this process: the selection and combination of aggregates to meet the limits of the specification being used and the determination of the optimum asphalt content.

17–5. Selection and Combination of Aggregates. Relatively little will be said here about the selection of the aggregates to be used in a given paving mixture, although this is a very important phase of the over-all design process. In the normal procedure, both coarse and fine aggregates available in the vicinity of the proposed work are sampled and carefully examined for compliance with the individual specifications for these materials. In some cases more than one suitable aggregate of each type will be found, and the question of selection of the particular ones to be used may then be largely a matter of economics. In other cases no suitable single aggregate may be locally available, in which case aggregates from several different sources may have to be blended to meet the requirements of the specifications, or imported aggregates may have to be used. Materials for use as mineral filler must be similarly examined and selected. Regardless of what the selection of aggregates may involve on a given project, all possible aggregates should be carefully examined and those selected which are suitable to the purpose and economically available. The aggregates finally selected for use must then be combined to provide an aggregate mixture which will comply with the limits established as to the composition of the mixture. See reference *244* for complete information about procedures to be used in combining aggregates to meet a given specification.

The procedure used in determining the proportions of coarse aggregate, fine aggregate, and mineral filler (and the optimum asphalt content) may best be illustrated by a discussion of the procedure to be followed in the design of a definitely specified asphaltic concrete mixture. This procedure will be followed in the subsequent discussion. It is assumed that the composition of the mixture which is to be designed (for which a job-mix formula is to be established) must be within the limits of the specifications in Table 17–4.

TABLE 17–4

MINERAL AGGREGATE AND MIX COMPOSITION

Passing Sieve Designation	Retained on Sieve Designation	Per Cent (by weight)
¾-inch	½-inch	0–6
½-inch	⅜-inch	9–40
⅜-inch	No. 4	9–45
No. 4	No. 10	8–27
Total Coarse Aggregate	No. 10	50–65
No. 10	No. 40	6–22
No. 40	No. 80	8–27
No. 80	No. 200	5–17
No. 200	–	5–8
Total Fine Aggregate and Filler	(Passing No. 10)	35–50
Total Mineral Aggregate	–	100
Total Mix		
Total Mineral Aggregate		92–95
Asphalt Cement		5–8
Total Mix		100

It will be assumed that the aggregates which are available for use in this mixture have been thoroughly tested and examined for compliance with the specifications relative to these materials. Sieve analyses of the aggregates which can be most economically used in this case, as determined by AASHO methods of tests T–27 and T–37 (mineral filler), are shown in Table 17–5.

The problem now is to determine the proportions of the separate aggregates which must be used in order to give an aggregate combination which will meet the requirements of the specification previously given.

Not only must the proportions selected be within the specification but they must be far enough from its extremes to provide room for the job-mix tolerance, so that when it is added or subtracted the mixture will not be outside the original specification master range. This is done by trial. Two factors are to be considered in the first trial combination: First, it is obvious that the fine aggregate and coarse aggregate only, in any combination, cannot meet the requirements of the specification for total mineral aggregate, as the specification requires from 5 to 8

TABLE 17–5

SIEVE ANALYSIS OF AGGREGATES

Per Cent (by weight)		Aggregate Type		
Passing Sieve Designation	Retained on Sieve Designation	Coarse Aggregate	Fine Aggregate	Mineral Filler
¾-inch	½-inch	5	–	–
½-inch	⅜-inch	32	–	–
⅜-inch	No. 4	37	–	–
No. 4	No. 10	22	7	–
No. 10	No. 40	4	28	–
No. 40	No. 80	–	39	5
No. 80	No. 200	–	24	30
No. 200	–	–	2	65
Total		100	100	100

per cent of material passing a No. 200 sieve and the coarse aggregate contains none of this material, while the fine aggregate contains only 2 per cent. Obviously, some mineral filler must be used in the mixture; for the first trial, the amount of mineral filler is arbitrarily set at 8 per cent. The total coarse aggregate in the mix must be from 50 to 65 per cent, and this figure is again more or less arbitrarily set at 52 per cent. The remaining 40 per cent must be fine aggregate.

Using the indicated proportions, calculations necessary in determining the sieve analysis of the combined aggregates are shown in Table 17–6.

A comparison of the figures in the last column of this table with the requirements of the specification will show that this combination of aggregates meets the stipulated requirements. This combination will therefore be judged satisfactory, and no additional trials will be made here. Additional trials would have had to be made if the combination had not been satisfactory. Various graphical methods have been devised which make the blending process somewhat simpler than the arithmetic process used here. Quite frequently no possible combination of the aggregates initially selected will prove to be satisfactory, in which case others must be selected or the available ones improved by blending. This situation can be avoided with experience. Additional trials might also be made in an effort to improve the gradation of the total aggregate; these also will not be considered here. The combination selected will be used in the preparation of the trial mixes required for the determination of the optimum asphalt content.

TABLE 17-6

CALCULATIONS FOR SIEVE ANALYSIS

Passing Sieve Size	Retained on Sieve Size	Per Cent (by weight)			
		Coarse Aggregate	Fine Aggregate	Mineral Filler	Total Aggregate
¾-inch	½-inch	0.52 × 5 = 2.6	–	–	2.6
½-inch	⅜-inch	0.52 × 32 = 16.6	–	–	16.6
⅜-inch	No. 4	0.52 × 37 = 19.2	–	–	19.2
No. 4	No. 10	0.52 × 22 = 11.4	0.40 × 7 = 2.8	–	14.2
No. 10	No. 40	0.52 × 4 = 2.2	0.40 × 28 = 11.2	–	13.4
No. 40	No. 80	–	0.40 × 39 = 15.6	0.08 × 5 = 0.4	16.0
No. 80	No. 200	–	0.40 × 24 = 9.6	0.08 × 30 = 2.4	12.0
No. 200	–	–	0.40 × 2 = 0.8	0.08 × 65 = 5.2	6.0
Total		52.0	40.0	8.0	100.0

17-6. Determination of Optimum Asphalt Content. The steps remaining in the process of determining the job-mix formula for a given asphaltic concrete surface course are directed toward the determination of the optimum asphalt content to be used in the mix. The asphalt content established as a part of the job-mix formula must, of course, be within the limits of the given specification relative to composition of the mixture.

The laboratory procedure used in determining the optimum asphalt content involves the preparation of trial mixtures using the selected aggregates and various percentages of asphalt, within the limits of the mix specification. Each trial mixture is prepared in a manner which is intended to secure a very high density. Densities which are secured in this fashion generally represent the ultimate densities which are practically attainable either in the laboratory or in the field. The density, stability, and other properties of each trial mixture are then determined and the results tabulated. Applicable criteria may then be applied to determine the optimum asphalt content.

Design criteria relative to the laboratory compacted specimens frequently require that the density of satisfactory mixes be not less than a certain percentage of the "theoretical maximum density." By "theoretical maximum density" is meant the density, usually expressed in terms of the bulk specific gravity, of a voidless mixture composed of the same ingredients in the same proportion as used in the actual mixture. Design criteria for density, stability, and other properties which are used by a given organization have been established by extensive laboratory and field experience and investigation and are generally applicable only to mixtures which are to be used as wearing surfaces, being regarded as unnecessary to the design of base and binder course mixtures in many cases. Although we are discussing asphaltic concrete, the same or similar design criteria may be applied to the design of the other high-type mixtures described in this chapter.

Two mix design methods are in widespread use: the Marshall method (Art. 17–7) and the Hveem method (Art. 17–11). The Hubbard-Field method has been widely used in the past, and is described briefly in Art. 17–12. Additional information relative to these test methods is given in Reference *118*.

In order to bring the discussion from the general down to the specific, the design of the hypothetical dense-graded asphaltic concrete mixture for which the aggregate combination has been determined will be continued in order to illustrate a method which may be used in selecting the optimum asphalt content and the final determination of the job-mix formula. It will be assumed that the stability determinations are to be performed by the use of the Marshall method. The procedure used in preparing and testing the laboratory specimens is more or less standardized, although some deviation may be seen in the practice of various laboratories.

17–7. Marshall Method of Design. This method of design was originally developed by Bruce Marshall, formerly of the Mississippi Highway Department, and is currently used by the Corps of Engineers in the design of paving mixtures for airfields. It is also used by several state highway departments. The method is applicable only to hot mixtures using penetration grades of asphalt cement and containing aggregates with maximum size of 1 inch or less.

The Marshall method uses standard test specimens 2.5 inches high and 4 inches in diameter. They are prepared by the use of a standard procedure for heating, mixing, and compacting the asphalt-aggregate mixtures. Density-voids determinations and calculations are made on the compacted specimens, using methods and equations presented in Art. 17–8.

17–8. Preparation of Test Specimens (Marshall Method). In preparing test specimens for the Marshall Method, it is essential that the gradation of the aggregate and the quantity of asphalt be very carefully controlled. Considerable care must be exercised in the entire conduct of laboratory testing operations so that test results will be reliable and reproducible. Tests should be planned on the basis of ½ per cent increments of asphalt content, with at least two asphalt contents above optimum and two below. In the example given, it will be assumed that, since the possible range in asphalt content is from 5 to 8 per cent of the total mix, trial mixtures will be prepared with asphalt contents over a portion of this range at intervals of ½ per cent. In other words, work will be begun on the assumption that mixtures will be prepared which contain 5.0, 5.5, 6.0, 6.5, and 7.0 per cent of asphalt.

As a first step in the procedure, the aggregates with the proper gradation are thoroughly dried and heated. Sufficient mixture is generally prepared at each asphalt content to form two or three specimens 4 inches in diameter and 2½ inches in height. Each specimen will require approximately 1200 grams of mixture.

The asphalt and the aggregates are heated separately and then mixed. The mixing and compaction temperatures are established by determining the temperature to which the asphalt must be heated to produce viscosities, respectively, of 85 ± 10 and 140 ± 15 seconds, Saybolt Furol.

The mixture is placed in the mold, spaded with a trowel, and compacted. A compactive effort of 35, 50, or 75 blows is specified according to the design traffic category. The compactive effort is applied to each side of the specimen. The test specimen is then tested for stability and flow after a density determination has been made.

17–9. Density Determination and Accompanying Calculations. The density (G) of the compacted mixture is generally determined at room temperature by determining the weight of a test specimen in air and the weight of the same specimen in water. The same method may be used to determine the density of a sample cut or cored from a completed pavement. With very open or porous mixtures, the sample may be coated with paraffin before the density determination. The density of the compacted mixture (uncoated specimen) may be expressed as follows:

$$G = \frac{W_a}{W_a - W_w} \tag{17–1}$$

where G = density (bulk specific gravity) of compacted mixture
W_w = weight of test specimen suspended in water (grams)
W_a = weight of test specimen in air (grams)

For example, assume that a test specimen of the specific mixture which is being discussed contains 5.0 per cent asphalt and weighs 3041.2 grams in air and 1713.2 grams in water. The specific gravity of the specimen (compacted mixture) then would be

$$G = \frac{3041.2}{3041.2 - 1713.2} = \frac{3041.2}{1328.0} = 2.290$$

This value of G means that the unit weight of the compacted mixture would be 2.29(62.4) = 142.9 pounds per cubic foot.

The theoretical maximum density of an asphaltic concrete mixture (G_t), or any mixture for that matter, has previously been defined as the theoretical density of a voidless mixture composed of the same ingredients in the same proportions. The theoretical maximum density of an asphaltic concrete mixture may be expressed as follows:

$$G_t = \frac{100}{\dfrac{W_1}{g_1} + \dfrac{W_2}{g_2} + \dfrac{W_3}{g_3} + \dfrac{W_4}{g_4}} \tag{17-2}$$

where G_t = theoretical maximum density
W_1 = per cent, by weight, of coarse aggregate
W_2 = per cent, by weight, of fine aggregate
W_3 = per cent, by weight, of mineral filler
W_4 = per cent, by weight, of asphalt cement
g_1 = specific gravity of coarse aggregate
g_2 = specific gravity of fine aggregate
g_3 = specific gravity of mineral filler
g_4 = specific gravity of asphalt cement

In asphalt paving technology, three different types of specific gravity are discussed. These are apparent, bulk, and effective specific gravities. Water absorption is normally used to determine the quantity of permeable voids in the aggregate, for use in specific gravity computations (ASTM Methods C127 and C128). The principal difference between bulk and apparent specific gravities is that the volume of the permeable voids are included in the volume of the aggregate for bulk specific gravity and are excluded in the volume of the aggregate for apparent specific gravity. In all cases, the bulk specific gravity is less than the apparent specific gravity. Recognizing that the absorption of asphalt by the aggregate is different from that of water, various organizations have devoted considerable effort to the development of methods of determining the effective specific gravity, for use in asphalt mix calculations. The effective specific gravity is normally somewhere between the bulk and apparent values. Methods for determining effective specific gravity are not yet standardized and reference *118* contains additional discussion of this problem. Regardless of the specific gravity used, the density and voids calculations are the same.

The theoretical maximum density of the mixture under discussion, assuming it to contain 5.0 per cent of asphalt cement by weight of the total mixture, may be computed as follows:

It is first necessary to determine the values of W_1, W_2, etc., for this mixture. This may be done in the following fashion. Of a total mixture weighing 100 grams, 5 grams will be asphalt cement, therefore $W_4 = 5.0$. The remaining 95.0 grams will be aggregate divided as follows: coarse aggregate, 52.0 per cent; fine aggregate, 40.0 per cent; and mineral filler, 8.0 per cent. Therefore there will be 95.0(0.52) = 49.4 grams of coarse aggregate in the total mix; 95.0 (0.40) = 38.0 grams of fine aggregate; and 95.0(0.08) = 7.6 grams of mineral filler. Therefore $W_1 = 49.4$; $W_2 = 38.0$; and $W_3 = 7.6$. Assume that the specific gravity of the coarse aggregate is 2.60, that of the fine aggregate 2.65, and that of the mineral filler 2.80. The specific gravity of the asphalt cement is 1.04. The theoretical maximum density of this mixture then will be

$$G_t = \frac{100}{\frac{49.4}{2.60} + \frac{38.0}{2.65} + \frac{7.6}{2.80} + \frac{5.0}{1.04}}$$

$$= \frac{100}{19.00 + 14.34 + 2.71 + 4.81} = \frac{100}{40.86}$$

$$= 2.447$$

The per cent (air) voids (V) in a compacted mixture (per cent voids total mix) may be obtained from the following equation:

$$V = \frac{G_t - G}{G_t} \times 100 \qquad (17\text{–}3)$$

For the specific mixture being considered,

$$V = \frac{(2.447 - 2.290)100}{2.447} = \frac{(0.157)100}{2.447} = 6.42\%$$

Expressed in another way, the density of the compacted mixture then would be 93.6 per cent of the theoretical maximum density. The per cent of theoretical density may be computed directly, being equivalent to $G/G_t(100)$.

The calculation for per cent voids in the compacted mineral aggregate is performed similarly, as follows. Let G_a = density of the compacted mineral aggregate in the mixture, as given by the following equation:

$$G_a = G\left(1 - \frac{W_4}{100}\right) \tag{17-4}$$

and the theoretical maximum density of the mineral aggregate alone (G_{ta}) then would be

$$G_{ta} = \frac{100}{\dfrac{W_5}{g_1} + \dfrac{W_6}{g_2} + \dfrac{W_7}{g_3}} \tag{17-5}$$

where the symbols W_5, W_6, and W_7 refer to the weights (percentages) of the different aggregates in the aggregate combination alone. Then the per cent voids in the compacted mineral aggregate, frequently designated as V.M.A. (%), is given by

$$\text{V.M.A. (\%)} = \frac{G_{ta} - G_a}{G_{ta}} \times 100 \tag{17-6}$$

If the calculations indicated above are followed through for the particular mixture which is being used, it will be found that the per cent voids in the compacted mineral aggregate is 17.4. This figure may be used only as general information, as a design criterion, or in estimating the approximate percentage of voids in the aggregate filled by a single per cent of asphalt, which information may be useful in adjusting the asphalt content of the mixture toward the desired goal. It may also be used in computing the per cent voids in the total mixture which are filled with asphalt, which figure is used as an additional design criterion by some agencies.

$$\text{\% voids filled with asphalt} = \frac{\text{V.M.A.(\%)} - \text{V(\%)}}{\text{V.M.A.(\%)}} \tag{17-7}$$

Following the density determinations and the calculation of the various items given above, the stability and flow determinations are made upon the compacted specimens.

17–10. Stability-Flow Test (Marshall Method). The Marshall stability is the maximum load resistance in pounds which the standard test specimen will develop at 140°F when subjected to load by a standardized procedure. (In 1966, the Asphalt Institute suggested that base course mixtures intended for use 4 inches or more below the pavement surface could be tested at 100°F, instead of 140°F.) Loading is done with the specimen lying on its side, in a split loading head; loading is perpendicular to the cylindrical axis of the specimen. The flow value is the total deformation, in units of 1/100 inch, occurring in the specimen

FIG. 17–2. The Marshall stability test. (Courtesy the Asphalt Institute.)

between no load and maximum load during the stability test. A typical setup for the Marshall stability test is shown in Fig. 17–2.

When testing is complete, five plots are prepared, as illustrated in Fig. 17–3.

Design criteria for the Marshall method are contained in Table 17–7 (*118*).

From curves like those of Fig. 17–3, asphalt contents are determined which correspond to the following.

1. Maximum stability.
2. Maximum unit weight.
3. Median of limits given in Table 17–7 for per cent voids, total mix.

The optimum asphalt content is the average of the asphalt contents thus determined.

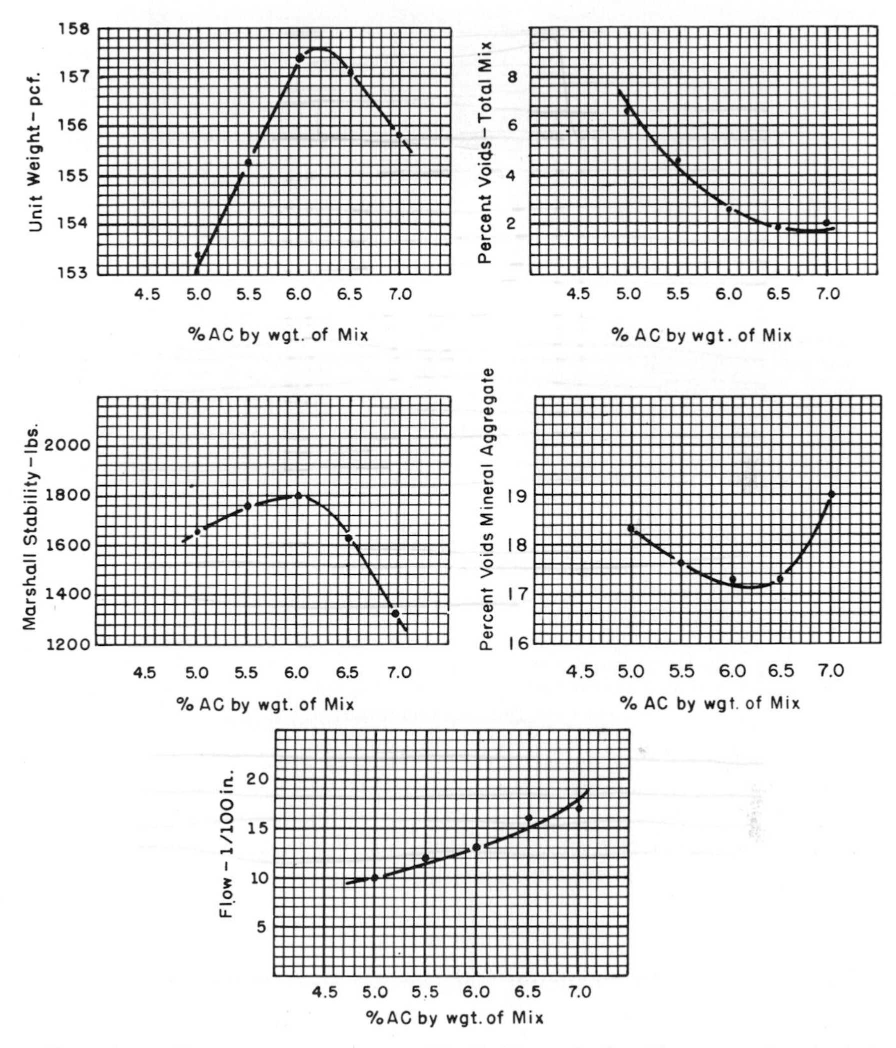

FIG. 17-3. Test property curves, Marshall method. (Courtesy the Asphalt Institute.)

For the specific mix shown in Fig. 17-3, specific values which correspond to the listings above are

1. 6.0 per cent
2. 6.2 per cent
3. 5.6 per cent

The optimum asphalt content equals the average of the three, or 5.9 per cent.

TABLE 17–7

MARSHALL DESIGN CRITERIA (*118*)

Traffic Category	Heavy and Very Heavy		Medium		Light	
No. of Compaction Blows Each End of Specimen	75		50		35	
Test Property	Min.	Max.	Min.	Max.	Min.	Max.
Stability, all mixtures	750	–	500	–	500	–
Flow, all mixtures	8	16	8	18	8	20
% Air voids						
Surfacing or leveling	3	5	3	5	3	5
Sand or stone sheet	3	5	3	5	3	5
Sand asphalt	5	8	5	8	5	8
Binder or base	3	8	3	8	3	8
% Voids in mineral aggregate	(See Fig. 17–4)		(See Fig. 17–4)		(See Fig. 17–4)	
Surfacing or leveling	–		–		–	
Sand or stone sheet	–		–		–	
Sand asphalt	–		–		–	
Binder or base	–		–		–	

Note:
1. Laboratory compactive efforts should closely approach the maximum density obtained in the pavement under traffic.
2. The flow value refers to the point where the load begins to decrease.
3. The portion of the asphalt cement lost by absorption into the aggregate particles must be allowed for when calculating per cent air voids.
4. Per cent voids in the mineral aggregate is to be calculated on the basis of the ASTM bulk specific gravity for the aggregate.

To judge the satisfactoriness of the mix at the optimum asphalt content, assume that the mix is asphaltic concrete and is to be designed for medium traffic. From Fig. 17–3, the following values are determined, at the optimum asphalt content.

Stability	1800
Flow	13
Per cent voids, total mix	2.8
Per cent voids, mineral aggregate	17

By comparing these values with the limits given in Table 17–7, it is seen that the mixture is satisfactory, except that the per cent voids in the total mix is too low. Adjustments would then be made in the mixture until satisfactory results were achieved.

17–11. Hveem Method of Design. The Hveem method of design of paving mixtures was developed under the direction of Francis N. Hveem, formerly Materials and Research Engineer, California Division of Highways (*118*), and is the standard method used in that state. It is gen-

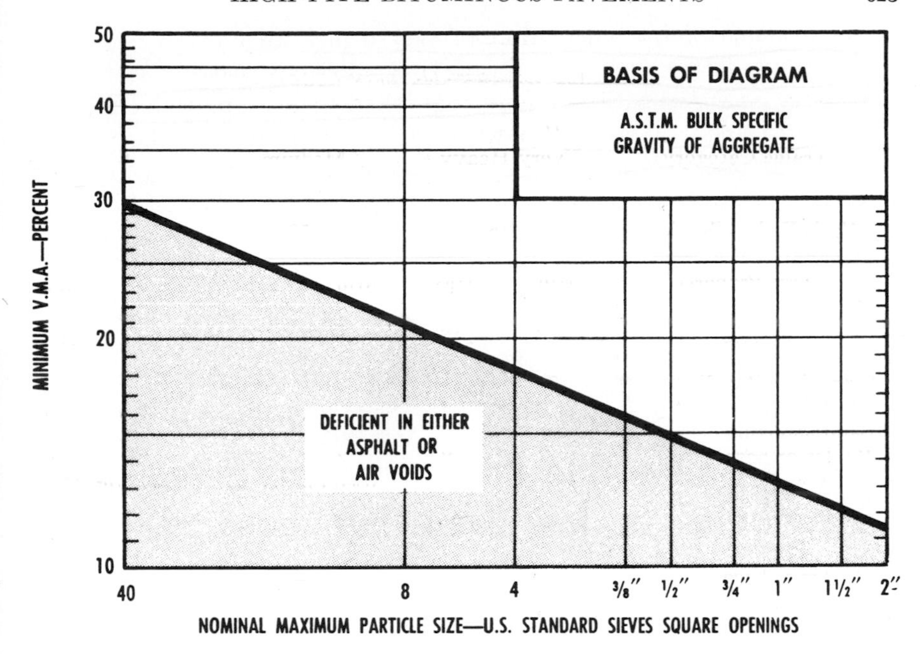

FIG. 17–4. Relationship between minimum VMA (per cent) and nominal maximum particle size of aggregate for compacted dense-graded paving mixtures. (Courtesy the Asphalt Institute.)

erally applicable to paving mixtures using either penetration grades of asphalt cement or liquid asphalts and which contain aggregates up to 1-inch maximum size. Therefore, it is applicable not only to the high-type paving mixtures described in this chapter, but also to the intermediate mixes of Chapter 16; it has been used principally in the design of dense-graded asphaltic mixtures. The method has been standardized as ASTM D1560 and D1561.

The first step in the Hveem method is to determine the "estimated optimum" asphalt content through determination of the Centrifuge Kerosene Equivalent. Factors determined from the C.K.E. test are combined with calculated surface area. The surface area of a given aggregate, or blend of aggregates, is calculated by use of the gradation and a table of surface area factors. Surface area factors to be applied to total per cents passing the various sizes or sieves are as follows:

No. 4	2
No. 8	4
No. 16	8
No. 30	14
No. 50	30
No. 100	60
No. 200	160

An example of the calculation of the surface area of an aggregate of known gradation is given in Table 17–8. Per cent passing each sieve (expressed as a decimal) is simply multiplied by the applicable surface area factor and the products added. It is assumed that the surface area of the material coarser than the No. 4 sieve is 2 square feet per pound.

TABLE 17–8

EXAMPLE OF CALCULATION OF SURFACE AREA OF AGGREGATE OF KNOWN GRADATION, HVEEM METHOD OF DESIGN

Sieve Size	Per Cent Passing	Surface Area Factor	Surface Area
¾-inch	100	–	–
⅜-inch	90	–	2.0
No. 4	75	2	1.5
No. 8	60	4	2.4
No. 16	45	8	3.6
No. 30	35	14	4.9
No. 50	25	30	7.5
No. 100	18	60	10.8
No. 200	10	160	16.0
	Surface area = 48.7 square feet per pound		48.7

In determining the Centrifuge Kerosene Equivalent in the laboratory, a special centrifuging device is used. A representative sample of 100 grams of aggregate passing the No. 4 sieve is saturated with kerosene and subjected to a centrifugal force of 400 times gravity for two minutes. The amount of kerosene retained, expressed as a percentage of the dry aggregate, is the C.K.E.

The "surface capacity" of the coarse aggregate portion of the mixture is determined by first placing 100 grams of dry aggregate which passes the ⅜-inch sieve and is retained on the No. 4 into a glass funnel. The sample and funnel are immersed in S.A.E. No. 10 lubricating oil at room temperature for five minutes. The funnel and sample are removed from the oil and drained for 15 minutes at a temperature of 140°F. The sample is then weighed and the amount of oil retained computed as a percentage of the dry aggregate weight.

To determine the estimated asphalt content, the following procedure is used.

1. Using the C.K.E. value and the chart of Fig. 17–5, the surface constant for fine material (K_f) is determined.

2. Using the per cent oil retained in the surface capacity test for

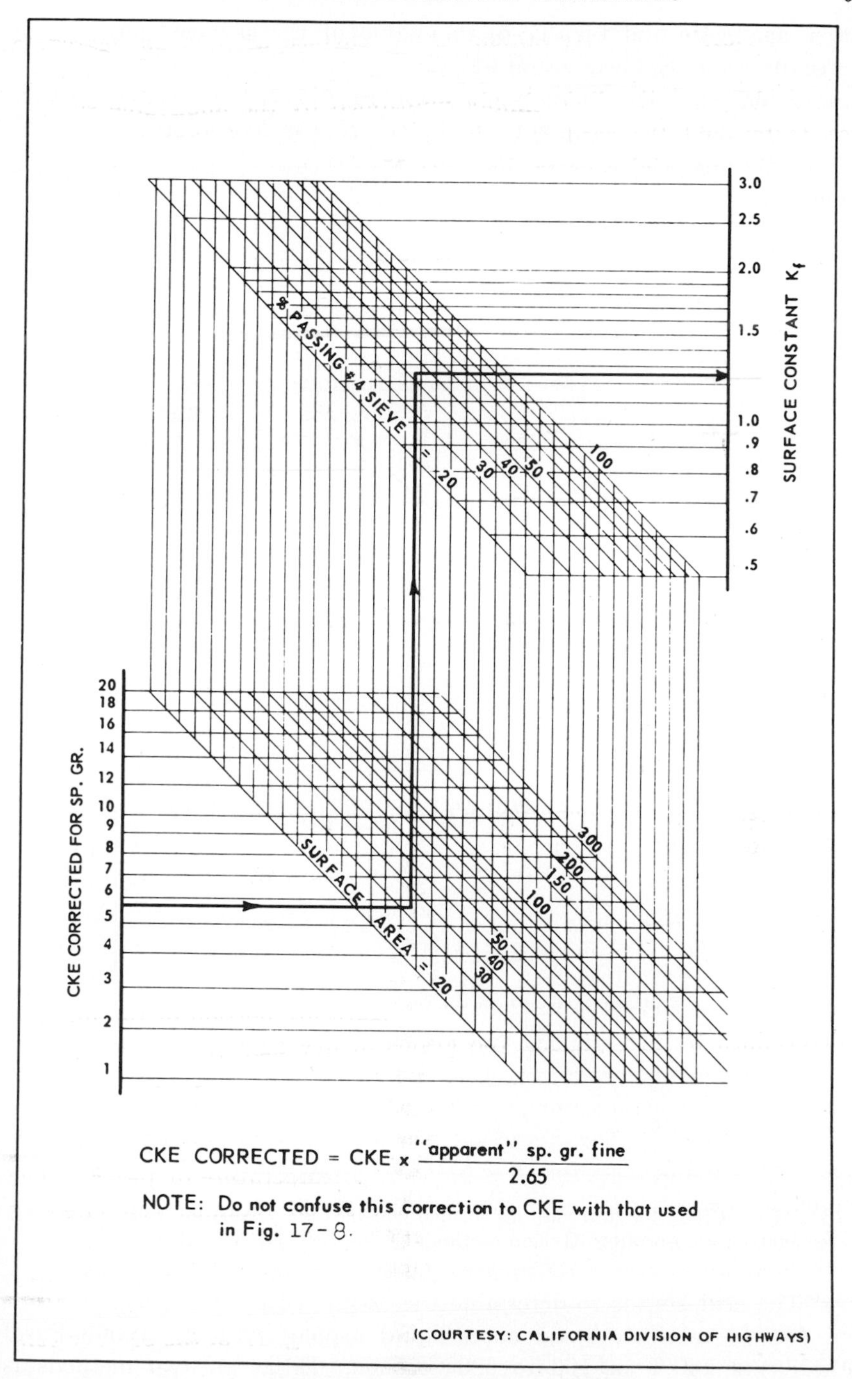

FIG. 17-5. Chart for determining surface constant for fine material (K_f) from C.K.E., Hveem method. (Courtesy the Asphalt Institute.)

coarse aggregate and Fig. 17–6, the value of the surface constant for coarse material (K_c) is determined.

3. Using the values of K_f and K_c and Fig. 17–7, the surface constant for the fine and coarse aggregate combined (K_m) is determined.

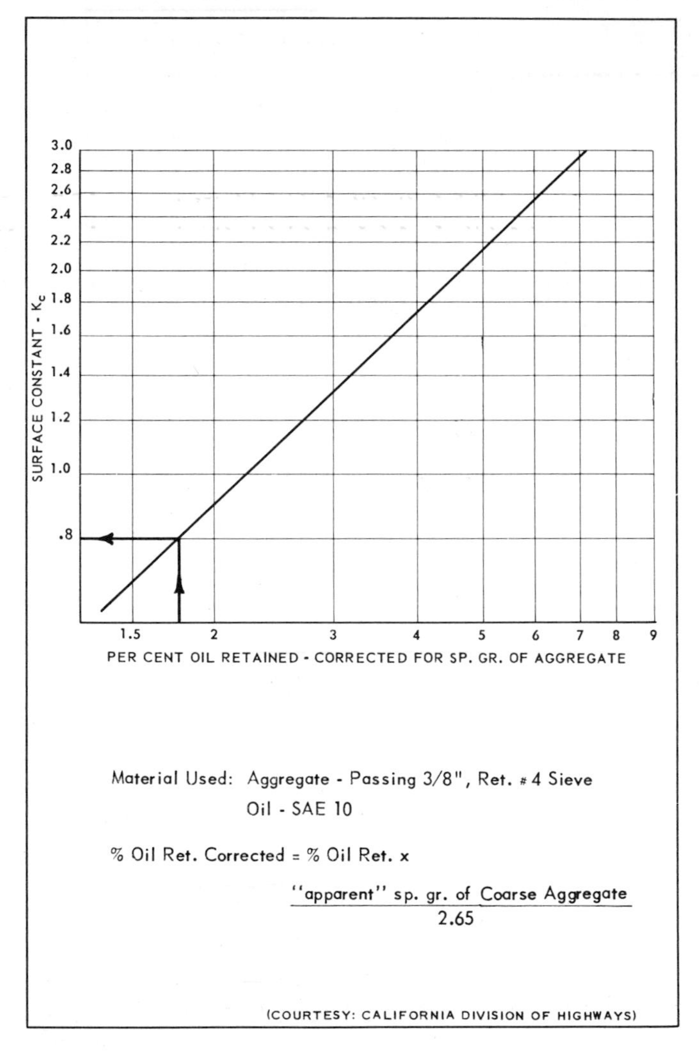

FIG. 17–6. Chart for determining surface constant for coarse material (K_c) from coarse aggregate absorption, Hveem method. (Courtesy the Asphalt Institute.)

4. The next step is to determine the estimated optimum asphalt content for the mixture, based on a liquid asphalt with a Saybolt-Furol viscosity at 140°F of 100 to 500 seconds. Either of two methods is used, referred to as Case I and Case II.

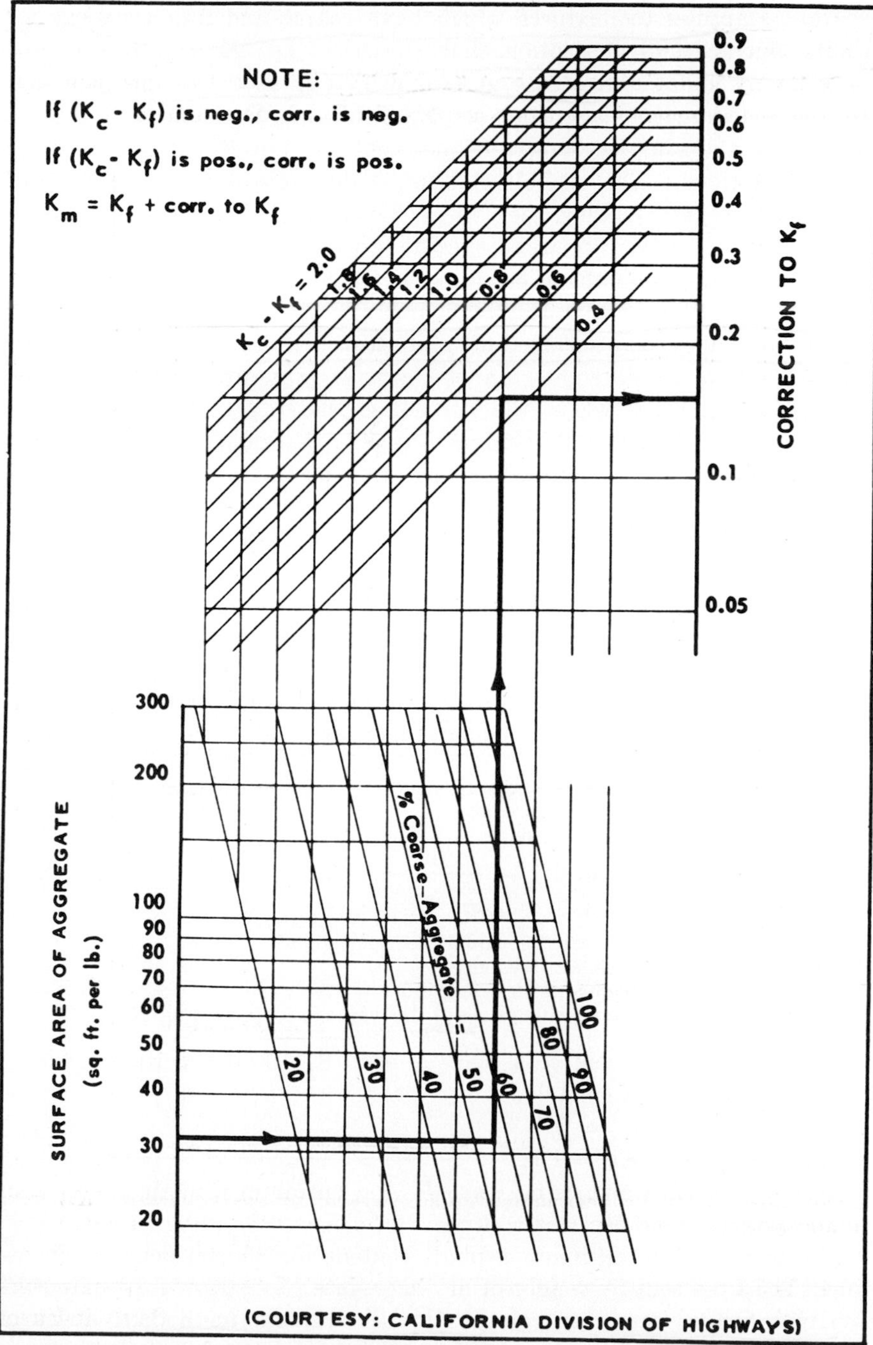

Fig. 17–7. Chart for combining K_f and K_c to determine surface constant for combined aggregate (K_m), Hveem method. (Courtesy the Asphalt Institute.)

Case I applies to mixtures which have coarse and fine aggregate of similar surface and absorption characteristics; i.e., K_f and K_c are approximately equal. This case is used primarily as a field method, and is used in the central laboratory of the California Division of Highways only when the sample contains 85 per cent or more of material passing the No. 4 sieve. The asphalt content is determined by using the per cent of aggregate passing the No. 4 sieve, the C.K.E. value adjusted for specific gravity (apparent specific gravities are used) and scales A, B, and C of the alignment chart of Fig. 17–8.

Case II applies to mixtures of coarse aggregates and fine aggregates which have dissimilar surface and absorption characteristics, i.e., K_f and K_c are markedly different. In the central laboratory, this approach is used in all cases where the percentage passing the No. 4 sieve is less than 85. The asphalt content based on a liquid asphalt is determined by using the computed surface area of the sample, the average apparent specific gravity of the aggregates, the surface constant K_m, and scales D to B on the alignment chart of Fig. 17–8.

5. The asphalt content of the mixture for the particular grade of asphalt to be used is found from Fig. 17–9, knowing the surface area of the sample, the grade of asphalt to be used, and the asphalt content from Fig. 17–8.

In order to illustrate the procedure, consider a mixture containing aggregates which have the following properties:

Specific gravity, coarse aggregate	2.45
Specific gravity, fine aggregate	2.64
Per cent passing No. 4 sieve	45

$$\text{Average specific gravity} = \frac{100}{\frac{55}{2.45} + \frac{45}{2.64}} = 2.53$$

Surface area of aggregate, square feet per pound	32.4
Centrifuge kerosene equivalent	5.6

Per cent oil retained, coarse aggregate = 1.9
(Corrected for specific gravity, this value is 1.76; see Fig. 17–6.)

The mixture is to contain asphalt cement, 85–100 penetration grade.

From Fig. 17–5, $K_f = 1.25$.

From Fig. 17–6, $K_c = 0.8$.

From Fig. 17–7, $K_m = 1.1$.

From Fig. 17–8 and using Case II, the optimum asphalt content for liquid asphalt is 4.6 per cent.

From Fig. 17–9, optimum asphalt content for 85–100 penetration asphalt is 5.7 per cent by weight of dry aggregate.

In designing a paving mixture by the Hveem method, a series of stabilometer test specimens are prepared for a range of asphalt contents, both above and below the estimated optimum asphalt content indicated

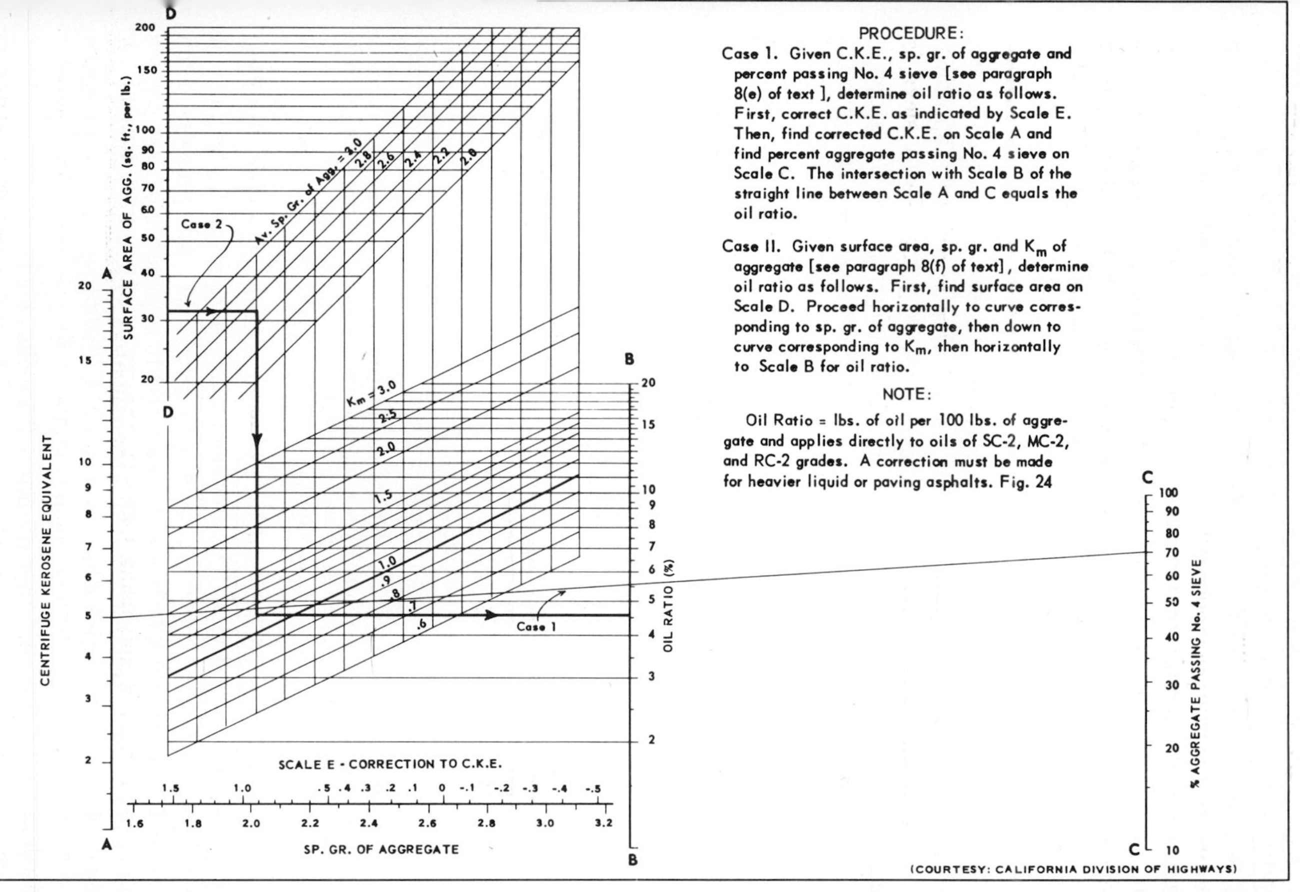

FIG. 17–8. Chart for computing oil ratio for dense graded bituminous mixtures, Hveem method. (Courtesy the Asphalt Institute.)

by the C.K.E. procedure. In addition, two swell test specimens are prepared at the estimated optimum asphalt content.

Samples are prepared by mixing heated aggregates and asphalt in accordance with a standardized procedure. As a part of the procedure, the mixture is cured for 15 hours in an oven at 140°F. After curing, the mixture is heated to 230°F preparatory to compaction. The compaction of the specimens is accomplished by use of a mechanical compactor

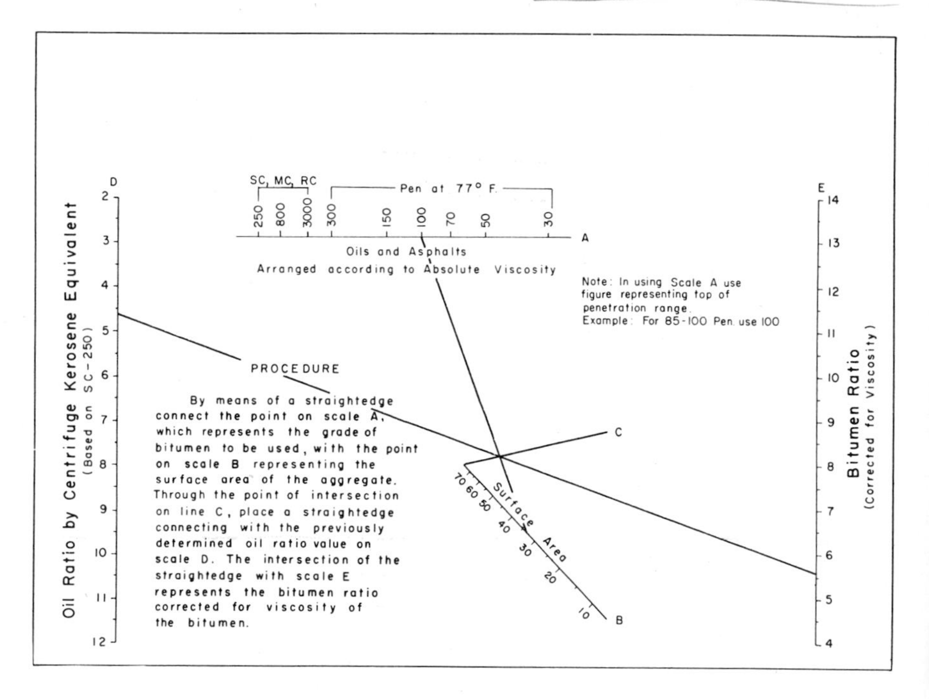

Fig. 17–9. Chart for correcting bitumen requirement due to increasing viscosity or lower penetration of asphalt, Hveem method of design. (Courtesy the Asphalt Institute.)

which consolidates the material by means of a kneading action resulting from a series of individual impressions made with a ram, the face of which is shaped as a sector of a 4-inch circle. At full pressure, each application of the ram subjects the material to a pressure of 500 pounds per square inch over an area of approximately 3.1 square inches. A detailed compaction procedure is followed, including final application of a static load to the specimen. Cylindrical specimens are formed, 4 inches in diameter and approximately 2.5 inches high.

The compacted test specimens are used in the performance of four tests: swell test, stabilometer test, bulk density determination, and cohesiometer test.

In the swell test, the mold and compacted specimen are placed in a deep aluminum pan. A tripod arrangement is used to record the change in height of the specimen during the test. The upper portion of the mold is partially filled with water, which is allowed to remain on top of the specimen for 24 hours. At the end of that time, the change in height of the specimen, or *swell*, is determined. A measurement is also made of the amount of water which moves through the specimen during the 24-hour period, or the *permeability* of the sample.

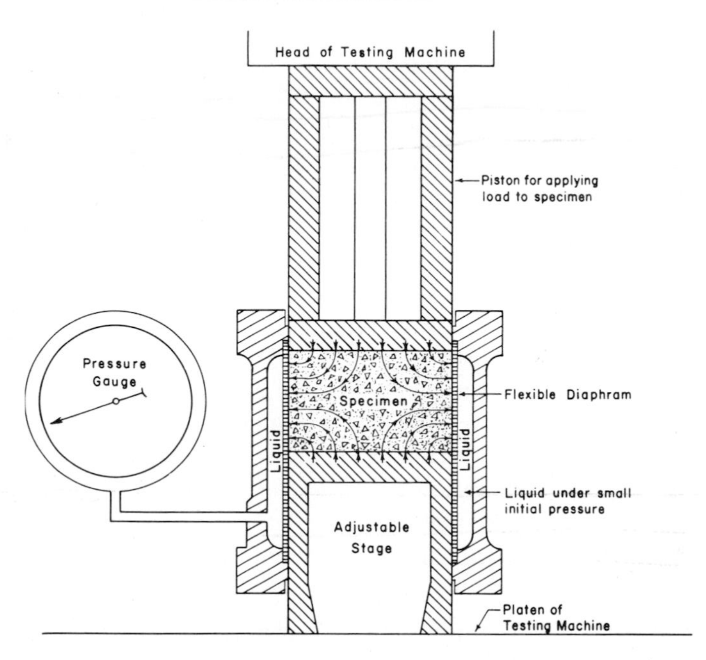

FIG. 17–10. Diagrammatic sketch of Hveem Stabilometer. (Courtesy the Asphalt Institute.)

Stabilometer test specimens are heated to 140°F before testing. A schematic drawing of the Hveem Stabilometer is shown in Fig. 17–10; the device was described briefly in Art. 12–19. In testing asphaltic mixtures, the stability value resulting from the test is dependent upon the ratio between the horizontal and vertical pressures at 400 pounds per square inch.

Bulk density determinations are made on the stabilometer test specimens in a fashion which is essentially the same as that previously described. The same specimens are used in the cohesiometer test; this device was described briefly in Art. 12–19. Samples are tested at 140°F. Cohesiometer values are expressed as grams per inch of width.

Design criteria for hot-mix asphaltic paving mixtures by the Hveem method are as follows:

	Light Traffic	Medium Traffic	Heavy Traffic
Stabilometer value	30+	35+	37+
Cohesiometer value	50+	50+	50+
Swell	Less than 0.030 inch		

An effort is also made to provide a minimum per cent voids in the total mix of approximately 4 per cent. In the application of these criteria, the optimum asphalt content for the mix should be the highest percentage of asphalt the mix will accommodate without loss of stability. Generally, the voids in the total mix should approach the 4 per cent minimum; however, in cases where the mixture shows a marked decrease in stability for slight increases in asphalt content, it is advisable to adjust the aggregate grading or reduce the asphalt content to provide a less critical mixture with more air voids.

If the cohesiometer value is relatively low, an increased value can usually be obtained by adjustment of the aggregate grading. For well-graded mixtures that are otherwise satisfactory, an increased cohesiometer value normally may be obtained by increasing the percentage of filler or by using a low penetration grade of asphalt cement.

A moisture vapor susceptibility test is also made. This test indicates the extent to which stabilometer values of bituminous mixtures are affected by moisture in the vapor state, entering the mixture from a wet subgrade or some other source.

17–12. Hubbard-Field Method of Mix Design. The Hubbard-Field method of design depends upon the results of an extrusion shear test performed on trial mix samples at various asphalt contents. The Hubbard-Field stability test was developed many years ago, and was the standard mix design method until recent years; it is being supplanted by the Marshall and California methods, but is still used by some agencies in the design of dense-graded asphaltic mixtures.

The standard method involves the formation of compacted, 2-inch-diameter specimens at asphalt contents both above and below the estimated optimum. A tamping procedure is used in conjunction with a static load to form each test specimen. The sample is then heated to 140°F and forced through an extrusion ring that is slightly smaller in diameter than the test sample, as shown in Fig. 17–11. The maximum load on the specimen during extrusion is the stability, in pounds.

Selection of the optimum asphalt content is based on the stability and a density-voids analysis. The Asphalt Institute recommends (*118*) these stability values: for heavy and very heavy traffic, a minimum of 2000 pounds; for medium and light traffic, a minimum of 1200 and a maximum of 2000 pounds. They suggest a minimum of 2 per cent air voids and a maximum of 5 per cent, regardless of traffic.

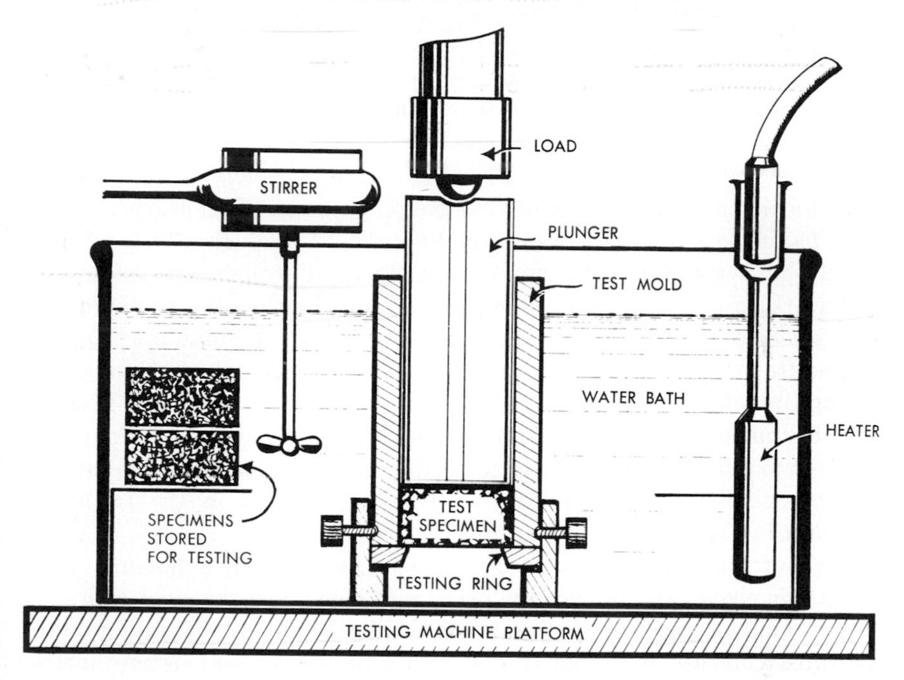

Fig. 17–11. The Hubbard-Field stability test. (Courtesy the Asphalt Institute.)

17–13. Construction Methods (Hot-Mix, Hot-Laid Asphaltic Concrete). All the mixtures described in this chapter are, of course, proportioned and mixed in central mixing plants. Construction procedures discussed here are intended to be generally applicable to all the hot-mix, hot-laid pavements described in this chapter, and are also largely applicable to the cold mixtures which are discussed. The necessary construction steps are described in some detail in this section: deviations from the basic procedure indicated will be mentioned in subsequent sections of this chapter. The specific procedures described here are primarily intended to be applicable to the construction of wearing surfaces, although the same procedures are also generally applicable to the construction of base and leveling courses. The fundamental steps in the construction of a high-type bituminous pavement may be listed as follows:

1. Preparation of the mixture
2. Preparation of base or leveling course
3. Transportation and placing of the surface-course mixture

4. Joint construction
5. Compaction and final finishing

17–14. Preparation of Mixture. The essential elements of a central mixing plant have already been described in Chapter 16. Plants used in the preparation of high-type hot mixtures are essentially the same as those which have been previously described, except that certain additional units may be added to the plant and the entire setup geared to the necessity for volume production of uniform hot mixtures with very close control over the proportioning and mixing steps.

Typical central mixing plants for the preparation of high-type bituminous paving mixtures are shown in Figs. 17–12 through 17–16. Figure 17–12 shows the flow chart for one of Barber-Greene's continuous-flow plants. The gradation unit shown is a four-bin element, providing for four sizes of aggregate. In Fig. 17–13 is shown a plant of this type in operation. Fig. 17–14 is a cutaway view of the heart of a batch-type hot-mix plant. Operation of this plant, as with certain plants produced by other manufacturers, is almost completely automatic. A Cedarapids batch-type plant is shown in Fig. 17–15. The automatic control system for this plant is shown in Fig. 17–16; it is housed in the trailer on the right in Fig. 17–15.

Two of the units which are frequently included in a hot-mix plant and which have not previously been mentioned are the dust-collecting system and a device for feeding mineral filler or "dust" into the mixture. The dust collector is generally operated adjacent to and in conjunction with the aggregate drier and is necessary for efficient plant operation in many cases. The collector serves to eliminate or abate the dust nuisance which exists around the plant and is especially important in plants which are located in or adjacent to municipalities. Modern dust-collection systems are highly efficient; frequently the fine material collected, if uniform in gradation, is fed back into the plant at the entrance to the hot elevator. To meet air pollution control requirements, many urban plants have "wet wash" systems in which exhaust from the dust collector enters a tower and passes through a series of water sprays that remove the remaining dust.

Mineral filler which is added to the mixture is not normally passed through the aggregate drier, this material being fed by a separate device directly into the mixing unit in some cases or into the aggregate batching unit. Separate feeder units are also used in the preparation of cold mixtures to supply liquefier, other fluxes, or hydrated lime to the mix. Central mixing plants for high-type mixtures are quite complex in nature and no further detailed description is deemed desirable in this text. The description of the basic units given in the preceding chapter and the

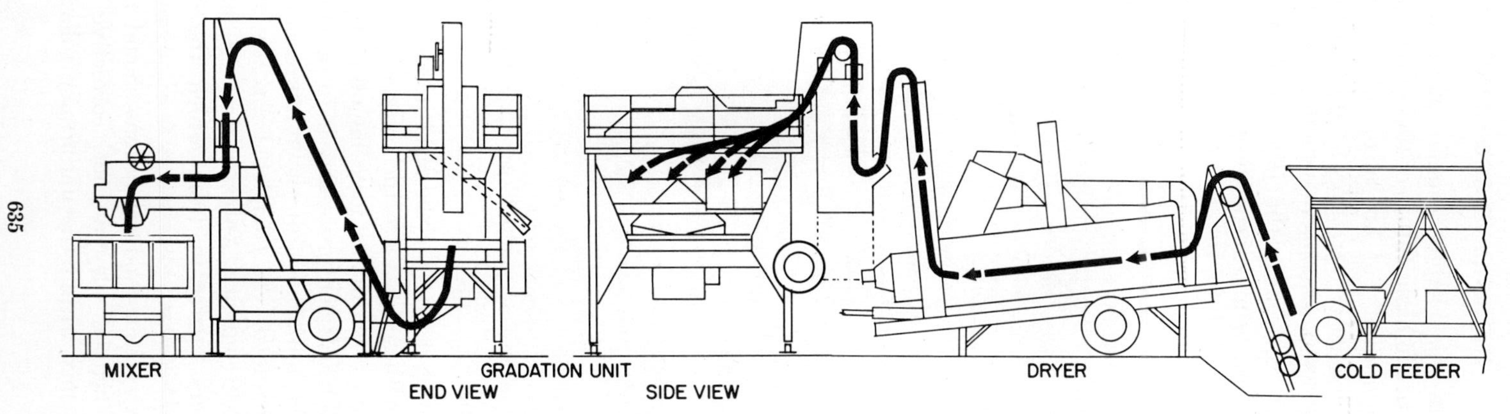

Fig. 17-12. Flow diagram for single-aggregate, continuous-flow hot-mix plant. (Courtesy Barber-Greene Company.)

information given in this chapter should suffice to give the reader a good understanding of the basic nature of these plants.

Control of the uniformity of hot mixes of this type is important, as any appreciable change or variation in gradation or asphalt content will be reflected in a change in some other characteristic of the mix. The control of the plant to produce the job mix chosen is a cooperative effort of the plant operator and inspector and requires continual inspection and careful control.

FIG. 17-13. A continuous-flow, multiple-aggregate hot-mix plant in operation. (Courtesy Barber-Greene Company.)

Sampling and testing are among the most important functions in plant control. Data from these tests are the tools with which to control the quality of the product. Samples may be obtained at various points in the plant to establish if the processing is in order up to those points. A final check of the uniformity of the mixture is made by obtaining a representative sample of the completed mixture and determining its gradation and asphalt content. An extraction test measures the asphalt content and provides aggregate from which the gradation may be determined.

The results of extractions and gradations should fall within the job-mix tolerance specified; if they do not, corrective measures must be taken to bring the mix within the uniformity tolerance. Adjustments

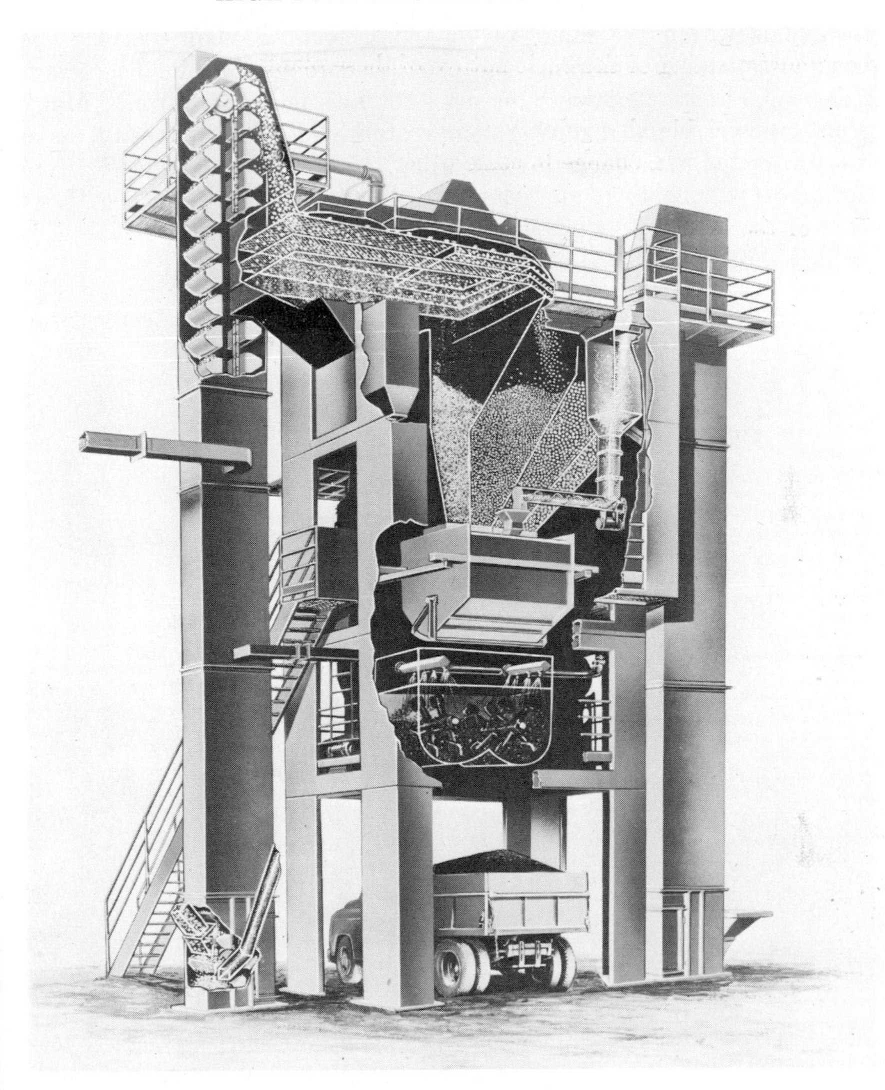

FIG. 17–14. Flow of material through a batch-type hot-mix plant. (Courtesy Barber-Greene Company.)

are frequently necessary, particularly during the early stages of plant operation or if changes occur in the aggregates. For a complete discussion of this phase of construction, see reference *244*.

The Asphalt Institute recommends that the temperature-viscosity relationship for the asphalt being used be used to arrive at a proper mixing temperature, since asphalts vary as to source or type and grade. The most effective temperature for plant mixing is that at which the asphalt viscosity is within the range of 75–150 seconds, Saybolt Furol. The higher viscosities in this range are usually more suitable for coarse aggregate mixes and the lower viscosities for fine aggregate mixes. Many agencies now specify a mixing temperature, based on viscosity, as a part of the job-mix formula and allow a variation of ±20°F from that figure.

Fig. 17–15. A batch-type hot-mix plant with automatic controls. (Courtesy Iowa Manufacturing Company.)

It is likely that a majority of mixtures of this type are now prepared at temperatures somewhere between 250 and 275°F, and the over-all range is perhaps 225 to 325°F. Modern plants incorporate recording temperature gages which provide a continuous record of asphalt and aggregate temperatures.

17–15. Temperatures and Mixing Times. As has been indicated, the aggregate and bituminous material are handled separately in the plant, prior to their combination in the mixing unit. The bituminous material is uniformly heated to a specified temperature in a tank or "kettle," while the aggregates are heated and dried in the drier unit. The temperature of the asphalt is carefully controlled in order to prevent overheating or "burning" of the material, with consequent destruction of certain desirable qualities. The temperature of the asphalt prior to its entrance into the mixer is generally from 225 to 350°F, and many modern specifications place the upper limit of this temperature at 300 or 325°F.

FIG. 17–16. Automatic control system for a hot-mix plant. (Courtesy Iowa Manufacturing Company.)

In batch-type plants, the dry aggregates are proportioned into the mixer unit and thoroughly manipulated for a brief time prior to the introduction of the bituminous material. Time for this dry-mixing operation generally would be from 15 to 20 seconds. The bituminous material is then introduced into the mixer and the materials mixed for a period which generally is a minimum of 45 seconds. Total time in the mixer in a batching plant would then generally be from 55 to 70 seconds, and may be longer if necessary to produce a uniform and homogeneous mix. Similar times are required in continuous mixing units. Time in the mixer should be held to a minimum consistent with proper mixing.

17–16. Automatic Control Systems. Several states now require automatic controls on hot-mix plants used to produce mixtures on state highway contracts, and contractors in other states have installed automatic controls, even though they are not required. In 1965, New York State began requiring all hot-mix plants to have automatic recording systems also, and this practice probably will spread.

Key operations in automatically controlled plants are automatic batching and cycling (*230*). The automatic systems retain the dial scales of a manually operated batch plant. However, instead of an operator's controlling the flow of material from a hot aggregate bin, a sensor determines when the pointer is approaching the index point and cuts off the flow, compensating for the material still in midair.

After the midair material reaches the weigh bucket, but before the gate on the next bin opens, over and under sensors determine if the weight is within the specified tolerance. If it is, the gate for the next bin opens. If not, the automatic cycle shuts down until excess material is wasted or the variance is approved by an inspector.

The operation described is for one bin. Actually, the entire batching and mixing operation is automated. A "memory" stores the weights of aggregate and asphalt for any given mix. The memory can be set into the system by dials on the instrument panel, by punched cards, or by a batch plug (several potentiometers enclosed in a casing).

The operator pushes a button to start the production cycle. The automatic equipment draws aggregate from each bin in sequence, dumps this into a pugmill for a dry mixing period, if required, dumps the asphalt, continues the mixing for the specified time, and then dumps the batch from the mixer into a truck. The cycle repeats itself for the number of batches required to load the truck, then shuts itself off.

Some plant owners have extended automatic controls to the feeding of cold aggregates, and to operation of the drier burner.

The principal advantage of automatic controls is the greater uniformity of the mixture. The contractor's costs drop, because labor requirements are lessened, and the production rate generally stays the same or rises slightly. If a recording system is used, the inspection effort is reduced.

17–17. Transportation of Mixture. The mixture is discharged from the plant into dump trucks or trailers for transportation to the job site. The vehicles which are used are required to have tight and smooth metal beds which have previously been cleaned of all foreign material. The vehicle bed may be sprayed with a light coat of lime-water, soap solution, or some similar substance to prevent adherence of the mixture. Fuel oils should not be used as they will have a detrimental effect on the mixture. Sometimes the vehicle is required to be insulated against excessive heat loss in the mixture during hauling and frequently is covered with a canvas to protect the mix from the weather.

17–18. Preparation of Base. Surface courses of asphaltic concrete are frequently placed upon new or existing bases which require very little preparation prior to the placing of the new surface other than thorough sweeping and cleaning to remove loose dirt and other foreign materials. In other cases, the existing base or surface upon which the mixture is to be placed may require extensive corrective measures. Quite frequently, when the existing surface is disintegrated, broken, or irregular in nature, specific defects may be corrected by the application of "patches" of asphaltic concrete. Excessive joint sealing compound and fatty areas should also be removed. In certain cases it may be desirable to place a leveling course of asphaltic concrete to correct irregularities in the existing surface. At points where the asphaltic concrete mixture is to be placed in contact with surfaces of manholes, curbs, gutters, and such, these surfaces are generally painted with a light coat of hot asphalt cement or liquid asphaltic material.

17–19. Placing of Mixture. Placing of the asphaltic concrete mixture is permitted only when the base is dry and under favorable weather conditions. Placing of hot mixtures is usually suspended when the air temperature becomes less than 40°F, except under unusual circumstances. Base, leveling, and surface courses are placed and compacted in separate operations. In certain cases, unusually thick layers composed of the same mixture may be placed in two or more courses.

The steps of spreading and finishing are, in the large majority of cases, accomplished by the use of mechanical spreading and finishing machines (pavers) such as were described in the preceding chapter. These machines generally process one lane width at a time and must be capable of producing a smooth, even surface which has the desired elevation, grade, and cross-section. This is best accomplished if the speed of laying is adjusted to the output of the plant, thus avoiding frequent long stops. Fig. 17–17 shows a Blaw-Knox Paver with a 19-foot screed engaged in the spreading and finishing of an asphaltic concrete wearing surface. The surface is checked during rolling and any irregularities, "fat" spots, or similar defects are corrected. In areas where the mechanical paver cannot operate satisfactorily, the mixture is placed on dump boards, spread, and raked by hand.

In placing the asphaltic concrete mixture, special attention must be given to the construction of joints between old and new surfaces or between successive days' work. It is essential that a proper bond be secured at longitudinal and transverse joints between the newly placed mixture and an existing surface, regardless of its nature, and special procedures, generally performed by hand, are utilized to ensure the formation of adequate joints.

The best longitudinal joint is formed when the material in the edge being laid against is still warm enough for effective compaction. This means that two finishers working in echelon can produce the most satisfactory joint. When one finisher is used a procedure of moving from one lane to the other may be used so that the edge of the previously placed lane is still hot when the new lane is placed. Various other procedures are used when laying against a cold edge, such as cutting the cold edge back and painting with bituminous material. Some agencies allow the use of an infrared joint heater attached to the front of

FIG. 17–17. Bituminous paver with 19-ft screed. (Courtesy Blaw-Knox Company.)

the finisher, which heats the cold material along the edge prior to placing a new lane.

17–20. Compaction of the Mixture. When the spreading and finishing operations described above have been completed, and while the mixture is still hot, rolling is begun.

Rolling may be carried out by steel-wheel or pneumatic-tired rollers, or by a combination of the two. Steel-wheel rollers may be of three types: three-wheel rollers of 10 to 12 tons in weight, two-axle tandem rollers of 8 to 12 tons, and three-axle tandem rollers of 12 to 18 tons.

Pneumatic-tired rollers are becoming more popular as they provide a closely knit surface by kneading aggregate particles together. They will achieve, during the rolling operation, a more uniform density than usually develops after vehicular traffic has used the asphalt surface for some time. The contact pressure exerted by a pneumatic-tire compactor is dependent on tire size, ply rating, wheel load, and tire inflation pressure. Tire contact pressures generally range from 40 to 90 psi. Use of a self-propelled pneumatic-tired roller for this purpose is shown in Fig. 17–18.

Rolling of the bulk of the pavement is done in a longitudinal direction, beginning at the edges and gradually progressing toward the center except on superelevated curves, where rolling begins on the low side and progresses toward the high side; successive trips of the roller overlap

FIG. 17–18. Pneumatic-tired roller on a Kansas paving project. (Courtesy *Engineering News-Record.*

to ensure complete and uniform coverage. Rolling progresses at a uniform, slow rate of speed, with the drive roll or wheel nearest the paver.

Rolling procedures vary with the properties of the mixture, thickness of layer, and other factors. In modern practice, rolling is divided into three phases, which follow closely behind one another; initial or "breakdown" rolling, intermediate rolling, and finish rolling. The breakdown and intermediate phases provide needed density and the final rolling gives the required final smoothness. A procedure used by most organizations calls for the use of three-wheel or tandem rollers for breakdown, pneumatic-tired rollers for intermediate rolling, and tandem rollers for finish rolling. Use of a steel-wheel roller in compaction of a hot-mix asphaltic pavement is illustrated in Fig. 17–19.

Experiments by the Washington State Highway Commission (*249*) showed the effectiveness of compaction done when the asphalt mixture was still hot (above 200°F), with thick lifts (up to 6 inches or more loose thickness), and using high-pressure (90 psi) pneumatic-tired rollers operating directly behind the paver.

Standards applied to the completed wearing surface are as high as is deemed practically possible. Requirements relative to the finished surface frequently stipulate that the surface shall be smooth, even, and true to the desired grade and cross-section, with no deviation of more than one-eighth of an inch from a line established by a template or a 10-foot straightedge being permitted. This seems to be a very high

Fig. 17–19. Three-axle tandem roller compacting an asphaltic mixture. (Courtesy Galion Iron Works & Manufacturing Company.)

standard, but such surfaces are readily attainable in most cases with modern construction equipment. Deviations from the specified thickness of surface course are generally permitted to be no more than one-quarter of an inch. The density to be obtained in the completed wearing surface may be stipulated in terms of a percentage of the theoretical maximum density or a percentage of the density of the laboratory-compacted mixture. A typical requirement is that the density of the completed surface course should be not less than 95 per cent of the laboratory compacted density for the same mixture. Requirements relative to the density of the completed pavement are very important and must be such as to guarantee, on the one hand, the attainment of a satisfactory density, and on the other that a density is not specified which cannot be economically secured or which is impossible of attainment without crushing the aggregates.

The density of the compacted mixture is determined on samples cored or sawed from the completed mat by the procedure previously discussed in Art. 17–8. Many agencies, in addition to determining density on compacted mixtures, determine for control purposes the per cent voids and voids filled and obtain large enough samples in order to determine the gradation and asphalt content of the compacted mixture. Experimental use is being made of nuclear density gages and air-flow meters (air permeability measuring devices) to measure density and control the field rolling process.

Rolling completes the construction of most asphaltic concrete pavements, and traffic may be permitted upon the surface as soon as the compacted mixture has adequately cooled. In some cases, also, a seal coat is applied to the newly completed pavement, as described in a previous chapter.

Stone-Filled Sheet Asphalt

Stone-filled sheet asphalt occupies an intermediate position between the asphaltic concrete mixtures which have been described and sheet asphalt, which is described in the following section. Stone-filled sheet asphalt is distinguished by the inclusion of a much greater percentage of fine material than is usually found in asphaltic concrete, with this type of mixture generally containing more than 60 per cent of material passing a No. 10 sieve. That is, stone-filled sheet-asphalt mixes contain less than 40 per cent of coarse aggregate. From Fig. 17–1, the Asphalt Institute defines "stone sheet" as a mixture which contains from 65 to 80 per cent fine aggregate (passing No. 8 sieve). The percentage of coarse aggregate is so small as to preclude interlocking of this aggregate material in many mixtures of this type.

Stone-filled sheet-asphalt mixtures have been more widely used in the surfacing of city streets than for rural highways. They are frequently laid directly upon the base course to a compacted thickness of 2 inches, although they may be built upon a binder course, as in the case of sheet-asphalt mixtures. They are mixed and placed at elevated temperatures. Construction methods employed in the preparation and placing of mixtures of this sort are the same as those used for the asphaltic concrete mixes previously described. Requirements relative to density and stability may also generally be the same. Since the aggregate is not quite as well graded as that used in dense-graded asphaltic concrete mixtures, the compacted aggregate will have a slightly higher percentage of voids and will require slightly more asphalt cement than comparable coarse dense-graded asphaltic concrete mixtures.

Asphalt cement is again used as the binder, and the mixture may contain coarse aggregate, fine aggregate, and mineral filler as before. Requirements relative to the quality of the aggregate materials are generally the same as previously given. Typical requirements relative to the composition of the mixture are as given in Table 17–9 (*122*).

TABLE 17–9

COMPOSITION OF THE MIXTURE, STONE-FILLED SHEET ASPHALT

Mix No.	VIa	VIb
Recommended use	Surface	Surface, leveling
Compacted depth for individual courses	1–2 inches	1–2 inches
Sieve Sizes	Per Cent	Passing (by Weight)
¾-inch	–	100
½-inch	100	–
⅜-inch	85–100	85–100
No. 4	–	–
No. 8	65–80	65–80
No. 16	50–70	47–68
No. 30	35–60	30–55
No. 50	25–48	20–40
No. 100	15–30	10–25
No. 200	6–12	3–8

Note 1: Normal asphalt content is from 4.5 to 8.5 per cent by weight of total mix. Upper limit may be raised when using absorptive aggregate.

Sheet Asphalt

Sheet asphalt is a type of paving mixture which is composed of carefully graded fine aggregate (sand), mineral filler, and asphalt cement. It is used almost exclusively as a surface mixture, being laid upon a binder course which is generally from 1 to 1.5 inches in thickness, or directly upon a prepared base course, such as an old concrete or brick pavement. The sheet-asphalt pavement itself may be from ½ to 3 inches in thickness, although the majority of these surfaces are from 1 to 1.5 inches thick. A binder course used in conjunction with a sheet-asphalt surface is generally composed of a mixture which closely resembles an open-graded asphaltic concrete mixture, as no mineral filler is generally used in the binder course.

At the present time, sheet asphalt is principally used in the surfacing of city streets, and many of the principal streets in the larger cities of the United States are paved with this material. Properly constructed, it is capable of withstanding the effects of very heavy wheel loads and very heavy volumes of traffic without damage and with very low maintenance costs. Surfaces of this type are frequently preferred in cities because of their added virtue of being extremely smooth, noiseless riding surfaces which are waterproof and easily cleaned. Sheet-asphalt surfaces are somewhat more difficult to design, and their construction is slightly more difficult to control than that of asphaltic concrete mixtures of comparable stability and durability. One of the principal faults of sheet asphalt is its susceptibility to contraction cracking in cold weather.

TABLE 17–10

Composition Requirements for Sheet Asphalt

Sieve Designation		Per Cent (by Weight)	
Passing	Retained On	Binder Course	Surface Course
1¼-inch	–	100	–
¾-inch	½-inch	20–40	–
½-inch	No. 10	25–45	–
No. 4	–	–	100
No. 10	–	15–30	–
No. 10	No. 40	–	10–40
No. 40	No. 80	–	20–48
No. 80	No. 200	–	12–36
No. 200	–	–	10–14
Bitumen	–	4–7	9.0–11.5

Note: Bituminous material used in these mixtures is either a petroleum asphalt or a blend of petroleum asphalt and native asphalt, penetration grade 50–60.

Sheet-asphalt mixtures are distinguished from the asphaltic concrete mixtures which have been discussed by the complete lack of coarse aggregate and very close control of the gradation of the sands used. More mineral filler is used in these mixes than in somewhat comparable asphaltic concretes, and the percentage of mineral filler frequently approaches 16 in a sheet-asphalt mix. The asphalt cement which is used may be slightly harder than that used in asphaltic concrete, asphalt cements having penetration ranges of 50 to 60 and 60 to 70 having been quite commonly used. Asphalt contents are also generally somewhat higher than for dense-graded asphaltic concrete mixtures. Construction procedures relative to the preparation and placing of sheet-asphalt mixtures are practically the same as those which have been described for asphaltic concrete pavements. Somewhat higher temperatures are sometimes used in heating the sand and asphalt cement which are to be used in a sheet-asphalt mix. As a final step, the surface of a sheet-asphalt pavement may be lightly brushed with portland cement or limestone dust immediately after the completion of rolling.

Many different organizations have specifications for sheet-asphalt wearing courses and the accompanying binder courses. Less variation is evident in specifications for this type of mixture than in specifications for some other types of bituminous mixtures, particularly with regard to the surface course. In Table 17–10 are the requirements relative to mix composition contained in the specification for Sheet Asphalt—Federal Classification J contained in the "Standard Specifications" of the State Highway Department of Delaware, dated July 1, 1960.

Sand Asphalt

In some sections of the country, a type of hot plant mixture which is of importance is sand asphalt. "Sand asphalt" may be defined as a surface or base course formed from a mixture of sand and asphalt cement, with or without mineral filler. This type of mixture has been extensively used in areas in which suitable sands are economically available, on highways which carry moderate amounts of traffic. The construction of sand-asphalt pavements is accomplished by methods which closely resemble those used in building sheet-asphalt surfaces, the chief difference between the two being that sand-asphalt mixtures are not as carefully specified or controlled, and consequently are usually somewhat less dense and less stable than comparable sheet asphalts.

The usual thickness of a base course of this type is from 3 inches to 4 inches, while a surface course would generally be about 2 inches in compacted thickness. Surface courses of sand asphalt usually contain from 5 to 10 per cent of asphalt, and base courses may contain slightly less. Requirements relative to the composition of sand-asphalt mixtures vary somewhat with different organizations, as might be expected, since it is usually desired to use local sand in a mixture of this kind. Some specifications are quite open in nature, while others quite closely control the grading of the sand used. A requirement relative to density and stability is also frequently included as a part of a specification for sand asphalt, with requirements in this respect being generally somewhat less severe than for sheet-asphalt or asphaltic concrete mixtures. For example, values specified for the Hubbard-Field stability test results may range from 800 to 1500 pounds at 140°F.

The variation in specifications of this sort makes it undesirable to attempt to include a "typical" specification for this type of mix. One specification has been selected, however, which is fairly representative of those in current use. Table 17–11 gives the requirements relative to the composition of the mixture for Class F—Sand-Asphalt Surface Course (Type–F1), as contained in the "Standard Specifications for Roads and Bridges" of the State Highway Commission of the State of North Carolina.

TABLE 17–11

COMPOSITION REQUIREMENTS FOR SAND ASPHALT

Passing Sieve Designation	Per Cent (by weight)		
	Grading A	Grading B	Grading C
½-inch	100	100	100
No. 4	95–100	95–100	90–100
No. 10	80–100	80–100	50–100
No. 40	50–80	50–95	30–70
No. 80	10–30	5–30	5–30
No. 200	4–8	4–8	4–8
Bitumen	6.5–8.0	6.0–8.5	6.0–8.5

Note 1: These mixtures utilize asphalt cement with penetration from 85 to 100.

Note 2: The amount between any two consecutive sieves except the ½ inch and the No. 4 shall be not less than 4%.

Note 3: The amount of material finer than the No. 200 sieve in the blended aggregate (exclusive of added mineral filler) before drying shall not exceed 8.0% and shall be determined by AASHO Method T-11 using a detergent (sodium hexanetophosate buffer with sodium carbonate).

A somewhat different type of sand-asphalt mix has been used by the Virginia Department of Highways in building skid-resistant pavements. A fine silica sand is used, with all the material passing a No. 10 sieve; it is required to contain at least 95 per cent silica particles. The binder is 85 to 100 penetration asphalt cement to which hydrated lime is added in the mixing process. In resurfacing, the hot mixture is applied at the rate of 15 to 25 pounds per square yard. A similar mixture has been used by the Port of New York Authority in resurfacing bridge pavements.

Hot-Tar Concrete

A type of bituminous concrete which is used in some areas utilizes the heavier grades of road tar as a binder material. This type of mixture is prepared in a central mixing plant and is mixed and laid at elevated temperatures. The temperatures used are considerably less than those normally used in similar asphaltic concretes, with the temperature of the final plant mixture generally being from 150 to 225°F. Road tars of grades RT–9, 10, 11, and 12 are all used in these mixtures, with heavier grades being generally preferred.

Both two-course and one-course construction is used in making tar-concrete pavements. If a binder course is used, its thickness is usually from 1.5 to 3 inches, with the top surface course being from ½ to 1 inch

Fig. 17–20. Placing a tar-rubber paving mixture at Ent Air Force Base, Colorado. (Courtesy Colorado Fuel & Iron Corporation.)

thick. A seal coat is frequently applied to the top course in this type of construction. The placing and finishing of this type of mixture are generally the same as for similar asphaltic concrete mixtures. Fig. 17–20 shows the placing and finishing of a hot tar–rubber paving mixture on a jet aircraft parking apron. If desired, lighter rollers may be used in compacting the hot-tar mixture than are used for similar asphalt mixtures.

An example of this type of mixture is seen in the specification for Hot Bituminous Concrete Pavement of the Bureau of Public Roads (*169*). The specification covers both asphaltic and tar concrete; tar,

TABLE 17–12

REQUIREMENTS FOR COMPOSITION OF HOT BITUMINOUS CONCRETE PAVEMENT (*169*)

Sieve Designation	Per Cent Passing (by Weight)
1-inch	100
¾-inch	70–100
½-inch	55–90
⅜-inch	40–80
No. 4	30–55
No. 10	22–47
No. 20	16–38
No. 40	12–32
No. 80	8–20
No. 200	4–8
Bitumen, per cent total mix (soluble in CS_2)	5.0–8.0

Note 1: For the binder course, percentage of material passing the No. 200 shall be from 0 to 5 per cent, and no mineral filler need be added.

Note 2: The fraction actually retained between any two consecutive sieve sizes shall be not less than 4 per cent of the total and, for the wearing course, at least one-half the fraction passing the No. 200 sieve shall be mineral filler.

when used, must be grade RT–11 or RT–12. Composition of this mixture is shown in Table 17–12.

Typical Cross-Sections

Fig. 17–21 shows a number of typical cross-sections in which the hot-mixed, hot-laid asphaltic surfaces which have been described have been used. They illustrate typical arrangements of wearing surface, binder course, and base which may be used in the construction of high-type bituminous pavements. All thicknesses shown are compacted measurements.

Cold-Mixed, Cold-Laid Pavements

The types of paving mixtures which are described in this section were developed because of a desire to place bituminous concrete in regions which were far removed from central plants or when it was desired to stock-pile the mixtures for future use. To meet this need, several processes were developed by which the paving mixture could be prepared in a central plant, shipped a considerable distance if necessary, and

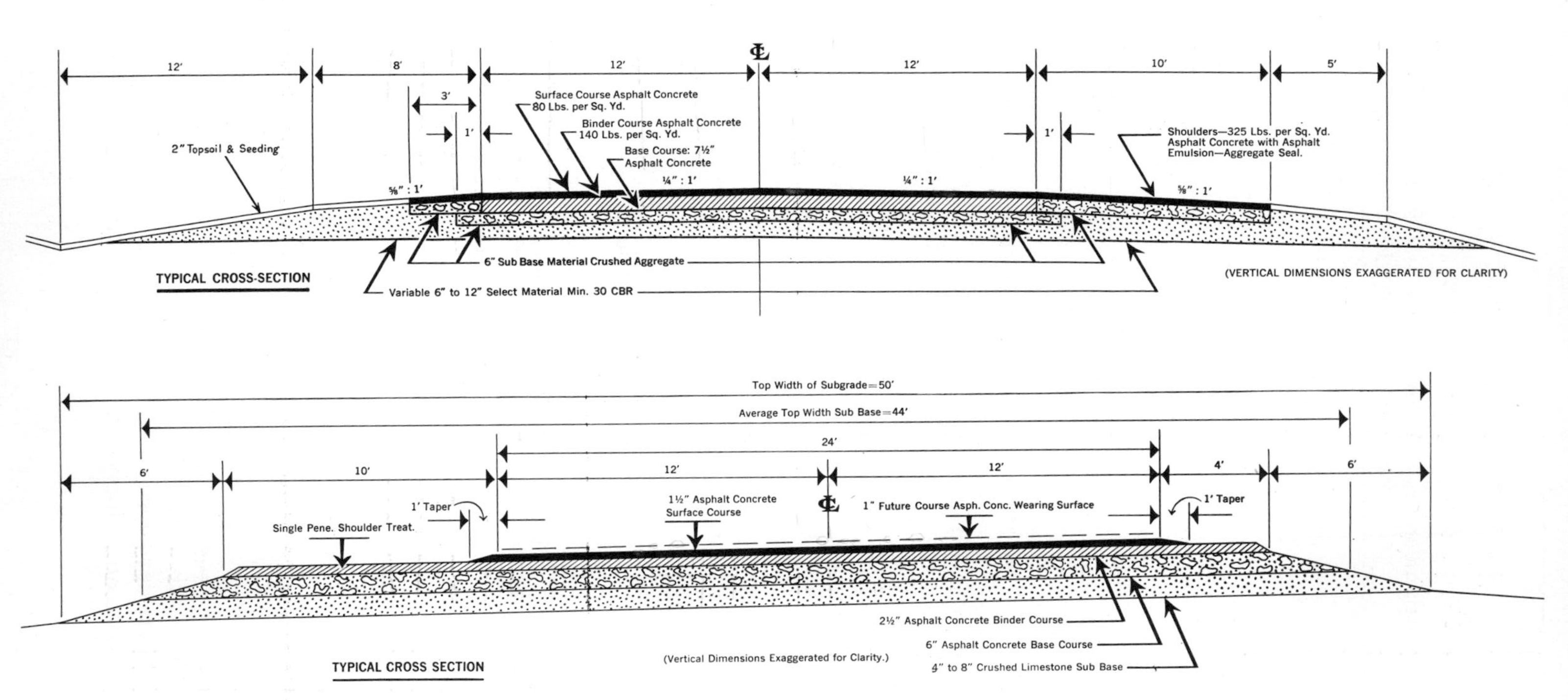
12′
8′
12′
12′
10′
5′
3′
1′
1′
Surface Course Asphalt Concrete 80 Lbs. per Sq. Yd.
Binder Course Asphalt Concrete 140 Lbs. per Sq. Yd.
Base Course: 7½″ Asphalt Concrete
2″ Topsoil & Seeding
⅝″ : 1′
¼″ : 1′
¼″ : 1′
⅝″ : 1′
Shoulders—325 Lbs. per Sq. Yd. Asphalt Concrete with Asphalt Emulsion—Aggregate Seal.
6″ Sub Base Material Crushed Aggregate
Variable 6″ to 12″ Select Material Min. 30 CBR
(VERTICAL DIMENSIONS EXAGGERATED FOR CLARITY)
Top Width of Subgrade=50′
Average Top Width Sub Base=44′
24′
6′
10′
12′
12′
4′
6′
1′ Taper
1′ Taper
1½″ Asphalt Concrete Surface Course
1″ Future Course Asph. Conc. Wearing Surface
Single Pene. Shoulder Treat.
2½″ Asphalt Concrete Binder Course
6″ Asphalt Concrete Base Course
4″ to 8″ Crushed Limestone Sub Base
(Vertical Dimensions Exaggerated for Clarity.)

TYPICAL CROSS-SECTION

TYPICAL CROSS SECTION

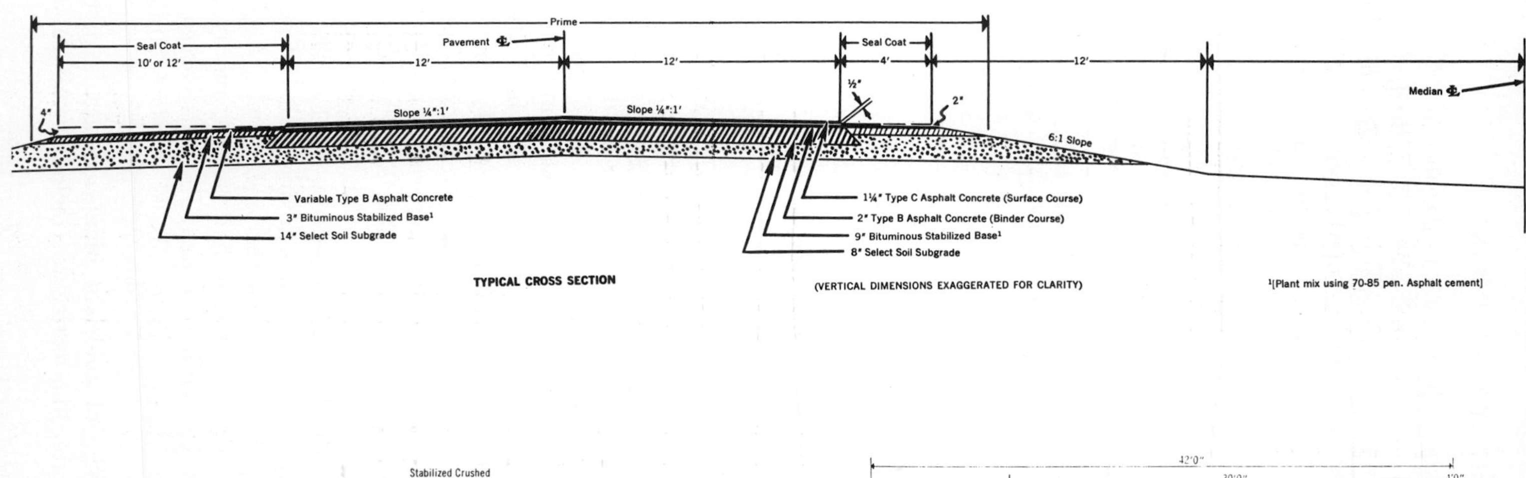

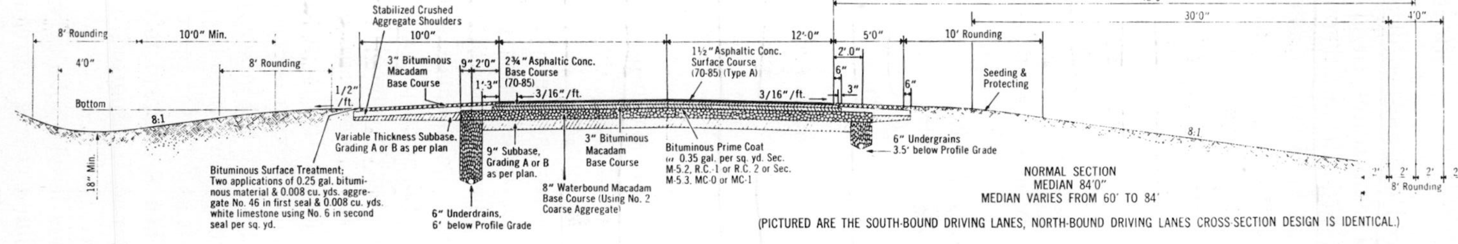

FIG. 17–21. Typical flexible pavement cross-sections. (Courtesy the Asphalt Institute.)

then placed and finished in the same fashion as decribed for the hot-mix types. The mixtures could also be stock-piled, if desired, for future use in resurfacing or other maintenance operations. Most of the processes used in preparing cold-laid mixtures of the type being discussed here were originally patented procedures. However, many of the basic patents have long since expired, and the mixtures described in this section are nonpatented types. The term "cold-mixed" may be somewhat of a misnomer when applied to the mixtures described here, as some heat may be used in their preparation. However, temperatures used are relatively lower than those used in the preparation of the hot-mix types, and they are designed to be placed at normal air temperatures.

17–21. Asphaltic Types. Two principal types of cold-mixed, cold-laid asphaltic concrete pavements will be discussed here. These may be designated as follows:

1. Liquefier type
2. Powdered asphalt type

Both binder and surface courses are manufactured from mixtures of the liquefier type. The materials included in a mix of this type are coarse aggregate, fine aggregate, asphalt cement, liquefier, and hydrated lime. The asphalt cement used is of paving grade, being generally in the 85–100 penetration range. The liquefier used is a petroleum distillate (naphtha), and usually conforms to the requirements for Liquefier, AASHO Designation M83. The aggregates used are quite coarse and the aggregate mixture is "open graded." The percentage of asphalt cement used varies somewhat, being generally from 3.5 to perhaps 6.0 per cent for binder courses, and from 4.5 to 7.0 per cent for surface courses. The amounts of liquefier and hydrated lime which are used are generally quite small, being from 0.5 to 2.0 per cent for the former, and 0.4 to 1.0 per cent for the latter. In the preparation of these mixtures, the asphalt cement is heated as in the preparation of hot mixes, generally to a temperature between 200 and 300°F. The aggregates are also heated to dry them, and their permissible temperature range at the time of mixing is generally from 45 to perhaps 140°F. The materials are fed into the mixer in sequence, with a mixing period of perhaps 20 seconds following each application of another ingredient. A typical sequence is as follows: aggregate, liquefier, asphalt cement, and hydrated lime. When a uniform mix has been produced, it is discharged from the plant and transported to the job site or placed in a stock pile for future use.

The powdered asphalt type of cold-laid plant mixture is prepared by combining coarse aggregate, fine aggregate (including mineral filler, if desired), powdered asphalt, and flux oil. The powdered asphalt is a very hard asphalt cement, usually with a penetration of less than 3 at 77°F, which has been pulverized in an impact mill until 95 per cent of it will pass a No. 20 sieve and not less than 50 per cent will pass the No. 80 sieve. The flux oil used is a viscous material which, for example, is free from water, has a specific gravity which is not less than 0.97, a Furol viscosity of 600 to 800 seconds at 122°F, and a loss on heating (50 grams, 325°F, five hours) of not more than 5 per cent. In the manufacture of this kind of mixture, the dried aggregates are introduced into the mixer, either singly or in combination, and thoroughly manipulated, after which the flux oil is added and these components are mixed long enough for the aggregate particles to be thoroughly coated. The powdered asphalt is then introduced into the mixer, and processing is continued until a uniform mixture has been secured. The mix may be discharged from the mixer at a temperature varying from about 90 to 175°F, depending on the detailed procedure which is being used. Both binder and surface courses are produced in this fashion. This type of mixture has many advantages when used for patching.

17–22. Cold-Laid Tar Concrete. The preparation of the mixtures which are included under this heading closely resembles the method which was described in the previous chapter for intermediate mixtures using tar as a binder. No special treatment is used, with grades RT–8 through RT–11 being used as a binder by different organizations, and the components of the mixture being moderately heated in the preparation of the mix. The gradings used to control the aggregates used in this type of mixture would place them in the bituminous concrete class, and the construction procedure utilized in placing and compacting the mixture is practically the same as that for other bituminous concrete pavements.

Patented and Miscellaneous Mixtures

As has been previously indicated, patented mixtures were once widely used in the construction of high-type bituminous pavements. Basic patents protecting many of these mixtures have expired, and the large majority of high-type pavements are now built through the use of non-patented mixtures. Patented mixtures which were formerly used, and on which the patents have expired, include those known as Warrenite, Amiesite, and Macasphalt pavements. Three modern types of patented mixtures which are in use in some areas are Colprovia, Laykold, and asphaltic concrete manufactured by the steam dispersion process. Another patented cold mixture makes use of Gilsapave, a mineral combination incorporating a native asphalt, Gilsonite, which is found only in the Uintah basin of Utah.

A number of experimental sections have been constructed in this country in which a small amount of powdered natural rubber was added to the asphalt normally used as a binder material. Advantages claimed for these "rubber-asphalt" mixtures include reduced maintenance costs, longer life, and excellent nonskid characteristics.

Hot asphaltic mixtures that contain a small amount of asbestos fiber are being used for thin overlays of existing surfaces, bridge deck surfaces, and curbs. Advantages claimed for the asbestos mixtures are greater toughness than conventional mixtures, higher asphalt contents, and higher laying temperatures.

PROBLEMS

17–1. Given below are the requirements of a specification relative to the grading of the mineral aggregates in an asphaltic concrete mixture, and the sieve analyses of two aggregates (A and B) which are economically available for this use. Determine the range of blends of aggregates A and B which will produce a combined aggregate which will meet the limits of the specification, and give the gradings of the aggregate combinations selected.

Sieve Designation	Per Cent Passing (by Weight)		
	Mix	Aggregate A	Aggregate B
1-inch	100	100	–
¾-inch	95–100	95	100
⅜-inch	60–75	72	76
No. 4	40–55	54	52
No. 8	30–40	43	38
No. 30	12–22	32	16
No. 200	3–6	8	5

17–2. A design is being prepared for an asphaltic concrete paving mixture. The following ingredients are to be used in the preparation of a trial mixture.

Coarse aggregate	55%	Specific gravity = 2.60
Fine aggregate	31%	Specific gravity = 2.69
Mineral filler	7%	Specific gravity = 3.10
Asphalt cement	7%	Specific gravity = 1.03

(a) Calculate the theoretical maximum density for this mixture.

(b) A laboratory-compacted specimen of this mixture has a weight in air of 1207.4 grams, while the weight of the same sample suspended in water is 670.3 grams. Calculate the per cent voids in the sample.

17–3. For the mixture given in Problem 17–2, compute the per cent voids in the compacted mineral aggregate (V.M.A.%) and the per cent voids filled with asphalt. Estimate the reduction in voids in the total mixture which would accompany a change in asphalt content from 7 to 6 per cent.

17–4. A sample cut from the completed pavement constructed by using the mix of Problem 17–2 weighs 3600 grams in air and 1900 grams in water. Compare the density obtained in the field with that of a laboratory-compacted specimen of the same mixture. Has the rolling process achieved a satisfactory degree of density in this case?

17–5. Given the following results of tests conducted in the laboratory in order to select the optimum asphalt content for an asphaltic concrete mixture:

Asphalt Content (per cent)	Hubbard-Field Stability 140 F (lb)	Voids, Compacted Mixture (per cent)
4.5	1750	8.5
5.0	1990	6.4
5.5	2260	4.4
6.0	2375	3.7
6.5	2230	3.2
7.0	2020	2.6
7.5	1800	1.9

Select the optimum asphalt content for this mixture, assuming that the applicable design criteria include a minimum of 2000 pounds for the Hubbard-Field stability value at 140°F, and air voids between 2 and 5 per cent.

17–6. An asphaltic concrete surface course mixture is being designed by the Marshall method. The aggregate combination in a trial mixture contains 40 per cent coarse aggregate, 52 per cent fine aggregate, and 8 per cent mineral filler; specific gravities of these materials are 2.68, 2.72, and 2.83, respectively. Asphalt cement to be used has a specific gravity of 1.02.

Results of tests on specimens prepared from this aggregate combination and various percentages of asphalt (by weight of total mixture) are as follows:

Per Cent Asphalt	Specimen Weight, Grams		Stability (pounds)	Flow (1/100 inch)
	In Air	In Water		
4.5	1308.7	765.0	1722	8
5.0	1314.2	772.3	1765	9
5.5	1318.9	779.7	1788	11
6.0	1323.3	782.1	1632	14
6.5	1328.6	782.9	1406	18

The mixture is to be designed for very heavy traffic. Determine the optimum asphalt content for this mixture. Is the mixture satisfactory?

17–7. A paving mixture is to be designed by the Hveem method. Gradation of the aggregate is as follows:

Sieve	Per Cent Passing (by Weight)
¾-inch	100
½-inch	94
⅜-inch	78
No. 4	62
No. 8	45
No. 16	34
No. 30	25
No. 50	18
No. 100	14
No. 200	8
No. 270	5

The Centrifuge Kerosene Equivalent for the aggregate is 3.8. Specific gravity of the coarse aggregate is 2.62; fine aggregate, 2.72. The per cent oil retained by the coarse aggregate is 2.5 per cent.

Estimate the quantity of asphalt required for this mixture, assuming that the binder is to be (a) a liquid asphalt, grade MC–800; (b) an asphalt cement, penetration grade 85–100.

Suppose the stabilometer value of the mixture of (b), using the computed quantity of asphalt, is 42; is the mixture satisfactory for heavy traffic? What other requirements pertain to the mixture?

17–8. Visit a plant in your area which is producing hot-mix asphaltic concrete or a similar high-type mixture. Prepare a report describing the controls which are applied to insure production of a mixture of uniformly high quality. Include such factors as aggregate gradings and proportions, quantity of bituminous material, temperatures, mixing times, and so on.

17–9. Describe in detail the laydown procedure used for a surface of asphaltic concrete, or a similar type of mixture, in your area. Give facts relative to temperature of the mix, spreader, rolling equipment and procedure, variations in the surface of the compacted layer, density achieved by rolling, and so on.

CHAPTER 18

DESIGN OF CONCRETE PAVEMENTS

The many and varied uses of portland cement concrete in the United States have made this material such an integral part of our everyday lives that it hardly seems necessary to define it. However, "portland cement concrete" may be defined as a plastic and workable mixture composed of mineral aggregate such as sand, gravel, crushed stone, or slag, interspersed in a binding medium of cement and water. When first combined, the materials listed form a plastic, workable mass which may be easily handled and shaped into any desired form. A short time after mixing, the concrete begins to stiffen or "set" because of chemical action between the cement and water in the mixture, and in a relatively short time forms a dense, hard mass which possesses considerable compressive and flexural strength.

The first use of portland cement concrete for paving in the United States occurred in 1891 when a strip 10 feet wide and 220 feet long was placed in the city of Bellefontaine, Ohio. During the period from 1891 to 1909, additional streets were paved with portland cement concrete in various cities in the United States and Canada. In 1909 the first sections of rural road were surfaced with this material, with surfaces of this type being constructed in Los Angeles County, California, and Wayne County, Michigan. Since that time the use of concrete in the paving of roads and streets has increased at a rapid rate. The Portland Cement Association had tabulated a total of 200,000 miles of rural roads and some 100,000 miles of city streets (and alleys) which had been completed in the United States by 1965. These figures are expressed in terms of miles of pavement 20 feet wide, having been converted to this basis from square yards of pavement. Information released by this organization shows that this type of surface had been built in every state in the Union and the District of Columbia.

When properly designed and constructed, concrete roads and streets are capable of carrying almost unlimited amounts of any type of traffic with ease, comfort, and safety. Surfaces of this type are smooth, dust-free, and skid-resistant, having a high degree of visibility for both day and night driving and generally having low maintenance costs. They are economical in many locations because of their low cost of maintenance and their relative permanency. They are, of course, classed as high-type pavements. The principal use of surfaces of this type has been in the construction of heavily traveled roads and city streets, including those in residential, business, and industrial areas. It is the standard material for urban expressways, even in states where asphalt surfaces are widely used. A wearing surface of portland cement concrete usually consists of a single layer of uniform cross-section which has a thickness of from 6 to 11 inches and which may not require a separate base course, often being constructed directly upon a prepared subgrade or subbase. A new concrete base course may be constructed to serve as a support for a brick or block pavement or one of the several types of bituminous wearing surfaces. Old concrete pavements have been extensively used as bases for new bituminous wearing surfaces in many areas.

A further distinction may be noted between the surfaces and bases which are discussed in this chapter and the so-called flexible pavements which have previously been described in this book. Concrete surfaces and bases are frequently classed as "rigid" pavements, the term "rigid" implying that pavements constructed of this material possess a certain degree of "beam strength" which permits them to span or "bridge over" some minor irregularities in the subgrade or subbase upon which they rest. Thus minor defects or irregularities in the supporting foundation layer may not be reflected in the surface course, although, of course, defects of this type are certainly not desirable, as they may lead to failure of the pavement through cracking, breaking, or similar distress.

The design and construction of concrete pavements is a fairly complex subject and it seems desirable to list the major topics which will be covered in this and the following chapter. Major subjects which will be discussed in subsequent sections include materials, proportioning of concrete mixtures, and the structural design of concrete pavements. The construction of portland cement concrete pavements is covered in detail in the next chapter.

In the interests of simplification, in the remainder of this text the term "concrete" will be taken to be synonymous with "portland cement concrete."

Materials

The materials included in concrete, as generally used in highway construction, are coarse aggregate, fine aggregate, water, cement, and one or more admixtures. Each of these materials will be discussed in turn in the following sections, along with a discussion of typical requirements of specifications relative to each.

18–1. Coarse Aggregate. Coarse aggregates most frequently used in portland cement concrete include crushed stone, gravel, and blast-furnace slag. Other similar inert materials may also be used, and the listed materials may be used singly or in combination with one another. Specific requirements relative to coarse aggregates to be used for this purpose may be divided into five groups, as follows: deleterious substances, percentage of wear, soundness, weight per cubic foot (slag), and grading.

Cleanliness of the aggregate is generally guaranteed by the inclusion in the specifications of requirements relative to the maximum percentages of various deleterious substances which are permitted to be present. Specific requirements in this respect vary somewhat with different agencies. The requirements in Table 18–1 relative to deleterious substances are indicative of those in current use (*135*).

TABLE 18–1

LIMITS FOR DELETERIOUS SUBSTANCES IN COARSE AGGREGATE FOR CONCRETE (ASTM C33)

Item	Maximum, Per Cent by Weight of Total Sample
Clay lumps	0.25
Soft particles	5.0
Chert that will readily disintegrate (soundness test, five cycles, or freezing-thawing test, 50 cycles), or that has a specific gravity less than 2.35	1.0 (severe exposure)
	5.0 (mild exposure)
Material finer than No. 200 sieve	1.0[a]
Coal and lignite	0.5

[a] In the case of crushed aggregates, if the material finer than the No. 200 sieve consists of the dust of fracture essentially free from clay or shale, this percentage may be increased to 1.5.

In certain areas, considerable difficulty has been experienced with aggregates which contain deleterious substances which react harmfully with the alkalies present in the cement. Detrimental alkali-aggregate reactions generally result in abnormal expansion of the concrete. Methods have been devised (ASTM Methods C227 and C289) for detecting aggregates with these harmful characteristics, and suitable stipulations are included in typical specifications (for example, ASTM C33).

The ability of the coarse aggregate to resist abrasion is generally controlled by the inclusion in the specifications of a maximum permissible percentage of wear in the Los Angeles Abrasion Test. Maximum permissible percentages of wear in this test (AASHO Designation T96) range from as low as 30 to as high as 65 in the specifications of the various state highway departments. On the average, however, the maximum permissible loss varies from 40 to 50 per cent for all three of the principal types of coarse aggregate, although many specifications do not contain a requirement relative to the percentage of wear of slag. Several state highway departments still use the Deval test for this purpose.

A number of states, particularly those in the northern portion of the country, include requirements relative to soundness, as measured by the use of sodium (or magnesium) sulfate, in their specifications

TABLE 18–2

GRADINGS FOR COARSE AGGREGATES IN PORTLAND CEMENT CONCRETE (ASTM C33)

	Per Cent Passing (by Weight)		
	Aggregate Designation		
Sieve Designation	2-in. to No. 4 (357)	1½-in. to No. 4 (467)	1-in. to No. 4 (57)
2½-inch	100	–	–
2-inch	95–100	100	–
1½-inch	–	95–100	100
1-inch	35–70	–	95–100
¾-inch	–	35–70	–
½-inch	10–30	–	25–60
⅜-inch	–	10–30	–
No. 4	0–5	0–5	0–10
No. 8	–	–	0–5

for coarse aggregate. The soundness test used is an accelerated weathering test. The weighted loss in the sodium sulfate soundness test (AASHO Designation T104) may generally be said to have a maximum permissible value ranging from 10 to 15 per cent, as measured at the end of five cycles of alternate wetting and drying. A small number of states use a freezing and thawing test for the same purpose. In states in which slag is used as a coarse aggregate, specifications generally require that the unit weight of this material be not less than 70 (or 75) pounds per cubic foot.

As would be expected, requirements relative to the grading of the coarse aggregates used in portland cement concrete vary considerably with different organizations. The material is generally required to be well graded from coarse to fine, with the maximum size of coarse aggregate permitted by most highway organizations being from 3 to 2 inches. In this work the division between coarse and fine aggregate is made on the No. 4 sieve. That is, material which is retained on a No. 4 sieve is classed as coarse aggregate. Grading requirements are generally quite open, permitting the use of a large number of locally available aggregates which have proved to be satisfactory for use in portland cement concrete.

Three more or less typical gradings for coarse aggregate are given in Table 18–2. These gradings are a part of ASTM C33 (*135*).

Gradings which are the same as, or similar to, those given above are currently in use by a number of state highway departments. A number of departments require coarse aggregates which are to be used in concrete paving operations to be shipped in separate fractions, with the division frequently being made on the 1-inch sieve.

18–2. Fine Aggregate. The fine aggregate which is most generally used in concrete is sand (either natural or manufactured) composed of tough, hard, and durable grains.

The cleanliness of the sand used may be controlled by the inclusion of requirements relative to the maximum amounts of various kinds of deleterious substances which may be present, in a fashion similar to that indicated above for coarse aggregate. Particular emphasis is frequently placed upon the maximum per cent of "silt" (material passing the No. 200 sieve) which may be contained in the fine aggregate. This amount must generally not exceed from 2 to 5 per cent of the total. An additional cleanliness requirement is generally imposed upon sands, in that excess amounts of organic impurities must not be present. The amount of organic material is generally controlled by subjecting the material to the colorimetric test for organic impurities (AASHO Designation T21).

Further requirements of specifications relative to sands may include stipulations as to soundness. The soundness of these materials is controlled by specifying the maximum loss which will be permitted in five cycles of alternate wetting and drying in the sodium sulfate soundness test. A typical requirement is that the loss in the sodium sulfate soundness test may not exceed 10 per cent. Modern specifications also frequently include stipulations intended to control the potential alkali reactivity of fine aggregates.

The grading requirements of the various highway agencies relative to fine aggregate are somewhat more uniform than similar requirements relative to coarse aggregate. Many of the state highway departments utilize grading requirements which are the same as, or very similar to, those given in the "Standard Specifications for Fine Aggregate for Portland Cement Concrete," AASHO Designation M6 (see Table 18-3).

Some specifications contain additional requirements relating to the portions of the fine aggregate which pass a No. 8 and a No. 30 sieve.

TABLE 18-3

GRADING OF FINE AGGREGATE

(AASHO Designation M6-48)

Sieve Designation	Per Cent Passing (by Weight)
3/8-inch	100
No. 4	95-100
No. 16	45-80
No. 50	10-30
No. 100	2-10

Fine aggregate which is obtained from a single source for use on a specific project is also frequently subjected to uniformity control by specifications relative to the variation in fineness modulus which will be permitted. The "fineness modulus" may be defined as the quantity which is secured by adding together the percentages by weight retained on the following sieves: 3-inch, 1½-inch, ¾-inch, 3/8-inch, No. 4, No. 8, No. 16, No. 30, No. 50, and No. 100, and dividing this total by 100. Specifications may require that the fineness modulus of the fine aggregate actually used on the job shall not vary more than 0.2 from the value obtained for an average sample of the material selected for use on the project.

18-3. Water. Almost any normal source of water may be used in the manufacture of portland cement concrete. Specifications for this material frequently require simply that water to be used in the mixture is to be suitable for drinking, although other water may be used in some instances if it has been demonstrated by laboratory test or field experience that it is suitable for this purpose. Water which is used must be free from an excess of alkalies, acids, oil, or organic matter. Detergents in water will cause high entrained air content. Sea water has been successfully used in some concrete mixtures, although it is not generally used in paving operations.

18-4. Cement. Portland cement may be defined in a number of different ways, with possible descriptions varying from involved technical descriptions relative to the chemical make-up of the material and its physical properties, to simple statements regarding its physical appearance and behavior when combined with water. In a more or less nontechnical sense, portland cement may be defined as a chemical combination of lime, silica, alumina, and iron oxide, together with small amounts of other materials, and gypsum, which is added in the final grinding process to control the setting time of the cement. The material is very finely ground, generally being of such a size that at least 90 per cent of it will pass a No. 200 sieve. Raw materials commonly used in the manufacture of portland cement include calcareous substances such as limestone, oyster shells, and marl, which are used in combination with argillaceous materials like clay, slate, shale, or blast-furnace slag. The manufacture of portland cement from these (or other) raw materials involves a series of complex operations in which the various processes and the proportions of the material used are carefully controlled in order to produce a material which will meet the rather exacting requirements of the specifications of the various agencies which use this material. In a typical process, the raw materials are successively crushed, mixed, and calcined to form clinker, which is then ground and combined with a small amount of gypsum to form cement. For air-entraining cements, the necessary material to impart air-entraining properties is added at the time the clinker is ground.

The specifications for cement which are most commonly used by highway agencies are those of the American Society for Testing Materials and the American Association of State Highway Officials. The specifications of these organizations cover a number of different types of cement, not all of which are of importance in highway work. The appropriate specifications of each of these organizations are indicated in the following tabulation, together with a listing of the types of cement which are most frequently used in highway construction:

1. Standard specifications for portland cement
   ASTM Designation C150
   AASHO Designation M85
   Types I, II, and III
2. Standard specifications for air-entraining portland cement
   ASTM Designation C175
   AASHO Designation M134
   Types IA, IIA, and IIIA

Detailed requirements of these specifications and detailed methods of test involved in their examination will not be discussed here. For this information the interested student is referred to the appropriate publications of the organizations listed. A general description of each of the listed types is, however, given below.

Type I is what might be termed standard or "normal" portland cement and is intended for use in general concrete construction where the cement is not required to have certain special properties. Type I cement is normally supplied by the manufacturer, unless one of the other types is specified.

Type II cement is also regarded as a standard type of portland cement and is used for general concrete construction. It is specifically recommended for use in situations in which the concrete will be exposed to moderate sulfate action or where a moderate heat of hydration is required. Both Types I and II are widely used by highway agencies at the present time in paving and structural work.

Type III cement is "high early strength" cement. It differs from the standard types described in that concrete made from it attains, in a much shorter period of time, compressive and flexural strengths which are comparable to those attained by concrete in which the same amount of one of the standard types is used. It is useful, for example, in paving operations in which the completed surface must be opened to traffic very soon after construction is complete.

Specifications for air-entraining cements cover three types which are comparable to Types I, II, and III, above, except that they contain air-entraining agents which are interground with the cement at the mill. The properties of air-entrained concretes made by using cements of these types will be discussed more fully in a later section, but the use of such a cement may be said to result in the trapping of a small percentage of air which is present in the form of small, disconnected bubbles distributed uniformly throughout the mix. Air-entrained concretes are, among other things, more durable than comparable concrete mixtures which do not contain air. A number of state highway departments permit (or require) the use of air-entraining cement, while others secure air-entrainment by the use of an admixture which is incorporated into the concrete during the mixing operation.

In addition to the types described, portland blast-furnace slag and portland-pozzolan cements are now available in some areas. Portland blast-furnace slag cement is covered by ASTM C205; there are two types, IS and IS-A (air entraining). In these cements, granulated blast-furnace slag of selected quality is interground with portland cement clinker. The ASTM specifications require the slag constituent to be from 25 to 65 per cent of the total weight of the cement. Cements designated as portland-pozzolan cement are covered by ASTM C340; the two types are designated IP and IP-A (air entraining). In these cements pozzolans, consisting of siliceous, or siliceous and aluminous material, are blended with ground portland cement clinker. Portland-pozzolan cement is normally used as a blend with one of the other three types of portland cement, as are certain types of natural cement (ASTM C10) and slag cement (ASTM C358).

In 1963, Connecticut's highway department built an experimental section of "self-stressing" concrete pavement in which an expansive cement, a mixture of portland cement and bauxite, was used. Expansion of the cement placed the pavement's reinforcing steel in tension, thus producing a prestress in the concrete. Results were inconclusive. In 1964, Wisconsin built a 1600-foot-long section of reinforced two-lane pavement with expansive cement; joints were spaced 157.5 and 300 feet apart.

Common units of measurement of cement quantity are the sack, which weighs 94 pounds and has an approximate loose volume of one cubic foot, and the barrel, which is equivalent to four sacks or 376 pounds of cement. Shipment of cement is commonly made in sacks or bags or in bulk quantities.

18–5. Admixtures. By an "admixture" is meant, in this case, any substance other than aggregate, water, or portland cement which may be added to a concrete mixture. A large number of admixtures may be used in conjunction with the standard ingredients of portland cement concrete for various purposes and in various ways. Only a few of these admixtures are of importance in highway construction, and certain of these will be discussed in the following paragraphs.

At the present time the most important admixtures which may be added to concretes used in highway construction are those used to produce air-entrained concrete. Numerous materials can be used as air-entraining agents, including natural wood resins, fats, various sulphonated hydrocarbons, and oils. Some of these materials are insoluble in water and must be saponified before they can be used as admixtures. ASTM C233 gives a program of testing for the evaluation of materials proposed for use as air-entraining agents, while ASTM C260 is a tentative specification for these materials.

Another admixture which is of some consequence is calcium chloride, which is generally used as an accelerating agent. That is, calcium chloride materially decreases the time of hardening of the cement and thus leads to increased early strength of the concrete mixture. The amount of calcium chloride used need not be large, being generally less than 2 per cent by weight of standard cement. Other advantages claimed for the use of small amounts of calcium chloride include improved curing of the concrete by reducing the loss of moisture during the early period of hydration, slight increases in the ultimate strength of the concrete, improved workability of the plastic mixture, increased resistance to abrasion, and decreased shrinkage. Caution should be exercised to avoid excessive use of calcium chloride, which could reduce the durability of concrete. The same properties obtained by the use of calcium chloride usually can be obtained by using Type III cement or by increasing the cement factor.

Powdered materials such as diatomaceous earth, pumice, fly ash, and hydrated lime are occasionally used as admixtures in concrete, primarily as workability agents. Research done in several states on the use of flyash in concrete was reported by the Highway Research Board in 1965 (*74*); in general, researchers reported such advantages as better protection against alkali-aggregate reaction and increased workability and strength from the use of this material. Widespread use is being made of various admixtures in casting concrete bridge decks to control the setting time, thus controlling dead load deflections and helping to ensure composite action between the deck and the underlying steel girders. Silicones have been used experimentally to increase the freeze-thaw resistance of concrete (*30*).

Admixtures may also be used to produce colored concrete. For example, red iron oxide has been used in concrete in some states for the purpose of marking traffic lanes. Similarly, the center strip in a concrete pavement is sometimes formed from concrete into which black iron oxide has been mixed.

Proportioning of Concrete Mixtures

As has been indicated, a concrete mixture contains four basic ingredients: coarse aggregate, fine aggregate, cement, and water, plus any desired admixture. Before concreting operations may begin on any given project, the proportions of these materials must be established. The procedure of determining the amount of each of these materials which must be used to produce a concrete mixture of the desired characteristics is known as the "design of the mix." In the brief discussion which follows, no admixtures will be considered, and the design of air-entrained mixtures will be given mention in a later portion of this chapter. The objective of any procedure which may be utilized for the design of a given mixture is to determine the most practical and economical combination of materials which will produce a workable concrete which will have the desired properties of durability and strength after it has hardened.

Certain basic factors govern the design of satisfactory concrete mixtures, regardless of the detailed method which may be used. Detailed procedures which may be used may be quite complex and lengthy, involving considerable laboratory and field testing, or they may be very simple in nature, depending largely upon the size of the project, the desired qualities of the hardened mixture, and the degree of experience which has been attained by the individual designer or organization in the use of materials suitable for concrete in a given area. The procedure outlined here is largely based upon that contained in a standard of the American Concrete Institute (*200*). It involves the selection of a trial mixture to be used as a starting point for field concreting operations or for a detailed laboratory study. The procedure presented here is not necessarily one which is typical of the present practice of a majority of highway agencies, but it is given here because of its exposition of the fundamental factors involved and because of its universal applicability. A procedure of this type may be regarded as being especially desirable when concrete of a given flexural strength is desired.

Concrete should be placed with the minimum quantity of mixing water consistent with proper handling. Proportions should be selected to produce concrete:

1. Of the stiffest consistency (lowest slump) which can be placed efficiently to provide a homogeneous mass.
2. With the maximum size of aggregate economically available and consistent with adequate placement.
3. Of adequate durability to withstand satisfactorily the weather and other destructive agencies to which it may be exposed.
4. Of the strength required to withstand the loads to be imposed without danger of failure.

A basic step to be taken in the design of a trial mixture is to select the water–cement ratio which is consistent with the desired durability and compressive strength desired in the hardened mixture. The water–cement ratio is a quantity of fundamental importance in the design of concrete mixtures and is defined as the ratio of the volume of water to the volume of a sack of cement. It is normally expressed as the number of gallons of water per sack of cement. For example, the water–cement ratio selected may be 6 gallons per sack of cement. It may also be expressed in terms of weight. A water–cement ratio of 6 gallons per sack of cement is equivalent to a water–cement ratio of 0.53 by weight, calculated as follows: $w/c = 6$ gal/sack $= 6(8.34)/94 = 50/94 = 0.53$. Generally speaking, both the durability and compressive strength of a given concrete mixture increase with a *decrease* in the water–cement ratio; that is, the lower the water–cement ratio, the higher will be the compressive strength and the greater the durability of a given concrete mixture, all other things being the same and assuming that a workable, homogeneous mixture is secured. Maximum values of the water–cement ratio to be used in designing concrete using average materials for a desired compressive strength and to resist known conditions of exposure are well established. State highway department specifications relative to concrete to be used in paving establish maximum permissible water–cement ratios varying from 5.0

to 6.5 gallons of water per sack of cement, depending upon the experience of the organization concerned. The average value used is probably between 5.5 and 6.0 gallons per sack. They also require a minimum of 1.25 to 1.50 barrels of cement per cubic yard, feeling that the use of less cement than this will result in surface abrasion. The relationship between water–cement ratio and flexural strength is not so clearly established and should be determined by laboratory or field testing, although the flexural strength of a given mixture may generally be expected to be from 15 to 20 per cent of the compressive strength for comparable conditions of curing and age. Compressive strengths of from 3000 to 3500 pounds per square inch (28-day curing) are commonly specified in highway work, while a typical required flexural strength is 650 pounds per square inch (28-day curing, 3rd point loading). Lower minimum flexural strengths may be specified for opening to traffic.

Another step required in the design of the mix, as outlined in the publication cited above, is to select the desired consistency of the concrete to permit proper handling and placing of the mixture under the anticipated job conditions. The consistency of the fresh concrete mixture may be measured by the slump test (AASHO Designation T119). The slump is measured by placing the fresh concrete mixture into a galvanized metal mold formed into the shape of the frustum of a cone, which is 8 inches in diameter at the base, 4 inches in diameter at the top, and 12 inches high. The concrete is placed in the mold in three layers, each of which is rodded by 25 strokes of a ⅝-inch diameter rod 24 inches long. The top layer is struck off even with the top of the mold and the mold raised vertically away from the concrete. The concrete mass subsides, and its height after subsidence is measured immediately. The difference between this height and the original height of the mass (12 inches) is the "slump." The slump test is illustrated in Fig. 18–1. The desired slump for concrete to be used in pavements is from 1 to 3 inches. Consistency may also be measured in the field by observing the penetration of a 6-inch, 30-pound metal ball into the fresh concrete (ASTM C360).

Another step which must be taken in the general design procedure which is being explained is the selection of the maximum size of aggregate which may be used. The maximum size of aggregate used in paving generally varies from 2.5 to 1.5 inches in slabs which are of normal thickness and which contain relatively small amounts of reinforcing. The maximum size of aggregate may be considerably less when the slab is to be heavily reinforced. From this step onward it will be assumed that the aggregates which are economically available for use in the mixture have been selected and found to meet the requirements which have previously been stated as applicable to these materials. No further consideration will be given in this portion of the discussion to variations in the aggregates which are to be used.

Additional variables which must be established in the design of a trial concrete mixture are the unit water content of the mixture and

FIG. 18-1. Slump test. (Courtesy Soiltest, Inc.)

the proportion of coarse aggregate to be used in the aggregate combination. Various methods have been presented for determining these quantities, and detailed design procedures now in use by the various organizations show considerable deviation at this point in the design process. The general method being discussed defines the unit water content as the amount of water required per cubic yard of concrete.

The quantity of water per unit volume of concrete necessary to produce a mixture of the desired consistency is influenced by the maximum size, particle shape, and grading of the aggregate, and by the amount of entrained air. It is relatively unaffected by the quantity of cement.

The approximate quantities of mixing water for different slumps and maximum sizes of aggregates are as follows:

Max. Size of Aggregate (inches)	Water, gal. per cu. yd.			
	1	1½	2	3
Slump	Non-Air-Entrained Concrete			
1–2	36	33	31	29
3–4	39	36	34	32
	Air-Entrained Concrete			
1–2	31	29	27	25
3–4	34	32	30	28

These quantities may be used with sufficient accuracy for preliminary estimates of proportions. They are maximum quantities which should be expected for fairly well-shaped, angular aggregates graded within the usual limits. If aggregates which are otherwise suitable lead to higher water requirements than given in the tabulation, the cement content should be increased to maintain the desired water-cement ratio.

The minimum amount of mixing water and maximum strength will result for given aggregates when the largest quantity of coarse aggregate is used consistent with desired workability. Precise quantity of coarse aggregate for a given mixture is best determined by laboratory investigation, with later adjustment in the field. However, in the absence of laboratory data, an estimate of the proper proportions can be made for aggregates of conventional grading from empirical relationships shown in Table 18–4 (*200*).

Volumes are based on aggregates in dry-rodded condition as described in Method of Test for Unit Weight of Aggregate (ASTM Designation C29).

These volumes are selected from empirical relationships to produce concrete with a degree of workability suitable for usual reinforced construction. For less workable concrete, such as required for concrete pavement construction, they may be increased about 10 per cent.

The information presented has been taken from the publication of the American Concrete Institute which has previously been cited. Additional information of this type is contained in that publication, and expanded tables of this sort are also to be found in the "Design and Control of Concrete Mixtures," tenth edition, published by the Portland Cement Association.

No attempt will be made here to illustrate the specific use of the fundamental data which have been presented, because of a lack of space. Suffice it to say that various trial mixtures may be designed and examined in the laboratory, following the determination of certain basic "design properties" of the cement and aggregates involved. The investigation may be quite complex or very simple in nature, depending on the situation and the extent of knowledge relative to the materials being used. The mix may be designed for a given flexural strength, compressive strength, or for both, and for the desired consistency. Some adjustment must generally be made in the selected mix on the basis of field conditions in order to secure concrete of satisfactory quality and economical proportions.

TABLE 18–4

VOLUME OF COARSE AGGREGATE PER UNIT OF VOLUME OF CONCRETE

Maximum Size of Aggregate (inches)	Volume of Dry-Rodded Coarse Aggregate per Unit Volume of Concrete for Different Fineness Moduli of Sand			
	2.40	2.60	2.80	3.00
1	0.70	0.68	0.66	0.64
1½	0.76	0.74	0.72	0.70
2	0.79	0.77	0.75	0.73
3	0.84	0.82	0.80	0.78

18–6. Air-Entrained Concrete Mixtures. Air-entrained concrete, as has been indicated, contains a small amount of entrapped air which is present in the form of small, disconnected air bubbles which are uniformly distributed throughout the mass. The desired amount of air is generally from 4 to 8 per cent of the total mix. The chief advantage accruing from the inclusion of this amount of air in the mixture is increased durability, including resistance to calcium chloride and other salts which are widely used in northern states for ice control and which have frequently resulted in surface scaling and spalling of concrete pavements. It also increases resistance to sulfate action and to freezing and thawing. Air-entrained concrete mixes have a high degree of workability when compared with similar regular mixtures, being generally somewhat sticky and plastic in nature and showing little tendency to segregation. They may be placed at slightly lower slumps than regular mixtures.

Air-entrained concrete mixtures should also be designed by the trial-mix method. It is necessary to determine and control the amount of air which is actually incorporated into the mixture. Several methods have been devised to measure the air content of air-entrained concrete mixtures, including the gravimetric method (AASHO T121 and ASTM C138), the volumetric method (ASTM C173), and the pressure method (AASHO T152 and ASTM C231). The last of these is most widely used and is based on the principle of Boyle's Law ($P_1V_1 = P_2V_2$); a known pressure is applied to a known volume of concrete and the change

FIG. 18–2. Determination of air content by the pressure method. (Courtesy Soiltest, Inc.)

in volume determined. A pressure device for determining the air content is shown in Fig. 18–2.

Structural Design of Concrete Pavements

The following several sections of this chapter will be devoted to what has been called the "structural design" of concrete pavements. Included in these sections is the design of the various elements of the pavement structure itself, including considerations relative to the various types of joints and their spacing, the use of reinforcing steel, and the thickness required in the slab cross-section.

Joints and Joint Spacing

Joints, except for construction joints, which are discussed separately, are installed in concrete pavements to control the stresses induced by volume changes in the concrete. Volume changes which are of greatest consequence in concrete pavements are those resulting from temperature change, as evidenced by the contraction or expansion of the slab due to a more or less uniform temperature change, and the warping of the slab due to the existence of a temperature differential between the top and bottom of the slab. Stresses induced by temperature may, under varying conditions, be direct tension or compression, or flexural in nature. Under certain conditions, stresses due to the existence of varying moisture conditions (a "moisture differential") in the depth of the slab may also be of appreciable magnitude.

Several different general types of joints are used in concrete pavements, and various considerations will be presented in this section relative to their design and use. The following is a brief outline of the subjects which will be covered, in sequence, in the following pages.

1. General considerations relative to volume changes and accompanying stresses
2. Longitudinal joints
3. Transverse contraction joints
4. Transverse expansion joints
5. Construction joints
6. Pumping

18–7. General Considerations. In this section concepts will be presented relative to the stresses which may be produced in a concrete slab because of its contraction due to a uniform temperature drop, expansion due to a uniform temperature increase, temperature warping, and, briefly, the effects of moisture.

18–8. Contraction. As do most construction materials, concrete changes in volume with a change in temperature. Thus a concrete slab, if it is free to move, contracts with a drop in temperature. However, if the contraction or "movement" of the slab is wholly or partially prevented, tensile stresses are developed. In a pavement, resistance to movement of the slab is caused by friction between the bottom of the slab and the subgrade. Consider first a plain concrete slab which has a length of L feet, a width of b feet, and a uniform depth of d inches. If the weight of the concrete, in pounds per cubic foot, is w, the coefficient of friction between the slab and the subgrade is C, and the tensile stress in the concrete is f_c pounds per square inch, then the following equation may be written:

Total frictional resistance

= total stress in the concrete (at the center of the slab)

$$Cw\left(\frac{d}{12}\right)b\left(\frac{L}{2}\right) = f_c(12b)d$$

and if the unit weight of the concrete (w) is taken to be 144 pounds per cubic foot, then

$$C(144)\left(\frac{d}{12}\right)b\left(\frac{L}{2}\right) = f_c(12b)\,d$$

or

$$f_c = \frac{CL}{2} \qquad (18\text{–}1)$$

If the tensile stress in the concrete exceeds a certain allowable value, the slab may be expected to crack transversely. Thus if it is desired to prevent the occurrence of a contraction crack, then the allowable length of slab (or spacing between contraction joints) may be determined from Eq. (18–1), as follows, considering only the effect of a uniform drop in temperature

$$L = \frac{2f_c}{C} \qquad (18\text{–}2)$$

A similar expression may be developed for a slab which contains reinforcing steel. Following the line of reasoning indicated above, and representing the area of steel by A_s (square inches), the following expression may be written:

$$C(144)\left(\frac{d}{12}\right)b\left(\frac{L}{2}\right) = f_c\left(\frac{144bd}{12} + nA_s\right)$$

when n = modular ratio = E_s/E_c, with E_s being the modulus of elasticity of the steel and E_c the modulus of elasticity of the concrete, both in pounds per cubic inch. This equation may be solved for L, the limiting length of slab or spacing between contraction joints for a slab which contains reinforcing steel and in which the steel in the concrete is assumed to take all the tensile stress caused by resistance of the subgrade to shortening of the slab due to a drop in temperature. In this case it would be assumed by the designer that the concrete would crack, and, usually, that the cracks would be large in number and small in width (hair cracks) and would not impair the structural efficiency of the slab. Again, only the effect of a uniform drop in temperature is being considered. The forces involved then would be, simply,

$$C(144)\left(\frac{d}{12}\right)b\left(\frac{L}{2}\right) = f_sA_s$$

and

$$L = \frac{f_sA_s}{6Cdb} \tag{18-3}$$

where f_s = the allowable tensile stress in the steel.

18-9. Expansion. If a concrete pavement slab is subjected to uniform increase in temperature, the slab will increase in length. Under extreme combinations of circumstances it may be imagined that a long length of concrete slab may buckle or "blow up" if this expansion is prevented. In order to prevent blow-ups, relief may be provided by the installation of transverse expansion joints. These joints, which are described in more detail later, generally provide a space an inch or so in width in which expansion may take place without damage to the slab.

However, friction between the slab and the subgrade prevents much of this expansion, and the compressive stresses created by this restraint are generally quite small as compared with the compressive strength of the concrete. This fact, coupled with the fact that concrete shrinks somewhat during hardening, has led many modern designers to the conclusion that expansion joints may be spaced at very great intervals, provided only that adequate contraction joints are provided and that the concrete does not have unusual expansion qualities or was placed at exceedingly low temperatures.

18-10. Temperature Warping. Very frequently a differential in temperature exists between the top and bottom of a concrete pavement. For example, during the day the temperature of the top portion of the slab may be considerably greater than that of the bottom. Under this circumstance the top portion of the slab tends to expand more than the bottom. This expansion, and its effect in producing a slightly convex slab surface, is resisted by the weight of the slab, with the consequence that fibers in the top portion of the slab are placed in compression and those in the bottom in tension. Stresses in the pavement from this case are termed "temperature warping stresses." Conversely, at night the top of the slab is frequently cooler than the bottom, with the result that the top of the slab is in tension and the bottom in compression.

Temperature warping is of importance in the design of concrete pavements in two ways. First, the flexural stresses due to this cause may, under certain conditions, be of considerable magnitude in themselves, without regard to their combination with stresses due to traffic loads. Secondly, the warping which does take place due to a temperature differential may partially destroy the subgrade support beneath portions of the slab, with the result that stresses due to traffic loads may be considerably increased over those which would exist if the pavement received uniform subgrade support.

Various investigators (*220*) have made extensive study of the nature, occurrence, and magnitude of temperature warping stresses. No specific data will be presented here with regard to the magnitude of temperature warping stresses. However, it may be noted that flexural stresses resulting from temperature warping may reach a magnitude of 200 pounds per square inch or more in a slab which is 10 feet wide, and may be much greater in wider slabs. Since the modulus of rupture of concrete in pavement slabs may be on the order of magnitude of 650 pounds per square inch or so, it may be seen that these stresses are of consequence and may result in cracking of the concrete if not controlled. The desire to control temperature warping stresses is the major factor behind the almost universal provision of longitudinal joints which divide concrete pavements into separate lanes, generally 11 or 12 feet in width. In recent years one-way ramps not more than 18 feet wide have been built successfully without longitudinal joints. The spacing of contraction joints is also greatly influenced by the existence of temperature warping stresses.

18–11. Moisture Effects. Little quantitative information is available as to the effects of moisture changes and the possible existence of a moisture differential upon the behavior of concrete pavements. Some general facts may be noted, however. For example, it is known that concrete generally contracts or "shrinks" with a decrease in moisture content and expands to some extent with an increase in moisture. Since it would normally be expected that a concrete slab would decrease in moisture content with the passing of time, the slab might be expected to contract slightly with a resultant opening of transverse joints. This circumstance would somewhat tend to offset later expansion of the slab due to temperature change, and thus further tend to substantiate the long spacing between expansion joints. Expansion of the slab due to an increase in moisture content would generally be substantially less than the contraction resulting from a similar decrease. If a difference in moisture content exists between the top and the bottom of the slab, the effect is somewhat similar to that produced by a temperature differential. In perhaps the majority of cases the moisture content in the bottom of the slab is greater than in the top, with the result that the bottom tends to expand, resulting in compression of the bottom fibers and somewhat offsetting stresses resulting from temperature warping during the daytime when the temperature differential might be expected to be greatest.

Both temperature and moisture warping stresses should be given consideration in the development of equations intended to evaluate the total stresses in a concrete pavement subjected to known traffic loads.

18–12. Current Practice, Spacing of Transverse Joints. After many years of laboratory and field studies, practice relative to the spacing of transverse joints has at last reached some degree of uniformity over the country. Details vary, but the following discussion reflects current practice for both plain and reinforced concrete pavements.

For both types of pavements, most designers agree with the views developed by the Portland Cement Association (*125*), in that except at intersections and structures no expansion joints need be provided in a concrete pavement when

1. The pavement is constructed of materials which have normal expansion characteristics.
2. The pavement is constructed during those periods of the year when normal construction temperatures prevail.
3. The pavement is divided into relatively short panels by contraction joints so spaced as to prevent the formation of intermediate cracks.
4. The contraction joints are properly maintained to prevent the infiltration of relatively incompressible material, such as soil fines.

If concrete pavements are built in cold weather, or of materials which have high coefficients of expansion, expansion joints may be necessary at intervals of several hundred feet. Otherwise, they are required only adjacent to structures and intersections.

For plain concrete pavements, the spacing of contraction joints must be close enough to control cracking. Spacing is dependent upon local experience and often is 15 or 20 feet. If the spacing is above 30 feet, it is recommended (*125*) that load transfer devices be used at the joints.

California has built a very large mileage of unreinforced concrete
California has built a very large mileage of unreinforced concrete
pavements with contraction joints at close spacings and no dowels. Many such recent pavements have transverse joints that are skewed slightly with respect to a line at right angles to the centerline of the pavement (4 or 5 feet in 24 feet) and are at random spacings (13, 19, 18, and 12 feet) to avoid rhythmic effects on automobiles associated with spacings that are exact multiples of 7.5 feet.

Distance between transverse contraction joints may be somewhat greater when the pavement is reinforced, since the distributed steel aids in controlling cracking. Joint spacing is dependent upon several factors, including the quantity of steel and the thickness of slab. Aaron (*92*) has recommended slab lengths from 50 to 100 feet; he believes that the most economical length of reinforced slab frequently will be from 60 to 80 feet. The most economical length may also depend on the cost of local labor and materials; a study in one state showed about 47 feet to be the most economical.

18–13. Longitudinal Joints. A longitudinal joint in a concrete pavement is, of course, a joint running continuously the length of the pavement. The joint divides, for example, a two-lane pavement into two sections, the width of each being the width of a traffic lane. The purpose of longitudinal joints is simply to control the magnitude of temperature warping stresses in such a fashion that longitudinal cracking of the pavement will not occur. Longitudinal cracking has been almost completely eliminated in concrete pavements by the provision of adequate longitudinal joints. In two-lane pavements, the two slabs are generally tied together by means of steel tie bars extending transversely across the joint and spaced at intervals along the length of the joint. The purpose of the tie bars in a longitudinal joint is to prevent movement of one slab with respect to the other, since, in addition to serving the function of relieving temperature warping stresses, the joint functions in transmitting a portion of the load carried by one slab to the other, thus reducing the stresses in the loaded slab. Adequate "load transfer" is dependent upon maintaining close contact between the two slabs (or portions of the joint) across the longitudinal joint. In multiple-lane pavements, no more than four lanes should be tied together.

18–14. Types of Longitudinal Joints. Longitudinal joints used in concrete pavements are of several different types, including both patented

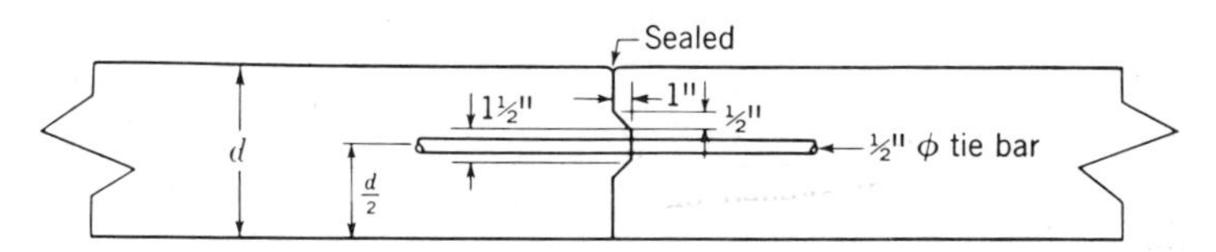

FIG. 18–3. Typical longitudinal joint.

and nonpatented varieties. Two of the most commonly used types are those designated as "deformed" or "keyed" and "weakened-plane" joints. A typical longitudinal joint of the former type is shown in Fig. 18–3. This type of joint often is used when the pavement is built one lane at a time; a weakened-plane joint generally is used if two or more lanes are paved at one time. Joints of this type are generally staked in position, prior to placing of the concrete, by steel pins driven at intervals through the joint assembly along its length. In this type of joint, concrete in adjacent slabs is separated by the material used in the joint which may, for example, be sheet metal or asphalt mastic board. The weakened-plane type of joint is formed by grooving, sawing, or the insertion of a premolded strip and is described in detail in the portion of the discussion relating to contraction joints.

Steel tie bars are used to tie the two slabs together and are firmly "bonded" to the concrete; the behavior of these bars should not be confused with that of the dowel bars mentioned in later sections. Tie bars which are used in longitudinal joints are generally designed to withstand the entire stress created by contraction of the slab in a transverse direction. There is some variation in the size of steel bars used for this purpose, with ½-inch round deformed bars being in common use. Tie bars for two-lane pavements are frequently spaced 30 inches on centers. They are generally 30 inches in length to provide the necessary length of embedment and bond.

18–15. Contraction Joints. Transverse contraction joints, as has been indicated, are used for two purposes; to control cracking of the slab resulting from contraction and to relieve temperature warping stresses. The spacing of contraction joints, as indicated by current practice of the various state highway departments, varies all the way from 15 feet to 100 feet. In unreinforced pavements in which no dowel bars are to be used at the joints for load transfer, the transfer of load is accomplished by aggregate interlock. When aggregate interlock is depended upon for load transfer, the spacing should be from 15 to 30 feet so that adjacent slabs will be maintained in close contact.

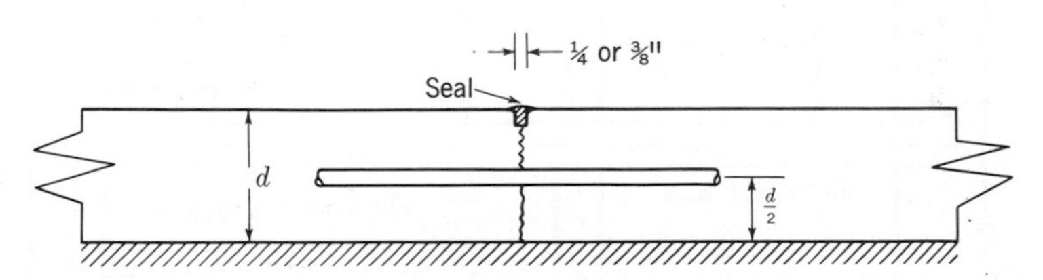

Fig. 18–4. Weakened plane or dummy contraction joint.

The type of contraction joint which is most widely used at the present time is the weakened-plane joint, one design of which is shown in Fig. 18–4. As contraction occurs, the lower portion of the slab cracks at the weakened plane and the structural integrity of the joint is maintained by the dowel bars and aggregate interlock. This type of joint is also called a "dummy" contraction joint.

Details of the designs of contraction joints vary to some extent, with bar diameters varying from ⅞ to 1¼ inches being used with lengths generally varying from 14 to 18 inches and spacings generally being 12–15 inches, center to center of the bars. The groove in the top of the pavement is frequently either of constant $^3/_{16}$- or ¼-inch width, as shown in Fig. 18–4. This groove is most often formed by sawing, but may be formed by use of a premolded strip inserted in the fresh concrete. The groove is later filled with one of a number of materials, including poured rubber and rubber-asphalt compounds and two-component, cold-applied elastomeric polymers. New York State has reported success with a preformed neoprene joint sealer which is under compression at all times (*39*).

Dowel bars used in this type of joint (and in expansion joints) are not bonded to the concrete on one side of the weakened plane, and freedom of movement is assured by painting or lubricating one end of the dowel, by enclosing one end in a sleeve, or by other similar methods. It is essential that freedom of movement be assured in the design and placing of the joint, since the purpose of the joint will be largely destroyed if movement is prevented. The dowels must be carefully placed so that they are horizontal and parallel to the center line of the slab. They must also be designed so as to ensure their proper functioning under the stresses imposed upon them in their capacity as load-transfer devices.

18–16. Expansion Joints. Expansion joints which are used at the present time are usually from ¾ to 1 inch in width and extend the

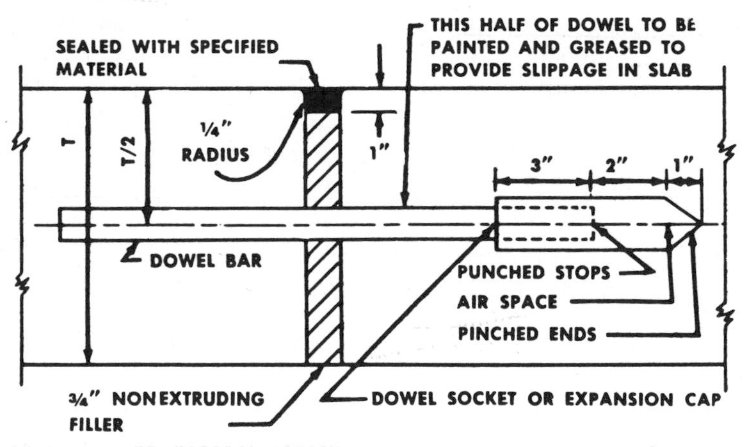

Fig. 18–5. Typical doweled expansion joint.

full depth of the slab. The joint space is filled with some compressible, elastic, nonextruding material. A variety of materials have been used as expansion joint fillers, including bituminous material, cork, rubber, cork-rubber compounds, bitumined fabrics, wood, and many others. Dowel bars are normally used in expansion joints, and their diameter, spacing, and length would generally be similar to that indicated for contraction joints. The dowels must be adequate to perform their load-transfer function and they must be designed and placed so as not to interfere with proper functioning of the joint. Fig. 18–5 shows a somewhat typical design of a transverse expansion joint.

18–17. Construction Joints. An additional type of joint which may be used in a concrete pavement may be called a "construction joint." Transverse construction joints may be placed at the end of a day's "run," or when work ceases due to some other interruption. If construction is stopped at the location of a transverse joint, the joint assembly may be installed as usual, concrete placed on one side of the joint, and the other side of the joint protected in an appropriate fashion until concreting operations can be resumed. If a transverse construction joint falls within the middle third of the regular joint interval, it should be a tied joint. Longitudinal construction joints may be made in a similar fashion when, for example, the pavement is constructed with a width of one lane. Joints of this type may also be formed by the use of steel forms which incorporate a keyway or by fastening a tongue-and-groove premolded or metal strip on the inside of the forms; tie bars may be bent back out of the way or tie bolts used. All formed longitudinal joints should be keyed regardless of whether they are tied or free.

18–18. Pumping of Joints. Special attention is given in this section to a problem which may be encountered in the design (and maintenance) of joints in concrete pavements. This problem is "pumping." The phenomenon of pumping may be defined as follows. With a certain combination of factors present, the movement of slab ends under traffic loads causes the extrusion or "pumping" of a portion of the subgrade material at joints, in cracks, and along the edges of the pavement. Pumping through joints will only occur under the following circumstances: (1) frequent occurrence of heavy wheel loads, (2) the existence of a surplus of water in the subgrade soil, and (3) the presence of a subgrade soil which is susceptible to pumping. All three of these elements must be present in order for pumping to occur. The amount of soil removed by pumping may be sufficient to cause a sizable reduction in subgrade support for the slab and may result in eventual failure of the pavement. The problem was especially severe after World War II in connection with expansion joints in existing pavements located on subgrade soils which contained high percentages (45 per cent or more) of silt and clay.

Design procedures intended to prevent pumping will be briefly discussed here, while maintenance operations intended to correct pumping which occurs in existing pavements will be discussed in a later chapter. Since much of the difficulty in this respect has been encountered at expansion joints, an obvious solution is to minimize the number of these joints or eliminate them altogether; this consideration has doubtless been a contributing factor in the present trend to minimize the use of expansion joints. Also, since pumping is associated with fine-grained soils, another obvious solution is to replace or improve these so-called "pumping soils." This may be done, for example, by the use of a nominal thickness of granular material or a selected sandy soil, or by the stabilization of the existing subgrade soil through the use of bituminous material or portland cement.

In modern practice, a granular subbase from 3 to 6 inches thick is employed by many organizations in locations where the subgrade soil is such as to be susceptible to pumping. Much thicker subbases may be used in areas subject to frost action or on very weak soils. Requirements for subbase materials vary; a suitable specification is AASHO M147 (Table 13–1). A desire to prevent pumping is one factor in the increasing use of cement-treated base courses beneath concrete pavements. Some states now use asphaltic plant mixtures under concrete pavements for the same purpose.

Use of Distributed Steel Reinforcing

Distributed steel reinforcing is used primarily to control cracking of a concrete pavement and to maintain the structural integrity of the slab between transverse joints. In accordance with principles previously stated, some organizations believe that reinforcing steel is unnecessary where contraction joints are spaced at intervals of 15 to 30 feet. If the distance between contraction joints is appreciably greater than this, it may generally be regarded as desirable, with the amount of steel required being a direct function of the spacing between joints. Cracking of the slab will occur even though the steel is present, but the steel serves to hold the edges of the cracks close together, thus preventing the progressive opening of these cracks and accompanying detrimental effects. Distributed steel, as commonly used in concrete pavements, is not held to contribute to the flexural strength of the slab. By holding cracks tightly closed, however, it maintains the shearing resistance of the slab, and, consequently, its load carrying capacity. These properties are greatly reduced if cracks are permitted to open.

Distributed reinforcing is used in the form of welded wire "fabric" or bar mats, both of which contain both longitudinal and transverse elements. Different sizes and spacing of the elements of the reinforcing "mat" are used by the various highway agencies in order to conform with requirements relative to the area of steel required across the width and length of the slab. The steel is usually placed in the slab at a distance of from 2 to 3 inches below the top of the slab. The amount of steel required in a given case may be determined through the use of the "subgrade drag theory" utilized in developing the equations presented earlier in this chapter under the discussion of joints. In fact, Eq. 18–3 may be applied directly to this case, with the length L being the predetermined spacing of transverse joints; *w, d,* and *b* being those applicable to the given case; and satisfactory values of f_s and C being assumed. The remaining unknown in this equation, A_s could then be determined and the size and spacing of the wires or bars selected. High yield point steel is generally used for this purpose, and working stresses used in the design (f_s) may generally be from 30,000 to 50,000 pounds per square inch.

According to the Wire Reinforcement Institute, the value A_s given by Eq. 18–3 may be inadequate for short slabs (40 feet or so). In such cases, these rules should prevail: The cross-sectional area of the longitudinal steel should not be less than 0.1 per cent of the cross-sectional area of the slab; longitudinal wires should not be smaller than No. 2 gage, with spacing of not more than 6 inches; and transverse wires should not be smaller than No. 4 gage, with spacing of not more than 12 inches.

Aaron (*92*) has presented recommendations relative to the use of welded wire fabric as reinforcement for different thicknesses and lengths of pavement slabs. These recommendations are contained in Table 18–5,

TABLE 18–5

REINFORCEMENT RECOMMENDATIONS USING 6 × 12 WELDED WIRE FABRIC

		Welded Wire Fabric Details		
Slab Thickness (inches)	Slab Length[1] (feet)	Style	Weight (lbs. per 100 sq. ft.)	Area per ft. Width[2] (sq. in.)
7	50	6 × 12—2/4[3]	54	.108
	60	6 × 12—2/4[3]	54	.108
	80	6 × 12—1/4	61	.126
8	50	6 × 12—2/4[3]	54	.108
	60	6 × 12—1/4	61	.126
	80	6 × 12—0/4	69	.148
9	50	6 × 12—2/4[3]	54	.108
	60	6 × 12—1/4	61	.126
	80	6 × 12—0/4	69	.148
	100	6 × 12—00/4	79	.172
10	50	6 × 12—2/4[3]	54	.108
	60	6 × 12—1/4	61	.126
	80	6 × 12—0/4	69	.148
	100	6 × 12—00/4	79	.172

[1] Widths may range from 20 to 24 feet.
[2] Longitudinal wires.
[3] Minimum sizes recommended for pavements.

which covers a range of thicknesses from 7 to 10 inches and slab lengths from 50 to 100 feet. Recommendations are intended to apply to either rural highways or urban expressways.

18–19. Continuously Reinforced Pavement. Following World War II, interest was focused upon the possibility of building continuously reinforced pavements without joints, with the construction of pavements of this type in Illinois and New Jersey. Longitudinal steel was used continuously over the length of a mile in the New Jersey experiment and 3500 feet in Illinois. The amount of steel was varied in different sections of these test roads, ranging from 0.3 to 1.0 per cent. Subsequently, test sections of continuously reinforced concrete pavement were built in Texas, California, Pennsylvania, and several other states.

Beginning in 1959, when there were less than 100 miles of continuously reinforced concrete highway pavement in the United States, use of this type of pavement has grown rapidly. By 1965, there were over 1800 miles of such pavements in 21 states. Most of the mileage is in three states—Texas (nearly 1,200 miles), Mississippi, and Illinois.

The required amount of longitudinal reinforcing steel in this kind of pavement is not in direct proportion to the length of the slab. Observations of long (a mile or more) continuously reinforced slabs indicate that only 400 or 500 feet of the pavement at each end is subject to longitudinal movement; for all practical purposes, the long central section is fully restrained. Seasonal movements at the slab ends total less than 2 inches, regardless of the slab length.

The Wire Reinforcement Institute states that the optimum quantity of reinforcing steel is that which results in a crack spacing of 3 to 10 feet (*276*). Crack intervals in this range can be gotten with from 0.5 to 0.7 per cent steel. The area of transverse steel required can be computed by Eq. 18–3.

Recommended slab thickness varies from 6 to 9 inches, depending on traffic conditions. A thickness of 8 inches is satisfactory for most primary rural highways and urban expressways.

Both welded wire fabric and bar mats are used for continuous reinforcement. For wire fabric, spacing of the longitudinal wires should be not less than 3 inches nor more than 6 inches; spacing of transverse wires should not exceed 16 inches; and the clearance between the longitudinal edge wire, and between the edge wire and the center joint, should be not less than 1 inch, nor more than 6 inches.

Since the great rise in the use of prestressed concrete for bridges following World War II, highway designers have toyed with the idea of prestressed concrete for highway pavements. Work thus far in the United States has been experimental. One such experimental installation is a 5-inch by 12-foot, by 530-foot prestressed slab in Pittsburgh.

And work on prestressed slabs has been done at the Portland Cement Association's research and development laboratories at Skokie, Illinois. Although the concept may hold some promise for the future, high labor costs and slow rates of production have discouraged its use.

Dimensions of the Cross-Section (Thickness Design)

It is the purpose of this section to discuss the determination of the design thickness of a concrete pavement. Several different factors enter into the thickness determination, as will be outlined in following sections. Basically, however, the pavement must have a thickness which is adequate to support the loads which will be applied to it during its service life, and the design must be economical.

Many organizations and individuals have given their attention to the problem of thickness design of concrete pavements during the last 50 years or so. Among the individuals who have been outstanding in their contributions to this branch of highway engineernig during this period are Clifford Older, of the Illinois Department of Highways; A. T. Goldbeck, of the National Crushed Stone Association; M. G. Spangler and Glenn Murphy, of Iowa State College; R. D. Bradbury, of the Wire Reinforcement Institute; H. M. Westergaard, of Harvard University; Frank Sheets of the Portland Cement Association; E. F. Kelley, L. W. Teller, and E. C. Sutherland, of the Bureau of Public Roads; Gerald Pickett, of Kansas State College; and many others. Many organizations have conducted experimental projects in order to further develop the fundamental concepts underlying the thickness design of concrete pavements. Outstanding among these were the early Bates Test Road, in Illinois, and the later Arlington (Virginia) tests of the Bureau of Public Roads. Of all the investigations which have been conducted to date, probably the most widely known are the theoretical analyses of Westergaard and the results of the Arlington experiments. Analysis of the results of studies at the AASHO Test Road located near Ottawa, Illinois, threw further light on the subject, largely substantiating the Westergaard Analyses.

Space will not permit the chronicling of the many contributions which have been made by these investigators and others. Rather, it has been decided to present the essentials of only one design method, in the belief that this method is typical of modern approaches to this subject. The method presented is that contained in the publication, "Thickness Design for Concrete Pavements," published by the Portland Cement Association in 1966 (*133*).

According to PCA engineers, design considerations that are vital to satisfactory performance and long life of a concrete pavement are reasonably uniform support for the pavement; the elimination of pumping by the use of a thin treated or untreated base course; adequate joint design; and a thickness that will keep load stresses within safe limits.

General objective of the design procedure is to determine the minimum thickness that will give the least annual cost.

Since direct compressive stresses in concrete pavements due to wheel loads are very small in relation to the compressive strength of the material, they can be ignored. Flexural strength is a key factor in thickness design, since the flexural stress produced by a heavy wheel load often is more than half the flexural strength.

Flexural strength is measured by modulus of rupture tests on beams subjected to third-point loading. PCA recommends the use of 28-day to 90-day strengths for roads and streets, since very few stress repetitions occur during the first 90 days of pavement life, compared with the millions of repetitions that occur after that time. Also, the strength of concrete increases with age.

Another major factor in thickness design is subgrade or subbase support, which is measured by the Westergaard modulus of subgrade reaction, k. To determine k, a plate bearing test generally similar to that described in Art. 12–14 is performed, generally using a 30-inch plate. Typical values of k for the groups of the AASHO soil classification system are given in Table 18–6.

TABLE 18–6

TYPICAL VALUES OF THE MODULUS OF SUBGRADE REACTION, k

AASHO Soil Group	k, lb. per cu. in.
A–1–a	400 plus
A–1–b	250 plus
A–2–4, A–2–5	300 plus
A–2–6, A–2–7	175–325
A–3	200–325
A–4	100–300
A–5	50–175
A–6	50–225
A–7–5, A–7–6	50–225

The Portland Cement Association (PCA) recommends that k values used in design be those determined in normal summer or fall weather (not during the "spring breakup" or when the subgrade is frozen). The k value can be increased by using a treated or untreated subbase, although use of a subbase for this purpose alone seldom would be economical. For a subgrade with a k value of 100, a 4-inch cement-treated base course should increase the k value to at least 300, according to PCA designers.

Concrete is subject to fatigue, as are other construction materials. In the PCA design concept, a fatigue failure occurs when a material ruptures under continued repetitions of loads that cause stress ratios of less than 1. The stress ratio is the ratio of flexural stress to the modulus of rupture.

TABLE 18–7

STRESS RATIOS AND ALLOWABLE LOAD REPETITIONS

(Portland Cement Association)

Stress[1] Ratio	Allowable Repetition	Stress[1] Ratio	Allowable Repetition
0.51[2]	400,000	0.69	2,500
0.52	300,000	0.70	2,000
0.53	240,000	0.71	1,500
0.54	180,000	0.72	1,100
0.55	130,000	0.73	850
0.56	100,000	0.74	650
0.57	75,000	0.75	490
0.58	57,000	0.76	360
0.59	42,000	0.77	270
0.60	32,000	0.78	210
0.61	24,000	0.79	160
0.62	18,000	0.80	120
0.63	14,000	0.81	90
0.64	11,000	0.82	70
0.65	8,000	0.83	50
0.66	6,000	0.84	40
0.67	4,500	0.85	30
0.68	3,500		

[1] Load stress divided by modulus of rupture.
[2] Unlimited repetitions for stress ratios of 0.50 or less.

Flexural fatigue research on concrete has shown that the number of stress repetitions to failure increases as the stress ratio decreases; studies show that, if the stress ratio is less than 0.55, concrete will withstand virtually unlimited stress repetitions without loss in load-bearing capacity. To be conservative, designers reduced this ratio to 0.50.

Allowable load repetitions for stress ratios from 0.50 to 0.85 are given in Table 18–7 (*133*).

In the past, thickness design methods for concrete pavements included an allowance for impact of moving loads. The PCA method discards this concept, substituting "load safety factors" in the design. The following load safety factors are recommended: Interstate and other multiple-lane projects where there will be an uninterrupted flow of heavy truck traffic, 1.2; highways and streets with moderate volumes of truck traffic, 1.1; and streets and highways that carry small volumes of truck traffic, 1.0.

Lane widths have important effects on the stresses caused by truck traffic, and on the thicknesses needed to control those stresses. In the 1920's, lanes were usually 9 feet wide, and virtually all trucks moved along the outside edges of the slab. Critical stresses occurred when truck wheels moved over the corners formed by transverse joints and the outside edges of the pavement.

Engineers used formulas based on corner loadings for many years to design concrete pavements in normal service. However, with general increases in lane widths from 9 to 12 feet, traffic has moved inward from outside corners and edges.

As a consequence, the critical stress location has moved from the outside corner to the transverse joint edge. At this critical location, designers positioned the axle loads laterally at the point of greatest load repetition. Maximum flexural stresses occur at the bottom of the slab, parallel to the joint edge.

Various load positions considered in the analysis are shown in Fig. 18–6; Case I is the critical one. PCA engineers did exhaustive analyses of several load positions for both single- and tandem-axle loads, using influence lines developed by Gerald Pickett and Gordon Ray and based on theoretical analyses by Westergaard. They then prepared design charts for the critical load position for both single and tandem axles. Fig. 18–7 applies to single-axle loads of from 10 to 50 kips. Fig. 18–8 is a similar chart for tandem-axle loads of from 20 to 100 kips. Both charts cover k values of from 50 to 600 pci, stresses of from 250 to 600 psi, and thicknesses of from 4 to 12 inches. The dashed, arrowed lines show the use of the charts.

Procedures recommended by PCA are based on a design life of 40 years.

Another major factor in designing a concrete pavement is, of course, the traffic to which it will be subjected during its service life. The number and magnitude of heavy vehicle loads are critical.

The PCA manual (*133*) details two methods for determining traffic volumes and the number and magnitude of heavy wheel load repetitions; neither method is explained here. One approach is to use figures based upon traffic volume and classification counts, plus loadometer surveys such as those explained in Chapter 4. Such information is generally

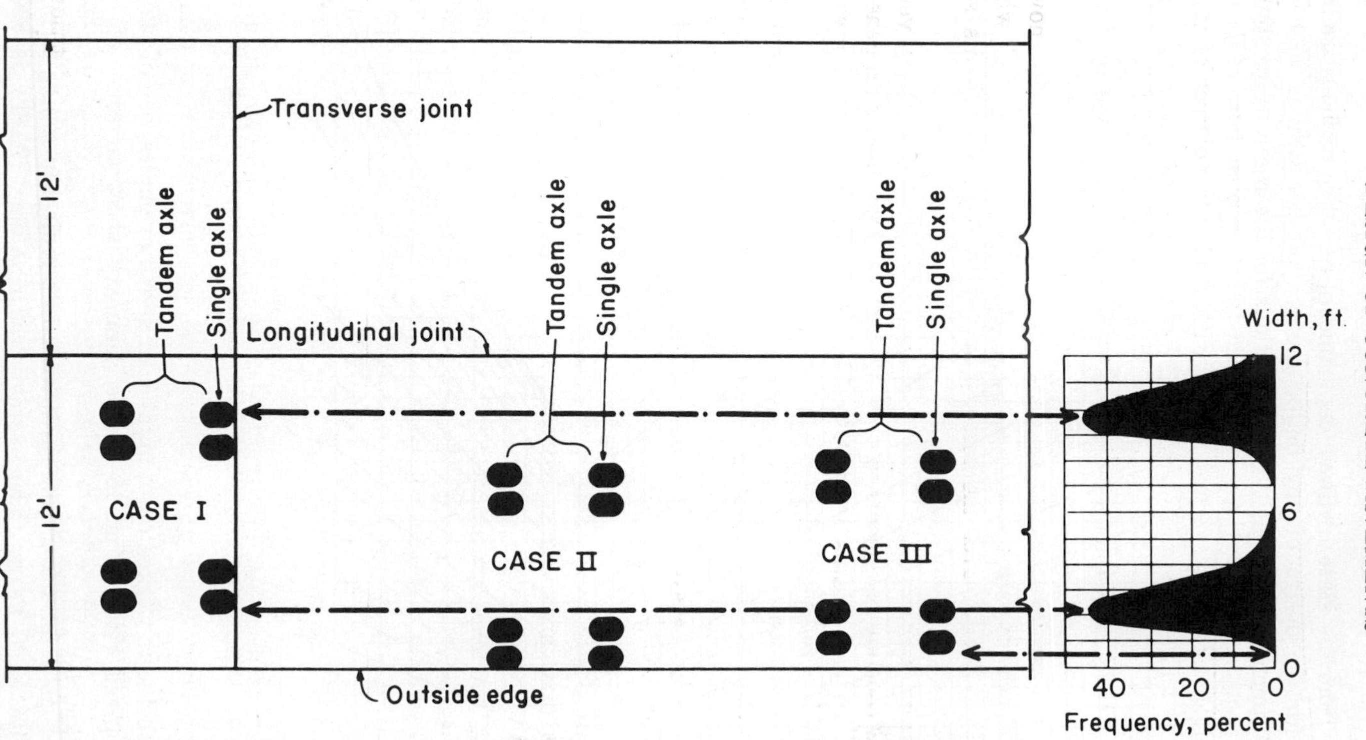

FIG. 18-6. Truck load positions for stress analysis. (Courtesy Portland Cement Association.)

available from the state highway departments or other traffic planning agencies for the design of any major route. Another way is to base the design upon traffic capacity considerations for the route in question; capacity is discussed in Art. 5–6.

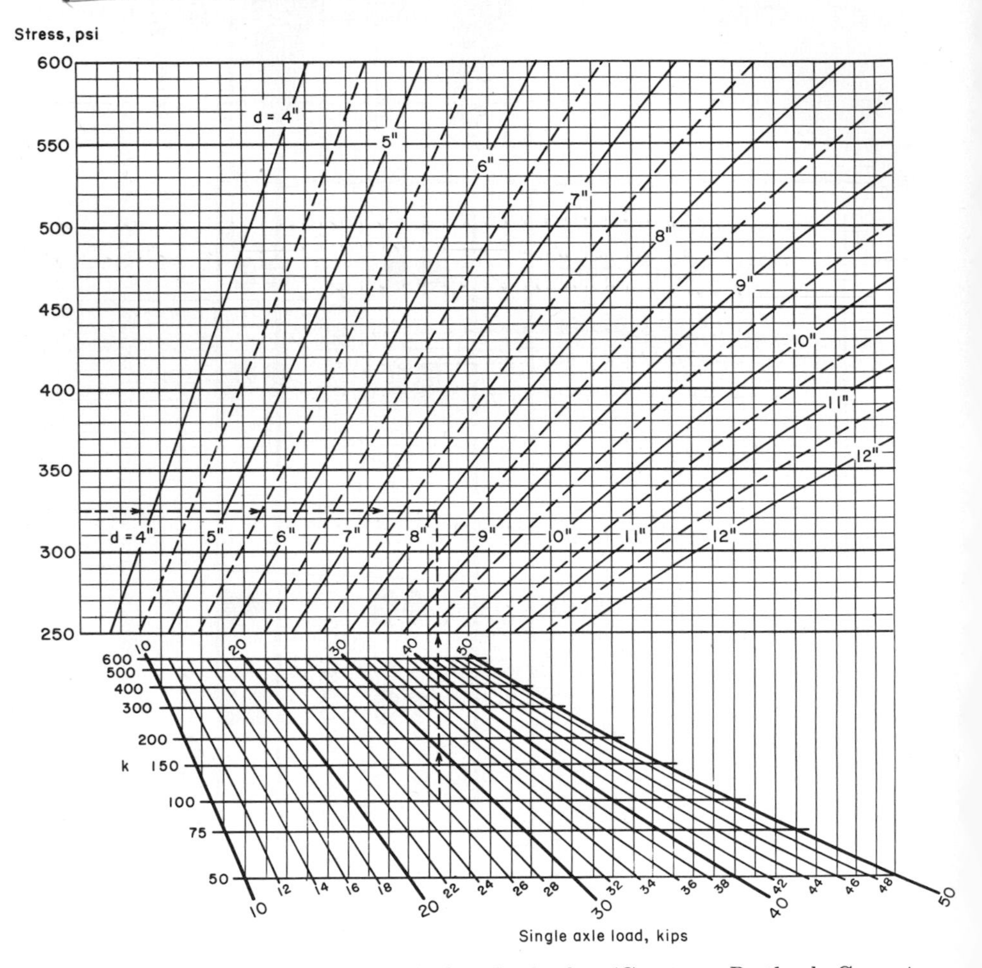

FIG. 18–7. Design chart for single-axle loads. (Courtesy Portland Cement Association.)

PCA engineers worked out a number of design examples, one of which is summarized below; Fig. 18–9 is a summary sheet on some of the calculations involved.

The design is for a four-lane rural Interstate project in rolling terrain. Design life is 40 years. The current traffic volume (ADT) is 8500, and the weighted average projection factor for 40 years is 2.2 (this corresponds to an average yearly growth of 3.5 per cent, using compound interest tables). Designers estimate that truck traffic volume (ADTT) is 13 per cent of the total.

The design ADT is $8500 \times 2.2 = 18{,}700$ vehicles per day, and the ADTT is 18,700 (0.13) = 2430. Truck traffic each way is $2430/2 = 1215$. The average hourly traffic volume in each direction

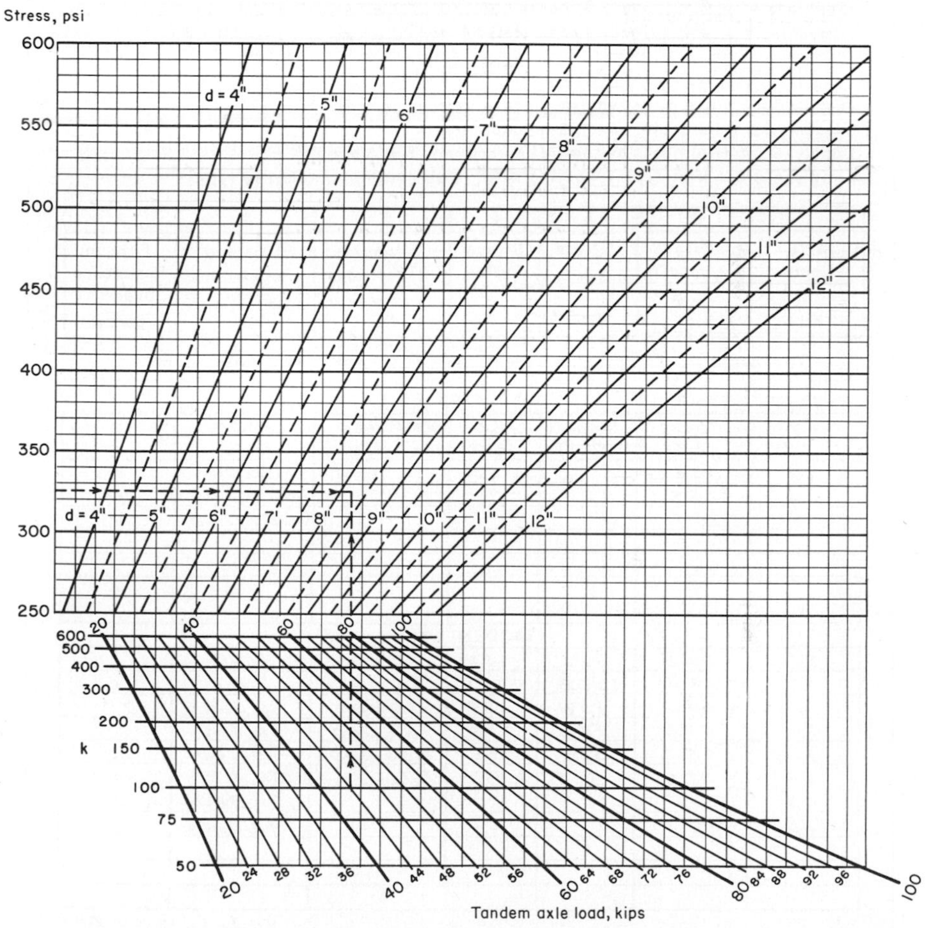

FIG. 18–8. Design chart for tandem-axle loads. (Courtesy Portland Cement Association.)

is 390. Data not shown here reveal that, at this hourly volume, 92 per cent of the trucks will be in the right lane. Thus, for this lane and a design life of 40 years, there will be $1215 \times 0.92 \times 365 \times 40 = 16.32$ million trucks.

PCA engineers evaluated three designs (1A, 1B, and 1C). For design 1A, the k of the subgrade was taken to be 100 pounds per cubic inch,

CALCULATION OF CONCRETE PAVEMENT THICKNESS
(Use with Case I Single & Tandem Axle Design Charts)

Project DESIGN ONE - A
Type Rural Interstate-Rolling Terrain No. of Lanes 4
Subgrade k 100 pci., Subbase 4-in. Granular Untreated
Combined k 130 pci., Load Safety Factor 1.2 (L.S.F.)

PROCEDURE

1. Fill in Col. 1,2 and 6, listing axle loads in decreasing order.
2. Assume 1st trial depth. Use 1/2-in. increments.
3. Analyze 1st trial depth by completing columns 3,4,5 and 7.
4. Analyze other trial depths, varying M.R*, slab depth and subbase type.**

1	2	3	4	5	6	7
Axle Loads kips	Axle Loads X/1.2 L.S.F. kips	Stress psi	Stress Ratios	Allowable Repetitions No.	Expected Repetitions No.	Fatigue Resistance Used*** percent

Trial depth 9.0 in. M.R.* 650 psi k 130 pci

SINGLE AXLES

1	2	3	4	5	6	7
30	36.0	340	.52	300,000	3100	1
28	33.6	325	.50	Unlimited	3100	0
26	31.2		<.50	"	6200	0
24	28.8		"	"	163,200	0
22	26.4		"	"	639,740	0

TANDEM AXLES

1	2	3	4	5	6	7
54	64.8	382	.59	42,000	3100	7
52	62.4	368	.57	75,000	3100	4
50	60.0	358	.55	130,000	30,360	23
48	57.6	348	.54	180,000	30,360	17
46	55.2	333	.51	400,000	48,140	12
44	52.8	318	<.50	Unlimited	150,470	0
42	50.4		"	"	171,360	0
40	48.0		"	"	248,060	0

Total = 64

* M.R. Modulus of Rupture for 3rd pt. loading.

** Cement-treated subbases result in greatly increased combined k values.

*** Total fatigue resistance used should not exceed about 125 percent.

FIG. 18–9. Summary sheet for design example, concrete pavement. (Courtesy Portland Cement Association.)

with a 4-inch untreated granular subbase. Combined k for the subgrade and subbase was taken to be 130. The load safety factor is 1.2, with a modulus of rupture of the concrete of 650 psi.

For design 1A, the engineers assumed a trial depth of 9 inches. Key calculations are shown in Fig. 18–9. Stresses shown in column 3 came from Figs. 18–7 and 18–8. Each stress ratio is the stress divided by the modulus of rupture. Allowable repetitions came from Table 18–7. The "fatigue resistance used" values were obtained by dividing expected repetitions by allowable repetitions and then expressing the quotients as percentages.

The design thickness of 9 inches is satisfactory, since it uses but 64 per cent of available fatigue resistance. A trial thickness of 8.5 inches was not satisfactory, because of "excessive fatigue consumption."

A second design calculation (1B) showed that 8.5 inches would be a satisfactory thickness if the modulus of rupture could be increased to 700 psi. That design used only 69 per cent of available fatigue resistance. However, the higher modulus of rupture would not allow a reduction in thickness to 8.0 inches.

A third design (1C) was tried by reducing the thickness to 8.0 inches, using a modulus of rupture of 650 psi and a cement-treated base with a k value of 300 pounds per cubic inch. Calculations showed this design to be more than adequate, since it used only 15 per cent of available fatigue resistance. This design, of course, might cost more. A reduction in thickness to 7.5 inches also would be satisfactory, according to PCA design criteria.

18–20. Present Status of Design. It should be noted at this stage that, although design principles and methods such as are illustrated above are in existence at the present time, these methods are not commonly applied as a part of the routine operations of a majority of highway agencies, including the state highway departments. Thicknesses of concrete pavements, as used in this country, commonly vary from 7 to 10 inches, and each organization normally has developed "standard designs" which are satisfactory for its purposes. Usually a concession is made to the principles outlined in the preceding section by the use of greater thicknesses for heavily traveled routes and in areas of poor subgrade support.

18–21. Composite Pavements. In broad terms, a composite pavement is one which combines dissimilar pavement types, i.e., "flexible" and "rigid" pavements. Usually, a composite pavement comprises a concrete- or cement-treated base course with a wearing surface of asphaltic concrete.

Comparatively few composite pavements have been designed and built as new pavements, although New York City has for many years built city streets with a portland cement concrete base, an asphaltic concrete binder course, and a sheet asphalt wearing surface. Other cities build similar pavement structures.

Many composite pavements have, of course, been produced by stage construction or by resurfacing operations. Overlays of existing pavements are discussed in Art. 20–15.

Relatively little design information is available on composite pavements, although some experimental installations have been built and the Corps of Engineers has extended methods used to design composite pavements for airfields to apply to roads and streets. The California method described in Art. 12–19 is applicable to pavement structures that include cement-treated base courses. Various aspects of composite pavement design are discussed in a publication of the Highway Research Board (*14*).

PROBLEMS

18–1. Using the equations in the text, determine the theoretical limiting distance between contraction joints for a slab which is 9 inches in uniform thickness and contains distributed reinforcing steel in the form of welded wire fabric containing 0.148 square inch of steel per foot of width. Assume that the steel carries all the tensile stresses caused by a uniform drop in temperature. The slab is 24 feet wide, the allowable stress in the steel 40,000 psi, and the coefficient of friction between the slab and subgrade 1.50.

18–2. Determine the necessary spacing of tie bars which are to be placed across a longitudinal center joint if half-inch round bars are to be used. The slab is 24 feet wide, the coefficient of friction between the slab and the subgrade may be assumed to be 1.25, and the slab is 9 inches in uniform thickness. The concrete may be taken to weigh 150 pounds per cubic foot, and the allowable stress in the tie bars is 36,000 pounds per square inch. What is the approximate length of bar required to develop the necessary bond?

18–3. Determine the present practice of your state highway department with regard to the following:

(a) Joints and joint spacing
(b) Use of distributed reinforcing steel
(c) Thickness of slabs

18–4. Determine the total area of reinforcing steel (wire mesh) required for a continuously reinforced concrete pavement that is 8 inches thick and 12 feet wide and contains the percentage of steel recommended by the Wire Reinforcement Institute. See if you can determine suitable sizes of deformed wire that would meet this requirement, if the wires are spaced (a) 3 inches center-to-center and (b) 4 inches center-to-center; consider wires only in the D-14 to D-20 size range. (See ASTM standard specification A496 for dimensions of deformed steel wire.)

18–5. A concrete pavement is to be built on a subgrade with a k value of 50 pci. The combined k of the subgrade and a 4-inch-thick untreated granular base is 75 pci. The design life is 40 years. The location of the loads is the critical one shown in Fig. 18–6. Over the design period the critical truck loads are as tabulated below:

Axle Load, Kips	Number of Expected Repetitions
	Single Axle
30	4,000
28	4,000
26	8,500
24	210,000
22	805,000
	Tandem Axle
54	4,000
52	4,000
50	39,600
48	39,600
46	61,800
44	183,500
42	183,500
40	183,500

(a) Determine the minimum thickness of slab that can be used safely for this urban Interstate route if the modulus of rupture of the concrete is 650 psi (28 days). Use the PCA method and a summary sheet like the one shown in Fig. 18–9.

(b) Can a thinner slab be used for this pavement if the modulus of rupture of the concrete is increased to 700 psi at 28 days?

18–6. Solve Problem 18–5 if a cement-treated base with a design k of 400 pci is substituted for the untreated granular base.

18–7. From current literature, prepare a report describing one of the experimental prestressed concrete paving slabs built in recent years. What do you think the future holds for this type of pavement construction in this country?

18–8. Follow the instructions of Problem 18–7, except for a recent project involving continuously reinforced concrete pavement.

CHAPTER 19

CONSTRUCTION OF CONCRETE PAVEMENTS

The construction of portland cement concrete pavements is marked by the use of a large number of specially designed machines, each of which performs a specific function in the construction process. Once paving operations are begun, the various steps in the construction procedure are arranged in the form of a continuing series of separate operations which are planned and coordinated so that the construction proceeds with a minimum loss of time and effort. Each of the separate steps must be done carefully and precisely so that the completed pavement will meet the exacting standards for structural strength and smoothness which are applied to it. The exact methods and machines used in the construction process vary somewhat from job to job, and no attempt will be made here to discuss all the possible variations in job methods and procedures. However, the following is the sequence of separate steps on typical projects:

1. Preparation and preliminary finishing of the subgrade
2. Placing of forms (where used)
3. Final finishing of the subgrade
4. Installation of joints
5. Batching of aggregates and cement
6. Mixing and placing concrete
7. Spreading and finishing concrete
8. Slipform paving
9. Curing

19–1. Preparation and Preliminary Finishing of the Subgrade. The preparation of the subgrade upon which a concrete pavement is to rest is, of course, a very important step in the over-all construction process. It is essential that uniform subgrade support be provided for the completed pavement during its service life, and that the pavement be free from other detrimental effects associated with unsatisfactory subgrade soils. Included among the most commonly encountered problems which are related to the character and condition of the subgrade soil are pumping and frost action, while other difficulties may become evident when soils of inadequate shearing strength, high-volume-change soils, organic soils, alkali soils, soils which are difficult to drain, and others are encountered. Provision must be made for the elimination of these defects in the subgrade soil before the pavement is placed.

In some cases the existing subgrade soil may be entirely suitable as a foundation and thus require only compaction and adequate drainage prior to the placing of a concrete slab. In other instances a layer of imported soil may be used between the existing soil and the slab, or the existing soil may be stabilized, as with portland cement, bituminous material, or granular materials. It is common practice to provide a blanket or "insulation" course of suitable granular material between the existing soil and the pavement itself. Various materials are used to form this layer or "subbase," including gravel, sand, sand and gravel, crushed rock, and so on. The depth of treatment varies from a minimum of 3 inches to 6 inches or so if the primary purpose is to prevent pumping; very much thicker layers of granular material may be used to prevent frost action in northern states.

In recent years many states, particularly those in the Far West and Southwest, have gone to the use of treated materials for subbases beneath concrete pavements. Most frequently used is a cement-treated base course, described more fully in Art. 19–14, but some states now use asphaltic mixtures for this purpose.

Whatever the solution used in a specific case, the layer immediately beneath the slab must be brought to a high degree of density and stability and must be adequately drained. Preliminary finishing of this layer will generally be accomplished by the use of motor graders or similar equipment which shape the layer to the desired cross-section, elevation, and line in what have previously been termed "finegrading operations." In most cases the subgrade layer will be brought very close to final shape before construction associated with the slab itself actually begins.

19–2. Placing of Forms. Steel forms are used in the construction of concrete pavements, unless a slip-form paver is used (Art. 19–9). They must be very carefully placed and secured in position so that the desired position, width, elevation, and grade may be secured in the final slab. Forms commonly used in highway work are straight 10-foot sections which are aligned both vertically and horizontally by slip joints, and which are held in position by three or more steel stakes driven at intervals at the back of the form. Forms of this type vary somewhat in size and weight, being available in heights of from 6 to 12 inches, with corresponding base widths over a similar range. One frequently used size, for example, has a height of 9 inches and a base width of 8 inches. Special forms are available for curbs, curb and gutter sections, and for the construction of sharp curves, including both rigid and flexible-type curved forms.

Several machines are frequently used in connection with the placing of the forms. One of these is the so-called "form grader" which cuts a trench of exact size in the correct position to receive the forms, as shown in Fig. 19–1. With this machine the trench is cut with a hydraulic cutter, which is adjustable hydraulically by the operator. Forms are then generally aligned and placed by hand, with the steel pins being driven either by hand or with the aid of an air hammer. A second machine may then be used to tamp the forms securely in place. The placing of the forms on most jobs is characterized by extreme care,

Fig. 19–1. Cutting of trench to exact line and grade, prior to placing of forms. (Courtesy Cleveland Formgrader Co.)

as many of the machines used in later operations ride on the forms and it is imperative that the forms be securely and carefully aligned and placed. Their elevation and alignment are, of course, carefully checked prior to continuance of construction operations. Concrete pavements generally are constructed in two-lane widths; forms are used on both sides. Sufficient forms are generally required to be available on the job to permit the forms to remain in place during preliminary curing of the slab and to provide several hundred feet of forms in place ahead of the paver.

19–3. Final Finishing of the Subgrade. The next step in the indicated sequence of operations at the job site is the final shaping of the subgrade to the exact dimensions required by the plans and specifications. This operation is generally accomplished by a machine called a "subgrader" or "fine-grader," which rides on the forms and cuts off the subgrade to the exact shape desired. The subgrade is generally left a little high prior to this step, and the earth (or subbase material) excavated by this machine is thrown outside the forms. Extensive use also is now being made of electronically controlled fine-graders without forms, particularly when slipform pavers are used. A machine of this type is shown in Fig. 19–2. This machine operates from a taut wire. Trimming is done by cutting blades on a rotating drum. In some instances, trimming of the subgrade is accompanied by final rolling with steel

Fig. 19–2. Cutting blades on rotating drum trim subgrade to close tolerances. (Courtesy Wire Reinforcement Institute.)

rollers, particularly where a granular subbase is being built. A scratch template is generally employed at this stage to check the final finish of the subgrade.

19–4. Installation of Joints. The installation of the various types of joints which may be used in a concrete pavement is also a very important step in the construction process. A portion of the process of construction of the required joints, which we may call the "installation of joint assemblies," normally takes place between the final finishing of the subgrade and the beginning of actual concreting operations. Such assemblies may not be required, as for a plain concrete pavement in which transverse joints are sawed and dowel bars are not used. Sawing of joints is covered in Art. 19–11. Extreme care must be used in all operations accompanying the construction of joints if they are to function properly. The faces of transverse joints must be straight, at right angles to the centerline of the pavement, and perpendicular to the surface of the finished slab. Dowel bars, as commonly used in transverse joints, must be carefully placed and aligned parallel to the centerline so that they will not inhibit free movement of the slab ends in a longitudinal direction and so that they will properly perform their load-transfer functions.

Joint assemblies which are used to facilitate the construction of joints are quite varied in nature, depending upon the exact type of joint which is to be used. Typical joint assemblies are shown in Fig. 19–3, which

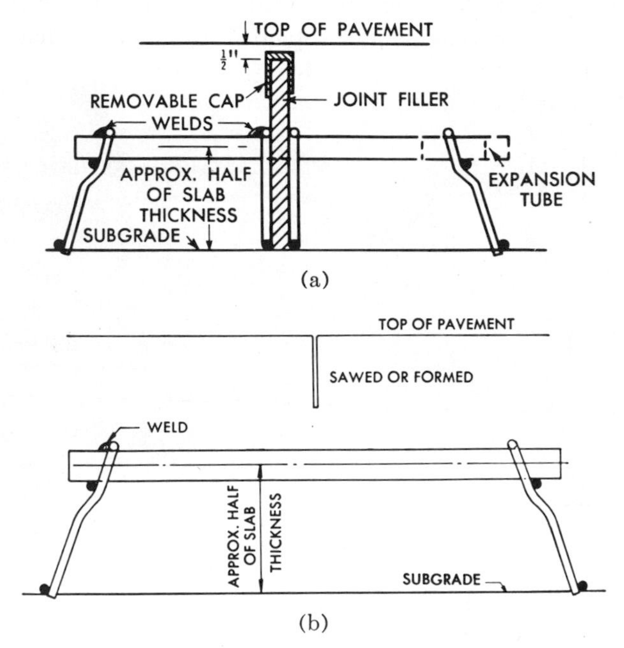

Fig. 19-3. Typical dowel assemblies. (Courtesy Bethlehem Steel Co.)

illustrates the use of the dowel unit manufactured by the Bethlehem Steel Company. The removable cap shown in Fig. 19–3(a) is removed after the concrete has been placed and subjected to preliminary finishing. The slot shown in Fig. 19–3(b) may be made just after the concrete has been finished by pressing a thin steel plate of appropriate dimensions into the concrete at the desired location or by sawing the hardened concrete. The spaces above the joint filler or the surface slot in a weakened plane joint, Fig. 19–3(b), are filled with one of several sealers previously discussed after the concrete has hardened. The filler is forced into the joint space by special machines. Edges of the surface groove are usually finished to a curve which has a ¼-inch radius. It should be noted that dowel bars used in transverse joints are commonly treated with paint, grease, or other materials on one side of the joint to prevent bonding of the bar to the concrete; the other end of the bar will, of course, be bonded. In transverse joints in which dowel bars are not used, which are commonly contraction joints of the weakened-plane type, the only steps involved in the construction are the formation of the surface groove, subsequent edging of the joint, and the installation of joint sealer. Longitudinal joints are installed in a similar fashion, except that the tie bars are bonded to the concrete on both sides of the joint.

Special machines are also available for placing dowel and tie bars and for forming the necessary surface groove through the concrete after it has been finished and while it is still plastic. One machine of this type is the Flex-Plane joint machine, which uses vibrating plungers and hydraulic lift controls to accomplish this operation. Tie bars across the longitudinal joint often are installed by a rotating wheel at the rear of the finishing machine or slipform paver. It may be further noted here that the prepared subgrade is usually sprinkled immediately prior to the placing of the concrete, and the forms are then oiled.

19–5. Batching of Aggregates and Cement. At this point in our discussion of the construction of concrete pavements, all preparations have been made at the job site for the placing of concrete between the forms. It is now necessary to examine the steps of preparing and mixing the concrete itself in some detail. Two general situations may exist. One of these involves batching of the aggregates and cement, with the concrete mixed at the site in a boom-and-bucket paver; this procedure is described in this article and Art. 19–6. The other is central mixing of the concrete (Art. 19–7).

On many highway jobs in rural locations, aggregates and cement are batched "dry" at a convenient location some distance away from the job site, brought to the job in trucks, and combined and mixed with water on the job, usually in a boom-and-bucket paver.

The aggregate and cement batching facilities may be located at the quarry or pit where the aggregates are produced or at a convenient railroad siding or dock, if shipment of the materials is made by water. Aggregate and cement are batched by weight. Facilities are provided for the weigh batching of cement and two, three, or four sizes of aggregate, depending on the specifications under which the work is being done. Aggregates and cement are generally stored and handled separately, either in separate batching units or in two different units of the plant. Semiportable batch plants are generally used on highway projects. Many different arrangements of plant components are possible.

In the batching of aggregates, from a simplified viewpoint, the essential units include a large storage bin with the required number of separate compartments. Compartments may be provided for sand and coarse aggregate, sand and two sizes of coarse aggregate, or sand and three

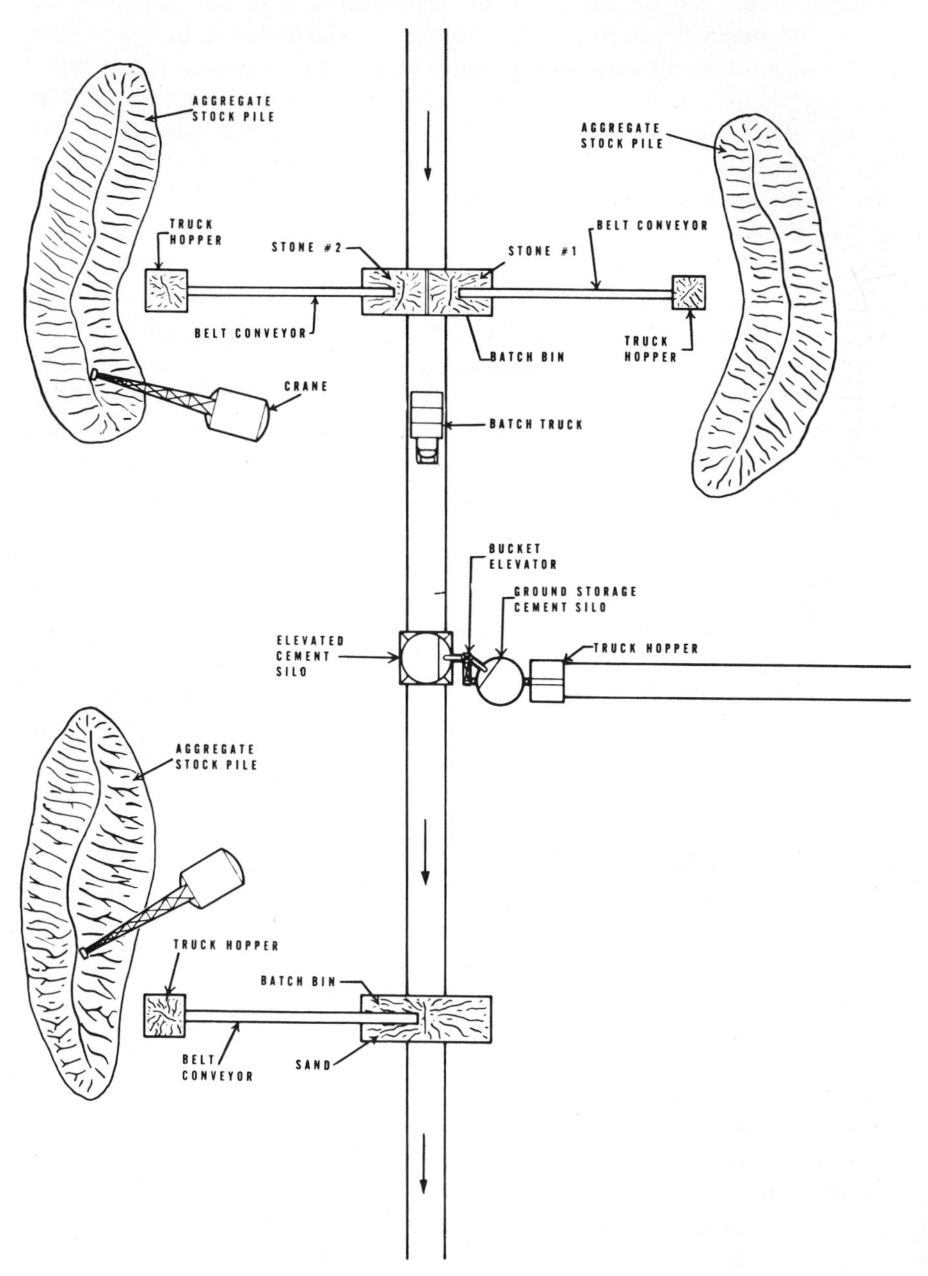

Fig. 19–4. Typical batching plant

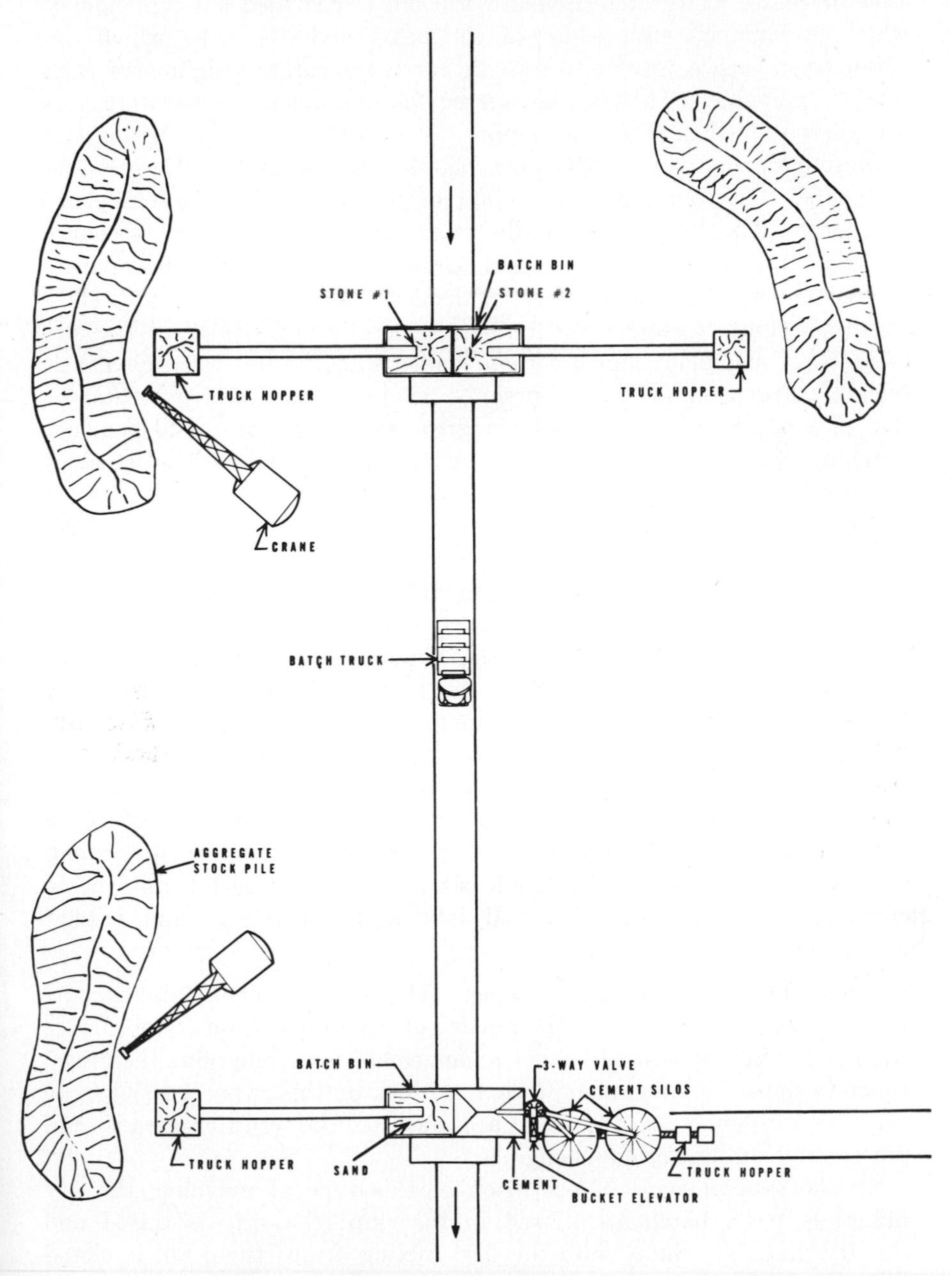

layouts. (Courtesy The C. S. Johnson Co.)

sizes of coarse aggregate. Beneath the bin is provided a weigh hopper which is equipped with scales of the beam and dial type which are arranged in such a way as to provide for the accurate weighing of each size of aggregate. That is, the aggregates are weighed separately, but are all contained in the same hopper. Control devices may be manual or automatic in nature, and some plants are equipped with recording devices so that continuous inspection of the weighing operation is not necessary. The hopper is usually mounted high enough to discharge directly into trucks which may handle from one to six batches in separate compartments. Aggregates may be loaded from railroad cars directly into the aggregate storage bin but are generally stock-piled at the site of the batching plant and are then loaded into the bin by a clamshell bucket. Provision must of course be made in the batch weights for moisture which is present in the aggregates. Aggregate (and cement) batching plants are frequently fitted with dual hoppers so that two compartments of a truck may be loaded at the same time.

Cement batching plants are similar to those that were just described. Bulk cement is generally used on highway projects, and facilities must be provided for the unloading of cement cars (both hopper-bottom and box-type cars are used), for the initial storage of the cement, and for the elevation of the cement into the bin of the cement batcher. The batching itself is done by weight, as indicated above. Cement may be discharged into the same truck compartment which holds the aggregate for a batch, or into a separate metal container mounted on the truck, as required by some state highway department specifications. In some cases it may be desirable to use one of the compartments of a multiple-compartment aggregate storage bin for the batching of cement. Cement batch units are of variable size, 300- to 600-barrel units being common. Fig. 19–4 shows typical layouts for aggregate and cement batching plants.

19–6. On-Site Mixing and Placing. The most frequently used model of concrete paver is the 34E paver, of the dual-drum type. It is equipped with a boom-and-bucket arrangement for discharging the mixed concrete onto the subgrade. Two machines of this type are shown in Fig. 19–5. Pavers of this sort are mounted on crawler treads and powered by gasoline or diesel engines.

In the general cycle of operation of this type of machine, the dry materials for a batch are placed in the skip; the skip is raised and the dry materials chuted into the first mixing drum; the drum is closed and the required amount of water added, after which the materials are mixed for approximately one-half the total mixing time. The mixture is then discharged into the second drum and mixed for the remainder of the specified time, after which the final mixture is discharged into the bucket. Control of the cycle of operation of the machine is very accurate and almost completely automatic in most machines of this type. The cycle is adjustable to any desired time sequence. The addition of the specified amount of water is also very carefully controlled, and measurement may be either by weight or by volume. Water is generally brought to the paver in large tank trucks, or it may be pumped from near-by sources. A separate device may be employed to add an air-entraining agent when required, or it may be done by hand. The total time of mixing generally varies from 60 to 90 seconds.

Also available is a triple-drum paver of the same type; its productivity is greater than that of the dual-drum unit.

FIG. 19–5. Two dual-drum, boom-and-bucket pavers place concrete for an Ohio highway. (Courtesy *Engineering News-Record.*)

When the bucket has been filled, it may be run out along the boom and emptied at any desired spot. The boom on a 34E paver is long enough that the machine may work either inside or outside the forms with equal facility. The paver usually operates outside the forms so as not to disturb the prepared subgrade. The paver is the key machine in this type of concrete paving operation, and all other operations and equipment should be coordinated with it in such a way that the paver is working at near maximum capacity all the time.

19–7. Central Mixing of Concrete. In this method, all ingredients of the concrete, including the water, are batched and mixed at a central location. The fresh concrete is then transported to the paving site in transit mixers or in trucks with agitating or non-agitating bodies.

Central mixing is the standard method on urban highway and bridge projects. It has become very popular on rural highway projects, too. Its principal advantages are closer control of the mixture, more uniform concrete, and higher production rates.

Central mixing plants vary widely. In urban areas, the plant often is a highly sophisticated, stationary "ready-mix" plant capable of producing automatically many mixtures of varying composition for many different purposes. Other plants are portable or semi-portable; they are automatically controlled, can be moved from one location to another quite easily, and have production rates up to about 500 cubic yards per hour with a single mixer.

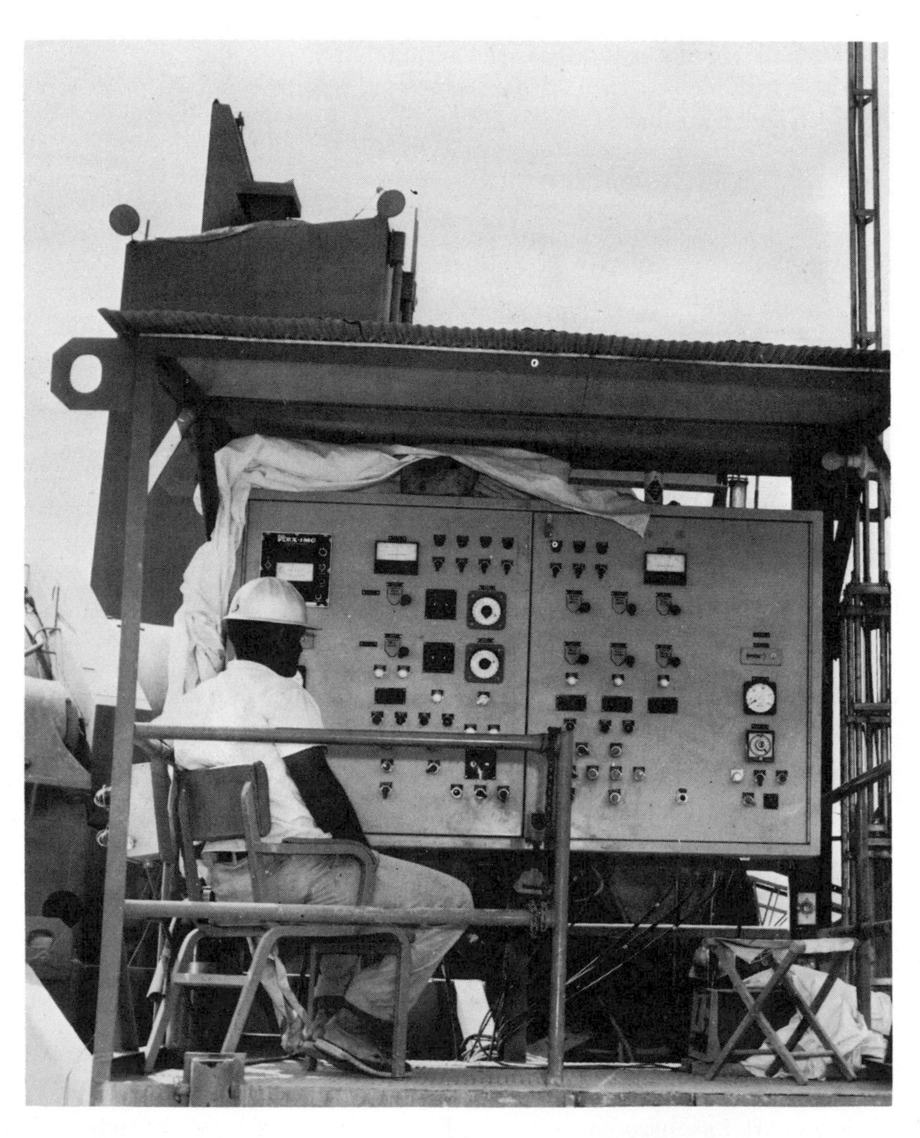

(a)

Fig. 19–6. One man at electronic console (a) controls operation of dual-drum concrete mixing plant (b). (Courtesy *Engineering News-Record.*)

(b)

FIG. 19-6. (*Continued*)

Nearly all modern concrete plants are fully automatic and equipped with recording systems. A single operator at a control panel performs simple steps that control all the functions of charging, mixing and discharging. Ingredient proportions can be controlled by preset dials, tapes or batch plugs. The control system incorporates an automatic moisture sensing device in the fine aggregate; adjustments are made automatically in batch quantities to correct for changes in moisture content. A plant also may be equipped with a "slump meter," which measures the current required to turn the mixer; the meter is calibrated with slump tests and gives an immediate check on the consistency of the concrete.

Permanent plants have more elaborate control systems that sense and govern the flow of all materials from the aggregate stockpiles through the plant to discharge of the concrete; ready-mix trucks often are dispatched from a central location by two-way radio.

Fig. 19-6(a) shows the electronic console that governs all operations in the plant of Fig. 19-6(b). This plant incorporates two tilting-drum mixers; the mixers discharge into trucks with side-dump bodies.

Fig. 19–7. Tilting-drum mixer discharges concrete into agitating body truck. (Courtesy Wire Reinforcement Institute.)

Fig. 19–7 shows a close-up view of a tilting-drum mixer discharging concrete into an elevating discharge, agitating truck body on an airfield paving project in Texas.

19–8. Spreading and Finishing. Spreading of the fresh concrete at the job site and finishing the slab to required smoothness are important operations that are done in a variety of ways. The number and kind of machines used vary from state to state, and among contractors.

If a boom-and-bucket paver is used, the bucket dumps the concrete on the subbase ahead of a spreader. The most commonly used spreader is one which incorporates a long screw or auger across the front of the machine; the screw distributes the concrete uniformly across the subbase between the forms. The spreader may carry vibrators, often has a transverse screed at the rear to give the slab a preliminary strike-off.

In a conventional paving train, the spreader is followed by a finishing machine that incorporates two transverse, oscillating screeds. Vibrators may be mounted on the front of this machine. This unit often pulls along a long wheelbase, float-type finisher. The float, which is suspended from a frame, operates independently of the side forms to iron out any irregularities left by the transverse screeds.

Fig. 19–8. Agitating body trucks discharge concrete into hopper-type spreader. (Courtesy *Engineering News-Record.*)

Also available are "one-pass" finishing machines that ride on the side forms, combine in one unit a spreader, vibrators, transverse screeds, and longitudinal float. These units have been described as "slipform pavers riding on side forms".

Contractors often use special machines to place central mix concrete between forms. These machines use hoppers to receive the concrete. In one machine of this type, the hopper travels across the grade to deposit the fresh concrete between the forms. Another machine of this type is shown in Fig. 19–8; on this project, two agitator-body trucks dumped concrete simultaneously into hoppers on either side of the grade. A transverse screw moved the concrete across the grade.

Simple box-type spreaders often are used with slipform pavers (Art. 19–9).

The placing of distributed steel reinforcing, either in the form of bar mats or wire mesh, complicates the spreading operation, but generally leaves the finishing operations unchanged. One way of placing the steel is to lay one course of concrete with a spreader; workmen then place the sheets or mats of steel on top of the fresh concrete. The spreader then makes a second pass over the steel (or a second spreader may be used).

A second method, and one that is growing in popularity, is to spread the concrete full depth, lay the steel on top, and then use another machine (a mesh installer) to force the steel into the concrete to the required depth. This operation is shown in Fig. 19–9.

Fig. 19–9. Blades of mesh placer push and vibrate reinforcing steel into fresh concrete. (Courtesy Wire Reinforcement Institute.)

Special methods are used to place continuous reinforcement.

The pavement is generally checked during final finishing operations by means of long wooden, steel, or aluminum straightedges, and any irregularities removed.

The final finish is applied to the surface by belting, brooming, or use of a burlap drag. Belting is done by the use of a narrow canvas or rubber belt which is moved longitudinally along the surface with a slight transverse motion; two men handle the belt, one on either side of the slab. In securing a broomed finish, which is considered to have slightly greater skid resistance than a belted surface, long-handled fiber brooms from 18 to 24 inches wide are used. The finish is obtained by placing the broom at one edge of the pavement and drawing it transversely across the surface in such a way that corrugations are produced which are not more than ⅛ of an inch in depth. Mechanical brooms are now common. A burlap drag is pulled longitudinally along the surface of the pavement, thus producing shallow surface corrugations; this is the method most commonly used in modern practice. The joints and edges of the pavement may be given a final finish, generally by the use of hand tools, with workmen operating from bridges as required.

The surface of the pavement is generally checked again after the concrete has hardened, with surface irregularities up to one-eighth of an inch (as measured by a 10-foot straightedge) being permitted. High spots detected at this stage may be removed by grinding. Use is also being made of recently developed "profilometers" or "road roughness indicators," which measure surface variations over a length of pavement by an accumulative measurement of deflections of a test tire. Roughness is usually expressed in terms of inches per mile.

19–9. Slipform Paving. Introduction of the first production models of slipform pavers in 1954 revolutionized concrete paving in the United States (*75*). Slipform highway paving now has been done in more than half the states; it has been used for unreinforced, conventionally reinforced, and continuously reinforced pavements. By 1966, over 80 per cent of the concrete paving done in California (all unreinforced) was being done by slipform machines; such machines in that state have laid many miles of three-lane (36-foot-wide) slabs in a single pass and a 48-foot-wide paver has been used on some projects.

Principal advantage of the slipform paver is the fact that one machine, under the control of a single operator, replaces the several machines in a conventional paving train. Hand finishing is held to a minimum. Since there are no side forms used, the labor of setting and handling the forms is eliminated. Pavements of outstanding smoothness have been built by the slipform method.

Several manufacturers produce slipform pavers, which vary in details of design and operation. All operate on the same general principle, however, to combine several operations in the one machine. Each of the machines has an electronic guidance system that operates from a taut wire to maintain line and grade; the taut wire often is the same one used to guide the subgrader (Art. 19–3) during final finishing of the subbase.

Concrete is dumped on the subbase ahead of the paver and spread to a reasonably uniform depth, then struck off by an oscillating screed. The concrete is then vibrated heavily and forced through the space formed by the traveling side forms that are part of the paver, the subbase, and a heavy transverse pan or beam (or beams) at the rear of the machine. The beam forms the surface of the slab and can be adjusted to give the desired cross-section. Some machines of this type have trailing side forms of variable length behind the machine to restrain the fresh concrete while the machine moves slowly forward; others carry no trailing forms. Careful control of the consistency (usual slump is 1 to 1½ inches) and entrained air content is essential for successful results.

Fig. 19–10 shows a Guntert & Zimmerman slipform paver on a highway project in northern Colorado; conventional dump trucks supply concrete to the paver, which has no trailing forms. The edge of the slab behind the end of the trailing forms (on another make of paver) is shown in Fig. 19–11; a mason is touching up the slab edge.

Fig. 19–10. Guntert & Zimmerman slipform paver at work on an Interstate project in Colorado. (Courtesy Colorado Department of Highways.)

Finishing operations vary, but handwork is held to a minimum. A burlap drag often gives the slab its final texture. Fig. 19–12 shows a long aluminum tube used to float the surface of the slab on some projects; the keyed joint is for later construction of an access road.

19–10. Curing. Curing of a concrete paving slab is necessary in order that the concrete may harden properly. It should be noted that water is absolutely necessary for the proper hydration of the cement and that the hardening of the concrete is not a drying-out process. Steps then must be taken to prevent loss of moisture from the concrete during the curing period. A large number of different curing methods are available, and the specifications of highway agencies relative to this phase of concrete construction may permit several different alternate procedures to be used.

Earth or straw may be spread over the surface of the pavement and kept constantly wet during the curing period. Another "wet-curing"

FIG. 19–11. Pavement edge left by trailing form of slipform paver. (Courtesy Wire Reinforcement Institute.)

FIG. 19–12. Aluminum tubular float finishing a concrete pavement. (Courtesy Wire Reinforcement Institute.)

process involves the spreading of burlap, felt, or cotton mats over the surface; the mats are sprinkled and kept constantly wet. Waterproof paper may be placed over the slab to retain the moisture and is widely used in some areas. Burlap, cotton mats, and paper are commonly furnished in rolls which cover the entire width of the pavement.

By far the most popular method in current use involves the spray application of light colored fluid to the entire area of the wet concrete. This is the commonly used "membrane" method. The fluid forms a film over the pavement which prevents moisture loss. The color in the fluid disappears after the passage of time. Membrane fluid is commonly applied by special spray machines which ride on the forms, or on rubber tires if a slipform paver is used, and ensure a uniform application of material over the entire area.

Preliminary wet curing is required by some highway departments, and the period of preliminary curing generally varies from one to three days. Total curing time, as evidenced by the period which must elapse before the pavement may be opened to traffic, commonly varies from 7 to 14 days. Opening of the pavement to traffic is very frequently based upon the attainment of a certain minimum flexural strength of the hardened concrete rather than on an arbitrary time period.

19–11. Joint Sawing. Most states in which a substantial mileage of concrete pavements is being built either permit or require the sawing of transverse joints; longitudinal joints are also being sawed in some cases. Sawing is done by the use of self-propelled, manually guided single-blade concrete saws or multi-blade saws which ride directly on the pavement or the side forms. Fig. 19–13 shows three self-propelled saws with diamond blades operating in tandem on a North Dakota highway project. To make joint sealing more effective, some states now require a step-down joint with a wider groove at the surface; such joints are cut by two or more saws operating in tandem.

Joints are sawed a short time after the concrete has been given its final finish. The time at which sawing is done is critical. On the one hand, the sawing must be done before random cracking occurs; however, premature sawing causes excessive spalling, water erosion, and excessive blade wear. The best way to determine the best time for sawing probably is to make a short trial cut a few hours after final finishing operations; appearance of the cut is then evaluated. Time of sawing varies from as little as 4 to as much as 24 hours after placing.

19–12. Other Considerations. Forms may generally be removed from the hardened concrete after 12 to 24 hours. Special devices are usually used to pull the forms so that the pavement will not be damaged in this process. Edges of the slab may be given their final finish at this time. At the end of the curing period and before opening the road to traffic, the surface grooves in longitudinal and transverse joints are cleaned and filled with joint sealing compound or some other material, like the neoprene compression seal described in Art. 18–15.

Some mention may be made of concreting in cold weather. Most highway organizations suspend concreting operations when the air temperature is 40°F and falling (or less), and resume operations when the temperature is 35°F and rising. In many sections of the country concrete construction is completely suspended during the winter months. When

FIG. 19–13. Three concrete saws with diamond blades cut a longitudinal joint. (Courtesy N. W. Ayer & Sons, Inc.)

concrete must be placed in cold weather, the heating of aggregates and water to moderate temperatures is quite common practice. In such circumstances the temperature of the concrete prior to placing may generally be required to be from 50 to 80°F. Calcium chloride frequently is used as an accelerator, to hasten the initial set of concrete placed in cold weather.

19–13. Concrete Base Courses. Concrete base courses are used occasionally to provide a support for a high-type bituminous pavement or a brick or block wearing surface (as in a vehicular tunnel). Their principal use at the present time is in urban areas. They are generally built of uniform thickness, with thicknesses of 6 to 9 inches being common.

Concrete bases have been built of both plain and reinforced concrete, with and without joints to control cracking of the base. Authorities differ as to the exact structural behavior of concrete bases which support a bituminous or brick wearing surface. The combined thickness of the base and wearing surface must, of course, be adequate to support the loads placed on the pavement structure. Many designers believe that a bituminous wearing surface of nominal thickness contributes but little to the structural strength of the pavement. It is customary in some organizations, however, to use concrete in the base course which has slightly less strength than that required of concrete for paving. A requirement of a flexural strength of 450 pounds per square inch is perhaps typical of such practice. This decrease in strength is generally justified by the fact that temperature warping stresses should be somewhat less in concrete bases than in pavements.

Another consideration relative to structural behavior of the base is the fact that extensive cracking which occurs in the base will generally be reflected in the wearing surface over a period of time. With this fact in mind, many designers feel that enough joints and reinforcing steel should be used in the base to effectively control cracking. Again because of lessened temperature warping stresses, longitudinal joints may not be necessary in a concrete base which has a width of 20 feet or so. Contraction joints of the weakened-plane type are generally provided in concrete bases to control cracking, while transverse expansion joints are less commonly used. The American Concrete Institute (*202*) has suggested that transverse joints in concrete base courses be placed at short intervals (15 to 20 feet) in order to prevent excessive opening of the joints and resultant cracking of the surface course.

Methods used in the construction of concrete base courses are very similar to those used in the construction of concrete pavements. Requirements relative to the surface finish of the completed base may be somewhat less severe than similar requirements relative to pavements. If a brick or block wearing surface is to be placed upon the base, the surface of the concrete is usually given a smooth finish, while a roughened, broomed finish is generally used if a bituminous wearing surface is to be placed. The base course must be cured as indicated previously for concrete pavements.

19–14. Cement-Treated Bases. Cement-treated bases are used widely beneath concrete pavements; in 1966 the Portland Cement Association reported that this type of mixture had been used in 29 states. CTB mixtures are also used beneath asphaltic wearing surfaces, although less frequently. Principal purposes of a CTB are to provide a stable base for the construction of the concrete slab, particularly when a slipform paver is used, and to prevent pumping.

A cement-treated base course mixture is very similar to the soil-cement mixtures described in Chapter 13. In many cases, however, there are two major differences; the material with which the cement is mixed is a crushed or uncrushed granular material of moderate to low plasticity, and cement contents are relatively low (from 2½ to 6 per cent by dry weight of aggregate in California, where bases of this type have been built for many years).

Specifications for materials vary. For Class A or B cement-treated base, California specifies (*192*) a clean, graded aggregate with a maximum size of 1 inch and from 2 to 15 per cent passing a No. 200 sieve, and a Sand Equivalent (Art. 14–10) of not less than 20. In addition, the state requires (*193*), beneath an 8-inch concrete slab, that the CTB be 0.35 foot thick, and that the cement content be sufficient to give a minimum compressive strength of 400 psi at 7 days. If the slab is 9 inches thick, the CTB is 6 inches thick with a minimum compressive strength of 750 psi at 7 days.

The Portland Cement Association recommends (*132*) that enough cement be mixed with the granular materials to develop an unconfined compressive strength of not less than 300 psi at 7 days.

Construction of a cement-treated base course is by either road mixing or plant mixing, using the general methods described in Chapter 13. The mixture is compacted at or near optimum moisture to maximum density. It is cured as is soil-cement, often by applying a bituminous material to the top surface.

CHAPTER 20

HIGHWAY MAINTENANCE

When a highway construction project has been completed, accepted from the contractor, and final payment has been made to him, a new facility is available for use by the traveling public. By the same token, a new responsibility is created to preserve the new investment and to serve and protect the interests of the traveling public. This is true not only of newly constructed highways but also for all highways of our road system. Sudden failures, damage by storms, gradual deterioration, and unexpected obstructions can cause personal injury, death, or delay. "Highway maintenance" is defined as the function of protecting the highway structure and keeping it in condition for safe use.

Public roads include state highways, county and local roads, and city streets. Most of the work of maintaining these highways is usually performed by various governmental agencies with their own labor and equipment. Thus state roads may be maintained by a state highway department; county roads may be maintained by a county road commission; and local roads may be maintained by local governmental agencies. Most cities, particularly the larger ones, have organizations for street and highway maintenance. Some highway departments now carry out certain maintenance operations by the contract method, particularly on heavily traveled routes. It is expected that this trend will continue in the future.

20–1. State Highway Maintenance. As indicated in the organization charts of Chapter 2, each highway department has a maintenance division. This division is headed by a maintenance engineer who directs the maintenance of the highways throughout an entire state. A state is generally divided into several geographical districts, the number often being determined by the land area and the number of miles of road within the area. Each district is made up of a number of counties, and all the work is under the direction of a district maintenance engineer. The state of Iowa, for example, has six districts or divisions in its 99 counties; each is headed by a district maintenance engineer to whom four or five resident maintenance engineers report. There are 25 highway districts in Texas, with a district maintenance engineer in each; each district is divided into maintenance sections, each of which includes about 180 miles of highways. In some states, the work in each county of a district is under the direction of a county superintendent. Often each county has maintenance garages where maintenance equipment and materials are stored and which serve as offices for maintenance superintendents.

While a majority of state highway departments maintain their state highways with their own labor and equipment, some states contract for this service with county governmental agencies that are qualified to do this work. Proponents of this method claim an economy due to the elimination of duplicate supervision, garages, and equipment, and to other savings brought about by combined operations. Under this arrangement counties benefit by an increase in their operating budgets, personnel, and equipment.

20-2. County and Local Road Maintenance. County road maintenance organizations are generally similar to those of the state highway departments but, naturally, they operate on a smaller scale. The type of maintenance performed on county roads may vary a great deal in a given state. Many counties compare favorably with the state groups, both from an organizational standpoint and as to a high degree of maintenance performance. A large number of other counties lack efficient personnel and funds for the purchase of equipment necessary for adequate maintenance. In recent years many counties in rural states have pooled their resources to obtain the services of a maintenance engineer and permit the purchase of maintenance equipment which they could not otherwise afford. The results have proved most gratifying.

20–3. City Street Maintenance. Trunk highways passing through cities and villages are usually maintained by the state highway departments. Some of the larger cities have large maintenance organizations and enough equipment for contracts to be made between the state highway department and the municipality for the maintenance of these trunk highways. City street organizations are, of course, responsible for the maintenance of the large majority of the streets located within the city limits.

20–4. Maintenance Budgets. A certain portion of the funds available for highway purposes is set aside for highway maintenance and may be referred to as the "maintenance budget." The amount of this budget is usually determined by making a study of cost records of previous years and estimating the amount of money needed to maintain the highways throughout the coming year. These estimates are made on an annual basis. This is generally true for all maintenance organizations, whether state, county, or municipal. When the annual budget for a state is determined, these funds are apportioned to the various districts.

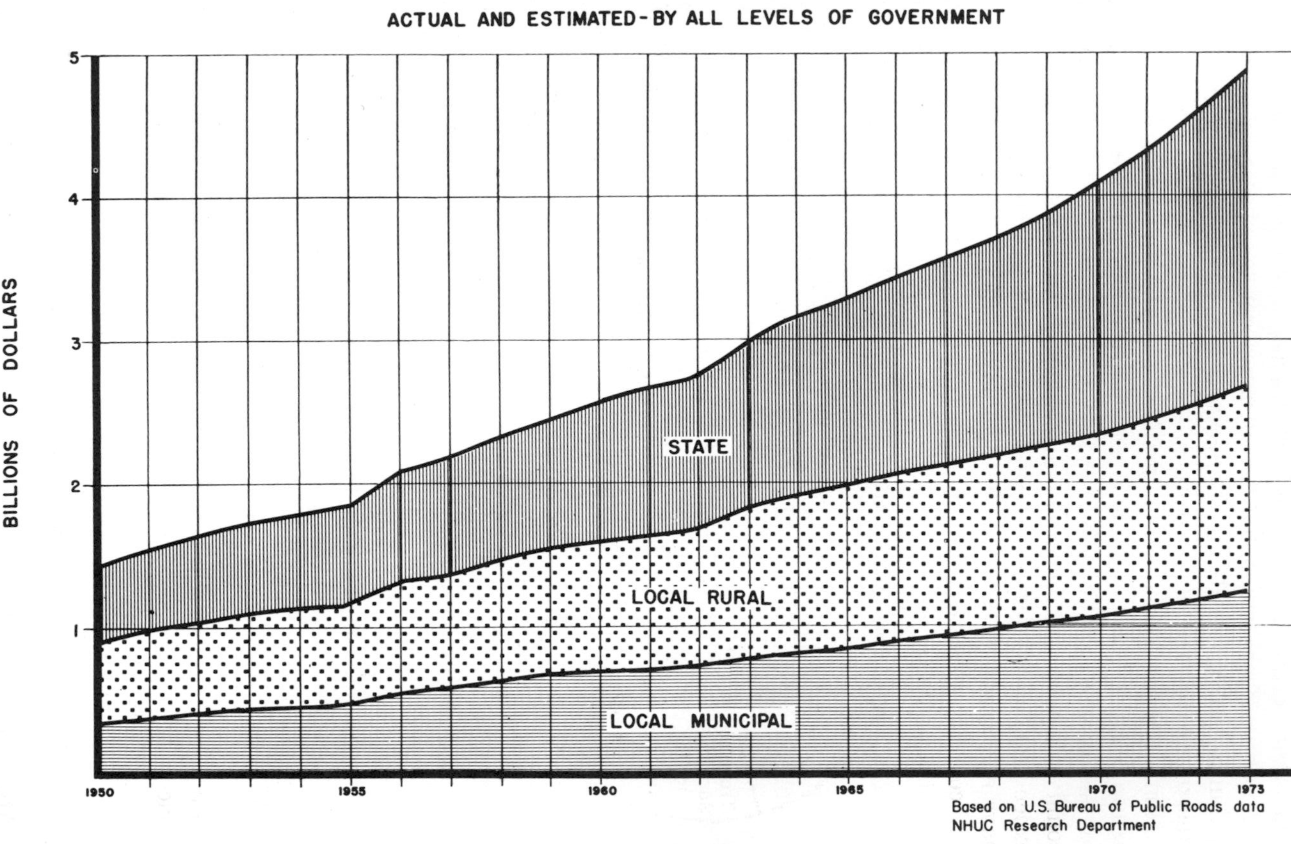

FIG. 20–1. Actual and estimated annual highway-maintenance expenditures, all levels of government. (Courtesy National Highway Users Conference, Inc.)

Allocation of district funds is based on the road mileage, types of road, and snow removal and other cost studies. The district maintenance engineer then allocates the funds of his district budget to the counties within his district on the basis of county mileage and his detailed knowledge of the roads within his district.

Annual expenditures for maintenance constitute a very sizable portion of the over-all budget. State maintenance expenditures were estimated to be about $1.1 billion in 1966, while maintenance expenditures on county, city, and town roads were estimated to be $1.85 billion—a total expenditure of $3 billion on all roads and streets. Maintenance costs have increased steadily over the years, and it is expected that this trend will continue (Fig. 20–1).

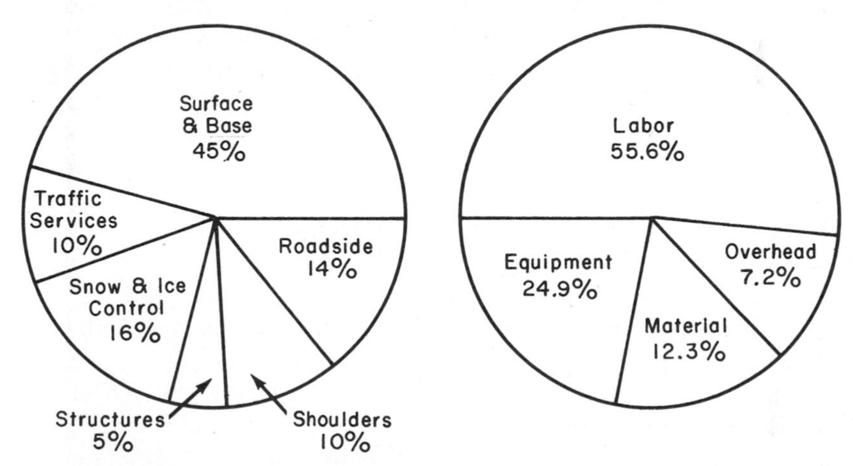

Fig. 20–2. The maintenance dollar, showing approximate services and cost distribution for a typical year. (Courtesy National Highway Users Conference, Inc.)

Fig. 20–2 shows the approximate services and cost distribution for a typical year, on a national basis. Snow removal, bridge repairs, and pavement repairs take nearly two-thirds of the annual maintenance budget on the New York State Thruway in a typical year. In Iowa, in a typical year, the breakdown is surfaces, 25.5 per cent; shoulders and approaches, 13.9 per cent; roadside and drainage, 12.8 per cent; traffic service, 11.8 per cent; snow and ice control, 33.9 per cent; and miscellaneous, 2.1 per cent.

Maintenance Operations

Maintenance work in most of the organizations follows the same general pattern. Much of it consists of maintaining the various types of road surfaces or that portion of the highway used by the moving vehicle.

Estimates based on a large volume of data from many highway departments show that almost half of the total state maintenance budget is used for this purpose. The remainder is spent on maintaining shoulders, drainage structures, roadsides, traffic service, snow and ice control, bridge repairs, and special services.

Maintenance of Road Surfaces

20–5. Soil-Aggregate Roads. This classification of roads is taken to include those constructed from natural earth, sand-clay mixes, and various coarse graded aggregate materials which generally require similar maintenance operations. Sand-clay roads are those built with artificial or natural mixtures of selected soils to produce a stabilized road. The sand-clay roads of the South Atlantic States are an example of this type. The coarse graded aggregate surfaces consist of a wide range of mineral aggregates such as gravel, crushed stone, crushed slag, chert, *caliche,* or similar substances combined with clay, stone dust, or other binder material to produce stability. Water-bound and traffic-bound macadam surfaces are included in the coarse graded aggregate group. The construction of roads of this general class was described in Chapters 13 and 14.

To maintain this type of road properly, it is necessary that the surface be kept smooth, firm, and free from excess loose material, with the proper crown for adequate drainage. The use of a dust palliative is frequently necessary in order to minimize the loss of material and to eliminate as much as possible the hazards due to dust. Maintenance of this type of road can include (1) patching, (2) blading, (3) scarifying and resurfacing, (4) stabilization and dust control, and sometimes, (5) application of a bituminous surface treatment.

20–6. Patching of Soil-Aggregate Roads. Failures of soil-aggregate roads are principally due to improper drainage, poorly mixed materials, or an inadequate foundation. A careful investigation should be made to determine which of these may be the cause of failure. Surface drainage can be ensured through the use of the proper amount of crown. Excessive moisture in the subgrade may be eliminated by lateral or side drains to intercept free water or to lower the water table. Perforated pipe underdrains may also be installed.

When road failure is due to a poor mixture or gradation of materials, it is usually necessary to remove the unsuitable material and replace it with suitable materials of the proper gradation. The edges of the area removed from the traveled roadway should be squared, and the patching material placed and thoroughly tamped, adding water when necessary to obtain the proper compaction. When the area to be repaired is large, or where the surface is worn away, it may be better to scarify, reshape the surface, and add the necessary additional material.

20–7. Blading. Blading is a general operation carried out on all soil and stabilized surfaces. Its purpose is to fill in the ruts and smooth out any irregularities in the surface.

Blading is done by a motor grader. The versatile motor grader also can do many other maintenance operations such as shaping of shoulders, cleaning and shaping of ditches, and similar work. Blading also may be done by trucks with underbody blades. This equipment is usually attached to four-wheel drive trucks and hydraulically operated.

Blading should be done as soon as practicable after a rain, when the surface materials are moist. The work is accomplished by dragging or blading the surface materials from the edges toward the center and then drawing the loosened material back again depositing it in depressions in the surface. Some blading may be required during dry weather in order to remove surplus loose materials from the traveled roadway. This material is bladed to the side of the road and bladed back during wet weather or when the material is moist. Blading operations are illustrated in Fig. 20–3.

Fig. 20–3. Blading a gravel road. (Courtesy Michigan Department of State Highways.)

Maintaining a crown is necessary on this type of road surface. The crown may vary from ¼ inch to ½ inch per foot, depending on the steepness of the grade and whether the section is in cut or fill. Water drains off the surface more readily on steep grades. A flat **A** crown is generally used on pervious surfaces except on curves where superelevation is provided.

20–8. Scarifying and Resurfacing. When the areas to be patched are numerous, or when they extend over a considerable area, it is often more efficient to recondition the entire surface. The usual procedure is to scarify the surface of the roadway to the full depth of the surface material, proper care being exercised so that any undesirable material from the shoulder or the subgrade is not mixed with the surface-course materials. When additional material must be added, it should be selected with care. A visual inspection and a knowledge of local materials may aid in the selection of the necessary materials. However, if the section being repaired is of considerable extent, laboratory and field testing of the separate materials and their combinations may be necessary.

When the scarifying of the old road surface has been completed and any necessary material added and mixed, a blade grader is used to shape the road surface to proper crown and thickness. The necessary compaction may be accomplished by permitting traffic to use the surface while continuing blading or dragging operations, or by rolling and watering if necessary until the surface is thoroughly compacted.

A familiar failure on gravel roads is the formation of corrugations running transverse to the traveled way. This "washboard" effect can be temporarily corrected by blading and dragging. Usually the construction of a more suitable wearing surface is the only permanent solution to the problem. The addition of a fine binder where the surface mixture is too coarse will tend to lessen the formation of corrugations. Likewise, where an excess of fine material exists, the addition of angular coarse material and a reduction of clay binder tend to alleviate this condition.

The spacing of these corrugations has been found to be about 31 inches from crest to crest, with a depth of 1.5 inches from crest to trough. This spacing is apparently related to some vibration frequency of the motor vehicle, but the frequency is much higher than the natural frequency of the motor vehicle spring system.

20–9. Stabilization and Dust Treatment with Palliatives. Stabilization of roads with the use of binder soil such as clay, the chlorides, or other additives is often done by maintenance forces. The methods of stabilization are similar to those discussed in Chapter 13. The use of dust palliatives serves two purposes: (1) the elimination of dust, and (2) the preservation of the soil-aggregate surface. Tests have indicated that the annual loss of material on gravel roads is equivalent to ½ to 1 inch of surface. In the case of an 18-foot surface, this would be 150 to 300 cubic yards of material per mile of road. The use of palliatives reduces this loss and at the same time aids in the stabilization of the surface. A comparison can readily be made between the cost of adding surface materials and the application of a dust palliative. The reduction of the dust hazard for the free flow of traffic is also a valuable contribution to driving safety.

The materials most commonly used as dust layers are calcium chloride, sodium chloride, and bituminous substances. Calcium chloride is most generally used because of its "hygroscopic" nature, in that it attracts moisture, and because it is "deliquescent," that is, it dissolves itself in the moisture which it attracts. When calcium chloride is scattered over the road, it absorbs and holds from 2 to 5 times its weight in water and then serves to keep the surface damp and dustless for prolonged periods.

In dry weather, calcium chloride will absorb moisture from the air during periods of high humidity, such as at night, in sufficient amounts to maintain a dustless surface during the day. Calcium chloride is susceptible to dilution by rain and gradually leaches into the soil where, if present in sufficient concentration, it tends to reduce the severity of frost heaves.

The calcium chloride is available in flake and granular form; and best results are obtained when mechanical spreaders are used. Either a lime-drill type or a centrifugal spreader may be used. Overlapping of the spreads in the middle of the road will safeguard against leaving an untreated portion in the middle of the roadway.

The best time to apply the material is directly after a rain. The first application is usually made in the spring when dust begins to appear, followed by periodic applications during the summer months when needed. The initial application is made at a rate of from ¾ to 1½ pounds per square yard. Later application requirements will vary. Some roads require as little as 1½ pounds per square yard through the season, and others require a total of 2 to 2½ pounds per square yard, depending on the texture of the surface, previous amount of stabilization, dryness of the season, and the amount of traffic. Shaded areas require less amounts of palliatives than do areas exposed to the sun. The same is true for low-lying sections of road. Frequent applications may be desirable on curves, grades, intersections, and other places where excessive wear occurs.

Sodium chloride, or common salt, is sometimes used as a dust layer.

It will retard evaporation but will not absorb moisture from the air. When rock salt is used it is usually mixed in the upper 2 inches of material at the rate of 2.5 to 3 pounds per square yard. Water is also usually added during the mixing process. Magnesium chloride is also used, the method of application being similar to that used for calcium chloride.

Use is also made of various bituminous materials for laying dust on soil-aggregate roads. The surface to be treated must be bladed to proper crown before the material is added, and any loose material must be removed. Materials used may be tars RT–1 and RT–2; the slow-curing road oils SC–70 and SC–250; or the medium-curing cut-back MC–70. Asphalt emulsions are also used.

The roadway should be fairly dry and free from dust when the bituminous materials are applied, except when an emulsion is used, when the presence of moisture is desirable. Emulsions require a certain period for breaking and curing. If traffic must follow soon after the use of the emulsion, the surface is covered with a light coating of clean coarse sand.

The amount of bituminous material for dust-laying ranges from 0.10 to 0.25 gallon per square yard. Emulsions often are diluted, one part emulsion to three parts of water. Application should be made with a pressure distributor.

20–10. Bituminous Surface Treatment. Bituminous surface treatment operations, which are sometimes done by maintenance forces, have been described in Chapter 15. As traffic increases, the road surface may be improved in various ways. The line of demarcation between maintenance and a capital improvement or betterment is sometimes not well defined. Care should be exercised to segregate maintenance costs from those of betterments as much as possible.

20–11. Maintenance of Bituminous Surfaces. The various types of bituminous surfaces and their construction have been previously described in Chapters 14, 15, 16, and 17. Weaknesses develop in these surfaces due to weathering, the action of alternate freezing and thawing, or failure of the base or subgrade. The maintenance of these surfaces may be divided into five distinct operations: (1) patching, (2) paint patching, (3) scarifying, (4) resealing, and (5) nonskid surface treatment. These operations are described in a "Policy on Maintenance of Roadway Surfaces" of the American Association of State Highway Officials, (*102*), from which much of the following discussion has been abstracted. This also applies to the maintenance of concrete surfaces, which is described later.

Before maintenance repairs are made to the surface, it is important to make the usual investigation for the cause of failure. If failure indicates a poor subgrade or faulty drainage, these items should be corrected before any repairs are made on the surface. Some states are now using road roughness measurements and the pavement serviceability concepts developed at the AASHO road test (Art. 12–8) as an aid in setting up maintenance programs. Methods and equipment used are described in reference *13*.

Patching may consist simply of an application of a thin coat of bituminous material to an area where cracking or raveling appears and spreading a coarse mineral aggregate over the affected area. Care must be exercised in the application of the proper amount of bituminous material in order to prevent bleeding, with the resulting slipperiness of the repaired surface.

In areas where the surface shows raveling and disintegration, a bituminous mixture may be placed over the weakened area and then compacted. Both hot and cold mixtures like those described in Chapter 17 are used, with asphaltic concrete and sheet asphalt being the most common. Where the surface has disintegrated to a depth of 2 or 3 inches, the failed area should be cut out, all loose material removed, the base course made true and firm, and the bituminous patching material placed in the hole. The edges of the hole should be cut vertically and should receive a coat of bituminous material to seal out any moisture which may enter the subgrade. The final patch should conform to the grade and contour of the existing surface.

Patches may also be made by a penetration method which consists of filling the hole with aggregate, compacting the aggregate, and then applying the bituminous material. Crushed stone is then applied and the patch is rolled. Many different patching methods are used with good results. The method used will depend on the type of bituminous surface and current practice prevailing in the particular locality.

Paint patching is essentially a preventive maintenance operation. Its purpose is to seal the surface to prevent penetration of water and to strengthen the wearing surface. The work consists of applying a thin coating of bitumen over the surface and placing fine stone chips, pea stone, or coarse sand over the patched area. The patch is then broomed and rolled with a light roller.

On low-type bituminous surfaces, such as bituminous treated gravel, where large areas are in poor condition, the entire surface should be scarified and reworked. This can be accomplished by the use of a motor grader with a scarifier attachment, or with a scarifier and a mechanical pulverizer or mixer. A combination of a ripper (not shown) and the cutter-crusher-compactor shown in Fig. 20–4 also is being used to salvage old bituminous roads and streets. The material is reworked, using additional bituminous material as required. It is then bladed to proper crown and grade and compacted. The compacted surface is then sealed with a bituminous material.

The seal coats and surface treatments described in Chapter 15 are widely used to rejuvenate old bituminous surfaces or to correct a slippery condition. Slipperiness is usually caused by an excess of bitumen on the road surface. This may be due to the application of too much

Fig. 20–4. Compaction-cutter-crusher attachment processes an old bituminous surface. (Courtesy American Tractor Equipment Company.)

bituminous material or an improper gradation of aggregate, or it may be caused by the kneading action of heavy traffic pushing the asphalt binder to the surface. Skid-proofing may be achieved by application of a surface treatment using sharp, angular aggregate such as blast furnace slag or by disking the surface. Slurry seal coats (Chapter 15) are widely used for this purpose. In severe cases the complete surface may have to be reworked in order to correct slipperiness.

A somewhat different type of maintenance operation has been carried out in Texas on bituminous pavement which has considerable surface cracking. In this method, medium-curing cut-back asphalt diluted with kerosene is poured into the cracks from a hand pouring pot. The material is then worked into the cracks with a rubber squeegee, after which sand is spread to blot or absorb the surplus asphalt. The sand is later removed. This process may also be carried out prior to seal coating.

20–12. Resurfacing of Old Bituminous Pavements. When maintenance operations become excessively expensive, it is difficult to maintain an adequately smooth surface, the pavement has to be widened, or for other reasons, the old pavement may be resurfaced. Before undertaking a resurfacing project, a careful study of its economic justification should be made. A general rule of thumb is that resurfacing will probably be economical when from 2 to 3 per cent of the total pavement area requires patching each year. Both the availability of money and the priority of the project will influence the decision. A comprehensive study of the condition of the existing surface and related factors is desirable before undertaking a resurfacing project. Included among the factors which must be given detailed consideration are the following:

1. History of pavement
2. Existing physical condition.
3. Character of existing and anticipated traffic.
4. Relation of geometric design to safe and efficient operation of anticipated traffic.

Corrective measures which may be required in the preparation of an old surface for resurfacing may be very considerable in extent or somewhat minor, depending on the type and condition of the existing surface and its adequacy to serve its new function as a base course. For example, major reconstruction measures may be required when substantial changes in grade, profile, or crown are called for in the new surface. Recommended reductions in grade and crown may be accomplished by scarifying and recompacting old surfaces such as water-bound macadam or stabilized soils; higher types of surfaces may require removal and reconstruction to the desired grade and section. Changes in width are generally accomplished by duplicating the original construction as closely as possible, with particular attention being paid to the development of adequate supporting power in the widened section so that this section will not subside under service to a lower elevation than exists in the original pavement. Space does not permit coverage of all the possible situations which may arise when the old surface must be reconstructed in whole or in part before resurfacing. Measures used are in general dependent upon the individual nature of the project concerned and are based upon the experience of an organization in the performance of similar work.

The old surface, regardless of its type, may be rough, with waves and bumps which are too considerable in magnitude to be compensated for by a relatively thin wearing surface. These bumps and waves may be removed by scarifying and recompacting certain surface types. When old bituminous wearing surfaces have these defects and are to be resurfaced, the bumps and waves may be removed by cutting or burning away the old surface with a heater planer. Burners in a big, insulated furnace heat and soften the surface, which is then scarified and remixed if desired, with additional bituminous material added as needed. Sometimes it may be more desirable to correct existing surface irregularities by the construction of a special leveling course, which may be the same bituminous mixture used in the original construction or a new mix. The construction of a leveling course is a major operation which usually is quite similar to the placing of a bituminous wearing surface. The course will have a variable thickness and will be constructed to the desired section and grade. The leveling course will have a lower bitumen content than the wearing course.

Holes and depressions which occur in the existing surface must also be corrected before resurfacing (or the placing of a leveling course) may begin. The holes are filled with the same mixture used in the original construction or with a special patching mixture. Hand methods are extensively used in patching operations. The infrared heater shown in Fig. 20–5 does an effective job in softening materials for removal or reworking during patching operations.

In many cases the preparation of the surface may consist simply of the removal of loose material on the surface by the use of rotary brooms or power blowers. If excessively dusty or unconsolidated areas remain after this treatment, they may have to be primed before construction may proceed.

The foregoing paragraphs have served only to illustrate the corrective measures which may be necessary before a new bituminous surface may be applied. Other corrective measures may be needed, and the remedial procedure to be followed in a given case is a matter for individual decision both as to the general and specific nature of the procedure used. In any event, the objective is to prepare the old surface in such fashion that it will adequately support the new wearing surface.

When preparations are complete, the new surface is constructed by methods which are very similar to those described previously for new surfaces. Multiple-lift surface treatments and hot asphaltic paving mixtures, particularly asphaltic concrete and sheet asphalt, are used most frequently. Road-mix surfaces of various types are used to some extent.

Many different types of mixtures have been used for resurfacing. Two tubes of the Lincoln Tunnel under the Hudson River between New York City and New Jersey were resurfaced with a hot sand-asbestos-asphalt mixture. The mixture contained 22 per cent ¼-inch crushed stone, 63.5 per cent silica sand, 4 per cent mineral filler, 2.5 per cent asbestos fiber, and 8 per cent asphalt cement (85–100 penetration grade). The contractor leveled the existing surface with a heater-planer, applied a tack coat of 85–100 asphalt cement, then placed the overlay with a conventional paver. Many miles of the original Pennsylvania Turnpike have been resurfaced with 3 inches of asphaltic concrete, 2 inches of binder course, and 1 inch of surface. Alabama, Florida and Mississippi

Fig. 20–5. Patching a bituminous pavement with an infrared heater (background). Trailer in foreground stores patching mixture kept hot by infrared heaters fueled by propane gas. (Courtesy United Suffolk Corporation.)

have resurfaced many miles of highways with a cold mixture incorporating crushed blast furnace slag, penetration grade asphalt cement, and liquefier (kerosene or No. 3 grade fuel oil).

In Oklahoma, "drill-lime stabilization" has been used to stabilize subgrades beneath bituminous pavements before patching or resurfacing. The procedure calls for hydrated lime to be introduced into a wet, plastic clay subgrade through holes drilled in the pavement; action of the lime is like that described in Art. 13–32.

20–13. Maintenance of Portland Cement Concrete Surfaces. The maintenance of concrete pavements consists for the most part of (1) filling and sealing joints and cracks in the pavement surface; (2) repairing spalled, scaled, and map-cracked areas; (3) patching areas where failure has occurred; (4) repairing areas damaged by settlement or pumping; and (5) treating buckled pavements.

The purpose of filling and sealing joints and cracks is to prevent the seepage of moisture to the subgrade and to preserve the original joint space. At the time the joint is sealed, the concrete must be dry and the joint space thoroughly cleaned of all scale, dirt, dust, and other foreign matter, including old joint sealer. Power cutters are used to cut and groove the joints prior to resealing; a mechanically driven wire

Fig. 20–6. Application of hot-applied joint sealing compound. (Courtesy Cutler Engineering.)

brush is used to thoroughly clean the joint. Immediately before sealing, the joint is blown out by the use of a jet of compressed air. Hot-applied rubber-asphalt sealing compounds are often used; they are generally heated and melted in a portable (rubber-tired) melting kettle of the double-boiler, indirect heating type which utilizes oil or other suitable material as a heat-transfer medium. The recommended type of equipment for applying the sealer is a mechanical pressure-type applicator, such as the one shown in Fig. 20–6. Mechanical applicators of the gravity type are also used, as are hand pouring pots. Cold-applied, two-component elastomeric polymers also are used as joint sealers. This maintenance operation is carried on throughout the year. The most favorable time, however, is during the warm summer months, when working temperatures are higher and better results can be obtained.

The repair of spalled, scaled, and map-cracked surfaces is a common maintenance problem for this type of pavement. Map-cracking is distinguished by irregular cracking over the pavement surface. Spalling is a chipping or splintering of sound pavement and usually occurs along the joints or cracks in the pavement. Scaling is caused by the deterioration or disintegration of the concrete and may occur anywhere on the pavement surface. The spalled, scaled, and map-cracked areas are cleaned of all loose and foreign material and sealed with one or more applications of bituminous material such as rapid-curing cut-back asphalt, quick-breaking emulsion or light tar, and immediately covered with coarse sand, pea gravel, or finely crushed stone chips. When excessive spalling and scaling has occurred to a depth of from 1.5 to 3 inches, it is often necessary to place a base course of penetration macadam or a bituminous mixture. A wearing course of bituminous material and aggregate cover is then placed over the repaired area.

An extensively repaired surface often results in an undesirable mottled appearance which is sometimes corrected by the application of a seal coat over the entire area.

Failed areas in concrete pavement should be repaired with cement concrete. The failed areas are marked out by a man qualified for this type of work. A crew then follows with jack hammers or other mechanical equipment, breaks and removes the broken concrete, and prepares the area for the new surface. At the edges of the patch the old slab of concrete should be undercut and the patch placed as shown in Fig. 20–7. High early strength cement should be used in the new concrete. The thickness of the patch should never be less than that of the existing slab, and the use of reinforcing steel is very often desirable. When patches are made adjacent to expansion joints, the expansion joint should be replaced. Contraction or construction joints may be omitted, however, if the patch extends across the entire width of pavement.

Surface irregularities can be removed from concrete pavements by grinding or cutting the surface. The machine shown in Fig. 20–8 is a "bumpcutter," a self-propelled unit that has a cutting head with diamond blades. Machines like this ground more than 1 million square feet of pavement in rehabilitating 16 miles of freeway in California. The blades produce a series of shallow, parallel grooves in the surface, giving it a non-skid texture.

Highway departments in some of the northern states are spraying a 50-50 mixture of boiled linseed oil and mineral spirits on concrete pavements and bridge decks to protect them against the deteriorating effects of deicing chemicals. It is recommended (*45*) that the mixture be applied in two coats, a first application of 0.025 gallon per square yard, followed by one of 0.015 gallon. The treatment can be applied to concrete of any age, but is most effective on new concrete after 28 days of curing.

Fig. 20–7. Concrete patching. (Courtesy Bureau of Public Roads.)

20–14. Pumping of Concrete Pavements. One of the major problems in the maintenance of concrete surfaces is correcting settlement of the slab caused by pumping. Pumping is usually indicated by (1) spalling of the pavement near the centerline and a transverse joint or crack, (2) the ejection of water from joints and cracks, (3) discoloration of pavement surface by subgrade soil, (4) the presence of mud boils at the pavement edge, and (5) breakage of the pavement.

Pumping of concrete pavements may be prevented by maintaining adequate drainage, correcting faulty drainage, and sealing joints and cracks. Where pumping has progressed to an appreciable degree, this condition is corrected by mud-jacking or subsealing. This procedure consists of drilling holes into the slab and forcing in a suitable slurry to fill the voids between the subgrade and the slab.

The usual method involves the drilling of one hole in the lowest portion of the settled pavement. This hole is about 1.5 inches in diameter

FIG. 20-8. Bumpcutter grooving pavement surface to increase skid resistance. (Courtesy California Division of Highways.)

and usually placed about 12 inches from the point. If needed, an additional hole is drilled about 30 inches from the edge. Water and sand are blown out of the area through the drilled hole by the use of compressed air. This is necessary to avoid dilution of the slurry to be forced under the pavement. As the slurry is forced into the drilled hole, pumping is continued until all voids are filled and the settled slab is raised to its proper position. The holes are then plugged.

The slurry mixtures used vary a great deal, but for the most part they consist of a combination of loam, cement, and water, or of asphaltic materials. Two widely used mixtures are as follows (*43*):

1. Soil 60 to 84 per cent, cement 16 to 40 per cent.
2. Soil 77 per cent, cement 16 per cent, and cut-back asphalt (SC–250), (MC–70), (RC–250), 7 per cent.

The mix used should be fluid enough to spread rapidly and flow into small cavities.

Several states have experimented with the use of asphaltic materials and special coal-tar pitch as material to be forced under concrete pavements at joints and cracks that show evidence of pumping. A low-penetration asphaltic cement with a high melting point has been used most often. North Carolina is subsealing concrete pavements with a cold slurry of asphalt emulsion (RS–2) and filler (*44*); the slurry sets up to form a mastic beneath the slab.

The buckling or blow-up of old concrete pavements is a continuing problem; it is usually caused by longitudinal expansion and the failure of transverse expansion joints to function properly. When this occurs, it is often necessary to remove the damaged portion of the pavement and replace it with a patch of concrete or bituminous material. Buckling or blow-ups may be prevented, when evidence of extreme compression is noted in the pavement, by cutting a wider expansion joint if one already exists or providing a new joint across the pavement.

20–15. Resurfacing of Old Concrete Pavements. When justified, old concrete pavements may be completely resurfaced. Three materials are in common use: asphaltic concrete, asphaltic concrete reinforced with welded wire fabric, and portland cement concrete.

In preparation for resurfacing with asphaltic concrete, joints and cracks may be cleaned and filled with asphaltic concrete. Bituminous patches which show an excess of bituminous material must be removed. Major defects must be corrected, and the cause of the failure must be corrected before the new surface is placed. A common problem with this type of resurfacing is that, over a period of time, joints and cracks in the old concrete pavement show up in the new wearing surface; this phenomenon is called "reflection cracking." In attempts to prevent reflection cracking, several different approaches have been used; these include: covering cracks and joints with metal plates, paper, or wire mesh reinforcing; construction of a layer of gravel or crushed stone several inches thick upon the old pavement; use of an open-graded bituminous leveling course of considerable thickness; use of rubber in the asphaltic mixture to provide additional flexibility in the pavement structure; and seating of the old slab with heavy rollers and impact breakers. None of these methods have been entirely satisfactory. When preparations have been completed, the asphaltic concrete is generally placed in two layers—a binder course and a surface course—by conventional methods.

Reflection cracking in asphaltic concrete mixtures laid on old concrete base courses has been prevented in a number of cases by the incorporation of a layer of welded wire fabric (like that used in reinforced concrete pavements) in the pavement structure (*277*). The style of fabric most commonly used is 3 × 6—10/10, which has a mesh opening of 3 by 6 inches with No. 10 gage wire. The sheets are placed so that the 6-inch spaced wires are parallel to the centerline. The entire pavement area normally is covered with mesh, but some success has been reported with the mesh laid in strips over the joints and cracks. The fabric is placed either directly on the existing pavement or on a leveling course. Sheets of fabric may be lapped for a short distance or butted up against one another. Sheets are placed in position, and the asphaltic concrete placed with a conventional paver equipped with a special hold-down device. Fine, dense-graded asphaltic concrete mixtures, with a maximum size of aggregate of 1 inch, have given the best results.

Old concrete pavements may be resurfaced with portland cement concrete. Considerable success has been achieved in recent years with thin overlays, from 1 to 3 inches in thickness. The old pavement must be thoroughly cleaned and all unsound material removed; the bumpcutter described earlier may be used to scarify the surface. The surface then may be washed down with detergent and acid baths (muriatic acid is often used). Best results are achieved if a thin grout layer is used to provide bond between the old pavement and the new surface. A 1:1 sand-cement grout is used; it is brushed onto the pavement just before the new concrete is applied. In some cases, the grout has been placed pneumatically. Concrete must be of high quality and resistant to freezing and thawing, particularly where it will be subjected to salt applications for ice control. In the northern states, air-entrained concrete is normally used. To obtain adequate bond, thorough compaction of the fresh concrete is essential. This is usually done by the use of special vibratory screeds. Whenever the new surface extends over a joint, or butts against an expansion joint in the old pavement, a joint will be necessary in the new concrete. Joints may be made by tooling the plastic concrete, or by sawing. Adequate curing of the new surface is essential. In some cases, particularly where the old pavement is excessively cracked, wire mesh reinforcing may be used in the new surface.

20–16. Maintenance of Block Pavements. Block pavements may consist of brick, shaped stone, asphalt blocks, or wood blocks. The maintenance of these surfaces usually consists of repairs to the foundation or base course on which the block rests, or the renewal of the cushion or bedding course. In replacing a base course or a sand cushion or bedding course, materials similar to those of the original pavement should be used. When it is necessary to reset old blocks, they should be carefully cleaned, turned over from the original position when placed, and rolled to a smooth surface. It is often desirable to place blocks that were removed from the center of the road toward the sides, and those removed from the sides near the center of the roadway. When blocks have to be replaced, care should be taken to see that they match in color and size as much as possible.

It is often necessary to fill the spaces between the blocks by means of a joint filler. Various forms of joint filler are used for this purpose, such as a sand filler, a sand-cement filler, and bituminous joint-filler mixtures.

Block pavements very often serve as base courses when it is uneconomical to maintain them and they are otherwise stable. A resurfacing or wearing course is added, the most common being a dense-graded bituminous concrete having a thickness of at least 1 inch.

20–17. Shoulder Maintenance. The importance of properly maintaining road shoulders becomes apparent when they are considered as a

Fig. 20–9. Shoulder erosion. (Courtesy Bureau of Public Roads.)

continuation of the traveled surface which must be used in cases of emergency. Adequate shoulders are also necessary if the full capacity of the roadway surface is to be utilized. A dangerous condition caused by shoulder erosion is shown in Fig. 20–9.

Shoulders may comprise many types of bases and surfaces including soil aggregate, soil surfaces which are capable of supporting vegetation and which may be seeded, and various bituminous surfaces. It is now almost standard practice to pave shoulders on heavily traveled roads; often the shoulders differ in color and texture from the travel lanes. Each of these approaches presents a slightly different type of maintenance problem.

A well-graded gravel shoulder will provide a satisfactory all-weather shoulder for light traffic; the maintenance of shoulders of this type will follow a pattern similar to that of soil-aggregate surfaces. This will include the dragging and blading of the shoulder surface for the proper slope and the filling of any ruts. It may be necessary to replace worn-out and lost materials at periodic intervals. Consideration must be given to the many mailbox turnouts which exist in most rural areas to prevent rutting and to provide an all-weather approach to the mailbox. This work can be done with a light motor grader.

Turf shoulders require the filling of holes and ruts, rolling or blading down high spots, and seeding, sodding, and fertilizing to maintain a satisfactory turf. It is also necessary to mow and clean the shoulders and provide some means of weed control or weed eradication. Mowing is usually accomplished with light equipment manufactured for this purpose. Treated bituminous and paved shoulders require maintenance that is similar to that for surfaces made of these same materials, which were previously described, including patching, surface treatment, and so on.

Closely related to shoulder maintenance is that of road approaches. Road approaches should be surfaced with materials of sufficient stability to support traffic loads. They require maintenance similar to that for other road surfaces.

Mention may also be made of the maintenance of sidewalks that are built along rural highways. Maintenance may therefore include patching, resurfacing, and controlling vegetation that may cause the sidewalk to become overgrown.

20–18. Roadside Maintenance. Roadside maintenance in general may include the care of the area between the traveled surface and the limits of the right-of-way. This will include any median strip or landscaped areas on divided or dual-lane highways; roadside parks and picnic and recreational areas; and various appurtenances, such as right-of-way fences, picnic tables, and the like. These activities may be catalogued as follows:

1. Mowing and cleaning of roadsides.
2. Control and eradication of weeds.
3. Control of erosion by seeding, sodding, or the planting of vegetation.
4. Care of roadside parks and picnic and recreational areas.
5. Care of trees and shrubs.
6. Removal of debris and rubbish.

The description of methods for maintaining roadsides which follows has been largely abstracted from the publication, "A Policy on Maintenance of Roadsides," of the American Association of State Highway Officials.

The mowing of roadsides is done to provide better sight distances as well as a better appearance. It also promotes better growth of the turf and aids in proper drainage. Mowing should be started as soon as the grass or weeds are high enough to be cut and it should be continued at periodic intervals when necessary, throughout the growing season. Mowing can be accomplished with machines or by hand. Some heavy-duty mowing machines are available which are capable of mowing roadside brush up to 2.5 inches in diameter. Cleaning of roadsides after mowing can be accomplished by the use of hand or power rakes. Grass cuttings can be left as a mulch, but brush cuttings should be removed or destroyed by burning.

There are three major categories of herbicides used to control weeds and brush along roadsides (*264*):

1. Ester and amine forms of 2,4-D are used commonly to control and eliminate broadleaf weeds in turf areas and in seldom-mowed areas in which the presence of tall weeds is objectionable.
2. Various compounds are used to control and eliminate weeds beneath guardrails; they include sterilants, combinations of systemic herbicides (e.g., 2,4-D and Dalapon), and contact herbicides, which rapidly kill all exposed plant parts.
3. Retardants are used in some cases to inhibit the growth of grass to a certain height.

Seeding, sodding, and the planting of vegetation are important maintenance operations for the prevention of erosion. Seeding may be done on relatively flat areas, while sodding is necessary on steeper slopes. On steep slopes where seeding or sodding is not practical, the ground surface may be protected by the planting of vines or similar ground cover. The planting of dune grasses on sandy slopes to prevent wind erosion has been quite successful (see Fig. 20–10).

Special roadside areas which have been developed for picnic and recreational areas, such as roadside parks and turnouts, should be maintained in a manner necessary to meet minimum health and aesthetic standards (see Fig. 20–11). Tables, benches, and refuse containers should be placed at convenient locations. The grounds should be kept cleaned, trees and shrubs trimmed, and stoves be kept in good condition. Toilet facilities should be kept in a sanitary condition, and any drinking water should be frequently tested for purity. The parking areas and access roads must also be properly maintained.

The care of trees and shrubs along the roadsides may include planting, trimming, fertilizing, spraying, and the construction of tree wells. Major tree surgery and the removal of broken limbs caused by storms are often necessary. A large number of highway departments have forestry departments that take care of all trees and shrubs of the state trunk-line system.

The removal of rubbish and debris (litter) is one of the vexing and expensive problems of roadside maintenance. Debris such as fallen branches, rocks, landslides, and articles that have fallen from trucks

FIG. 20–10. Planting beach grass to prevent erosion. (Courtesy Bureau of Public Roads.)

should be removed immediately in order to protect the traveling public. The remains of animals killed by motor vehicles should also be removed promptly and buried at a convenient location. Garbage and trash dumped within the highway right-of-way must be removed because of its unsightliness and for reasons of health.

20–19. Drainage and Drainage Structures. All drainage structures and appurtenances have to be kept in good working condition so as to provide free and unobstructed flow. Generally, maintenance operations include those relating to (1) surface drainage, (2) ditches, and (3) culverts. Modern roads are generally well developed for drainage, but many of the older ones lack adequate drainage. Observation of conditions at times of heavy rains and high water will give the best indications of the ability of the drainage system to perform its function properly without erosion or damage. The entire system must be so maintained as to handle heavy rainfalls and to function properly under all conditions, even under winter conditions of ice and snow.

The first objective of surface drainage is to remove the water quickly so that it will not interfere with the use of the road. To accomplish

Fig. 20–11. A roadside park. (Courtesy Michigan Department of State Highways.)

this, the roads are crowned and elevated. On the high-type road surfaces, these crowns are more permanent than on the lower-type of road surfaces. Thus, on gravel roads and similar types, it is necessary to maintain a proper crown by blading.

Wide shallow ditches are preferable for maintenance and are less dangerous when vehicles have to leave the road. Certain types of roadside ditches may be kept clean and the proper slope maintained by the use of a power grader. Care should be exercised to preserve seeded and sodded areas and shrubs during maintenance operations. The original line or grade of the ditch should also be maintained. When ditches become eroded owing to excessive grade, handling of large volumes of water, or a combination of both, it may be necessary to construct ditch checks to retard the flow or to line them, as described in Chapter 7.

All culverts should be checked regularly to see that they are free from obstructions. Inlets and outlets should be kept open and free from refuse, and culverts should be cleaned and repaired when necessary.

20–20. Snow and Ice Control. Considering the public's complete dependence upon the uninterrupted use of highways throughout the year, winter maintenance has become one of the most essential activities of many state highway departments. This is particularly true in the northern states, where a planned program for snow removal and ice control has to be considered when the annual maintenance budget is prepared. The amount of money expended for these services will vary a great deal in the many states, and it is dependent in large measure upon the length and intensity of storms and the amount of snowfall in a particular area. For example, in Iowa, in fiscal 1964–1965, 33.9 per cent of the maintenance budget went for snow and ice control; in Texas less than 2.8 per cent was expended for this purpose.

Measures preparing for snow removal and ice control are taken before the start of the winter season. The most important of these are (1) the preparation of the necessary equipment; (2) stock-piling of ice-control chemicals and abrasives, if used; (3) placing of snow fences, snow guides, and containers for ice control abrasives; (4) organization arrangements.

The necessary equipment needed for snow-removal operations consists of large trucks, various types of snowplows, power graders, bank-slicers, and sanders. Some trucks are equipped with underbody blades, others with "Vee" plows or with bank-slicers. All equipment should be put into working order before the winter season starts. An ample supply of snowplow shoes and cutting edges should be on hand for each type of plow. Tire chains, flags, blinker lights, tow chains, and other safety devices should also be checked and placed, ready for use.

Ice-control chemicals and abrasives should be stock-piled previous to the first storm in the fall—the quantities to be stock-piled may be estimated on the basis of past experience in a particular district or division. Stock piles should be located with care at convenient places along the road or at central locations, such as main or branch garage yards, gravel or sand pits, or at other places where overhead bins or mechanical loaders are available. Virginia is storing ice-control chemicals for use on rural highways in structures that look like farm silos and, thus, blend into the countryside. Generally, a coarse heavy sand is the best abrasive. Cinders or other abrasives, however, may be desirable, depending on availability and cost. Calcium chloride is frequently incorporated and thoroughly mixed with the abrasive at the time the stock piles are made. Fifty to 100 pounds of calcium chloride per cubic yard are used, and a light application of calcium chloride is usually spread over the surface of the pile to prevent freezing.

Snow fences are generally placed parallel to the road at a distance from 75 to 125 feet from the centerline of the road in areas where observations show that drifts constantly form. A snow fence usually consists of fabricated wooden slats mounted on steel posts, which can be removed,

FIG. 20–12. A natural snow fence. (Courtesy Michigan Department of State Highways.)

rolled in bundles, and stored when not in use. The purpose of a snow fence is to change the velocity of the snow-laden wind so that the snow is deposited before it reaches the roadway. As the snow accumulates near the fence, the fence is raised in order that it may effectively gather snow. Extension posts are provided when this is necessary. In certain areas where right-of-way is available, natural windbreaks or snow fences may be formed by planting evergreens parallel to the road (see Fig. 20–12).

Snow guides may be made from peeled saplings approximately 12 feet long with a maximum top diameter of 2 inches. They are sometimes painted with alternate bands of black and white paint or some other suitable color, and they are often fitted with reflectorized buttons. They are used for marking the roadway or road hazards in areas of heavy snowfall. Guides marking the locations of guardrails, culverts, headwalls, and curbs are essential during plowing operations when these structures are covered with snow.

The placing of containers for abrasives at strategic locations has proved very beneficial to the traveling public, who can avail themselves of these materials if needed before maintenance crews arrive. The containers are usually barrels equipped with a utensil, such as a gallon can, for removing the abrasive.

To properly carry on the work of snow removal and ice control, good organization is necessary. Many states maintain a weather reporting service, with all districts of a division reporting adverse weather or road conditions to a central office. These give such information as temperature, snowfall, and other adverse weather conditions, such as rain, fog, mist, and smog. Road conditions are then rebroadcast to serve as a source of information for the traveling public. Maintenance crews are generally subject to call at all hours, and provisions are usually made for dual crews to operate for alternate 12-hour periods during emergencies. The use of radio communication has facilitated work during snow-removal and ice-control operations. Mobile and fixed station units assist in the rapid dispatching of equipment, the reporting of adverse conditions, and the general control of operations. Use is made of both one-way and two-way radio communication systems for other maintenance operations as well.

The procedure followed to keep trunk highways clear of snow varies. One method is to start plowing with trucks equipped with underbody blades after one half to an inch of snow has fallen, using a conventional curved cutting edge to remove the snow before it becomes packed by traffic. These units are generally kept working during a continuous snowfall up to the time when the snow becomes too deep for their effective operation. When this occurs, light trucks equipped with side delivery plows start operating and keep in operation until the storm has subsided or until sufficient snow has accumulated to require heavier trucks equipped with side delivery plows or light trucks with "Vee" plows. When this equipment proves inadequate, heavy "Vee" plows or "snogos" are brought into operation. When the storm has abated, snow removal operations are continued until the snow is removed to a point beyond the outer edge of the shoulder. On trunk lines in urban areas, use is often made of mechanical loaders to speed snow removal. Use is sometimes made of heavy rotary plows for opening up roads after storms of such intensity that snow-removal operations have to be temporarily suspended (see Fig. 20–13). Care must be exercised at railroad crossings to remove snow between the rails so that ice does not form.

Ice control means the application of quick-action emergency measures to counteract slippery conditions as early as possible, and the removal

FIG. 20–13. Snow removal—a rotary plow. (Courtesy Michigan Department of State Highways.)

of all ice from the pavement after skid control has been established. These operations may consist of removing sleet or thin ice, thick ice or frozen slush, and hard-packed snow from the surface of the roadway. Salt, calcium chloride, a mixture of the two, and abrasives mixed with chlorides are widely used for ice control in urban and rural areas. Many organizations use a mixture of 1 part calcium chloride and 2 parts sodium chloride, by volume. Application rates vary from 200 to 600 pounds of mixture per mile of two-lane roadways (Fig. 20–14). They depend largely on straight chemicals in urban areas, because abrasives may clog storm sewers.

To remove sleet or thin ice from the roadway, treated abrasive may be placed on the surface at the rate of 2 to 3 cubic yards per mile, as the ice forms. Chlorides may also be used alone. First consideration in ice-control treatment is given to approaches to railroad crossings, road intersections, and dangerous horizontal and vertical curves. This kind of work is usually done by light trucks in advance of other ice-control measures.

FIG. 20–14. Sanding an icy pavement. (Courtesy Bureau of Public Roads.)

Sleet or thin ice on gravel roads should first be serrated. This process forms grooves, and in many cases will bring up enough gravel to control skidding. If this does not occur, chloride may be added. Thick ice or frozen slush is also scored with a serrated blade in order to concentrate the chemical on the road. When the ice or slush has been loosened, it is bladed off the road.

Abrasives treated with calcium or sodium chloride, or chlorides alone, should be used as sparingly as possible on portland cement concrete pavements. This is because repeated freezing and thawings of concrete in contact with these salts are conducive to scaling. It has been found *(57)* that unscaled pavements more than four years old are less susceptible to damage, and that pavements made from air-entrained concrete, or those that have been given a protective treatment, are generally free from damage. On urban expressways, the overwhelming need for getting a bare pavement quickly leads to excessive use of salts, with attention being focused on building pavements that can withstand this treatment.

Some tests are being carried out in and near urban areas for controlling ice conditions on the highways by the use of coils placed under the pavements, the heat being supplied by an electric current. Installations of this nature are beneficial at railroad crossings, intersections, ramps, bridge decks, and other points where hazardous conditions exist. Expense of installing and operating such systems has limited their use to only a few critical locations.

20–21. Maintenance of Traffic-Control Devices. Signs, signals, and markings for the direction, warning, and regulation of traffic are essential to traffic safety and must be properly maintained. This work includes the installation, repair, and painting of signs, and the care of pavement markings. Because of their location, many pavement markings must necessarily be renewed practically every year. Many departments have mechanized marking equipment for the application of the painted strip (see Fig. 20–15). Large machines of this type can apply simultaneously as many as three stripes, either solid or broken, and use two colors of paint, if desired. Reflectorized glass beads may be added to the painted surface during the striping process. The beads have proved beneficial for night driving. Thermoplastic compounds are used for pavement markings, also, as are raised plastic disks cemented to the road surface. Many highway departments maintain shops for the fabrication and repair of signs and other traffic control devices. Automatic traffic signals have to be checked, repaired, and often readjusted to traffic conditions. Highway departments are continually standardizing and modernizing traffic signs and signals, and much of this work is done by maintenance forces. During the winter season in the northern states, special attention is given to warning signs to keep them free from snow and visible as much as possible.

20–22. Bridge Maintenance. Highway departments continually inspect bridges and elevated structures, then take any necessary steps to minimize deterioration or repair damage caused by accidents, floods, or other unforeseen events.

Bridge superstructures require periodic maintenance to preserve and protect the investment made in them. For instance, steel bridges (unless they are made of special alloys or have been galvanized) are cleaned and painted regularly to prevent corrosion.

A serious problem for the maintenance engineer is the deterioration

FIG. 20-15. Multiple-line paint striping machine used by the Wyoming Highway Department. (Courtesy *Public Works*.)

of concrete bridge decks, particularly on heavily traveled routes in northern states, where chlorides are used extensively for snow and ice control. The chlorides, of course, can be very damaging to concrete.

Highway agencies use a wide variety of methods to repair bridge deck concrete. The Port of New York Authority uses two basic methods (*45*): They install a patching compound consisting of a coal tar epoxy resin binder and clean sharp sand, mixed in proportions of 1:4, as a surface patch with a limiting depth of about 1 inch. Before placing the patch, they sawcut its perimeter and paint the area with liquid epoxy coal tar. They use portland cement concrete for all other deck repairs, varying the mix as necessary to fit design requirements or construction limitations.

Practice on the New Jersey Turnpike is to remove and replace any seriously damaged deck areas, patch any spalled areas, then seal and surface the entire deck (*45*). The sealing material is a coal tar epoxy. The new surface is a special neoprene-asphalt mixture laid to a thickness of 1½ inches. One of the reports of the National Cooperative Highway Research Program contains a comprehensive discussion of methods used to replace deteriorated concrete in structures (*21*).

Bridge seats, rollers, rockers, and expansion joints should be inspected frequently and maintained as necessary to assure their proper functioning.

Bridge inspections should be complete enough to discover any undermining of the bridge footings or any damage to the substructure. Timber bulkheads should be repaired as soon as evidence of failure is shown. Piling which shows deterioration caused by erosion, corrosion, or attack by organisms should be replaced when necessary.

The stream bed should be kept clean and free from debris in order that the free flow of water may be maintained. Provision for the installation of some sort of barrier to prevent floating debris from damaging the bridge floor during periods of high water must sometimes be made.

The maintenance of pumping facilities at highway underpasses is also a very important operation.

The maintenance of bridges is usually carried on by a special bridge crew trained and skilled in this type of work. Some state highway departments have heavy equipment that is necessary for all phases of bridge maintenance, including mobile shops.

20–23. Regulation and Control of Highway Uses. The control of certain uses of the highway right-of-way or the area immediately adjacent to it is usually delegated to the maintenance division. This regulation and control usually extends to (1) permits regulating travel, (2) commercial advertising, (3) encroachments, (4) obstructions, (5) vendors, (6) damage to trees, shrubs, or other highway appurtenances, and (7) public utilities.

State highway departments generally require permits for moving unusual vehicles or loads over the highway system. This applies to oversize or overweight vehicles or loads and to the moving of buildings where the use of the highway is involved. Applicants for permits are generally required to list the gross weights of loads, size of vehicle or load, and the route of travel. In granting permits, the governing factors include the size and nature of vehicle or load, the traffic density on the proposed route, and the distance to be traveled. The route can be checked for the proper clearance and permissible loading of the structures to be encountered.

Most states have restrictions which limit the maximum loads permitted on any one axle, and they maintain permanent weighing stations for their control. Emergency weight restrictions limiting wheel loads permitted during the spring break-up in northern states are applied as necessary to protect pavement structures.

Permits are sometimes required by state highway departments when abutting property owners desire a driveway, culvert, or the placing of service pipes and the like within the right-of-way limits.

The regulation and control of commercial advertising present a problem to most highway organizations. The most desirable policy is to exclude all commercial advertising from the right-of-way. Attention was focused on billboard control on Interstate and primary routes with the passage of the Highway Beautification Act of 1965. Billboard control presents complicated legal and administrative problems about which little generalization is possible.

An encroachment may be considered as a structure of permanent or semipermanent nature, illegally erected within the right-of-way and attached to the land, such as a fence, building, or gasoline pump. Responsibility for initial action toward the removal of the encroachment lies usually with the maintenance engineer. When failure to remove the encroachment as requested occurs, legal action is usually necessary.

Obstructions are casual objects lying within the right-of-way which are not attached to the land, such as logs, cordwood, junked cars, scrap, and the like. Arrangements must be made for their removal.

The roadside vendor usually contributes to a traffic hazard, unsightly roadside conditions, and possible interference with maintenance operations. A vendor in this case may be defined as one who sells merchandise of any kind from a stand, counter, table, or vehicle located on a trunk line. Procedures for their control are initiated by the maintenance division.

The regulation and control of public utilities within the highway right-of-way are necessary to protect the existing highway structure, the traveling public, and the appearance of the roadside. Definite policies for the location of pole lines and underground conduits have been established in all highway departments. The engineering divisions of most utilities generally appreciate the problems involved and cooperate accordingly.

Maintenance forces are often required to investigate and report damage to highway appurtenances. This includes the cutting or removal of trees and shrubs from the highway right-of-way, damage to guardrail, culverts, bridge abutments, traffic devices, and so forth.

20–24. Maintenance Management. Modern management methods must be applied to maintenance operations (as well as to other operations of a highway agency) in order to cope with rising traffic volumes and public demands for increased services.

Attention focused on maintenance operations in the early 1960's, as maintenance expenditures and unit costs rose rapidly. During the period from 1956 through 1964, highway maintenance expenditures by all levels of government rose from $2.1 billion to $3.1 billion (see Fig. 20–1). In 1950, the average annual expenditure for maintenance per mile of road or street was $430; in 1964, it was almost double that amount. The highway maintenance and operation cost trend index prepared by the Bureau of Public Roads rose from 71 in 1950 to 123 in 1965.

Maintenance costs are made up principally of machinery costs (own-

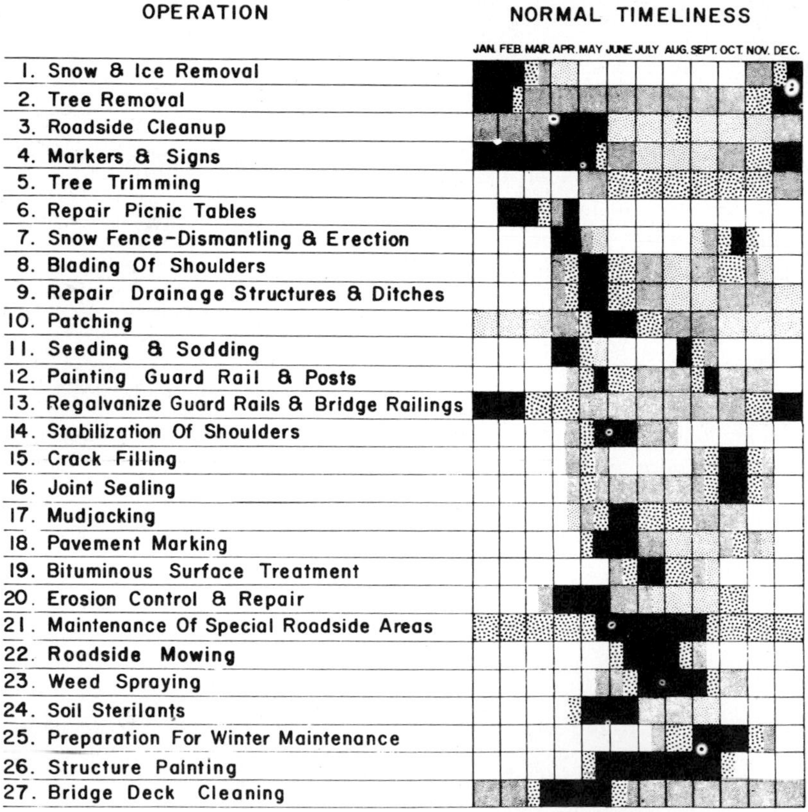

OPERATION

NORMAL TIMELINESS

JAN. FEB. MAR. APR. MAY JUNE JULY AUG. SEPT. OCT. NOV. DEC.

1. Snow & Ice Removal
2. Tree Removal
3. Roadside Cleanup
4. Markers & Signs
5. Tree Trimming
6. Repair Picnic Tables
7. Snow Fence-Dismantling & Erection
8. Blading Of Shoulders
9. Repair Drainage Structures & Ditches
10. Patching
11. Seeding & Sodding
12. Painting Guard Rail & Posts
13. Regalvanize Guard Rails & Bridge Railings
14. Stabilization Of Shoulders
15. Crack Filling
16. Joint Sealing
17. Mudjacking
18. Pavement Marking
19. Bituminous Surface Treatment
20. Erosion Control & Repair
21. Maintenance Of Special Roadside Areas
22. Roadside Mowing
23. Weed Spraying
24. Soil Sterilants
25. Preparation For Winter Maintenance
26. Structure Painting
27. Bridge Deck Cleaning

Fig. 20–16. Seasonal schedule of maintenance operations; depth of tone indicates timeliness of operation. (Courtesy Michigan Department of State Highways.)

ership and upkeep) and labor. Hence, efficient management that lowers these costs automatically lowers maintenance costs. As a result, many organizations are conducting studies aimed at establishing maintenance standards; measuring and increasing productivity of men and machines; and planning, scheduling, and controlling maintenance operations. Most of these studies are not complete, but they already have lead to improved maintenance management methods.

Potential savings are quite large. In 1966, B.P.R. chief engineer Francis C. Turner estimated that a saving equivalent to a one-cent increase in gasoline tax could be achieved by every state highway department by increasing the productivity of labor and equipment used in maintenance operations.

An example of a seasonal schedule of maintenance operations is shown in Fig. 20–16.

20–25. Maintenance and Highway Design. The importance of highway maintenance in preserving the highway investment is readily apparent. Of additional importance is its relation to highway design. Accurate maintenance costs are effective in determining the selection of the types of surfaces and other features of design. For an example, on a section of road built on a low grade line, the annual cost of snow removal was $296.56 per mile (*56*). On a similar section, a highway was built to a modern grade line under the same climatic conditions, and the annual cost of snow removal was $19.53 per mile. The annual savings in snow removal cost was $277.03 per year.

The experience of the maintenance engineer is also beneficial to the designer. The maintenance engineer knows the weaknesses of previous designs and can improve those of the future. Many states require that the maintenance engineer review and criticize highway construction plans before they are approved.

BIBLIOGRAPHY

Publications of the Highway Research Board, Washington, D.C.

1. The AASHO Road Test: History and Description of Project. *Special Report No. 61A* (1961).
2. The AASHO Road Test: Pavement Research. *Special Report No. 61E* (1962).
3. Airport Runway Evaluation in Canada. N. W. McLeod. *Research Report No. 4B* (1947) and *Supplement* (1948).
4. Aluminum Highway Culverts and Bridges. *Bulletin No. 361* (1962).
5. Application of the Classifications and Group Index in Estimating Desirable Subbase and Total Pavement Thicknesses. D. J. Steele. *Proceedings,* Vol. 25 (1945).
6. Application of Triaxial Test Results to the Calculation of Flexible Pavement Thickness. E. S. Barber. *Proceedings,* Vol. 26 (1946).
7. Cement-treated Soil Mixtures. *Highway Research Record No. 36* (1963).
8. Chemical Soil Stabilization and Soil Aggregate Stabilization. *Bulletin No. 357* (1962).
9. Classification of Soils and Subgrade Materials for Highway Construction. Report of Committee on Classification of Materials for Subgrades and Granular Type Roads. *Proceedings,* Vol. 25 (1945).
10. Color Air Photos, Soil Properties and Tests. *Highway Research Record No. 63* (1964).
11. Compaction and Correlation between Compaction and Classification Data. *Bulletin No. 325* (1962).
12. Compaction of Embankments, Subgrades and Bases. *Bulletin No. 58* (1954).
13. Comparison of Different Methods of Measuring Pavement Condition—Interim Report. *National Cooperative Highway Research Report No. 7* (1964).
14. Composite Pavement Design. *Highway Research Record No. 37* (1963).
15. Construction on Marsh Deposits. *Highway Research Record No. 57* (1964).
16. Cost Allocation Revisited. John Rapp. *Highway Research News, No. 17* (February, 1965).
17. Depressed Curb Inlets. *Highway Research Record No. 58* (1964).
18. Economic Cost of Traffic Accidents. *Bulletin No. 263* (1964).
19. Economic Analysis in Highway Programming, Location and Design. *Special Report No. 56* (1960).
20. Engineering Economy—1963. *Highway Research Record No. 77* (1965).
21. Evaluation of Methods of Replacement of Deteriorated Concrete in Structures. *National Cooperative Highway Research Report No. 1* (1964).
22. Factors Influencing Compaction Test Results. *Bulletin No. 319* (1962).
23. Factors That Influence Field Compaction of Soils. *Bulletin No. 272* (1960).
24. The Factors Underlying the Rational Design of Pavements. F. N. Hveem and R. M. Carmany. *Proceedings,* Vol. 28 (1948).
25. Flexible Pavement Design. *Highway Research Record No. 13* (1963).
26. Flexible Pavement Design—1963 and 1964. *Highway Research Record No. 71* (1965).
27. Flexible Pavement Design Correlation Study. *Bulletin No. 133* (1956).
28. Flexible Pavement Design in Four States. *Bulletin No. 136* (1956).
29. Foundations for Flexible Pavements. O. J. Porter. *Proceedings,* Vol. 22 (1942).
30. Freezing and Thawing of Concrete and Use of Silicones. *Highway Research Record No. 18* (1963).

31. Geophysical Methods and Statistical Soil Surveys in Highway Engineering. *Highway Research Record No. 81* (1965).
32. Highway Capacity Manual—1965. *Special Report No. 87* (1966).
33. Highway Financing. *Highway Research Record No. 20* (1963).
34. Highway Sufficiency Ratings. *Bulletin No. 53* (1952).
35. Indirect and Sociological Effects of Highway Location and Improvement. *Highway Research Record No. 75* (1965).
36. Investigational Concrete Pavements—Progress Reports of Cooperative Research Projects on Joint Spacing. *Research Report No. 3B* (1945).
37. Iowa State Maintenance Study: Time Utilization, Productivity, Methods and Management—1959–60. *Special Report No. 65* (1961).
38. Joint Spacing in Concrete Pavements. *Research Report No. 17-B* (1956).
39. Joints and Sealants—A Symposium and Other Papers. *Highway Research Record No. 80* (1965).
40. Laboratory Analysis of Soils. *Bulletin No. 95* (1953).
41. Legal, Administrative and Financial Aspects of Urban Parking Surveys. D. R. Levin. *Bulletin No. 19* (1949).
42. Lives of Highway Surfaces—Half Century Trends. G. D. Gronberg and N. B. Blosser. *Proceedings,* Vol. 35 (1956).
43. Maintenance Methods for Preventing and Correcting the Pumping Action of Concrete Pavement Slabs. *Current Road Problems No. 4R* (1947).
44. Maintenance Practices. *Highway Research Record No. 11* (1963).
45. Maintenance Practices 1963—Administration, Methods and Materials. *Highway Research Record No. 61* (1964).
46. Methods and Costs of Peat Displacement in Highway Construction. J. W. Cushing and O. L. Stokstad. *Proceedings,* Vol. 14 (1934).
47. Motor Vehicle Operating Costs, Road Roughness and Slipperiness of Various Bituminous and Portland Cement Concrete Surfaces. R. A. Moyer. *Proceedings,* Vol. 22 (1942).
48. Nuclear Excavation. *Highway Research Record No. 50* (1964).
49. Nuclear Measurements. *Highway Research Record No. 66* (1963).
50. Nuclear Testing of Asphaltic Concrete Pavement and Soil Subgrades. *Bulletin No. 360* (1962).
51. Objectives and Findings of Highway Needs Studies. J. P. Buckley and Carl E. Fritts. *Proceedings,* Vol. 28 (1948).
52. Parking. *Bulletin No. 19* (1949).
53. Pavement Performance Concepts. *Bulletin No. 250* (1960).
54. Photogrammetry: Developments and Applications, 1960. *Bulletin No. 283* (1961).
55. Progress Report of Subcommittee on Methods of Measuring Strength of Subgrade Soil—Review of Methods of Design of Flexible Pavements. Committee on Flexible Pavement Design. *Proceedings,* Vol. 25 (1945).
56. Progress Report of the Project Committee on Maintenance Costs. Department of Maintenance. *Proceedings,* Vol. 29 (1949).
57. Recommended Practice for Snow Removal and Treatment of Icy Pavements. *Current Road Problems Bulletin No. 9-2R* (1948).
58. Report of Department of Highway Finance. . . . T. H. MacDonald. *Proceedings,* Vol. 18 (1938).
59. Resurfacing and Patching Concrete Pavements with Bonded Concrete. E. J. Felt. *Proceedings,* Vol. 35 (1956).
60. Roadside Development. *Highway Research Record No. 93* (1965).
61. Sand-Equivalent Test for Control of Materials During Construction. F. N. Hveem. *Proceedings,* Vol. 32 (1953).
62. Soil and Slope Stabilization and Moisture and Density Determination Developments. *Bulletin No. 309* (1962).
63. Soil-Bituminous Roads. *Current Road Problems Bulletin No. 12* (1946).
64. Soil-Cement Stabilization. *Highway Research Record No. 86* (1965).

65. Soil Compaction. *Highway Research Record No. 22* (1963).
66. Soil Density Control Methods. *Bulletin No. 159* (1957).
67. Soil-Testing Methods—Moisture, Density, Classification, Soil-Cement. *Bulletin No. 122* (1956).
68. State Highway Administrative Organizations. *Special Report No. 51* (1959).
69. Studies of Fill Construction over Mud Flats. . . . O. J. Porter. *Proceedings,* Part II, vol. 18 (1938).
70. Studies in Highway Engineering Economy. *Bulletin No. 306* (1961).
71. Surface Drainage. *Research Report No. 11-B* (1950).
72. Surface Treatments, Bituminous Mixtures and Pavements. *Highway Research Record No. 104* (1965).
73. Symposium on Asphalt Paving Mixtures, Corps of Engineers, Department of the Army. *Research Report No. 7B* (1949).
74. Symposium on Fly Ash in Concrete. *Highway Research Record No. 73* (1965).
75. Symposium on Slip-Form Paving. *Highway Research Record No. 98* (1965).
76. Theoretical Analysis of Structural Behavior of Road Test Flexible Pavements. *National Cooperative Highway Research Program Report No. 10* (1964).
77. Thin Bituminous Concrete Retreatments Used in Alabama, Florida and Mississippi. *Highway Research News, No. 4* (May, 1963).
78. Traffic Assignment. *Bulletin No. 61* (1952).
79. Traffic Assignment by Mechanical Methods. *Bulletin No. 130* (1956).
80. Vertical Sand Drains for Stabilization of Embankments. *Bulletin No. 115* (1955).

Publications of the American Road Builders' Association, Washington, D.C.

81. Committee Report—Flexible Pavement Design. *Technical Bulletin No. 119* (1947).
82. Continuous Steel Reinforcement in Concrete Pavements. Wayne R. Woolley. *Technical Bulletin No. 223* (1957).
83. Handling and Placing Welded Wire Fabric in Concrete Pavement. Henry Aaron. *Technical Bulletin No. 260* (1965).
84. The Highway Program Begins a New Era. C. M. Upham. *Bulletin No. 96* (1945).
85. Practical Density Limits for Highway Subgrades and Bases. A. W. Johnson. *Technical Bulletin No. 203* (1954).
86. Pumping of Subgrade Through Paving Joints and Cracks; and Other Titles. *Technical Bulletin No. 103* (1946).
87. Recent Developments in Portland Cement Concrete Pavement Design and Construction. Gordon K. Ray. *Technical Bulletin No. 256* (1965).
88. Report of Committee on Calcium Chloride Soil Stabilization. *Technical Bulletin No. 127* (1947).
89. Skid-Proofing City Streets, County and State Highways Using Blast Furnace Slag. E. W. Bauman. *Technical Bulletin No. 241* (1959).
90. Stabilization of Soil with Asphalt. *Technical Bulletin No. 200* (1953).
91. The Use of Linseed Oil for the Protection of Portland Cement Concrete Surfaces. C. E. Morris. *Technical Bulletin No. 257* (1965).
92. Welded Wire Fabric in Portland Cement Concrete Pavements. Henry Aaron. *Technical Bulletin No. 217* (1956).

Publications of the American Association of State Highway Officials, Washington, D.C.

93. Electronic Computer Methods of Forecasting and Assigning Traffic to Large-Scale Urban and Rural Systems. S. E. Ridge. Paper presented at annual convention, San Francisco, Calif., December, 1958.
94. Guide Specifications for Highway Construction (1963).
95. Historic American Highways (1953).
96. An Informational Guide for Physical Maintenance of Pavements (1963).
97. Manual of Highway Construction Practice and Methods (1958).

98. Manual of Uniform Traffic Control Devices for Streets and Highways (1961).
99. Policy on Arterial Highways in Urban Areas (1957).
100. A Policy on Design Standards—Interstate System (revised May, 1965).
101. A Policy on Geometric Design of Rural Highways (1966).
102. Policy on Maintenance of Roadway Surfaces (1965).
103. Policy on Maintenance of Shoulders, Road Approaches and Sidewalks (1958).
104. Policy on Maximum Dimensions and Weights of Motor Vehicles to be Operated over the Highways of the United States (1963).
105. Road User Benefit Analyses for Highway Improvements (1959).
106. Standard Specifications for Highway Bridges (8th ed., 1961).
107. Standard Specifications for Highway Materials and Methods of Sampling and Testing. Parts I and II (8th ed., 1961).

Publications of The Asphalt Institute, College Park, Md.

108. Analysis of Asphaltic Types. . . . *Construction Series No. 49* (1939).
109. The Asphalt Handbook. *Construction Series No. 81* (1947).
110. Asphalt Mixed-in-Place (Road Mix) Manual. *Manual Series No. 14* (1965).
111. Asphalt Paving Manual. *Manual Series No. 8* (1960).
112. Asphalt Overlays. *Asphalt* (July, 1966).
113. Asphalt Plant Manual. *Manual Series No. 3* (1959).
114. Asphalt Surface Treatments and Penetration Macadam. *Manual Series No. 13* (1964).
115. Drainage of Asphalt Pavements. *Manual Series No. 15* (1966).
116. Flexible Pavement Reaction Under Field Load Bearing Tests. Prevost Hubbard. *Research Series No. 9* (1943).
117. Manual on Design and Construction of Asphaltic Roads and Streets. Pacific Coast Division (1952).
118. Mix Design Methods for Asphaltic Concrete. *Manual Series No. 2* (1962).
119. The Rational Design of Asphalt Paving Mixtures. *Research Series No. 1* (1935).
120. Required Thickness of Asphalt Pavement in Relation to Subgrade Support. Prevost Hubbard and F. C. Field. *Research Series No. 8* (1941).
121. Soil Manual. *Manual Series No. 10* (1961).
122. Specifications and Construction Methods for Asphalt Concrete. *Specification Series No. 1* (1964).
123. Thickness Design, Asphalt Pavement Structures for Highways and Streets. *Manual Series No. 1* (1963).

Publications of the Portland Cement Association, Chicago, Ill.

124. A Charted Summary of Concrete Road Specifications used by State Highway Departments (1963).
125. Concrete Pavement Design (1951).
126. Concrete Pavement Inspector's Manual (1959).
127. Design and Control of Concrete Mixtures (10th ed.).
128. New Developments in Concrete Pavement Construction. Gordon K. Ray and Harold J. Halm (1965).
129. Soil-Cement Construction Handbook (1956).
130. Soil-Cement Laboratory Handbook (1959).
131. Soil Primer (1962).
132. Subgrades, Subbases and Shoulders for Concrete Pavement (1960).
133. Thickness Design for Concrete Pavements (1966).

Publications of the American Society for Testing Materials, Philadelphia, Pa.

134. The Significance of Tests of Petroleum Products. Committee D2 on Petroleum Products and Lubricants (1946).
135. Standards, Part 10—Concrete and Mineral Aggregates (October, 1965).

136. Standards, Part 11—Bituminous Materials; Soils; and Skid Resistance (March, 1966).
137. Standards on Bituminous Materials for Highway Construction, Waterproofing, and Roofing. Committee D-4 on Road and Paving Materials, Committee D-8 on Waterproofing and Roofing Materials (9th ed., 1962).
138. Symposium on Accelerated Durability Testing of Bituminous Materials. *Special Technical Publication No. 94* (1949).
139. Symposium on Load Tests of Bearing Capacity of Soils. *Special Technical Publication No. 79* (1948).

Publications of the Federal Government

140. The Administration of Federal Aid for Highways and Other Activities of the Bureau of Public Roads. Bureau of Public Roads, Washington, D.C. (1957).
141. Analyses of Direct Costs and Frequencies of Illinois Motor-Vehicle Accidents, 1958. Charles M. Billingsley and Dayton P. Jorgenson. *Public Roads* (August, 1963).
142. Capacity Charts for the Hydraulic Design of Culverts. *Hydraulic Engineering Circular No. 10*. Bureau of Public Roads (1965).
143. Concrete Manual. U.S. Bureau of Reclamation, Denver (6th ed., 1956).
144. Control of Soils in Military Construction. *TM5-541*. Department of the Army. Washington, D.C. (1954).
145. Debris-Control Structures. *Hydraulic Engineering Circular No. 9*. Bureau of Public Roads (1964).
146. Design of Roadside Drainage Channels. *Hydraulic Design Series No. 4*. Bureau of Public Roads (1965).
147. Design Charts for Open-Channel Flow. *Hydraulic Design Series No. 3*. Bureau of Public Roads (1961).
148. Earth Manual (1st ed., revised). Bureau of Reclamation (1963).
149. Estimate of Receipts and Disbursements for Highways, 1963–70. Bureau of Public Roads (June, 1963).
150. Federal-Aid Financing and the Highway Trust Fund. Bureau of Public Roads (1964).
151. Final Report of the Highway Cost Allocation Study. *House Document No. 72*, 87th Congress, 1st Session. U.S. Government Printing Office (1961).
152. Further Developments and Application of the Immersion-Compression Test. J. T. Pauls and J. F. Goode. *Public Roads* (December, 1948).
153. Geophysical Methods of Subsurface Exploration in Highway Construction. R. W. Moore. *Public Roads* (August, 1950).
154. Highways and Economic and Social Changes. Bureau of Public Roads (1964).
155. Highways and The Nation's Economy. Joint Committee on the Economic Report, 81st Congress, 2nd Session. U.S. Government Printing Office. Washington, D.C. (1950).
156. Highway Practice in the United States of America. Public Roads Administration (1949).
157. Hydraulic Charts for the Selection of Highway Culverts. *Hydraulic Engineering Circular No. 5*. Bureau of Public Roads (1964).
158. Hydrology of a Highway Stream Crossing. *Hydraulic Engineering Circular No. 3*. Bureau of Public Roads (1961).
159. Interregional Highways. *House Document No. 397,* 78th Congress, 2nd Session. U.S. Government Printing Office (1944).
160. Manual of Procedures for Metropolitan Area Traffic Studies. Public Roads Administration (1946).
161. The Origin, Distribution and Airphoto Identification of U.S. Soils. *Technical Development Report No. 52,* Civil Aeronautics Administration, Department of Commerce (1946).

162. Peak Rate of Runoff from Small Watersheds. *Hydraulic Design Series No. 2.* Bureau of Public Roads (1961).
163. A Quarter Century of Financing Municipal Highways, 1937–61. Bureau of Public Roads (1964).
164. Rainfall-Frequency Atlas of the United States. *Technical Paper No. 40.* U.S. Weather Bureau (1961).
165. Rainfall Intensity-Duration-Frequency Curves for Selected Stations. *Technical Paper No. 25.* U.S. Weather Bureau (1955).
166. Report of Western Conference on Increasing Highway Engineering Productivity. H. A. Radzikowski. Bureau of Public Roads (March, 1957).
167. Roads and Airfields. *TM5-250.* Department of the Army. Washington, D.C. (August, 1957).
168. The Role of Aerial Surveys in Highway Engineering. W. T. Pryor. Bureau of Public Roads (1960).
169. Standard Specifications for Roads and Bridges on Federal Highway Projects, FP-61. Bureau of Public Roads (1961).
170. State Highway-User Taxes Paid in 1954 and 1955 on Vehicles of Various Type and Weight Groups. *Public Roads* (February, 1958).
171. A Study of Viscosity-Graded Asphalt Cements. J. Y. Wellborn, E. R. Oglio, and J. A. Zenewitz. *Public Roads* (June, 1966).
172. Supplementary Report of the Highway Cost Allocation Study. *House Document No. 124,* 89th Congress, 1st Session. U.S. Government Printing Office (1965).
173. A Test for Determining the Effect of Water on Bituminous Mixtures. J. T. Pauls and H. M. Rex. *Public Roads* (July, 1945).
174. Transition Curves for Highways. Joseph Barnett. Public Roads Administration (1938).
175. The Unified Soil Classification System (with appendixes). *Technical Memorandum No. 3–357.* Waterways Experiment Station, Vicksburg, Miss. (1953).
176. Utilization of the Planning Survey. H. S. Fairbank. Bureau of Public Roads, paper presented at the Highway Engineering and Highway Safety Conference, Ann Arbor, Mich. (1938).

Publications of the Various State Highway Departments

177. Acquisition and Analysis of Traffic Data in Georgia. Division of Highway Planning, Georgia State Highway Department. Atlanta. (September, 1965).
178. The Centrifuge Kerosene Equivalent as Used in Establishing the Oil Content for Dense Graded Bituminous Mixtures. F. N. Hveem. Division of Highways, Department of Public Works, State of California. Sacramento (1946).
179. Construction and Material Specifications. Department of Highways, State of Ohio. Columbus (January 1, 1957).
180. Continuous Station and Short Count Analysis. R. J. Paquette. Michigan State Highway Department. Lansing (1937).
181. Field Manual of Soil Engineering. Michigan State Highway Department. Lansing (4th ed., 1960).
182. Method of Test for Determination of the Resistance "R" Value of Treated and Untreated Bases by the Stabilometer Test Method. *Test Method No. 301-F.* Division of Highways, State of California. Sacramento (September, 1964).
183. Parking Survey of the Downtown Area—City of Baltimore. Vol. IV. State Roads Commission. Baltimore (1946).
184. Public Works Specifications. Department of Public Works, State of New York. Albany (1962).
185. Standard Specifications. State Highway Department of Delaware. Dover (July 1, 1960).
186. Standard Specifications for Highways and Bridges. State of Maine. Augusta (June, 1965).

187. Standard Specifications for Highway Construction. Department of Roads. State of Nebraska. Lincoln (1965).

188. Standard Specifications for Road and Bridge Construction. Department of Highways. State of Kentucky. Frankfort (1956).

189. Standard Specifications for Road and Bridge Construction. Texas Highway Department. Austin (1962).

190. Standard Specifications for Roads and Bridges. Division of Highways. Department of Public Works and Buildings. State of Illinois. Springfield (January 2, 1958).

191. Standard Specifications for Roads and Structures. State Highway and Public Works Commission. State of North Carolina. Raleigh (April 1, 1959).

192. Standard Specifications of the Division of Highways. State of California. Sacramento (July, 1964).

193. Structural Design of the Roadbed. *Planning Manual,* Section 7-601.1. Division of Highways, State of California. Sacramento (August, 1964).

194. Summary of Treatments for Highway Embankments on Soft Foundations. Lyndon H. Moore. Department of Public Works, State of New York. Albany (1966).

195. Traffic Density Study—Manual of Instructions. State Highway Department of South Carolina. Columbia (1948).

196. Transportation Needs—Baltimore Metropolitan Area. Vol. I, State Roads Commission. Baltimore (1946).

Other Publications

197. Abbett, R. W. Engineering Contracts and Specifications. John Wiley & Sons, Inc., New York (2nd ed., 1948).

198. Abraham, Herbert. Asphalts and Allied Substances (2 vols, 6th ed.). D. Van Nostrand Co., Inc., Princeton, N.J. (1960).

199. American Bitumuls & Asphalt Co. Specification B-4 for Emulsified Asphalt Sand-Mix Pavement. San Francisco.

200. American Concrete Institute. ACI Standards, 1957. Detroit.

201. American Concrete Institute. Manual of Concrete Inspection. Detroit (1957).

202. American Concrete Institute. Specifications for Concrete Pavements and Concrete Bases. *ACI Standard 617-58.* Detroit (1958).

203. American Concrete Pipe Association. Concrete Pipe Handbook. Chicago (1965).

204. American Society of Civil Engineers. Construction Practice for Rigid Pavement. *Journal,* Aero-Space Transport Division, May, 1964. New York.

205. American Society of Civil Engineers. Development of CBR Flexible Pavement Design Method for Airfields—A Symposium. *Proceedings,* January, 1949. New York.

206. American Society of Civil Engineers. Systems Engineering in Highway Engineering. *Conference Preprint No. 232,* May, 1965. New York.

207. Armco Drainage and Metal Products, Inc. Drainage Handbook. Middletown, Ohio.

208. Automobile Manufacturers Association. Automobile Facts and Figures, 1966. Detroit.

209. Automotive Safety Foundation. Multiproject Scheduling for Highway Programs. Washington, D.C. (1964).

210. Baker, Robert F. Factors Influencing the Choice of Compaction Equipment. *Public Works,* January, 1957. Ridgewood, N.J.

211. The Barber Asphalt Co. Trinidad and Bermudez Lake Asphalts. New York.

212. Barber-Greene Co. Bituminous Construction Handbook. Aurora, Ill. (1957).

213. The Barrett Division, Allied Chemical & Dye Corp. Barrett Road Book. New York.

214. Belcher, D. J., Gregg, L. E., and Woods, K. B. The Formation, Distribution

and Engineering Characteristics of Soils. *Research Series No. 78.* Engineering Experiment Station, Purdue University, Lafayette, Ind. (1943).

215. Benson, Fred J. Surface Treatments for Highways. *Public Works,* May, 1957. Ridgewood, N.J.
216. Bethlehem Steel Co. Solving Drainage Problems. Bethlehem, Pa. (1958).
217. Bethlehem Steel Co. Steel for Highways. Bethlehem, Pa.
218. Betz, M. J. Interpretation of Costs. American Society of Civil Engineers. *Conference Preprint No. 207,* May, 1965. New York.
219. Bixby, Howard M. Construction and Inspection Practices on Graded Aggregate Base Courses in the Midwest. *Crushed Stone Journal,* December, 1957.
220. Bradbury, R. D. Reinforced Concrete Pavements. Wire Reinforcement Institute. Washington, D.C. (1938).
221. Bros, Inc. Handbook of In-Place Soil Stabilization. Minneapolis (1964).
222. Calcium Chloride Institute. Calcium Chloride for Stabilization of Bases and Wearing Surfaces. *Manual SM-1.* Washington, D.C.
223. Carstens, R. H., and Csanyi, L. H. Economic Analysis for Highway Improvements in Under-Developed Countries. *Engineering Report No. 44.* Iowa Engineering Experiment Station. Ames (1964).
224. Casagrande, A. Classification and Identification of Soils. *Bulletin No. 432.* Graduate School of Engineering, Harvard University, Cambridge, Mass. (1947).
225. Chevron Asphalt Co. Paving Handbook. San Francisco (1965).
226. Clark, G. H. Rock Asphalts of Alabama and Their Use in Paving. *Special Report No. 13.* Geological Survey of Alabama (1925).
227. Compton, F. E., & Co. Roads and Streets. Compton's Pictured Encyclopedia and Fact-finding Index (Vol. 12). Chicago (1947).
228. Eichler, J. O., and Tubis, H. Photogrammetry Kit (3rd ed., 1958). Margate, N.J.
229. Engineering Experiment Station, Purdue University. Bituminous Surface Treatment. *Research Series No. 82.* Lafayette, Ind. (1941).
230. *Engineering News-Record.* Modern Construction Operations: The Takeover of Automation. December 9, 1965.
231. Gray, B. E. Design and Construction of Bituminous Pavements. Annual Meeting of Highway Engineers and Commissioners of Michigan, Houghton (1934).
232. Gray, Joseph E. Crushed Stone Base Courses. *Crushed Stone Journal,* June, 1956.
233. Green, Roy M. Cold-Mixed Limestone Rock Asphalt Pavement Wearing Surfaces. 12th Annual Short Course in Highway Engineering. A & M College of Texas, College Station (1936).
234. Huang, E. Y. Manual of Current Practice for the Design, Construction and Maintenance of Soil-Aggregate Roads. University of Illinois, Urbana (June, 1959).
235. Joint Fact-Finding Committee on Highways, Streets, and Bridges, California Legislature. Engineering Facts and a Future Program—A Study for the California Legislature. Sacramento (1946).
236. Joint Sealer Manufacturers Association. Joint Sealing Manual. Chicago (1957).
237. Jonas, Douglas L. Reducing Equipment Needs with CPM. *Western Construction,* November, 1964.
238. Kansas Highways Fact-Finding and Research Committee. Highway Needs of Kansas. Topeka (1948).
239. Kentucky Rock Asphalt Institute. Kentucky Natural Sandstone Rock Asphalt—Specifications and Designs. Louisville (1944).
240. Leisch, J. E. Interchange Design—Reprints and Papers. DeLeuw, Cather & Co. Chicago. November, 1962.

241. Leisch, J. E. New Techniques in Alignment Design and Stakeout. American Society of Civil Engineers. New York. May, 1965.
242. Marshall Consulting and Testing Laboratory. The Marshall Method for the Design and Control of Bituminous Paving Mixtures. Jackson, Miss. (1949).
243. Martin, George E. How To Make Old Roads Fit for Today's Traffic. *Public Works.* Ridgewood, N.J.
244. Martin, J. Rogers, and Wallace, Hugh A. Design and Control of Asphalt Pavements. McGraw-Hill Book Co., Inc., New York (1958).
245. McLeod, N. W. Flexible Pavement Thickness Requirements. *Proceedings.* Association of Asphalt Paving Technologists. (1956).
246. Meyer, Carl F. Route Surveying. International Textbook Co., Scranton, Pa. (1959).
247. Michigan Good Roads Federation, Highway Study Committee. Highway Needs in Michigan—An Engineering Analysis. Lansing (1948).
248. Michigan, University of. International Conference on the Structural Design of Asphalt Pavements. *Preprint Volume (Supplement).* Ann Arbor (1962).
249. Minor, Carl E. Placing and Compacting Thick Lifts of Asphalt Concrete. Paper presented at the 11th annual convention of the National Asphalt Paving Association, February, 1966.
250. Moffitt, F. H. Photogrammetry. International Textbook Co., Scranton, Pa., 2nd ed. (1956).
251. National Committee on Urban Transportation. Better Transportation for Your City. Public Service Administration. Chicago (1958).
252. National Lime Association. Lime Stabilization Construction Manual. Washington, D.C. (1965).
253. National Slag Association. Composition of Asphaltic Concrete—Hot and Cold Mixed Types—Recommended Design and Quantities. Washington, D.C.
254. Owen, Wilfred. Automotive Transportation Trends and Problems. The Brookings Institution. Washington, D.C. (1949).
255. Power Crane and Shovel Association. The Functional Design, Job Application and Job Analysis of Power Cranes and Shovels. *Technical Bulletin No. 1.* New York.
256. Preston, E. M., & Associates. A Manual for Applying the Critical Path Method to Highway Department Engineering and Administration. Columbus, Ohio (1963).
257. Proctor, R. R. Fundamental Principles of Soil Compaction. *Engineering News-Record,* August 31, September 7, 21, 28, 1933.
258. Public Administration Service. Traffic Engineering Functions and Administration. *Publication No. 100.* Chicago (1948).
259. *Public Works.* Street and Highway Manual and Catalog File. Ridgewood, N.J. (1966).
260. Rice, James M. A Graphical Method of Combining Sizes of Aggregates for a Specification. *Crushed Stone Journal,* December, 1956.
261. Ritter, J. B., and Schaffer, L. R. Blending Natural Earth Deposits for Least Cost. American Society of Civil Engineers. *Journal,* Construction Division, March, 1961.
262. Ritter, L. J. Soil Stabilization. *Proceedings.* First Annual Florida Highway Conference. University of Florida, Gainesville (1947).
263. Ritter, L. J. What You Should Know About Soil Engineering. *Public Works.* Ridgewood, N.J. (1953).
264. *Rural and Urban Roads.* Herbicides Spray-Down Brush, Weeds and Grass. May, 1964.
265. Salt Institute. Salt for Ice and Snow Removal. Chicago.
266. Salt Institute. Salt for Road Stabilization. Chicago (1966).
267. Smith, Wilbur, & Associates. Future Highways and Urban Growth. New Haven, Conn. February, 1961.

268. Smith, Wilbur, & Associates. Parking and the Central City. New Haven, Conn. May, 1965.
269. Swineford, F. E. The Modern Macadam Pavement. *Crushed Stone Journal,* March, 1948.
270. Tallamy, B. D. Control of Highway Access: Experiences in New York. American Society of Civil Engineers. *Journal,* Highway Division, January, 1956. New York.
271. Texas Transportation Institute. A Laboratory and Field Evaluation of Lightweight Aggregates as Coverstone for Seal Coats and Surface Treatments. *Research Report No. 51-2.* College Station, Texas (April, 1966).
272. Tyler, O. R. Adhesion of Bituminous Films to Aggregates. *Research Series No. 62.* Engineering Experiment Station, Purdue University. Lafayette, Ind. (1938).
273. Tyler, O. R., Goetz, W. H., and Slesser, C. Natural Sandstone Rock Asphalt. *Research Series No. 78.* Engineering Experiment Station, Purdue University, Lafayette, Ind. (1941).
274. U.S. Waterways Experiment Station. The California Bearing Ratio as Applied to the Design of Flexible Pavements for Airports. *Technical Memorandum No. 213-1.* Vicksburg, Miss. (1945).
275. Uvalde Rock Asphalt Co. Handbook. San Antonio, Texas.
276. Wire Reinforcement Institute. Continuously Reinforced Concrete Pavement. Washington, D.C. (1964).
277. Wire Reinforcement Institute. Reinforced Bituminous Concrete Overlays. Washington, D.C. (1962).
278. Wolfard, N. E. Native Road Materials and Highway Maintenance. *Circular No. 20.* Oklahoma Geological Survey, Norman (1929).
279. Woods, K. B., *et al.* Highway Engineering Handbook. McGraw-Hill Book Co., Inc., New York (1960).

INDEX